Birkhäuser

Progress in Mathematics

Volume 340

Progress in Mathematics is a series of books intended for professional mathematicians and scientists, encompassing all areas of pure mathematics. This distinguished series, which began in 1979, includes research level monographs, polished notes arising from seminars or lecture series, graduate level textbooks, and proceedings of focused and refereed conferences. It is designed as a vehicle for reporting ongoing research as well as expositions of particular subject areas.

More information about this series at https://link.springer.com/bookseries/4848

Anton Alekseev • Edward Frenkel • Marc Rosso
Ben Webster • Milen Yakimov
Editors

Representation Theory, Mathematical Physics, and Integrable Systems

In Honor of Nicolai Reshetikhin

Editors
Anton Alekseev
Section de Mathématiques
Université de Genève
Genève 4, Switzerland

Edward Frenkel
Department of Mathematics
University of California at Berkeley
Berkeley, CA, USA

Marc Rosso
Mathematics Department
Institut de Mathématiques de Jussieu
Paris Cedex 13, France

Ben Webster
Department of Pure Mathematics
University of Waterloo
Waterloo, ON, Canada

Milen Yakimov
Department of Mathematics
Northeastern University
Boston, MA, USA

ISSN 0743-1643 ISSN 2296-505X (electronic)
Progress in Mathematics
ISBN 978-3-030-78150-7 ISBN 978-3-030-78148-4 (eBook)
https://doi.org/10.1007/978-3-030-78148-4

Mathematics Subject Classification: 81T13, 82B23, 17B37, 14N35, 60F10

This book is published under the imprint Birkhäuser, www.birkhauser-science.com by the registered company Springer Nature Switzerland AG
The registered company address is: Gewerbestrasse 11, 6330 Cham, Switzerland

Nicolai Reshetikhin

Preface

Anton Alekseev, Edward Frenkel, Marc Rosso,

Ben Webster, and Milen Yakimov

It is a pleasure to present this volume dedicated to Nicolai Reshetikhin, mathematician and friend we admire. Kolya, as he is affectionately known, has made a number of groundbreaking contributions in representation theory, integrable systems, and topology. His ideas have profoundly influenced the evolution of these fields in the past 40 years and will certainly continue to do so for many years to come. This book is a collection of chapters by distinguished mathematicians and physicists, many of them Kolya's students and collaborators, who develop further the themes of his research. Some of these chapters are based on the talks given by their authors at a conference in honor of Kolya's 60th birthday held at CIRM, Luminy, in June 2018.

In this preface, we present a brief summary of some of Kolya's discoveries.

1 Quantum Groups

Kolya's scientific career began in St. Petersburg, then known as Leningrad, in the late 1970s. He was a member of the famous Leningrad School led by Ludvig Faddev. Kolya joined the Leningrad School at an opportune moment, when Faddeev, Sklyanin, and Takhtajan were developing the quantum inverse scattering method and applying it to statistical models such as the Heisenberg XYZ spin chain [51, 52]. Their work, synthesizing the classical inverse scattering method used in soliton theory and the results of Bethe, Baxter, and others on the exactly solved models of statistical mechanics, heralded a revolution in quantum integrable systems. New powerful algebraic structures were emerging, including what came to be known as *quantum groups*.

Kolya was at the forefront of this research from the start. In fact, one of his first scientific papers [32], joint with Kulish, introduced the first quantum group, now known as $U_q(sl_2)$, as the algebraic structure behind the higher spin generalization of the quantum models such as the sine-Gordon and the Heisenberg XXZ spin chain. Soon after that Sklyanin endowed the algebra $U_q(sl_2)$ with a Hopf algebra structure [55], and just a few years later Drinfeld [12] and Jimbo [24] generalized the construction from sl_2 to an arbitrary Kac–Moody algebra. Thus, quantum groups were born.

Since then, they have become as ubiquitous as Lie groups and Lie algebras in many areas of mathematics and mathematical physics, far beyond the theory of integrable systems where they originated.

Kolya played a big role in these developments. His works, such as the influential papers [13] with Faddeev and Takhtajan and [44] with Semenov-Tian-Shansky, elucidated the algebraic structure of quantum groups. And he also pioneered many exciting applications of quantum groups in other fields: integrable systems, representation theory, combinatorics, and topology. We discuss some of these works in the next sections.

After obtaining his doctorate degree in 1984, Kolya joined the Leningrad Branch of the Steklov Institute of the Academy of Sciences, where he worked for 5 years. In 1989, he came to Harvard University as a winner of the Harvard Prize Fellowship. Two years later, he joined the faculty at University of California, Berkeley. In 2021, after 30 years of service, he retired from Berkeley and joined the Yau Center for Mathematical Sciences at Tsinghua University, Beijing. Kolya has also held numerous visiting positions, such as Niels Bohr Visiting Professorship at the Aarhus University and Humboldt Visiting Professorship at the Technical University of Berlin and the Max Planck Institute for Gravitational Physics. He is currently affiliated with the KdV Institute of the University of Amsterdam and the St. Petersburg State University.

2 Quantum Integrable Systems

The quantum inverse scattering method was initially applied to models related to the Lie algebra sl_2, such as the XXX and XXZ model. As far as we know, Kolya was the first to systematically extend this method to quantum models associated to other Lie algebras. In modern language, these models correspond to finite-dimensional representations of the Yangians $Y(\mathfrak{g})$, where $\mathfrak{g}$ is a simple Lie algebra, and quantum affine algebras $U_q(\widehat{\mathfrak{g}})$, where $\widehat{\mathfrak{g}}$ is an affine Kac–Moody algebra (including the twisted ones). The commuting Hamiltonians acting on these representations come from a family of commuting transfer-matrices in these algebras. Thus, one gets a vast collection of quantum integrable systems associated to these Lie algebras.

In the case the XXX and XXZ models, which correspond to $Y(sl_2)$ and $U_q(\widehat{sl}_2)$, respectively, the spectra of these Hamiltonians can be computed using the algebraic Bethe Ansatz method, introduced by Bethe and developed further by the Leningrad School. The problem of generalizing this method to the integrable systems associated to other Lie algebras is highly non-trivial. In a series of papers [36–38], Kolya proposed an analogue of this method, which he dubbed *analytic Bethe Ansatz*, for the quantum integrable systems associated to the Yangians and quantum affine algebras. A crucial role in it was played by an elegant generalization of the famous Baxter relations observed in the XXX and XXZ model.

The resulting Bethe Ansatz equations [38, 50] have been widely used in the subject, even though the method itself remained something of a mystery. It was finally put on a firm foundation with the advent of the theory of *q-characters* developed by Kolya and Edward Frenkel [20] (see Sect. 4 below). Using this theory, Frenkel and Hernandez [17] proved the generalized Baxter relations and gained new insights into the analytic Bethe Ansatz.

The quantum integrable systems associated to $U_q(\widehat{\mathfrak{g}})$ have a limit, in which the symmetry algebra becomes the affine Kac–Moody algebra $\widehat{\mathfrak{g}}$ itself. Together with Feigin and E. Frenkel, Kolya showed how to obtain the quantum Hamiltonians of the corresponding integrable system, called the *Gaudin model*, from the center of the completed enveloping algebra of $\widehat{\mathfrak{g}}$ at the critical level [16]. Moreover, they were able to describe the spectrum of the Hamiltonians of the Gaudin model in terms of the geometric objects called *opers*. An interesting aspect of this construction is that the opers are associated not to $\mathfrak{g}$ but to the Langlands dual Lie algebra ${}^L\mathfrak{g}$ of $\mathfrak{g}$. The appearance of the *Langlands duality* here is important. It manifests a deep connection between the Gaudin model (and its generalizations) and the geometric Langlands correspondence developed by Beilinson and Drinfeld [3]. As far as we know, the paper [16] was the first case study of the Langlands duality in quantum integrable systems. Since then, dualities of this kind have been extensively studied and connected to various dualities of quantum field theories.

Around the same time, together with Varchenko, Kolya established a link between the Gaudin model and the critical level limit of the solutions of the Knizhnik–Zamolodchikov (KZ) equations [49].

Closely related to this topic is another groundbreaking work [22], in which Kolya and Igor Frenkel introduced the celebrated *qKZ equations* (in a special case, this equation was also introduced by Fedor Smirnov, another alumnus of the Leningrad School [56]). The qKZ equations have roughly the same relationship with the quantum integrable models associated to $U_q(\widehat{\mathfrak{g}})$ as the KZ equations with the Gaudin models. In particular, the critical level limits of the solutions of the qKZ equations give rise to the eigenvectors of the quantum integrable systems associated to $U_q(\widehat{\mathfrak{g}})$, see [40]. Together with Jasper Stokman and Bart Vlaar, Kolya explored similar structures for the so-called boundary qKZ equations [45, 46].

3 Topology

One of the most striking and influential applications of quantum groups is low-dimensional topology. In the mid-1980s, Vaughan Jones and others defined stunning new invariants of knots—most famously, the Jones polynomial, HOMFLYPT polynomial, and Kauffman polynomial. In a groundbreaking work [39, 57], Kolya and Vladmir Turaev showed that these invariants were special cases of a much more general, categorical construction that attached a knot invariant to each representation of the quantum group associated to a simple Lie algebra. These are now called *Reshetikhin–Turaev invariants*.

This work supplied an enormous family of new powerful invariants, and its novel categorical approach opened the door to defining invariants of 3-manifolds. This was done by Kolya with Turaev in [48]. Among other things, this gave a mathematical interpretation of the invariants that Edward Witten had proposed earlier in the context of the Chern-Simons theory [58] (from Witten's perspective, the invariant arises as the partition function of the Chern-Simons theory on the corresponding 3-manifold). These *Witten–Reshetikhin–Turaev invariants* have had tremendous influence in 3-manifold theory, topological quantum field theory (including topological quantum computing), tensor categories, and beyond. This is one of Kolya's best known works.

Since then, Kolya continued making new strides in topology, making new advances such as defining new invariants of links and tangles in 3-manifolds with the additional datum of a flat connection in the complement [25], the recursion formulas for knot invariants [10], and the extension of the Reshetikhin–Turaev invariants to braided categories with weaker properties than the ribbon property usually required [4].

4 Representation Theory and Combinatorics

Kolya has made a number of important contributions to representation theory of quantum groups. For example, in [29], he and Anatol Kirillov introduced a family of what came to be known as *Kirillov–Reshetikhin modules* over the Yangians $Y(\mathfrak{g})$. These modules have been used extensively in representation theory and combinatorics. Another important result, a bijection between semi-standard Young tableaux and rigged configurations, was obtained by Kolya with Kerov and Kirillov from the study of asymptotic completeness of the Bethe Ansatz equations [26]. This led Kolya and Kirillov to new formulas for the Kostka polynomials [27, 28].

Another direction concerns *deformations of $\mathcal{W}$-algebras*. According to the results of the paper [16] mentioned earlier, quantum Hamiltonians of the Gaudin model could be obtained from the center of the enveloping algebra of $\widehat{\mathfrak{g}}$ at the critical level. The center, however, is not just a commutative algebra. It also has a Poisson algebra structure due to the possibility of deforming the level away from the critical level. It is known from [14] that this Poisson algebra is isomorphic to the classical $\mathcal{W}$-algebra associated to ${}^L\mathfrak{g}$ (and this is in fact the root of the Langlands duality mentioned earlier). Now consider the $U_q(\widehat{\mathfrak{g}})$-analogue of the XXZ model, which is a q-deformation of the Gaudin model. It turns out that the quantum Hamiltonians of this model may also be obtained from the center at the critical level, but now of the quantum affine algebra $U_q(\widehat{\mathfrak{g}})$. And this q-deformed center also has a Poisson algebra structure. This way, Kolya and E. Frenkel discovered in [18] q-deformations of the classical $\mathcal{W}$-algebras, in particular, a q-deformation of the classical Virasoro algebra (corresponding to $\mathfrak{g} = sl_2$). Together with Semenov-Tian-Shansky, they were able to obtain this Poisson algebra by means of a q-deformed Drinfeld–Sokolov reduction [21] (this construction was generalized to other Lie algebras in [53]).

Two years later, Frenkel and Reshetikhin quantized these Poisson algebras [19]. Namely, they introduced the deformed $\mathcal{W}$-algebra $\mathcal{W}_{q,t}(\mathfrak{g})$ for an arbitrary simple Lie algebra $\mathfrak{g}$ (in the case of $\mathfrak{g} = sl_n$ this algebra was constructed earlier in [2, 15, 54]). The deformed $\mathcal{W}$-algebra depends on two parameters, q and t, and one recovers the center of $U_q(\widehat{\mathfrak{g}})$ at the critical level in the limit $t \to 1$. Other limits also give rise to interesting algebras. Frenkel and Reshetikhin used the $t \to 1$ limit in [20] to introduce the theory of *q-characters* of finite-dimensional representations of quantum affine algebras. The q-characters proved to be a powerful tool in the study of these representations and beyond.

Recently, the deformed $\mathcal{W}$-algebras and related structures appeared in the study of four-dimensional gauge theories associated to Nakajima quiver varieties [1, 30].

5 Poisson Algebras, BV Theory, and Quantization

Poisson geometry, quantization, and Batalin–Vilkovisky (BV) structures are closely related topics. Their connection was established in the work of Kontsevich on quantization of Poisson manifolds using Feynman path integral [31] and in the work of Cattaneo–Felder [5] which explained the Feynman rules used by Kontsevich in terms of a BV structure of the Poisson σ-model.

Kolya's interest in the topic dates back to his joint work with Takhtajan [47], where they gave a simple formula for quantization of Kähler manifolds in terms of Feynman graphs. Recently, in collaboration with Alberto Cattaneo and Pavel Mnev, Kolya put forward an ambitious *Cattaneo–Mnev–Reshetikhin program* which bridges the BV quantization and low dimensional topology. More precisely, the aim of this program is to upgrade low dimensional algebraic topology to quantum algebraic topology in order to define and compute quantum invariants of manifolds in dimensions 2, 3, and 4 by gluing them from simple pieces (e.g., tetrahedra or cubes).

This program is still under development, but significant progress has already been achieved. The first important result is a better formulation of quantum field theory (QFT) on manifolds with boundary. While BV formalism is the standard tool of treating symmetries in the bulk, the Batalin–Fradkin–Vilkovisky formalism (BFV) is needed to understand the symmetry at the boundary. The combined BV-BFV theory was stated at the classical level in [6], and the quantum version was addressed in [7]. Interesting partial results include applications to integrable systems [8] and an example of a BF theory verifying Atiyah–Segal type gluing axioms [9].

6 Other Works

The scope of Kolya's research has been extremely broad, and he has made substantial contributions in numerous directions in addition to those described in the previous sections.

Throughout his career, he has actively worked on the semiclassical structures behind quantum groups. He has made key contributions to the geometry of *Poisson–Lie groups* and the dynamics of related integrable systems, such as the ones in [23] on the symplectic foliation of the standard Poisson–Lie groups and the associated Coxeter–Toda lattices. He introduced powerful representation theoretic methods to study degenerate integrability on non-Coxeter symplectic leaves [42] and spin Calogero–Moser integrable systems [43].

Kolya has done extensive research on quantum groups at roots of unity. In his solo paper [41], he constructed quasitriangular structures on the unrestricted quantum groups at roots of unity. In a joint paper with De Concini et al. [11], he described the tensor product structure of the finite-dimensional representations of these algebras and developed a general axiomatic setting of Cayley–Hamilton Hopf algebras.

He was also very much involved in the study of *random matrices, random processes, and random surfaces*. Jointly with Andrei Okounkov, he obtained integral

Reshetikhin's mathematical descendants at the Luminy conference. From left to right: David Keating, Noah Snyder, Sevak Mkrtchyan, Qingtao Chen, Aaron Brookner, Alexander Shapiro, Ben Webster, Peter Tingley, Nicolai Reshetikhin, Theo Johnson-Freyd, Olya Mandelshtam, Harold Williams, Kurt Trampel, Meredith Shea, Kent Vashaw, Raeez Lorgat, Gus Schrader, and Kai Chieh Chen

representations of the correlation functions of the Schur process and used them to obtain explicit formulas for the asymptotic correlation functions for 3D Young diagrams in the bulk limit [33]. This paper inspired much subsequent research on determinantal processes and their asymptotic behavior. In another paper [34], Kolya and Okounkov carried out a detailed study of random skew 3D partitions and their various asymptotics expressed in terms of Airy and Pearcey kernels. These ideas led to a remarkable duality that was proposed by Kolya, Okounkov, and Vafa between the topological A-model in string theory and a classical statistical mechanical model of crystal melting [35].

Kolya has brought up many outstanding graduate students. According to the Mathematics Genealogy Project, as of the summer of 2020, Kolya had 20 students and 33 descendants. He has said that the opportunity to work with graduate students is one of the joys of being in academia.

A conference to mark Kolya's 60th birthday took place at Centre International de Rencontres Mathématiques, Luminy, in June 2018. It was funded by the National Science Foundation grant DMS-1803265 and the European Research Council project MODFLAT.

In conclusion, we wish Kolya many more years of health, productivity, success, and inspiration.

Genève 4, Switzerland — Anton Alekseev

Berkeley, CA, USA — Edward Frenkel

Paris Cedex 13, France — Marc Rosso

Waterloo, ON, USA — Ben Webster

Boston, MA, USA — Milen Yakimov

References

1. M. Aganagic, E. Frenkel, and A. Okounkov, *Quantum q-Langlands Correspondence*, Trans. Moscow Math. Soc. **79** (2018) 1–83.
2. H. Awata, H. Kubo, S. Odake, and J. Shiraishi, *Quantum $\mathcal{W}_N$ algebras and Macdonald polynomials,* Comm. Math. Phys. **179** (1996) 401–416.
3. A. Beilinson and V. Drinfeld, *Quantization of Hitchin's integrable system and Hecke eigensheaves*, Preprint.
4. C. Blanchet, N. Geer, B. Patureau-Mirand, and N. Reshetikhin. *Holonomy braidings, biquandles and quantum invariants of links with* $\mathsf{SL}_2(\mathbb{C})$*-flat connections.* Selecta Math. **26**, no. 2 (2020) 1–58.
5. A. Cattaneo, G. Felder, *A path integral approach to the Kontsevich quantization formula*, Comm. Math. Phys. **212**, no. 3 (2000) 591–611.

6. A. Cattaneo, P. Mnev, N. Reshetikhin. *Classical BV theories on manifolds with boundary*, Comm. Math. Phys. **332**, no. 2 (2014) 535–603.
7. A. Cattaneo, P. Mnev, N. Reshetikhin. *Perturbative quantum gauge theories on manifolds with boundary*, Comm. Math. Phys. **357**, no. 2 (2018) 631–730.
8. A. Cattaneo, P. Mnev, N. Reshetikhin, *Poisson sigma model and semiclassical quantization of integrable systems*, Rev. Math. Phys. **30**, no. 6 (2018) 1840004, 26 pp.
9. A. Cattaneo, P. Mnev, N. Reshetikhin, *A Cellular Topological Field Theory*, Comm. Math. Phys. **374**, no. 2 (2020) 1229–1320.
10. Q. Chen, N. Reshetikhin, *Recursion formulas for HOMFLY and Kauffman invariants*, J. Knot Theory Ramifications **23**, no. 5 (2014) 1450024, 23 pp.
11. C. De Concini, C. Procesi, N. Reshetikhin, and M. Rosso, *Hopf algebras with trace and representations*, Invent. Math. **161**, no. 1 (2005) 1–44.
12. V.G. Drinfeld, *Hopf algebras and the quantum Yang-Baxter equation*, Soviet Math. Dokl. **32** (1985) 254–258.
13. L.D. Faddeev, N.Yu. Reshetikhin, and L.A. Takhtajan, *Quantization of Lie Groups and Lie Algebras*, Leningrad Math. J. **1** (1990) 193–225.
14. B. Feigin and E. Frenkel, *Affine Kac-Moody algebras at the critical level and Gelfand-Dikii algebras*, in *Infinite Analysis*, eds. A. Tsuchiya, T. Eguchi, M. Jimbo, Adv. Ser. in Math. Phys. **16**, 197–215, World Scientific, 1992.
15. B. Feigin and E. Frenkel, *Quantum $\mathcal{W}$–algebras and elliptic algebras*, Comm. Math. Phys. **178** (1996) 653–678.
16. B. Feigin, E. Frenkel, and N. Reshetikhin, *Gaudin model, Bethe ansatz and critical level*, Comm. Math. Phys. **166** (1994) 27–62.
17. E. Frenkel and D. Hernandez, *Baxter's Relations and Spectra of Quantum Integrable Models*, Duke Math. J. **164** (2015) 2407–2460.
18. E. Frenkel and N. Reshetikhin, *Quantum affine algebras and deformations of the Virasoro and $\mathcal{W}$-algebras*, Comm. Math. Phys. **178** (1996) 237–264.
19. E. Frenkel and N. Reshetikhin, *Deformations of $\mathcal{W}$–algebras associated to simple Lie algebras*, Comm. Math. Phys. **197** (1998) 1–32.
20. E. Frenkel and N. Reshetikhin, *The q–characters of representations of quantum affine algebras and deformations of $\mathcal{W}$–algebras*, in Recent Developments in Quantum Affine Algebras and Related Topics, N. Jing and K. Misra (eds.), Contemporary Mathematics **248**, pp. 163–205, AMS, 1999.
21. E. Frenkel, N. Reshetikhin, and M.A. Semenov-Tian-Shansky, *Drinfeld-Sokolov reduction for difference operators and deformations of $\mathcal{W}$-algebras I.*, Comm. Math. Phys. **192** (1998) 605–629.
22. I.B. Frenkel and N.Yu. Reshetikhin, *Quantum affine algebras and holonomic difference equations*, Comm. Math. Phys. **146** (1992) 1–60.
23. T. Hoffmann, J. Kellendonk, N. Kutz, and N. Reshetikhin, *Factorization dynamics and Coxeter-Toda lattices*, Comm. Math. Phys. **212**, no. 2 (2000) 297–321.
24. M. Jimbo, *A q-difference analogue of $U(g)$ and the Yang-Baxter equation*, Lett. Math. Phys. **10** (1985) 63–69.

25. R. Kashaev and N. Reshetikhin, *Invariants of tangles with flat connections in their complements*, in Graphs and patterns in mathematics and theoretical physics, pp. 151–172, Proc. Sympos. Pure Math. **73**, AMS, 2005.
26. S.V. Kerov, A.N. Kirillov, and N.Yu. Reshetikhin, *Combinatorics, the Bethe ansatz and representations of the symmetric group*, J. Soviet Math. **41**, no. 2 (1988) 916–924.
27. A.N. Kirillov and N.Yu. Reshetikhin, *The Bethe ansatz and the combinatorics of Young tableaux*, J. Soviet Math. **41**, no.2 (1988) 925–955.
28. A.N. Kirillov and N.Yu. Reshetikhin, *Multiplicity of weights in irreducibile tensor representations of a complete linear superalgebra*, Funct. Anal. Appl. **22**, no. 4 (1988) 328–330.
29. A.N. Kirillov and N.Yu. Reshetikhin, *Representations of Yangians and multiplicities of occurrence of the irreducible components of the tensor product of representations of simple Lie algebras*, J. Soviet Math. **52** (1990) 3156–3164.
30. T. Kimura and V. Pestun, *Quiver W-algebras*, Lett. Math. Phys. **108** (2018) 1351–1381.
31. M. Kontsevich, *Deformation quantization of Poisson manifolds*, Lett. Math. Phys. **66**, no. 3 (2003) 157–216.
32. P.P. Kulish and N.Yu. Reshetikhin, *Quantum linear problem for the sine-Gordon equation and higher representations*, Zapiski Nauchn. Sem. LOMI **101** (1981) 101–110. English translation: J. Soviet Math. **23** (1983) 2435–2441.
33. A. Okounkov and N. Reshetikhin, *Correlation function of Schur process with application to local geometry of a random 3-dimensional Young diagram*, J. Amer. Math. Soc. **16**, no. 3 (2003) 581–603.
34. A. Okounkov and N. Reshetikhin, *Random skew plane partitions and the Pearcey process*, Comm. Math. Phys. **269**, no. 3 (2007) 571–609.
35. A. Okounkov, N. Reshetikhin, and C. Vafa, *Quantum Calabi-Yau and classical crystals*, in The unity of mathematics, Progr. Math. **244**, pp. 597–618, Birkhüser Boston, 2006.
36. N.Yu. Reshetikhin, *A Method Of Functional Equations In The Theory Of Exactly Solvable Quantum Systems*, Lett. Math. Phys. **7** (1983) 205–213.
37. N.Yu. Reshetikhin, *Integrable Models of Quantum One-dimensional Magnets With $O(N)$ and $Sp(2k)$ Symmetry*, Theor. Math. Phys. **63** (1985) 555–569.
38. N.Yu. Reshetikhin, *The spectrum of the transfer matrices connected with Kac–Moody algebras*, Lett. Math. Phys. **14** (1987) 235–246.
39. N.Yu. Reshetikhin, *Quantized universal enveloping algebras, the Yang-Baxter equation and invariants of links I., II.* LOMI Preprints E–4–87, E–17–87 (1987).
40. N. Reshetikhin, *Jackson-type integrals, Bethe vectors, and solutions to a difference analog of the Knizhnik-Zamolodchikov system*, Lett. Math. Phys. **26** (1992) 153–165.
41. N. Reshetikhin, *Quasitriangularity of quantum groups at roots of 1*, Comm. Math. Phys. **170**, no. 1 (1995) 79–99.

42. N. Reshetikhin, *Integrability of characteristic Hamiltonian systems on simple Lie groups with standard Poisson Lie structure*, Comm. Math. Phys. **242**, no. 1 (2003) 1–29.
43. N. Reshetikhin, *Degenerate integrability of the spin Calogero-Moser systems and the duality with the spin Ruijsenaars systems*, Lett. Math. Phys. **63** (2003), no. 1, 55–71.
44. N.Yu. Reshetikhin and M.A. Semenov-Tian-Shansky, *Quantum R-matrices and factorization problems*, J. Geom. Phys. **5**, no. 4 (1988) 533–550.
45. N. Reshetikhin, J. Stokman, and B. Vlaar, *Boundary quantum Knizhnik-Zamolodchikov equations and Bethe vectors*, Comm. Math. Phys. **336** (2015) 953–986.
46. N. Reshetikhin, J. Stokman, and B. Vlaar, *Integral solutions to boundary quantum Knizhnik-Zamolodchikov equations*, Adv. in Math. **323** (2018) 486–528.
47. N. Reshetikhin and L.A. Takhtajan, *Deformation quantization of Kähler manifolds*, L.D. Faddeev seminar on Mathematical Physics, Amer. Math. Soc. Transl. Ser. 2, **201**, pp. 257–276, AMS, 2000.
48. N. Reshetikhin and V. G. Turaev. *Invariants of 3-manifolds via link polynomials and quantum groups*, Invent. Math. **103**, no. 1 (1991) 547–597.
49. N. Reshetikhin and A. Varchenko, *Quasiclassical asymptotics of solutions to the KZ equations*, in Geometry, topology, and physics, Conf. Proc. Lect. Notes Geom. Topology, IV, pp. 293–322, Int. Press, Cambridge, MA, 1995.
50. N. Reshetikhin and P. Wiegmann, *Towards the classification of completely integrable quantum field theories (the Bethe-Ansatz associated with Dynkin diagrams and their automorphisms)*, Physics Letters B **189** (1987) 125–131.
51. E.K. Sklyanin, L.A. Takhtajan, and L.D. Faddeev, *Quantum inverse problem method. I.*, Theor. Math. Phys. **40** (1979) 688–706.
52. L.A. Takhtajan and L.D. Faddeev, *The quantum method of the inverse problem and the Heisenberg XYZ model*, Russ. Math. Surv. **34**, no. 5 (1979) 11–68.
53. M.A. Semenov-Tian-Shansky and A.V. Sevostyanov, *Drinfeld-Sokolov reduction for difference operators and deformations of $\mathcal{W}$-algebras. II.*, Comm. Math. Phys. **192** (1998) 631–647.
54. J. Shiraishi, H. Kubo, H. Awata, and S. Odake, *A quantum deformation of the Virasoro algebra and the Macdonald symmetric functions*, Lett. Math. Phys. **38** (1996) 33–51.
55. E.K. Sklyanin, *On an algebra generated by quandratic relations*, Uspekhi Mat. Nauk **40**, no. 2 (1985) 214.
56. F. Smirnov, *Form Factors in Completely Integrable Models of Quantum Field Theory*, Advanced Series in Mathematical Physics **14**, World Scientific, 1992.
57. V. G. Turaev, *The Yang-Baxter equation and invariants of links*, Invent. Math. **92**, no. 3 (1988) 527–554.
58. E. Witten, *Quantum field theory and the Jones polynomial*, Communications in Mathematical Physics **121**, no. 3 (1989) 351–399.

Contents

Examples of Finite-Dimensional Pointed Hopf Algebras in Positive Characteristic

Nicolás Andruskiewitsch, Iván Angiono, and István Heckenberger

To Nikolai Reshetikhin on his 60th birthday with admiration.

Abstract We present new examples of finite-dimensional Nichols algebras over fields of positive characteristic. The corresponding braided vector spaces are not of diagonal type, admit a realization as Yetter-Drinfeld modules over finite abelian groups, and are analogous to braidings over fields of characteristic zero whose Nichols algebras have finite Gelfand-Kirillov dimension.

We obtain new examples of finite-dimensional pointed Hopf algebras by bosonization with group algebras of suitable finite abelian groups.

2010 Mathematics Subject Classification 16T20, 17B37

1 Introduction

1.1 Overview

This is a contribution to the classification of finite-dimensional pointed Hopf algebras in positive characteristic. Beyond the classical theme of cocommutative Hopf algebras—see, for instance, [CF] and the references therein—the problem was considered in several recent works [CLW, HW, NW, NWW1, NWW2, W]. As in various of these papers, the focus of our work is on finite-dimensional Nichols algebras over finite abelian groups. Let $\Bbbk$ be an algebraically closed field

N. Andruskiewitsch (✉) · I. Angiono
FaMAF-CIEM (CONICET), Universidad Nacional de Córdoba, Córdoba, Republica Argentina
e-mail: nicolas.andruskiewitsch@unc.edu.ar; ivan.angiono@unc.edu.ar

I. Heckenberger
Philipps-Universität Marburg, Fachbereich Mathematik und Informatik, Marburg, Germany
e-mail: heckenberger@mathematik.uni-marburg.de

A. Alekseev et al. (eds.), *Representation Theory, Mathematical Physics, and Integrable Systems*, Progress in Mathematics 340,
https://doi.org/10.1007/978-3-030-78148-4_1

of characteristic $p \geq 0$. When $p = 0$, such Nichols algebras are necessarily of diagonal type and their classification was achieved in [H]. When $p > 0$, finite-dimensional Nichols algebras of diagonal type of rank 2 and 3 were classified in [HW, W]. Notice that there are more examples than in characteristic 0: indeed, 1 in the diagonal is no longer excluded.

Example 1.1 Assume that $p > 0$. Given $\theta \in \mathbb{N}$, we set $\mathbb{I}_\theta = \{1, 2, \dots, \theta\}$. Let $\mathbf{q} = (q_{ij})_{i,j \in \mathbb{I}_\theta} \in \Bbbk^{\theta \times \theta}$ be a matrix with $q_{ii} = 1 = q_{ij}q_{ji}$ for all $i \neq j \in \mathbb{I}_\theta$. Let$(V, c)$ be a braided vector space of dimension θ, of diagonal type with matrix $\mathbf{q}$ with respect to a basis $(x_i)_{i \in \mathbb{I}_\theta}$, that is $c : V \otimes V \to V \otimes V$ is given by $c(x_i \otimes x_j) = q_{ij} x_j \otimes x_i$. Then the corresponding Nichols algebra is

$$\mathscr{B}(V) = \mathfrak{s}_{\mathbf{q}}(V) := T(V)/\langle x_i^p,\ i \in \mathbb{I}_\theta, \quad x_i x_j - q_{ij} x_j x_i,\ i < j \in \mathbb{I}_\theta \rangle.$$

Clearly, $\dim \mathfrak{s}_{\mathbf{q}}(V) = p^\theta$.

Furthermore, if $p > 0$, then there are finite-dimensional Nichols algebras over abelian groups that are *not* of diagonal type, a remarkable example being the Jordan plane that has dimension p^2 [CLW] (it gives rise to pointed Hopf algebras of order p^3, see [NW]), in contrast with characteristic 0, where it has Gelfand-Kirillov dimension 2. In fact, Nichols algebras over abelian groups *with finite Gelfand-Kirillov dimension and assuming* $p = 0$ were the subject of the recent papers [AAH1, AAH2]. Succinctly, the main relevant results in loc. cit. are the following:

- It was conjectured in [AAH1] that finite GK-dimensional Nichols algebras of diagonal type have arithmetic root system; the conjecture is true in rank 2 and also in affine Cartan type [AAH2].
- A class of braided vector spaces arising from abelian groups was introduced in [AAH1]; they are decomposable with components being points and blocks. Assuming the validity of the above Conjecture, the finite GK-dimensional Nichols algebras from this class were classified in [AAH1].

Beware that there are finite GK-dimensional Nichols algebras over abelian groups that do not belong to the referred class, see [AAH1, Appendix].

The braided vector spaces in the class alluded to above can be labeled with flourished Dynkin diagrams. The main result of [AAH1] says that the Nichols algebra of a braided vector space in the class has finite Gelfand-Kirillov dimension if and only if its flourished Dynkin diagram is admissible.

From now on we assume that $p > 2$. (The case $p = 2$ has to be treated separately.) In the present paper, we show, adapting arguments from [AAH1], that the Nichols algebras of many braided vector spaces of admissible flourished Dynkin diagrams are finite-dimensional. This result extends Example 1.1 and the Jordan plane [CLW] and is reminiscent of a familiar phenomenon in Lie algebras in positive characteristic. By bosonization we obtain many new examples of finite-dimensional pointed Hopf algebras.

1.2 The Main Result

To describe more precisely our main Theorem we need first to discuss blocks.

For $k < \ell \in \mathbb{N}_0$, we set $\mathbb{I}_{k,\ell} = \{k, k+1, \dots, \ell\}$, $\mathbb{I}_\ell = \mathbb{I}_{1,\ell}$.

A *block* $\mathcal{V}(\epsilon, \ell)$, where $\epsilon \in \Bbbk^\times$ and $\ell \in \mathbb{N}_{\geq 2}$, is a braided vector space with a basis $(x_i)_{i \in \mathbb{I}_\ell}$ such that for $i, j \in \mathbb{I}_\ell$, $1 < j$:

$$c(x_i \otimes x_1) = \epsilon x_1 \otimes x_i, \qquad c(x_i \otimes x_j) = (\epsilon x_j + x_{j-1}) \otimes x_i. \tag{1.1}$$

In characteristic 0, the only Nichols algebras of blocks with finite GKdim are the Jordan plane $\mathscr{B}(\mathcal{V}(1,2))$ and the super Jordan plane $\mathscr{B}(\mathcal{V}(-1,2))$; both have GKdim $= 2$. In our context with $p > 2$, the Jordan plane $\mathscr{B}(\mathcal{V}(1,2))$ has dimension p^2 [CLW]; see Lemma 3.1. Our starting result is that the super Jordan plane $\mathscr{B}(\mathcal{V}(-1,2))$ has dimension $4p^2$, see Proposition 3.2. For simplicity a block $\mathcal{V}(\epsilon, 2)$ of dimension 2 is called an ϵ-block. We also prove that a block $\mathcal{V}(\epsilon, 2)$ has finite-dimensional Nichols algebra only when $\epsilon = \pm 1$, see Proposition 3.3.

The braided vector spaces in this paper belong to the class analogous to the one considered in [AAH1]. Briefly, (V, c) belongs to this class if

$$V = V_1 \oplus \cdots \oplus V_t \oplus V_{t+1} \oplus \cdots \oplus V_\theta, \tag{1.2}$$

$$c(V_i \otimes V_j) = V_j \otimes V_i,\ i, j \in \mathbb{I}_\theta, \tag{1.3}$$

where V_h is a ϵ_h-block, with $\epsilon_h^2 = 1$, for $h \in \mathbb{I}_t$; and $\dim V_i = 1$ with braiding determined by $q_{ii} \in \Bbbk^\times$ (we say that i is a point), $i \in \mathbb{I}_{t+1,\theta}$; the braiding between points i and j is given by $q_{ij} \in \Bbbk^\times$ while the braiding between a point and block, respectively two blocks, should have the form as in (4.1), respectively (6.1). For convenience, we attach to (V, c) a flourished graph $\mathcal{D}$ with θ vertices, those corresponding to a 1-block decorated with $\boxplus$, those to -1-block decorated with $\boxminus$ and the point i with $\overset{q_{ii}}{\circ}$. If $i \neq j$ are points, and there is an edge between them decorated by $\widetilde{q}_{ij} := q_{ij}q_{ji}$ when this is $\neq -1$, or no edge if $\widetilde{q}_{ij} = 1$. If h is a block and j is a point, then there is an edge between h and j decorated either by $\mathscr{G}_{hj}$ if the interaction is weak and $\mathscr{G}_{hj} \neq 0$ is the ghost, cf. (4.2), or by $(-, \mathscr{G}_{hj})$ if the interaction is mild and $\mathscr{G}_{hj}$ is the ghost; but no edge if the interaction is weak and $\mathscr{G}_{hj} = 0$. There are no edges between blocks and we assume that the diagram is connected by a well-known reduction argument.

This class of braided vector spaces together with those of diagonal type does not exhaust that of Yetter-Drinfeld modules arising from abelian groups; there are still those containing a pale block as in [AAH1, Chapter 8]. Synthetically our main result is the following.

Theorem 1.2 *Let V be a braided vector space as in the following list, then* $\dim \mathscr{B}(V) < \infty$.

(a) *V has braiding* (4.1) *and is listed in Table 1, or*
(b) *V has braiding* (5.1) *and is listed in Table 2, or*
(c) *V has braiding* (6.1), *or*
(d) *V has braiding* (7.1) *and is listed in Table 3.*

By bosonization with suitable abelian groups, we get examples of finite-dimensional pointed Hopf algebras in positive characteristic.

Concrete examples of such Hopf algebras are described in Sects. 3.3, 4.5, 5.3, 6.1, and 7.3. We also give a presentation by generators and relations of the Nichols algebras; references to this information and the dimensions are also given in the Tables.

All braided vector spaces in this Theorem belong to the class described above except those in (d) that contain a pale block.

Table 1 Finite-dimensional Nichols algebras of a block and a point

V	diagram	q_{22}	$\mathscr{G}$	$\mathscr{B}(V)$	$\dim K$	$\dim \mathscr{B}(V)$
$\mathfrak{L}(1,\mathscr{G})$	$\boxplus \overset{\mathscr{G}}{\text{———}} \overset{1}{\bullet}$	1	discrete	§4.3.1	$p^{\mathfrak{r}+1}$	$p^{\mathfrak{r}+3}$
$\mathfrak{L}(-1,\mathscr{G})$	$\boxplus \overset{\mathscr{G}}{\text{———}} \overset{-1}{\bullet}$	-1	discrete	§4.3.2	$2^{\mathfrak{r}+1}$	$2^{\mathfrak{r}+1}p^2$
$\mathfrak{L}(\omega,1)$	$\boxplus \overset{1}{\text{———}} \overset{\omega}{\bullet}$	$\in \mathbb{G}'_3$	1	§4.3.5	3^3	3^3p^2
$\mathfrak{L}_-(1,\mathscr{G})$	$\boxminus \overset{\mathscr{G}}{\text{———}} \overset{1}{\bullet}$	1	discrete	§4.3.3	$2^{\frac{\mathfrak{r}}{2}}p^{\frac{\mathfrak{r}}{2}+1}$	$2^{\frac{\mathfrak{r}}{2}+2}p^{\frac{\mathfrak{r}}{2}+3}$
$\mathfrak{L}_-(-1,\mathscr{G})$	$\boxminus \overset{\mathscr{G}}{\text{———}} \overset{-1}{\bullet}$	-1	discrete	§4.3.4	$2^{\frac{\mathfrak{r}}{2}+1}p^{\frac{\mathfrak{r}}{2}}$	$2^{\frac{\mathfrak{r}}{2}+3}p^{\frac{\mathfrak{r}}{2}+2}$
$\mathfrak{C}_1$	$\boxminus \overset{(-1,1)}{\text{———}} \overset{-1}{\bullet}$	-1	1	§4.4	16	$64p^2$

1.3 Contents of the Paper

Section 2 is devoted to preliminaries. The next Sections contain the examples of finite-dimensional Nichols algebras and some realizations over abelian groups; each Section describes a family of braided vector spaces with a certain decomposition as we describe now. In Sect. 3 we compute Nichols algebras of a block. In Sect. 4 we present examples of Nichols algebras corresponding to one block and one point, while in Sect. 5 we consider the case one block and several points. Section 6 is devoted to examples of several blocks and one point. Finally in Sect. 7 we give examples of finite-dimensional Nichols algebras whose braided vector spaces decompose as one pale block and one point.

Table 2 Finite-dimensional Nichols algebras of a block and several points, $\omega \in \mathbb{G}_3'$

V	diagram	$\mathscr{B}(V)$	$dim\mathscr{B}(V)$
$\mathfrak{L}(A_{\theta-1})$, $\theta > 2$	$\boxplus \overset{1}{\text{——}} \overset{-1}{\bullet} \overset{-1}{\text{——}} \overset{-1}{\circ} \dots \overset{-1}{\circ} \overset{-1}{\text{——}} \overset{-1}{\circ}$ $\theta - 1$ vertices	§5.2.7	$p^2 2^6$ $p^2 2^{(\theta-1)(\theta-2)}$
$\mathfrak{L}(A_2, 2)$	$\boxplus \overset{2}{\text{——}} \overset{-1}{\bullet} \overset{-1}{\text{——}} \overset{-1}{\circ}$	§5.2.6	$p^2 2^{12}$
$\mathfrak{L}(A(1\|0)_2; \omega)$	$\boxplus \overset{1}{\text{——}} \overset{-1}{\bullet} \overset{\omega}{\text{——}} \overset{-1}{\circ}$	§5.2.2	$p^2 2^7 3^4$
$\mathfrak{L}(A(1\|0)_1; \omega)$	$\boxplus \overset{1}{\text{——}} \overset{-1}{\bullet} \overset{\omega^2}{\text{——}} \overset{\omega}{\circ}$	§5.2.1	$p^2 2^4 3^2$
$\mathfrak{L}(A(1\|0)_3; \omega)$	$\boxplus \overset{1}{\text{——}} \overset{\omega}{\bullet} \overset{\omega^2}{\text{——}} \overset{-1}{\circ}$	§5.2.3	$p^2 2^7 3^4$
$\mathfrak{L}(A(1\|0)_1; r)$	$\boxplus \overset{1}{\text{——}} \overset{-1}{\bullet} \overset{r^{-1}}{\text{——}} \overset{r}{\circ}$, $r \in \mathbb{G}_N'$, $N > 3$	§5.2.1	$p^2 2^4 N^2$
$\mathfrak{L}(A(2\|0)_1; \omega)$	$\boxplus \overset{1}{\text{——}} \overset{-1}{\bullet} \overset{\omega}{\text{——}} \overset{\omega^2}{\circ} \overset{\omega}{\text{——}} \overset{\omega^2}{\circ}$	§5.2.4	$p^2 2^8 3^9$
$\mathfrak{L}(D(2\|1); \omega)$	$\boxplus \overset{1}{\text{——}} \overset{-1}{\bullet} \overset{\omega}{\text{——}} \overset{\omega^2}{\circ} \overset{\omega^2}{\text{——}} \overset{\omega}{\circ}$	§5.2.5	$p^2 2^8 3^9$

Table 3 Finite-dimensional Nichols algebras of a pale block and a point

V	ϵ	$\widetilde{q}_{12}$	q_{22}	$\mathscr{B}(V)$	dim K	dim $\mathscr{B}(V)$
$\mathfrak{E}_p(q)$	1	1	-1	Sect. 7.1	2^p	$2^p p^2$
$\mathfrak{E}_+(q)$	-1	1	1	Sect. 7.2	$2p$	$2^3 p$
$\mathfrak{E}_-(q)$	-1	1	-1	Sect. 7.2	$2p$	$2^3 p$
$\mathfrak{E}_\star(q)$	-1	-1	-1	Sect. 7.2	$2^4 p^2$	$2^6 p^2$

2 Preliminaries

2.1 Conventions

The $\mathtt{q}$-numbers are the polynomials

$$(n)_{\mathtt{q}} = \sum_{j=0}^{n-1} \mathtt{q}^j, \qquad (n)^!_{\mathtt{q}} = \prod_{j=1}^{n} (j)_{\mathtt{q}}, \qquad \binom{n}{i}_{\mathtt{q}} = \frac{(n)^!_{\mathtt{q}}}{(n-i)^!_{\mathtt{q}} (i)^!_{\mathtt{q}}} \in \mathbb{Z}[\mathtt{q}],$$

$n \in \mathbb{N}$, $0 \le i \le n$. If $q \in \Bbbk$, then $(n)_q$, $(n)^!_q$, $\binom{n}{i}_q$ denote the evaluations of $(n)_{\mathtt{q}}$, $(n)^!_{\mathtt{q}}$, $\binom{n}{i}_{\mathtt{q}}$ at $\mathtt{q} = q$.

Let $\mathbb{G}_N$ be the group of N-th roots of unity, and $\mathbb{G}_N'$ the subset of primitive roots of order N; $\mathbb{G}_\infty = \bigcup_{N \in \mathbb{N}} \mathbb{G}_N$. All the vector spaces, algebras, and tensor products are over $\Bbbk$.

All Hopf algebras have bijective antipode.

2.2 Yetter-Drinfeld Modules

Let Γ be an abelian group. We denote by $\widehat{\Gamma}$ the group of characters of Γ. The category ${}^{\Bbbk\Gamma}_{\Bbbk\Gamma}\mathcal{YD}$ of Yetter-Drinfeld modules over the group algebra $\Bbbk\Gamma$ consists of Γ-graded Γ-modules, the Γ-grading being denoted by $V = \oplus_{g\in\Gamma} V_g$; that is, $hV_g = V_g$ for all $g, h \in \Gamma$. If $g \in \Gamma$ and $\chi \in \widehat{\Gamma}$, then the one-dimensional vector space $\Bbbk_g^{\chi}$, with action and coaction given by g and χ, is in ${}^{H}_{H}\mathcal{YD}$. Let $W \in {}^{\Bbbk\Gamma}_{\Bbbk\Gamma}\mathcal{YD}$ and $(w_i)_{i\in I}$ a basis of W consisting of homogeneous elements of degree g_i, $i \in I$, respectively. Then there are skew-derivations ∂_i, $i \in I$, of $T(W)$ such that for all $x, y \in T(W)$, $i, j \in I$

$$\partial_i(w_j) = \delta_{ij}, \qquad \partial_i(xy) = \partial_i(x)(g_i \cdot y) + x\partial_i(y). \tag{2.1}$$

For a definition of Yetter-Drinfeld modules over arbitrary Hopf algebras we refer, e.g., to [R, 11.6].

2.3 Nichols Algebras

Nichols algebras are graded Hopf algebras $\mathscr{B} = \oplus_{n\geq 0}\mathscr{B}^n$ in ${}^{H}_{H}\mathcal{YD}$ coradically graded and generated in degree one. They are completely determined by $V := \mathscr{B}^1 \in {}^{H}_{H}\mathcal{YD}$ and it is customary to denote $\mathscr{B} = \mathscr{B}(V)$. If $W \in {}^{\Bbbk\Gamma}_{\Bbbk\Gamma}\mathcal{YD}$ as in Sect. 2.2, then the skew-derivations ∂_i induce skew-derivations on $\mathscr{B}(W)$. Moreover, an element $w \in \mathscr{B}^k(W)$, $k \geq 1$, is zero if and only if $\partial_i(w) = 0$ in $\mathscr{B}(W)$ for all $i \in I$. A pre-Nichols algebra of V is a graded Hopf algebra in ${}^{H}_{H}\mathcal{YD}$ generated in degree one, with the one-component isomorphic to V.

Example 2.1 Let V be of dimension 1 with braiding $c = \epsilon\,\mathrm{id}$. Let N be the smallest natural number such that $(N)_\epsilon = 0$. Then $\mathscr{B}(V) = \Bbbk[T]/\langle T^N\rangle$, or $\mathscr{B}(V) = \Bbbk[T]$ if such N does not exist.

A braided vector space V is of diagonal type if there exists a basis $(x_i)_{i\in\mathbb{I}_\theta}$ of V and $\mathbf{q} = (q_{ij})_{i,j\in\mathbb{I}_\theta} \in \Bbbk^{\theta\times\theta}$ such that $q_{ij} \neq 0$ and $c(x_i \otimes x_j) = q_{ij}x_j \otimes x_i$ for all $i, j \in \mathbb{I} = \mathbb{I}_\theta$. Given a braided vector space V of diagonal type with a basis (x_i), we denote in $T(V)$, or $\mathscr{B}(V)$, or any intermediate Hopf algebra,

$$x_{ij} = (\mathrm{ad}_c\, x_i)\, x_j, \qquad x_{i_1 i_2 \dots i_M} = (\mathrm{ad}_c\, x_{i_1})\, x_{i_2 \dots i_M}, \tag{2.2}$$

for $i, j, i_1, \dots, i_M \in \mathbb{I}$, $M \geq 2$. A braided vector space V of diagonal type is of Cartan type if there exists a generalized Cartan matrix $\mathbf{a} = (a_{ij})$ such that $q_{ij}q_{ji} = q_{ii}^{a_{ij}}$ for all $i \neq j$.

Theorem 2.2 *If V is of Cartan type with matrix* $\mathbf{a}$ *that is not finite, then* $\dim \mathscr{B}(V) = \infty$.

Proof The argument in [AAH2, Proposition 3.1] is characteristic-free and applies here because there are infinite real roots in the root system of **a**. □

3 Blocks

We consider braided vector spaces $\mathcal{V}(\epsilon, 2)$ with braiding (1.1), $\epsilon^2 = 1$.

3.1 The Jordan Plane

Here we deal with $\mathcal{V} = \mathcal{V}(1, 2)$. In characteristic 0, $\mathscr{B}(\mathcal{V})$ is the well-known algebra presented by x_1 and x_2 with the relation (3.1). In positive characteristic, $\mathscr{B}(\mathcal{V})$ is a truncated version of that algebra.

Lemma 3.1 ([CLW]) *$\mathscr{B}(\mathcal{V})$ is presented by generators x_1, x_2 and relations*

$$x_2x_1 - x_1x_2 + \frac{1}{2}x_1^2, \tag{3.1}$$

$$x_1^p, \tag{3.2}$$

$$x_2^p. \tag{3.3}$$

Also $\dim \mathscr{B}(\mathcal{V}) = p^2$ *and* $\{x_1^a x_2^b : 0 \le a, b < p\}$ *is a basis of* $\mathscr{B}(\mathcal{V})$. □

In characteristic 2, the relations of $\mathscr{B}(\mathcal{V})$ are different.

Let $\Gamma = \mathbb{Z}/p = \langle g \rangle$. We realize $\mathcal{V}$ in ${}^{\Bbbk\Gamma}_{\Bbbk\Gamma}\mathcal{YD}$ by $g \cdot x_1 = x_1$, $g \cdot x_2 = x_2 + x_1$, $\deg x_i = g$, $i \in \mathbb{I}_2$. Thus the Hopf algebra $\mathscr{B}(\mathcal{V})\#\Bbbk\Gamma$ has dimension p^3.

3.2 The Super Jordan Plane

Let $\mathcal{V} = \mathcal{V}(-1, 2)$ be the braided vector space with basis x_1, x_2 and braiding

$$c(x_i \otimes x_1) = -x_1 \otimes x_i, \quad c(x_i \otimes x_2) = (-x_2 + x_1) \otimes x_i, \quad i \in \mathbb{I}_2. \tag{3.4}$$

Let g be a generator of the cyclic group $\mathbb{Z}$. We realize $\mathcal{V}(-1, 2)$ in ${}^{\Bbbk\mathbb{Z}}_{\Bbbk\mathbb{Z}}\mathcal{YD}$ by $g \cdot x_1 = -x_1$, $g \cdot x_2 = -x_2 + x_1$, $\deg x_i = g$, $i \in \mathbb{I}_2$. As in (2.2),

$$x_{21} = (\operatorname{ad}_c x_2)\, x_1 = x_2x_1 + x_1x_2. \tag{3.5}$$

The Nichols algebra $\mathscr{B}(\mathcal{V}(-1,2)) = T(\mathcal{V}(-1,2))/\mathcal{J}(\mathcal{V}(-1,2))$ (called the super Jordan plane) was studied in [AAH1, 3.3] over fields of characteristic 0. Assuming $p > 2$, the basic features of $\mathscr{B}(\mathcal{V}(-1,2))$ are summarized here:

Proposition 3.2 *The defining ideal $\mathcal{J}(\mathcal{V}(-1,2))$ is generated by*

$$x_1^2, \tag{3.6}$$

$$x_2x_{21} - x_{21}x_2 - x_1x_{21}, \tag{3.7}$$

$$x_{21}^p, \tag{3.8}$$

$$x_2^{2p}. \tag{3.9}$$

The set $B = \{x_1^a x_{21}^b x_2^c : a \in \mathbb{I}_{0,1}, b \in \mathbb{I}_{0,p-1}, c \in \mathbb{I}_{0,2p-1}\}$ is a basis of $\mathscr{B}(\mathcal{V})$ and $\dim \mathscr{B}(\mathcal{V}) = 4p^2$.

Proof Since $\partial_2(x_{21}) = 0$, we have that $\partial_2(x_{21}^n) = 0$ for every $n \in \mathbb{N}$. Also $\partial_1(x_{21}) = x_1$ and $g \cdot x_{21} = x_{21}$. Both (3.6) and (3.7) are 0 in $\mathscr{B}(\mathcal{V})$ being annihilated by ∂_1 and ∂_2, cf. (2.1). From (3.6) and (3.7) we see that in $\mathscr{B}(\mathcal{V})$

$$x_2^2x_1 = x_1(x_2^2 + x_{21}), \tag{3.10}$$

$$x_{21}x_1 = x_1x_2x_1 = x_1x_{21}. \tag{3.11}$$

By the preceding, we have

$$\partial_1(x_{21}^n) = \sum_{1\le i\le n} x_{21}^{i-1}\partial_1(x_{21})x_{21}^{n-i} = nx_1x_{21}^{n-1}.$$

Hence $x_{21}^p = 0$, i.e., (3.8) holds. Next we prove (3.9). Clearly $\partial_1(x_2^n) = 0$ for every $n \in \mathbb{N}$. We observe that

$$g \cdot x_2^2 = (-x_2 + x_1)^2 = x_2^2 - x_{21}, \qquad \partial_2(x_2^2) = g \cdot x_2 + x_2 = x_1.$$

Setting for simplicity $a := x_{21}$ and $b := x_2^2$, we have for any $n \in \mathbb{N}$:

$$\partial_2(x_2^{2n}) = \sum_{1\le i\le n} x_2^{2(i-1)}\partial_2(x_2^2)\left(g \cdot x_2^{2(n-i)}\right) = \sum_{1\le i\le n} b^{i-1}x_1\,(b-a)^{n-i}.$$

By (3.7), (3.10), and (3.11) we have

$$ax_1 = x_1a, \qquad bx_1 = x_1(b+a), \qquad ba = a(a+b), \tag{3.12}$$

hence

$$\partial_2(x_2^{2n}) = x_1 \sum_{1\le i\le n} (b+a)^{i-1}(b-a)^{n-i}.$$

We prove recursively that for all $n \in \mathbb{N}$

$$(b-a)^n = b^n - nab^{n-1}, \quad A_n := \sum_{i\in\mathbb{I}_n} (b+a)^{i-1}(b-a)^{n-i} = nb^{n-1}. \tag{3.13}$$

The case $n = 1$ is evident. We start with the first identity:

$$\begin{aligned}(b-a)^{n+1} &= (b-a)(b-a)^n = (b-a)(b^n - nab^{n-1}) \\ &= b^{n+1} - nbab^{n-1} - ab^n + na^2b^{n-1} = b^{n+1} - (n+1)ab^n\end{aligned}$$

as desired. For the second identity we use the first:

$$A_{n+1} = (b-a)^n + (b+a)A_n = b^n - nab^{n-1} + (b+a)nb^{n-1} = (n+1)b^n.$$

The claim is proved; summarizing we have

$$\partial_2(x_2^{2n}) = nx_1x_2^{2(n-1)}. \tag{3.14}$$

In particular, this implies (3.9).

We now argue as in [AAH1, 3.3.1]. The quotient $\widetilde{\mathcal{B}}$ of $T(\mathcal{V})$ by (3.6)–(3.9) projects onto $\mathscr{B}(\mathcal{V})$ and the subspace I spanned by B is a left ideal of $\widetilde{\mathcal{B}}$, by (3.7), (3.11). Since $1 \in I$, $\widetilde{\mathcal{B}}$ is spanned by B. To prove that $\widetilde{\mathcal{B}} \simeq \mathscr{B}(\mathcal{V})$, we just need to show that B is linearly independent in $\mathscr{B}(\mathcal{V})$. We claim that this is equivalent to prove that $B' = \{x_2^c x_{21}^b x_1^a : a \in \{0, 1\}, b \in \mathbb{I}_{0,p-1}, c \in \mathbb{I}_{0,2p-1}\}$ is linearly independent. Indeed, $\widetilde{\mathcal{B}}$ is spanned by B' since the subspace spanned B' is also a left ideal; if B' is linearly independent, then the dimension of $\widetilde{\mathcal{B}}$ is $4p^2$, so B should be linearly independent and vice versa. Suppose that there is a non-trivial linear combination of elements of B' in $\mathscr{B}(\mathcal{V})$ of minimal degree. As

$$\partial_1(x_2^c x_{21}^b) = b\, x_2^c x_{21}^{b-1} x_1, \qquad \partial_1(x_2^c x_{21}^b x_1) = x_2^c x_{21}^b, \tag{3.15}$$

such linear combination does not have terms with a or b greater than 0. We claim that the elements x_2^c, $c \in \mathbb{I}_{0,2p-1}$, are linearly independent, yielding a contradiction. By homogeneity it is enough to prove that they are $\neq 0$. If c is even this follows from (3.14). If $c = 2n+1$ with $n < p$, then

$$\partial_2(x_2^{2n+1}) = \partial_2(x_2^{2n})g \cdot x_2 + x_2^{2n} = -nx_1x_2^{2n-1} + x_2^{2n}.$$

Again a degree argument gives the desired claim. □

Let $\Gamma = \mathbb{Z}/2p$. We may realize $\mathcal{V}(-1, 2)$ in ${}^{\Bbbk\Gamma}_{\Bbbk\Gamma}\mathcal{YD}$ by the same formulas as above; thus $\mathscr{B}(\mathcal{V})\#\Bbbk\Gamma$ is a pointed Hopf algebra of dimension $8p^3$.

3.3 Realizations

Let H be a Hopf algebra. A *YD-pair* for H is a pair $(g, \chi) \in G(H) \times \operatorname{Hom}_{\text{alg}}(H, \Bbbk)$ such that

$$\chi(h)\, g = \chi(h_{(2)})h_{(1)}\, g\, \mathcal{S}(h_{(3)}), \qquad h \in H. \tag{3.16}$$

Let $\Bbbk_g^\chi$ be a one-dimensional vector space with H-action and H-coaction given by χ and g respectively; then (3.16) says that $\Bbbk_g^\chi \in {}^H_H\mathcal{YD}$.

If $\chi \in \operatorname{Hom}_{\text{alg}}(H, \Bbbk)$, then the space of (χ, χ)-derivations is

$$\operatorname{Der}_{\chi,\chi}(H, \Bbbk) = \{\eta \in H^* : \eta(h\ell) = \chi(h)\eta(\ell) + \chi(\ell)\eta(h)\, \forall h, \ell \in H\}.$$

A *YD-triple* for H is a collection (g, χ, η) where (g, χ) is a YD-pair for H, $\eta \in \operatorname{Der}_{\chi,\chi}(H, \Bbbk)$, $\eta(g) = 1$ and

$$\eta(h)g_1 = \eta(h\_2)h\_1g_2\mathcal{S}(h\_3), \qquad h \in H. \tag{3.17}$$

Given a YD-triple (g, χ, η) we define $\mathcal{V}_g(\chi, \eta) \in {}^H_H\mathcal{YD}$ as the vector space with a basis $(x_i)_{i\in\mathbb{I}_2}$, whose H-action and H-coaction are given by

$$h \cdot x_1 = \chi(h)x_1, \quad h \cdot x_2 = \chi(h)x_2 + \eta(h)x_1, \quad \delta(x_i) = g \otimes x_i, \quad h \in H,\ i \in \mathbb{I}_2;$$

the compatibility is granted by (3.16), (3.17). As a braided vector space, $\mathcal{V}_g(\chi, \eta) \simeq \mathcal{V}(\epsilon, 2)$, $\epsilon = \chi(g)$.

Consequently, if H is finite-dimensional and $\epsilon^2 = 1$, then $\mathscr{B}(\mathcal{V}_g(\chi, \eta))\#H$ is a Hopf algebra satisfying

$$\dim\left(\mathscr{B}(\mathcal{V}_g(\chi, \eta))\#H\right) = \begin{cases} p^2 \dim H, & \text{when } \epsilon = 1, \\ 4p^2 \dim H, & \text{when } \epsilon = -1. \end{cases} \tag{3.18}$$

3.4 Exhaustion in Rank 2

We recall some facts from [AAH1, §3.4].

Let H be a Hopf algebra with bijective antipode and $V \in {}^H_H\mathcal{YD}$. Let $0 = V_0 \subsetneq V_1 \cdots \subsetneq V_d = V$ be a flag of Yetter-Drinfeld submodules with $\dim V_i = \dim V_{i-1} + 1$ for all i. Then $V^{\text{diag}} := \operatorname{gr} V$ is of diagonal type. If $\mathscr{B}$ is a pre-Nichols algebra of V, then it is a graded filtered Hopf in ${}^{H}_{H}\mathcal{YD}$ and $\mathcal{B}^{\text{diag}} := \operatorname{gr} \mathscr{B}$ is a pre-Nichols algebra of V^{diag}.

Proposition 3.3 *Let $\epsilon \in \Bbbk^{\times}$. If $\dim \mathscr{B}(\mathcal{V}(\epsilon, 2)) < \infty$, then $\epsilon^2 = 1$.*

Proof Let $\mathcal{V} = \mathcal{V}(\epsilon, 2)$; it has a flag as above and $\mathcal{V}^{\text{diag}}$ is the braided vector space of diagonal type with matrix $(q_{ij})_{i,j\in\mathbb{I}_2}$, $q_{ij} = \epsilon$ for all $i, j \in \mathbb{I}_2$. Hence

$$\dim \mathscr{B}(\mathcal{V}^{\text{diag}}) \leq \dim \mathscr{B}(\mathcal{V}(\epsilon, 2)). \tag{3.19}$$

Step 1 If $\epsilon \notin \mathbb{G}_\infty$, then $\dim \mathscr{B}(\mathcal{V}(\epsilon, 2)) = \infty$.

Proof Here $\dim \mathscr{B}(\mathcal{V}^{\text{diag}}) = \infty$ by Example 2.1 and (3.19) applies. □

Step 2 If $\epsilon \in \mathbb{G}'_N$, $N \geq 4$, then $\dim \mathscr{B}(\mathcal{V}(\epsilon, \ell)) = \infty$ for all $\ell \geq 2$.

Proof Here $\mathcal{V}^{\text{diag}}$ is of Cartan type with Cartan matrix $\begin{pmatrix} 2 & 2-N \\ 2-N & 2 \end{pmatrix}$. Thus Theorem 2.2 and (3.19) apply. □

Step 3 Let $\epsilon \in \mathbb{G}'_3$. Then $\dim \mathscr{B}(\mathcal{V}(\epsilon, 2)) = \infty$.

Proof The proof of [AAH1, §3.5 – Step 3] holds verbatim. □

The Proposition is proved. □

4 One Block and One Point

4.1 The Setting and the Statement

Let $(q_{ij})_{1\leq i,j\leq 2}$ be a matrix of invertible scalars and $a \in \Bbbk$. We assume that $\epsilon := q_{11}$ satisfies $\epsilon^2 = 1$. Let V be a braided vector space with a basis $(x_i)_{i\in\mathbb{I}_3}$ and a braiding given by

$$(c(x_i \otimes x_j))_{i,j\in\mathbb{I}_3} = \begin{pmatrix} \epsilon x_1 \otimes x_1 & (\epsilon x_2 + x_1) \otimes x_1 & q_{12} x_3 \otimes x_1 \\ \epsilon x_1 \otimes x_2 & (\epsilon x_2 + x_1) \otimes x_2 & q_{12} x_3 \otimes x_2 \\ q_{21} x_1 \otimes x_3 & q_{21}(x_2 + a x_1) \otimes x_3 & q_{22} x_3 \otimes x_3 \end{pmatrix}. \tag{4.1}$$

Let $V_1 = \langle x_1, x_2 \rangle$ (the block) and $V_2 = \langle x_3 \rangle$ (the point). Let $\Gamma = \mathbb{Z}^2$ with canonical basis g_1, g_2. We realize (V, c) in ${}^{\Bbbk\Gamma}_{\Bbbk\Gamma}\mathcal{YD}$ as $\mathcal{V}_{g_1}(\chi_1, \eta) \oplus \Bbbk^{\chi_2}_{g_2}$ with suitable χ_1, χ_2, and η, where $V_1 = \mathcal{V}_{g_1}(\chi_1, \eta)$, while $V_2 = \Bbbk^{\chi_2}_{g_2}$. Thus $V_1 \simeq \mathcal{V}(\epsilon, 2)$; thus we use the notations and results from Sect. 3.2.

The *interaction* between the block and the point is $q_{12}q_{21}$; it is

weak if $q_{12}q_{21} = 1$, mild if $q_{12}q_{21} = -1$, strong if $q_{12}q_{21} \notin \{\pm 1\}$.

In characteristic 0, we introduced a normalized version of a called the ghost, which is discrete when it belongs to $\mathbb{N}$. In our context, $p > 2$, we need a variant of this notion. First we say that V has *discrete ghost* if $a \in \mathbb{F}_p^\times$. When this is the case, we pick a representative $\mathtt{r} \in \mathbb{Z}$ of $2a$ by imposing

$$\mathtt{r} \in \begin{cases} \{1-p, \dots, -1\}, & \epsilon = 1, \\ \{1, \dots, 2p-1\} \cap 2\mathbb{Z}, & \epsilon = -1; \end{cases} \qquad \text{set } \mathscr{G} := \begin{cases} -\mathtt{r}, & \epsilon = 1, \\ \mathtt{r}, & \epsilon = -1. \end{cases} \tag{4.2}$$

Then $\mathscr{G}$ is called the *ghost*. In this Section we consider the following braided vectors spaces with braiding (4.1), where the ghost is discrete and $q_{22} \in \mathbb{G}_\infty$:

$\mathfrak{L}(q_{22}, \mathscr{G})$:	weak interaction,	$\epsilon = 1$;
$\mathfrak{L}_-(q_{22}, \mathscr{G})$:	weak interaction,	$\epsilon = -1$;
$\mathfrak{C}_1$:	mild interaction,	$\epsilon = q_{22} = -1, \quad \mathscr{G} = 1$.

In this Section, we shall prove part (a) of Theorem 1.2.

Theorem 4.1 *Let V be a braided vector space with braiding* (4.1). *If V is as in Table 1, then* $\dim \mathscr{B}(V) < \infty$.

To prove the Theorem, we consider $K = \mathscr{B}(V)^{\operatorname{co} \mathscr{B}(V_1)}$. By [HS, Proposition 8.6], $\mathscr{B}(V) \simeq K \# \mathscr{B}(V_1)$ and K is the Nichols algebra of

$$K^1 = \operatorname{ad}_c \mathscr{B}(V_1)(V_2). \tag{4.3}$$

Now $K^1 \in {}^{\mathscr{B}(V_1)\#\Bbbk\Gamma}_{\mathscr{B}(V_1)\#\Bbbk\Gamma}\mathcal{YD}$ with the adjoint action and the coaction given by

$$\delta = (\pi_{\mathscr{B}(V_1)\#\Bbbk\Gamma} \otimes \operatorname{id})\Delta_{\mathscr{B}(V)\#\Bbbk\Gamma}. \tag{4.4}$$

In order to describe K^1, we set

$$z_n := (\operatorname{ad}_c x_2)^n x_3, \qquad n \in \mathbb{N}_0. \tag{4.5}$$

4.2 Weak Interaction

Here $q_{12}q_{21} = 1$. In general,

$$c^2_{|V_1 \otimes V_2} = \operatorname{id} \iff q_{12}q_{21} = 1 \text{ and } a = 0. \tag{4.6}$$

If $a = 0$, then

$$\mathscr{B}(V) \simeq \mathscr{B}(\mathcal{V}(\epsilon, 2))\underline{\otimes}\mathscr{B}(\Bbbk x_3). \tag{4.7}$$

Here $\underline{\otimes}$ denotes the *braided* tensor product of Hopf algebras (the structure of Hopf algebra in ${}^{\Bbbk G}_{\Bbbk G}\mathcal{YD}$).

From now on we assume that the ghost is discrete, in particular $\neq 0$. We follow the exposition in [AAH1, §4.2].

Lemma 4.2 *The following formulae hold in* $\mathscr{B}(V)$ *for all* $n \in \mathbb{N}_0$*:*

$$\begin{aligned} g_1 \cdot z_n &= \epsilon^n q_{12} z_n, & x_1 z_n &= \epsilon^n q_{12} z_n x_1, & x_{21}^n x_2 &= (n\epsilon x_1 + x_2) x_{21}^n, \\ g_2 \cdot z_n &= q_{21}^n q_{22} z_n, & x_{21} z_n &= q_{12}^2 z_n x_{21}, & x_2 z_n &= \epsilon^n q_{12} z_n x_2 + z_{n+1}. \end{aligned} \tag{4.8}$$

Proof The proof of [AAH1, Lemma 4.2.1] is valid in any characteristic. □

Let $(\mu_n)_{n\in\mathbb{N}_0}$ be the family of elements of $\Bbbk$ defined recursively by

$$\mu_0 = 1, \quad \mu_{2k+1} = -(a + k\epsilon)\mu_{2k}, \quad \mu_{2k} = (a + k + \epsilon(a + k - 1))\mu_{2k-1}.$$

This can be reformulated as

$$\mu_{n+1} = \begin{cases} (2a + n)\mu_n & \text{if } n \text{ is odd}, \\ -\frac{2a+n}{2}\mu_n & \text{if } n \text{ is even}, \end{cases} \qquad \text{when } \epsilon = 1; \tag{4.9}$$

$$\mu_{n+1} = \begin{cases} \mu_n & \text{if } n \text{ is odd}, \\ -(a - \frac{n}{2})\mu_n & \text{if } n \text{ is even}, \end{cases} \qquad \text{when } \epsilon = -1. \tag{4.10}$$

Thus $\mu_n = 0 \iff n > |\mathtt{r}|$.

Lemma 4.3 *For all* $k \in \{1, \dots, p-1\}$, $\partial_1(z_k) = \partial_2(z_k) = 0$,

$$\partial_3(z_{2k}) = \mu_{2k} x_{21}^k, \qquad \partial_3(z_{2k+1}) = \mu_{2k+1} x_1 x_{21}^k. \tag{4.11}$$

Therefore, $z_n = 0 \iff n > |\mathtt{r}|$.

Proof The proof of [AAH1, Lemma 4.2.2] is valid in any characteristic, taking into the account the conventions on $\mathtt{r}$. □

If $\epsilon = 1$, then (4.11) says that

$$\partial_3(z_{2k}) = \frac{(-1)^k}{2^k}\mu_{2k} x_1^{2k}, \qquad \partial_3(z_{2k+1}) = \frac{(-1)^k}{2^k}\mu_{2k+1} x_1^{2k+1}. \tag{4.12}$$

Recall that $K = \mathscr{B}(V)^{\operatorname{co}\mathscr{B}(V_1)} \simeq \mathscr{B}(K^1)$, $K^1 = \operatorname{ad}\mathscr{B}(V_1)(V_2)$.

Lemma 4.4 *The family* $(z_n)_{0\leq n\leq |\mathfrak{r}|}$ *is a basis of* K^1.

Proof The family is linearly independent, because the z_n's are homogeneous of distinct degrees and are $\neq 0$ by (4.11). We have for all $n \in \mathbb{N}_0$

$$\operatorname{ad}_c x_1(z_n) = x_1 z_n - g_1 \cdot z_n x_1 \overset{(4.8)}{=} \epsilon^n q_{12} z_n x_1 - \epsilon^n q_{12} z_n x_1 = 0, \tag{4.13}$$

$$\operatorname{ad}_c x_{21}(z_n) = \operatorname{ad}_c x_2 \operatorname{ad}_c x_1(z_n) - \epsilon \operatorname{ad}_c x_1 \operatorname{ad}_c x_2(z_n) = 0. \tag{4.14}$$

The Lemma follows. □

If $\epsilon = 1$, then we define recursively $\nu_{k,n}$ as follows: $\nu_{n,n} = 1$,

$$\nu_{0,n+1} = -\big(\frac{n}{2} + a\big)\nu_{0,n}, \quad \nu_{k,n+1} = \nu_{k-1,n} - \left(\frac{n+k}{2} + a\right)\nu_{k,n}, \quad 1 \leq k \leq n.$$

Lemma 4.5 *The coaction* (4.4) *on* z_n, $0 \leq n \leq |\mathfrak{r}|$, *is given by* (4.15), *when* $\epsilon = 1$, *and by* (4.16), (4.17), *when* $\epsilon = -1$:

$$\delta(z_n) = \sum_{k=0}^{n} \nu_{k,n}\, x_1^{n-k} g_1^k g_2 \otimes z_k. \tag{4.15}$$

$$\delta(z_{2n}) = \sum_{k=1}^{n} k\binom{n}{k}\mu_{k,n}\, x_1 x_{21}^{n-k} g_1^{2k-1} g_2 \otimes z_{2k-1} \tag{4.16}$$
$$+ \sum_{k=0}^{n}\binom{n}{k}\mu_{k,n} x_{21}^{n-k} g_1^{2k} g_2 \otimes z_{2k},$$

$$\delta(z_{2n+1}) = \sum_{k=0}^{n}\binom{n}{k}\mu_{k,n+1}\, x_1 x_{21}^{n-k} g_1^{2k} g_2 \otimes z_{2k} \tag{4.17}$$
$$+ \sum_{k=0}^{n}\binom{n}{k}\mu_{k+1,n+1} x_{21}^{n-k} g_1^{2k+1} g_2 \otimes z_{2k+1}.$$

Proof The proof of [AAH1, Lemma 4.2.1] can be adapted since $\binom{n}{k} \neq 0$ by assumption. □

We are ready to prove the finite-dimensionality in the case of weak interaction. We claim that the braided vector space K^1 is of diagonal type with braiding matrix

$$(\mathbf{p}_{ij})_{0\leq i,j\leq |\mathfrak{r}|} = (\epsilon^{ij} q_{12}^i q_{21}^j q_{22})_{0\leq i,j\leq |\mathfrak{r}|}.$$

Hence, the corresponding generalized Dynkin diagram has labels

$$\mathbf{p}_{ii} = \epsilon^i q_{22}, \qquad \mathbf{p}_{ij}\mathbf{p}_{ji} = q_{22}^2, \qquad i \neq j \in \mathbb{I}_{0,|\mathfrak{r}|}.$$

Indeed, by Lemma 4.4 it is enough to compute

$$c(z_i \otimes z_j) = g_1^i g_2 \cdot z_j \otimes z_i = \epsilon^{ij} q_{12}^i q_{21}^j q_{22} z_j \otimes z_i,$$

by Lemmas 4.5 and 4.2, (4.13) and (4.14). We proceed then case by case.

Case 1 $q_{22}^2 = 1$.

Here the Dynkin diagram of K^1 is totally disconnected with vertices $i \in \mathbb{I}_{0,|\mathfrak{r}|}$ labeled with $\epsilon^i q_{22}$. The vertices with label 1, respectively -1, contribute with p, respectively 2, to $\dim \mathscr{B}(K^1)$.

Case 2 $\epsilon = 1$, $q_{22} \in \mathbb{G}_3'$, $|\mathfrak{r}| = 1$ (provided that $p > 3$).

The Dynkin diagram is of Cartan type A_2, so $\dim \mathscr{B}(K^1) < \infty$.

4.3 The Presentation by Generators and Relations

We still assume that the interaction is weak. We start by some general Remarks that are proved exactly as in [AAH1, §4.3].

Remark 4.6 Let

$$y_{2k} = x_{21}^k, \qquad y_{2k+1} = x_1 x_{21}^k, \qquad k \in \mathbb{N}_0.$$

By Lemma 4.2

$$\partial_3(z_t) = \mu_t y_t, \qquad z_t y_n = \epsilon^{nt} q_{21}^n y_n z_t, \qquad t, n \in \mathbb{N}_0. \tag{4.18}$$

Lemma 4.7 *Assume that* $\epsilon^2 = q_{22}^2 = 1$. *In* $\mathscr{B}(\mathfrak{L}(q_{22}, \mathscr{G}))$, *or correspondingly* $\mathscr{B}_-(\mathfrak{L}(q_{22}, \mathscr{G}))$,

$$z_{|\mathfrak{r}|+1} = 0, \tag{4.19}$$

$$z_t z_{t+1} = q_{21} q_{22} z_{t+1} z_t \qquad t \in \mathbb{N}_0, t < |\mathfrak{r}|, \tag{4.20}$$

$$z_t^2 = 0 \qquad t \in \mathbb{N}_0,\ \epsilon^t q_{22} = -1. \tag{4.21}$$

$$\partial_3(z_t^{n+1}) = \mu_t q_{21}^{nt} q_{22}^n n\, y_t z_t^n, \qquad n, t \in \mathbb{N}_0, \epsilon^t q_{22} = 1. \tag{4.22}$$

Lemma 4.8 *Let $\mathscr{B}$ be a quotient algebra of $T(V)$. Assume that $x_1x_3 = q_{12}x_3x_1$, and either*
(a) (3.1), *or else*
(b) (3.7), $x_{21}x_3 = q_{12}^2 x_3x_{21}$

hold in $\mathscr{B}$. Then for all $n \in \mathbb{N}_0$,

$$x_1 z_n = \epsilon^n q_{12} z_n x_1 \tag{4.23}$$

$$x_{21} z_n = q_{12}^2 z_n x_{21}. \tag{4.24}$$

Lemma 4.9 *Let $\mathscr{B}$ be a quotient algebra of $T(V)$, $\epsilon^2 = q_{22}^2 = 1$.*

(i) *Assume that* (4.20) *and* (4.21) *hold in $\mathscr{B}$. Then for $0 \le t < k \le |\mathtt{r}|$,*

$$z_t z_k = \epsilon^{tk} q_{21}^{k-t} q_{22} z_k z_t. \tag{4.25}$$

(ii) *Assume that $z_t^2 = 0$ in $\mathscr{B}$ for $t \in \mathbb{N}_0$ such that $\epsilon^t q_{22} = -1$. Then $z_t z_{t+1} = q_{21}q_{22}z_{t+1}z_t$ in $\mathscr{B}$.*

In other words, **(ii)** says that (4.21) for a specific t implies (4.20) for t.

4.3.1 The Nichols Algebra $\mathscr{B}(\mathfrak{L}(1, \mathscr{G}))$

Proposition 4.10 *Let $\mathscr{G} \in \mathbb{I}_{p-1}$. The algebra $\mathscr{B}(\mathfrak{L}(1, \mathscr{G}))$ is presented by generators x_1, x_2, x_3 and relations* (3.1)–(3.3), *together with*

$$x_1x_3 = q_{12}\, x_3x_1, \tag{4.26}$$

$$z_{1+\mathscr{G}} = 0, \tag{4.27}$$

$$z_t z_{t+1} = q_{12}^{-1}\, z_{t+1} z_t, \qquad 0 \le t < \mathscr{G}, \tag{4.28}$$

$$z_t^p = 0, \qquad 0 \le t \le \mathscr{G}. \tag{4.29}$$

The dimension of $\mathscr{B}(\mathfrak{L}(1, \mathscr{G}))$ is $p^{\mathscr{G}+3}$, since it has a PBW-basis

$$B = \{x_1^{m_1} x_2^{m_2} z_{\mathscr{G}}^{n_{\mathscr{G}}} \dots z_1^{n_1} z_0^{n_0} : m_i, n_j \in \mathbb{I}_{0,p}\}.$$

□

4.3.2 The Nichols Algebra $\mathscr{B}(\mathfrak{L}(-1, \mathscr{G}))$

Proposition 4.11 *Let $\mathscr{G} \in \mathbb{I}_{p-1}$. The algebra $\mathscr{B}(\mathfrak{L}(-1, \mathscr{G}))$ is presented by generators x_1, x_2, x_3 and relations* (3.1)–(3.3), (4.26), (4.27) *and*

$$z_t^2 = 0, \qquad 0 \le t \le \mathscr{G}. \tag{4.30}$$

The dimension of $\mathscr{B}(\mathfrak{L}(1, \mathscr{G}))$ is $2^{\mathscr{G}+1}p^2$, since it has a PBW-basis

$$B = \{x_1^{m_1} x_2^{m_2} z_{\mathscr{G}}^{n_{\mathscr{G}}} \dots z_1^{n_1} z_0^{n_0} : n_i \in \{0, 1\}, m_j \in \mathbb{I}_{0,p-1}\}.$$

□

4.3.3 The Nichols Algebra $\mathscr{B}(\mathfrak{L}_-(1, \mathscr{G}))$

Proposition 4.12 *Let $\mathscr{G} \in \mathbb{I}_{2p-1} \cap 2\mathbb{Z}$. The algebra $\mathscr{B}(\mathfrak{L}_-(1, \mathscr{G}))$ is presented by generators x_1, x_2, x_3 and relations* (3.6)–(3.9), (4.26) *and*

$$z_{1+\mathscr{G}} = 0, \tag{4.31}$$

$$x_{21} z_0 = q_{12}^2 z_0 x_{21}, \tag{4.32}$$

$$z_{2k+1}^2 = 0, \qquad 0 \le k < \mathscr{G}/2, \tag{4.33}$$

$$z_{2k} z_{2k+1} = q_{12}^{-1} z_{2k+1} z_{2k}, \qquad 0 \le k < \mathscr{G}/2. \tag{4.34}$$

The dimension of $\mathscr{B}(\mathfrak{L}(1, \mathscr{G}))$ is $2^{\frac{\mathscr{G}}{2}+2} p^{\frac{\mathscr{G}}{2}+3}$, since it has a PBW-basis

$$B = \{x_1^{m_1} x_{21}^{m_2} x_2^{m_3} z_{\mathscr{G}}^{n_{\mathscr{G}}} \dots z_1^{n_1} z_0^{n_0} : m_1, n_{2k+1} \in \{0, 1\}, \\ m_3 \in \mathbb{I}_{0,2p-1}, m_2, n_{2k} \in \mathbb{I}_{0,p-1}\}.$$

□

4.3.4 The Nichols Algebra $\mathscr{B}(\mathfrak{L}_-(-1, \mathscr{G}))$

Proposition 4.13 *Let $\mathscr{G} \in \mathbb{I}_{2p-1} \cap 2\mathbb{Z}$. The algebra $\mathscr{B}(\mathfrak{L}_-(-1, \mathscr{G}))$ is presented by generators x_1, x_2, x_3 and relations* (3.6)–(3.9), (4.26), (4.28) *and*

$$z_{2k}^2 = 0, \qquad 0 \le k \le \mathscr{G}/2, \tag{4.35}$$

$$z_{2k-1} z_{2k} = -q_{12}^{-1} z_{2k} z_{2k-1}, \qquad 0 < k \le \mathscr{G}/2. \tag{4.36}$$

The dimension of $\mathscr{B}(\mathfrak{L}(1, \mathscr{G}))$ *is* $2^{\frac{\mathscr{G}}{2}+3}p^{\frac{\mathscr{G}}{2}+2}$, *since it has a PBW-basis*

$$B = \{x_1^{m_1} x_{21}^{m_2} x_2^{m_3} z_{\mathscr{G}}^{n_{\mathscr{G}}} \dots z_1^{n_1} z_0^{n_0} : m_1, n_{2k} \in \{0, 1\},$$
$$m_3 \in \mathbb{I}_{0,2p-1}, m_2, n_{2k-1} \in \mathbb{N}_0\}.$$

□

4.3.5 The Nichols Algebra $\mathscr{B}(\mathfrak{L}(\omega, 1))$

Proposition 4.14 *Let* $\omega \in \mathbb{G}_3'$. *The algebra* $\mathscr{B}(\mathfrak{L}(\omega, 1))$ *is presented by generators* x_1, x_2, x_3 *and relations* (3.1)–(3.3), (4.26),

$$z_2 = 0, \tag{4.37}$$

$$z_0^3 = 0, \tag{4.38}$$

$$z_1^3 = 0, \tag{4.39}$$

$$z_{1,0}^3 = 0. \tag{4.40}$$

The dimension of $\mathscr{B}(\mathfrak{L}(\omega, 1))$ *is* $3^3 p^2$, *since it has a PBW-basis*

$$B = \{x_1^{m_1} x_2^{m_2} z_1^{n_1} z_{1,0}^{n_2} z_0^{n_3} : m_i \in \mathbb{I}_{0,p-1}, n_j \in \mathbb{I}_{0,2}\}.$$

□

4.4 Mild Interaction

We assume in this Subsection that $q_{12}q_{21} = -1 = \epsilon$, $a = 1$, $q_{22} = -1$. The corresponding braided vector space is denoted $\mathfrak{C}_1$, as above. We proceed as above but now the elements $z_n = (\mathrm{ad}_c\, x_2)^n x_3$ are not enough to describe K^1 and we need $f_n = \mathrm{ad}_c\, x_1(z_n)$, $n = 0, 1$. Then

$$\begin{aligned} x_1 z_0 &= f_0 + q_{12} z_0 x_1, \\ x_1 z_1 &= f_1 - q_{12} z_1 x_1 + q_{12} f_0 x_1, \\ x_2 z_0 &= z_1 + q_{12} z_0 x_2. \end{aligned} \tag{4.41}$$

Proposition 4.15 *The Nichols algebra $\mathscr{B}(\mathfrak{C}_1)$ is presented by generators x_1, x_2, x_3 and relations* (3.6)–(3.9),

$$\begin{aligned} x_2z_1 + q_{12}z_1x_2 &= \frac{1}{2}f_1 + q_{12}f_0x_2, \\ x_2f_1 = q_{12}f_1x_2, &\quad x_2f_0 + q_{12}f_0x_2 = -f_1, \\ z_0^2 = 0, \quad f_0^2 = 0, &\quad z_1^2 = 0, \qquad f_1^2 = 0. \end{aligned} \tag{4.42}$$

The dimension of $\mathscr{B}(\mathfrak{C}_1)$ is $64p^2$, since it has a PBW-basis

$$B = \{x_1^{m_1}x_{21}^{m_2}x_2^{m_3}f_1^{n_1}f_0^{n_2}z_1^{n_3}z_0^{n_4} : m_1, n_i \in \{0, 1\}, m_2, m_3 \in \mathbb{I}_p\}.$$

□

4.5 Realizations

Let H be a Hopf algebra, (g_1, χ_1, η) a YD-triple and (g_2, χ_2) a YD-pair for H, see Sect. 3.3. Let (V, c) be a braided vector space with braiding (4.1). Then $\mathcal{V}_{g_1}(\chi_1, \eta) \oplus \Bbbk_{g_2}^{\chi_2} \in {}^H_H\mathcal{YD}$ is a *principal realization* of (V, c) over H if

$$q_{ij} = \chi_j(g_i), \qquad i, j \in \mathbb{I}_2; \qquad a = q_{21}^{-1}\eta(g_2).$$

Thus $(V, c) \simeq \mathcal{V}_{g_1}(\chi_1, \eta) \oplus \Bbbk_{g_2}^{\chi_2}$ as braided vector space. Hence, if H is finite-dimensional and (V, c) is as in Table 1, then $\mathscr{B}(\mathcal{V}_{g_1}(\chi_1, \eta) \oplus \Bbbk_{g_2}^{\chi_2})\#H$ is a finite-dimensional Hopf algebra. Examples of finite-dimensional pointed Hopf algebras $A = \mathscr{B}(\mathcal{V}_{g_1}(\chi_1, \eta) \oplus \Bbbk_{g_2}^{\chi_2})\#\Bbbk\Gamma$ like this are listed in Table 4. In all cases Γ is a product of two cyclic groups, $g_1 = (1, 0)$, $g_2 = (0, 1)$ and $\chi_j(g_i) = 1$ if $i \neq j$; hence it remains to fix the value of q_{12}.

5 One Block and Several Points

5.1 The Setting and the Main Result

Let $\theta \in \mathbb{N}_{\geq 3}$, $\mathbb{I}_{2,\theta} = \mathbb{I}_\theta - \{1\}$, $\mathbb{I}_\theta^\dagger = \mathbb{I}_\theta \cup \{\frac{3}{2}\}$. Let $\lfloor i \rfloor$ be the largest integer $\leq i$. We start from the data

$$(q_{ij})_{i,j\in\mathbb{I}_\theta} \in (\Bbbk^\times)^{\theta\times\theta}, \quad q_{11}^2 = 1; \qquad (a_2, \dots, a_\theta) \in \Bbbk^{\mathbb{I}_{2,\theta}}.$$

Table 4 Pointed Hopf algebras K from a block and a point

V	diagram	Γ	q_{12}	$\dim A$
$\mathfrak{L}(1,\mathscr{G})$	$\boxplus \overset{\mathscr{G}}{\text{———}} \overset{1}{\bullet}$	$\mathbb{Z}/p \times \mathbb{Z}/p$	1	$p^{\mathtt{r}+5}$
$\mathfrak{L}(-1,\mathscr{G})$	$\boxplus \overset{\mathscr{G}}{\text{———}} \overset{-1}{\bullet}$	$\mathbb{Z}/p \times \mathbb{Z}/2p$	1	$2^{\mathtt{r}+2}p^4$
$\mathfrak{L}(\omega,1)$	$\boxplus \overset{1}{\text{———}} \overset{\omega}{\bullet}$	$\mathbb{Z}/p \times \mathbb{Z}/3p$	1	3^4p^4
$\mathfrak{L}_-(1,\mathscr{G})$	$\boxminus \overset{\mathscr{G}}{\text{———}} \overset{1}{\bullet}$	$\mathbb{Z}/2p \times \mathbb{Z}/p$	1	$2^{\frac{\mathtt{r}}{2}+3}p^{\frac{\mathtt{r}}{2}+5}$
$\mathfrak{L}_-(-1,\mathscr{G})$	$\boxminus \overset{\mathscr{G}}{\text{———}} \overset{-1}{\bullet}$	$\mathbb{Z}/2p \times \mathbb{Z}/2p$	± 1	$2^{\frac{\mathtt{r}}{2}+5}p^{\frac{\mathtt{r}}{2}+4}$
$\mathfrak{C}_1$	$\boxminus \overset{(-1,1)}{\text{———}} \overset{-1}{\bullet}$	$\mathbb{Z}/2p \times \mathbb{Z}/2p$	± 1	$256p^4$

We assume that $q_{11} = 1 =: a_1$. Let (V, c) be the braided vector space of dimension $\theta + 1$, with a basis $(x_i)_{i\in\mathbb{I}_\theta^\dagger}$ and braiding given by

$$c(x_i \otimes x_j) = \begin{cases} q_{\lfloor i\rfloor j} x_j \otimes x_i, & i \in \mathbb{I}_\theta^\dagger,\ j \in \mathbb{I}_\theta; \\ q_{\lfloor i\rfloor 1}(x_{\frac{3}{2}} + a_{\lfloor i\rfloor} x_1) \otimes x_i, & i \in \mathbb{I}_\theta^\dagger,\ j = \frac{3}{2}. \end{cases} \tag{5.1}$$

We say that the block and the points have *discrete ghost* if $a_j \in \mathbb{F}_p^{\mathbb{I}_{2,\theta}}$, $(a_j) \neq 0$. When this is the case, we pick the representative $\mathtt{r}_j \in \mathbb{Z}$ of $2a_j$ by imposing $\mathtt{r}_j \in \{1-p, \dots, -1, 0\}$, and set $\mathscr{G}_j = -\mathtt{r}_j$. The *ghost* between the block and the points is the vector $\mathscr{G} = (\mathscr{G}_j)_{j\in\mathbb{I}_{2,\theta}}$ given by

$$\mathscr{G} = -(\mathtt{r}_j)_{j\in\mathbb{I}_{2,\theta}} \in \mathbb{N}_0^{\mathbb{I}_{2,\theta}}. \tag{5.2}$$

The braided subspace V_1 spanned by $x_1, x_{\frac{3}{2}}$ is $\simeq \mathcal{V}(1,2)$, while V_{diag} spanned by $(x_i)_{i\in\mathbb{I}_{2,\theta}}$ is of diagonal type. Obviously,

$$V = V_1 \oplus V_{\text{diag}}. \tag{5.3}$$

Let $\mathcal{X}$ be the set of connected components of the Dynkin diagram of the matrix $\mathbf{q} = (q_{ij})_{i,j\in\mathbb{I}_{2,\theta}}$. If $J \in \mathcal{X}$, then we set $J' = \mathbb{I}_{2,\theta} - J$,

$$V_J = \sum_{j\in J} \Bbbk_{g_j}^{\chi_j}, \qquad \mathscr{G}_J = (\mathscr{G}_j)_{j\in J}.$$

We shall use the results and notations from the preceding Sections, but with $\frac{3}{2}$ replacing 2 when appropriate, e. g. $x_{\frac{3}{2}1} = x_{\frac{3}{2}}x_1 - x_1x_{\frac{3}{2}}$. Let

$$K = \mathscr{B}(V)^{\mathrm{co}\,\mathscr{B}(V_1)} \text{ and } \qquad K^1 = \mathrm{ad}_c\,\mathscr{B}(V_1)(V_{\mathrm{diag}}) \in {}^{\mathscr{B}(V_1)\#\Bbbk\Gamma}_{\mathscr{B}(V_1)\#\Bbbk\Gamma}\mathcal{YD},$$

$$\text{so that} \qquad \mathscr{B}(V) \simeq K\#\mathscr{B}(V_1), \quad K \simeq \mathscr{B}(K^1)$$

Let

$$z_{j,n} := (ad_c x_{\frac{3}{2}})^n x_j, \qquad j \in \mathbb{I}_{2,\theta}, \qquad n \in \mathbb{N}_0. \tag{5.4}$$

For all $i, j \in \mathbb{I}_{2,\theta}$, $n \in \mathbb{N}_0$, we have as in [AAH1, §5.2.1] that

$$g_1 \cdot z_{j,n} = q_{1j} z_{j,n}, \qquad \text{(by Lemma 4.2)} \tag{5.5}$$

$$g_i \cdot z_{j,n} = q_{i1}^n q_{ij} z_{j,n}, \tag{5.6}$$

Lemma 5.1 *The braided vector space K^1 is of diagonal type in the basis*

$$(z_{j,n})_{j\in\mathbb{I}_{2,\theta}, 0\le n\le \mathscr{G}_j} \tag{5.7}$$

with braiding matrix

$$(\mathbf{p}_{im,jn})_{\substack{i,j\in\mathbb{I}_{2,\theta},\\ 0\le m\le\mathscr{G}_i,\, 0\le n\le\mathscr{G}_j}} = (q_{i1}^n q_{1j}^m q_{ij})_{\substack{i,j\in\mathbb{I}_{2,\theta},\\ 0\le m\le\mathscr{G}_i,\, 0\le n\le\mathscr{G}_j}}.$$

Hence, the corresponding generalized Dynkin diagram has labels

$$\mathbf{p}_{im,im} = q_{ii}, \qquad \mathbf{p}_{im,jn}\mathbf{p}_{jn,im} = q_{ij}q_{ji}, \qquad (i,m) \neq (j,n).$$

Proof The proof in [AAH1, Lemma 7.2.5] applies as the combinatorial numbers appearing there are not zero. □

Let K_J be the braided vector subspace of K^1 spanned by $(z_{j,n})_{\substack{j\in J,\\ 0\le n\le\mathscr{G}_j}}$.

Corollary 5.2 *The braided subspaces corresponding to the connected components of the Dynkin diagram of K^1 are K_J, $J \in \mathcal{X}$. Hence*

$$\dim K = \dim \mathscr{B}(K^1) = \prod_{J\in\mathcal{X}} \dim \mathscr{B}(K_J). \qquad \square \tag{5.8}$$

Observe that if $\mathscr{G}_J = 0$, then $K_J = V_J$.

In this Section, we shall prove part (b) of Theorem 1.2.

Theorem 5.3 *Let V be a braided vector space with braiding* (5.1). *Assume that for every $J \in \mathcal{X}$, either $\mathscr{G}_J = 0$, or else* $\dim V_J = 1$ *and $V_1 \oplus V_j$ is as in Table 1, or else V_J is as in Table 2. Then*

$$\dim \mathscr{B}(V) = p^2 \prod_{J \in \mathcal{X}} \dim \mathscr{B}(K_J) < \infty. \tag{5.9}$$

Proof By Corollary 5.2 we reduce to connected components in $\mathcal{X}$. If $J \in \mathcal{X}$ has weak interaction and $\mathscr{G}_J = 0$, then $\mathscr{B}(V) \simeq \mathscr{B}(V_1 \oplus V_{J'}) \underline{\otimes} \mathscr{B}(V_J)$, hence $\dim \mathscr{B}(V) = \dim \mathscr{B}(V_1 \oplus V_{J'}) \dim \mathscr{B}(V_J)$.

If $J \in \mathcal{X}$ is a point, then Theorem 4.1 applies. We need to analyze those J with $|J| \geq 2$ and $\mathscr{G}_J \neq 0$.

Below we denote $\imath = \sqrt{-1}$.

$\mathfrak{L}(A_2)$, $\mathscr{G}_J = (1, 0)$: Here K_J is of Cartan type A_3 and $\dim \mathscr{B}(K_J) = 2^6$.

$\mathfrak{L}(A_j)$, $j > 2$, $\mathscr{G}_J = (1, 0)$: Here K_J is of Cartan type D_{j-1} and $\dim \mathscr{B}(K_J) = 2^{j(j+1)}$.

$\mathfrak{L}(A_2, 2)$: Here K_J is of Cartan type D_4 and $\dim \mathscr{B}(K_J) = 2^{12}$.

$\mathfrak{L}(A(1|0)_2; \omega)$, provided that $p > 3$: Here K_J is of diagonal type in the basis $1 = z_{i,0}$, $2 = z_{i,1}$, $3 = z_{j,0}$ with diagram

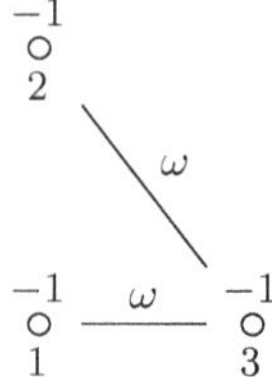

Then $\dim \mathscr{B}(K_J) < \infty$ by [H, Table 2, row 15]; it has the same root system as $\mathfrak{g}(2, 3)$ and dimension $2^7 3^4$, see [AA, 8.3.4].

$\mathfrak{L}(A(1|0)_1; \omega)$ and $\mathfrak{L}(A(1|0)_3; \omega)$, provided that $p > 3$: Here K_J is of diagonal type in the basis $1 = z_{i,0}$, $2 = z_{j,1}$, $3 = z_{j,0}$, respectively $1 = z_{i,0}$, $2 = z_{i,1}$, $3 = z_{j,0}$, and its diagram is

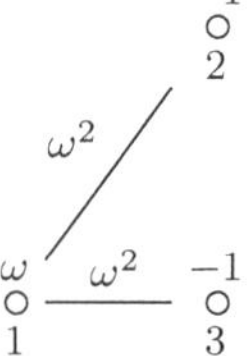

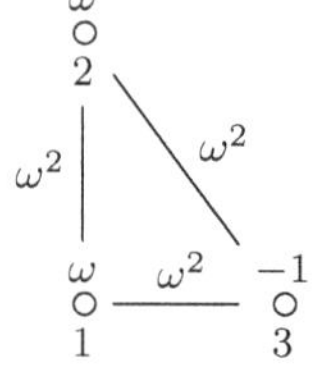

Then $\dim \mathscr{B}(K_J) < \infty$ by [H, Table 2, row 8], respectively [H, Table 2, row 15]. In the first case it has the same root system as $\mathfrak{sl}(2|2)$ and dimension $2^4 3^2$, see [AA, 5.1.8]. In the second case it has the same root system as $\mathfrak{g}(2, 3)$ and dimension $2^7 3^4$, see [AA, 8.3.4].

$\mathrm{R}\mathfrak{L}(A(1|0)_1; r)$: analogous to $\mathfrak{L}(A(1|0)_1; \omega)$; same root system as $\mathfrak{sl}(2|2)$ and dimension $2^4 N^2$, see [AA, 5.1.8].

$\mathfrak{L}(A(2|0)_1; \omega)$ and $\mathfrak{L}(D(2|1); \omega)$, $p > 3$: In both cases, $\dim \mathscr{B}(K_J) = 2^8 3^9$; it has the same root system as $\mathfrak{g}(3, 3)$, see [H, Table 3, row 18], [AA, 8.4.5]. □

5.2 *The Presentation of the Nichols Algebras*

We give defining relations and an explicit PBW-basis of $\mathscr{B}(V)$, for all V as in Theorem 5.3, assuming that the Dynkin diagram of V_{diag} is connected, i.e., $V_{\text{diag}} = V_J$, where $J = \mathbb{I}_{2,\theta}$. Essentially the relations are the same as in [AAH1] up to adding the suitable p-powers; we omit the proofs as they are minor variations of those in *loc. cit.* The passage from connected V_{diag} to the general case is standard, just add the quantum commutators between points in different components. Since the case $|J| = 1$ was treated in Sect. 4, we also suppose that $|J| > 1$. These braided vector spaces have names given in [AAH1], see Table 2. The braided vector subspace $V_1 \oplus \Bbbk x_2$ of such V is of type

- $\mathfrak{L}(-1, 2)$ when V is of type $\mathfrak{L}(A_2, 2)$,
- $\mathfrak{L}(\omega, 1)$ when V is of type $\mathfrak{L}(A(1|0)_3; \omega)$, or
- $\mathfrak{L}(-1, 1)$ for all the other cases.

Thus the subalgebra generated by $V_1 \oplus \Bbbk x_2$ is a Nichols algebra. We recall its relations up to the change of index with respect to Sect. 4; the 2 and 3 there are now $\frac{3}{2}$ and 2. As in (2.2), we set $x_{i_1 i_2 \dots i_M} = \operatorname{ad}_c x_{i_1} x_{i_2 \dots i_M}$. Also, we have now $z_n = (\operatorname{ad}_c x_{\frac{3}{2}})^n x_2$, $n \in \mathbb{N}_0$.

First, the defining relations of $\mathscr{B}(\mathfrak{L}(-1, 1))$ are

$$x_{\frac{3}{2}} x_1 - x_1 x_{\frac{3}{2}} + \frac{1}{2} x_1^2, \tag{5.10}$$

$$x_1^p, \tag{3.2}$$

$$x_{\frac{3}{2}}^p, \tag{5.11}$$

$$x_1 x_2 - q_{12}\, x_2 x_1, \tag{5.12}$$

$$(\operatorname{ad}_c x_{\frac{3}{2}})^2 x_2, \tag{5.13}$$

$$x_2^2,\ x_{\frac{3}{2}2}^2. \tag{5.14}$$

Second, the defining relations of $\mathscr{B}(\mathfrak{L}(-1,2))$ are (5.10), (3.2), (5.11), (5.12), (5.14) and

$$(\operatorname{ad}_c x_{\frac{3}{2}})^3 x_2, \tag{5.15}$$

$$x_{\frac{3}{2}\frac{3}{2}2}^2. \tag{5.16}$$

Third, the defining relations of $\mathscr{B}(\mathfrak{L}(\omega,1))$ are (5.10), (3.2), (5.11), (5.12), (5.13) and

$$x_2^3, \tag{5.17}$$

$$x_{\frac{3}{2}2}^3, \tag{5.18}$$

$$[x_{\frac{3}{2}2}, x_2]_c^3. \tag{5.19}$$

We also observe that, since $q_{1j}q_{j1} = 1$ and $\mathscr{G}_j = 0$, we have

$$x_1 x_j = q_{1j} x_j x_1, \qquad x_{\frac{3}{2}} x_j = q_{1j} x_j x_{\frac{3}{2}}, \qquad j \in \mathbb{I}_{3,\theta}. \tag{5.20}$$

5.2.1 The Nichols Algebra $\mathscr{B}(\mathfrak{L}(A(1|0)_1; r))$, $r \in \mathbb{G}'_N$, $N \geq 3$

Let

$$\text{ц}_{12} = x_{\frac{3}{2}23}, \qquad \text{ц}_{123} = [x_{\frac{3}{2}2}, x_{23}]_c. \tag{5.21}$$

Proposition 5.4 *The algebra $\mathscr{B}(\mathfrak{L}(A(1|0)_1; r))$ is presented by generators x_i, $i \in \mathbb{I}_3^\dagger$, and relations* (5.10), (3.2), (5.11)–(5.14) (5.20), *and*

$$(\operatorname{ad}_c x_3)^2 x_2 = 0, \qquad x_3^N = 0, \tag{5.22}$$

$$\text{ц}_{123}^N = 0. \tag{5.23}$$

The set

$$B = \{x_1^{m_1} x_{\frac{3}{2}}^{m_2} x_{\frac{3}{2}2}^{n_1} \text{ц}_{12}^{n_2} \text{ц}_{123}^{n_3} x_3^{n_4} x_{23}^{n_5} x_2^{n_6} : \\ 0 \leq n_1, n_2, n_5, n_6 < 2,\ 0 \leq n_3, n_4 < N,\ 0 \leq m_1, m_2 < p\}$$

is a basis of $\mathscr{B}(\mathfrak{L}(A(1|0)_1; r))$ and $\dim \mathscr{B}(\mathfrak{L}(A(1|0)_1; r)) = p^2 2^4 N^2$. □

5.2.2 The Nichols Algebra $\mathscr{B}(\mathfrak{L}(A(1|0)_2; \omega))$

Proposition 5.5 *The algebra $\mathscr{B}(\mathfrak{L}(A(1|0)_2; \omega))$ is presented by generators x_i, $i \in \mathbb{I}_3^{\dagger}$, and relations* (5.10), (3.2), (5.11)–(5.14), (5.20), *and*

$$x_2^2 = 0, \qquad x_3^2 = 0, \qquad x_{23}^3 = 0, \tag{5.24}$$

$$\text{ц}_{12}^3 = 0, \qquad \text{ц}_{123}^6 = 0, \qquad [\text{ц}_{123}, x_3]_c^3 = 0. \tag{5.25}$$

The set

$$\begin{aligned} B = \big\{ & x_1^{m_1} x_{\frac{3}{2}}^{m_2} x_{\frac{3}{2}2}^{n_1} \text{ц}_{12}^{n_2} [\text{ц}_{12}, [\text{ц}_{12}, \text{ц}_{123}]_c]_c^{n_3} [\text{ц}_{12}, \text{ц}_{123}]_c^{n_4} \text{ц}_{123}^{n_5} \\ & [\text{ц}_{123}, x_3]_c^{n_6} [\text{ц}_{123}, x_{32}]_c^{n_7} x_3^{n_8} x_{32}^{n_9} x_2^{n_{10}} : 0 \le m_1, m_2, < p, \\ & 0 \le n_2, n_6, n_9 < 3,\ 0 \le n_5 < 6,\ 0 \le n_1, n_3, n_4, n_7, n_8, n_{10} < 2 \big\} \end{aligned}$$

is a basis of $\mathscr{B}(\mathfrak{L}(A(1|0)_2; \omega))$ and $\dim \mathscr{B}(\mathfrak{L}(A(1|0)_2; \omega)) = p^2 2^7 3^4$. □

5.2.3 The Nichols Algebra $\mathscr{B}(\mathfrak{L}(A(1|0)_3; \omega))$

Proposition 5.6 *The algebra $\mathscr{B}(\mathfrak{L}(A(1|0)_3; \omega))$ is presented by generators x_i, $i \in \mathbb{I}_3^{\dagger}$, and relations* (5.10), (3.2), (5.11)–(5.13), (5.17)–(5.19) (5.20),

$$x_2^3 = 0, \qquad x_3^2 = 0, \qquad x_{223} = 0, \tag{5.26}$$

$$x_{\frac{3}{2}2} x_{\frac{3}{2}23} + q_{13} q_{23} x_{\frac{3}{2}23} x_{\frac{3}{2}2} = 0, \qquad \text{ц}_{123}^6 = 0. \tag{5.27}$$

The set

$$\begin{aligned} B = \big\{ & x_1^{m_1} x_{\frac{3}{2}}^{m_2} x_{\frac{3}{2}2}^{n_1} \text{ц}_{12}^{n_2} [\text{ц}_{12}, \text{ц}_{13}]_c^{n_3} \text{ц}_{123}^{n_4} [\text{ц}_{123}, \text{ц}_{13}]_c^{n_5} [\text{ц}_{13}, x_{32}]_c^{n_6} \\ & \text{ц}_{13}^{n_7} x_3^{n_8} x_{32}^{n_9} x_2^{n_{10}} : 0 \le m_1, m_2 < p,\ 0 \le n_2, n_3, n_5, n_6, n_8, n_9 < 2, \\ & 0 \le n_1, n_7, n_{10} < 3,\ 0 \le n_4 < 6 \big\} \end{aligned}$$

is a basis of $\mathscr{B}(\mathfrak{L}(A(1|0)_3; \omega))$ and $\dim \mathscr{B}(\mathfrak{L}(A(1|0)_3; \omega)) = p^2 2^7 3^4$. □

5.2.4 The Nichols Algebra $\mathscr{B}(\mathfrak{L}(A(2|0)_1; \omega))$

Proposition 5.7 *The algebra* $\mathscr{B}(\mathfrak{L}(A(2|0)_1; \omega))$ *is presented by generators* x_i, $i \in \mathbb{I}_4^\dagger$, *and relations* (5.10), (3.2), (5.11)–(5.14), (5.20), *and*

$$x_{24} = 0, \quad x_{332} = 0, \quad x_{334} = 0, \quad x_{443} = 0, \tag{5.28}$$

$$x_2^2 = 0, \quad x_3^3 = 0, \quad x_{34}^3 = 0, \quad x_4^3 = 0, \tag{5.29}$$

$$\begin{aligned} [x_{\frac{3}{2}23}, x_2]_c{}^3 &= 0, & [[x_{\frac{3}{2}23}, x_2]_c, [x_{\frac{3}{2}23}, x_{24}]_c]_c{}^3 &= 0, \\ [x_{\frac{3}{2}23}, x_{24}]_c{}^3 &= 0, & [[x_{\frac{3}{2}23}, x_2]_c, [[x_{\frac{3}{2}23}, x_{24}]_c, x_2]_c]_c{}^3 &= 0, \\ [[x_{\frac{3}{2}23}, x_{24}]_c, x_2]_c{}^3 &= 0, & [[x_{\frac{3}{2}23}, x_{24}]_c, [[x_{\frac{3}{2}23}, x_{24}]_c, x_2]_c]_c{}^3 &= 0. \end{aligned} \tag{5.30}$$

The set

$$\begin{aligned} B = \big\{ & x_1^{m_1} x_{\frac{3}{2}}^{m_2} x_{\frac{3}{2}2}^{n_1} \text{ц}_{12}^{n_2} [\text{ц}_{12}, \text{ц}_{1234}]_c^{n_3} \text{ц}_{123}^{n_4} [\text{ц}_{123}, \text{ц}_{1234}]_c^{n_5} [\text{ц}_{123}, [\text{ц}_{1234}, x_3]_c]_c^{n_6} \\ & \text{ц}_{1234}^{n_7} [\text{ц}_{1234}, [\text{ц}_{1234}, x_3]_c]_c^{n_8} [\text{ц}_{1234}, x_3]_c^{n_9} [\text{ц}_{1234}, x_{32}]_c^{n_{10}} x_{\frac{3}{2}234}^{n_{11}} x_3^{n_{12}} x_{32}^{n_{13}} \\ & x_{324}^{n_{14}} x_{34}^{n_{15}} x_2^{n_{16}} x_4^{n_{17}} \,:\, 0 \le n_1, n_2, n_3, n_{10}, n_{11}, n_{13}, n_{14}, n_{15} < 2, \\ & 0 \le m_1, m_2 < p,\ 0 \le n_4, n_5, n_6, n_7, n_8, n_9, n_{12}, n_{16}, n_{17} < 3 \big\} \end{aligned}$$

is a basis of $\mathscr{B}(\mathfrak{L}(A(2|0)_1; \omega))$ *and* $\dim \mathscr{B}(\mathfrak{L}(A(2|0)_1; \omega)) = p^2 2^8 3^9$. □

5.2.5 The Nichols Algebra $\mathscr{B}(\mathfrak{L}(D(2|1); \omega))$

Proposition 5.8 *The algebra* $\mathscr{B}(\mathfrak{L}(D(2|1); \omega))$ *is presented by generators* x_i, $i \in \mathbb{I}_4^\dagger$, *and relations* (5.10), (3.2), (5.11)–(5.14), (5.20), *and*

$$\begin{aligned} & [[[x_{\frac{3}{2}23}, x_{24}]_c, x_3]_c, x_3]_c^3 = 0, \quad [[x_{\frac{3}{2}23}, x_{24}]_c, x_3]_c^3 = 0, \\ & [[x_{\frac{3}{2}23}, x_{24}]_c, x_{334}]_c^3 = 0, \qquad [x_{\frac{3}{2}23}, x_{24}]_c^3 = 0, \qquad [x_{\frac{3}{2}23}, x_2]_c^3 = 0, \end{aligned} \tag{5.31}$$

$$x_{24} = 0, \quad x_{332} = 0, \quad x_{443} = 0, \quad [[x_{234}, x_3]_c, x_3]_c = 0, \tag{5.32}$$

$$x_2^2 = 0, \quad x_3^3 = 0, \quad x_{34}^3 = 0, \quad x_{334}^3 = 0, \quad x_4^3 = 0. \tag{5.33}$$

The set

$$
\begin{aligned}
B = \{ & x_1^{m_1} x_{\frac{3}{2}}^{m_2} x_{\frac{3}{2}2}^{n_1} \text{ц}_{12}^{n_2} \text{ц}_{123}^{n_3} \text{ц}_{1234}^{n_4} [\text{ц}_{1234}, x_3]_c^{n_5} [[\text{ц}_{1234}, x_3]_c, x_3]_c^{n_6} [\text{ц}_{1234}, x_{334}]_c^{n_7} \\
& x_{\frac{3}{2}234}^{n_8} [x_{\frac{3}{2}234}, x_3]_c^{n_9} x_3^{n_{10}} x_{32}^{n_{11}} x_{324}^{n_{12}} [x_{324}, x_3]_c^{n_{13}} x_{334}^{n_{14}} x_{34}^{n_{15}} x_2^{n_{16}} x_4^{n_{17}} : \\
& 0 \le m_1, m_2 < p,\ 0 \le n_1, n_2, n_8, n_9, n_{11}, n_{12}, n_{13}, n_{16} < 2, \\
& \qquad\qquad 0 \le n_3, n_4, n_5, n_6, n_7, n_{10}, n_{14}, n_{15}, n_{17} < 3\}
\end{aligned}
$$

is a basis of $\mathscr{B}(\mathfrak{L}(D(2|1); \omega))$ *and* $\dim \mathscr{B}(\mathfrak{L}(D(2|1); \omega)) = p^2 2^8 3^9$. □

5.2.6 The Nichols Algebra $\mathscr{B}(\mathfrak{L}(A_2, 2))$

Proposition 5.9 *The algebra* $\mathscr{B}(\mathfrak{L}(A_2, 2))$ *is presented by generators* x_i, $i \in \mathbb{I}_3^{\dagger}$, *and relations* (5.10), (3.2), (5.11), (5.12), (5.14)–(5.16), (5.20), *and*

$$
x_{\frac{3}{2}}^2 = 0, \qquad x_{\frac{3}{2}}^2 = 0, \qquad x_{32}^2 = 0. \tag{5.34}
$$

$$
\begin{aligned}
& [x_{\frac{3}{2}\frac{3}{2}23}, x_2]_c, x_{\frac{3}{2}2}]_c^2 = 0, && x_{\frac{3}{2}\frac{3}{2}23}^2 = 0, \quad [x_{\frac{3}{2}\frac{3}{2}23}, x_{\frac{3}{2}2}]_c^2 = 0, \\
& [x_{\frac{3}{2}\frac{3}{2}23}, x_2]_c^2 = 0, && x_{3\frac{3}{2}2}^2 = 0, \quad [x_{32}, x_{\frac{3}{2}2}]_c^2 = 0, \\
& [[x_{\frac{3}{2}\frac{3}{2}23}, x_2]_c, x_{\frac{3}{2}2}]_c, x_3]_c^2 = 0.
\end{aligned} \tag{5.35}
$$

The set $B =$

$$
\begin{aligned}
\{ & x_1^{m_1} x_{\frac{3}{2}}^{m_2} x_{\frac{3}{2}\frac{3}{2}2}^{n_1} x_{\frac{3}{2}\frac{3}{2}23}^{n_2} [x_{\frac{3}{2}\frac{3}{2}23}, x_{\frac{3}{2}2}]_c^{n_3} [[x_{\frac{3}{2}\frac{3}{2}23}, x_2]_c, x_{\frac{3}{2}2}]_c^{n_4} [[[x_{\frac{3}{2}\frac{3}{2}23}, x_2]_c, x_{\frac{3}{2}2}]_c, x_3]_c^{n_5} \\
& [x_{\frac{3}{2}\frac{3}{2}23}, x_2]_c^{n_6} x_3^{n_7} x_{3\frac{3}{2}2}^{n_8} [x_{32}, x_{\frac{3}{2}2}]^{n_9} x_{32}^{n_{10}} x_{\frac{3}{2}2}^{n_{11}} x_2^{n_{12}} : m_1, m_2 \in \mathbb{I}_{0,p-1}, n_i \in \mathbb{I}_{0,1}\}.
\end{aligned}
$$

is a basis of $\mathscr{B}(\mathfrak{L}(A_2, 2))$ *and* $\dim \mathscr{B}(\mathfrak{L}(A_2, 2)) = p^2 2^{12}$. □

5.2.7 The Nichols Algebra $\mathscr{B}(\mathfrak{L}(A_{\theta-1}))$

For details of the following result we refer to [AAH1, §5.3.8]. In particular, one defines x_{ij}, $2 \le i \le j$, as usual, $\text{я}_{1\ell} = [x_{\frac{3}{2}2}, x_{3\ell}]_c$ and $\text{я}_{k\ell} = [\text{я}_{1\ell}, x_{2k}]$, $k > 1$.

Proposition 5.10 *The algebra* $\mathscr{B}(\mathfrak{L}(A_{\theta-1}))$ *is presented by generators* x_i, $i \in \mathbb{I}_\theta^\dagger$, *and relations* (5.10), (3.2), (5.11)–(5.14), (5.20), *and*

$$x_{ij}^2 = 0, \qquad 2 \le i \le j \le \theta, \tag{5.36}$$

$$[x_{k-1\,k\,k+1}, x_k] = 0, \qquad 3 \le k < \theta. \tag{5.37}$$

$$x_{1j}^2, \qquad я_{k\ell}^2, \qquad j, k, \ell \in \mathbb{I}_\theta, k < \ell. \tag{5.38}$$

Furthermore there is a PBW-basis in terms of the positive roots of the root system of type D_θ *and* $\dim \mathscr{B}(\mathfrak{L}(A_{\theta-1})) = p^2 2^{\theta(\theta-1)}$. □

5.3 Realizations

Let H be a Hopf algebra, (g_1, χ_1, η) a YD-triple and (g_j, χ_j), $j \in \mathbb{I}_{2,\theta}$, a family of YD-pairs for H, see Sect. 3.3. Let (V, c) be a braided vector space with braiding (5.1). Then

$$\mathcal{V} := \mathcal{V}_{g_1}(\chi_1, \eta) \oplus \left(\oplus_{j \in \mathbb{I}_{2,\theta}} \Bbbk_{g_j}^{\chi_j} \right) \in {}^H_H\mathcal{YD} \tag{5.39}$$

is a *principal realization* of (V, c) over H if

$$q_{ij} = \chi_j(g_i), \qquad i, j \in \mathbb{I}_\theta; \qquad a_j = q_{j1}^{-1} \eta(g_j), \ j \in \mathbb{I}_{2,\theta}.$$

Thus $(V, c) \simeq \mathcal{V}$ as braided vector space. Consequently, if H is finite-dimensional and (V, c) is as in Table 2, then $\mathscr{B}(\mathcal{V})\#H$ is a finite-dimensional Hopf algebra. Examples of finite-dimensional pointed Hopf algebras $A = \mathscr{B}(V)\#\Bbbk\Gamma$ with Γ abelian like this are listed in Table 5 where the interaction is weak, $\epsilon = 1$ and $\omega \in \mathbb{G}_3'$. As in Sect. 4.5, Γ is a product of θ cyclic groups, g_i is the i-th canonical generator, $\chi_j(g_i) = 1$ if $i \neq j$; we set $q_{ij} = 1$, $i < j$.

6 Several Blocks, One Point

Let $t \ge 2$ and $\theta = t + 1$. We shall use the following notation:

$$\mathbb{I}_k^\ddagger = \{k, k + \tfrac{1}{2}\}, \qquad k \in \mathbb{I}_t; \qquad \mathbb{I}^\ddagger = \mathbb{I}_1^\ddagger \cup \cdots \cup \mathbb{I}_t^\ddagger \cup \{\theta\}.$$

Table 5 Pointed Hopf algebras from a block and several points

V_J	$\mathscr{G}_J$	Γ	$dim A$
$\overset{-1}{\circ}\overset{-1}{\text{———}}\overset{-1}{\circ}\dots\overset{-1}{\circ}\overset{-1}{\text{———}}\overset{-1}{\circ}$	$(1,0,\dots,0)$	$\mathbb{Z}/p\times\mathbb{Z}/2p\times\mathbb{Z}/2$	2^8p^4
		$\mathbb{Z}/p\times\mathbb{Z}/2p\times(\mathbb{Z}/2)^{\theta-2}$	$2^{(\theta-1)^2}p^4$
$\overset{-1}{\circ}\overset{-1}{\text{———}}\overset{-1}{\circ}$	$(2,0)$	$\mathbb{Z}/p\times\mathbb{Z}/2p\times\mathbb{Z}/2$	$2^{14}p^4$
$\overset{-1}{\circ}\overset{\omega}{\text{———}}\overset{-1}{\circ}$	$(1,0)$	$\mathbb{Z}/p\times\mathbb{Z}/2p\times\mathbb{Z}/6$	$2^93^5p^4$
$\overset{-1}{\circ}\overset{\omega^2}{\text{———}}\overset{\omega}{\circ}$	$(1,0)$	$\mathbb{Z}/p\times\mathbb{Z}/2p\times\mathbb{Z}/3$	$2^53^3p^4$
	$(0,1)$	$\mathbb{Z}/p\times\mathbb{Z}/6\times\mathbb{Z}/3p$	$2^83^6p^4$
$\overset{-1}{\circ}\overset{r^{-1}}{\text{———}}\overset{r}{\circ}$, $r\in\mathbb{G}'_N, N>3$	$(1,0)$	$\mathbb{Z}/p\times\mathbb{Z}/2p\times\mathbb{Z}/N$	$2^5N^3p^4$
$\overset{-1}{\circ}\overset{\omega}{\text{———}}\overset{\omega^2}{\circ}\overset{\omega}{\text{———}}\overset{\omega^2}{\circ}$	$(1,0,0)$	$\mathbb{Z}/p\times\mathbb{Z}/2p\times\mathbb{Z}/3\times\mathbb{Z}/3$	$2^93^{11}p^4$
$\overset{-1}{\circ}\overset{\omega}{\text{———}}\overset{\omega^2}{\circ}\overset{\omega^2}{\text{———}}\overset{\omega}{\circ}$	$(1,0,0)$	$\mathbb{Z}/p\times\mathbb{Z}/2p\times\mathbb{Z}/3\times\mathbb{Z}/3$	$2^93^{11}p^4$

The Poseidon braided vector space $\mathfrak{P}(\mathbf{q},\mathscr{G})$ depends on a datum

- $\mathbf{q}\in\Bbbk^{\theta\times\theta}$ such that $\epsilon_i := q_{ii} = \pm 1$, $q_{ij}q_{ji} = 1$ for all $i\neq j\in\mathbb{I}_\theta$.
- $\mathscr{G} = (\mathscr{G}_j)\in\mathbb{N}^t$, (a_j) such that $\mathscr{G}_j = \begin{cases} -\mathtt{r}_j, & \epsilon_j = 1,\\ \mathtt{r}_j, & \epsilon_j = -1;\end{cases}$ cf. (4.2);

it has a basis $(x_i)_{i\in\mathbb{I}_\theta^\ddagger}$ and braiding

$$c(x_i\otimes x_j) = \begin{cases} q_{ij}\, x_j\otimes x_i, & \lfloor i\rfloor\leq t,\ \lfloor i\rfloor\neq\lfloor j\rfloor,\\ \epsilon_j\, x_j\otimes x_i, & \lfloor i\rfloor = j\leq t,\\ (\epsilon_j\, x_j + x_{\lfloor j\rfloor})\otimes x_i, & \lfloor i\rfloor\leq t,\ j = \lfloor i\rfloor + \frac{1}{2},\\ q_{\theta j}\, x_j\otimes x_\theta, & i=\theta,\ j\in\mathbb{I}_\theta,\\ q_{\theta j}\,(x_j + a_j x_{\lfloor j\rfloor})\otimes x_\theta, & i=\theta,\ j\notin\mathbb{I}_\theta.\end{cases} \tag{6.1}$$

We shall use the elements

$$y_j^{\langle n\rangle} := \begin{cases} x_j x^m_{j+\frac{1}{2}\, j}, & n = 2m+1 \text{ odd};\\ x^m_{j+\frac{1}{2}\, j}, & n = 2m \text{ even}.\end{cases} \qquad j\in\mathbb{I}_t,\ n\in\mathbb{N}_0; \tag{6.2}$$

$$\text{Ш}_{\mathbf{n}} := (\operatorname{ad}_c x_{\frac{3}{2}})^{n_1}\dots(\operatorname{ad}_c x_{t+\frac{1}{2}})^{n_t}x_\theta, \qquad \mathbf{n} = (n_1,\dots,n_t)\in\mathbb{N}_0^t. \tag{6.3}$$

Let $\mathcal{A} := \{\mathbf{n} \in \mathbb{N}_0^t : 0 \leq \mathbf{n} \leq \mathbf{a} = (|\mathfrak{r}_1|, \dots, |\mathfrak{r}_t|)\}$, ordered lexicographically. For $\mathbf{m}, \mathbf{n} \in \mathcal{A}$ we set

$$\mathbf{p}_{\mathbf{m},\mathbf{n}} := \epsilon_\theta \prod_{i,j \in \mathbb{I}_t} q_{ij}^{m_i n_j} q_{i\theta}^{m_i} q_{\theta j}^{n_j} \qquad \epsilon_{\mathbf{n}} := \mathbf{p}_{\mathbf{n},\mathbf{n}}$$

We also need the following notation:

$$t_+ := \{i \in \mathbb{I}_t : \epsilon_i = 1\}, \qquad t_- = t - t_+,$$
$$M_+ := \{\mathbf{m} \in \mathcal{A} : \epsilon_{\mathbf{m}} = 1\}, \qquad M_- = |\mathcal{A}| - M_+.$$

The main result of this Section is the following.

Proposition 6.1 *The algebra $\mathscr{B}(\mathfrak{P}(\mathbf{q}, \mathscr{G}))$ is presented by generators x_i, $i \in \mathbb{I}^{\ddagger}$, and relations*

$$\left.\begin{aligned} &x_{i+\frac{1}{2}}^p = 0, \quad x_i^p = 0, \\ &x_{i+\frac{1}{2}} x_i - x_i x_{i+\frac{1}{2}} + \frac{1}{2} x_i^2 = 0, \end{aligned}\right\} \quad i \in \mathbb{I}_t,\ \epsilon_i = 1; \tag{6.4}$$

$$\left.\begin{aligned} &x_i^2 = 0, \quad x_{i+\frac{1}{2} i}^{2p} = 0, \quad x_{i+\frac{1}{2}}^p = 0, \\ &x_{i+\frac{1}{2}} x_{i+\frac{1}{2} i} - x_{i+\frac{1}{2} i} x_{i+\frac{1}{2}} - x_i x_{i+\frac{1}{2} i} = 0, \end{aligned}\right\} \quad i \in \mathbb{I}_t,\ \epsilon_i = -1; \tag{6.5}$$

$$x_i x_j = q_{ij}\, x_j x_i, \qquad \lfloor i \rfloor \neq \lfloor j \rfloor \in \mathbb{I}_t; \tag{6.6}$$

$$x_i x_\theta = q_{i\theta}\, x_\theta x_i, \qquad i \in \mathbb{I}_t; \tag{6.7}$$

$$(\operatorname{ad}_c x_{i+\frac{1}{2}})^{1+|\mathfrak{r}_i|}(x_\theta) = 0, \qquad i \in \mathbb{I}_t, \tag{6.8}$$

$$ш_{\mathbf{m}} ш_{\mathbf{n}} = \mathbf{p}_{\mathbf{m},\mathbf{n}}\, ш_{\mathbf{n}} ш_{\mathbf{m}} \qquad \mathbf{m} \neq \mathbf{n} \in \mathcal{A}; \tag{6.9}$$

$$ш_{\mathbf{n}}^2 = 0, \qquad \mathbf{n} \in \mathcal{A},\ \epsilon_{\mathbf{n}} = -1, \tag{6.10}$$

$$ш_{\mathbf{n}}^p = 0, \qquad \mathbf{n} \in \mathcal{A},\ \epsilon_{\mathbf{n}} = 1. \tag{6.11}$$

A basis of $\mathscr{B}(\mathfrak{P}(\mathbf{q}, \mathscr{G}))$ is given by

$$B = \Big\{ y_1^{\langle m_1 \rangle} x_{\frac{3}{2}}^{m_2} \dots y_t^{\langle m_{2t-1} \rangle} x_{t+\frac{1}{2}}^{m_{2t}} \prod_{\mathbf{n} \in \mathcal{A}} ш_{\mathbf{n}}^{b_{\mathbf{n}}} : 0 \leq b_{\mathbf{n}} < 2 \text{ if } \epsilon_{\mathbf{n}} = -1,$$
$$0 \leq b_{\mathbf{n}}, m_i < p \text{ if } \epsilon_{\mathbf{n}} = 1,\ i \in \mathbb{I}_t \Big\}.$$

Hence $\dim \mathscr{B}(\mathfrak{P}(\mathbf{q}, \mathscr{G})) = 2^{2t_- + M_-} p^{2t_+ + M_+}$. □

6.1 Realizations

Let H be a Hopf algebra, (g_i, χ_i, η_i), $i \in \mathbb{I}_t$, a family of YD-triples and (g_θ, χ_θ) a YD-pair for H, see Sect. 3.3. Let (V, c) be a braided vector space with braiding (6.1). Then

$$\mathcal{V} := \Big(\oplus_{i\in\mathbb{I}_t} \mathcal{V}_{g_i}(\chi_i, \eta_i)\Big) \oplus \Bbbk_{g_\theta}^{\chi_\theta} \in {}^H_H\mathcal{YD} \tag{6.12}$$

is a *principal realization* of (V, c) over H if

$$q_{ij} = \chi_j(g_i), \qquad i, j \in \mathbb{I}_\theta; \qquad a_j = q_{j1}^{-1}\eta(g_j),\ j \in \mathbb{I}_t.$$

Thus $(V, c) \simeq \mathcal{V}$ as braided vector space. Consequently, if H is finite-dimensional, then $\mathscr{B}(\mathcal{V})\#H$ is a finite-dimensional Hopf algebra.

If $q_{ij} = 1$ for $i \neq j$, then we may choose $H = \Bbbk\Gamma$, where

$$\Gamma = \mathbb{Z}/n_1 \times \dots \mathbb{Z}/n_\theta, \qquad n_i = \begin{cases} p & \text{if } q_{ii} = 1, \\ 2p & \text{if } q_{ii} = -1. \end{cases}$$

Thus $\mathscr{B}(\mathfrak{P}(\mathbf{q}, \mathscr{G}))\#\Bbbk\Gamma$ is a Hopf algebra of dimension $2^{3t_- + M_- + \delta_{\epsilon_\theta, -1}} p^{3t+M_+ + 1}$.

7 A Pale Block and a Point

Let V be a braided vector space of dimension 3 with braiding given in the basis $(x_i)_{i\in\mathbb{I}_3}$ by

$$(c(x_i \otimes x_j))_{i,j\in\mathbb{I}_3} = \begin{pmatrix} \epsilon x_1 \otimes x_1 & \epsilon x_2 \otimes x_1 & q_{12}x_3 \otimes x_1 \\ \epsilon x_1 \otimes x_2 & \epsilon x_2 \otimes x_2 & q_{12}x_3 \otimes x_2 \\ q_{21}x_1 \otimes x_3 & q_{21}(x_2 + x_1) \otimes x_3 & q_{22}x_3 \otimes x_3 \end{pmatrix}. \tag{7.1}$$

Let $V_1 = \langle x_1, x_2\rangle$, $V_2 = \langle x_3\rangle$. Let $\Gamma = \mathbb{Z}^2$ with a basis g_1, g_2. We realize V in ${}^{\Bbbk\Gamma}_{\Bbbk\Gamma}\mathcal{YD}$ by $V_1 = \mathcal{V}_{g_1}$, $V_2 = \mathcal{V}_{g_2}$, $g_1 \cdot x_1 = \epsilon x_1$, $g_2 \cdot x_1 = q_{21}x_1$, $g_1 \cdot x_2 = \epsilon x_2$, $g_2 \cdot x_2 = q_{21}(x_2 + x_1)$, $g_i \cdot x_3 = q_{i2}x_3$.

As usual, let $\widetilde{q}_{12} = q_{12}q_{21}$; in particular the Dynkin diagram of the braided subspace $\langle x_1, x_3\rangle$ is $\overset{\epsilon}{\circ}\overset{\widetilde{q}_{12}}{\text{------}}\overset{q_{22}}{\circ}$.

As for other cases, we consider $K = \mathscr{B}(V)^{\operatorname{co}\mathscr{B}(V_1)}$; then $K = \oplus_{n\geq 0}K^n$ inherits the grading of $\mathscr{B}(V)$; $\mathscr{B}(V) \simeq K\#\mathscr{B}(V_1)$ and K is the Nichols algebra of $K^1 = \operatorname{ad}_c \mathscr{B}(V_1)(V_2)$. Now $K^1 \in {}^{\mathscr{B}(V_1)\#\Bbbk\Gamma}_{\mathscr{B}(V_1)\#\Bbbk\Gamma}\mathcal{YD}$ with the adjoint action and the coaction given by (4.4), i.e., $\delta = (\pi_{\mathscr{B}(V_1)\#\Bbbk\Gamma} \otimes \operatorname{id})\Delta_{\mathscr{B}(V)\#\Bbbk\Gamma}$. Next we introduce $\text{ш}_{m,n} =$

$(\mathrm{ad}_c\, x_1)^m (\mathrm{ad}_c\, x_2)^n x_3$; we distinguish two cases:

$$w_m = (\mathrm{ad}_c\, x_1)^m x_3 = \text{ш}_{m,0}, \qquad z_n = (\mathrm{ad}_c\, x_2)^n x_3 = \text{ш}_{0,n}.$$

By direct computation,

$$g_1 \cdot \text{ш}_{m,n} = q_{12}\epsilon^{m+n} \text{ш}_{m,n}, \qquad g_2 \cdot w_m = q_{21}^m q_{22} w_m, \tag{7.2}$$

$$z_{n+1} = x_2 z_n - q_{12}\epsilon^n z_n x_2, \qquad \text{ш}_{m+1,n} = x_1 \text{ш}_{m,n} - q_{12}\epsilon^{m+n} \text{ш}_{m,n} x_1, \tag{7.3}$$

$$\partial_1(\text{ш}_{m,n}) = 0, \qquad \partial_2(\text{ш}_{m,n}) = 0, \tag{7.4}$$

$$\partial_3(w_m) = \prod_{0\le j\le m-1} (1 - \epsilon^j \widetilde{q}_{12}) x_1^m. \tag{7.5}$$

7.1 The Block Has $\epsilon = 1$

Here $\mathscr{B}(V_1) \simeq S(V_1)$ is a polynomial algebra, so that x_1 and x_2 commute, and

$$(\mathrm{ad}_c\, x_2)^s \text{ш}_{m,n} = \text{ш}_{m,n+s} \qquad \text{for all } m, n, s \in \mathbb{N}_0. \tag{7.6}$$

Thus $\text{ш}_{m,n}$, $m, n \in \mathbb{N}_0$ generate K^1. As in [AAH1, §8.1], we have that

$$g_2 \cdot \text{ш}_{m,n} = q_{21}^{m+n} q_{22} \sum_{0\le j\le n} \binom{n}{j} \text{ш}_{m+j,n-j}. \tag{7.7}$$

For $q \in \Bbbk^\times$, let $\mathfrak{E}_p(q) = V$ be the braided vector space as in (7.1) under the assumptions that $\epsilon = 1$, $q_{12} = q = q_{21}^{-1}$, $q_{22} = -1$. We call $\mathscr{B}(\mathfrak{E}_p(q))$ and the Nichols algebras $\mathscr{B}(\mathfrak{E}_\pm(q))$, $\mathscr{B}(\mathfrak{E}_\star(q))$ studied in Propositions 7.2–7.4 the *Endymion algebras*.

Proposition 7.1 *The algebra $\mathcal{B}(\mathfrak{E}_p(q))$ is presented by generators x_1, x_2, x_3 and relations*

$$x_1^p = 0, \quad x_2^p = 0, \quad x_1 x_2 = x_2 x_1, \tag{7.8}$$

$$x_1 x_3 = q x_3 x_1, \tag{7.9}$$

$$z_t^2 = 0, \qquad t \in \mathbb{I}_{0,p-1}. \tag{7.10}$$

The dimension of $\mathscr{B}(\mathfrak{E}_p(q))$ is $2^p p^2$, since it has a PBW-basis

$$B = \{x_1^{m_1} x_2^{m_2} z_{p-1}^{n_{p-1}} \dots z_0^{n_0} : n_i \in \{0, 1\},\ m_j \in \mathbb{I}_{0,p-1}\}.$$

Proof We claim that $\dim K^1 = p$ and $K \simeq \Lambda(K^1)$. In fact, by (7.4), (7.5) and the hypothesis $\widetilde{q}_{12} = 1$, $w_m = 0$ for all $m > 0$; thus $\text{ш}_{m,n} = 0$ for all $m > 0$ by (7.6). Using this fact and (7.7),

$$g_2 \cdot z_n = q_{21}^n q_{22} z_n, \qquad n \in \mathbb{N}_0. \tag{7.11}$$

We have that for all $n \in \mathbb{N}_0$:

$$\partial_3(z_n) = (-1)^n x_1^n, \tag{7.12}$$

$$\delta(z_n) = \sum_{0 \le j \le n} (-1)^{n+j} \binom{n}{j} x_1^{n-j} g_1^j g_2 \otimes z_j. \tag{7.13}$$

Indeed the proof follows as in [AAH1, Lemma 8.1.4]. As $x_1^p = 0$, we have $z_p = 0$. Thus the set $(z_n)_{n \in \mathbb{I}_{0,p-1}}$ is a basis of K^1. The braiding is

$$c(z_n \otimes z_s) = \sum_{0 \le j \le n} (-1)^{n+j} \binom{n}{j} \operatorname{ad}_c(x_1^{n-j} g_1^j g_2) z_s \otimes z_j = q_{12}^n q_{21}^s q_{22} z_s \otimes z_n.$$

That is, K^1 is of diagonal type and the Dynkin diagram consists of disconnected p points labeled with -1. Hence $\mathscr{B}(K^1) = \Lambda(K^1)$. Moreover, B is a basis of $\mathscr{B}(V)$ since $\mathscr{B}(V) \simeq K \# \mathscr{B}(V_1)$.

The presentation follows as in [AAH1, Proposition 4.3.7].

7.2 *The Block Has* $\epsilon = -1$

Here $\mathscr{B}(V_1) \simeq \Lambda(V_1)$ is an exterior algebra and consequently $\text{ш}_{m,n}$, $m, n \in \{0, 1\}$ generates K^1. By direct computation,

$$g_2 \cdot z_1 = q_{21} q_{22}(z_1 + w_1), \qquad \partial_3(z_1) = (1 - \widetilde{q}_{12}) x_2 - \widetilde{q}_{12} x_1, \tag{7.14}$$

$$\delta(z_1) = g_1 g_2 \otimes z_1 + \big((1 - \widetilde{q}_{12}) x_2 - \widetilde{q}_{12} x_1\big) g_2 \otimes x_3. \tag{7.15}$$

7.2.1 Case 1: $\widetilde{q}_{12} = 1$

Here $w_1 = 0$ by (7.4) and (7.5), so

$$\text{ш}_{1,1} = -(\operatorname{ad}_c x_2) w_1 = 0.$$

Thus $z_0 = x_3$ and z_1 form a basis of K^1 and the braiding of K^1 is given by

$$\begin{aligned} c(x_3 \otimes x_3) &= q_{22} x_3 \otimes x_3, & c(x_3 \otimes z_1) &= q_{21} q_{22} z_1 \otimes x_3, \\ c(z_1 \otimes x_3) &= q_{12} q_{22} x_3 \otimes z_1, & c(z_1 \otimes z_1) &= -q_{22} z_1 \otimes z_1. \end{aligned} \tag{7.16}$$

That is, K^1 is of diagonal type with Dynkin diagram $\overset{q_{22}}{\circ} \xrightarrow{\;q_{22}^2\;} \overset{-q_{22}}{\circ}$.

For $q \in \Bbbk^\times$, let $\mathfrak{E}_\pm(q) = V$ be the braided vector space as in (7.1) under the assumptions that $\epsilon = -1$, $q_{12} = q = q_{21}^{-1}$, $q_{22} = \pm 1$.

Proposition 7.2 *The algebra $\mathcal{B}(\mathfrak{E}_+(q))$ is presented by generators x_1, x_2, x_3 and relations*

$$x_1^2 = 0, \quad x_2^2 = 0, \quad x_1 x_2 = -x_2 x_1, \tag{7.17}$$

$$(x_2 x_3 - q x_3 x_2)^2 = 0, \quad x_3^p = 0, \tag{7.18}$$

$$x_3(x_2 x_3 - q x_3 x_2) = q^{-1}(x_2 x_3 - q x_3 x_2) x_3, \tag{7.19}$$

$$x_1 x_3 = q x_3 x_1. \tag{7.20}$$

Let $z_1 = x_2 x_3 - q x_3 x_2$. Then $\mathcal{B}(\mathfrak{E}_+(q))$ has a PBW-basis

$$B = \{x_1^{m_1} x_2^{m_2} x_3^{m_3} z_1^{m_4} : m_1, m_2, m_4 \in \{0, 1\},\ m_3 \in \mathbb{I}_{0,p-1}\};$$

hence $\dim \mathcal{B}(\mathfrak{E}_+(q)) = 2^3 p$.

Proof Notice that $x_3^p = 0$ since x_3 is a point labeled with $q_{22} = 1$ in K^1. Also, B is a basis thanks to the isomorphism $\mathscr{B}(\mathfrak{E}_+(q)) \simeq \mathscr{B}(K^1)\#\mathscr{B}(V_1)$. The rest of the proof follows as in [AAH1, Proposition 8.1.6]. □

Proposition 7.3 *The algebra $\mathcal{B}(\mathfrak{E}_-(q))$ is presented by generators x_1, x_2, x_3 and relations* (7.17), (7.20),

$$x_3^2 = 0, \quad (x_2 x_3 - q x_3 x_2)^p = 0, \tag{7.21}$$

$$x_3(x_2 x_3 - q x_3 x_2) = -q^{-1}(x_2 x_3 - q x_3 x_2) x_3. \tag{7.22}$$

Let $z_1 = x_2 x_3 - q x_3 x_2$. Then $\mathcal{B}(\mathfrak{E}_-(q))$ has a PBW-basis

$$B = \{x_1^{m_1} x_2^{m_2} x_3^{m_3} z_1^{m_4} : m_1, m_2, m_3 \in \{0, 1\},\ m_4 \in \mathbb{I}_{0,p-1}\};$$

hence $\dim \mathcal{B}(\mathfrak{E}_-(q)) = 2^3 p$.

Proof Notice that $z_1^p = 0$ since z_1 is a point labeled with 1 in K^1. Also, B is a basis thanks to the isomorphism $\mathscr{B}(\mathfrak{E}_-(q)) \simeq \mathscr{B}(K^1)\#\mathscr{B}(V_1)$. The rest of the proof follows as in [AAH1, Proposition 8.1.6]. □

7.2.2 Case 2: $\widetilde{q}_{12} = -1$

We consider now a fixed choice of q_{22} and $\widetilde{q}_{12}$, which is the corresponding one to the example of finite GKdim over a field of characteristic 0.

For $q \in \Bbbk^{\times}$, let $\mathfrak{E}_{\star}(q) = V$ be the braided vector space as in (7.1) under the assumptions that $\epsilon = -1$, $q_{22} = -1$, $q_{12} = q$, $q_{21} = -q^{-1}$. Recall (2.2).

Proposition 7.4 *The algebra $\mathcal{B}(\mathfrak{E}_{\star}(q))$ is presented by generators x_1, x_2, x_3 and relations* (7.17),

$$x_3^2 = 0, \quad x_{31}^2 = 0, \tag{7.23}$$

$$x_2[x_{23}, x_{13}]_c - q^2[x_{23}, x_{13}]_c x_2 = q\, x_{13}x_{213}, \tag{7.24}$$

$$x_{23}^{2p} = 0, \quad [x_{23}, x_{13}]_c^p = 0, \quad x_{213}^2 = 0. \tag{7.25}$$

Moreover $\mathcal{B}(\mathfrak{E}_{\star}(q))$ has a PBW-basis

$$B = \{x_2^{m_1} x_{23}^{m_2} x_{213}^{m_3} [x_{23}, x_{13}]_c^{m_4} x_1^{m_5} x_{13}^{m_6} x_3^{m_7} : \\ m_1, m_3, m_5, m_6, m_7 \in \{0, 1\},\ m_2 \in \mathbb{I}_{0,2p-1}, m_4 \in \mathbb{I}_{0,p-1}\};$$

hence $\dim \mathcal{B}(\mathfrak{E}_{\star}(q)) = 2^6 p^2$.

Proof Relations (7.17) are 0 in $\mathcal{B}(\mathfrak{E}_{\star}(q))$ because $\mathscr{B}(V_1) \simeq \Lambda(V_1)$; (7.23) are 0 since x_1, x_3 generate a Nichols algebra of Cartan type A_2 at -1.

Notice that $[x_{23}, x_{13}]_c = x_{23}x_{13} + x_{13}x_{23}$. By (7.17) and (7.23),

$$x_2x_{23} = -q\, x_{23}x_2, \qquad x_2x_{213} = q\, x_{213}x_2, \tag{7.26}$$

$$x_{23}x_3 = -q\, x_3x_{23}, \qquad x_{23}x_1 = -q^{-1}x_1x_{23} - q^{-1}x_{213}, \tag{7.27}$$

$$x_{213}x_1 = q^{-1}x_1x_{213}, \qquad [x_{23}, x_{13}]_c x_1 = -q^{-2}x_1[x_{23}, x_{13}]_c, \tag{7.28}$$

$$x_1x_{13} = -q\, x_{13}x_1, \qquad x_{213}x_3 = q[x_{23}, x_{13}]_c - q^2x_3x_{213}, \tag{7.29}$$

$$x_{13}x_3 = -q\, x_3x_{13}, \qquad x_{13}[x_{23}, x_{13}]_c = [x_{23}, x_{13}]_c x_{13}, \tag{7.30}$$

$$x_{213}x_{13} = q\, x_{13}x_{213}, \qquad [x_{23}, x_{13}]_c x_3 = q^2x_3[x_{23}, x_{13}]_c. \tag{7.31}$$

As in [AAH1, Proposition 8.1.8] we check that

$$\partial_3(x_{213}) = 4x_2x_1 \neq 0, \qquad \partial_3([x_{23}, x_{13}]_c) = 2q^{-1}x_1x_{13} \neq 0.$$

Now we prove that (7.25) holds in $\mathscr{B}(V)$. We check that ∂_i annihilates these terms for $i = 1, 2, 3$. To simplify the notation, let $u = [x_{23}, x_{13}]_c$. As ∂_1, ∂_2 annihilate x_{23},

x_{213} and u, it remains the case $i = 3$. Using (7.26)–(7.31),

$$\partial_3(x_{213}^2) = 4\big(q^{-2}x_2x_1x_{213} + x_{213}x_2x_1\big) = 0,$$

$$\partial_3(u^p) = 2q^{-1}\sum_{k=0}^{p-1}(-q)^{2+2k-p}u^k x_1x_{13}u^{p-1-k} = 2q^{-1}p\,u^{p-1}x_1x_{13} = 0.$$

For the remaining relation, we check that $\partial_3(x_{23}^2) = q^{-1}x_{213} - x_{13}(2x_2 + x_1)$ and the following equalities hold (using (7.26)–(7.31)):

$$\begin{aligned} x_{213}x_{23}^2 &= q(x_{23}^2 + u)x_{213}, & x_{213}u &= q^2\,ux_{213},\\ x_{13}x_{23}^2 &= ux_{13} + x_{23}^2x_{13}, & x_{13}u &= ux_{13},\\ x_1x_{23}^2 &= q^2x_{23}^2x_1 - qx_{13}x_{213}, & x_1u &= q^2\,ux_1,\\ x_2x_{23}^2 &= q^2x_{23}^2x_2, & x_2u &= q^2\,ux_2 + qx_{13}x_{213}. \end{aligned}$$

Using the previous computations and $x_{13}^2 = 0$,

$$\begin{aligned} \big(q^{-1}x_{213} - 2x_{13}x_2 - x_{13}x_1\big)(x_{23}^2 + u) &= q(x_{23}^2 + u)x_{213} - 2q^2x_{13}x_{23}^2x_2\\ &\quad - q^2x_{13}x_{23}^2x_1 + q\,ux_{213} - 2q^2x_{13}ux_2 - q^2x_{13}ux_1\\ &= qx_{23}^2x_{213} + q\,ux_{213} - 2q^2(ux_{13} + x_{23}^2x_{13})x_2\\ &\quad - q^2(ux_{13} + x_{23}^2x_{13})x_1 + q\,ux_{213} - 2q^2\,ux_{13}x_2 - q^2\,ux_{13}x_1\\ &= q^2(x_{23}^2 + 2\,u)\big(q^{-1}x_{213} - 2x_{13}x_2 - x_{13}x_1\big). \end{aligned}$$

We apply this equality to compute

$$\begin{aligned} \partial_3(x_{23}^{2p}) &= \sum_{k=0}^{p-1} q^{2k+2-2p}x_{23}^{2k}\partial_3(x_{23}^2)(x_{23} + x_{13})^{2p-2-2k}\\ &= \sum_{k=0}^{p-1} q^{2k+2-2p}x_{23}^{2k}\big(q^{-1}x_{213} - x_{13}(2x_2 + x_1)\big)(x_{23}^2 + u)^{p-1-k}\\ &= \sum_{k=0}^{p-1} x_{23}^{2k}(x_{23}^2 + 2u)^{p-1-k}\big(q^{-1}x_{213} - x_{13}(2x_2 + x_1)\big). \end{aligned}$$

Using (7.24), (7.29), and (7.30) we get $ux_{23}^2 = (u + x_{23}^2)u$. Hence $a = x_{23}^2 + u$ and $b = u$ satisfy the last equation of (3.12), so (3.13) applies and we have

$$\partial_3(x_{23}^{2p}) = p(x_{23}^2 + u)^{p-1}\big(q^{-1}x_{213} - x_{13}(2x_2 + x_1)\big) = 0.$$

The rest of the proof follows as in [AAH1, Proposition 8.1.8]. □

7.3 Realizations

Here we present a realization of a braided vector space as in (7.1) over a group algebra $H = \Bbbk\Gamma$, with Γ a finite abelian group. We consider $V_1 = \langle x_1, x_2 \rangle$, $V_2 = \langle x_3 \rangle$. We realize V in ${}^{\Bbbk\Gamma}_{\Bbbk\Gamma}\mathcal{YD}$ by $V_1 = V_{g_1}$, $V_2 = V_{g_2}$, $g_1 \cdot x_1 = \epsilon x_1$, $g_2 \cdot x_1 = q_{21} x_1$, $g_1 \cdot x_2 = \epsilon x_2$, $g_2 \cdot x_2 = q_{21}(x_2 + x_1)$, $g_i \cdot x_3 = q_{i2} x_3$. In all the cases Γ will be a product of two cyclic groups, $g_1 = (1, 0)$, $g_2 = (0, 1)$. Examples of finite-dimensional pointed Hopf algebras $A = \mathscr{B}(V_{g_1} \oplus V_{g_2}) \# H$ are listed in Table 6.

Table 6 Pointed Hopf algebras K from a pale block and a point

V	$(\epsilon, \widetilde{q}_{12}, q_{22})$	Γ	q_{12}	$\dim A$
$\mathfrak{E}_p(q)$	$(1, 1, -1)$	$\mathbb{Z}/p \times \mathbb{Z}/2p$	1	$2^{p+1} p^4$
$\mathfrak{E}_+(q)$	$(-1, 1, 1)$	$\mathbb{Z}/2p \times \mathbb{Z}/p$	1	$2^4 p^3$
$\mathfrak{E}_-(q)$	$(-1, 1, -1)$	$\mathbb{Z}/2p \times \mathbb{Z}/2p$	1	$2^5 p^3$
$\mathfrak{E}_\star(q)$	$(-1, -1, -1)$	$\mathbb{Z}/2p \times \mathbb{Z}/2p$	± 1	$2^8 p^4$

Acknowledgments The work of N. A. and I. A. was partially supported by CONICET, Secyt (UNC). The work of N. A. and I. A., respectively I. H., was partially done during visits to the University of Marburg, respectively Córdoba, supported by the Alexander von Humboldt Foundation through the Research Group Linkage Programme.

References

[AA] N. Andruskiewitsch and I. Angiono. *On Finite dimensional Nichols algebras of diagonal type*. Bull. Math. Sci. **7** 353–573 (2017).

[AAH1] N. Andruskiewitsch, I. Angiono and I. Heckenberger. *On finite GK-dimensional Nichols algebras over abelian groups*. Mem. Amer. Math. Soc. Vol. **271**, No. 1329 (2021).

[AAH2] N. Andruskiewitsch, I. Angiono and I. Heckenberger. *On finite GK-dimensional Nichols algebras of diagonal type*. Contemp. Math. **728** (2019), 1–23.

[CF] Chang, H., Farnsteiner, R. *Finite group schemes of p-rank* $\leq$ 1. Math. Proc. Camb. Philos. Soc. **166** (2), 297–323. doi:10.1017/S0305004117000834 (2019)

[CLW] C. Cibils, A. Lauve, S. Witherspoon, *Hopf quivers and Nichols algebras in positive characteristic*, Proc. Amer. Math. Soc. 137(12) (2009) 4029–4041.

[H] I. Heckenberger, *Classification of arithmetic root systems*. Adv. Math. **220** (2009), 59–124.

[HS] I. Heckenberger and H.-J Schneider, *Yetter–Drinfeld modules over bosonizations of dually paired Hopf algebras*, Adv. Math. **244** (2013), 354–394.

[HW] I. Heckenberger and J. Wang: *Rank 2 Nichols algebras of diagonal type over fields of positive characteristic*, SIGMA, Symmetry Integrability Geom. Methods Appl. **11**, Paper 011, 24 p. (2015).

[NW] V. C. Nguyen, X. Wang, *Pointed p^3-dimensional Hopf algebras in positive characteristic*. Algebra Colloq. **25** 399–436 (2018).

[NWW1] V. C. Nguyen, L. Wang and X. Wang, *Classification of connected Hopf algebras of dimension p^3 I*, J. Algebra, **424** (2015), 473–505.

[NWW2] V. C. Nguyen, L. Wang and X. Wang, *Primitive deformations of quantum p-groups*. Algebr. Represent. Theor. **22** (2019), 837–865.

[W] J. Wang: *Rank three Nichols algebras of diagonal type over arbitrary fields*. Isr. J. Math. **218**, 1–26 (2017).

[R] Radford, D. E., *Hopf algebras*, Series on Knots and Everything 49. Hackensack, NJ: World Scientific. xxii, 559 p. (2012).

Poisson Vertex Algebra Cohomology and Differential Harrison Cohomology

Bojko Bakalov, Alberto De Sole, Victor G. Kac, and Veronica Vignoli

To Nikolai Reshetikhin on his 60-th birthday.

Abstract We construct a canonical map from the Poisson vertex algebra cohomology complex to the differential Harrison cohomology complex, which restricts to an isomorphism on the top degree. This is an important step in the computation of Poisson vertex algebra and vertex algebra cohomologies.

Keywords Poisson vertex algebra · Harrison cohomology · Chiral operad · Classical operad · Poisson vertex algebra cohomology · Variational Poisson cohomology

1 Introduction

The present paper is a next step in the development of the cohomology theory of vertex algebras started in [BDSHK19, BDSHK20]. Recall (see, e.g., [BDSHK19]) that, to any linear symmetric (super)operad $\mathcal{P}$ over a field $\mathbb{F}$, one canonically associates a $\mathbb{Z}$-graded Lie superalgebra

B. Bakalov
Department of Mathematics, North Carolina State University, Raleigh, NC, USA
e-mail: bojko_bakalov@ncsu.edu

A. De Sole · V. Vignoli
Dipartimento di Matematica, Sapienza Università di Roma, Rome, Italy
e-mail: desole@mat.uniroma1.it; vignoli@mat.uniroma1.it
www.mat.uniroma1.it/~desole

V. G. Kac (✉)
Department of Mathematics, MIT, Cambridge, MA, USA
e-mail: kac@math.mit.edu

A. Alekseev et al. (eds.), *Representation Theory, Mathematical Physics, and Integrable Systems*, Progress in Mathematics 340,
https://doi.org/10.1007/978-3-030-78148-4_2

$$W(\mathcal{P}) = \bigoplus_{k=-1}^{\infty} W^k(\mathcal{P}), \quad \text{where} \quad W^k(\mathcal{P}) = \mathcal{P}(k+1)^{S_{k+1}}. \tag{1.1}$$

An odd element $X \in W^1(\mathcal{P})$, satisfying $[X, X] = 0$, defines a cohomology complex $(W(\mathcal{P}), \operatorname{ad} X)$, which is a differential graded Lie superalgebra.

A cohomology theory of vertex algebras is constructed by considering the operad $\mathcal{P}_{\mathrm{ch}}(V)$, attached to a vector superspace V with an even endomorphism ∂. In order to describe this construction, let, for $n \in \mathbb{Z}_{\geq 0}$,

$$V_n = V[\lambda_1, \dots, \lambda_n] / \langle \partial + \lambda_1 + \cdots + \lambda_n \rangle,$$

where the indeterminates λ_i have even parity and $\langle \Phi \rangle$ stands for the image of the endomorphism Φ, and let

$$\mathcal{O}_n^{\star, T} = \mathbb{F}[z_i - z_j, (z_i - z_j)^{-1}]_{1 \leq i < j \leq n}.$$

The superspace $\mathcal{P}_{\mathrm{ch}}(V)(n)$ is defined as the set of all linear maps

$$Y \colon V^{\otimes n} \otimes \mathcal{O}_n^{\star, T} \to V_n, \qquad v_1 \otimes \cdots \otimes v_n \otimes f \mapsto Y_{\lambda_1, \dots, \lambda_n}(v_1 \otimes \cdots \otimes v_n \otimes f), \tag{1.2}$$

satisfying the following two sesquilinearity properties ($1 \leq i \leq n$):

$$Y_{\lambda_1, \dots, \lambda_n}(v_1 \otimes \cdots \otimes (\partial + \lambda_i) v_i \otimes \cdots \otimes v_n \otimes f) = Y_{\lambda_1, \dots, \lambda_n}\Big(v_1 \otimes \cdots \otimes v_n \otimes \frac{\partial f}{\partial z_i}\Big), \tag{1.3}$$

and

$$Y_{\lambda_1, \dots, \lambda_n}(v_1 \otimes \cdots \otimes v_n \otimes (z_i - z_j) f) = \Big(\frac{\partial}{\partial \lambda_j} - \frac{\partial}{\partial \lambda_i}\Big) Y_{\lambda_1, \dots, \lambda_n}(v_1 \otimes \cdots \otimes v_n \otimes f). \tag{1.4}$$

In [BDSHK19] we also defined the action of S_n on $\mathcal{P}_{\mathrm{ch}}(V)(n)$ and the $\circ_i$-products, making $\mathcal{P}_{\mathrm{ch}}(V)$ an operad.

As a result, we obtain the Lie superalgebra

$$W_{\mathrm{ch}}(V) = W(\mathcal{P}_{\mathrm{ch}}(V)) = \bigoplus_{k=-1}^{\infty} W_{\mathrm{ch}}^k(V),$$

see (1.1). We show in [BDSHK19] that odd elements $X \in W_{\mathrm{ch}}^1(\Pi V)$, such that $[X, X] = 0$, correspond bijectively to vertex algebra structures on the $\mathbb{F}[\partial]$-module V, such that ∂ is the translation operator (where Π stands for the reversal of parity). This leads to the vertex algebra cohomology complex $(W_{\mathrm{ch}}(\Pi V), \operatorname{ad} X)$, with coefficients in the adjoint V-module. The cohomology with coefficients in an arbitrary V-module M is obtained by a simple reduction procedure.

Now, suppose that the $\mathbb{F}[\partial]$-module V has an increasing $\mathbb{Z}_{\geq 0}$-filtration by $\mathbb{F}[\partial]$-submodules. Taking the increasing filtration of $\mathcal{O}_n^{\star,T}$ by the number of divisors, we obtain an increasing filtration of $V^{\otimes n} \otimes \mathcal{O}_n^{\star,T}$. This filtration induces a decreasing filtration of the superspace $\mathcal{P}_{\mathrm{ch}}(V)(n)$ (see [BDSHK19]). The associated graded spaces $\operatorname{gr} \mathcal{P}_{\mathrm{ch}}(V)(n)$ form a graded operad.

On the other hand, in [BDSHK19] we introduced the closely related operad $\mathcal{P}_{\mathrm{cl}}(V)$, which "governs" the Poisson vertex algebra (PVA) structures on the $\mathbb{F}[\partial]$-module V. Let $\mathbb{F}\mathcal{G}(n)$ be the vector space (with even parity) spanned by the set $\mathcal{G}(n)$ of labeled oriented graphs with n vertices. The vector superspace $\mathcal{P}_{\mathrm{cl}}(V)(n)$ is the space of linear maps (cf. (1.2))

$$Y : \mathbb{F}\mathcal{G}(n) \otimes V^{\otimes n} \to V_n \,, \qquad \Gamma \otimes v \mapsto Y^{\Gamma}(v) \,, \tag{1.5}$$

satisfying the sesquilinearity conditions (3.28) and (3.29) in Sect. 3.7, which are the "classical" analogs of (1.3) and (1.4). The corresponding $\mathbb{Z}$-graded Lie superalgebra

$$W_{\mathrm{cl}}(\Pi V) = \bigoplus_{k=-1}^{\infty} W_{\mathrm{cl}}^{k}(\Pi V)$$

is such that odd elements $X \in W_{\mathrm{cl}}^{1}(\Pi V)$ with $[X, X] = 0$ parametrize the PVA structures on the $\mathbb{F}[\partial]$-module V by

$$ab = (-1)^{p(a)} X^{\bullet \to \bullet}(a \otimes b) \,, \qquad [a_\lambda b] = (-1)^{p(a)} X^{\bullet \quad \bullet}_{\lambda, -\lambda-\partial}(a \otimes b) \,. \tag{1.6}$$

When V is endowed with an increasing $\mathbb{Z}_{\geq 0}$-filtration by $\mathbb{F}[\partial]$-submodules, we have a canonical linear map of graded operads

$$\operatorname{gr} \mathcal{P}_{\mathrm{ch}}(V) \to \mathcal{P}_{\mathrm{cl}}(\operatorname{gr} V) \,. \tag{1.7}$$

It is proved in [BDSHK19] that the map (1.7) is injective. The main result of [BDSHK20] is that this map is an isomorphism, provided that the filtration of V is induced by a grading by $\mathbb{F}[\partial]$-modules. If, in addition, this filtration of V is such that $\operatorname{gr} V$ inherits from the vertex algebra structure of V a PVA structure, then, as a result, the vertex algebra cohomology is majorized by the classical PVA cohomology:

$$\dim H^n_{\mathrm{ch}}(V) \leq \dim H^n_{\mathrm{cl}}(V) \,. \tag{1.8}$$

Unfortunately, (1.8) is not a "practical" inequality, since the direct computation of $H_{\mathrm{cl}}(V, M)$ may be very hard. However, if the superspace V is endowed with the structure of a commutative associative superalgebra with an even derivation ∂, there exists a much smaller $\mathbb{Z}$-graded Lie superalgebra, constructed in [DSK13]:

$$W_{\mathrm{PV}}(\Pi V) = \bigoplus_{k=-1}^{\infty} W_{\mathrm{PV}}^{k}(\Pi V)\,.$$

It has the property that PVA structures on the differential algebra V correspond bijectively to odd elements $X \in W_{\mathrm{PV}}^{1}(\Pi V)$ such that $[X, X] = 0$, cf. (1.6). This produces the variational Poisson cohomology complex $(W_{\mathrm{PV}}(\Pi V), \operatorname{ad} X)$. It is easy to see that this complex is a subcomplex of the complex $(W_{\mathrm{cl}}(\Pi V), \operatorname{ad} X)$, corresponding to the graphs without edges.

The present paper is the first paper towards proving that the inclusion of complexes

$$(W_{\mathrm{PV}}(\Pi V), \operatorname{ad} X) \hookrightarrow (W_{\mathrm{cl}}(\Pi V), \operatorname{ad} X) \tag{1.9}$$

induces an isomorphism in cohomology, under some assumptions on the differential algebra V. In this case, we may replace H_{cl} by H_{PV} in (1.8). Since there are by now well-developed tools for computing variational Poisson cohomology, see [DSK13] and [BDSK20], this allows one to get a hold of the vertex algebra cohomology.

In the present paper, given a vector superspace V with an even endomorphism ∂, we construct a canonical map of complexes

$$(W_{\mathrm{cl}}(\Pi V), \operatorname{ad} X) \xrightarrow{\varphi} (C_{\partial,\mathrm{Har}}(V), d)\,, \tag{1.10}$$

which restricts to a bijective linear map on the top degree:

$$\operatorname{gr}^{n-1} W_{\mathrm{cl}}^{n-1}(\Pi V) \xrightarrow{\sim} C_{\partial,\mathrm{Har}}^{n}(V)\,,$$

see Theorem 4.1 in Sect. 4. Here $X \in W_{\mathrm{cl}}^{1}(\Pi V)$ corresponds to the PVA structure on V, given by (1.6). In particular, V is endowed with a structure of a commutative associative superalgebra with an even derivation ∂, hence we may consider the differential Harrison complex

$$\Big(C_{\partial,\mathrm{Har}}(V) = \bigoplus_{n\in\mathbb{Z}_{\geq 0}} C_{\partial,\mathrm{Har}}^{n}(V)\,,\ d\Big)\,.$$

Here $C_{\partial,\mathrm{Har}}^{n}(V)$ is the subspace of $\operatorname{Hom}_{\mathbb{F}[\partial]}(V^{\otimes n}, V)$ satisfying Harrison's conditions (2.11) in Sect. 2.3. It was shown in [Har62, GS87] that the subspace $C_{\partial,\mathrm{Har}}(V)$ of $\bigoplus_{n\in\mathbb{Z}_{\geq 0}} \operatorname{Hom}_{\mathbb{F}[\partial]}(V^{\otimes n}, V)$ is invariant with respect to the Hochschild differential d (which commutes with ∂). Finally, the map φ in (1.10) maps $Y \in W_{\mathrm{cl}}^{n-1}(\Pi V)$ to Y^{Λ_n}, where Λ_n is the labeled oriented graph

$$\Lambda_n = \underset{1}{\bullet}\!\longrightarrow\!\underset{2}{\bullet}\!\longrightarrow \cdots \longrightarrow\!\underset{n}{\bullet}$$

In our next paper [BDSHKV21] we show that, if V is an algebra of differential polynomials in finitely many variables, then $H^n_{\partial,\mathrm{Har}}(V,d) = 0$ for $n > 1$ and the inclusion (1.10) is a quasi-isomorphism, hence in this case the inequality (1.8) turns into the inequality

$$\dim H^n_{\mathrm{ch}}(V) \leq \dim H^n_{\mathrm{PV}}(V) . \tag{1.11}$$

We refer to the Ph.D. thesis [Vig19] for examples and more details.

Throughout the paper, the base field $\mathbb{F}$ has characteristic 0, and, unless otherwise specified, all vector spaces, their tensor products, and Hom's are over $\mathbb{F}$.

2 Differential Harrison Cohomology Complex

In this section, we recall the definition of the Harrison cohomology complex and we introduce the differential Harrison cohomology complex.

2.1 Hochschild Cohomology Complex

First, we review the Hochschild cohomology complex, of which Harrison's is a subcomplex, see [Hoc45] and [Har62]. We use the original Harrison's definition. For other definitions see [GS87, Lod13].

Let A be an associative algebra over the base field $\mathbb{F}$, and M be an A-bimodule. We will write $A^{\otimes n}$ for the n-fold tensor product $A \otimes \cdots \otimes A$. The Hochschild cohomology complex is defined as follows. The space of n-cochains is

$$\mathrm{Hom}(A^{\otimes n}, M) , \tag{2.1}$$

and the differential $d\colon \mathrm{Hom}(A^{\otimes n}, M) \to \mathrm{Hom}(A^{\otimes n+1}, M)$ is defined by

$$\begin{aligned}(df)(a_1 \otimes \cdots \otimes a_{n+1}) &= a_1 f(a_2 \otimes \cdots \otimes a_{n+1}) \\ &+ \sum_{i=1}^{n} (-1)^i f(a_1 \otimes \cdots \otimes a_{i-1} \otimes a_i a_{i+1} \otimes a_{i+2} \otimes \cdots \otimes a_{n+1}) \\ &+ (-1)^{n+1} f(a_1 \otimes \cdots \otimes a_n) a_{n+1} .\end{aligned} \tag{2.2}$$

Then $d^2 = 0$, and we get the Hochschild cohomology complex

$$0 \longrightarrow M \xrightarrow{d} \mathrm{Hom}(A, M) \xrightarrow{d} \mathrm{Hom}(A^{\otimes 2}, M) \xrightarrow{d} \cdots . \tag{2.3}$$

If A is an associative algebra with a derivation $\partial : A \to A$, and M is a differential bimodule over A (i.e., the action of ∂ is compatible with the bimodule structure), we may consider the *differential Hochschild cohomology complex* by taking the subspace of n-cochains

$$\mathrm{Hom}_{\mathbb{F}[\partial]}(A^{\otimes n}, M) . \tag{2.4}$$

It is clear by the definition (2.2) that the differential d maps $\mathrm{Hom}_{\mathbb{F}[\partial]}(A^{\otimes n}, M)$ to $\mathrm{Hom}_{\mathbb{F}[\partial]}(A^{\otimes n+1}, M)$. Hence, we have a cohomology subcomplex.

Remark 2.1 It is straightforward, using the Koszul–Quillen rule, to extend the definition of the Hochschild complex to the case when A is an associative superalgebra, as well as all other definitions and results of the paper. We restricted here to the purely even case for the simplicity of the exposition.

2.2 Monotone Permutations

Consider the symmetric group S_n. Using Harrison's notation in [Har62] (see also [GS87]), we have the following definition:

Definition 2.2 A permutation $\pi \in S_n$ is called *monotone* if, for each $i = 1, \ldots, n$, one of the following two conditions holds:

(a) $\pi(j) < \pi(i)$ for all $j < i$;
(b) $\pi(j) > \pi(i)$ for all $j < i$.

(Not necessarily the same condition (a) or (b) holds for every i.) When (b) holds, we call i a *drop* of π. Also, $\pi(1) = k$ is called the *start* of π (and we say that π *starts* at k).

We denote by $\mathcal{M}_n \subset S_n$ the set of monotone permutations, and by $\mathcal{M}_n^k \subset \mathcal{M}_n$ the set of monotone permutations starting at k.

Here is a simple description of all monotone permutations starting at k. Let us identify the permutation $\pi \in S_n$ with the n-tuple $[\pi(1), \ldots, \pi(n)]$. To construct all $\pi \in \mathcal{M}_n^k$, we let $\pi(1) = k$. Then, for every choice of $k - 1$ positions in $\{2, \ldots, n\}$ we get a monotone permutation π as follows. In the selected positions we put the numbers 1 to $k - 1$ in decreasing order from left to right; in the remaining positions we write the numbers $k + 1$ to n in increasing order from left to right. (The selected positions are the drops of π.)

According to the above description, we have a bijective correspondence

$$\mathcal{M}_n^k \xrightarrow{\sim} \left\{ D \subset \{2, \ldots, n\} \,\middle|\, |D| = k - 1 \right\} , \tag{2.5}$$

associating the monotone permutation $\pi \in \mathcal{M}_n^k$ to the set $D(\pi)$ of drops of π, which are

$$\pi^{-1}(k-1) < \pi^{-1}(k-2) < \cdots < \pi^{-1}(1) \in \{2, \ldots, n\}.$$

Example 2.3 The only monotone permutation starting at 1 is the identity, while the only monotone permutation starting at n is

$$\sigma_n = [n \;\; n-1 \;\; \cdots \;\; 2 \;\; 1]. \tag{2.6}$$

Example 2.4 Let $n = 5$ and $k = 3$. The monotone permutations starting at 3 are

$$\begin{aligned} &[3 \;\; \underline{2} \;\; \underline{1} \;\; 4 \;\; 5], \quad [3 \;\; \underline{2} \;\; 4 \;\; \underline{1} \;\; 5], \quad [3 \;\; \underline{2} \;\; 4 \;\; 5 \;\; \underline{1}], \\ &[3 \;\; 4 \;\; \underline{2} \;\; \underline{1} \;\; 5], \quad [3 \;\; 4 \;\; \underline{2} \;\; 5 \;\; \underline{1}], \quad [3 \;\; 4 \;\; 5 \;\; \underline{2} \;\; \underline{1}], \end{aligned}$$

where we underlined the positions of the drops.

Given a monotone permutation π, we denote by $\mathrm{dr}(\pi)$ the sum of all the drops with respect to π. According to the previous description, we can easily see that

$$(-1)^{\mathrm{dr}(\pi)} = (-1)^{k-1} \operatorname{sign}(\pi), \tag{2.7}$$

if k is the start of π.

Note that the description (2.5) of $\mathcal{M}_n^k$ in terms of positions of drops allows us to count the number of elements in $\mathcal{M}_n^k$, for fixed n and k. We have

$$|\mathcal{M}_n^k| = \binom{n-1}{k-1}. \tag{2.8}$$

Remark 2.5 Let us denote by $\mathcal{M}_n^{k,k-1} \subset \mathcal{M}_n^k$ the subset of all monotone permutations π starting at k with $\pi(2) = k-1$, and by $\mathcal{M}_n^{k,k+1} \subset \mathcal{M}_n^k$ the subset of all monotone permutations π starting at k with $\pi(2) = k+1$. Then we have

$$\mathcal{M}_n^k = \mathcal{M}_n^{k,k-1} \sqcup \mathcal{M}_n^{k,k+1}.$$

Lemma 2.6 *There are natural identifications*

$$\mathcal{M}_n^{k,k-1} \simeq \mathcal{M}_{n-1}^{k-1} \quad (\textit{resp. } \mathcal{M}_n^{k,k+1} \simeq \mathcal{M}_{n-1}^k),$$

mapping $\pi \in \mathcal{M}_n^{k,k-1}$ to $\bar{\pi} \in \mathcal{M}_{n-1}^{k-1}$ (resp. $\pi \in \mathcal{M}_n^{k,k+1}$ to $\bar{\pi} \in \mathcal{M}_{n-1}^k$), given by $\bar{\pi}(1) := k-1$ (resp. k), and, for $i = 2, \ldots, n-1$,

$$\bar{\pi}(i) := \begin{cases} \pi(i+1), & \textit{if } \pi(i+1) < k \\ \pi(i+1) - 1, & \textit{if } \pi(i+1) > k \end{cases}. \tag{2.9}$$

Moreover,

$$(-1)^{\mathrm{dr}(\bar{\pi})} = (-1)^{\mathrm{dr}(\pi)+k} \quad (\textit{resp. } (-1)^{\mathrm{dr}(\pi)+k-1}) .$$

Proof Straightforward; see [Vig19] for details. □

Remark 2.7 Observe that, given a monotone permutation π, either $\pi(n) = 1$ or $\pi(n) = n$. Denote by ${}^1\mathcal{M}_n^k \subset \mathcal{M}_n^k$ the set of all the monotone permutations π starting at k with $\pi(n) = 1$, and by ${}^n\mathcal{M}_n^k \subset \mathcal{M}_n^k$ the set of all the monotone permutations π starting at k with $\pi(n) = n$. As in Remark 2.5, we have

$$\mathcal{M}_n^k = {}^1\mathcal{M}_n^k \sqcup {}^n\mathcal{M}_n^k .$$

Lemma 2.8 *There are natural identifications*

$${}^1\mathcal{M}_n^k \simeq \mathcal{M}_{n-1}^{k-1} \quad (\textit{resp. } {}^n\mathcal{M}_n^k \simeq \mathcal{M}_{n-1}^k) ,$$

mapping $\pi \in {}^1\mathcal{M}_n^k$ *to* $\tilde{\pi} \in \mathcal{M}_{n-1}^{k-1}$ *(resp.* $\pi \in {}^n\mathcal{M}_n^k$ *to* $\tilde{\pi} \in \mathcal{M}_{n-1}^k$*), given by* $\tilde{\pi}(i) := \pi(i) - 1$ *(resp.* $\tilde{\pi}(i) := \pi(i)$*), for* $i = 1, \dots, n-1$. *Moreover,*

$$(-1)^{\mathrm{dr}(\tilde{\pi})} = (-1)^{\mathrm{dr}(\pi)+n} \quad (\textit{resp. } (-1)^{\mathrm{dr}(\pi)}) .$$

Proof Straightforward; see [Vig19] for details. □

2.3 Differential Harrison Cohomology Complex

Let us now recall Harrison's original definition of his cohomology complex [Har62]. Let A be a commutative associative algebra, and M be a symmetric A-bimodule, i.e., such that $am = ma$, for all $a \in A$ and $m \in M$. For every $1 < k \leq n$ define the following endomorphism on the space $\mathrm{Hom}(A^{\otimes n}, M)$:

$$(L_k F)(a_1 \otimes \cdots \otimes a_n) := \sum_{\pi \in \mathcal{M}_n^k} (-1)^{\mathrm{dr}(\pi)} F(a_{\pi(1)} \otimes \cdots \otimes a_{\pi(n)}) . \tag{2.10}$$

A *Harrison n-cochain* is defined as a Hochschild n-cochain $F \in \mathrm{Hom}(A^{\otimes n}, M)$ fixed by all operators L_k:

$$L_k F = F , \quad \text{for every } 2 \leq k \leq n . \tag{2.11}$$

We will denote by

$$C^n_{\mathrm{Har}}(A, M) \subset \mathrm{Hom}(A^{\otimes n}, M) \tag{2.12}$$

the space of Harrison n-cochains.

Furthermore, if A is a differential algebra with a derivation $\partial : A \to A$, and M is a symmetric differential bimodule, we may consider the space of *differential Harrison n-cochains*

$$C^n_{\partial,\mathrm{Har}}(A, M) \subset \mathrm{Hom}_{\mathbb{F}[\partial]}(A^{\otimes n}, M) , \tag{2.13}$$

again defined by Harrison's conditions (2.11).

Proposition 2.9

(a) *The Harrison complex $(C_{\mathrm{Har}}(A, M), d)$ is a subcomplex of the Hochschild complex.*
(b) *If A is a differential algebra, with a derivation $\partial : A \to A$, the differential Harrison complex $(C_{\partial,\mathrm{Har}}(A, M), d)$ is a subcomplex of the differential Hochschild complex.*

Proof The proof of (a) is in [Har62, GS87]. Part (b) is straightforward. □

Remark 2.10 Clearly, $H^0_{\partial,\mathrm{Har}}(A, M) = M$ and $H^1_{\partial,\mathrm{Har}}(A, M) = \mathrm{Der}_{\mathbb{F}[\partial]}(A, M)$. It follows from [GS87] that $H^n_{\partial,\mathrm{Har}}(A, M)$ is a direct summand of the differential Hochschild cohomology $HH^n_\partial(A, M)$, for $n \geq 2$.

3 The Classical Operad and PVA Cohomology

In this section, we recall some basic notions that will be used throughout the paper and review the construction of the PVA cohomology complex as described in [BDSHK19].

3.1 Symmetric Group Actions

There is a natural left action of S_n on an arbitrary n-tuple of objects $(\lambda_1, \dots, \lambda_n)$:

$$\sigma(\lambda_1, \dots, \lambda_n) = (\lambda_{\sigma^{-1}(1)}, \dots, \lambda_{\sigma^{-1}(n)}) , \quad \sigma \in S_n . \tag{3.1}$$

Also, given $V = V_{\bar{0}} \oplus V_{\bar{1}}$ a vector superspace with parity p, we have a linear left action of the symmetric group S_n on the tensor product $V^{\otimes n}$ ($\sigma \in S_n$, $v_1, \dots, v_n \in V$):

$$\sigma(v_1 \otimes \cdots \otimes v_n) := \epsilon_v(\sigma)\, v_{\sigma^{-1}(1)} \otimes \cdots \otimes v_{\sigma^{-1}(n)} , \tag{3.2}$$

where, following the Koszul–Quillen rule,

$$\epsilon_v(\sigma) := \prod_{i<j\,:\,\sigma(i)>\sigma(j)} (-1)^{p(v_i)p(v_j)} . \tag{3.3}$$

In particular, if V is purely even $\epsilon_v(\sigma) = 1$, while if V is purely odd $\epsilon_v(\sigma) = \text{sign}(\sigma)$. The corresponding right action of S_n on the space $\text{Hom}(V^{\otimes n}, V)$ is given by ($f \in \text{Hom}(V^{\otimes n}, V)$, $\sigma \in S_n$):

$$f^\sigma(v_1 \otimes \cdots \otimes v_n) = f(\sigma(v_1 \otimes \cdots \otimes v_n)) . \tag{3.4}$$

3.2 Composition of Permutations and Shuffles

Let $n \geq 1$ and $m_1, \ldots, m_n \geq 0$. We introduce the following notation:

$$M_0 = 0 \quad \text{and} \quad M_i = \sum_{j=1}^{i} m_j , \quad i = 1, \ldots, n . \tag{3.5}$$

Given $\sigma \in S_n$ and $\tau_1 \in S_{m_1}, \ldots, \tau_n \in S_{m_n}$, we describe the *composition*

$$\sigma(\tau_1, \ldots, \tau_n) \in S_{M_n}$$

by saying how it acts on the tensor power $V^{\otimes M_n}$ of a vector space V:

$$(\sigma(\tau_1, \ldots, \tau_n))(v_1 \otimes \cdots \otimes v_{M_n}) = \sigma(\tau_1(v_1 \otimes \cdots \otimes v_{M_1}) \otimes \cdots \otimes \tau_n(v_{M_{n-1}+1} \otimes \cdots \otimes v_{M_n})) . \tag{3.6}$$

Definition 3.1 A permutation $\sigma \in S_{m+n}$ is called an (m, n)-*shuffle* if

$$\sigma(1) < \cdots < \sigma(m) , \quad \sigma(m+1) < \cdots < \sigma(m+n) .$$

The subset of (m, n)-shuffles is denoted by $S_{m,n} \subset S_{m+n}$.

Observe that, by definition, $S_{0,n} = S_{n,0} = \{1\}$ for every $n \geq 0$. If either m or n is negative, we set $S_{m,n} = \emptyset$ by convention.

3.3 n-Graphs

For an oriented graph Γ, we denoted by $V(\Gamma)$ the set of vertices of Γ, and by $E(\Gamma)$ the set of edges. We call Γ an *n-graph* if $V(\Gamma) = \{1, \ldots, n\}$. Denote by $\mathcal{G}(n)$ the set of all n-graphs without tadpoles, and by $\mathcal{G}_0(n)$ the set of all acyclic n-graphs.

An n-graph L will be called an *n-line*, or simply a *line*, if its set of edges is of the form $\{i_1 \to i_2, i_2 \to i_3, \ldots, i_{n-1} \to i_n\}$, where $\{i_1, \ldots, i_n\}$ is a permutation of $\{1, \ldots, n\}$.

We have a natural left action of S_n on the set $\mathcal{G}(n)$: for the n-graph Γ and the permutation σ, the new n-graph $\sigma(\Gamma)$ is defined to be the same graph as Γ but with the vertex which was labeled as i relabeled as $\sigma(i)$, for every $i = 1, \dots, n$. So, if the n-graph Γ has an oriented edge $i \to j$, then the n-graph $\sigma(\Gamma)$ has the oriented edge $\sigma(i) \to \sigma(j)$. Note that S_n permutes the set of n-lines.

Let us recall the cocomposition of n-graphs, as described in [BDSHK19]. Given an n-tuple $(m_1, \dots, m_n)$ of positive integers, let M_i be as in (3.5). If $\Gamma \in \mathcal{G}(M_n)$, define $\Delta_i^{m_1,\dots,m_n}(\Gamma) \in \mathcal{G}(m_i)$, $i = 1, \dots, n$, to be the subgraph of Γ associated with the set of vertices $\{M_{i-1} + 1, \dots, M_i\}$, relabeled as $\{1, \dots, m_i\}$. Define also $\Delta_0^{m_1,\dots,m_n}(\Gamma)$ to be the graph obtained from Γ by collapsing the vertices and the edges of each $\Delta_i^{m_1,\dots,m_n}(\Gamma)$ into a single vertex, relabeled as i. Then the cocomposition map is the map

$$
\begin{aligned}
\Delta^{m_1,\dots,m_n} : \mathcal{G}(M_n) &\to \mathcal{G}(n) \times \mathcal{G}(m_1) \times \cdots \times \mathcal{G}(m_n)\,, \\
\Gamma &\mapsto \big(\Delta_0^{m_1,\dots,m_n}(\Gamma),\, \Delta_1^{m_1,\dots,m_n}(\Gamma), \dots, \Delta_n^{m_1,\dots,m_n}(\Gamma)\big)\,.
\end{aligned} \tag{3.7}
$$

Example 3.2 Let $n = 3$, $(m_1, m_2, m_3) = (3, 1, 4)$, and $\Gamma \in \mathcal{G}(8)$ be the following graph

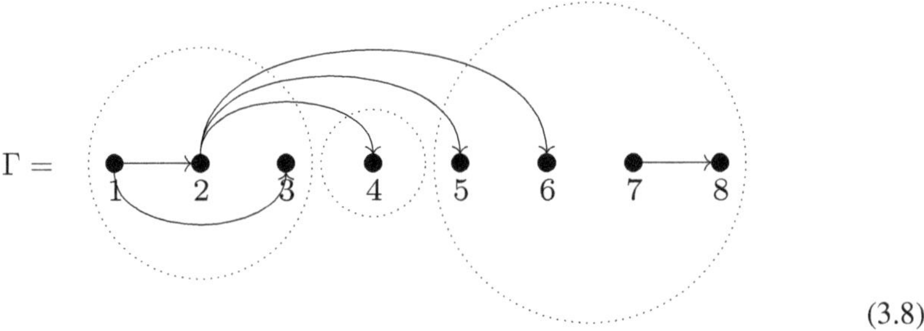

(3.8)

The cocomposition $\Delta^{3,1,4}(\Gamma) = \big(\Delta_0^{3,1,4}(\Gamma), \Delta_1^{3,1,4}(\Gamma), \Delta_2^{3,1,4}(\Gamma), \Delta_3^{3,1,4}(\Gamma)\big)$ is given by the following graphs. $\Delta_1^{3,1,4}(\Gamma)$ is the subgraph of Γ generated by the first three vertices:

$\Delta_1^{3,1,4}(\Gamma) =$ 1 → 2 3 $\in \mathcal{G}(3)$;

$\Delta_2^{3,1,4}(\Gamma)$ is the subgraph of Γ associated with the fourth vertex:

$\Delta_2^{3,1,4}(\Gamma) =$ 1 $\in \mathcal{G}(1)$;

$\Delta_3^{3,1,4}(\Gamma)$ is the subgraph of Γ associated with the last four vertices:

$$\Delta_3^{3,1,4}(\Gamma) = \quad \underset{1}{\bullet} \quad \underset{2}{\bullet} \quad \underset{3}{\bullet} \longrightarrow \underset{4}{\bullet} \quad \in \mathcal{G}(4)\,;$$

and, finally, $\Delta_0^{3,1,4}(\Gamma)$ is

$$\Delta_0^{3,1,4}(\Gamma) = \quad \underset{1}{\bullet} \quad \underset{2}{\bullet} \quad \underset{3}{\bullet} \qquad \in \mathcal{G}(3)\,.$$

From the construction of $\Delta_i^{m_1,\dots,m_n}(\Gamma)$, it is easy to see that there is a natural bijective correspondence

$$\Delta\colon E(\Gamma) \xrightarrow{\sim} E\big(\Delta_0^{m_1\dots m_n}(\Gamma)\big) \sqcup E\big(\Delta_1^{m_1\dots m_n}(\Gamma)\big) \sqcup \dots \sqcup E\big(\Delta_n^{m_1\dots m_n}(\Gamma)\big)\,. \tag{3.9}$$

Definition 3.3 Let $k \in \{1, \dots, M_n\}$ and $j \in \{1, \dots, n\}$. We say that j is *externally connected* to k (via the graph Γ and its cocomposition $\Delta^{m_1\dots m_n}(\Gamma)$) if there is an unoriented path (without repeating edges) of $\Delta_0^{m_1\dots m_n}(\Gamma)$ joining j to i, where $i \in \{1, \dots, n\}$ is such that $k \in \{M_{i-1}+1, \dots, M_i\}$, and the edge out of i is the image, via the map Δ in (3.9), of an edge which has its head or tail in k. Given a set of variables $x_1, \dots, x_n$, we denote

$$X(k) = \sum_{\substack{j \text{ externally} \\ \text{connected to } k}} x_j\,. \tag{3.10}$$

Example 3.4 For the graph (3.8), we have

$$X(1) = 0,\ X(2) = x_1 + x_2 + x_3,\ X(3) = 0,\ X(4) = x_1 + x_3,\ X(5) = x_1 + x_2 + x_3,$$
$$X(6) = x_1 + x_2 + x_3,\ X(7) = 0,\ X(8) = 0.$$

3.4 Lie Conformal Algebras and Poisson Vertex Algebras

Definition 3.5 A Lie conformal (super)algebra is a vector (super)space V, endowed with an even endomorphism $\partial \in \mathrm{End}(V)$ and a bilinear (over $\mathbb{F}$) λ-bracket $[\cdot\,_\lambda\,\cdot]\colon V \times V \to V[\lambda]$ satisfying sesquilinearity ($a, b \in V$):

$$[\partial a_\lambda b] = -\lambda[a_\lambda b]\,, \qquad [a_\lambda \partial b] = (\lambda + \partial)[a_\lambda b]\,, \tag{3.11}$$

skew symmetry ($a, b \in V$):

$$[a_\lambda b] = -(-1)^{p(a)p(b)}[b_{-\lambda-\partial}a]\,, \tag{3.12}$$

and the Jacobi identity ($a, b, c \in V$):

$$[a_\lambda[b_\mu c]] - (-1)^{p(a)p(b)}[b_\mu[a_\lambda, b]] = [[a_\lambda b]_{\lambda+\mu}c]\,. \tag{3.13}$$

Definition 3.6 A Poisson vertex (super)algebra (PVA) is a commutative associative (super)algebra V endowed with an even derivation ∂ and a Lie conformal (super)algebra λ-bracket $[\cdot_\lambda\cdot]$ that satisfies the left Leibniz rule

$$[a_\lambda bc] = [a_\lambda b]c + (-1)^{p(a)p(b)}b[a_\lambda c]\,. \tag{3.14}$$

3.5 Operads

Recall that a (linear, unital, symmetric) *superoperad* $\mathcal{P}$ is a collection of vector superspaces $\mathcal{P}(n)$, $n \geq 0$, with parity p, endowed, for every $f \in \mathcal{P}(n)$ and $m_1, \dots, m_n \geq 0$, with a *composition* parity preserving linear map,

$$\begin{aligned} \mathcal{P}(n) \otimes \mathcal{P}(m_1) \otimes \cdots \otimes \mathcal{P}(m_n) &\rightarrow \mathcal{P}(M_n)\,, \\ f \otimes g_1 \otimes \cdots \otimes g_n &\mapsto f(g_1 \otimes \cdots \otimes g_n)\,, \end{aligned} \tag{3.15}$$

where M_n is as in (3.5), satisfying the following associativity axiom:

$$f\big((g_1 \otimes \cdots \otimes g_n)(h_1 \otimes \cdots \otimes h_{M_n})\big) = \big(f(g_1 \otimes \cdots \otimes g_n)\big)(h_1 \otimes \cdots \otimes h_{M_n}) \in \mathcal{P}\Big(\sum_{j=1}^{M_n} \ell_j\Big)\,, \tag{3.16}$$

for every $f \in \mathcal{P}(n)$, $g_i \in \mathcal{P}(m_i)$ for $i = 1, \dots, n$, and $h_j \in \mathcal{P}(\ell_j)$ for $j = 1, \dots, M_n$. In the left-hand side of (3.16) the linear map

$$g_1 \otimes \cdots \otimes g_n \colon \bigotimes_{j=1}^{M_n} \mathcal{P}(\ell_j) \rightarrow \bigotimes_{i=1}^{n} \mathcal{P}\Big(\sum_{j=M_{i-1}+1}^{M_i} \ell_j\Big)$$

is the tensor product of composition maps applied to

$$h_1 \otimes \cdots \otimes h_{M_n} = (h_1 \otimes \cdots \otimes h_{M_1}) \otimes (h_{M_1+1} \otimes \cdots \otimes h_{M_2}) \otimes \cdots \otimes (h_{M_{n-1}+1} \otimes \cdots \otimes h_{M_n})\,.$$

We assume that $\mathcal{P}$ is endowed with a *unit* element $1 \in \mathcal{P}(1)$ satisfying the following unity axioms:

$$f(1 \otimes \cdots \otimes 1) = 1(f) = f\,, \quad \text{for every} \quad f \in \mathcal{P}(n)\,. \tag{3.17}$$

Furthermore, we assume that, for each $n \geq 1$, $\mathcal{P}(n)$ has a right action of the symmetric group S_n, denoted f^σ, for $f \in \mathcal{P}(n)$ and $\sigma \in S_n$, satisfying the following equivariance axiom ($f \in \mathcal{P}(n)$, $g_1 \in \mathcal{P}(m_1), \dots, g_n \in \mathcal{P}(m_n)$, $\sigma \in S_n$, $\tau_1 \in S_{m_1}, \dots, \tau_n \in S_{m_n}$):

$$f^\sigma(g_1^{\tau_1} \otimes \cdots \otimes g_n^{\tau_n}) = \big(f(\sigma(g_1 \otimes \cdots \otimes g_n))\big)^{\sigma(\tau_1,\dots,\tau_n)}\,, \tag{3.18}$$

where the left action of $\sigma \in S_n$ on the tensor product of vector superspaces was defined in (3.2), and the composition $\sigma(\tau_1, \dots, \tau_n)$ is described in (3.6).

For simplicity, from now on, we will use the term operad in place of superoperad. Given an operad $\mathcal{P}$, one defines, for each $i = 1, \dots, n$, the $\circ_i$-product $\circ_i : \mathcal{P}(n) \otimes \mathcal{P}(m) \to \mathcal{P}(n+m-1)$ by insertion in position i, i.e.,

$$f \circ_i g = f(\overbrace{1 \otimes \cdots \otimes 1}^{i-1} \otimes \overset{i}{g} \otimes \overbrace{1 \otimes \cdots \otimes 1}^{n-i})\,. \tag{3.19}$$

Example 3.7 The simplest example of an operad is $\mathcal{P} = \mathcal{H}om$. Given a vector superspace V, $\mathcal{H}om = \mathcal{H}om(V)$ is defined as the collection of

$$\mathcal{H}om(n) := \operatorname{Hom}(V^{\otimes n}, V)\,, \qquad n \geq 0\,,$$

endowed with the composition maps ($f \in \mathcal{H}om(n)$, $g_i \in \mathcal{H}om(m_i)$ for $i = 1, \dots, n$, $v_j \in V$ for $j = 1, \dots, M_n$)

$$(f(g_1 \otimes \cdots \otimes g_n))(v_1 \otimes \cdots \otimes v_{M_n}) := f((g_1 \otimes \cdots \otimes g_n)(v_1 \otimes \cdots \otimes v_{M_n}))\,,$$

where M_n is as in (3.5). $\mathcal{H}om$ is a unital operad with unity $1 = \mathbb{1}_V \in \operatorname{End}(V)$, and the right action of S_n on $\mathcal{H}om(n)$ is given by (3.4).

3.6 The $\mathbb{Z}$-graded Lie Superalgebra Associated with an Operad

Recall that, given an operad $\mathcal{P}$, one can construct the associated $\mathbb{Z}$-graded Lie superalgebra $W(\mathcal{P})$. It is defined, as a $\mathbb{Z}$-graded vector superspace

$$W(\mathcal{P}) = \sum_{n \geq -1} W^n(\mathcal{P}) = \sum_{n \geq -1} \mathcal{P}(n+1)^{S_{n+1}}\,, \tag{3.20}$$

with the following Lie bracket. For $f \in W^n(\mathcal{P})$ and $g \in W^m(\mathcal{P})$, their $\Box$-product is defined by

$$f\Box g = \sum_{\sigma\in S_{m+1,n}} (f \circ_1 g)^{\sigma^{-1}} \in W^{m+n}(\mathcal{P}) \tag{3.21}$$

and the Lie bracket on $W(\mathcal{P})$ is given by

$$[f, g] = f\Box g - (-1)^{p(f)p(g)} g\Box f\,. \tag{3.22}$$

See, e.g., [BDSHK19, Sec. 3] for details.

3.7 The Classical Operad $\mathcal{P}_{\mathrm{cl}}$ [BDSHK19]

Let $V = V_{\bar{0}} \oplus V_{\bar{1}}$ be a vector superspace with parity p, endowed with an even endomorphism $\partial \in \operatorname{End} V$. For $n \geq 0$, define $\mathcal{P}_{\mathrm{cl}}(n)$ as the vector superspace of all maps

$$f : \mathcal{G}(n) \times V^{\otimes n} \longrightarrow V[\lambda_1, \dots, \lambda_n]/\langle \partial + \lambda_1 + \cdots + \lambda_n\rangle\,, \tag{3.23}$$

which are linear in the second factor, mapping the n-graph $\Gamma \in \mathcal{G}(n)$ (by definition $p(\Gamma) = \bar{0}$) and the monomial $v_1 \otimes \cdots \otimes v_n \in V^{\otimes n}$ to the polynomial

$$f^{\Gamma}_{\lambda_1,\dots,\lambda_n}(v_1 \otimes \cdots \otimes v_n)\,, \tag{3.24}$$

satisfying the cycle relations and the sesquilinearity conditions described as follows.

The *cycle relations* say that

$$\text{if } \Gamma \notin \mathcal{G}_0(n)\,, \quad \text{then} \quad f^{\Gamma} = 0\,, \tag{3.25}$$

and

$$\text{if } C \subset E(\Gamma) \ \text{ is an oriented cycle of } \Gamma\,, \quad \text{then} \quad \sum_{e\in C} f^{\Gamma\backslash e} = 0\,, \tag{3.26}$$

where $\Gamma\backslash e$ is the graph obtained from Γ by removing the edge e. Observe that for oriented cycles of length 2, the cycle relation (3.26) means that changing orientation of a single edge of the n-graph $\Gamma \in \mathcal{G}(n)$ amounts to a change of sign of f^{Γ}.

The *sesquilinearity conditions* are as follows. Let $\Gamma = \Gamma_1 \sqcup \cdots \sqcup \Gamma_s$ be the decomposition of Γ as a disjoint union of its connected components, and let $I_1, \dots, I_s \subset \{1, \dots, n\}$ be the sets of vertices associated with these connected components. For a graph $\tilde{\Gamma}$ and its set of vertices $\tilde{I} \subset \{1, \dots, n\}$, we write

$$\lambda_{\tilde{\Gamma}} = \sum_{i \in \tilde{I}} \lambda_i , \qquad \partial_{\tilde{\Gamma}} = \sum_{i \in \tilde{I}} \partial_i , \tag{3.27}$$

where ∂_i denotes the action of ∂ on the i-th factor in the tensor product $V^{\otimes n}$. Then, for every $\alpha = 1, \dots, s$,

$$\frac{\partial}{\partial \lambda_i} f^{\Gamma}_{\lambda_1,\dots,\lambda_n}(v_1 \otimes \cdots \otimes v_n) \text{ is the same for all } i \in I_\alpha \tag{3.28}$$

and

$$f^{\Gamma}_{\lambda_1,\dots,\lambda_n}(\partial_{\Gamma_\alpha}(v_1 \otimes \cdots \otimes v_n)) = -\lambda_{\Gamma_\alpha} f^{\Gamma}_{\lambda_1,\dots,\lambda_n}(v_1 \otimes \cdots \otimes v_n) . \tag{3.29}$$

Observe that the second sesquilinearity condition (3.29) implies

$$f^{\Gamma}_{\lambda_1,\dots,\lambda_n}(\partial v) = -\sum_{i=1}^{n} \lambda_i \, f^{\Gamma}_{\lambda_1,\dots,\lambda_n}(v) = \partial\big(f^{\Gamma}_{\lambda_1,\dots,\lambda_n}(v)\big), \qquad v \in V^{\otimes n} . \tag{3.30}$$

Remark 3.8 When the graph Γ is connected, the first sesquilinearity condition (3.28) implies that $f^{\Gamma}_{\lambda_1,\dots,\lambda_n}(v_1 \otimes \cdots \otimes v_n)$ is a polynomial of $\lambda_1 + \cdots + \lambda_n$. Hence, it is an element of

$$V[\lambda_1 + \cdots + \lambda_n] \big/ \langle \partial + \lambda_1 + \cdots + \lambda_n \rangle \simeq V .$$

In this case, we will omit the subscripts of f^{Γ}.

The classical operad $\mathcal{P}_{\mathrm{cl}}(V)$ is defined as the collection of the vector superspaces $\mathcal{P}_{\mathrm{cl}}(n)$, $n \geq 0$, endowed, for every $f \in \mathcal{P}_{\mathrm{cl}}(n)$ and $m_1, \dots, m_n \geq 0$, with the composition parity preserving linear map

$$\begin{aligned} \mathcal{P}_{\mathrm{cl}}(n) \otimes \mathcal{P}_{\mathrm{cl}}(m_1) \otimes \cdots \otimes \mathcal{P}_{\mathrm{cl}}(m_n) &\to \mathcal{P}_{\mathrm{cl}}(M_n) , \\ f \otimes g_1 \otimes \cdots \otimes g_n &\mapsto f(g_1, \dots, g_n) , \end{aligned}$$

described as follows. Let M_i be as in (3.5), and

$$\Lambda_i = \sum_{j=M_{i-1}+1}^{M_i} \lambda_j , \qquad i = 1, \dots, n . \tag{3.31}$$

If $\Gamma \in \mathcal{G}(M_n)$, then

$$(f(g_1, \dots, g_n))^{\Gamma} : V^{\otimes M_n} \to V[\lambda_1, \dots, \lambda_{M_n}] \big/ \langle \partial + \lambda_1 + \cdots + \lambda_{M_n} \rangle$$

is defined by the formula:

$$
\begin{aligned}
&(f(g_1,\dots,g_n))^{\Gamma}_{\lambda_1,\dots,\lambda_{M_n}}(v_1\otimes\cdots\otimes v_{M_n})\\
&= f^{\Delta_0^{m_1\dots m_n}(\Gamma)}_{\Lambda_1,\dots,\Lambda_n}\Bigg(\Bigg(\Big(\Big|_{x_1=\Lambda_1+\partial}(g_1)^{\Delta_1^{m_1\dots m_n}(\Gamma)}_{\lambda_1+X(1),\dots,\lambda_{M_1}+X(M_1)}\Big)\otimes\cdots\\
&\cdots\otimes\Big(\Big|_{x_n=\Lambda_n+\partial}(g_n)^{\Delta_n^{m_1\dots m_n}(\Gamma)}_{\lambda_{M_{n-1}+1}+X(M_{n-1}+1),\dots,\lambda_{M_n}+X(M_n)}\Big)\Bigg)(v_1\otimes\cdots\otimes v_{M_n})\Bigg),
\end{aligned}
\tag{3.32}
$$

where $\Delta^{m_1,\dots,m_n}(\Gamma)$ is the cocomposition of Γ described in Sect. 3.3, $X(1),\dots\ \dots,X(M_n)$ are the variables as in (3.10), and the notation is as follows. For given graphs $\Gamma_1\in\mathcal{G}(m_1),\ \dots,\Gamma_n\in\mathcal{G}(m_n)$, we have

$$
\begin{aligned}
&\Big((g_1)^{\Gamma_1}_{\lambda_1,\dots,\lambda_{M_1}}\otimes\cdots\otimes(g_n)^{\Gamma_n}_{\lambda_{M_{n-1}+1},\dots,\lambda_{M_n}}\Big)(v_1\otimes\cdots\otimes v_{M_n})\\
&:=(-1)^{\sum_{i<j}p(g_j)(p(v_{M_{i-1}+1})+\cdots+p(v_{M_i}))}(g_1)^{\Gamma_1}_{\lambda_1,\dots,\lambda_{M_1}}(v_1\otimes\cdots\otimes v_{M_1})\otimes\cdots\\
&\qquad\cdots\otimes(g_n)^{\Gamma_n}_{\lambda_{M_{n-1}+1},\dots,\lambda_{M_n}}(v_{M_{n-1}+1}\otimes\cdots\otimes v_{M_n}),
\end{aligned}
\tag{3.33}
$$

and for polynomials $P(\lambda)=\sum_m p_m\lambda^m$ and $Q(\mu)=\sum_n q_n\mu^n$ with coefficients in V, we write

$$
\big(\big|_{x=\partial}P(\lambda+y)\big)\otimes\big(\big|_{y=\partial}Q(\mu+x)\big)=\sum_{m,n}((\mu+\partial)^n p_m)\otimes((\lambda+\partial)^m q_n). \tag{3.34}
$$

For each $n\geq 1$, $\mathcal{P}_{\mathrm{cl}}(n)$ has a natural right action of the symmetric group S_n, which is given by ($f\in\mathcal{P}_{\mathrm{cl}}(n)$, $\Gamma\in\mathcal{G}(n)$, $v_1,\dots,v_n\in V$):

$$
(f^{\sigma})^{\Gamma}_{\lambda_1,\dots,\lambda_n}(v_1\otimes\cdots\otimes v_n)=f^{\sigma(\Gamma)}_{\sigma(\lambda_1,\dots,\lambda_n)}(\sigma(v_1\otimes\cdots\otimes v_n)), \tag{3.35}
$$

where $\sigma(\lambda_1,\dots,\lambda_n)$ is defined by (3.1), $\sigma(v_1\otimes\cdots\otimes v_n)$ is defined by (3.2), and $\sigma(\Gamma)$ is defined in Sect. 3.3.

On the space $\mathcal{P}_{\mathrm{cl}}(n)$ we can also define a grading:

$$
\mathcal{P}_{\mathrm{cl}}(n)=\bigoplus_{r\geq 0}\mathrm{gr}^r\,\mathcal{P}_{\mathrm{cl}}(n), \tag{3.36}
$$

where $\mathrm{gr}^r\,\mathcal{P}_{\mathrm{cl}}(n)$ is the subspace of all maps in $\mathcal{P}_{\mathrm{cl}}(n)$ vanishing on graphs with a number of edges not equal to r. Then $\mathcal{P}_{\mathrm{cl}}$ is a graded operad, i.e., the compositions and the actions of the symmetric groups are compatible with the grading.

3.8 PVA Cohomology [BDSHK19]

Given a vector superspace V with parity p, and an even endomorphism $\partial \in \mathrm{End}(V)$, let ΠV be the same vector space with reversed parity $\bar{p} = 1 - p$, and consider the corresponding classical operad $\mathcal{P}_{\mathrm{cl}}(\Pi V)$ from Sect. 3.7. The associated $\mathbb{Z}$-graded Lie superalgebra is $W_{\mathrm{cl}}(\Pi V) := W(\mathcal{P}_{\mathrm{cl}}(\Pi V))$, with Lie bracket defined by (3.22).

Theorem 3.9 ([BDSHK19, Theorem 10.7]) *We have a bijective correspondence between the odd elements $X \in W^1_{\mathrm{cl}}(\Pi V)$, such that $X \square X = 0$, and the Poisson vertex superalgebra structures on V, defined as follows. The commutative associative product and the λ-bracket of the Poisson vertex superalgebra V corresponding to X are given by*

$$ab = (-1)^{p(a)} X^{\bullet \!\longrightarrow\! \bullet}(a \otimes b) \ , \quad [a_\lambda b] = (-1)^{p(a)} X^{\bullet \ \bullet}_{\lambda, -\lambda-\partial}(a \otimes b) \, . \tag{3.37}$$

Thanks to the Jacobi identity for the Lie superalgebra $W_{\mathrm{cl}}(\Pi V)$, if $X \in W^1_{\mathrm{cl}}(\Pi V)_{\bar{1}}$ satisfies $X \square X = 0$, then $(\mathrm{ad}\, X)^2 = 0$. In view of Theorem 3.9, this means that we have a cohomology complex

$$(W_{\mathrm{cl}}(\Pi V), \mathrm{ad}\, X) \, ,$$

called the *PVA cohomology complex*, where $X \in W^1(\Pi V)_{\bar{1}}$ is given by (3.37).

4 Relation Between PVA and Differential Harrison Cohomology Complexes

4.1 Main Theorem

Let V be a Poisson vertex algebra. By Theorem 3.9, we have an odd element $X \in W_{\mathrm{cl}}(\Pi V)$ such that $[X, X] = 0$, which is associated with the PVA structure of V by (3.37). Thus, there is the PVA cohomology complex

$$(W_{\mathrm{cl}}(\Pi V), \mathrm{ad}\, X) \, . \tag{4.1}$$

A classical n-cochain is an element $Y \in W^{n-1}_{\mathrm{cl}}(\Pi V)$, namely a map

$$Y \colon \mathcal{G}(n) \times (\Pi V)^{\otimes n} \longrightarrow (\Pi V)[\lambda_1, \dots, \lambda_n] / \langle \partial + \lambda_1 + \dots + \lambda_n \rangle \, , \tag{4.2}$$

satisfying relations (3.25), (3.26), (3.28), (3.29), and the following symmetry property (by definition (3.20)):

$$Y^\sigma = Y \, , \quad \forall \sigma \in S_n \, . \tag{4.3}$$

Recall the grading of the superoperad $\mathcal{P}_{\text{cl}}(\Pi V)$ from (3.36): $\text{gr}^r\ W^{n-1}_{\text{cl}}(\Pi V)$ is the set of maps Y as in (4.2) such that

$$Y^\Gamma = 0 \quad \text{unless} \quad |E(\Gamma)| = r\,.$$

Note that if $\Gamma \in \mathcal{G}(n)$ has $|E(\Gamma)| \geq n$, then necessarily Γ contains a cycle. Hence, by the cycle relation (3.25), $Y^\Gamma = 0$. Therefore the top degree in $\text{gr}\, W^{n-1}_{\text{cl}}(\Pi V)$ is $n-1$, i.e.,

$$\text{gr}^r\ W^{n-1}_{\text{cl}}(\Pi V) = 0 \quad \text{if} \quad r \geq n\,. \tag{4.4}$$

Note that, if $\Gamma \in \mathcal{G}_0(n)$, then $|E(\Gamma)| = n-1$ if and only if Γ is connected. By Remark 3.8, the top degree subspace $\text{gr}^{n-1}\ W^{n-1}_{\text{cl}}(\Pi V)$ consists of all collections of maps

$$Y^\Gamma : (\Pi V)^{\otimes n} \longrightarrow (\Pi V)\,, \qquad \text{for } \Gamma \in \mathcal{G}_0(n),\ \ |E(\Gamma)| = n-1\,, \tag{4.5}$$

satisfying (3.25), (3.26), (4.3), and $Y^\Gamma(\partial(v_1 \otimes \cdots \otimes v_n)) = \partial Y^\Gamma(v_1 \otimes \cdots \otimes v_n)$. If Γ is not connected, then $Y^\Gamma = 0$.

In addition, as explained in Sect. 2.3, there is another cohomology complex associated with V, viewed as a differential algebra, namely the differential Harrison complex

$$(C_{\partial,\text{Har}}(V), d)\,, \tag{4.6}$$

where $C^n_{\partial,\text{Har}}(V) \subset \text{Hom}_{\mathbb{F}[\partial]}(V^{\otimes n}, V)$ is defined by Harrison's conditions (2.11) and d is the Hochschild differential (2.2).

The main result of this paper is the following:

Theorem 4.1 *Let V be a Poisson vertex algebra. One has a surjective morphism of cochain complexes*

$$(W_{\text{cl}}(\Pi V), \text{ad}\, X) \to (C_{\partial,\text{Har}}(V), d)\,, \tag{4.7}$$

mapping $Y \in W^{n-1}_{\text{cl}}(\Pi V)$ to Y^{Λ_n}, where Λ_n is the standard n-line

$$\Lambda_n = \underset{1}{\bullet}\!\longrightarrow\!\underset{2}{\bullet}\!\longrightarrow \cdots \longrightarrow\!\underset{n}{\bullet} \tag{4.8}$$

Moreover, the morphism (4.7) *restricts to a bijective linear map on the top degree:*

$$\text{gr}^{n-1}\ W^{n-1}_{\text{cl}}(\Pi V) \xrightarrow{\sim} C^n_{\partial,\text{Har}}(V)\,. \tag{4.9}$$

We will prove Theorem 4.1 in Sect. 4.6. For that, we will need some preliminary results.

4.2 Lines

We say that a graph $\Gamma \in \mathcal{G}(n)$ is a non-connected line if it has the following form:

$$\Gamma = \underset{i_1^1}{\bullet}\!\to\!\underset{i_2^1}{\bullet}\!\to\!\cdots\!\to\!\underset{i_{k_1}^1}{\bullet} \quad \underset{i_1^2}{\bullet}\!\to\!\underset{i_2^2}{\bullet}\!\to\!\cdots\!\to\!\underset{i_{k_2}^2}{\bullet} \quad \cdots \quad \underset{i_1^s}{\bullet}\!\to\!\underset{i_2^s}{\bullet}\!\to\!\cdots\!\to\!\underset{i_{k_s}^s}{\bullet} = L_1 \sqcup L_2 \sqcup \cdots \sqcup L_s\,, \tag{4.10}$$

where $1 \leq k_1 \leq \cdots \leq k_s$ are such that $k_1 + \cdots + k_s = n$, and the set of indices $\{i_b^a\}$ is a permutation of $\{1, \ldots, n\}$ such that

$$i_1^l = \min\{i_1^l, \ldots, i_{k_l}^l\} \quad \forall\, l = 1, \ldots, s\,. \tag{4.11}$$

If $k_l = k_{l+1}$, we also assume that $i_1^l < i_1^{l+1}$. In particular, the connected lines are all of the form

$$\sigma(\Lambda_n)\,, \quad \sigma \in S_n \text{ such that } \sigma(1) = 1\,, \tag{4.12}$$

where Λ_n is the n-line (4.8). Let $\mathcal{L}(n)$ be the set of n-graphs that are non-connected lines. Let also $\mathbb{F}\mathcal{G}(n)$ be the vector space with basis the set of graphs $\mathcal{G}(n)$.

Definition 4.2 The *cycle relations* in $\mathbb{F}\mathcal{G}(n)$ are the following elements:

(i) all $\Gamma \in \mathcal{G}(n) \setminus \mathcal{G}_0(n)$ (i.e., graphs containing a cycle);
(ii) all linear combinations $\sum_{e\in C} \Gamma \setminus e$, where $\Gamma \in \mathcal{G}(n)$ and $C \subset E(\Gamma)$ is an oriented cycle.

Denote by $R(n) \subset \mathbb{F}\mathcal{G}(n)$ the subspace spanned by the cycle relations (4.2) and (4.2).

Note that reversing an arrow in a graph $\Gamma \in \mathcal{G}(n)$ gives us, modulo cycle relations, the element $-\Gamma \in \mathbb{F}\mathcal{G}(n)$.

Example 4.3 For $n = 3$, a cycle relation of type (4.2) is

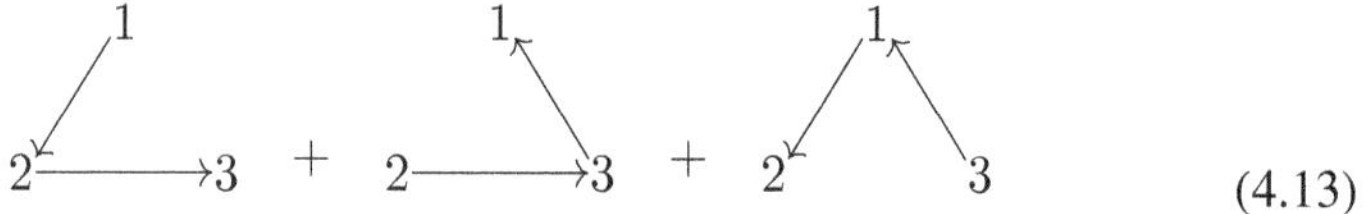

(4.13)

Remark 4.4 The cycle relations (3.25) and (3.26) on $Y \in \mathcal{P}_{\mathrm{cl}}$ can be restated by saying that $Y^{\Gamma} = 0$ for all $\Gamma \in R(n)$.

Theorem 4.5 ([BDSHK20, Theorem 4.7]) *The set $\mathcal{L}(n)$ is a basis for the quotient space $\mathbb{F}\mathcal{G}(n)/R(n)$.*

By Theorem 4.5 and (4.12), we can write every connected graph $\Gamma \in \mathcal{G}(n)$, uniquely, modulo cycle relation, as follows:

$$\Gamma \equiv \sum_{\substack{\sigma\in S_n\\ \sigma(1)=1}} c_\sigma^\Gamma\, \sigma\Lambda_n\,, \tag{4.14}$$

where $c_\sigma^\Gamma \in \mathbb{F}$ and the action of the symmetric group on graphs is defined in Sect. 3.3. Here and further, by $\equiv$ we mean equivalence modulo cycle relations, i.e., equality in the quotient space $\mathbb{F}\mathcal{G}(n)/R(n)$.

4.3 Connected Lines

Lemma 4.6 *For every n, the following identity on n-lines holds:*

$$\underset{1}{\bullet}\!\to\!\underset{2}{\bullet}\!\to\cdots\to\!\underset{n}{\bullet} + \underset{2}{\bullet}\!\to\!\underset{1}{\bullet}\!\to\!\underset{3}{\bullet}\!\to\cdots\to\!\underset{n}{\bullet} + \cdots + \underset{2}{\bullet}\!\to\!\underset{3}{\bullet}\!\to\cdots\to\!\underset{n-1}{\bullet}\!\to\!\underset{1}{\bullet}\!\to\!\underset{n}{\bullet} + \underset{2}{\bullet}\!\to\!\underset{3}{\bullet}\!\to\cdots\to\!\underset{n}{\bullet}\!\to\!\underset{1}{\bullet} \equiv 0\,. \tag{4.15}$$

Proof Let us consider the first two terms in the left-hand side of (4.15). Reversing the edges $2 \to 1$ and $1 \to 3$ in the second graph, and using (4.13), we have

$$\begin{smallmatrix} 1 & & & & \\ \swarrow & & & & \\ 2\to & 3 & \to\cdots\to & n \end{smallmatrix} + \begin{smallmatrix} & 1 & & & \\ \swarrow & & \nwarrow & & \\ 2 & & 3 & \to 4\to\cdots\to & n \end{smallmatrix} \equiv - \begin{smallmatrix} & 1 & & \\ & & \nwarrow & \\ 2\to & & 3 & \to\cdots\to n \end{smallmatrix} \equiv \begin{smallmatrix} & 1 & & \\ & \swarrow & & \\ 2\to & 3 & \to 4\to\cdots\to & n \end{smallmatrix} \tag{4.16}$$

Adding (4.16) to the third term appearing in the left-hand side of (4.15), and applying again (4.13), we obtain

$$\begin{smallmatrix} & 1 & & \\ & \swarrow & & \\ 2\to & 3 & \to 4\to\cdots\to & n \end{smallmatrix} + \begin{smallmatrix} & & 1 & & \\ & \swarrow & & \nwarrow & \\ 2\to & 3 & & 4 & \to\cdots\to n \end{smallmatrix} \equiv - \begin{smallmatrix} & 1 & & \\ & & \nwarrow & \\ 2\to 3\to & & 4 & \to\cdots\to n \end{smallmatrix} \equiv \begin{smallmatrix} & 1 & \\ & \swarrow & \\ 2\to 3\to & 4 & \to\cdots\to n \end{smallmatrix} \tag{4.17}$$

We proceed in the same way, up to

$$\begin{smallmatrix} & 1 & \\ & \swarrow & \\ 2\to 3\to\cdots\to & (n-1) & \to n \end{smallmatrix} + \begin{smallmatrix} & & 1 & & \\ & \swarrow & & \nwarrow & \\ 2\to 3\to\cdots\to & (n-1) & & n \end{smallmatrix} + \begin{smallmatrix} & 1 & \\ & & \nwarrow \\ 2\to 3\to\cdots\to (n-1)\to & & n \end{smallmatrix} \equiv 0\,. \tag{4.18}$$

This completes the proof of the lemma. $\square$

Remark 4.7 Observe that equation (4.15) can be viewed as a "local" identity: Lemma 4.6 holds even if we attach the same graph Γ at any vertex of every graph appearing in the identity.

Lemma 4.8 *Let Λ_n be as in (4.8). For every $k \in \{2, \ldots, n\}$, the following identity holds:*

$$\Lambda_n + (-1)^k \sum_{\pi \in \mathcal{M}_n^k} \pi \Lambda_n \equiv 0\,, \tag{4.19}$$

where the sum is over all monotone permutations π starting at k and the action of S_n on graphs is described in Sect. 3.3.

Proof If $k = 2$, then (4.19) is equivalent to (4.15). Suppose that $k > 2$. Recall the description (2.5) of the monotone permutations $\pi \in \mathcal{M}_n^k$ in terms of the set $D(\pi)$ of drops. Given the set of drops $D = \{d_1, \ldots, d_{k-1}\}$ such that $2 \le d_{k-1} < \cdots < d_1 \le n$, the corresponding monotone permutation $\pi^D \in \mathcal{M}_n^k$ is uniquely determined by $\pi^{-1}(i) = d_i$, $\forall i = 1 \ldots, k-1$. Hence:

$$\pi^D(\Lambda_n) = \quad k \to (k+1) \to \cdots \to \underset{[d_{k-1}]}{\underline{(k-1)}} \to \cdots \to \underset{[d_2]}{\underline{2}} \to \cdots \to \underset{[d_1]}{\underline{1}} \to \cdots \to n$$

where the underlined positions correspond to drops, while all other positions have vertices in increasing order from $k+1$ to n.

We then have

$$\sum_{\pi \in \mathcal{M}_n^k} \pi \Lambda_n = \sum_{2 \le d_{k-1} < \cdots < d_1 \le n} \quad k \to (k+1) \to \cdots \to \underset{[d_{k-1}]}{\underline{(k-1)}} \to \cdots \to \underset{[d_2]}{\underline{2}} \to \cdots \to \underset{[d_1]}{\underline{1}} \to \cdots \to n \,.$$

By Lemma 4.6, the sum over $d_1 \in \{d_2 + 1, \ldots, n\}$ gives

$$\sum_{d_1 = d_2 + 1}^{n} \quad k \to (k+1) \to \cdots \to \underset{[d_{k-1}]}{\underline{(k-1)}} \to \cdots \to \underset{[d_2]}{\underline{2}} \to \cdots \to \underset{[d_1]}{\underline{1}} \to \cdots \to n$$

$$= -\; k \to (k+1) \to \cdots \to \underset{[d_{k-1}]}{\underline{(k-1)}} \to \cdots \to \underset{[d_2]}{\overset{\overset{1}{\downarrow}}{\underline{2}}} \to \cdots \to n \,.$$

Similarly, using again Lemma 4.6 (cf. Remark 4.7), the sum over $d_2 \in \{d_3 + 1, \ldots, n\}$ is

$$-\sum_{d_2 = d_3 + 1}^{n} \quad k \to (k+1) \to \cdots \to \underset{[d_{k-1}]}{\underline{(k-1)}} \to \cdots \to \underset{[d_3]}{\underline{3}} \to \cdots \to \underset{[d_2]}{\overset{\overset{1}{\downarrow}}{\underline{2}}} \to \cdots \to n$$

$$= (-1)^2 \; k \to (k+1) \to \cdots \to \underset{[d_{k-1}]}{\underline{(k-1)}} \to \cdots \to \underset{[d_3]}{\overset{\overset{\overset{1}{\downarrow}}{\overset{2}{\downarrow}}}{\underline{3}}} \to \cdots \to n \,.$$

Repeating the same argument k times, we conclude that

$$\sum_{\pi\in\mathcal{M}_n^k}\pi\Lambda_n=(-1)^{k-2}\sum_{d_{k-1}=2}^{n}\quad\begin{array}{c}\begin{array}{c}1\\ \downarrow\\ \vdots\\ \downarrow\\ (k-2)\\ \downarrow\end{array}\\ k\to(k+1)\to\cdots\to\underline{(k-1)}\to\cdots\to n\\ [d_{k-1}]\end{array}$$

$$=(-1)^{k-1}\quad\begin{array}{l}\begin{array}{c}1\\ \downarrow\\ \vdots\\ \downarrow\\ (k-1)\\ \downarrow\end{array}\\ k\to(k+1)\to\cdots\to n\end{array}$$

$$=(-1)^{k-1}\Lambda_n\,,$$

proving the lemma. □

4.4 Relation Between the Symmetry Property and Harrison's Conditions

Recall that every $Y\in W_{\mathrm{cl}}^{n-1}(\Pi V)$ satisfies the symmetry property (4.3).

Lemma 4.9 *If $Y\in W_{\mathrm{cl}}^{n-1}(\Pi V)$, then Y^{Λ_n} satisfies Harrison's relations* (2.11), *hence it lies in the differential Harrison cohomology complex:*

$$Y^{\Lambda_n}\in C^n_{\partial,\mathrm{Har}}(V)\,.$$

Conversely, given $F\in C^n_{\partial,\mathrm{Har}}(V)$, there exists a unique $Y\in\mathrm{gr}^{n-1}W_{\mathrm{cl}}^{n-1}(\Pi V)$, such that

$$Y^{\Lambda_n}=F\,.$$

Consequently, the linear map

$$\mathrm{gr}^{n-1}W_{\mathrm{cl}}^{n-1}(\Pi V)\stackrel{\sim}{\rightarrow}C^n_{\partial,\mathrm{Har}}(V)\,,\quad Y\mapsto Y^{\Lambda_n}$$

is bijective.

Proof First, we prove that, if $Y \in W_{\mathrm{cl}}^{n-1}(\Pi V)$ satisfies the symmetry relations (4.3), then $f = Y^{\Lambda_n}$ satisfies Harrison's conditions (2.11). By Lemma 4.8 (cf. Remark 4.4), we get

$$Y^{\Lambda_n} = (-1)^{k-1} \sum_{\pi \in \mathcal{M}_n^k} Y^{\pi(\Lambda_n)} .$$

Evaluating both sides of this identity on $v_1 \otimes \cdots \otimes v_n \in V^{\otimes n}$, the left side is simply $Y^{\Lambda_n}(v_1 \otimes \cdots \otimes v_n) = f(v_1 \otimes \cdots \otimes v_n)$. On the right-hand side, we have

$$\begin{aligned}
(-1)^{k-1} \sum_{\pi \in \mathcal{M}_n^k} Y^{\pi(\Lambda_n)}(v_1 \otimes \cdots \otimes v_n) &= (-1)^{k-1} \sum_{\pi \in \mathcal{M}_n^k} (Y^{\pi^{-1}})^{\pi(\Lambda_n)}(v_1 \otimes \cdots \otimes v_n) \\
&= (-1)^{k-1} \sum_{\pi \in \mathcal{M}_n^k} \operatorname{sign}(\pi) Y^{\Lambda_n}(v_{\pi(1)} \otimes \cdots \otimes v_{\pi(n)}) \\
&= L_k f(v_1 \otimes \cdots \otimes v_n) ,
\end{aligned}$$

by the definition (2.10) of L_k. Hence, f satisfies Harrison's conditions (2.11) as claimed.

We next turn to the second claim of the lemma. Let $F \in C^n_{\partial,\mathrm{Har}}(V)$, i.e., $F \colon V^{\otimes n} \to V$ is an $\mathbb{F}[\partial]$-module homomorphism satisfying Harrison's conditions (2.11). Then the corresponding $Y \in \mathrm{gr}^{n-1} W_{\mathrm{cl}}^{n-1}(\Pi V)$, such that $Y^{\Lambda_n} = F$, is defined as follows. For $\Gamma \in R(n)$, or if $\Gamma \in \mathcal{L}(n)$ is not connected, we set

$$Y^{\Gamma} = 0 . \tag{4.20}$$

For $\Gamma \in \mathcal{L}(n)$ connected, there exists a unique $\tau \in S_n$ such that $\tau(1) = 1$ and $\Gamma = \tau(\Lambda_n)$. We then set

$$Y^{\Gamma}(v_1 \otimes \cdots \otimes v_n) = \operatorname{sign}(\tau) F(v_{\tau(1)} \otimes \cdots \otimes v_{\tau(n)}) . \tag{4.21}$$

In particular, for $\Gamma = \Lambda_n$, we have $\tau = 1$ and $Y^{\Lambda_n} = F$.

Since, by Theorem 4.5, $\mathcal{L}(n)$ is a basis for the vector space $\mathbb{F}\mathcal{G}(n)/R(n)$, equations (4.20) and (4.21) determine a unique well-defined element $Y \in \mathrm{gr}^{n-1} \mathcal{P}_{\mathrm{cl}}(\Pi V)(n)$. Indeed, Y satisfies the cycle relations (3.25) and (3.26), by (4.20), Theorem 4.5 and Remark 4.4. The first sesquilinearity condition (3.28) is obvious, and the second condition (3.29) follows from (3.30).

To prove that $Y \in \mathrm{gr}^{n-1} W_{\mathrm{cl}}^{n-1}(\Pi V)$, it remains to show that Y satisfies the symmetry conditions (4.3). Obviously, the action of the symmetric group S_n preserves $R(n)$ and the set of non-connected lines. Hence, when we evaluate (4.3) on $\Gamma \in R(n)$ or on a non-connected n-line $\Gamma \in \mathcal{L}(n)$, we get $0 = 0$.

We are left to prove that (4.3) holds when evaluated on a connected line $\Gamma \in \mathcal{L}(n)$, which, as remarked above, can be obtained as $\Gamma = \tau(\Lambda_n)$, for a unique $\tau \in S_n$

such that $\tau(1) = 1$. The right-hand side of (4.3), when evaluated on such a Γ is given by (4.21). The left-hand side is, by (3.35),

$$\begin{aligned}(Y^\sigma)^\Gamma(v_1 \otimes \cdots \otimes v_n) &= Y^{\sigma\tau(\Lambda_n)}(\sigma(v_1 \otimes \cdots \otimes v_n)) \\ &= \operatorname{sign}(\sigma) Y^{\sigma\tau(\Lambda_n)}(v_{\sigma(1)} \otimes \cdots \otimes v_{\sigma_n}) . \end{aligned} \tag{4.22}$$

By Lemma 4.8, we have, modulo $R(n)$,

$$\sigma\tau(\Lambda_n) \equiv (-1)^{\tau^{-1}\sigma^{-1}(1)-1} \sum_{\pi \in \mathcal{M}_n^{\tau^{-1}\sigma^{-1}(1)}} \sigma\tau\pi(\Lambda_n) .$$

Hence, by Remark 4.4, the right-hand side of (4.22) becomes

$$\operatorname{sign}(\sigma)(-1)^{\tau^{-1}\sigma^{-1}(1)-1} \sum_{\pi \in \mathcal{M}_n^{\tau^{-1}\sigma^{-1}(1)}} Y^{\sigma\tau\pi(\Lambda_n)}(v_{\sigma^{-1}(1)} \otimes \cdots \otimes v_{\sigma^{-1}(n)}) . \tag{4.23}$$

Note that, if $\pi \in \mathcal{M}_n^{\tau^{-1}\sigma^{-1}(1)}$, then $\sigma\tau\pi(1) = 1$. Hence, we can apply formula (4.21) to get

$$Y^{\sigma\tau\pi(\Lambda_n)}(v_{\sigma^{-1}(1)} \otimes \cdots \otimes v_{\sigma^{-1}(n)}) = \operatorname{sign}(\sigma\tau\pi) F(v_{\tau\pi(1)} \otimes \cdots \otimes v_{\tau\pi(n)}) . \tag{4.24}$$

Combining (4.22)–(4.24), we get, by (2.7) and the definition (2.10) of L_k,

$$\begin{aligned}&(Y^\sigma)^\Gamma(v_1 \otimes \cdots \otimes v_n) \\ &= \operatorname{sign}(\tau)(-1)^{\tau^{-1}\sigma^{-1}(1)-1} \sum_{\pi \in \mathcal{M}_n^{\tau^{-1}\sigma^{-1}(1)}} \operatorname{sign}(\pi) F(v_{\tau\pi(1)} \otimes \cdots \otimes v_{\tau\pi(n)}) \\ &= \operatorname{sign}(\tau)(L_{\tau^{-1}\sigma^{-1}(1)} F)(v_{\tau(1)} \otimes \cdots \otimes v_\tau(n)) , \end{aligned}$$

which equals (4.21) by Harrison's conditions (2.11).

Hence, Y is a well-defined element of $\operatorname{gr}^{n-1} W_{\text{cl}}^{n-1}(\Pi V)$, such that $Y^{\Lambda_n} = F$, as required. The uniqueness of such a Y is obvious since, by Theorem 4.5, $Y \in \operatorname{gr}^{n-1} W_{\text{cl}}^{n-1}(\Pi V)$ is uniquely determined by its value on Λ_n. □

4.5 *Relation Between* ad X *and the Hochschild Differential*

Lemma 4.10 *For $Y \in W_{\text{cl}}^{n-1}(\Pi V)$ and $X \in W_{\text{cl}}^1(\Pi V)$ defined in* (3.37), *we have*

$$[X, Y]^{\Lambda_{n+1}}(v_1 \otimes \cdots \otimes v_{n+1}) = (-1)^{n+1} d(Y^{\Lambda_n})(v_1 \otimes \cdots \otimes v_{n+1}) ,$$

where Λ_n is as in (4.8) *and d is the Hochschild differential* (2.2).

Proof Recall that, by definition (3.22), the adjoint action of X on Y is given by

$$\begin{aligned} &= X\Box Y - (-1)^{n-1} Y\Box X \\ &= \sum_{\sigma\in S_{n,1}} (X\circ_1 Y)^{\sigma^{-1}} + (-1)^n \sum_{\tau\in S_{2,n-1}} (Y\circ_1 X)^{\tau^{-1}}, \end{aligned} \tag{4.25}$$

since $\bar{p}(X) = 1$ and $\bar{p}(Y) = n-1$. The elements of $S_{n,1}$ are

$$\begin{aligned} \sigma_k &= \left(\begin{array}{ccccccc|c} 1\ 2 \cdots k-1 & k & k+1 & \cdots & n & n+1 \\ 1\ 2 \cdots k-1 & k+1 & k+2 & \cdots & n+1 & k \end{array}\right) \\ &= (k\ \ k+1\ \ \cdots\ \ n+1)\,, \end{aligned} \tag{4.26}$$

for $1\le k\le n+1$. The elements of $S_{2,n-1}$ are

$$\tau_{i,j} = \left(\begin{array}{cc|ccccccc} 1 & 2 & 3 \cdots i+1 & i+2 \cdots & j & j+1 \cdots n+1 \\ i & j & 1 \cdots i-1 & i+1 \cdots & j-1 & j+1 \cdots n+1 \end{array}\right), \tag{4.27}$$

for $1\le i< j-1\le n$, and

$$\tau_{i,i+1} = \left(\begin{array}{cc|ccc} 1 & 2 & 3 \cdots i+1 & i+2 \cdots n+1 \\ i & i+1 & 1 \cdots i-1 & i+2 \cdots n+1 \end{array}\right), \tag{4.28}$$

for $1\le i\le n$.

Using (4.26), we evaluate

$$\begin{aligned} &\left((X\circ_1 Y)^{\sigma_k^{-1}}\right)^{\Lambda_{n+1}}(v_1\otimes\cdots\otimes v_{n+1}) \\ &= (-1)^{n+1-k}(X\circ_1 Y)^{\sigma_k^{-1}(\Lambda_{n+1})}(v_1\otimes\cdots\otimes v_{k-1}\otimes v_{k+1}\otimes\cdots\otimes v_{n+1}\otimes v_k)\,. \end{aligned} \tag{4.29}$$

For $k=1$, by the symmetric group's action described in Sect. 3.3, we have

$$\sigma_1^{-1}(\Lambda_{n+1}) = \underset{\sigma_1^{-1}(1)}{\bullet}\longrightarrow\underset{\sigma_1^{-1}(2)}{\bullet}\longrightarrow\cdots\longrightarrow\underset{\sigma_1^{-1}(n+1)}{\bullet}$$

$$= \underset{1}{\bullet}\longrightarrow\cdots\longrightarrow\underset{n}{\bullet}\qquad\underset{n+1}{\bullet}\,.$$

When we apply the cocomposition map $\Delta^{n,1}$ to this graph, we get

$$\Delta_0^{n,1}(\sigma_1^{-1}(\Lambda_{n+1})) = \underset{1}{\bullet}\longleftarrow\underset{2}{\bullet}$$

and

$$\Delta_1^{n,1}(\sigma_1^{-1}(\Lambda_{n+1})) \;=\; \underset{1}{\bullet}\longrightarrow\underset{2}{\bullet}\longrightarrow\cdots\longrightarrow\underset{n}{\bullet} \;=\; \Lambda_n\,.$$

Hence, by the definition (3.32) of the composition map, (4.29) becomes

$$\begin{aligned}
&\big((X\circ_1 Y)^{\sigma_1^{-1}}\big)^{\Lambda_{n+1}}(v_1\otimes\cdots\otimes v_{n+1})\\
&\quad=(-1)^n(X\circ_1 Y)^{\sigma_1^{-1}(\Lambda_{n+1})}(v_2\otimes\cdots\otimes v_{n+1}\otimes v_1)\\
&\quad=(-1)^n X^{\bullet\!\longleftarrow\!\bullet}(Y^{\Lambda_n}(v_2\otimes\cdots\otimes v_{n+1})\otimes v_1)\\
&\quad=(-1)^{n+1}Y^{\Lambda_n}(v_2\otimes\cdots\otimes v_{n+1})\,v_1\,.
\end{aligned}\tag{4.30}$$

Similarly, for $k=n+1$, we have $\sigma_{n+1}=1$, and applying the cocomposition map $\Delta^{n,1}$ to $\sigma_{n+1}^{-1}(\Lambda_{n+1})=\Lambda_{n+1}$, we get

$$\Delta_0^{n,1}(\sigma_{n+1}^{-1}(\Lambda_{n+1})) \;=\; \underset{1}{\bullet}\longrightarrow\underset{2}{\bullet}$$

and

$$\Delta_1^{n,1}(\sigma_{n+1}^{-1}(\Lambda_{n+1})) \;=\; \underset{1}{\bullet}\longrightarrow\underset{2}{\bullet}\longrightarrow\cdots\longrightarrow\underset{n}{\bullet} \;=\; \Lambda_n\,.$$

Then (4.29) becomes

$$\begin{aligned}
&\big((X\circ_1 Y)^{\sigma_{n+1}^{-1}}\big)^{\Lambda_{n+1}}(v_1\otimes\cdots\otimes v_{n+1})\\
&\quad=X^{\bullet\!\longrightarrow\!\bullet}(Y^{\Lambda_n}(v_1\otimes\cdots\otimes v_n)\otimes v_{n+1})\\
&\quad=Y^{\Lambda_n}(v_1\otimes\cdots\otimes v_n)\,v_{n+1}\,.
\end{aligned}\tag{4.31}$$

Furthermore, for $2\leq k\leq n$, we have

$$\sigma_k^{-1}(\Lambda_{n+1}) = \underset{\sigma_k^{-1}(1)}{\bullet}\longrightarrow\underset{\sigma_k^{-1}(2)}{\bullet}\longrightarrow\cdots\longrightarrow\underset{\sigma_k^{-1}(n+1)}{\bullet}$$

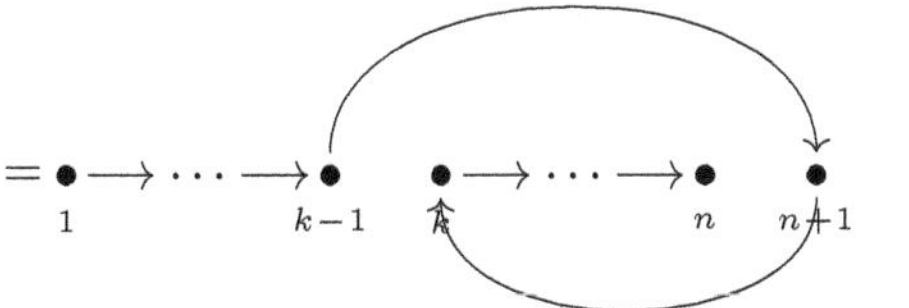

Hence, applying the cocomposition map $\Delta^{n,1}$ we get

$$\Delta_0^{n,1}(\sigma_k^{-1}(\Lambda_{n+1})) = \underset{1}{\bullet} \quad \underset{2}{\bullet}$$

which has a cycle. Therefore,

$$\left((X \circ_1 Y)^{\sigma_k^{-1}}\right)^{\Lambda_{n+1}} = X^{\bullet\ \bullet}(\cdots) = 0\,. \tag{4.32}$$

Next, we compute $\left((Y \circ_1 X)^{\tau_{i,j}^{-1}}\right)^{\Lambda_{n+1}}(v_1 \otimes \cdots \otimes v_{n+1})$ where $\tau_{i,j}$ is defined in (4.27). By the symmetric group's action described in Sect. 3.3 we have

$$\tau_{i,j}^{-1}(\Lambda_{n+1}) = \underset{\tau_{i,j}^{-1}(1)}{\bullet} \longrightarrow \underset{\tau_{i,j}^{-1}(2)}{\bullet} \longrightarrow \cdots \longrightarrow \underset{\tau_{i,j}^{-1}(n+1)}{\bullet}$$

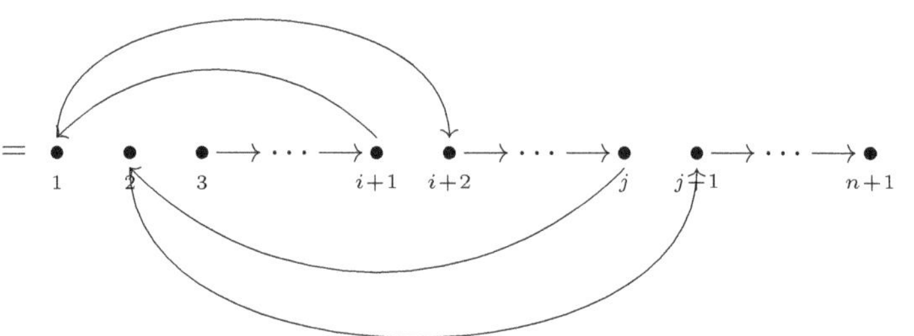

Hence, applying the cocomposition $\Delta^{2,1,\dots,1}$ we get

$$\Delta_0^{2,1,\dots,1}(\tau_{i,j}^{-1}(\Lambda_{n+1})) = \underset{1}{\bullet} \quad \underset{2}{\bullet} \longrightarrow \cdots \longrightarrow \underset{i}{\bullet} \quad \underset{i+1}{\bullet} \longrightarrow \cdots \longrightarrow \underset{j-1}{\bullet} \quad \underset{j}{\bullet} \longrightarrow \cdots \longrightarrow \underset{n}{\bullet}\,,$$

which has a cycle. Therefore,

$$\left((Y \circ_1 X)^{\tau_{i,j}^{-1}}\right)^{\Lambda_{n+1}} = Y^{\Delta_0^{2,1,\dots,1}(\tau_{i,j}^{-1}(\Lambda_{n+1}))}(\cdots) = 0\,. \tag{4.33}$$

Finally, we evaluate $\left((Y \circ_1 X)^{\tau_{i,i+1}^{-1}}\right)^{\Lambda_{n+1}}(v_1 \otimes \cdots \otimes v_{n+1})$ with $\tau_{i,i+1}$ defined in (4.28). In this case, we have

$$\tau_{i,i+1}^{-1}(\Lambda_{n+1}) = \underset{\tau_{i,i+1}^{-1}(1)}{\bullet} \longrightarrow \underset{\tau_{i,i+1}^{-1}(2)}{\bullet} \longrightarrow \cdots \longrightarrow \underset{\tau_{i,i+1}^{-1}(n+1)}{\bullet}$$

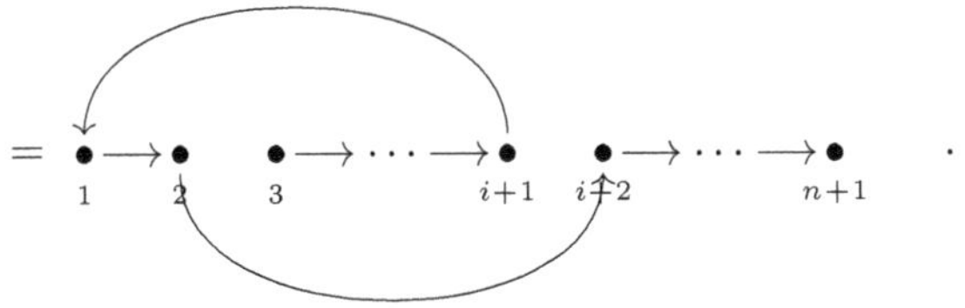

$$= \underset{1}{\bullet} \longrightarrow \underset{2}{\bullet} \quad \underset{3}{\bullet} \longrightarrow \cdots \longrightarrow \underset{i+1}{\bullet} \quad \underset{i+2}{\bullet} \longrightarrow \cdots \longrightarrow \underset{n+1}{\bullet} \ .$$

Hence, applying the cocomposition $\Delta^{2,1,\dots,1}$ we get

$$\Delta_0^{2,1,\dots,1}(\tau_{i,i+1}^{-1}(\Lambda_{n+1})) = \underset{1}{\bullet} \quad \underset{2}{\bullet} \longrightarrow \cdots \longrightarrow \underset{i}{\bullet} \quad \underset{i+1}{\bullet} \longrightarrow \cdots \longrightarrow \underset{n}{\bullet}$$

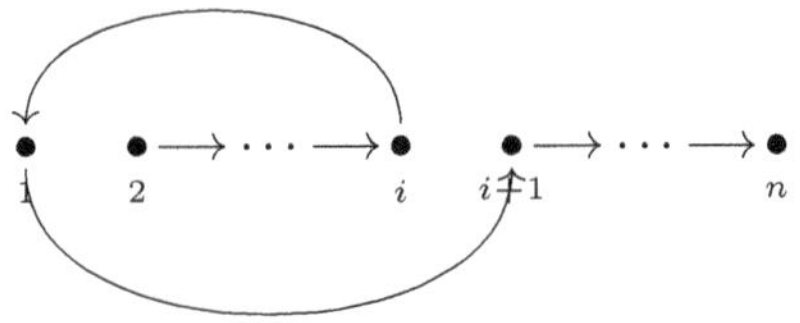

and

$$\Delta_1^{2,1,\dots,1}(\tau_{i,i+1}^{-1}(\Lambda_{n+1})) = \underset{1}{\bullet} \longrightarrow \underset{2}{\bullet} \ .$$

Therefore, by definition (3.32) of the composition map, we obtain

$$\begin{aligned}
&\left((Y \circ_1 X)^{\tau_{i,j}^{-1}}\right)^{\Lambda_{n+1}}(v_1 \otimes \cdots \otimes v_n) = \\
&= (Y \circ_1 X)^{\tau_{i,j}^{-1}(\Lambda_{n+1})}(v_i \otimes v_{i+1} \otimes v_1 \otimes \cdots \otimes v_{i-1} \otimes v_{i+2} \otimes \cdots \otimes v_{n+1}) \\
&= Y^{\Delta_0^{2,1,\dots,1}(\tau_{i,i+1}^{-1}(\Lambda_{n+1}))}\left(X^{\bullet\to\bullet}(v_i \otimes v_{i+1}) \otimes v_1 \otimes \cdots \otimes v_{i-1} \otimes v_{i+2} \otimes \cdots \otimes v_{n+1}\right) \\
&= Y^{\Delta_0^{2,1,\dots,1}(\tau_{i,i+1}^{-1}(\Lambda_{n+1}))}(v_i v_{i+1} \otimes v_1 \otimes \cdots \otimes v_{i-1} \otimes v_{i+2} \otimes \cdots \otimes v_{n+1}) \, .
\end{aligned} \tag{4.34}$$

Note that

$$\Delta_0^{2,1,\dots,1}(\tau_{i,i+1}^{-1}(\Lambda_{n+1})) = \sigma(\Lambda_n) \, ,$$

where $\sigma = (1\ 2 \cdots i) \in S_n$ is the i-cycle. Hence, by the symmetry property (4.3), we can replace Y by $Y^{\sigma^{-1}}$ in the right-hand side of (4.34) to get

$$(-1)^{i+1} Y^{\Lambda_n}(v_1 \otimes \cdots \otimes v_{i-1} \otimes v_i v_{i+1} \otimes v_{i+2} \otimes \cdots \otimes v_{n+1}) . \tag{4.35}$$

Combining equations (4.25) and (4.30)–(4.35), we conclude that

$$\begin{aligned}
&[X,Y]^{\Lambda_{n+1}}(v_1 \otimes \cdots \otimes v_n) \\
&\quad = (-1)^{n+1} v_1\, Y^{\Lambda_n}(v_2 \otimes \cdots \otimes v_{n+1}) + Y^{\Lambda_n}(v_1 \otimes \cdots \otimes v_n)\, v_{n+1} \\
&\qquad + (-1)^n \sum_{i=1}^{n} (-1)^{i+1} Y^{\Lambda_n}(v_1 \otimes \cdots \otimes v_{i-1} \otimes v_i v_{i+1} \otimes v_{i+2} \otimes \cdots \otimes v_{n+1}) \\
&\quad = (-1)^{n+1} d(Y^{\Lambda_n})(v_1 \otimes \cdots \otimes v_n) ,
\end{aligned}$$

completing the proof. □

Corollary 4.11 *Let X be defined in* (3.37) *and $Y \in W_{\mathrm{cl}}^{n-1}(\Pi V)$ be such that $[X, Y] = 0$. Then Y^{Λ_n} is a cocyle in the Hochschild cohomology.*

Proof Obvious, by Lemma 4.10. □

4.6 Proof of Theorem 4.1

By Lemma 4.9, given $Y \in W_{\mathrm{cl}}^{n-1}(\Pi V)$, Y^{Λ_n} is a cochain in the differential Harrison complex. Conversely, for any $F \in C^n_{\partial,\mathrm{Har}}(V)$, there is a unique $Y \in \mathrm{gr}^{n-1} W_{\mathrm{cl}}^{n-1}(\Pi V)$ such that $F = Y^{\Lambda_n}$. Hence, the following diagram is well-defined:

$$\begin{array}{ccccc}
Y \ni & W_{\mathrm{cl}}^{n-1}(\Pi V) & \xrightarrow{\mathrm{ad}\, X} & W_{\mathrm{cl}}^{n}(\Pi V) \\
\downarrow & \downarrow & & \downarrow \\
Y^{\Lambda_n} \ni & C^n_{\partial,\mathrm{Har}}(V) & \xrightarrow{d} & C^{n+1}_{\partial,\mathrm{Har}}(V)
\end{array} , \tag{4.36}$$

where the vertical maps are surjective and restrict to bijective maps on top degree: $\mathrm{gr}^{n-1} W_{\mathrm{cl}}^{n-1}(\Pi V) \xrightarrow{\sim} C^n_{\partial,\mathrm{Har}}(V)$. Lemma 4.10 says that, up to a sign, the diagram (4.36) is commutative. This completes the proof of the theorem.

Acknowledgments This research was partially conducted during the authors' visits to the University of Rome La Sapienza, to MIT, and to IHES. The first author was supported in part by a Simons Foundation grant 584741. The second author was partially supported by the national PRIN fund n. 2015ZWST2C_001 and the University funds n. RM116154CB35DFD3 and RM11715C7FB74D63. The third author was partially supported by the Bert and Ann Kostant fund and by a Simons Fellowship.

References

[BDSHK19] B. Bakalov, A. De Sole, R. Heluani and V.G. Kac, *An operadic approach to vertex algebra and Poisson vertex algebra cohomology*. Japan. J. Math. **14** (2019), 1–94.

[BDSHK20] B. Bakalov, A. De Sole, R. Heluani and V.G. Kac, *Chiral vs classical operad*. IMRN **19** (2020), 6463–6488.

[BDSHKV21] B. Bakalov, A. De Sole, R. Heluani, V.G. Kac, and V. Vignoli, *Classical and variational Poisson cohomology*, To appear in Japan J. Math. (2021), Preprint arXiv:2101.10939.

[BDSK20] B. Bakalov, A. De Sole, and V.G. Kac, *Computation of cohomology of Lie conformal and Poisson vertex algebras*. Selecta Math. (N.S.) **26** (2020) n. 4, 1–5.

[DSK13] A. De Sole and V.G. Kac, *Variational Poisson cohomology*. Japan. J. Math. **8** (2013), 1–145.

[GS87] M. Gerstenhaber and S.D. Schack, *A Hodge-type decomposition for commutative algebra cohomology*. J. Pure Appl. Algebra **48** (1987), 229–247.

[Har62] D.K. Harrison, *Commutative algebras and cohomology*. Trans. Amer. Math. Soc. **104** (1962), 191–204.

[Hoc45] G. Hochschild, *On the cohomology groups of an associative algebra*. Ann. of Math. (2) **46** (1945), 58–67.

[Lod13] J.L. Loday, *Cyclic homology*. Springer Science & Business Media. vol. 301, 2013.

[Vig19] V. Vignoli, *On Poisson vertex algebra cohomology*. Ph.D. thesis, University of Rome La Sapienza, 2019.

Theta Invariants of Lens Spaces via the BV-BFV Formalism

Alberto S. Cattaneo, Pavel Mnev, and Konstantin Wernli

Abstract The goal of this paper is to investigate the Theta invariant—an invariant of framed 3-manifolds associated with the lowest order contribution to the Chern-Simons partition function—in the context of the quantum BV-BFV formalism. Namely, we compute the state on the solid torus to low degree in $\hbar$ and apply the gluing procedure to compute the Theta invariant of lens spaces. We use a distributional propagator which does not extend to a compactified configuration space, so to compute loop diagrams we have to define a regularization of the product of the distributional propagators, which is done in an *ad hoc* fashion. Also, a polarization has to be chosen for the quantization process. Our results agree with results in the literature for one type of polarization, but for another type of polarization there are extra terms.

1 Introduction

The goal of this paper is to investigate the theta invariant—the first example of a perturbative Chern-Simons invariant—through the BV-BFV formalism,[1] which allows for perturbative quantization compatible with cutting and gluing [CMR17]. We will briefly discuss the BV-BFV formalism in Sect. 3.

[1]The letters B, F, and V here stand for Batalin, Fradkin, and Vilkovisky, who introduced what is now known as BV [BV77, BV81, BV83] and BFV [FV75, FV77, BF83, BF86] formalisms related to the quantization of gauge theories.

A. S. Cattaneo (✉) · K. Wernli
Institut für Mathematik, Universität Zürich, Zürich, Switzerland
e-mail: cattaneo@math.uzh.ch; konstantin.wernli@math.uzh.ch

P. Mnev
University of Notre Dame, Notre Dame, IN, USA
e-mail: pmnev@nd.edu

A. Alekseev et al. (eds.), *Representation Theory, Mathematical Physics, and Integrable Systems*, Progress in Mathematics 340,
https://doi.org/10.1007/978-3-030-78148-4_3

The concept of perturbative Chern-Simons invariants goes back to the idea of Schwarz [Sch78, Sch07] that physical quantities associated with topological quantum field theories should be topological invariants of the spacetime M on which the theory is defined. Among physical quantities of interest for topological quantum field theories is the partition function. If the space of fields of the theory is F_M, and its action functional is $S_M \colon F_M \to \mathbb{R}$, the first attempt at defining the partition function is

$$Z_M = \int_{F_M} e^{\frac{i}{\hbar} S_M[\Phi]}. \tag{1}$$

On the right-hand side one is supposed to integrate over the space of fields F_M. However, apart from the case $\dim M = 1$, there is hardly a case of interest where it has been possible to define a measure on F_M. Therefore, an alternative definition of Z_M is required. There are different ideas for this in the literature, one of them being *perturbative quantization*. Here one extrapolates the asymptotic behavior in the $\hbar \to 0$ limit of an integral of the form $Z[\hbar] = \int_F \exp(i/\hbar S)$, where F is a finite-dimensional manifold, to the infinite-dimensional case. In the finite-dimensional case, the result is a power series describing the asymptotics whose coefficients depend on the local behavior of S around critical points, and there is hope to define a similar power series in the infinite-dimensional case. Now, if the theory is topological, we expect the coefficients of this power series to be topological invariants of the spacetime. In Chern-Simons theory, this is not quite the case: It has been shown repeatedly [AS91, AS94, Kon94, KT99, BC98, Les04a] that the coefficients, starting with the lowest order, depend on choices that are made in the quantization process. However, it is possible to cancel the dependence on the choices, at the price of introducing a framing (i.e. a trivialization of the tangent bundle of the spacetime manifold). The result is an invariant of framed 3-manifolds known as Theta invariant. We review perturbative Chern-Simons quantization in more detail in Sect. 2.

In this paper we compute the contribution of the theta graph on lens spaces by the cutting and gluing method provided by the BV-BFV formalism. Since lens spaces admit a Heegaard splitting of genus 1, it is enough to compute the state—to the necessary order—on a solid torus. This is done in Sect. 4.1. In order to treat Chern-Simons as a perturbation of abelian BF theory, we make the assumption that the Lie algebra of coefficients admits a splitting into Lagrangian subspaces with respect to the invariant form used in the definition of the Chern-Simons form, see Sect. 4.1. This determines a polarization on the space of boundary fields. As a gauge fixing we choose the axial gauge propagator. We then compute the effective action in Sect. 5. Since the axial gauge propagator does not extend to a compactified configuration space, a different regularization has to be chosen for loop diagrams. In this note we restrict ourselves to a naive ad hoc regularization discussed in Sect. 5.4. This issue is addressed in more detail in [Wer18].

In Sect. 6 we explain how to compute the pairing of the states on two solid tori. In Sect. 7 we present the result for the effective action. Remarkably, the result—

both for the state on the solid torus and the weight of the Theta graph—depends on whether the subspaces are subalgebras or not. This seems similar to other results in the literature [Wei87, Haw08] where such dependence on the polarization is discussed. If both subspaces are subalgebras, the result can be compared with the results by Kuperberg-Thurston-Lescop (see Sect. 8). Finally, in Sect. 9, we present some conclusions and discuss further questions and related work.

2 Perturbative Quantization of Chern-Simons Theory

We briefly review the idea behind perturbative quantization of Chern-Simons theory.

2.1 Chern-Simons Theory

Chern-Simons is an example of a topological field theory that after the seminal paper by Witten [Wit89] received widespread attention throughout the mathematical physics and topology communities. It is a field theory defined on three-dimensional manifolds with the space of fields F_M the space of connections on the trivial principal G-bundle over M:

$$F_M = Conn(M \times G) \cong \Omega^1(M, \mathfrak{g}). \tag{2}$$

Here G is a Lie group and $\mathfrak{g}$ its Lie algebra, subject to assumptions depending on the context. We identify the space of connections with $\mathfrak{g}$-valued 1-forms on M. The action functional is

$$S_M[A] = \int_M \frac{1}{2}\langle A, \mathrm{d}A\rangle + \frac{1}{6}\langle A, [A, A]\rangle. \tag{3}$$

Here $\langle\cdot, \cdot\rangle$ is an invariant symmetric bilinear form on $\mathfrak{g}$. The action (3) is the integral of the Chern-Simons form defined in [CS74], whence the name. An important feature of Chern-Simons theory is its gauge invariance under infinitesimal gauge transformations[2] $A \to \mathrm{d}_A c$, where $c \in C^\infty(M, \mathfrak{g})$ (if $\partial M = \emptyset$ or appropriate boundary conditions are chosen). Here $\mathrm{d}_A c = \mathrm{d}c + [A, c]$. One can extend the theory to nontrivial bundles, but this requires more work. A thorough discussion of classical Chern-Simons theory can be found in the review papers by Freed [Fre95] for the case of trivializable bundles and [Fre02] for the nontrivializable case.

[2]Only the exponential of the Chern-Simons action, with the correct normalization, is invariant under large gauge transformations, but this is irrelevant for the perturbative treatment in this paper.

2.2 Perturbative Quantization of Gauge Theories

We remain very brief in this subsection. The interested reader is invited to study the references [Res10, Mne17] for a thorough introduction to the perturbative quantization of gauge theories from the mathematical viewpoint, or the short introductory paper [Pol05] (which is devoted exclusively to Chern-Simons theory).

The quantity of interest in this paper is the Chern-Simons partition function Z_M. Its naive definition would be

$$Z_M = \int_{F_M} e^{\frac{i}{\hbar} S_M},$$

but attempts at defining an appropriate measure on F_M have failed (see [GJ87] for a discussion of measures on field spaces in quantum field theory). In perturbative quantization, one tries to define Z_M by extrapolating the behavior of a finite-dimensional oscillatory integral $Z = \int_F e^{\frac{i}{\hbar} S}$ in the $\hbar \to 0$ limit. It is well known that if S has non-degenerate critical points, then the integral concentrates in a neighborhood of them, and one can derive a series in powers of $\hbar$ describing the asymptotic behavior. One can mimic the definition of this power series in the infinite-dimensional case if the critical points of S_M are non-degenerate. However, the critical points of functionals invariant under a symmetry, such as the Chern-Simons functional, are never non-degenerate, so one needs an additional method from physics, called *gauge fixing*. There are different variants of this method, the most commonly used being the Faddeev-Popov (FP) ghosts [FP67] and the BRST formalism.[3] The idea is to embed the space of fields F_M in the degree 0 part of a graded vector space[4] $\mathcal{F}_M$, and define a new functional $\mathcal{S}_M : \mathcal{F}_M \to \mathbb{R}$ such that $\mathcal{S}_M$ has non-degenerate critical points, and $\mathcal{S}_M|_{F_M} = S_M$.

Both the FP ghosts and the BRST formalism are subsumed in the BV formalism named after Batalin and Vilkovisky, who introduced it in [BV77, BV81, BV83]. We will briefly discuss the BV formalism and its adaptation to the case with boundary, the BV-BFV formalism, in the next section.

2.3 Perturbative Chern-Simons Theory and Invariants of 3-Manifolds and Links

There is a considerable number of invariants of 3-manifolds or rational homology spheres or possibly knots and links in 3-manifolds that are related or sup-

[3]Sometimes also BRS formalism, it was introduced independently by Becchi, Rouet and Stora [BRS75, BRS76] and Tyutin [Tyu76].

[4]In other examples the space of fields can be non-linear, but we will only deal with the linear case, which is also easier in view of quantization.

posedly related to perturbative Chern-Simons theory, including the Kontsevich integral [FK89, Kon93, BN95], the Le-Murakami-Ohtsuki invariant [LMO98], the Aarhus integral [BNGRT02], the Kontsevich-Kuperberg-Thurston-Lescop invariant of rational homology spheres[Kon94, KT99, Les04a, Les04b], the Axelrod-Singer invariants [AS91, AS94] associated with acyclic flat connections, the Bott-Cattaneo [BC98, BC99] invariants of rational homology spheres, the effective action approach by the first two authors [CM08] and of course the conjectural asymptotic expansion of the Reshetikhin–Turaev invariant [RT91]. The authors hope that the methods of cutting and gluing diagrams introduced in [CMR17], and used in this paper to compute weights of theta graphs, can help to shed light on the various known and conjectured relations between these invariants in the future.

3 Review of BV-BFV Quantization

The BV formalism is generally considered as the most powerful gauge-fixing formalism, as it can deal with a large class of gauge theories, especially also those where the gauge symmetries do not close off-shell. Detailed introductions can be found in [Mne08, Mne17, Fio03]. In view of functorial quantization, an important fact is that it admits a natural extension to manifolds with boundary, the BV-BFV formalism, that is compatible with cutting and gluing both at the classical level and the level of perturbative quantization [CMR14, CMR17]. We briefly review the main concepts. For details and proofs we refer to the references above.

3.1 BV Formalism

In the BV formalism one embeds the space of states F_M into an *odd symplectic* $\mathbb{Z}$-graded vector space $(\mathcal{F}_M, \omega)$, where the odd symplectic form ω is required to have $\mathbb{Z}$-degree -1. The $\mathbb{Z}$-degree is referred to as *ghost number*. One needs to find a BV action $\mathcal{S}_M$ such that $\mathcal{S}_M\big|_{F_M} = S_M$, which satisfies the Classical Master Equation (CME) $(\mathcal{S}, \mathcal{S}) = 0$, where $(\cdot, \cdot)$ is the odd Poisson bracket associated with ω. Notice that if $\mathcal{S}_M$ has a Hamiltonian vector field $\mathcal{Q}_M$, then $\mathcal{Q}_M^2 = 0$, from the CME. This leads to the following definition of BV theory [CMR14]:

Definition 3.1 A *BV vector space* is a quadruple $(\mathcal{F}, \omega, \mathcal{Q}, \mathcal{S})$ where $\mathcal{F}$ is a $\mathbb{Z}$-graded vector space, ω is a symplectic form of degree -1, $\mathcal{Q}$ is a degree $+1$ vector field, and $\mathcal{S}$ is a function of degree 0, such that $\mathcal{Q}^2 = 0$ and

$$\iota_{\mathcal{Q}}\omega = \delta\mathcal{S}. \tag{4}$$

A *d-dimensional (linear) BV theory* is an association of a *BV* vector space $(\mathcal{F}_M, \omega_M, \mathcal{Q}_M, \mathcal{S}_M)$ to every d-dimensional manifold M.

Here the de Rham differential on $\mathcal{F}$ is denoted by δ. Equation (4) says that $\mathcal{Q}$ is the Hamiltonian vector field of $\mathcal{S}$ and together with $\mathcal{Q}^2 = 0$ this implies the CME. We will only be interested in local theories. Roughly this means that $\mathcal{F}_M$ is given by sections of a sheaf over M, and the other objects are given by integrals of functions of jets of the fields.

In order to quantize one requires also the Quantum Master Equation (QME)

$$\Delta(e^{\frac{i}{\hbar}\mathcal{S}}) = 0 \Leftrightarrow \frac{1}{2}(\mathcal{S}, \mathcal{S}) + i\hbar\Delta\mathcal{S} = 0. \tag{5}$$

Here Δ is the BV Laplacian associated with ω, which in Darboux coordinates (x^i, p_i) is given by $\sum_i \pm\frac{\partial}{\partial x^i}\frac{\partial}{\partial p_i}$.[5,6] To define the partition function ψ_M one needs to choose a Lagrangian $\mathcal{L}_M \subset \mathcal{F}_M$, such that the partition function restricted to that Lagrangian has a unique critical point. This is the choice of gauge fixing. The partition function is then defined by the BV integral

$$\psi_M = \int_{\mathcal{L}_M} e^{\frac{i}{\hbar}\mathcal{S}}. \tag{6}$$

Under the assumption that the QME is satisfied, the partition function ψ_M defined by (6) is invariant under deformations of the gauge-fixing Lagrangian $\mathcal{L}_M$.

In the infinite-dimensional case usually at hand in quantum field theory, the BV Laplacian is ill-defined and has to be regularized, whereas the BV integral (6) has to be replaced by its formal asymptotic version, which yields a sum over Feynman graphs.

3.1.1 Effective Action and Residual Fields

In certain cases, it is not possible to find directly a Lagrangian which satisfies the requirement that the action restricted to it has a unique critical point. A good example is the case of abelian BF theory. In d dimensions, the BV space of fields is $\mathcal{F}_M = \Omega^\bullet(M)[1] \oplus \Omega^\bullet(M)[d-2]$, and the BV action functional is

$$\mathcal{S}_M[\mathsf{A}, \mathsf{B}] = \int_M \mathsf{B} \wedge \mathrm{d}\mathsf{A}. \tag{7}$$

The critical points are given by closed forms, and the gauge symmetries are given by shifting A, B by exact forms. Hence if the de Rham cohomology $H^\bullet(M)$ is non-

[5] As always, the formalism is developed for finite-dimensional F_M, and then needs to adapted to the infinite-dimensional case.

[6] In this paper we completely ignore the (important) distinction between functions and half-densities, see e.g. [Khu04].

trivial, there are critical points which are inequivalent under gauge transformations, also known as zero modes.

The solution to this problem is to choose a BV space of residual fields $\mathcal{V}_M$ and a splitting $F_M = \mathcal{V}_M \times \mathcal{Y}$, such that one can find a gauge-fixing Lagrangian $\mathcal{L}_M \subset \mathcal{Y}$. Elements of $\mathcal{V}_M$ are known as residual fields, zero modes, infrared fields, or slow fields, whereas the elements of $\mathcal{Y}$ are called fluctuations, fast fields, or ultraviolet fields. The partition function ψ_M gets replaced by an *effective action* $\psi_M(\mathsf{x})$, which is a function of the residual fields and formally defined via BV pushforward:

$$\psi_M(\mathsf{x}) = \int_{\xi \in \mathcal{L}_M \subset \mathcal{Y}} e^{\frac{i}{\hbar}\mathcal{S}[\mathsf{x},\xi]}. \tag{8}$$

In the case of abelian BF theory one can choose as residual fields representatives of the cohomology: $\mathcal{V}_M = H^\bullet(M)[1] \oplus H^\bullet(M)[d-2]$. One way to do this is to pick a Riemannian metric and use the harmonic representatives, a possible choice of gauge-fixing Lagrangian is then given by d*-exact forms. In the finite-dimensional case, the QME for the BV action implies that the effective action is $\Delta_{\mathcal{V}_M}$-closed, i.e. closed with respect to the BV Laplacian on residual fields, and changes by a Δ-exact term under a deformation of the gauge-fixing Lagrangian.

3.2 BV-BFV Extension on Manifolds with Boundary

A crucial feature of the BV formalism is that admits an extension to manifolds that is compatible with cutting and gluing, in the form of the BV-BFV formalism. The classical case is discussed in [CMR14]. An introduction can also be found in [Sch15]. The perturbative quantization was introduced in [CMR17]. Further discussions of the gauge theories on manifolds with boundary both at classical and quantum level can be found in [CMR11, CMR15]. For a brief review see also [CMR, CMW18]. We briefly review the main notions in this subsection.

3.2.1 Classical Case

First one needs the definition of BFV vector space (for background on the BFV complex see [Sta97, Sch10]).

Definition 3.2 A BFV vector space is a triple $(\mathcal{F}^\partial, \omega^\partial, \mathcal{Q}^\partial)$, where $\mathcal{F}^\partial$ is a $\mathbb{Z}$-graded vector space, ω^∂ is a symplectic form of degree 0 on $\mathcal{F}^\partial$, and $\mathcal{Q}^\partial$ is a degree $+1$ vector field which is symplectic and satisfies $(\mathcal{Q}^\partial)^2 = 0$.

By degree reasons $\mathcal{Q}^\partial$ is automatically Hamiltonian with Hamiltonian function $\mathcal{S}^\partial$.

Now the idea is to associate to the boundary of a manifold a BFV vector space, and to the bulk a suitable generalization of a BV vector space which is compatible with the BFV data associated with the boundary. The solution is the notion of BV-BFV vector space as introduced in [CMR14].

Definition 3.3 Let $\mathcal{F}^\partial = (\mathcal{F}^\partial, \omega^\partial = \delta\alpha^\partial, \mathcal{Q}^\partial)$ be a BFV vector space with exact symplectic form ω^∂. A BV-BFV vector space over $\mathcal{F}^\partial$ is a quintuple $(\mathcal{F}, \omega, \mathcal{Q}, \mathcal{S}, \pi)$, where $\mathcal{F}$ is a $\mathbb{Z}$-graded vector space, ω is a degree -1 symplectic form, $\mathcal{Q}$ is a degree $+1$ vector field, $\mathcal{S}$ is a degree 0 function on $\mathcal{F}$ and $\pi : \mathcal{F} \to \mathcal{F}^\partial$ is a surjective submersion such that $\mathcal{Q}^2 = 0$, $\delta\pi\,\mathcal{Q} = \mathcal{Q}^\partial$ and

$$\iota_Q \omega = \delta S + \pi^* \alpha^\partial. \tag{9}$$

Remark 3.1 Equation (9) should be thought of as a generalization of the CME. In fact, in the case $\mathcal{F}^\partial = \{0\}$ the definition of BV-BFV vector space reduces to that of an ordinary BV vector space.

We are now ready to define the notion of classical BV-BFV theory. Namely, a d-dimensional BV-BFV theory associates to closed $d-1$-dimensional manifold Σ an exact BFV vector space $\mathcal{F}^\partial_\Sigma$ and to a d-dimensional manifold M with boundary ∂M a BV-BFV vector space $\mathcal{F}_M$ over the BFV vector space $\mathcal{F}^\partial_{\partial M}$. Some remarks are in order:

Remark 3.2

(i) The exactness condition (like the linearity condition) can be relaxed. See [CMR14, Remark 3.3] for a short discussion.
(ii) We also require locality from BV-BFV theories. That means the vector spaces $\mathcal{F}, \mathcal{F}^\partial$ are given by sections of a sheaf and as such are typically infinite-dimensional (over $\mathbb{R}$ or $\mathbb{C}$). They can be equipped with natural Banach or Fréchet topologies depending on the situation.
(iii) With enough care, a BV-BFV theory yields a functor from a cobordism category (perhaps equipped with extra structure) to a category of vector spaces where composition is given by (homotopy) fibered product. Again, we refer to [CMR14, Section 4] for a more detailed discussion.

3.2.2 Quantization

Classical BV-BFV theories admit a perturbative quantization that is compatible with cutting and gluing [CMR17]. Roughly, the idea is to combine geometric quantization of $(\mathcal{F}^\partial, \omega)$ and perturbative quantization as in Eq. (8). We briefly outline the main steps.

1. Choose a polarization $\mathcal{P}$ on $(\mathcal{F}^\partial, \omega)$ with smooth leaf space $\mathcal{B}^\mathcal{P}$. In our case it will be enough to find a splitting $\mathcal{F}^\partial = \mathcal{B}^\mathcal{P}_1 \times \mathcal{B}^\mathcal{P}_2$, where both $\mathcal{B}^\mathcal{P}_1$ and $\mathcal{B}^\mathcal{P}_2$ are Lagrangian subspaces of $\mathcal{F}^\partial$. One can choose either $\mathcal{B}^\mathcal{P}_i$ as base space or fibers of the polarization, respectively.

2. If necessary, change α^∂ by an exact term such that it vanishes on the fibers of $\mathcal{P}$. To preserve Eq. (9) one has to change $\mathcal{S}$ by a boundary term.
3. Choose a section σ of $\mathcal{F} \to \mathcal{F}^\partial \to \mathcal{B}^\mathcal{P}$, and a splitting $\mathcal{F} = \sigma(\mathcal{B}^\mathcal{P}) \times \mathcal{Y}$ (subject to certain requirements discussed in [CMR17]). We immediately suppress σ from the notation.
4. Then, proceed to split $\mathcal{Y} = \mathcal{V} \times \mathcal{Y}'$ such that there is a Lagrangian $\mathcal{L} \subset \mathcal{Y}'$ on which the action $\mathcal{S}$ has a unique critical point. We now have a splitting $\mathcal{F} = \mathcal{B}^\mathcal{P} \times \mathcal{V} \times \mathcal{Y}'$, and accordingly we write $\mathsf{X} = \mathbb{X} + \mathsf{x} + \xi$ for $\mathsf{X} \in \mathcal{F}$.
5. Finally, define the state or partition function formally by

$$\psi(\mathbb{X}, \mathsf{x}) = \int_{\xi \in \mathcal{L}_M \subset \mathcal{Y}} e^{\frac{i}{\hbar}\mathcal{S}[\mathbb{X}, \mathsf{x}, \xi]}. \tag{10}$$

A short discussion is in order.

In the finite-dimensional case, Eq. (10) is an example of a "BV pushforward in families" introduced in [CMR17]. In the infinite-dimensional case, definition (10) has to be interpreted via the Feynman graphs and rules discussed in [CMR17] and [CMW17]. We will briefly review the ones relevant for this paper below.

The state (10) is a functional on both $\mathcal{V}$ and $\mathcal{B}^\mathcal{P}$. We think of it as an element of $\widehat{\mathcal{H}}^\mathcal{P} = \widehat{S}\mathcal{V}^* \otimes \mathcal{H}^\mathcal{P}$, where $\mathcal{H}^\mathcal{P}$ is a certain space of functionals on $\mathcal{B}^\mathcal{P}$ that should be thought of as a geometric quantization of $(\mathcal{F}^\partial, \omega)$, and $\widehat{S}$ denotes the formal completion of the symmetric algebra. Also the boundary action S^∂ can be quantized and yields a coboundary operator $\Omega^\mathcal{P}$ on $\mathcal{H}^\mathcal{P}$. The precise construction of this space and the coboundary operator Ω are not relevant for this paper, the interested reader is again referred to [CMR17, Section 4.1]. Also $\mathcal{V}$ carries a coboundary operator, the BV Laplacian $\Delta_\mathcal{V}$, and the state is a cocycle in the bicomplex $\mathcal{H}^\mathcal{P}$:

$$(\hbar^2 \Delta_\mathcal{V} + \Omega^\mathcal{P})\psi = 0. \tag{11}$$

Equation (11) is called the modified Quantum Master Equation (mQME).

An important feature of the quantum BV-BFV formalism is that the perturbative expansions associated with manifolds with boundary can be glued together using a form of the BKS[7] pairing discussed in [CMR17]. The gluing procedure is important to this paper and we will discuss it in more detail below in Sect. 6.2.

4 Split Chern-Simons Theory on the Solid Torus

We now apply the BV-BFV formalism to the case of split Chern-Simons theory, in the example of the solid torus. Partly this example was already discussed in [CMW17]. In this paper we use the axial gauge fixing to explicitly compute the

[7]For Blattner, Kostant, Sternberg. See [BW97].

Feynman diagrams in low orders on the solid torus and, via gluing, also on lens spaces.

4.1 Split Chern-Simons Theory

Split Chern-Simons theory as an example of a perturbation of abelian BF theory was first considered in [CMR17] and investigated more thoroughly in [CMW17], see also [Wer18]. Let $\mathfrak{g}$ be a Lie algebra with symmetric ad-invariant bilinear form $\langle\cdot,\cdot\rangle$. Consider the BV-extended Chern-Simons action functional

$$\mathcal{S}[\mathsf{C}] = \int_M \frac{1}{2}\langle \mathsf{C}, \mathrm{d}\mathsf{C}\rangle + \frac{1}{6}\langle \mathsf{C}, [\mathsf{C},\mathsf{C}]\rangle = \int_M \frac{1}{2} B_{ab}\mathsf{C}^a\mathsf{C}^b + \frac{1}{6} f_{abc}\mathsf{C}^a\mathsf{C}^b\mathsf{C}^c,$$

where $\mathsf{C} \in \mathcal{F}_M = \Omega^\bullet(M,\mathfrak{g})[1]$ is the superfield, i.e. an inhomogeneous Lie algebra-valued differential form, and $\langle\cdot,\cdot\rangle$ and $[\cdot,\cdot]$ denote extensions of the bilinear form and the Lie bracket to Lie algebra-valued differential forms. In the second expression we pick a basis e_a of $\mathfrak{g}$ and define $B_{ab} = \langle e_a, e_b\rangle$, $f_{abc} = \langle e_a, [e_b, e_c]\rangle$ and expand the field as $\mathsf{C} = \mathsf{C}^a e_a$. In [CMR14] it is shown that this is an example of a BV-BFV theory.

Now assume that $\mathfrak{g}$ admits a splitting $\mathfrak{g} = V \oplus W$ into maximal isotropic subspaces. We choose a basis ξ_i of V and a dual basis ξ^i of W. Then the space of fields splits as $\Omega^\bullet(M,\mathfrak{g}) = \Omega^\bullet(M,V)\oplus\Omega^\bullet(M,W)$ and the superfield C splits as $\mathsf{A}+\mathsf{B} = \mathsf{A}^i\xi_i + \mathsf{B}_i\xi^j$. Integrating by parts one can rewrite the action as

$$\mathcal{S}[\mathsf{A},\mathsf{B}] = \int_M \langle \mathsf{B}, \mathrm{d}\mathsf{A}\rangle + \frac{1}{6}\langle \mathsf{A}, [\mathsf{A},\mathsf{A}]\rangle + \frac{1}{2}\langle \mathsf{B}, [\mathsf{A},\mathsf{A}]\rangle + \frac{1}{2}\langle \mathsf{A}, [\mathsf{B},\mathsf{B}]\rangle + \frac{1}{6}\langle \mathsf{B}, [\mathsf{B},\mathsf{B}]\rangle \tag{12}$$

so it becomes a "BF-like" theory with interaction term

$$\begin{aligned}\mathcal{V}(\mathsf{A},\mathsf{B}) &= \frac{1}{6}\langle \mathsf{A}, [\mathsf{A},\mathsf{A}]\rangle + \frac{1}{2}\langle \mathsf{B}, [\mathsf{A},\mathsf{A}]\rangle + \frac{1}{2}\langle \mathsf{A}, [\mathsf{B},\mathsf{B}]\rangle + \frac{1}{6}\langle \mathsf{B}, [\mathsf{B},\mathsf{B}]\rangle \\ &= \frac{1}{6} g_{ijk}\mathsf{A}^i\mathsf{A}^j\mathsf{A}^k + \frac{1}{2} g^i_{jk}\mathsf{B}_i\mathsf{A}^j\mathsf{A}^k + \frac{1}{2} h_i^{jk}\mathsf{A}^i\mathsf{B}_j\mathsf{B}_k + \frac{1}{6} h^{ijk}\mathsf{B}_i\mathsf{B}_j\mathsf{B}_k,\end{aligned}$$

where we introduced the structure constants $g_{ijk}, g^i_{jk}, h_i^{jk}, h^{ijk}$ defined by applying $\langle\cdot,[\cdot,\cdot]\rangle$ to the basis vectors ξ_i, ξ^i. If $(\mathfrak{g}, V, W)$ form a quasi-Manin triple (i.e. V is a subalgebra), by isotropy the interaction term simplifies to

$$\mathcal{V}(\mathsf{A},\mathsf{B}) = \frac{1}{2}\langle \mathsf{B}, [\mathsf{A},\mathsf{A}]\rangle + \frac{1}{2}\langle \mathsf{A}, [\mathsf{B},\mathsf{B}]\rangle + \frac{1}{6}\langle \mathsf{B}, [\mathsf{B},\mathsf{B}]\rangle. \tag{13}$$

In the special case where $(\mathfrak{g}, V, W)$ is a Manin triple (i.e. V, W are subalgebras), by isotropy we get that the interaction term simplifies to

$$\mathcal{V}(\mathsf{A}, \mathsf{B}) = \frac{1}{2}\langle \mathsf{B}, [\mathsf{A}, \mathsf{A}]\rangle + \frac{1}{2}\langle \mathsf{A}, [\mathsf{B}, \mathsf{B}]\rangle. \tag{14}$$

We now specialize to the case of the solid torus $S^1 \times D$. We give it coordinates $x = (t, z) \in [0, 1) \times \{z \in \mathbb{C} | |z| \leq 1\}$.

Remark 4.1 Notice that by choosing global coordinates we also choose a trivialization of its tangent bundle—a framing. We will fix this framing on the solid torus. However, as we glue lens spaces from solid tori, the framing of the glued lens space will depend on the gluing diffeomorphism. See also Sect. 6.

4.2 Polarization

In split Chern-Simons theory, the space of boundary fields splits as

$$\mathcal{F}^{\partial} = \Omega^{\bullet}(\partial M, V) \oplus \Omega^{\bullet}(\partial M, W). \tag{15}$$

By the isotropy condition this is a splitting into Lagrangian subspaces, so we can use either of them as base or fibers of the polarization. The coordinate on the base is denoted by a blackboard bold letter $\mathbb{A}$ or $\mathbb{B}$, and we speak of $\mathbb{A}$- or $\mathbb{B}$-representation, respectively.[8] This terminology comes from the p- and q-representations in Quantum Mechanics.

Since $\partial M \cong S^1 \times S^1$ is connected, the only choice is between the $\mathbb{A}$- or the $\mathbb{B}$-representation on ∂M. For computations we will use the $\mathbb{A}$-representation, i.e. we will split the A-field as

$$\mathsf{A} = \widehat{\mathsf{A}} + \widetilde{\mathbb{A}},$$

where $\widetilde{\mathbb{A}}$ is an extension of $\mathbb{A} = \iota^*\mathsf{A}$ to the bulk. The computations for the $\mathbb{B}$-representation are exactly the same, one just needs some care in translating the results. See Sect. 6.3.

4.3 Residual Fields and Fluctuations

The minimal space of residual fields for this polarization is

$$\mathcal{V}_M = H^{\bullet}(M, \partial M) \otimes V \oplus H^{\bullet}(M) \otimes W \ni (\mathsf{a}, \mathsf{b}).$$

[8] If one thinks of Chern-Simons theory as an AKSZ theory, this amounts to lifting a target polarization instead of a source polarization.

As representatives of the cohomology we choose

$$\chi^1 = 1$$
$$\chi^2 = dt$$
$$\chi_1 = \mu dt$$
$$\chi_2 = \mu,$$

where $\mu = \frac{1}{2\pi i} d\bar{z}dz$ is unit volume form on the disk, dt is the coordinate differential on S^1, and we omit wedge symbols. Notice that

$$\int_M \chi^i \chi_j = \int_M \chi_j \chi^i = \delta^i_j,$$

i.e. χ^i is the dual basis to χ_j. We can then expand

$$\mathsf{a} = z^i \chi_i$$
$$\mathsf{b} = z_i^+ \chi^i,$$

where z^i, z_i are linear V(resp. W)-valued functions on $H^\bullet(M, \partial M)$ and $H^\bullet(M)$. The fluctuations α, β are then defined by

$$\widehat{\mathsf{A}} = \alpha + \mathsf{a}$$
$$\widehat{\mathsf{B}} = \beta + \mathsf{b}.$$

Notice that the fluctuations depend on the extension of $\mathbb{A}$, whereas the backgrounds a, b do not. As discussed in [CMR15] one now chooses a discontinuous extension of $\mathbb{A}$ which drops to zero outside the boundary, which we also denote by $\mathbb{A}$, abusing notation. This leads to a splitting of the action as

$$\mathcal{S} = \mathcal{S}_0 + \mathcal{S}^{back} + \mathcal{S}^{source} + \mathcal{S}^{int}, \tag{16}$$

where

$$\mathcal{S}_0 = \int_M \widehat{\mathsf{B}} d\widehat{\mathsf{A}}$$
$$\mathcal{S}^{back} = \int_{\partial M} \mathsf{b}\mathbb{A}$$
$$\mathcal{S}^{source} = \int_{\partial M} \beta\mathbb{A}$$
$$\mathcal{S}^{int} = \mathcal{V}(\mathsf{A}, \mathsf{B}).$$

4.4 Axial Gauge Propagator

Axial gauge propagators on product manifolds were discussed in [CMR15] (Appendix C.1), but the idea is older and goes back to works of Fröhlich [FK89].

The identity these distributional propagators satisfy is

$$\mathrm{d}\eta_M = \delta_M^{(d)}(x_1, x_2) + (-1)^{d-1} \sum_i (-1)^{d \cdot \deg \chi_i} \pi_1^* \chi_i \pi_2^* \chi^i. \tag{17}$$

On the solid torus, we have two choices for a distributional propagator. The horizontal propagator is

$$\eta^{hor}((z_1, t_1), (z_2, t_2)) = \eta_D(z_1, z_2) \delta_{S^1}^{(1)}(t_1, t_2) + \mu_1 \eta_{S^1}(t_1, t_2), \tag{18}$$

while the axial propagator is

$$\eta^{ax}((z_1, t_1), (z_2, t_2)) = \delta_D^{(2)}(z_1, z_2) \eta_{S^1}(t_1, t_2) + \eta_D(z_1, z_2)(dt_1 - dt_2), \tag{19}$$

where η_D and η_{S^1} are propagators on the disk and the circle, respectively.

Lemma 4.1 *These propagators satisfy*

$$\mathrm{d}\eta = \delta_M^{(3)}(x_1, x_2) + \sum_i (-1)^{\deg \chi_i} \pi_1^* \chi_i \pi_2^* \chi^i = \delta_M^{(3)}(x_1, x_2) - \mu_1 dt_1 + \mu_1 dt_2. \tag{20}$$

Proof First note that in the distributional sense we have

$$\begin{aligned} \mathrm{d}_D \eta_D &= \delta_D^{(2)}(z_1, z_2) - \mu_1 \\ \mathrm{d}_{S^1} \eta_{S^1} &= \delta_{S^1}^{(1)}(t_1, t_2) - (dt_1 - dt_2) \\ \mathrm{d}_{S^1} \delta_{S^1}^{(1)} &= 0, \end{aligned}$$

where the first two identities are (17) for D and S^1, respectively, and the third is for dimensional reasons. Using this, we evaluate (omitting the arguments)

$$\begin{aligned} \mathrm{d}\eta^{hor} &= \mathrm{d}\eta_D \delta_{S^1} + \mu_1 \mathrm{d}\eta_{S^1} \\ &= (\delta_D - \mu_1)\delta_{S^1} + \mu_1(\delta_{S^1} - (dt_1 - dt_2)) \\ &= \delta_D \delta_{S^1} - \mu_1(dt_1 - dt_2) \\ &= \delta_M - \mu_1 dt_1 + \mu_1 dt_2 \end{aligned}$$

and

$$\begin{aligned} d\eta^{ax} &= \delta_D(\delta_{S^1} - dt_1 + dt_2) + (\delta_D - \mu_1)(dt_1 - dt_2) \\ &= \delta_M - \mu_1 dt_1 + \mu_1 dt_2. \end{aligned}$$

□

We will now state some other desirable properties the propagators have.

Proposition 4.1 *Suppose* $\int_{D_2} \eta_{D,12}\mu_2 = 0$, $\int_{D_2} \eta_{D,12}\eta_{D,23} = 0$ *and* $\int_{S^1_2} \eta_{S^1,12} dt_2 = 0$, $\int_{\partial D_2} \eta_{D,12} = 1$. *Then the following identities hold for* $\eta \in \{\eta^{ax}, \eta^{hor}\}$*:*

(1) $\int_1 \eta_{12} = 0$,
(2) $\int_1 dt_1 \eta_{12} = 0$,
(3) $\int_2 \eta_{12} dt_2 = 0$,
(4) $\int_2 \eta_{12}\mu_2 = 0$,
(5) $\int_2 \eta_{12}\mu_2 dt_2 = 0$,
(6) $\int_2 \eta_{12}\eta_{23} = 0$,
(7) $\int_{2,\partial} \eta_{12} = 1$.

Proof All identities are direct applications of the assumptions together with some degree counting (notice a form needs to have bidegree (2,1) to contribute to the integral). □

4.5 The State

For reasons discussed below, from now on we only work with the horizontal propagator. It determines the BV gauge-fixing Lagrangian $\mathcal{L}$ and therefore the state by the formal BV integral

$$\widehat{\psi} = \int_{(\alpha,\beta)\in\mathcal{L}} e^{\frac{i}{\hbar}\mathcal{S}[\mathsf{A},\mathsf{B}]} \tag{21}$$

which is evaluated by Feynman graphs and rules as in [CMR17, CMW17], which we briefly recall below.

4.5.1 Feynman Graphs and Rules

In this case the Feynman graphs and rules are as follows. Admissible graphs are directed connected graphs with a trivalent and univalent vertices. The trivalent vertices can have any number (between 0 and 3) of outgoing and incoming half-edges. Outgoing half-edges represent A fields while incoming half-edges represent

B fields. Univalent vertices can have only incoming half-edges and are decorated by a boundary field $\mathbb{A}$. See Fig. 1. Incoming half-edges must either be connected to an outgoing half-edge or a b residual field (a half-edge ending in a residual field is thought of as a decoration of the vertex incident to the half-edge). Outgoing half-edges must either be connected to an incoming half-edge or an a residual field.

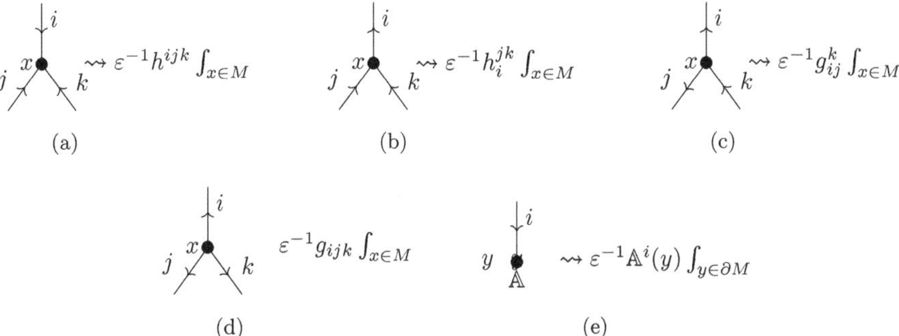

Fig. 1 Feynman graphs and rules on the solid torus: vertices. (**a**) BBB vertex. (**b**) ABB vertex. (**c**) AAB vertex. (**d**) AAA vertex. (**e**) Univalent vertex

A Feynman graph Γ is evaluated as follows. First, choose dual bases ξ_i of V and ξ^i of W and expand the fields accordingly: $\mathsf{A} = \mathsf{A}^i\xi_i = \mathbb{A}^i\xi_i + \mathsf{a}^i\xi_i + \alpha^i\xi_i$, and similarly for the B fields. Now one labels every trivalent vertex by a point $x \in M$, every univalent vertex by a point $y \in \partial M$ (for this reason trivalent vertices are called bulk vertices and univalent vertices are called boundary vertices), and all half-edges with an index i_k. Every half-edge also "inherits" the label of the vertex it is attached to. Now one computes a de Rham current as follows. For every trivalent vertex with half-edges labeled (i, j, k) one multiplies with the appropriate structure constants, see Fig. 1. If a half-edge labeled by (x, i) ends in the residual field a, one multiplies with $\mathsf{a}^i(x)$, and similarly for the other half-edges and the b residual fields. For an edge formed by half-edges (x, i) and (y, j) multiply[9] with $\delta^i_j\eta(x, y)$, see Fig. 2. Finally, for a boundary vertex labeled by (y, i) multiply by $\mathbb{A}^i$.

As for power counting, vertices come with a factor of $\varepsilon^{-1} = \frac{i}{\hbar}$, while edges carry a factor of $\varepsilon = (-i\hbar)$. In the next section we will see plenty of examples of Feynman graphs. There we depict the torus in a cross-section, placing boundary (univalent) vertices on the boundary and bulk (trivalent) vertices in the bulk. Notice that in the special case of the axial gauge, the weights will factor into integrals over the disk and integrals over the circle, since the residual fields and the propagator are products of forms on the disk and the circle.

[9] η is a de Rham current and one must be very careful taking products, see the paragraph below.

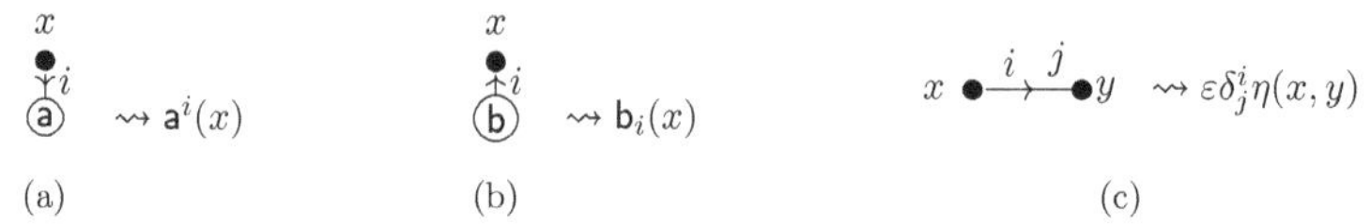

x $\bullet$ i ⓐ $\rightsquigarrow \mathsf{a}^i(x)$ (a)

x $\bullet$ i ⓑ $\rightsquigarrow \mathsf{b}_i(x)$ (b)

$x \bullet \xrightarrow{i \quad j} \bullet y \rightsquigarrow \varepsilon \delta^i_j \eta(x, y)$ (c)

Fig. 2 Feynman graphs and rules on the solid torus: decorations and edges. (**a**) A half-edge ending in an a residual field. (**b**) A half-edge coming from a b residual field. (**c**) An edge between vertices x and y

4.5.2 Regularization

Notice that since we work with the horizontal propagator, the integrals over the circle now contain distributions which cannot be extended to smooth forms on compactified configuration spaces as in [CMR17]. Hence one needs another way to define the product of distributions arising in Feynman graphs. We will discuss this question, and the question of the correct space of states and operators to state the mQME, in more detail and generality elsewhere. A regularization that is conjecturally equivalent to the one used here, and based on approximating the axial gauge by regular gauges, is discussed in the thesis of the third author [Wer18]. For the purpose of this note is sufficient that to say that for tree diagrams the wavefront sets of the involved distributions are transversal and the product can be defined. The loop diagrams which are relevant for the computation of the theta invariant will be discussed separately in Sect. 5.4. The choice of regularization is much easier for distributions on the circle, which is why we stick to the horizontal propagator.

5 Effective Action on the Solid Torus

We will now compute all the terms in the effective action contributing to the two-loop contribution on a glued lens space.

The integrals over the bulk vertices will factorize into contributions from the disk and contributions from the circle, both of which can be computed explicitly. The results are presented in the following way: The contribution of a graph Γ with m univalent boundary vertices to the state is a sum of functionals on boundary fields of the form

$$\psi_\Gamma(\mathbb{A}) = \int_{(\partial M)^m} \omega_\Gamma \mathbb{A}_1 \cdots \mathbb{A}_m = \sum_k P^k_\Gamma(z, z^+)_{i_1 i_2 \ldots i_m} \int_{(\partial M)^m} \omega^k_\Gamma \mathbb{A}^{i_1}_1 \cdots \mathbb{A}^{i_m}_m,$$

where

- $P^k_\Gamma(z, z^+)_{i_1 \ldots i_m}$ is a product of z and z^+ and the structure constants of the Lie algebra obtained by contracting the structure constants according to the graph, with $i_1, \ldots, i_m$ labeling the legs ending on the boundary, k labels the different products of z, z^+ appearing,

- $\omega_\Gamma^k := \pi_*(\widetilde{\omega}_\Gamma^k)$ is a distributional form (that we call "coefficient") obtained by integrating the product of propagators and representatives of cohomology over the bulk points in the graph, i.e. taking the pushforward along the fibers of the map $M^{\times N} \times (\partial M)^{\times m} \to (\partial M)^{\times m}$ that forgets the bulk points,
- $\mathbb{A}_i^j$ denotes the pullback of the ξ_j component of $\mathbb{A}$ under the i-th projection $(\partial M)^m \to \partial M$.

If we need to address individual coefficients we will usually distinguish them by form degree and denote a coefficient of form degree p by $\omega_\Gamma^{(p)}$.

5.1 Zero-Point Contribution

By our choice of polarization, the zero-point contribution—the effective action of the free part—just consists of the term (Fig. 3)

$$\psi_{\Gamma_0} = -\varepsilon^{-1} \int_{\partial_1 M} \mathsf{b}_k \mathbb{A}^k = \varepsilon^{-1} \left(-z_{1,k}^+ \int_{\partial_1 M} \mathbb{A}^k - z_{2,k}^+ \int_{\partial_1 M} dt \mathbb{A}^k \right).$$

Fig. 3 Single diagram contributing to zero-point effective action

5.2 One-Point Contribution

5.2.1 Possible Diagrams

If we have just one bulk vertex it can carry no β's due to the polarization. Also, notice that a^2=0. The following labeling survive: abb, bbb, $\mathsf{ab}\alpha$, $\mathsf{bb}\alpha$, $\mathsf{a}\alpha\alpha$, $\mathsf{b}\alpha\alpha$, $\alpha\alpha\alpha$. One can check that upon integration the ones with two b's vanish for degree reasons, with the exception of abb. Let us look at the remaining terms (Fig. 4).

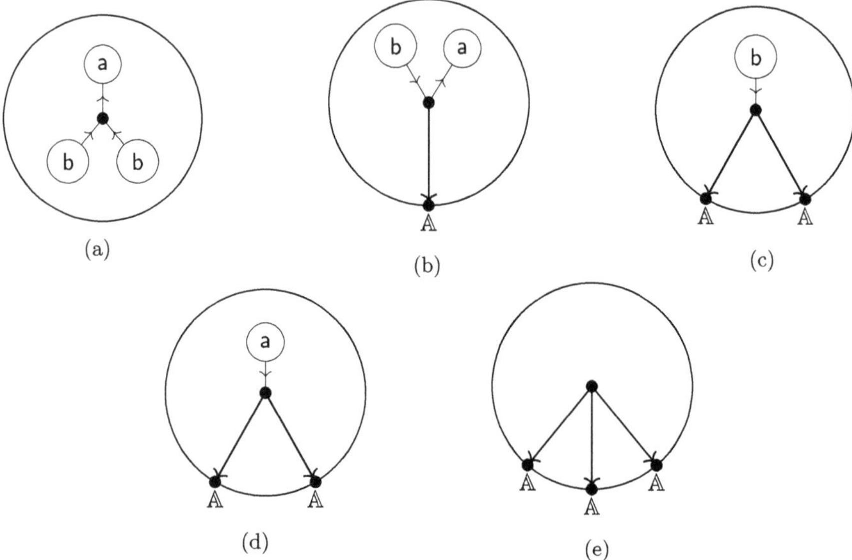

Fig. 4 Graphs in the solid torus (depicted in a cross-section) with 1 interaction vertex. A bullet denotes a point we integrate over, a long arrow denotes a propagator. (**a**) $\Gamma_{1,0}$. (**b**) $\Gamma_{1,1}$. (**c**) $\Gamma^{\mathsf{b}}_{1,2}$. (**d**) $\Gamma^{\mathsf{a}}_{1,2}$. (**e**) $\Gamma_{1,3}$

5.2.2 baa Term

The first one contains no propagator:

$$\Psi_{\Gamma_{1,0}} = \frac{\varepsilon^{-1}}{2}\int_M \langle \mathsf{a}, [\mathsf{b}, \mathsf{b}]\rangle = \frac{\varepsilon^{-1}}{2} g_i^{jk}(z^{1i} z^+_{1j} z^+_{1k} + 2z^{2i} z^+_{1j} z^+_{2k}) = \varepsilon^{-1}(P^1_{\Gamma_{1,0}} + P^2_{\Gamma_{1,0}}).$$

For the other terms explicit computations of the pushforwards over bulk vertices can be found in the thesis of the third author [Wer18].

5.2.3 baα Term

We have

$$\begin{aligned}\Psi_{\Gamma_{1,1}} &:= -\varepsilon^{-1}\int_{M\times\partial M} g^i_{jk}\mathsf{b}_{1,i}\mathsf{a}^j_1\eta_{12}\mathbb{A}^k_2 \\ &= -\varepsilon^{-1} g^i_{jk}(z^+_{1i}z^{1j} - z^+_{2i}z^{2j})\int_{\partial M}\pi_*(\mu_1 dt_1\eta_{12}))\mathbb{A}^k_2 - \varepsilon^{-1} g^i_{jk} z^+_{1i} z^{2j}\int_{\partial M}\pi_*(\mu_1\eta_{12})\,\mathbb{A}^k_2.\end{aligned}$$

5.2.4 b$\alpha\alpha$ Term

Let us turn to the next term

$$\Psi_{\Gamma^{\mathsf{b}}_{1,2}} := \frac{\varepsilon^{-1}}{2} \int_{M\times\partial M\times\partial M} g^{i}_{jk} b_{1,i} \eta_{12}\eta_{13}\mathbb{A}^{j}_{2}\mathbb{A}^{k}_{3}.$$

It evaluates to

$$\Psi_{\Gamma^{\mathsf{b}}_{1,2}} = \frac{\varepsilon^{-1}}{2} g^{i}_{jk} z^{+}_{1i} \int_{\partial M\times\partial M} \pi_{*}\left(\eta_{12}\eta_{13}\right) \mathbb{A}^{j}_{2}\mathbb{A}^{k}_{3} + \frac{\varepsilon^{-1}}{2} g^{i}_{jk} z^{+}_{2i} \int_{\partial M\times\partial M} \pi_{*}\left(dt_{1}\eta_{12}\eta_{13}\right) \mathbb{A}^{j}_{2}\mathbb{A}^{k}_{3}.$$

5.2.5 a$\alpha\alpha$ Term

The next term is

$$\Psi_{\Gamma^{\mathsf{a}}_{1,2}} := \frac{\varepsilon^{-1}}{2} \int_{M\times\partial M\times\partial M} f_{ijk} a^{i}_{1} \eta_{12}\eta_{13}\mathbb{A}^{j}_{2}\mathbb{A}^{k}_{3}.$$

It evaluates to

$$\Psi_{\Gamma^{\mathsf{a}}_{1,2}} = -\frac{\varepsilon^{-1}}{2} f_{ijk} z^{2,i} \int_{\partial M\times\partial M} \pi_{*}\left(\mu_{1}\eta_{12}\eta_{13}\right) \mathbb{A}^{j}_{2}\mathbb{A}^{k}_{3} - \frac{1}{2} f_{ijk} z^{1,i} \int_{\partial M\times\partial M} \pi_{*}\left(\mu_{1} dt_{1}\eta_{12}\eta_{13}\right) \mathbb{A}^{j}_{2}\mathbb{A}^{k}_{3}.$$

5.2.6 $\alpha\alpha\alpha$ Term

The next term is

$$\Psi_{\Gamma_{1,3}} := \frac{1}{6} \int_{M\times\partial M\times\partial M\times\partial M} f_{ijk} \eta_{12}\eta_{13}\eta_{14}\mathbb{A}^{i}_{2}\mathbb{A}^{j}_{3}\mathbb{A}^{k}_{4}.$$

It evaluates to

$$\Psi_{\Gamma_{1,3}} = -\frac{1}{6} f_{ijk} \int_{\partial M\times\partial M\times\partial M} \pi_{*}\left(\eta_{12}\eta_{13}\eta_{14}\right) \mathbb{A}^{i}_{2}\mathbb{A}^{j}_{3}\mathbb{A}^{k}_{4}.$$

Notice the last two terms are not present in the case where $(\mathfrak{g}, V, W)$ is a Manin triple.

5.3 2-Point Tree Contribution

In principle, tree diagrams with two points could contribute to the 2-point effective action after gluing. However, in this subsection we will argue that it is not so. The point is that in the gluing process 2-point graphs can only be paired with 0-point graphs on the other side, i.e. graphs from the free effective action. As explained below in Sect. 7.1.1, pairing against a 0-point diagram on the other side amounts to placing a linear combination of 1, dt and $d\theta$ at this point and integrating over it. The claim then follows from the following lemmata:

Lemma 5.1 *For $x_2 \in \partial M$, we have*

$$\int_2 \eta_{12} dt_2 = dt_1$$

$$\int_2 \eta_{12} d\theta_2 = -\psi = \frac{z d\bar{z} - \bar{z} dz}{4\pi i}.$$

Proof We have

$$\int_2 \eta_2 dt_2 = \int_2 (\eta_{D,12}\delta_{S^1,12} + \mu_1 \eta_{S^1,12}) dt_2 = \int_{\partial D,2} \eta_{D,12} \int_{S^1,2} \delta_{S^1,12} dt_2 = dt_1$$

since $\int_{\partial D,2} \eta_{D,12} = 1$ and

$$\int_2 \eta_2 d\theta_2 = (\eta_{D,12}\delta_{S^1,12} + \mu_1 \eta_{S^1,12}) d\theta_2 = \int_{\partial D,2} \eta_{D,12} d\theta_2$$

$$= \int_{\partial D,2} 2(\phi_{12} - \psi_1) d\theta_2 = -\psi_1$$

since $\int_{\partial D,2} \phi_{12} d\theta_2 = 0$, as can easily be checked by the residue theorem. □

Lemma 5.2 *Integrating η against dt, ψ or ψdt placed either at head or boundary vanishes.*

Proof This follows from the fact that η_{S^1} (resp. η_D) vanish when integrated against dt (resp. ψ). □

Now consider a two-point tree diagram as in Fig. 5a. It consists of a single arrow between the two points, some legs on the boundary, and maybe some residual fields. Now integrate all the legs against dt or $d\theta$. The result is a graph consisting of single arrow with a product γ_i of residual fields, dt's and $d\theta$'s on both ends (Fig. 5b). From the two Lemmata above together with Proposition 4.1 it now follows that the contribution of such a graph is zero after gluing.

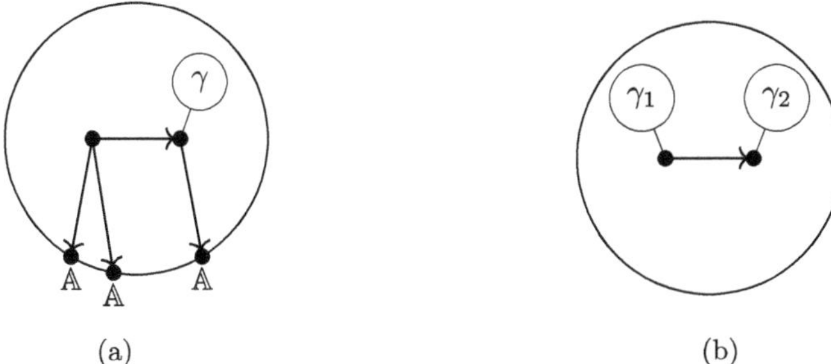

Fig. 5 Two-point tree diagrams. (**a**) Example of a two-point tree diagram. (**b**) After integrating all legs against dt, $d\theta$

5.4 *Loop Diagrams*

At two-point order one can also see loops appearing. Loop diagrams will contain products of distributions that cannot be defined without further choices, namely, the choice of a regularization procedure. Usually one extends the propagator smoothly to a compactification of the configuration space, this can be interpreted as a sort of point-splitting regularization. For the distributional propagators at hand such a regularization is not possible, and another normalization scheme is needed.

5.4.1 Regularization

For the purpose of this note, where we work with distributional forms on the circle, the choice of regularization is straightforward. Consider the loop given by two arrows between the two points (their directions do not matter, since the circle delta form and propagator are symmetric up to a sign). The ill-defined products of distributions that can appear are

$$\left(\delta^{(1)}_{S^1}(t_1 - t_2)\right)^2 \qquad \text{and} \qquad \delta^{(1)}_{S^1}(t_1 - t_2)\eta_{S^1}(t_1, t_2).$$

The only sensible value to attach to these two terms is zero.[10] The first one is the square of the one-form $\delta^{(1)}_{S^1}(t_1 - t_2) = \delta(t_1 - t_2)(dt_1 - dt_2)$. For the second one, notice that the circle propagator is antisymmetric with respect to the diagonal, so its diagonal value should be zero.[11] A slightly more rigorous approach is to smear

[10] In the thesis of the third author [Wer18], this is called the "universal" regularization of the axial gauge. Conjecturally, it can be obtained by approximating the axial gauge via Riemann-Hodge gauges.

[11] Here we are thinking of the propagator as a function on $S^1 \times S^1$, so we should define it on the diagonal.

out the delta function, which will give the above results, as long as one chooses a symmetric nascent delta function. Setting the circle propagator to 0 on the diagonal means that the product with the delta distribution vanishes. These two choices are enough to evaluate all the loop diagrams at two-point order, and, in fact, at any order.

5.4.2 Evaluation

Consider first a loop with both arrows pointing the same way as in Fig. 6a, call this a loop of type A. The contribution of such a loop is

$$(\eta_{12})^2 = \left(\eta_{D,12}\delta^{(1)}_{S^1,12} + \mu_1\eta_{S^1,12}\right)^2 = \left(\eta_{D,12}\delta^{(1)}_{S^1,12}\right)^2 + 2\mu_1\eta_{D,12}\delta^{(1)}_{S^1,12} + (\mu_1\eta_{S^1,12})^2 = 0$$

by our choice of regularization above, and since $\mu^2 = 0$. On the other hand, if the arrows point in opposite directions as in Fig. 6b (call this loop of type B), the loop contributes

$$\eta_{12}\eta_{21} = \left(\eta_{D,12}\delta^{(1)}_{S^1,12} + \mu_1\eta_{S^1,12}\right)\left(\eta_{D,21}\delta^{(1)}_{S^1,12} + \mu_2\eta_{S^1,21}\right) = -\mu_1\mu_2\eta^2_{S^1,12}$$

and again all other terms vanish due to the regularization. In particular, with this regularization the contribution of the theta graph is zero, since it always contains a type A loop. From the contribution of the type B loop we see that if we place an a residual field at either end the contribution vanishes (since $\mathsf{a} \wedge \mu = 0$). So, the contributing loop diagrams contain a type B loop with a b residual or propagator placed at the ends. The resulting loop diagrams can be seen in Fig. 7.

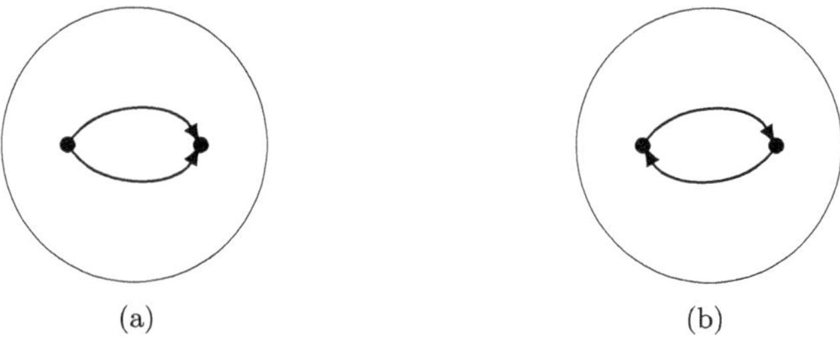

Fig. 6 (**a**) Loop with both arrows directed the same way ("type A"). (**b**) Loop with both arrows directed opposite ways ("type B")

Their contributions can then be evaluated to

$$\psi_{\Gamma_{2,0}} = \frac{\varepsilon}{2} z^+_{2,i} z^+_{2,j} h^{ik}_l h^{jl}_k \pi_*(-\mu_1 dt_1 \mu_2 dt_2 \eta^2_{S^1,12}) \tag{22}$$

$$\psi_{\Gamma_{2,1}} = \frac{\varepsilon}{12} z^+_{2,i} h^{ij}_k g^k_{jl} \int_{\partial M,1} \pi_*(\mu_1 dt_1 \mu_2 \eta_{23}) \mathbb{A}^l_3 \tag{23}$$

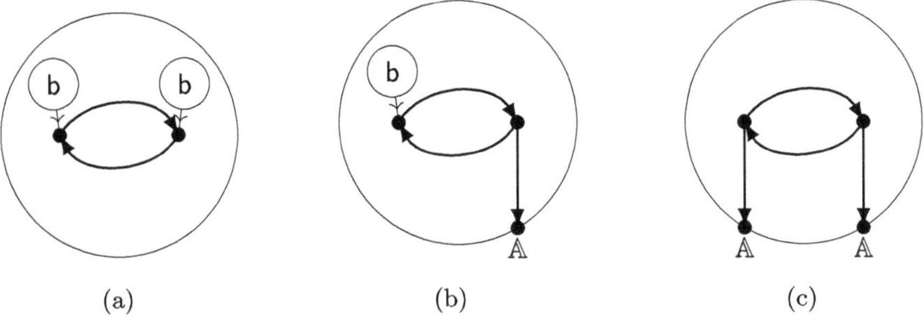

Fig. 7 Loop diagrams in the 2-point effective action. (**a**) $\Gamma_{2,0}$. (**b**) $\Gamma_{2,1}$. (**c**) $\Gamma_{2,2}$

$$\psi_{\Gamma_{2,2}} = \frac{\varepsilon}{2} g^k_{ij} g^j_{kl} \int_{\partial M \times \partial M} \pi_*(\mu_1 \mu_2 \eta^2_{S^1,12} \eta_{13} \eta_{24}) \mathbb{A}^i_3 \mathbb{A}^l_4. \tag{24}$$

Remark 5.1 (Regularization and Framing Contribution) One can say that imposing $(\delta^{(1)}_{S^1} = 0)^2$ (alternatively, using the axial gauge) is using the blackboard framing of the solid torus. As we will discuss below, the different choices of gluing diffeomorphism leading to the same diffeomorphism type of lens spaces are related by changing the framing on one of two solid tori, and this change is reflected through a changed value of the two-loop contribution. Thus one can argue that the regularization of the theory depends on the choice of a framing, an effect that is visible in other approaches to perturbative Chern-Simons theory: In the Axelrod-Singer approach one cancels dependence on the regularization through the introduction of a framing, in the Kontsevich-Kuperberg-Thurston approach the framing defines the regularization.

6 Gluing of Lens Spaces

In this section we will describe how to compute the 2-point effective action on lens spaces by applying the gluing procedure described in [CMR17]. The computation is carried out in the next section.

6.1 *Lens Spaces*

In this note we will apply the following conventions for lens spaces. Consider two solid tori $M_1 = S^1 \times D = M_2$. The boundary is $S^1 \times S^1$ with coordinates $(t, \theta) \in (\mathbb{R}/\mathbb{Z})^2$. Pick two coprime integers p and q. Since they are coprime, there exist m, n such that $mq - np = 1$. Let $\varphi \in \mathrm{Diff}(S^1 \times S^1)$ be defined by

$$\varphi\colon \begin{pmatrix} t \\ \theta \end{pmatrix} \mapsto \begin{pmatrix} m & p \\ n & q \end{pmatrix} \begin{pmatrix} t \\ \theta \end{pmatrix} = \begin{pmatrix} mt + p\theta \\ nt + q\theta \end{pmatrix}. \tag{25}$$

Then we define the lens space $L_{p,q}$ by

$$L_{p,q} = M_1 \cup_\varphi M_2. \tag{26}$$

Note that with this convention $L_{1,0} \cong S^3$ and $L_{0,1} \cong S^1 \times S^2$.

Remark 6.1 (Dependence of Framing on Choices) It is well known that the diffeomorphism type of $L_{p,q}$ is independent of the choice of m and n and also independent of the choice of q (mod p). Changing q by a multiple of p , to preserve the equation $mq - np = 1$, we have to change n by the same multiple of m. Let

$$T = \begin{pmatrix} 1 & 0 \\ 1 & 1 \end{pmatrix}.$$

Then the former change corresponds to multiplying φ with T^k from the right, while the latter corresponds to multiplying with T^k from the left. Since in the gluing we identify $\partial M_1 \ni x \sim \varphi(x) \in \partial M_2$, multiplying from the right with T^k corresponds to gluing after performing k Dehn twists around the longitude (given by $(t, 0)$) in ∂M_2, while the latter operation corresponds to performing k *inverse* Dehn twists around the longitude in ∂M_1. We can extend Dehn twists around the longitude to the bulk of the solid torus: Using polar coordinates on the disk a possible representation[12] is[13]

$$T\colon (t, r, \theta) \mapsto (t, r, \theta + t).$$

A Dehn twist changes the homotopy class of the framing on the solid torus by one generator. Hence both operations, i.e. shifting either m or q by p and performing the according shift of n, correspond to changes of framing of the resulting lens space. The first operation will change the framing of $L_{p,q}$ by $+k$ units, the second operation by $-k$ units. This is discussed in detail in [FG91, Appendix B].

Remark 6.2 (The case $p = 0$) If $p = 0$, then $qm = 1$, so we have $q = m = \pm 1$. That case needs to be considered separately: The resulting space $S^1 \times S^2$ is not a rational homology 3-sphere, unlike all lens spaces. In the following we will always assume $p \neq 0$ unless otherwise stated.

[12]Only the isotopy class of a Dehn twist is well-defined.

[13]The following formula can be extended to $r = 0$ by the identity.

6.2 Gluing Perturbative Expansions in BV-BFV

The gluing procedure discussed in [CMR17] amounts to the following prescription:

- Take a diagram Γ_1 in the $\mathbb{A}$-representation and a diagram Γ_2 in the $\mathbb{B}$-representation with the same number n of legs and multiply $\psi_{\Gamma_1} = \int_{(\partial M_1)^n} \omega_{\Gamma_1}\mathbb{A}_1 \cdots \mathbb{A}_n$ and $\psi_{\Gamma_2} = \int_{(\partial M_2)^n} \omega_{\Gamma_2}\mathbb{B}_1 \cdots \mathbb{B}_n$. Here the diagrams can be non-connected, since the state is the exponential of the effective action.
- Sum over all ways of contracting $\mathbb{A}$ and $\mathbb{B}$ fields to a delta form $\epsilon\delta^{(d-1)}_{\partial M}(x, \varphi(x))$ (the factor ε comes from the definition via path integrals as in [CMR17]).
- Perform the integration over $(\partial M_1)^n \times (\partial M_2)^n$.
- Reduce the residual fields.

Equivalently, the state glued from Ψ_{Γ_1} and Ψ_{Γ_2} can be defined as follows. Let $\sigma \in S_n$ be a permutation and denote $\Phi_\sigma : (\partial M)^n \to (\partial M)^n$ the map defined by

$$(x_1, \ldots, x_n) \mapsto (\varphi(x_{\sigma(1)}), \ldots, \varphi(x_{\sigma(n)})).$$

Then the above prescription results in

$$\psi_{\Gamma_1} * \psi_{\Gamma_2} := \sum_{\sigma \in S_n} \int_{(\partial M)^n} \omega_{\Gamma_1} \Phi^*_\sigma \omega_{\Gamma_2}.$$

In this integral only the top degree part survives. Sometimes it will be convenient to use the reformulation

$$\psi_{\Gamma_1} * \psi_{\Gamma_2} = \sum_{\sigma \in S_n} \int_{(\partial M)^n} (\Phi_\sigma^{-1})^*(\omega_{\Gamma_1} \Phi^*_\sigma \omega_{\Gamma_2}) = \sum_{\sigma \in S_n} \int_{(\partial M)^n} (\Phi_\sigma^{-1})^*(\omega_{\Gamma_1})\omega_{\Gamma_2}. \tag{27}$$

In all examples that we consider, the graphs are invariant with respect to permuting the boundary points. Hence, the sum is a constant times the pairing computed using Φ_{id}, which we will also denote φ, abusing notation.

Notice also that for graphs not depending on boundary fields the gluing procedure is trivial, i.e. the corresponding contributions are simply multiplied with the rest, or, equivalently, added to the effective action.

6.3 The Effective Action on M_2

On M_2 we will choose the opposite polarization, namely, $\partial M = \partial_2 M$ (the $\mathbb{B}$-representation). To avoid confusion, from now on we decorate objects with a superscript depending on which representation they are computed in, e.g. $\psi^{\mathbb{A}}$ (resp. $\psi^{\mathbb{B}}$) denotes the state in the $\mathbb{A}$-representation ($\mathbb{B}$-representation). The residual fields change roles: $\mathsf{a}^{\mathbb{B}} = z_1^{\mathbb{B}} 1 + z_2^{\mathbb{B}} dt$, $\mathsf{b}^{\mathbb{B}} = z_1^{+,\mathbb{B}} \mu dt + z_2^{+,\mathbb{B}} \mu$. Now the self-duality of Chern-Simons theory comes in handy. Let $\Gamma^{\mathbb{A}}$ be a diagram appearing in the $\mathbb{A}$-representation. Then there is a diagram $\Gamma^{\mathbb{B}}$ in the $\mathbb{B}$-representation obtained from $\Gamma^{\mathbb{A}}$ by reversing all arrows and exchanging $\mathsf{b}^{\mathbb{A}}$ (resp. $\mathsf{a}^{\mathbb{A}}$) with $\mathsf{a}^{\mathbb{B}}$ (resp. $\mathsf{b}^{\mathbb{B}}$) fields (and of course $\mathbb{A}$ and $\mathbb{B}$ fields). See Fig. 8. We will call this diagram $\Gamma^{\mathbb{B}}$ the *dual diagram* of $\Gamma^{\mathbb{A}}$. However, since we have

$$\eta^{\mathbb{A}}(x_1, x_2) = -\eta^{\mathbb{B}}(x_2, x_1),$$

and the form parts of $\mathsf{a}^{\mathbb{A}}$ and $\mathsf{b}^{\mathbb{B}}$ (resp. $\mathsf{a}^{\mathbb{B}}$ and $\mathsf{b}^{\mathbb{A}}$) are the same, the form parts of $\Gamma^{\mathbb{A}}$ and $\Gamma^{\mathbb{B}}$ are the same up to a sign. The structure constants and coordinates on the space of residual fields have to be replaced with their "dual" counterparts. This amounts to the following prescription. To compute the state $\psi_{\Gamma^{\mathbb{B}}}$ from the state $\psi_{\Gamma^{\mathbb{A}}}$,

- replace $z^{i,k,\mathbb{A}}$ by $z_{i,k}^{+,\mathbb{B}}$, $z_{i,k}^{+,\mathbb{A}}$ by $z_{i,k,\mathbb{B}}$,
- replace g^i_{jk} by h_i^{jk}, g_{ijk} by h^{ijk} and vice versa,
- replace every $\mathbb{A}^i$ by $\mathbb{B}_i$,
- multiply by $(-1)^{\#E(\Gamma_{\mathbb{A}})}$, where $E(\Gamma_{\mathbb{A}})$ denotes the set of edges of $\Gamma_{\mathbb{A}}$.

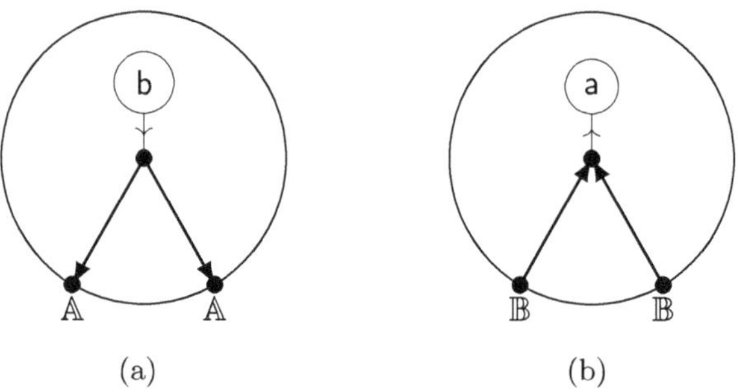

Fig. 8 Corresponding diagrams in $\mathbb{A}$- and in $\mathbb{B}$-representations. (**a**) $\Gamma^{\mathbb{A}}$. (**b**) Corresponding $\Gamma^{\mathbb{B}}$

6.4 *Reducing the Residual Fields*

Naively, after pairing the states ψ_{M_1}, ψ_{M_2}, the new state depends is a function on the direct sum of the spaces of residual fields $\widetilde{\mathcal{V}_M} = \mathcal{V}_{M_1} \oplus \mathcal{V}_{M_2}$. This produces a valid state in the sense that it satisfies the mQME. However, in most cases it is not the minimal possible space of residual fields and it is possible to reduce it using the methods of [CMR17]. We will discuss the reduction first since it will allow us to simplify the computations later.

In the case of lens spaces there is a significant difference between the cases $p \neq 0$ and the case $p = 0$, and we will discuss these separately.

6.4.1 Case $p \neq 0$

Let $M = L_{p,q}$ for $p \neq 0$. Recall that $M_1 = M_2 = D^2 \times S^1$, $\partial_1 M_1 = \partial_2 M_2 = S^1 \times S^1 =: \mathbb{T}^2$ and $\partial_2 M_1 = \partial_1 M_2 = \emptyset$. Let $\varphi\colon \mathbb{T}^2 \to \mathbb{T}^2$ be the diffeomorphism given by $t \mapsto mt + p\theta, \theta \mapsto nt + q\theta$ so that $M = M_1 \cup_\varphi M_2$. We have $H^\bullet_{Di}(M_i) = H^\bullet(M_i, \partial M_i) \cong H^\bullet(D^2, S^1) \otimes H^\bullet(S^1)$ and $H^\bullet(M_i) \cong H^\bullet(S^1)$ for $i \neq j$. The spaces of redshirt residual fields can be identified by $\int_\Sigma \mathsf{b}_1 \mathsf{a}_2 = \int_\Sigma \mathsf{b}_1^\times \mathsf{a}_2^\times$. The reduced state is then defined by

$$\check{\psi} = \int_{\mathcal{L}^\times} \widetilde{\psi}, \tag{28}$$

where $\mathcal{L}^\times$ is the zero section of $T^*[-1](L_1^\times \oplus L_2^\times)$. In our case, we have

$$L_1^\times = \langle dt \rangle, L_2^\times = \langle m dt + p d\theta \rangle.$$

Taking the pushforward of the zero section amounts to contracting a pair of $z_2^{+,\mathbb{A}}$ and $z^{2,\mathbb{B}}$ coordinates to the number

$$V = \Lambda^{-1} = \left(\int_{\mathbb{T}^2} dt \varphi^*(dt) \right)^{-1},$$

while setting $z^{2,\mathbb{A}} = z_2^{+,\mathbb{B}} = 0$. For $L_{p,q}$ we have $V = \frac{1}{p}$. The resulting representatives of the cohomology of $L_{p,q}$ are

$$\chi_1 = 1_{S^3}$$

$$\chi_2 = \begin{cases} \mu_1 \wedge dt_1 & \text{on } M_1 \\ 0 & \text{on } M_2 \end{cases}$$

with coordinates $z^1 = z^{1,\mathbb{B}}, z^2 = z^{1,\mathbb{A}}$ and dual basis (with respect to the Poincare pairing)

$$\chi^1 = \begin{cases} 0 & \text{on } M_1 \\ \mu_2 \wedge dt_2 & \text{on } M_2 \end{cases}$$

$$\chi^2 = 1_{S^3}$$

with coordinates $z_1^+ = z_1^{+,\mathbb{B}}, z_2^+ = z_1^{+,\mathbb{A}}$. We can use this to simplify the calculation of the state on $L_{p,q}$ by ignoring pairings of diagrams that would vanish after reducing residual fields.

6.4.2 The Case $p = 0$

In this case there are no redshirt residual fields, and we do not have to perform the reduction. The resulting manifold is $M = S^2 \times S^1$. Sticking to the same conventions as above, we will get representatives of cohomology

$$\chi_1 = 1$$
$$\chi_2 = dt$$
$$\chi_3 = \begin{cases} \mu & \text{on } M_1 \\ 0 & \text{on } M_2 \end{cases}$$
$$\chi_4 = \begin{cases} \mu \wedge dt & \text{on } M_1 \\ 0 & \text{on } M_2 \end{cases}$$

with coordinates $z^1 = z^{1,\mathbb{B}}$, $z^2 = z^{2,\mathbb{B}}$, $z^3 = z^{2,\mathbb{A}}$, $z^4 = z^{1,\mathbb{A}}$. The dual basis is

$$\chi^1 = \begin{cases} 0 & \text{on } M_1 \\ \mu \wedge dt & \text{on } M_2 \end{cases}$$
$$\chi^2 = \begin{cases} 0 & \text{on } M_1 \\ \mu & \text{on } M_2 \end{cases}$$
$$\chi^3 = dt$$
$$\chi^4 = 1$$

with coordinates $z_1^+ = z_1^{+,\mathbb{B}}$, $z_2^+ = z_2^{+,\mathbb{B}}$, $z_3^+ = z_1^{+,\mathbb{A}}$, $z_4^+ = z_2+, \mathbb{A}$.

7 The Effective Action on Lens Spaces

We now proceed to the central part of this note, the computation of the effective action on lens spaces up to two-loop order. We will consider separately the cases where (g, V, W) form a Manin triple and the one where it does not.

Since we are interested in the two-point effective action after gluing, we have to consider all pairs of diagrams with a total of at most two interaction vertices. Also, since we are interested in the state only after reduction of the residual fields, only pairings with no residual fields or the same number of $z_2^{+,\mathbb{A}}$ and $z^{2,\mathbb{B}}$ survive, all others can be ignored. We now compute all the relevant pairings for the case where $(\mathfrak{g}, V, W)$ form a Manin triple.

7.1 Case of a Manin Triple

In this case, the A^3 and B^3 vertices vanish and the number of diagrams to be considered is considerably reduced.

7.1.1 Pairing Against Order 0 Diagram

First we consider all pairings against the single order 0 diagram on M_2. Its contribution to the state, since M_2 is in the $\mathbb{B}$-representation, is

$$\Psi^{\mathbb{B}}_{\Gamma_0} = -z_1^{k,\mathbb{B}} \int_{\partial M} \mathbb{B}_k - z_2^{k,\mathbb{B}} \int_{\partial M} dt\mathbb{B}_k.$$

We have $\varphi^* dt = mdt + pd\theta$. Hence (as used already in Sect. 5.3) gluing a boundary point on the $\mathbb{A}$ side against the 0-point action up to constants corresponds to multiplying the corresponding form with $1, dt$ or $d\theta$ and integrating over that boundary point. Often, it is best to perform this integration first—the results are known from Lemma 5.1—and then compute the integral. Notice also that this diagram does not pair to diagrams with no boundary vertices.

We can pair the two 0-point terms on either side, see Fig. 9. The result is

$$\psi^{\mathbb{A}}_{\Gamma_0} * \psi^{\mathbb{B}}_{\Gamma_0} = \int_{\partial M} \mathsf{b}^{\mathbb{A}} \varphi^* \mathsf{a}^{\mathbb{B}} = z_{2,k}^{+,\mathbb{A}} z^{2,\mathbb{B}} \int_{\partial M} dt \varphi^* dt = p z_{2,k}^{+,\mathbb{A}} z^{2,\mathbb{B}}. \tag{29}$$

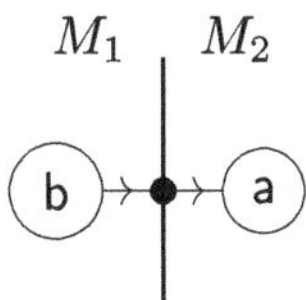

Fig. 9 Pairing order 0 diagrams on either side

These are precisely the fields that we ought to reduce in the reduction of the redshirt residual fields as discussed in Sect. 6.4. We will perform this reduction in Sect. 7.1.3.

Let us turn to the 1-point diagrams. The first one ($\Gamma^{\mathbb{A}}_{1,0}$) does not contain $\mathbb{A}$ fields and hence does not pair to $\Gamma^{\mathbb{B}}_0$. The two possible pairings are shown in Fig. 10. Using Lemma 5.1 one quickly sees that the pairing in Fig. 10b vanishes. Alternatively one can use that $\int_{S^1,i} \eta_{S^1}(t_1, t_2) dt_i = 0$ for $i = 1, 2$. Together with degree counting this implies the same. The remaining pairing shown in Fig. 10a evaluates to

$$\psi^{\mathbb{A}}_{1,1} * \psi^{\mathbb{B}}_{\Gamma_0} = g^i_{jk}(z_{1i}^{+,\mathbb{A}} z^{1j,\mathbb{A}} - z_{2i}^{+,\mathbb{A}} z^{2j\mathbb{A}}) z^{1k,\mathbb{B}} + m g^i_{jk} z_{1i}^{+,\mathbb{A}} z^{2j,\mathbb{A}} z_{1k,\mathbb{B}}. \tag{30}$$

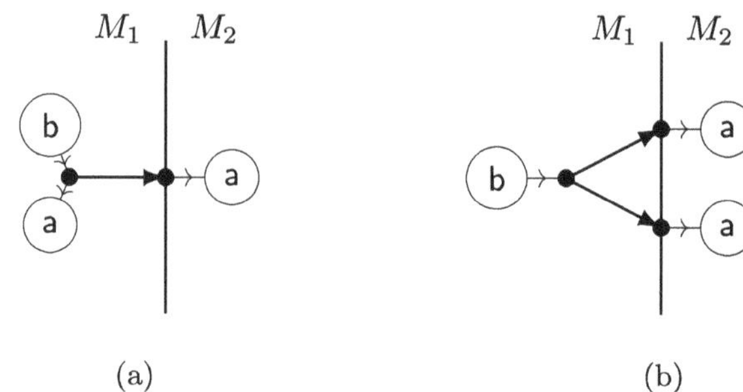

Fig. 10 Pairing 1-point diagrams on M_1 to 0-point diagram on M_2. (**a**) $\psi_{\Gamma^{\mathbb{A}}_{1,1}} * \psi_{\Gamma^{\mathbb{B}}_0}$. (**b**) $\psi_{\Gamma^{\mathbb{A}}_{1,1}} * (\psi_{\Gamma^{\mathbb{B}}_0})^2$

We can also pair against the loop diagrams, see Fig. 11 to obtain

$$\psi^{\mathbb{A}}_{\Gamma_{2,1}} * \psi^{\mathbb{B}}_{\Gamma_0} = \frac{m}{12} z^{+,\mathbb{A}}_{2i} z^{2,l,\mathbb{B}} h^{ij}_k g^k_{jl} \tag{31}$$

$$\psi^{\mathbb{A}}_{\Gamma_{2,2}} * \psi^{\mathbb{B}}_{\Gamma_0} = \frac{m^2}{24} z^{2,k,\mathbb{B}} z^{2,l,\mathbb{B}} h^{ij}_k g^k_{jl}. \tag{32}$$

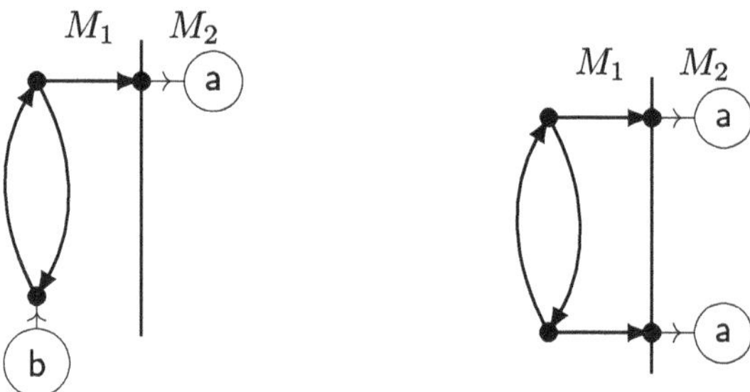

Fig. 11 Pairing 2-point diagrams on M_1 to 0-point diagram on M_2

7.1.2 Pairing the 1-Point Functions

Pairing the 1-point functions is computationally more intense, since we have to take the pullback of non-constant forms. The graph $\Gamma_{1,0}$ with no legs does not depend on boundary, hence its contribution and the one of its dual diagrams simply add to the effective action. Part of it survives after reducing residual fields, and in fact we will show this is the only one-point contribution to the effective action.

Under the Manin triple assumption, the other pairings are the ones described in Fig. 12.

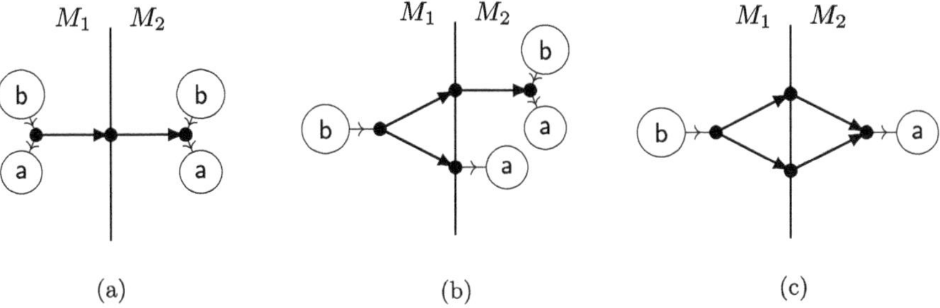

Fig. 12 Pairing 1-point diagrams on M_1 to 1-point diagrams on M_2. **(a)** $\psi_{\Gamma^{\mathbb{A}}_{1,1}} * \psi_{\Gamma^{\mathbb{B}}_{1,1}}$. **(b)** $\psi_{\Gamma^{\mathsf{b},\mathbb{A}}_{1,2}} * \left(\psi_{\Gamma^{\mathbb{B}}_{1,1}} \psi_{\Gamma^{\mathbb{B}}_0}\right)$. **(c)** $\psi_{\Gamma^{\mathsf{b},\mathbb{A}}_{1,2}} * \psi_{\Gamma^{\mathsf{b},\mathbb{B}}_{1,2}}$

The next diagram $\Gamma_{1,1}$ has only constant form coefficients and the pairings are (see [Wer18] for the computations)

$$\psi_{\Gamma^{\mathbb{A}}_{1,1}} * \psi_{\Gamma^{\mathbb{B}}_{1,1}} = -n g^i_{jk} h^{kl}_m z^{+,\mathbb{A}}_{1i} z^{2j,\mathbb{A}} z^{1m,\mathbb{B}} z^{+,\mathbb{B}}_{2l} \tag{33}$$

$$\left(\psi_{\Gamma^{\mathbb{A}}_{1,1}} \psi_{\Gamma^{\mathbb{A}}_0}\right) * \psi_{\Gamma^{\mathsf{b},\mathbb{B}}_{1,2}} = n g^i_{jk} h^{klm} z^{+,\mathbb{A}}_{1i} z^{2j,\mathbb{A}} z^{+,\mathbb{A}}_{1,l} z^{+,\mathbb{B}}_{2,m} \tag{34}$$

$$\psi_{\Gamma^{\mathsf{b},\mathbb{A}}_{1,2}} * \left(\psi_{\Gamma^{\mathbb{B}}_{1,1}} \psi_{\Gamma^{\mathbb{B}}_0}\right) = n h_{ijk} h^{kl}_m z^{2i,\mathbb{A}} z^{1j,\mathbb{B}} z^{1m,\mathbb{B}} z^{+,\mathbb{B}}_{2,l} \tag{35}$$

$$\psi_{\Gamma^{\mathsf{b},\mathbb{A}}_{1,2}} * \psi_{\Gamma^{\mathsf{b},\mathbb{B}}_{1,2}} = g^i_{jk} h^{jk}_l z^{+,\mathbb{A}}_{2i} z^{2l,\mathbb{B}} \begin{cases} \frac{-p}{2} s(q,p) & p \neq 0 \\ \frac{q}{12} & p = 0 \end{cases}. \tag{36}$$

7.1.3 Reducing the Residual Fields

If $p \neq 0$, we have to reduce the residual fields as discussed in Sect. 6.4. We recall that this amounts to pairing $z^{+,\mathbb{A}}_{2i}$ with $z^{2,j,\mathbb{B}}$ to $\delta^j_i \cdot 1/p$ and setting their conjugates variables $z^{2,i,\mathbb{A}} = z^{+,\mathbb{B}}_{2,j} = 0$. This eliminates many of the pairings above, namely, (32)–(35). We denote the resulting effective action with S^{MT}_{eff}, where MT stands for Manin triple. We have that

$$S^{MT}_{eff} = S^{MT,(1)}_{eff} + S^{MT,(2)}_{eff}, \tag{37}$$

where

$$S^{MT,(1)}_{eff} = g^i_{jk} \left(z^{+,\mathbb{A}}_{1i} z^{1j,\mathbb{A}} z^{1k,\mathbb{B}} + \frac{1}{2} z^{+,\mathbb{B}}_{1i} z^{1j,\mathbb{B}} z^{1k,\mathbb{B}} \right) + h^{jk}_i \left(z^{1i,\mathbb{B}} z^{+,\mathbb{A}}_{1j} z^{+,\mathbb{B}}_{1k} + \frac{1}{2} z^{+,\mathbb{A}}_{1i} z^{1j,\mathbb{A}} z^{1k,\mathbb{A}} \right) \tag{38}$$

and

$$S_{eff}^{MT,(2)} = g^i_{jk} h_i^{jk} \left(\frac{1}{2} s(q,p) + \frac{q+m}{12p} \right). \tag{39}$$

7.2 The General Case

At this order, the Manin triple condition amounts to ignoring diagrams $\Gamma^{\mathsf{a}}_{1,2}$ and $\Gamma_{1,3}$. Considering them amounts to computing the additional pairings described in Figs. 13 and 14, respectively. To simplify the computations we will now drop terms that vanish after reducing fields, i.e. we keep only terms that contain no $z^{2,i,\mathbb{A}}$- and $z^{+,\mathbb{B}}_{2,j}$-variables and exactly the same number of $z^{+,\mathbb{A}}_{2,i}$ and $z^{2,j,\mathbb{B}}$ variables. In Fig. 13 this eliminates the diagrams b, c, and d, together with degree counting: The dimension of the domain of integration is $3+2+2+3 = 10$ in all cases, but the corresponding form degrees are 12, 11, and 14, respectively. On the other hand, diagram 13a yields a contribution corresponding to the $\check{\mathsf{a}}^3$ vertex on the glued lens space, while its dual will contribute the $\check{\mathsf{b}}^3$ vertex.

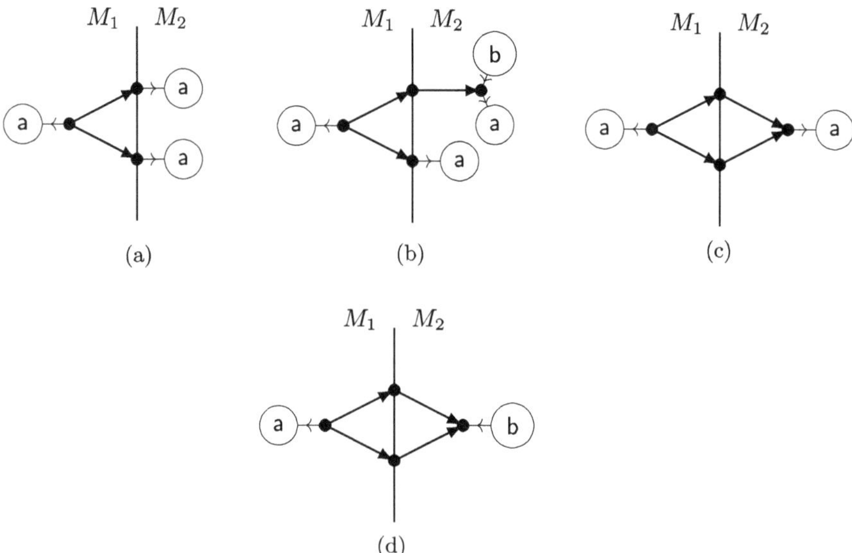

Fig. 13 Pairing diagram $\Gamma^{\mathsf{a},\mathbb{A}}_{1,2}$ on M_1 to diagrams on M_2. Here we excluded the diagrams including $\Gamma^{\mathbb{B}}_{1,3}$, we will get them from symmetry from the ones for $\Gamma^{\mathbb{A}}_{1,3}$. **(a)** $\psi_{\Gamma^{\mathsf{a},\mathbb{A}}_{1,2}} * \left(\psi_{\Gamma^{\mathbb{B}}_0}\right)^2$. **(b)** $\psi_{\Gamma^{\mathsf{a},\mathbb{A}}_{1,2}} * \left(\psi_{\Gamma^{\mathbb{B}}_{1,1}} \psi_{\Gamma^{\mathbb{B}}_0}\right)$. **(c)** $\psi_{\Gamma^{\mathsf{a},\mathbb{A}}_{1,2}} * \psi_{\Gamma^{\mathsf{b},\mathbb{B}}_{1,2}}$. **(d)** $\psi_{\Gamma^{\mathsf{b},\mathbb{A}}_{1,2}} * \psi_{\Gamma^{\mathsf{b},\mathbb{B}}_{1,2}}$

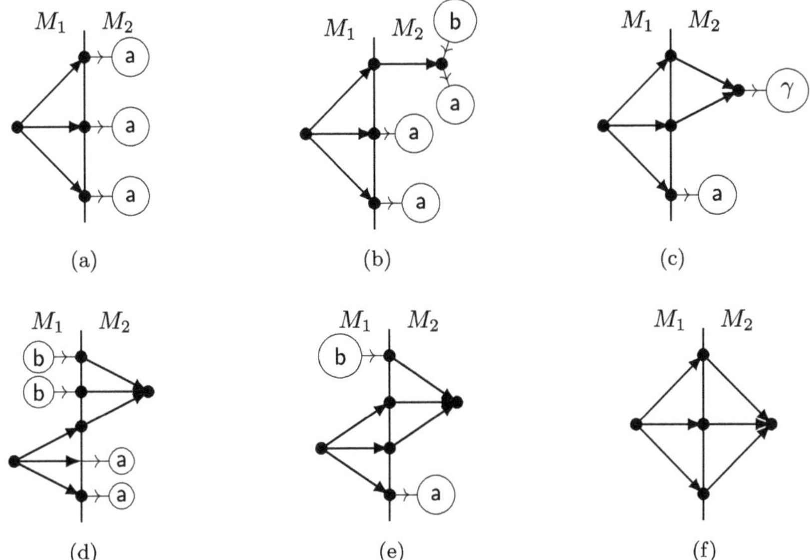

Fig. 14 Pairing diagram $\Gamma^{\mathbb{A}}_{1,3}$ on M_1 to diagrams on M_2. Here $\gamma \in \{\mathsf{a}, \mathsf{b}\}$. (**a**) $\psi_{\Gamma^{\mathbb{A}}_{1,3}} * \left(\psi_{\Gamma^{\mathbb{B}}_0}\right)^3$. (**b**) $\psi_{\Gamma^{\mathbb{A}}_{1,3}} * \left(\psi_{\Gamma^{\mathbb{B}}_{1,1}} (\psi_{\Gamma^{\mathbb{B}}_0})^2\right)$. (**c**) $\psi_{\Gamma^{\mathbb{A}}_{1,3}} * \left(\psi_{\Gamma^{\mathsf{a},\mathbb{B}}_{1,2}} \psi_{\Gamma^{\mathbb{B}}_0}\right)$. (**d**) $\psi_{\Gamma^{\mathbb{A}}_{1,3}} \psi^2_{\Gamma^{\mathbb{A}}_0} * \left(\psi_{\Gamma^{\mathbb{B}}_{1,3}} \psi^2_{\Gamma^{\mathbb{B}}_0}\right)$. (**e**) $\psi_{\Gamma^{\mathbb{A}}_{1,3}} \psi_{\Gamma^{\mathbb{A}}_0} * \left(\psi_{\Gamma^{\mathbb{B}}_{1,3}} \psi_{\Gamma^{\mathbb{B}}_0}\right)$. (**f**) $\psi_{\Gamma^{\mathbb{A}}_{1,3}} * \left(\psi_{\Gamma^{\mathbb{B}}_{1,3}}\right)$

Now let us look at the diagrams in Fig. 14. The first two diagrams in Fig. 14a, b, and also the one in Fig. 14d do not contribute, even if we do not reduce residual fields, this follows from the discussion on vanishing of two-point tree contributions after gluing. After reducing residual fields, diagram 14c vanishes for degree reasons: only the zero-form part of a survives, so the total form degree is 10, while integration is over a 12-dimensional space. By degree counting, the only nonzero term contains the one-form parts of $\mathsf{b}^{\mathbb{A}}$ and $\mathsf{a}^{\mathbb{B}}$. After reducing residual fields, this corresponds to part of a theta diagram for the glued propagator, together with the last diagram 14e. Below we list the weights of the glued graphs, we only list the ones yielding a nonzero contribution after reducing the residual fields.

$$\psi_{\Gamma^{\mathsf{a},\mathbb{A}}_{1,2}} * \left(\psi_{\Gamma^{\mathbb{B}}_0}\right)^2 = \frac{1}{2} g_{ijk} z^{1i\mathbb{A}} z^{1j\mathbb{B}} z^{1k\mathbb{B}} \tag{40}$$

$$\psi_{\Gamma^{\mathbb{A}}_{1,3}} \psi_{\Gamma^{\mathbb{A}}_0} * \left(\psi_{\Gamma^{\mathbb{B}}_{1,3}} \psi_{\Gamma^{\mathbb{B}}_0}\right) = h^{ijk} g_{ljk} z^{+,\mathbb{A}}_{2i} z^{2l,\mathbb{B}} \frac{1}{4} \int_{(\partial M)^4} (\omega_{\Gamma_{0,3}})_{123} dt_4 (\varphi^{\times 4})^* (\omega_{\Gamma_{0,3}})_{234} dt_1) \tag{41}$$

$$\psi_{\Gamma^{\mathsf{b},\mathbb{B}}_{1,3}} * \psi_{\Gamma^{\mathsf{b},\mathbb{A}}_{1,3}} = \frac{1}{6} h^{ijk} g_{ijk} \int_{(\partial M)^3} (\omega_{\Gamma_{0,3}}) (\varphi^{\times 3})^* (\omega_{\Gamma_{0,3}}). \tag{42}$$

The resulting effective action is

$$S_{eff} = S_{eff}^{MT} + S_{eff}^{NMT} = S_{eff}^{MT} + S_{eff}^{NMT,(1)} + S_{eff}^{NMT,(2)}, \tag{43}$$

where

$$S_{eff}^{NMT,(1)} = \frac{1}{2} f_{ijk} z^{1i\mathbb{A}} z^{1j\mathbb{B}} z^{1k\mathbb{B}} + \frac{1}{2} f^{ijk} z_{1i}^{+,\mathbb{B}} z_{1j}^{+,\mathbb{A}} z_{1k}^{+,\mathbb{A}} \tag{44}$$

and

$$S_{eff}^{NMT,(2)} = \frac{1}{2} f_{ijk} f^{ijk} \left(s(q,p) + \sum_{k=0}^{p-1} \eta_{S^1}(k/p) f(qk/p) + \eta_{S^1}(k/p) f(mk/p) \right), \tag{45}$$

where the function $f: S^1 \to \mathbb{R}$ is given by

$$f(\theta) = \cos(2\pi\theta)\eta_{S^1}(\theta) - \frac{1}{\pi} \sin 2\pi\theta \log 2|\sin \pi\theta|$$

for $\theta \notin \mathbb{Z}$, and $f(k) = 0$ for $k \in \mathbb{Z}$.

7.3 *Weights of Oriented Theta Graphs on Lens Spaces*

From this computation we can now extract the weights of oriented theta graphs on lens spaces by comparing with the state on the glued lens space computed with the reduced glued residual fields and the corresponding glued propagator. We will briefly describe the first diagrams contributing to the state on any rational homology sphere, and then use the formula for the reduced propagator to identify these with the diagrams arising from the gluing described in the previous paragraphs.

7.3.1 Low-Order Diagrams on Rational Homology Spheres

On orientable rational homology spheres M the space of residual fields is $(H^\bullet(M) \oplus H^\bullet(M))$ [1]. A choice of volume form $v \in H^3(M)$ gives a basis $\langle 1, v \rangle$ for $H^\bullet(M)$. Since there is no boundary, there are no source terms. Feynman diagrams can only be closed graphs, possibly with residual fields placed the vertices. Hence the 1-point contribution to the effective action is three zero modes contracted at a single vertex and integrated over M. This yields a numerical coefficient which is either 0 or 1, times a cubic polynomial in the coordinate on the space of residual fields, multiplied with the structure constants of the corresponding vertex.[14] The two-point contribution to the effective action consists of connected graphs with exactly two bulk vertices. If there is a single edge between them (and two residual fields on either side) the contribution vanishes since the propagator vanishes when integrated against residual fields. That leaves diagrams with two and three edges. For a diagram with two edges, we have to place a residual field at both vertices. Since residual fields are inhomogeneous forms concentrated in form degrees 0 and 3, and the propagator is a 2-form, but we integrate over the six-dimensional space $M \times M$, these contributions also vanish. Hence we only have the oriented theta graphs 16c and d. Notice that the Manin triple condition rules out $\Gamma_{3,al}$.

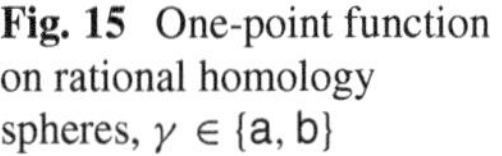

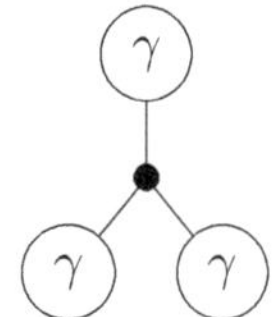

Fig. 15 One-point function on rational homology spheres, $\gamma \in \{\mathsf{a}, \mathsf{b}\}$

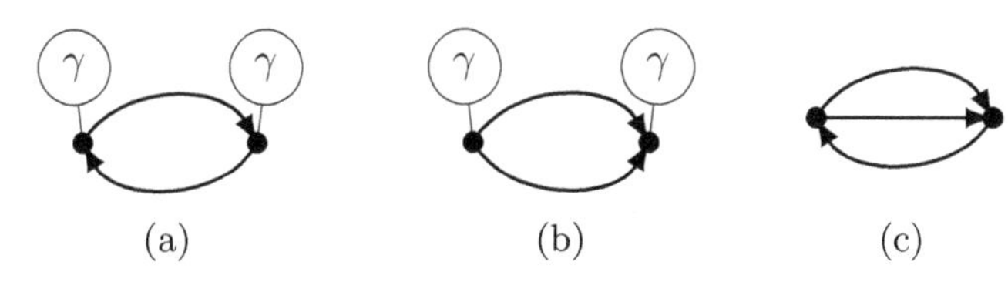

Fig. 16 Oriented two-point diagrams. $\gamma \in \{\mathsf{a}, \mathsf{b}\}$. **(a)** $\Gamma^{\gamma\gamma}_{2,op}$. **(b)** $\Gamma^{\gamma\gamma}_{2,al}$. **(c)** $\Gamma_{3,op}$. **(d)** $\Gamma_{3,al}$

7.3.2 Lens Spaces

On lens spaces one can use the decomposition described in Sect. 6 to compute the diagrams using the glued propagator for reduced residual fields computed in [CMR17]. The result is given in terms of diagrams described in Sect. 5. In particular, we can identify the weights of the oriented theta diagrams with the terms in effective actions (39) and (45), namely (ignoring the Lie algebra coefficients)

$$w_{\Gamma_{3,op}} = \frac{1}{2}s(q,p) + \frac{q+m}{12p} \tag{46}$$

$$w_{\Gamma_{3,al}} = \frac{1}{2}s(q,p) + \sum_{k=0}^{p-1} \eta_{S^1}(k/p)f(qk/p) + \eta_{S^1}(k/p)f(mk/p) + \frac{q+m}{2\pi^2}H_{1/p}. \tag{47}$$

[14]Notice that this is true on any closed manifold where the representatives of the residual fields are closed under wedge product.

8 Comparing to Existing Results

The theta invariant has been previously investigated by Axelrod and Singer [AS91, AS94] and following up on their method by Bott and Cattaneo in [BC98, BC99]. Following up on a preprint by Kontsevich [Kon94], slightly different methods were employed by Kuperberg-Thurston [KT99] and Lescop [Les04b, Les04a, Les15, Les16] to investigate the theta invariant. Axelrod and Singer use a Riemann-Hodge gauge fixing

$$K_g = d_g^* \circ (\Delta_g + P_g)^{-1}$$

associated with a particular metric g on M. In this formula, d_g^* is the codifferential, $\Delta_g = dd_g^* + d_g^* d$ is the Hodge-de Rham Laplacian, and P_g is the projection to harmonic forms. They analyze how the weight of the theta graph changes when the metric is varied continuously. Since in this paper we fix the axial gauge from the beginning, and this is a non-metric gauge, we cannot compare their results to ours directly. However, it is shown in the thesis [Wer18] of the third author that the axial gauge can be approximated arbitrarily well by Riemann-Hodge gauge fixings.

8.1 *Kuperberg-Thurston-Lescop Theta Invariants*

On the other hand, Lescop defines a theta invariant of *framed* rational homology spheres (M, τ) by

$$\Theta(M, \tau) = \int_{C_2(M)} \omega^3 \tag{48}$$

for a propagator $\omega \in \Omega^2(C_2(M, \tau))$ defined using the parallelization τ. Note that there is no factor of $1/6$ like in our normalization. It is further shown that

$$\Theta(M, g \cdot \tau) - \Theta(M, \tau) = \frac{1}{2} \deg(g),$$

where $g \colon M \to SO(3)$ and deg denotes the degree of the map, i.e. the invariant changes by a half-integer under a normalized change of framing. She concludes that

$$\Theta(M) := \Theta(M, \tau) - \frac{1}{4} p_1(\tau),$$

where $p_1(\tau)$ is a certain relative Pontryagin number associated with τ, is an invariant of rational homology 3-spheres, and observes that

$$\Theta(M) = 6\lambda(M),$$

where $\lambda(M)$ is the Casson-Walker invariant of rational homology 3-spheres, normalized to $1/2$ of Walker's original normalization [Wal92].

8.2 *Comparison with Weights of Oriented Theta Graphs*

In the Manin triple case, the weight of the oriented theta graph for the lens space $L(p, q)$ is

$$\frac{1}{2}s(q, p) + \frac{q+m}{12p}.$$

Note that the first term is precisely $\lambda(L_{p,q})$, since $\lambda_W(L_{p,q}) = s(q, p)$ [Wal92]. On the other hand, we can change q and m by a multiple of p by composing the gluing diffeomorphism with a Dehn twist on one of the solid tori, thus changing the 2-framing of the resulting lens space by ± 1 (see [FG91]). This change results in shifting the weight of the theta graph by $\pm 1/12$.

On the other hand, in the non-Manin triple case there is an irrational term in the weight of the theta graph, and if we change q or m, the weight changes by $\frac{1}{2\pi^2}H_{1/p}$ where H_x is the analytic continuation of the harmonic numbers to real values. Indeed these weights have no analogue in the literature known to the authors. Still, note that up to a shift by a multiple of an irrational number they only depend on p and $q (\mathrm{mod} p)$.

9 Conclusions and Outlook

In this note we have computed the lowest orders of the Chern-Simons state for a split Lie algebra on lens spaces using the quantum BV-BFV formalism. We fixed most of the choices, except for the choice of gluing diffeomorphism, and polarization of the Lie algebra of coefficients. We analyzed how the state changes as we vary these choices. In particular, if we choose a polarization that is compatible with the algebraic structure we obtain results that are very similar to the ones observed in the literature on perturbative Chern-Simons invariants, namely, that the two-loop part, up to a framing dependent term, equals the Casson-Walker invariant. If we choose a polarization which is incompatible with the algebraic structure, we obtain different weights that so far have not been observed in the literature.

It has often been observed that the state depends heavily on the polarization, a pedagogical example can be found in [Wei91]. It has been proposed that in the quantization of symplectic groupoids one should always use polarizations compatible with the algebraic structure [Haw08]. In our case we can observe that the map $\mathcal{F}^\partial \to \mathcal{B}$ from boundary fields to the base space of the polarization is a Lie algebra morphism if and only if we choose a Manin triple polarization.

It would be interesting to see whether we can extend these results to a larger class of 3-manifolds. A first difficulty is that on higher genus handlebodies there is no axial gauge, so we cannot do explicit computations as in this paper. Another approach based on theta functions that might generalize to higher genus easier is considered in [Wer18]. Also, on handlebodies of nonzero Euler characteristic there might be problems in the regularization of tadpoles.

Acknowledgments A. S. C. and K. W. acknowledge partial support of SNF Grant No. 200020-172498/1. This research was (partly) supported by the NCCR SwissMAP, funded by the Swiss National Science Foundation, and by the COST Action MP1405 QSPACE, supported by COST (European Cooperation in Science and Technology). P. M. acknowledges partial support of RFBR Grant No. 17-01-00283a. K. W. acknowledges partial support by the Forschungskredit of the University of Zurich, grant no. FK-16-093.

References

[AS91] S. Axelrod and I. M. Singer. Chern-Simons perturbation theory. In *Differential geometric methods in theoretical physics, Proceedings, New York*, volume 1, pages 3–45, 1991, arXiv:hep-th/9110056.

[AS94] S. Axelrod and I. M. Singer. Chern-Simons perturbation theory. II. *J. Diff. Geom.*, 39(1):173–213, 1994, arXiv:hep-th/9304087.

[BC98] R. Bott and A. S. Cattaneo. Integral invariants of 3-manifolds. *J. Diff. Geom.*, 48(1):91–133, 1998, arXiv:dg-ga/9710001.

[BC99] R. Bott and A. S. Cattaneo. Integral invariants of 3-manifolds II. *J. Diff. Geom.*, 53(1):1–13, 1999, arXiv:math/9802062.

[BF83] I. Batalin and E. Fradkin. A generalized canonical formalism and quantization of reducible gauge theories. *Phys. Lett. B*, 122(2):157–164, March 1983.

[BF86] I. A. Batalin and E. S. Fradkin. Operator quantization and abelization of dynamical systems subject to first-class constraints. *La Rivista Del Nuovo Cimento Series 3*, 9(10):1–48, oct 1986.

[BN95] D. Bar-Natan. On the Vassiliev knot invariants. *Topology*, 34(2):423–472, 1995.

[BNGRT02] D. Bar-Natan, S. Garoufalidis, L. Rozansky, and D. P. Thurston. The Århus integral of rational homology 3-spheres I: A highly non trivial flat connection on S_3. *Selecta Mathematica*, 8(3):315–339, sep 2002.

[BRS75] C. Becchi, A. Rouet, and R. Stora. Renormalization of the abelian Higgs-Kibble model. *Commun. Math. Phys.*, 42(2):127–162, 1975.

[BRS76] C. Becchi, A. Rouet, and R. Stora. Renormalization of gauge theories. *Ann. Phys.*, 98:287–321, 1976.

[BV77] I. A. Batalin and G. A. Vilkovisky. Relativistic S-matrix of dynamical systems with boson and fermion constraints. *Phys. Lett. B*, 69(3):309–312, aug 1977.

[BV81] I. A. Batalin and G. A. Vilkovisky. Gauge algebra and quantization. *Phys. Lett. B*, 102(1):27–31, June 1981.

[BV83] I. A. Batalin and G. A. Vilkovisky. Quantization of gauge theories with linearly dependent generators. *Phys. Rev. D*, 28(10):2567–2582, November 1983.

[BW97] S. Bates and A. Weinstein. *Lectures on the Geometry of Quantization (Berkeley Mathematical Lecture Notes; Vol 8)*. American Mathematical Society, 1997.

[CM08] A. S. Cattaneo and P. Mnev. Remarks on Chern-Simons Invariants. *Commun. Math. Phys.*, 293:803–836, November 2008, arXiv:0811.2045.

[CMR] A. S. Cattaneo, P. Mnev, and N. Reshetikhin. Perturbative BV theories with Segal-like gluing, arXiv:1602.00741v2.

[CMR11] A. S. Cattaneo, P. Mnev, and N. Reshetikhin. Classical and quantum Lagrangian field theories with boundary. In *Proceedings, 11th Hellenic School and Workshops on Elementary Particle Physics and Gravity (CORFU2011)*, volume CORFU2011, page 44. 2011, arXiv:1207.0239.

[CMR14] A. S. Cattaneo, P. Mnev, and N. Reshetikhin. Classical BV Theories on manifolds with boundary. *Commun. Math. Phys.*, 332(2):535–603, August 2014, arXiv:1201.0290.

[CMR15] A. S. Cattaneo, P. Mnev, and N. Reshetikhin. Semiclassical Quantization of Classical Field Theories. In *Mathematical Aspects of Quantum Field Theories*, D. Calaque and T. Strobl, editors, Mathematical Physics Studies, pages 275–324. Springer, Cham, 2015.

[CMR17] A. S. Cattaneo, P. Mnev, and N. Reshetikhin. Perturbative quantum gauge theories on manifolds with boundary. *Commun. Math. Phys.*, 357(2):631–730, dec 2017.

[CMW17] A. S. Cattaneo, P. Mnev, and K. Wernli. Split Chern–Simons theory in the BV-BFV formalism. In *Quantization, Geometry and Noncommutative Structures in Mathematics and Physics*, pages 293–324. Springer International Publishing, 2017.

[CMW18] A. S. Cattaneo, N. Moshayedi, and K. Wernli. Perturbative quantization of nonlinear AKSZ sigma models on manifolds with boundary. 2018, arXiv:1807.11782v1.

[CS74] S.-S. Chern and J. Simons. Characteristic forms and geometric invariants. *Ann. of Math. (2)*, 99(1):48–69, 1974.

[FG91] D. S. Freed and R. E. Gompf. Computer calculation of Witten's 3-manifold invariant. *Commun. Math. Phys.*, 141(1):79–117, 1991.

[Fio03] D. Fiorenza. An introduction to the Batalin-Vilkovisky formalism. In *Comptes Rendus des Rencontres Mathematiques de Glanon, Edition 2003*. 2003, arXiv:math/0402057.

[FK89] J. Fröhlich and C. King. The Chern-Simons theory and knot polynomials. *Commun. Math. Phys.*, 126(1):167–199, 1989.

[FP67] L. D. Faddeev and V. N. Popov. Feynman diagrams for the Yang-Mills field. *Phys. Lett. B*, 25(1):29–30, jul 1967.

[Fre95] D. S. Freed. Classical Chern-Simons theory, 1. *Advances in Mathematics*, 113(2):237–303, jul 1995.

[Fre02] D. S. Freed. Classical Chern-Simons Theory, Part 2. *Houston J. Math.*, 28:293–, 2002.

[FV75] E. S. Fradkin and G. A. Vilkovisky. Quantization of relativistic systems with constraints. *Phys. Lett. B*, 55(2):224–226, feb 1975.

[FV77] E. S. Fradkin and G. A. Vilkovisky. Quantization of Relativistic Systems with Constraints: Equivalence of Canonical and Covariant Formalisms in Quantum Theory of Gravitational Field. *CERN Preprint*, (CERN-TH-2332), 1977.

[GJ87] J. Glimm and A. Jaffe. *Quantum Physics*. Springer New York, 1987.

[Haw08] E. Hawkins. A groupoid approach to quantization. *J. Symplectic Geom.*, 6(1):61–125, 2008.

[Khu04] H. M. Khudaverdian. Semidensities on Odd Symplectic Supermanifolds. *Commun. Math. Phys.*, 247(2):353–390, may 2004.

[Kon93] M. Kontsevich. Vassiliev's knot invariants. *Adv. in Sov. Math*, 16(2):137–150, 1993.

[Kon94] M. Kontsevich. Feynman diagrams and low-dimensional topology. In *First European Congress of Mathematics Paris, July 6–10, 1992*, A. Joseph, F. Mignot, F. Murat, B. Prum, and R. Rentschler, editors, volume 120 of *Progress in Mathematics*, pages 97–121. Birkhäuser Basel, 1994.

[KT99] G. Kuperberg and D. P. Thurston. Perturbative 3-manifold invariants by cut-and-paste topology, 1999, arXiv:math/9912167.

[Les04a] C. Lescop. On the Kontsevich-Kuperberg-Thurston construction of a configuration-space invariant for rational homology 3-spheres. 2004, arXiv:math/0411088v1.

[Les04b] C. Lescop. Splitting formulae for the Kontsevich-Kuperberg-Thurston invariant of rational homology 3-spheres. 2004, arXiv:math/0411431v1.

[Les15] C. Lescop. A formula for the Θ-invariant from Heegaard diagrams. *Geom. Topol.*, 19(3):1205–1248, 2015, arXiv:1209.3219.

[Les16] C. Lescop. A combinatorial definition of the Θ-invariant from Heegaard diagrams. *North-W. Eur. J. of Math.*, 2:17–87, February 2016, arXiv:1402.2261.

[LMO98] T. T. Le, J. Murakami, and T. Ohtsuki. On a universal perturbative invariant of 3-manifolds. *Topology*, 37(3):539–574, may 1998.

[Mne08] P. Mnev. Discrete BF theory. September 2008, arXiv:0809.1160.

[Mne17] P. Mnev. Lectures on Batalin-Vilkovisky formalism and its applications in topological quantum field theory, 2017, arXiv:1707.08096.

[Pol05] M. Polyak. Feynman diagrams for pedestrians and mathematicians. *Proc. Symp. Pure Math.*, 73:15–42, 2005, arXiv:math/0406251.

[Res10] N. Reshetikhin. Lectures on quantization of gauge systems. In *New Paths Towards Quantum Gravity*, pages 125–190. Springer Berlin Heidelberg, 2010.

[RT91] N. Reshetikhin and V. G. Turaev. Invariants of 3-manifolds via link polynomials and quantum groups. *Invent. Math.*, 103(1):547–597, December 1991.

[Sch78] A. Schwarz. The partition function of degenerate quadratic functional and Ray-Singer invariants. *Lett. Math. Phys.*, 2(3):247–252, 1978.

[Sch07] A. Schwarz. Topological quantum field theories. 2007, arXiv:hep-th/0011260.

[Sch10] F. Schätz. Invariance of the BFV complex. *Pacific J. Math.*, 248(2):453–474, Dec 2010, arXiv:0812.2357.

[Sch15] M. Schiavina. *BV-BFV Approach to General Relativity*. PhD thesis, Universität Zürich, 2015.

[Sta97] J. Stasheff. Homological reduction of constrained Poisson algebras. *J. Diff. Geom.*, 45(1):221–240, 1997.

[Tyu76] I. V. Tyutin. Gauge invariance in field theory and statistical physics in operator formalism. *Preprints of P.N. Lebedev Physical Institute, No. 39*, December 1976, arXiv:0812.0580.

[Wal92] K. Walker. *An Extension of Casson's Invariant. (Am-126), Volume 126*. Princeton University Press, 1992.

[Wei87] A. Weinstein. Symplectic groupoids and Poisson manifolds. *Bulletin of the American Mathematical Society*, 16(1):101–105, jan 1987.

[Wei91] A. Weinstein. Symplectic groupoids, geometric quantization, and irrational rotation algebras. In *Mathematical Sciences Research Institute Publications*, A. Weinstein and P. Dazord, editors, pages 281–290. Springer US, 1991.

[Wer18] K. Wernli. *Perturbative Quantization of Split Chern-Simons Theory on Handle-bodies and Lens Spaces by the BV-BFV formalism*. PhD thesis, Universität Zürich, 2018.

[Wit89] E. Witten. Quantum field theory and the Jones polynomial. *Commun. Math. Phys.*, 121(3):351–399, 1989.

Generalized Demazure Modules and Prime Representations in Type D_n

Vyjayanthi Chari, Justin Davis, and Ryan Moruzzi

To Kolya Reshetikhin, on his 60th birthday.

Abstract The goal of this paper is to understand the graded limit of a family of irreducible prime representations of the quantum affine algebra associated with a simply laced simple Lie algebra $\mathfrak{g}$. This family was introduced by Hernandez and Leclerc (Duke Math J 154:265–341, 2010; Monoidal categorifications of cluster algebras of type A and D, Symmetries, Integrable Systems and Representations, Springer Proceedings in Mathematics & Statistics, vol. 40, pp. 175–193, 2013) in the context of monoidal categorification of cluster algebras. The graded limit of a member of this family is an indecomposable graded module for the current algebra $\mathfrak{g}[t]$; or equivalently a module for the maximal standard parabolic subalgebra in the affine Lie algebra $\widehat{\mathfrak{g}}$. In the case when $\mathfrak{g}$ is of type A_n the problem was studied by Brito et al. (J Inst Math Jussieu, 31 pp., 2015), where it was shown that the graded limit is isomorphic to a level two Demazure module. In this paper we study the case when $\mathfrak{g}$ is of type D_n. We show that in certain cases the limit is a generalized Demazure module, i.e., it is a submodule of a tensor product of level one Demazure modules. We give a presentation of these modules and compute their graded character (and hence also the character of the prime representations) in terms of Demazure modules of level two.

VC was partially supported by DMS-1719357, the Simons Fellows Program and by the Infosys Visiting Chair position at the Indian Institute of Science.

V. Chari (✉) · J. Davis
Department of Mathematics, University of California, Riverside, CA, USA
e-mail: chari@math.ucr.edu; jdavis@math.ucr.edu

R. Moruzzi Jr.
Department of Mathematics, Ithaca College, Ithaca, NY, USA
e-mail: rmoruzzi@ithaca.edu

A. Alekseev et al. (eds.), *Representation Theory, Mathematical Physics, and Integrable Systems*, Progress in Mathematics 340,
https://doi.org/10.1007/978-3-030-78148-4_4

1 Introduction

Kirillov–Reshetikhin modules (or KR-modules) are a family of finite-dimensional modules for the quantum affine algebra associated with a simple Lie algebra. These modules were introduced and originally studied by Kirillov and Reshetikhin in [22] in connection with the closely related Yangians. The interest in these representations arose from their study of solvable lattice models which allowed them to formulate an important conjecture on the character of the tensor products of these modules. Subsequently, with the introduction of q-characters in [14], which was further developed in [12, 13], it became more natural and more tractable to study the Kirillov-Reshetikhin modules for the quantum affine algebra associated with a simple Lie algebra. In [29, 30] the Kirillov–Reshetikhin conjecture was solved by geometric methods when $\mathfrak{g}$ is simply laced. The non-simply laced case was done in [18] using a q-character approach. A combinatorial version of this conjecture was proved in [11]. A crystal basis approach to these modules was first developed in [16, 17] and there is a vast literature on crystal bases for these modules (see, for instance, [24, 37]). We emphasize here that our references to the topics discussed in the introduction are nowhere near comprehensive that would require a separate volume!

One approach to understanding the character of KR-modules and more generally the character of finite-dimensional representations of quantum affine algebras is to study the classical ($q \to 1$) limit of these modules. The systematic study of this approach was begun in [9], where a necessary and sufficient condition for the existence of the limit was proved. All the interesting representations studied in the literature admit a classical limit. The limit, when it exists, is a finite-dimensional module for the corresponding affine Lie algebra and so also a module for current algebra $\mathfrak{g}[t]$ which is the Lie algebra of polynomial maps from $\mathbb{C} \to \mathfrak{g}$. The notion of the graded limit of a representation of a quantum affine algebra was developed in [5, 8]. We refer the reader to [21] for a discussion of the role of graded limits in the KR-conjecture. An interesting problem is to give a presentation of the graded limit and to compute its graded character; this is particularly so, because a presentation and the character of an irreducible finite-dimensional module for the quantum affine algebra is not known in general. This problem was studied for a class of representations called minimal affinizations in [27, 33, 34] when $\mathfrak{g}$ is of classical type and in [25, 28] for some exceptional types.

D. Hernandez and B. Leclerc introduced in [19, 20] the notion of a monoidal categorification of cluster algebras. They defined certain tensor subcategories of representations of the quantum affine algebra and showed that their Grothendieck ring was a cluster algebra of Dynkin type. The cluster variables are particular " prime" irreducible modules (which we shall refer to as HL-modules) of the quantum affine algebra and a cluster monomial is an irreducible tensor product of HL-modules. (Recall that a representation is said to be prime if it is not isomorphic to a tensor product of non-trivial representations.) The HL-modules include some of the well-known ones, such as certain KR-modules, but also include many new examples of non-isomorphic prime irreducible representations which are not well-understood.

We are interested in the graded limit of HL-modules. In the case of A_n this problem was studied in [3, 4] and we discuss this briefly. The results of those papers imply that the graded limit of an HL-module is isomorphic to a Demazure module occurring in a level two representation of the affine Lie algebra. The graded character of Demazure modules is known to be given by the Demazure character formula. In recent work [1], the graded characters of HL-modules in type A_n are given explicitly in terms of Macdonald polynomials. The appearance of level one Demazure modules as graded limits goes back to the work of [7, 15, 32] and we say no more about it here.

In this paper we are interested in the limits of HL-modules in type D_n. We restrict our attention to the modules which correspond to the roots of D_n in which the simple roots occur with multiplicity at most one. Even with this restriction, the situation is substantially more complicated than A_n. We prove that the graded limit of an HL-module is a generalized Demazure module, i.e., it occurs as a particular submodule of the tensor product of level one Demazure modules. The study of generalized Demazure modules is not very well-developed and, in particular, no presentation or character formula is known in general for these modules. Thus, in the first two sections we work entirely in the context of the current algebra $\mathfrak{g}[t]$ where $\mathfrak{g}$ is of type D_n and give a presentation and a graded character formula for a family of generalized Demazure modules. In the final section of the paper we explain briefly the connection with the work of Hernandez and Leclerc. These connections are similar to the A_n case which is explained in great detail in [3, 4]. There are difficulties which arise when we take the HL-modules associated with the remaining cluster variables; namely those which involve a simple root with multiplicity two. However, it looks likely that these will have a flag where the successive quotients are level two Demazure modules and we hope to study this elsewhere.

2 Generalized Demazure Modules

In this section we set up the basic notation and define the Demazure and generalized Demazure modules associated with an integrable highest weight representation of the untwisted affine Lie algebra $D_n^{(1)}$. In the latter part of the section we state and prove the main result assuming two key steps in the proof.

2.1 *The Simple Lie Algebra of Type D_n*

Throughout this paper $\mathfrak{g}$ will denote a simple Lie algebra of type D_n and $\mathfrak{h}$ a fixed Cartan subalgebra. Let $\kappa : \mathfrak{g}\times\mathfrak{g} \to \mathbb{C}$ be the Killing form on $\mathfrak{g}$ and $(\ ,\ ) : \mathfrak{h}^*\times\mathfrak{h}^* \to$

$\mathbb{C}$ be the induced symmetric bilinear form on $\mathfrak{h}^*$. Let R and W denote the set of roots and the Weyl group, respectively, of the pair $(\mathfrak{g}, \mathfrak{h})$. Fix a set of simple roots $\{\alpha_i : 1 \le i \le n\}$ where α_{n-1}, α_n are the spin nodes and α_{n-2} is the trivalent node and fix a set of fundamental weights $\{\omega_i : 1 \le i \le n\}$ satisfying $(\omega_i, \alpha_j) = \delta_{i,j}$. It will be convenient to set $\omega_0 = 0 = \omega_{n+1}$. As usual we let Q and P be the integer span of the simple roots and fundamental weights, respectively; the sets Q^+ and P^+ are defined in the obvious way and we set $R^+ = R \cap Q^+$.

Set,

$$\alpha_{i,j} = \alpha_i + \cdots + \alpha_j, \quad \alpha_{i,n} = \alpha_i + \cdots + \alpha_{n-2} + \alpha_n, \quad 1 \le i \le j \le n-1,$$
$$\beta_{i,j} = \alpha_i + \cdots + \alpha_{j-1} + 2(\alpha_j + \cdots + \alpha_{n-2}) + \alpha_{n-1} + \alpha_n, \quad 1 \le i < j \le n-1,$$
$$h_{i,j} = h_i + \cdots + h_j, \quad 1 \le i \le j \le n-1,$$
$$h_{i,n} = h_i + h_{i+1} + \cdots + h_{n-2} + h_n, \quad 1 \le i \le n-2.$$

We shall adopt the convention that $\alpha_{i,j} = h_{i,j} = 0$ if $i > j$ or if $(i, j) = (n-1, n)$. Note that

$$R^+ = \{\alpha_{i,j} : 1 \le i \le j \le n\} \sqcup \{\beta_{i,j} : 1 \le i < j \le n-1\}.$$

Define a partial order on P by $\lambda \le \mu$ iff $\mu - \lambda \in Q^+$. Finally, let $\{x_\alpha^\pm, h_i : \alpha \in R^+, i \in I\}$ be a Chevalley basis of $\mathfrak{g}$ and set $x_{\alpha_i}^\pm = x_i^\pm$. Let $\mathfrak{g} = \mathfrak{n}^- \oplus \mathfrak{h} \oplus \mathfrak{n}^+$ be the corresponding triangular decomposition.

2.2 *The Affine Lie Algebra D_n^1*

Let $\widehat{\mathfrak{g}}$ be the untwisted affine Lie algebra associated with $\mathfrak{g}$ and recall that it has the following explicit realization. Given an indeterminate t, let $\mathbb{C}[t, t^{-1}]$ be the algebra of Laurent polynomials and set

$$\widehat{\mathfrak{g}} = \mathfrak{g} \otimes \mathbb{C}[t, t^{-1}] \oplus \mathbb{C}c \oplus \mathbb{C}d.$$

Define the commutator on $\widehat{\mathfrak{g}}$ by requiring c to be central, and

$$[x \otimes t^r, y \otimes t^s] = [x, y] \otimes t^{r+s} + r\delta_{r+s,0}\kappa(x, y)c, \quad [d, x \otimes t^r] = r(x \otimes t^r), \quad x, y \in \mathfrak{g}, \quad r, s \in \mathbb{Z}.$$

Then

$$\widehat{\mathfrak{h}} = \mathfrak{h} \oplus \mathbb{C}c \oplus \mathbb{C}d$$

is a Cartan subalgebra of $\widehat{\mathfrak{g}}$ and we extend an element $\lambda \in \mathfrak{h}^*$ to an element of $\widehat{\mathfrak{h}}^*$ by setting $\lambda(c) = 0 = \lambda(d)$. Define elements $\Lambda_0, \delta, \alpha_0 \in \widehat{\mathfrak{h}}^*$ by

$$\Lambda_0(\mathfrak{h} \oplus \mathbb{C}d) = 0, \;\; \Lambda_0(c) = 1, \;\; \delta(\mathfrak{h} \oplus \mathbb{C}c) = 0, \;\; \delta(d) = 1, \;\; \alpha_0 = -\beta_{1,2} + \delta.$$

The set $\{\alpha_i : 0 \le i \le n\}$ is a set of simple roots for the pair $(\widehat{\mathfrak{g}}, \widehat{\mathfrak{h}})$ and the corresponding set of affine fundamental weights is given by $\{\Lambda_i : 0 \le i \le n\}$ where

$$\Lambda_i = \omega_i + a_i^\vee \Lambda_0, \quad a_i^\vee = 2, \;\; 2 \le i \le n-2, \;\; a_i^\vee = 1, \;\; i = 1, n-1, n.$$

Let $\widehat{W}$ be the affine Weyl group and note that it contains an isomorphic copy of W. Define the affine root lattice $\widehat{Q}$ and affine weight lattice $\widehat{P}$ and the subsets $\widehat{Q}^+$, $\widehat{P}^+$ and the set of affine positive roots in the natural way. Let

$$\widehat{\mathfrak{b}} = \widehat{\mathfrak{h}} \oplus (\mathfrak{n}^+ \otimes 1) \oplus (\mathfrak{g} \otimes t\mathbb{C}[t]), \;\; \widehat{\mathfrak{p}} = (\mathfrak{n}^- \otimes 1) \oplus \widehat{\mathfrak{b}}$$

be the Borel subalgebra and the standard maximal parabolic subalgebra of $\widehat{\mathfrak{g}}$ with respect to $\widehat{R}^+$. Let $\widehat{\mathfrak{n}}^+$ be the commutator subalgebra of $\widehat{\mathfrak{b}}$ and note that the commutator subalgebra of $\widehat{\mathfrak{p}}$ is just the current algebra $\mathfrak{g}[t] = \mathfrak{g} \otimes \mathbb{C}[t]$ associated with $\mathfrak{g}$. The action of d induces a canonical $\mathbb{Z}_+$- grading on $\mathfrak{g}[t]$ and on its universal enveloping algebra with the grade of an element $x \otimes t^r \in \mathfrak{g}[t]$ being r.

2.3 *Highest Weight Modules and Demazure Modules*

Given $\Lambda \in \widehat{P}^+$ let $\widehat{V}(\Lambda)$ be the integrable irreducible $\widehat{\mathfrak{g}}$-module generated by an element $\widehat{v}_\Lambda$ with relations

$$\widehat{\mathfrak{n}}^+ \widehat{v}_\Lambda = 0, \;\; h\widehat{v}_\Lambda = \Lambda(h)\widehat{v}_\Lambda, \;\; h \in \widehat{\mathfrak{h}},$$

$$(x_i^- \otimes 1)^{\Lambda(h_i)+1} \widehat{v}_\Lambda = 0, \;\; 1 \le i \le n, \;\; (x_{\beta_{1,2}}^+ \otimes t^{-1})^{\Lambda(h_0)+1} \widehat{v}_\Lambda = 0.$$

We have

$$\widehat{V}(\Lambda) = \bigoplus_{\Lambda' \in \widehat{P}} \widehat{V}(\Lambda)_{\Lambda'}, \;\; \widehat{V}(\Lambda)_{\Lambda'} = \{v \in \widehat{V}(\Lambda) : hv = \Lambda'(h)v, \;\; h \in \widehat{\mathfrak{h}}\}.$$

It is known that

$$\dim \widehat{V}(\Lambda)_{w\Lambda} = 1, \;\; \text{for all } w \in \widehat{W},$$

and we let $\widehat{V}_w(\Lambda)$ be the $\widehat{\mathfrak{b}}$-module generated by $\widehat{V}(\Lambda)_{w\Lambda}$. The modules $\widehat{V}_w(\Lambda)$ are finite-dimensional and are called the Demazure modules associated with Λ. Moreover $\widehat{v}_\Lambda \in \widehat{V}_w(\Lambda)$ for all $w \in \widehat{W}$. The following is now essentially immediate.

Lemma 2.1 *Suppose that* $w, w' \in \widehat{W}$ *and* $\Lambda \in \widehat{P}^+$. *Then*

$$\dim \operatorname{Hom}_{\widehat{\mathfrak{b}}}(\widehat{V}_w(\Lambda), \widehat{V}_{w'}(\Lambda)) \leq 1,$$

and any non-zero element of this space is injective. □

2.4 Generalized Demazure Modules

Given $\Lambda, \Lambda' \in \widehat{P}^+$ and $w, w' \in \widehat{W}$ we set

$$\widehat{V}_{w,w'}(\Lambda, \Lambda') = \mathbb{U}(\widehat{\mathfrak{b}})(\widehat{V}(\Lambda)_{w\Lambda} \otimes \widehat{V}(\Lambda')_{w'\Lambda'}).$$

These (along with an obvious extension to an arbitrary number of dominant integral weights and Weyl group elements) are called the generalized Demazure modules. Further, since

$$\dim \operatorname{Hom}_{\widehat{\mathfrak{g}}}(\widehat{V}(\Lambda) \otimes \widehat{V}(\Lambda'), \ \widehat{V}(\Lambda + \Lambda')) = 1$$

it is easy to see that

$$\widehat{V}(\Lambda)_{w\Lambda} \otimes \widehat{V}(\Lambda')_{w\Lambda'} \cong \widehat{V}(\Lambda + \Lambda')_{w(\Lambda+\Lambda')} \text{ and so } \widehat{V}_{w,w}(\Lambda, \Lambda') \cong \widehat{V}_w(\Lambda + \Lambda').$$

In this paper we shall be interested in certain special families of Demazure and generalized Demazure modules; namely those associated with pairs (w, Λ) such that the restriction of $w\Lambda$ to $\mathfrak{h}$ is in $(-P^+)$. This restriction guarantees that $\widehat{V}_w(\Lambda)$ is stable under $\widehat{\mathfrak{p}}^+$. Equivalently, we can (and do) regard these modules as $\mathbb{Z}$-graded modules for $\mathfrak{g}[t]$ and call them the stable Demazure modules. We recall a result proved independently in [23, 26].

Theorem 2.1 *Let* $\Lambda, \Lambda' \in \widehat{P}^+$ *and* $w, w' \in \widehat{W}$. *If* $\sigma \in \widehat{W}$ *is such that* $\sigma(w\Lambda + w'\Lambda') \in \widehat{P}^+$, *then there exists a projection of* $\widehat{\mathfrak{g}}$*-modules*

$$\widehat{V}(\Lambda) \otimes \widehat{V}(\Lambda') \longrightarrow \widehat{V}(\sigma(w\Lambda + w'\Lambda')) \to 0,$$

which induces an isomorphism of the one-dimensional spaces

$$\widehat{V}(\Lambda)_{w\Lambda} \otimes \widehat{V}(\Lambda')_{w'\Lambda'} \cong \widehat{V}(\sigma(w\Lambda + w'\Lambda'))_{w\Lambda+w'\Lambda'}.$$

□

The following corollary is immediate from the preceding theorem and is needed later in this section.

Corollary 2.1 *Retain the notation of the theorem. Assume that* $\widehat{V}_w(\Lambda)$ *and* $\widehat{V}_{w'}(\Lambda')$ *are stable Demazure modules. Then* $\widehat{V}_{\sigma^{-1}}(\sigma(w\Lambda+w'\Lambda'))$ *is also a stable Demazure module and there exists a surjective map of* $\widehat{\mathfrak{p}}$*-modules*

$$\widehat{V}_{w,w'}(\Lambda,\Lambda') \to \widehat{V}_{\sigma^{-1}}(\sigma(w\Lambda+w'\Lambda')) \to 0.$$

□

2.5 A Presentation of Stable Demazure Modules

Let $w_\circ$ be the longest element in W. Suppose that (w,Λ) and (w',Λ') are such that $\Lambda(c)=\Lambda'(c)=\ell$ and $\lambda=w_\circ w\Lambda|_{\mathfrak{h}}=w_\circ w'\Lambda'|_{\mathfrak{h}}\in P^+$. Then it is known [15, 32] that $V_w(\Lambda)$ is isomorphic to $V_{w'}(\Lambda')$ as $\mathfrak{g}[t]$-modules up to an overall shift of the action of d. In particular, we can regard the stable Demazure modules as being indexed by a pair $(\ell,\lambda)\in\mathbb{N}\times P^+$.

In what follows we prefer to regard the stable (generalized) Demazure modules as $\mathbb{Z}$-graded modules for the current algebra of $\mathfrak{g}[t]$, where the grading is given by the action of d. The maps between graded modules will be of degree zero. Given $s\in\mathbb{Z}$ and a graded $\mathfrak{g}[t]$-module we denote by τ_s^*V the graded $\mathfrak{g}[t]$-module obtained by shifting the grades by s.

We recall a presentation of the stable Demazure modules given in [10]. Let $D(\ell,\lambda)$ be the $\mathfrak{g}[t]$-module generated by an element w_λ with the following defining relations: for $1\le i\le n$, $\alpha\in R^+$, $h\in\mathfrak{h}$, we have

$$\mathfrak{n}^+[t]w_\lambda=0,\ \ (h\otimes t^r)w_\lambda=\delta_{r,0}\lambda(h)w_\lambda,\ \ (x_i^-\otimes 1)^{\lambda(h_i)+1}w_\lambda=0, \tag{2.1}$$

$$(x_\alpha^-\otimes t^{s_\alpha})w_\lambda=0,\ \ (x_\alpha^-\otimes t^{s_\alpha-1})^{m_\alpha+1}w_\lambda=0, \tag{2.2}$$

where we write $\lambda(h_\alpha)=\ell(s_\alpha-1)+m_\alpha$ with $s_\alpha\in\mathbb{N}$ and $0<m_\alpha\le\ell$. The $\mathbb{Z}$-grading on $D(\ell,\lambda)$ is given by requiring w_λ to have grade zero.

In the case $\ell=1,2$ the relations can be further streamlined. The following was proved when $\ell=1$ in [9] and when $\ell=2$ in [4].

Lemma 2.2 *If* $\ell=1$ *the relations in* (2.2) *are a consequence of the relations in* (2.1). *If* $\ell=2$, *then the second relation in* (2.2) *is a consequence of the others.* □

2.6 Stable Generalized Demazure Modules

In the language of $\mathfrak{g}[t]$-modules developed above, a generalized Demazure module corresponds to considering a tensor product $\tau_s^* D(\ell, \lambda) \otimes \tau_{s'}^* D(\ell', \lambda')$ and taking the $\mathfrak{g}[t]$-module through $w_\lambda \otimes w_{\lambda'}$. Presentations of generalized Demazure modules have not been much studied although there is some work on the subject. They play an important role in the study of classical limits of representations of quantum affine algebras and first arose in this context in [33, 34] (see [36] for other examples). In this paper we shall give a presentation of a particular family of generalized Demazure modules, namely those of the form

$$D(\lambda, \mu) := \mathbb{U}(\mathfrak{g}[t])(w_\lambda \otimes w_\mu) \subset D(1, \lambda) \otimes D(1, \mu),$$

with some restrictions on the pair $(\lambda, \mu) \in P^+ \times P^+$. Our motivations for considering this problem were discussed extensively in the introduction and further details can also be found in Sect. 4. We note the following reformulation of Corollary 2.1.

Lemma 2.3 *There exists a (unique up to scalars) map* $\eta_{\lambda,\mu} : D(\lambda, \mu) \to D(2, \lambda + \mu) \to 0$, *of* $\mathfrak{g}[t]$*-modules extending the assignment* $w_\lambda \otimes w_\mu \to w_{\lambda+\mu}$. □

2.7 The Modules $V(\lambda, \mu)$

Given $\lambda, \mu \in P^+$, let $V(\lambda, \mu)$ be the $\mathfrak{g}[t]$-module generated by an element $w_{\lambda,\mu}$ satisfying the following defining relations:

$$\mathfrak{n}^+[t] w_{\lambda,\mu} = 0, \quad (h \otimes t^r) w_{\lambda,\mu} = \delta_{r,0}(\lambda + \mu)(h) w_{\lambda,\mu}, \quad (x_i^- \otimes 1)^{(\lambda+\mu)(h_i)+1} w_{\lambda,\mu} = 0, \tag{2.3}$$

$$(x_\alpha^- \otimes t^{\max\{\lambda(h_\alpha), \mu(h_\alpha)\}}) w_{\lambda,\mu} = 0, \tag{2.4}$$

where $1 \le i \le n$, $h \in \mathfrak{h}$ and $\alpha \in R^+$. Define a grading on $V(\lambda, \mu)$ by requiring the grade of $w_{\lambda,\mu}$ to be zero. Clearly $V(\lambda, \mu)$ is a graded quotient of $D(1, \lambda + \mu)$ and $V(\lambda, \mu) \cong V(\mu, \lambda)$. Moreover Lemma 2.2 shows that if $\mu = 0$, then $V(\lambda, 0) \cong D(1, \lambda)$. The following is immediate from Eqs. (2.1)–(2.4) and Lemma 2.3.

Lemma 2.4 *The assignments* $w_{\lambda,\mu} \to w_{\lambda+\mu}$ *and* $w_{\lambda,\mu} \to w_\lambda \otimes w_\mu$ *define surjective maps of graded* $\mathfrak{g}[t]$*-modules*

$$\psi_{\lambda,\mu} : V(\lambda, \mu) \to D(2, \lambda + \mu), \quad \varphi_{\lambda,\mu} : V(\lambda, \mu) \to D(\lambda, \mu) \to 0,$$

and $\psi_{\lambda,\mu} = \eta_{\lambda,\mu} \circ \varphi_{\lambda,\mu}$. □

2.8 Interlacing Weights

Let

$$P^+(1) = \{\lambda \in P^+ : \lambda(h_i) \leq 1, \quad 1 \leq i \leq n\}.$$

It is obvious that any $\lambda \in P^+(1)$ can be written uniquely (up to order) as a sum $\lambda = \lambda_1 + \lambda_2$ where $\lambda_s \in P^+(1)$, $s = 1, 2$ and satisfy the following: for $1 \leq i \leq j \leq n$ and $(i, j) \neq (n-1, n)$

$$\lambda_r(h_i) = 1 = \lambda_r(h_j) \implies \lambda_p(h_s) = 1 \text{ for some } i < s < j, \quad \{r, p\} = \{1, 2\},$$

and

$$\lambda(h_{n-1} + h_n) > 0 \implies \lambda_p(h_{n-1} + h_n) = 0 \text{ for some } p \in \{1, 2\}.$$

It will be convenient to call $(\lambda_1, \lambda_2) \in P^+ \times P^+$ an *interlacing* pair if $\lambda_1 + \lambda_2 \in P^+(1)$ and the preceding two conditions hold.

Examples The pairs $(\omega_i, 0)$ for $0 \leq i \leq n$, $(\omega_{n-1} + \omega_n, 0)$ and the elements of the set $\{(\omega_i, \omega_j) : 0 \leq i \neq j \leq n \ \ (i, j) \neq (n-1, n)\}$ are interlacing. The pair $(\omega_2 + \omega_5, \omega_3 + \omega_6 + \omega_7)$ is interlacing in D_7 but not in D_8.

2.9 The Main Theorem

The following is the main result of this note.

Theorem 2.2 *Let $(\lambda_1, \lambda_2) \in P^+ \times P^+$ be an interlacing pair. For all $\nu \in P^+$ the map*

$$\varphi_{\lambda_1+\nu,\lambda_2+\nu} : V(\lambda_1 + \nu, \lambda_2 + \nu) \to D(\lambda_1 + \nu, \lambda_2 + \nu)$$

is an isomorphism. Moreover $D(\lambda_1 + \nu, \lambda_2 + \nu)$ admits a flag and the quotients are isomorphic to $\tau_s^* D(2, \mu)$ for some $s \in \mathbb{Z}_+$, $\mu \in P^+$.

Remark 2.1 One can give a precise formula for the graded character of $D(\lambda_1 + \nu, \lambda_2 + \nu)$ and this can be found in Sect. 2.15.

Remark 2.2 An analogous result was proved for A_n in [4] when $\nu = 0$. In fact in that case it was proved that

$$V(\lambda_1, \lambda_2) \cong D(\lambda_1, \lambda_2) \cong D(2, \lambda_1 + \lambda_2).$$

Our methods are different and work for A_n as well and prove the more general case when $\nu \neq 0$.

Remark 2.3 It is plausible that the map $\varphi_{\lambda,\mu} : V(\lambda, \mu) \to D(\lambda, \mu)$ is always an isomorphism. This is known to be true for $\mathfrak{sl}_2$ and examples exist in $\mathfrak{sl}_n$ (see, for instance, [2, 35]).

In the rest of the section we shall assume that (λ_1, λ_2) *is an interlacing pair and* $\lambda = \lambda_1 + \lambda_2$. *Moreover the property of interlacing pairs means that we can and will assume without loss of generality that if* $\lambda(h_{n-1} + h_n) > 0$, *then* $\lambda_1(h_{n-1}) > 0$. *This means that if* p *is maximal such that* $\lambda(h_{p+1}) = 1$, *then* $p \leq n - 2$ *and* $\lambda_1(h_{p+1}) = 1$ *and* $\lambda_2(h_{p+1}) = 0$.

2.10 A First Reduction

The theorem is proved in several steps. The first reduction is the following proposition which gives a condition for the generalized Demazure module to be isomorphic to a Demazure module.

Proposition 2.1 *If* $\lambda(h_{n-1} + h_n) = 1$ *or if* $\lambda = \omega_{i-1} + \omega_i + \delta_{i+1,n}\omega_n$ *for* $1 \leq i \leq n - 1$, *then for all* $\nu \in P^+$ *we have*

$$V(\lambda_1 + \nu, \lambda_2 + \nu) \cong D(2, 2\nu + \lambda) \cong D(\lambda_1 + \nu, \lambda_2 + \nu).$$

2.11 The Root β_λ

Suppose that $\lambda(h_{n-1}+h_n) \in \{0, 2\}$ and $\lambda \neq \omega_{i-1}+\omega_i+\delta_{i+1,n}\omega_n$ and let $p \leq n-2$ be maximal such that $\lambda_1(h_{p+1}) = 1 = \lambda(h_{p+1})$. Let $1 \leq p' \leq p \leq n - 2$ be maximal so that $(\lambda_1 - \lambda_2)(h_{p',p}) = 0$. Observe that if $p' < p$, then the definition of interlacing pairs forces $\lambda_2(h_p) = 1 = \lambda_1(h_{p'})$. Set

$$\beta_\lambda = \beta_{p',p+1} = \alpha_{p'} + \cdots + \alpha_p + 2(\alpha_{p+1} + \cdots + \alpha_{n-2}) + \alpha_{n-1} + \alpha_n$$

and note that $\lambda_1(h_{\beta_\lambda}) = 3 - \delta_{p',p}$ and $\lambda_2(h_{\beta_\lambda}) = (1 - \delta_{p'p})$. It is helpful to note that the assumption $\lambda(h_{n-1} + h_n) \in \{0, 2\}$ implies that $p = n - 2$ iff $\lambda(h_{n-1} + h_n) = 2$.

Lemma 2.5 *Suppose that* $\lambda(h_{n-1} + h_n) \in \{0, 2\}$ *and* $\lambda \neq \omega_{i-1} + \omega_i + \delta_{i+1,n}\omega_n$. *Then* $\lambda_1 - \beta_\lambda \in P^+$ *and there exists* $\nu_0 \in P^+$ *such that* $(\lambda_1 - \beta_\lambda - \nu_0, \lambda_2 - \nu_0)$ *is an interlacing pair.*

Proof It is simple to see that with our assumptions

$$\lambda_1 - \beta_\lambda = \lambda_1 - \omega_{p+1} - \delta_{p,n-2}\omega_n - (1 - \delta_{p',p})\omega_{p'} + \omega_{p'-1} + (1 - \delta_{p',p})\omega_p \in P^+.$$

Taking $\nu_0 = \lambda_2(h_{p'-1})\omega_{p'-1} + (1-\delta_{p',p})\lambda_2(h_p)\omega_p$ it is easy to check that $(\lambda_1 - \beta_\lambda - \nu_0, \lambda_2 - \nu_0)$ is an interlacing pair.

□

2.12 The Second Reduction

The second reduction is the following proposition which, in particular, gives an upper bound for the dimension of $V(\lambda_1+\nu, \lambda_2+\nu)$.

Proposition 2.2 *Suppose that* $\lambda(h_{n-1}+h_n) \in \{0,2\}$ *and* $\lambda \neq \omega_{i-1}+\omega_i+\delta_{i+1,n}\omega_n$. *There exists a right exact sequence of* $\mathfrak{g}[t]$*-modules*

$$\tau^*_{(\lambda_1+\nu)(h_{\beta_\lambda})-1}V(\lambda_1+\nu-\beta_\lambda, \lambda_2+\nu) \to V(\lambda_1+\nu, \lambda_2+\nu) \to D(2, \lambda+2\nu) \to 0, \tag{2.5}$$

with $w_{\lambda_1+\nu-\beta_\lambda,\lambda_2+\nu} \to (x^-_{\beta_\lambda} \otimes t^{(\lambda_1+\nu)(h_{\beta_\lambda})-1})w_{\lambda_1+\nu,\lambda_2+\nu}$.

2.13 Proof of Theorem 2.2

Assuming Propositions 2.1 and 2.2 we complete the proof of Theorem 2.2. The proof is by an induction with respect to the partial order on P^+. The minimal elements with respect to this order are $0, \omega_1, \omega_{n-1}, \omega_n$ and Proposition 2.1 shows that induction begins. Moreover the proposition also establishes the theorem if $\lambda = \omega_{i-1}+\omega_i+\delta_{i+1,n}\omega_n$ or $\lambda(h_{n-1}+h_n) = 1$. Hence we have only to prove the inductive step when $\lambda(h_{n-1}+h_n) \in \{0,2\}$ and $\lambda \neq \omega_{i-1}+\omega_i+\delta_{i+1,n}\omega_n$. We shall need the following result to complete the proof of the inductive step.

Lemma 2.6 *If* $\lambda(h_{n-1}+h_n) \in \{0,2\}$ *and* $\lambda \neq \omega_{i-1}+\omega_i+\delta_{i+1,n}\omega_n$, *there exists an inclusion of* $\mathfrak{g}[t]$*-modules*

$$\tau^*_{(\lambda_1+\nu)(h_{\beta_\lambda})-1}D(1, \lambda_1+\nu-\beta_\lambda) \hookrightarrow D(1, \lambda_1+\nu), \tag{2.6}$$

which sends to $w_{\lambda_1+\nu-\beta_\lambda} \to (x^-_{\beta_\lambda} \otimes t^{(\lambda_1+\nu)(h_{\beta_\lambda})-1})w_{\lambda_1+\nu}$.

Proof By Lemma 2.2 it suffices to prove that $w := (x^-_{\beta_\lambda} \otimes t^{(\lambda_1+\nu)(h_{\beta_\lambda})-1})w_{\lambda_1+\nu}$ satisfies the relations in (2.1). If $\beta_\lambda - \alpha_i \notin R^+$, then it is clear that $(x_i^+ \otimes t^r)w = 0$ for all $r \in \mathbb{Z}_+$. Otherwise we must have $i = p+1$, or $i = p'$ if $p' < p$ or $i = n$ if $p = n-2$ and we have to prove that

$$(x^-_{\beta_\lambda-\alpha_i} \otimes t^{(\lambda_1+\nu)(h_{\beta_\lambda})-1+r})w_{\lambda_1+\nu} = 0, \quad r \in \mathbb{Z}_+.$$

But this follows from (2.2) since in all cases $\lambda_1(h_i) = 1$ and so $(\lambda_1 + \nu)(h_{\beta_\lambda}) - 1 \geq (\lambda_1 + \nu)(h_{\beta_\lambda - \alpha_i})$. The second relation in (2.1) is trivial if $r = 0$ and if $r > 0$, then it follows from (2.2) since

$$(h \otimes t^r)w = (x^-_{\beta_\lambda} \otimes t^{(\lambda_1+\nu)(h_{\beta_\lambda})+r-1})w_{\lambda_1+\nu} = 0.$$

The third relation is immediate since the modules are all finite-dimensional.

□

2.14 *Proof of Theorem 2.2, the Final Step*

Lemma 2.3, Proposition 2.2, and the inductive hypothesis establish the following inequalities:

$$\dim D(\lambda_1 + \nu, \lambda_2 + \nu) \leq \dim V(\lambda_1 + \nu, \lambda_2 + \nu),$$
$$\dim V(\lambda_1 + \nu, \lambda_2 + \nu) \leq \dim D(2, 2\nu + \lambda) + \dim D(\lambda_1 + \nu - \beta_\lambda, \lambda_2 + \nu).$$

The inductive step follows if we prove that

$$\dim D(\lambda_1 + \nu, \lambda_2 + \nu) = \dim D(2, 2\nu + \lambda) + \dim D(\lambda_1 + \nu - \beta_\lambda, \lambda_2 + \nu). \quad (2.7)$$

For this, we observe that Lemma 2.6 gives an inclusion

$$0 \to D(1, \lambda_1 + \nu - \beta_\lambda) \otimes D(1, \lambda_2 + \nu) \to D(1, \lambda_1 + \nu) \otimes D(\lambda_2 + \nu),$$

which sends

$$w_{\lambda_1+\nu-\beta_\lambda} \otimes w_{\lambda_2+\nu} \to ((x^-_{\beta_\lambda} \otimes t^{(\lambda_1+\nu)(h_{\beta_\lambda})-1)})w_{\lambda_1+\nu}) \otimes w_{\lambda_2+\nu}.$$

Since $(\lambda_1 + \nu)(h_{\beta_\lambda}) - 1 \geq (\lambda_2 + \nu)(h_{\beta_\lambda})$ we see that (2.2) gives

$$(x^-_{\beta_\lambda} \otimes t^{(\lambda_1+\nu)(h_{\beta_\lambda})-1)}) \left(w_{\lambda_1+\nu} \otimes w_{\lambda_2+\nu}\right) = \left((x^-_{\beta_\lambda} \otimes t^{(\lambda_1+\nu)(h_{\beta_\lambda})-1})w_{\lambda_1+\nu}\right) \otimes w_{\lambda_2+\nu}.$$

It follows that we have an inclusion

$$\iota : D(\lambda_1 + \nu - \beta_\lambda, \lambda_2 + \nu) \hookrightarrow D(\lambda_1 + \nu, \lambda_2 + \nu)$$

and it suffices to prove that the corresponding quotient is isomorphic to $D(2, 2\nu + \lambda)$. By Lemma 2.4 we have the following surjective maps

$$V(\lambda_1 + \nu, \lambda_2 + \nu) \twoheadrightarrow D(\lambda_1 + \nu, \lambda_2 + \nu) \twoheadrightarrow D(2, 2\nu + \lambda).$$

These maps are all unique up to scalars and Proposition 2.2 shows that the kernel of the composite map is generated by the element $(x^-_{\beta_\lambda} \otimes t^{(\lambda_1+\nu)(h_{\beta_\lambda})-1)})w_{\lambda_1+\nu,\lambda_2+\nu}$. Hence the kernel of

$$D(\lambda_1 + \nu, \lambda_2 + \nu) \twoheadrightarrow D(2, 2\nu + \lambda)$$

is generated by $(x^-_{\beta_\lambda} \otimes t^{(\lambda_1+\nu)(h_{\beta_\lambda})-1)})(w_{\lambda_1+\nu} \otimes w_{\lambda_2+\nu})$. But this means that the latter kernel is precisely the image of ι and hence the corresponding quotient is isomorphic to $D(2, 2\nu + \lambda)$ as needed.

2.15 A Character Formula

Given a graded $\mathfrak{g}[t]$-module $V = \oplus_{s\in\mathbb{Z}} V[s]$ we have,

$$(\mathfrak{g} \otimes t^r)V[s] \subset V[r + s], \quad V_\mu = \bigoplus_{s\in\mathbb{Z}} V_\mu[s], \quad V_\mu[s] = V_\mu \cap V[s],$$

for all $r, s \in \mathbb{Z}$ and $\mu \in P$. Let $\mathbb{Z}[v, v^{-1}][P]$ be the group ring of P with coefficients in $\mathbb{Z}[v, v^{-1}]$ where v is an indeterminate. Let $e(\mu)$, $\mu \in P$ be a basis for the group ring and set

$$\mathrm{ch}_{\mathrm{gr}}\, V = \sum_{\mu\in P} \left(\sum_{s\in\mathbb{Z}} \dim V_\mu[s] v^s \right) e(\mu).$$

Notice that in the course of proving Theorem 2.2 we also established that the right exact sequence in Proposition 2.2 is exact. Hence we get an equality of graded characters

$$\mathrm{ch}_{\mathrm{gr}}\, D(\lambda_1 + \nu, \lambda_2 + \nu) = \mathrm{ch}_{\mathrm{gr}}\, D(2, 2\nu + \lambda) + v^{(\lambda_1+\nu)(h_{\beta_\lambda})-1}\, \mathrm{ch}_{\mathrm{gr}}\, D(\lambda_1 - \beta_\lambda + \nu, \lambda_2 + \nu). \tag{2.8}$$

Using this we can derive a formula for $\mathrm{ch}_{\mathrm{gr}}\, D(\lambda_1 + \nu, \lambda_2 + \nu)$ in terms of Demazure modules as follows.

Given $\mu \in P^+$ define elements $\lfloor \mu/2 \rfloor$ and $\mathrm{res}_2\, \mu$ of P^+ by,

$$\lfloor \mu/2 \rfloor(h_i) = \lfloor \mu(h_i)/2 \rfloor, \quad 1 \le i \le n, \quad \mathrm{res}_2\, \mu = \mu - 2\lfloor \mu/2 \rfloor.$$

Then $\mathrm{res}_2\, \mu \in P^+(1)$ and we set $\beta_\mu = \beta_{\mathrm{res}_2\, \mu}$ if $\mathrm{res}_2\, \mu$ satisfies the conditions of Sect. 2.11 and otherwise set $\beta_\mu = 0$. Let (μ_1, μ_2) be the interlacing pair corresponding to $\mathrm{res}_2\, \mu$ and let

$$r_\mu = \lfloor \mu/2 \rfloor(h_{\beta_\mu}) + \max\{\mu_1(h_{\beta_\mu}), \mu_2(h_{\beta_\mu})\} - 1.$$

Given $\mu \in P^+$ set

$$\mu^0 = \mu, \quad \mu^1 = \mu^0 - \beta_{\mu^0}, \ldots, \mu^s = \mu^{s-1} - \beta_{\mu^{s-1}},$$

where s is minimal so that $\beta_{\mu^s} = 0$. Setting $r_k = r_{\mu^k}$ a repeated application of (2.8) gives

$$\mathrm{ch}_{\mathrm{gr}}\, D(\lfloor \mu/2 \rfloor + \mu_1,\ \lfloor \mu/2 \rfloor + \mu_2) = \sum_{k=0}^{s} v^{r_k}\, \mathrm{ch}_{\mathrm{gr}}\, D(2, \mu^k).$$

3 Proof of Propositions 2.1 and 2.2

In this section we first establish some technical results on the properties of interlacing pairs and then prove Propositions 2.1 and 2.2.

Throughout this section we shall assume without further mention that (λ_1, λ_2) *is an interlacing pair and* $\lambda = \lambda_1 + \lambda_2$. *We shall also assume without loss of generality that there exists p maximal with* $\lambda(h_{p+1}) = 1$ *with* $p \le n-2$ *and* $\lambda_1(h_{p+1}) = 1$.

3.1 Elementary Properties of Interlacing Pairs

Lemma 3.1 *For all* $1 \le i \le j \le n$ *we have*

$$|(\lambda_1 - \lambda_2)(h_{i,j})| \le 1 \tag{3.1}$$

and hence $|(\lambda_1 - \lambda_2)(h_\alpha)| \le 2$ *for all* $\alpha \in R^+$. *Further if* $\lambda(h_{n-1} + h_n) = 1$, *then*

$$|(\lambda_1 - \lambda_2)(h_\alpha)| \le 1 \ \textit{for all } \alpha \in R^+.$$

Proof Equation (3.1) is proved by induction on $j - i$; induction obviously begins at $i = j$ since $\lambda \in P^+(1)$. If $j > i$ the inductive step is clear if $\lambda(h_i) = 0$ or $\lambda(h_j) = 0$. If $\lambda(h_i) = 1 = \lambda(h_j)$, then either $\lambda(h_s) = 0$ for all $i < s < j$ or there exists $i < p < j$ minimal such that $\lambda_2(h_p) = 1$. In the first case we have proved that $(\lambda_1 - \lambda_2)(h_{i,j}) = 0$ and in the second case we have proved that $(\lambda_1 - \lambda_2)(h_{i,j}) = (\lambda_1 - \lambda_2)(h_{p+1,j})$ and the inductive step follows. Since $\beta_{i,j} = \alpha_{i,n} + \alpha_{j,n-1}$ for $1 \le i < j \le n-1$ it is immediate that

$$|(\lambda_1 - \lambda_2)(\beta_{i,j})| \le 2 \ \text{ for all } 1 \le i < j \le n-1.$$

Moreover, if $\lambda(h_{n-1}+h_n) = 1$ our assumptions imply that $\lambda_1(h_{n-1}) = 1$ and $\lambda_2(h_{n-1}+h_n) = 0$ and so using (3.1) we have

$$-1 \le (\lambda_1-\lambda_2)(h_{i,n}) = (\lambda_1-\lambda_2)(h_{i,n-2}) = (\lambda_1-\lambda_2)(h_{i,n-1}) - 1 \le 0,$$

$$1 \ge (\lambda_1-\lambda_2)(h_{j,n-1}) = (\lambda_1-\lambda_2)(h_{j,n-2}) + 1 \ge 0.$$

Hence we have

$$-1 \le (\lambda_1-\lambda_2)(h_{\beta_{i,j}}) = (\lambda_1-\lambda_2)(h_{i,n}) + (\lambda_1-\lambda_2)(h_{j,n-1}) \le 1$$

and the proof of the lemma is complete. □

3.2 The Set $R(\lambda_1, \lambda_2)$

Recall our convention that $\alpha_{i,j} = 0$ if $i > j$. Set

$$R(\lambda_1,\lambda_2) = \{\beta_{i,j} : (\lambda_1-\lambda_2)(h_{\beta_{i,j}}) = \pm 2\}.$$

Note that Lemma 3.1 shows that $R(\lambda_1,\lambda_2) = \emptyset$ if $\lambda(h_{n-1}+h_n) = 1$ or $\lambda = \omega_{i-1}+\omega_i+\delta_{i,n-1}\omega_n$. The next result establishes the converse.

Lemma 3.2 *Suppose that $\lambda(h_{n-1}+h_n) \in \{0,2\}$ and $\lambda \ne \omega_{i-1}+\omega_i+\delta_{i,n-1}\omega_n$. Then $\beta_\lambda \in R(\lambda_1,\lambda_2)$ and more generally,*

$$\beta_{i,j} \in R(\lambda_1,\lambda_2) \iff \beta_{i,j} = \alpha_{i,p'-1} + \beta_\lambda + \alpha_{j,p} \text{ and } (\lambda_1-\lambda_2)(h_{i,p'-1}) = 0 = (\lambda_1-\lambda_2)(h_{j,p}) = 0.$$

Proof Recall from Sect. 2.11 that $\beta_\lambda = \beta_{p',p+1}$ where $1 \le p' \le p$ is maximal with $(\lambda_1-\lambda_2)(h_{p',p}) = 0$ and that $(\lambda_1-\lambda_2)(h_{\beta_\lambda}) = 2$. Hence a calculation shows that

$$\beta_{i,j} \in R(\lambda_1,\lambda_2) \implies i \le p' \text{ or } p' < j \le p+1,$$

which shows that $\beta_{i,j} = \alpha_{i,p'-1} + \beta_\lambda + \alpha_{j,p}$.

Since $\beta_{i,p+1} \in R^+$ if $i < p'$ we have

$$(\lambda_1-\lambda_2)(h_{\beta_{i,p+1}}) = (\lambda_1-\lambda_2)(h_{i,p'-1}) + 2$$

and Lemma 3.1 forces $(\lambda_1-\lambda_2)(h_{i,p'-1}) \in \{-1,0\}$. If $j < p+1$ we have $\beta_{p',j} = \beta_{p',p+1} + \alpha_{j,p}$ and hence a similar argument shows that $(\lambda_1-\lambda_2)(h_{j,p}) \in \{-1,0\}$. Together with hypothesis that $(\lambda_1-\lambda_2)(h_{\beta_{i,j}}) = \pm 2$ we get $(\lambda_1-\lambda_2)(h_{i,p'-1}) = 0 = (\lambda_1-\lambda_2)(h_{j,p})$ as needed. The converse is obvious. □

Remark 3.1 Notice that, in particular, we have established (with our conventions on (λ_1, λ_2)) that $\beta \in R(\lambda_1, \lambda_2)$ iff $(\lambda_1 - \lambda_2)(h_\beta) = 2$.

3.3 Proof of Proposition 2.1

Assume that $\lambda(h_{n-1} + h_n) = 1$ or that $\lambda = \omega_{i-1} + \omega_i + \delta_{i,n-1}\omega_n$. By Lemma 3.1 we have $(\lambda_1 - \lambda_2)(h_\alpha) \in \{-1, 0, 1\}$ for all $\alpha \in R^+$, and so

$$\max\{(\nu + \lambda_1)(h_\alpha), (\nu + \lambda_2)(h_\alpha)\} = \nu(h_\alpha) + \lceil(\lambda(h_\alpha)/2\rceil.$$

The defining relations give

$$V(\lambda_1 + \nu, \lambda_2 + \nu) \cong D(2, 2\nu + \lambda_1 + \lambda_2).$$

Since the maps in Lemmas 2.3 and 2.4 are unique up to scalars it now follows that the map

$$V(\lambda_1 + \nu, \lambda_2 + \nu) \twoheadrightarrow D(\lambda_1 + \nu, \lambda_2 + \nu) \twoheadrightarrow D(2, 2\nu + \lambda_1 + \lambda_2)$$

is an isomorphism and hence all the maps are isomorphisms. Proposition 2.1 is proved.

3.4 The Kernel of ψ

We turn to the proof of Proposition 2.2. We have $R(\lambda_1, \lambda_2) \neq \emptyset$ and our conventions imply that $\lambda_1(h_{\beta_\lambda}) = 3 - \delta_{p',p}$ and $\lambda_2(h_{\beta_\lambda}) = 1 - \delta_{p',p}$. Using the remark following Lemma 3.2 we have $\lambda_1(h_\beta) \geq \lambda_2(h_\beta)$ for all $\beta \in R(\lambda_1, \lambda_2)$. Hence for all $\beta \in R(\lambda_1, \lambda_2)$,

$$(\nu + \lambda_1)(h_\beta) = \max\{(\nu + \lambda_1)(h_\beta), (\nu + \lambda_2)(h_\beta)\} = \nu(h_\beta) + \lceil(\lambda(h_\beta)/2\rceil + 1.$$

An inspection of the defining relations of the modules shows that the kernel K of the canonical map

$$\psi_{\lambda_1+\nu,\lambda_2+\nu} : V(\lambda_1 + \nu, \lambda_2 + \nu) \to D(2, \lambda + \nu) \to 0, \quad w_{\lambda_1+\nu,\lambda_2+\nu} \to w_{2\nu+\lambda},$$

is generated by the elements

$$(x_\beta^- \otimes t^{(\nu+\lambda_1)(h_\beta)-1})w_{\lambda_1+\nu,\lambda_2+\nu}, \quad \beta \in R(\lambda_1, \lambda_2).$$

The proof of Proposition 2.2 is now shown in two steps. The first step is to show that

$$K = \mathbb{U}(\mathfrak{g}[t])(x^-_{\beta_\lambda} \otimes t^{(\nu+\lambda_1)(h_{\beta_\lambda})-1)})w_{\lambda_1+\nu,\lambda_2+\nu}.$$

For this, let $\beta \in R(\lambda_1, \lambda_2)$ and write $\beta = \alpha_{i,p'-1} + \beta_{p',p+1} + \alpha_{j,p}$ as in Lemma 3.2 and assume that $i \le p' - 1$ or $j \le p$ (otherwise $\beta = \beta_\lambda$ and there is nothing to prove). The defining relations

$$(x^-_{i,p'-1} \otimes t^{(\nu+\lambda_1)(h_{i,p'-1})})w_{\lambda_1+\nu,\lambda_2+\nu} = 0 = (x^-_{j,p} \otimes t^{(\nu+\lambda_1)(h_{j,p})})w_{\lambda_1+\nu,\lambda_2+\nu},$$

imply if $i \le p' - 1$ and $j \le p$

$$(x^-_{i,p'-1} \otimes t^{(\nu+\lambda_1)(h_{i,p'-1})})(x^-_{j,p} \otimes t^{(\nu+\lambda_1)(h_{j,p})})(x^-_{\beta_\lambda} \otimes t^{(\nu+\lambda_1)(h_{\beta_\lambda})-1})w_{\lambda_1+\nu,\lambda_2+\nu} =$$

$$[x^-_{i,p'-1}, [x^-_{j,p}, x^-_{\beta_\lambda}]] \otimes t^{(\nu+\lambda_1)(h_{\beta_{i,j}})-1})w_{\lambda_1+\nu,\lambda_2+\nu} = (x^-_{\beta_{i,j}} \otimes t^{(\nu+\lambda_1)(h_{\beta_{i,j}})-1})w_{\lambda_1+\nu,\lambda_2+\nu}.$$

An obvious modification works if $i \ge p'$ or $j \ge p + 1$ and the first step is proved.

The next step is to show that the element $(x^-_{\beta_\lambda} \otimes t^{(\nu+\lambda_1)(h_{\beta_\lambda})-1})w_{\lambda_1+\nu,\lambda_2+\nu}$ satisfies the defining relations of the element $w_{\lambda_1+\nu-\beta_\lambda,\lambda_2+\nu} \in V(\lambda_1 + \nu - \beta_\lambda, \lambda_2 + \nu)$. This establishes the existence of the map $V(\lambda_1 + \nu - \beta_\lambda, \lambda_2 + \nu) \to K \to 0$.

3.5 *Proof of Proposition 2.2*

We prove that $(x^-_{\beta_\lambda} \otimes t^{(\nu+\lambda_1)(h_{\beta_\lambda})-1})w_{\lambda_1+\nu,\lambda_2+\nu}$ satisfies the relations in (2.3) with (λ_1, λ_2) replaced by $(\lambda_1 - \beta_\lambda, \lambda_2)$. Since $(x^+_i \otimes t^r)w_{\lambda_1+\nu,\lambda_2+\nu} = 0$ for $1 \le i \le n$ and $r \in \mathbb{Z}_+$ the first relation in (2.3) is immediate if $\beta_\lambda - \alpha_i \notin R^+$. Otherwise we must prove that

$$(\dagger)\ \ (x^-_{\beta_\lambda-\alpha_i} \otimes t^{(\nu+\lambda_1)(h_{\beta_\lambda})-1+r})w_{\lambda_1+\nu,\lambda_2+\nu} = 0.$$

Since $\beta_\lambda = \beta_{p',p+1}$ it follows that either $i = p + 1$ and if $p' < p$, then we can also have $i = p'$ and if $p = n - 2$, then, in addition, we can have $i = n$. In the first two cases we have $\lambda_1(h_i) = 1$ and $\lambda_2(h_i) = 0$ (see Sect. 2.11). In particular,

$$(x^-_{\beta_\lambda-\alpha_i} \otimes t^{\nu(h_{\beta_\lambda-\alpha_i})+2-\delta_{p,p'}})w_{\lambda_1+\nu,\lambda_2+\nu} = 0$$

is a defining relation and $(\dagger)$ follows by applying $(h \otimes t^r)$ since

$$(\nu + \lambda_1)(h_{\beta_\lambda}) - 1 + r = \nu(h_{\beta_\lambda}) + 2 - \delta_{p,p'} + r \ge \nu(h_{\beta_\lambda-\alpha_i}) + 2 - \delta_{p,p'},\quad r \in \mathbb{Z}_+.$$

A similar argument establishes the case when $p = n - 2$ and $i = n$.

It is trivial to see that the second relation in (2.1) holds and the third is then immediate since $V(\nu+\lambda_1, \nu+\lambda_2)$ is finite-dimensional.

3.6 Proof of Proposition 2.2, the Second Step

We need the following result to prove that $(x^-_{\beta_\lambda} \otimes t^{(\nu+\lambda_1)(h_{\beta_\lambda})-1})w_{\lambda_1+\nu,\lambda_2+\nu}$ satisfies the relations in (2.4).

Lemma 3.3 *The relation* $(x^-_\alpha \otimes t^{\nu(h_\alpha)+\max\{\lambda_1(h_\alpha),\lambda_2(h_\alpha)\}})w_{\lambda_1+\nu,\lambda_2+\nu} = 0$ *holds in* $V(\lambda_1+\nu, \lambda_2+\nu)$ *for all* $\alpha \in R^+$ *iff it holds for all the elements of the set*

$$\{\alpha_{i,j} : 1 \le i \le j \le n,\ \ \lambda(h_i) = 1 = \lambda(h_j),\ \ \lambda(h_{i,j}) = 2 - \delta_{i,j}\}.$$

Proof The forward direction of the Lemma is obvious. For the reverse direction we proceed by induction with respect to the partial order on R^+ induced by the partial order $\le$ on P. If $\alpha = \alpha_i$ for some $1 \le i \le n$ there is nothing to prove. For the inductive step choose $1 \le r \le n$ so that $\alpha - \alpha_r \in R^+$. If $\lambda(h_r) = 0$, then the result follows from $[x^-_r, x^-_{\alpha-\alpha_r}] = x^-_\alpha$ and

$$(x^-_r \otimes t^{\nu(h_r)})w_{\lambda_1+\nu,\lambda_2+\nu} = 0 = (x^-_{\alpha-\alpha_r} \otimes t^{\nu(h_{\alpha-\alpha_r})+\max\{\lambda_1(h_\alpha),\lambda_2(h_\alpha)\}})w_{\lambda_1+\nu,\lambda_2+\nu},$$

where the second equality is an application of the inductive hypothesis. Otherwise, $\lambda(h_r) > 0$ for all $1 \le r \le n$ with $\alpha - \alpha_r \in R^+$, in which case α is one of the following:

- $\alpha_{i,j}$ with $\lambda(h_i) = 1 = \lambda(h_j)$,
- $\beta_{i,j}$ with $i < j-1$ and $\lambda(h_i) = \lambda(h_j) = 1$,
- $\beta_{j-1,j}$ with $\lambda(h_j) = 1$.

In the first two cases choose $i < s \le j$ minimal with $\lambda(h_s) = 1$ and $\lambda(h_{i,s}) = 2$. The result follows since Lemma 3.1 and the inductive hypothesis give

$$(x^-_{i,s} \otimes t^{\nu(h_{i,s})+1})w_{\lambda_1+\nu,\lambda_2+\nu} = 0 = (x^-_{\alpha-\alpha_{i,s}} \otimes t^{\nu(h_{\alpha-\alpha_{i,s}})+\max\{\lambda_1(h_\alpha)-1,\lambda_2(h_\alpha)-1\}})w_{\lambda_1+\nu,\lambda_2+\nu}.$$

In the third case if $\lambda(h_{j-1}) = 1$ or if $\lambda(h_{j-1}) = 0$ and there exists $s > j$ minimal with $\lambda(h_s) = 1$ then the result follows as before by working with the pairs $(\alpha_{j-1,j}, \alpha - \alpha_{j-1,j})$ and $(\alpha_{j-1,s}, \alpha - \alpha_{j-1,s})$ respectively. Otherwise we have $\max\{\lambda_1(h_\alpha), \lambda_2(h_\alpha)\} = 2$ and the result follows from

$$(x^-_{j-1,n-1} \otimes t^{\nu(h_{j-1,n-1})+1})w_{\lambda_1+\nu,\lambda_2+\nu} = 0 = (x^-_{j,n} \otimes t^{\nu(h_{j,n})+1})w_{\lambda_1+\nu,\lambda_2+\nu}.$$

□

3.7 Proof of Proposition 2.2, the Final Step

To complete the proof of the second step, we now show that (2.4) also holds. By the preceding Lemma it suffices to prove that

$$(*)\quad (x_\alpha^- \otimes t^{\nu(h_\alpha)+\max\{(\lambda_1-\beta_\lambda)(h_\alpha),\lambda_2(h_\alpha)\}})(x_{\beta_\lambda}^- \otimes t^{(\nu+\lambda_1)(h_{\beta_\lambda})-1})w_{\lambda_1+\nu,\lambda_2+\nu} = 0$$

for roots of the form $\alpha = \alpha_{i,j}$ with

$$(**)\quad (\lambda - \beta_\lambda)(h_i) = 1 = (\lambda - \beta_\lambda)(h_j),\quad (\lambda - \beta_\lambda)(h_s) = 0,\quad i < s < j,$$

and note that this forces $\beta_\lambda(h_i) \le 0$ and $\beta_\lambda(h_j) \le 0$. We claim that $(\beta_\lambda, \alpha) \le 0$. Otherwise we would have that $\beta_\lambda - \alpha \in R$ and since $\alpha - \beta_\lambda \notin R^+$ we must have $\beta_\lambda - \alpha \in R^+$. It is elementary to see that this is only possible if $\alpha_{i,j} = \alpha_{p',r}$ or $\alpha_{p+1,r}$ for some $r \ne p$ and $p' < r \le n$. Since $\beta_{p',p+1}(h_{p'}) = 1 - \delta_{p',p}$, and $\beta_{p',p+1}(h_{p+1}) = 1$ we see that $(**)$ forces $p' = p$ and $\alpha = \alpha_{p,r}$. However, if $p' = p$, then we have $\lambda(h_p) = 0$ by definition and $\beta_\lambda(h_p) = 0$ which again contradicts $(**)$ and the claim is proved.

If $\alpha + \beta_\lambda \notin R^+$, i.e., $(\alpha, \beta_\lambda) = 0$, then

$$(x_\alpha^- \otimes t^{\nu(h_\alpha)+\max\{(\lambda_1-\beta_\lambda)(h_\alpha),\lambda_2(h_\alpha)\}})w_{\lambda_1+\nu,\lambda_2+\nu} = 0$$

is a defining relation and $(*)$ follows since $[x_\alpha^-, x_{\beta_\lambda}^-] = 0$. If $(\alpha, \beta_\lambda) < 0$, then $\alpha + \beta_\lambda \in R^+$ and $[x_\alpha^-, x_{\beta_\lambda}^-] = x_{\alpha+\beta_\lambda}^-$. It suffices to show that

$$(x_{\alpha+\beta_\lambda}^- \otimes t^{\nu(h_\alpha)+\max\{(\lambda_1-\beta_\lambda)(h_\alpha),\lambda_2(h_\alpha)\}+(\nu+\lambda_1)(h_{\beta_\lambda})-1})w_{\lambda_1+\nu,\lambda_2+\nu} = 0.$$

But this holds since

$$\begin{aligned}\max\{\lambda_1(h_{\alpha+\beta_\lambda}), \lambda_2(h_{\alpha+\beta_\lambda})\} &= \max\{\lambda_1(h_\alpha) + 3 - \delta_{p',p}, \lambda_2(h_\alpha) + 1 - \delta_{p',p}\}\\ &\le \max\{\lambda_1(h_\alpha) + 3 - \delta_{p',p}, \lambda_2(h_\alpha) + 2 - \delta_{p',p}\}\\ &= \max\{\lambda_1(h_\alpha) + 1, \lambda_2(h_\alpha)\} + 2 - \delta_{p',p}.\end{aligned}$$

This completes the proof of Proposition 2.2.

4 Connection with the Representation Theory of Quantum Affine Algebras

In this section we explain very briefly the relationship between our results and the representation theory of quantum affine algebras. We also discuss the connection with the papers [19, 20] on monoidal categorification.

4.1 A Multiplicative Monoid of Weights

Let $\mathbb{C}(q)$ be the field of rational functions in an indeterminate q and set $\mathbb{A} = \mathbb{Z}[q, q^{-1}]$. Let $\mathbb{U}_q(\widehat{\mathfrak{g}})$ be the quantized enveloping algebra (defined over $\mathbb{C}(q)$) associated with $\widehat{\mathfrak{g}}$. Let $\mathbb{U}_{\mathbb{A}}(\widehat{\mathfrak{g}})$ be the $\mathbb{A}$-form of $\mathbb{U}_q(\widehat{\mathfrak{g}})$ and recall that it is a free $\mathbb{A}$-module and

$$\mathbb{U}_q(\widehat{\mathfrak{g}}) \cong \mathbb{U}_{\mathbb{A}}(\widehat{\mathfrak{g}}) \otimes_{\mathbb{A}} \mathbb{C}(q).$$

Regarding $\mathbb{C}$ as an $\mathbb{A}$-module by letting q act as 1 we have that $\mathbb{U}_{\mathbb{A}}(\widehat{\mathfrak{g}}) \otimes_{\mathbb{A}} \mathbb{C}$ is an algebra over $\mathbb{C}$ which has the universal enveloping algebra $\mathbb{U}(\widehat{\mathfrak{g}})$ as a canonical quotient. Finally, recall that $\mathbb{U}_q(\widehat{\mathfrak{g}})$ is a Hopf algebra and that $\mathbb{U}_{\mathbb{A}}(\widehat{\mathfrak{g}})$ is a Hopf subalgebra.

Let $\mathcal{P}_{\mathbb{Z}}^+$ be the free abelian monoid generated by elements $\{\boldsymbol{\omega}_{i,q^r} : 1 \le i \le n,\ r \in \mathbb{Z}\}$ and let $\mathrm{wt} : \mathcal{P}^+ \to P^+$ be the morphism of monoids given by $\mathrm{wt}\, \boldsymbol{\pi} = \sum_{i=1}^n (\deg \pi_i)\omega_i$.

Definition 4.1 Let $\mathcal{P}_{\mathbb{Z}}^+(1)$ be the subset of $\mathcal{P}_{\mathbb{Z}}^+$ containing the identity of the and elements of the form $\boldsymbol{\omega}_{i_1,a_1} \cdots \boldsymbol{\omega}_{i_k,a_k}$, where $1 \le i_1 < i_2 < \cdots < i_k \le n$, $a_j \in q^{\mathbb{Z}}$ for $1 \le k \le n$, and

$$a_j/a_{j+1} = q^{\pm(i_{j+1}-i_j+2)}, \quad k \ge 2,$$
$$a_j/a_{j+1} = q^{\pm(i_{j+1}-i_j+2)} \implies a_{j+1}/a_{j+2} = q^{\mp(i_{j+2}-i_{j+1}+2)},$$

where the second requirement holds for all $j \le k-2$ if $(i_{k-1}, i_k) \ne (n-1, n)$ and if $(i_{k-1}, i_k) = (n-1, n)$, then it holds for $j \le k-3$ and we require $a_{k-1} = a_k$. □

Clearly $\mathrm{wt}\, \mathcal{P}_{\mathbb{Z}}^+(1) = P^+(1)$.

4.2 Prime Representations

To each element $\boldsymbol{\pi} \in \mathcal{P}_{\mathbb{Z}}^+$ one can associate an (unique up to isomorphism) irreducible finite-dimensional representation $[\boldsymbol{\pi}]$ of $\mathbb{U}_q(\widehat{\mathfrak{g}})$. The trivial representation corresponds to the identity of the monoid. Given $\boldsymbol{\pi}, \boldsymbol{\pi}' \in \mathcal{P}_{\mathbb{Z}}^+(1)$, the tensor product $[\boldsymbol{\pi}] \otimes [\boldsymbol{\pi}']$ is generically irreducible and isomorphic to $[\boldsymbol{\pi}\boldsymbol{\pi}']$. However, necessary and sufficient conditions for this to hold are not known outside the case of $\mathfrak{sl}_2$ and motivates the interest in understanding the prime irreducible representations. Recall that a representation is prime if it cannot be written as a tensor product of two non-trivial representations. The following result is not hard to prove (see, for instance, [3, 4] for a similar statement for A_n).

Lemma 4.1 *The module* $[\boldsymbol{\pi}]$ *is prime for all* $\boldsymbol{\pi} \in \mathcal{P}_{\mathbb{Z}}^+(1)$. □

The results of [5, 9] show that the representation $[\pi]$ admits an $\mathbb{A}$-form denoted $[\pi_{\mathbb{A}}]$ and $[\pi_{\mathbb{A}}] \otimes_{\mathbf{A}} \mathbb{C}$ is an indecomposable and usually reducible module for the enveloping algebra $\mathbb{U}(\widehat{\mathfrak{g}})$ and hence also for the current algebra $\mathfrak{g}[t]$. Moreover if we pull-back this representation via the automorphism of $\mathfrak{g}[t] \to \mathfrak{g}[t]$ sending $x \otimes t^r \to x \otimes (t-1)^r$, $x \in \mathfrak{g}$, $r \in \mathbb{Z}_+$ we get a representation $[\pi_{\mathbb{C}}]$ of $\mathfrak{g}[t]$ and

$$(\mathfrak{g} \otimes t^N \mathbb{C}[t])[\pi_{\mathbb{C}}] = 0, \quad N >> 0.$$

4.3 Graded Limits

We now explain how to deduce the following result.

Theorem 4.1 *For $\pi \in \mathcal{P}_{\mathbb{Z}}^+(1)$, there exists an isomorphism of $\mathfrak{g}[t]$-modules*

$$[\pi_{\mathbb{C}}] \cong D(\lambda_1, \lambda_2),$$

where $\mathrm{wt}\,\pi = \lambda$ *and (λ_1, λ_2) is the interlacing pair associated with λ. In particular, $[\pi_{\mathbb{C}}]$ acquires the structure of a graded $\mathfrak{g}[t]$-module.*

The theorem can be proved by the same methods as the ones used in [4] for $\mathfrak{sl}_{n+1}$. It uses the following idea developed in [27] where similar questions were studied for a different family of irreducible representations.

Suppose that $\pi_1, \pi_2 \in \mathcal{P}_{\mathbb{Z}}^+$ are such that we have an injective map of $\mathbb{U}_q(\widehat{\mathfrak{g}})$-modules $[\pi_1\pi_2] \to [\pi_1] \otimes [\pi_2]$. Since $\mathbb{U}_{\mathbb{A}}(\widehat{\mathfrak{g}})$ is a Hopf subalgebra of $\mathbb{U}_q(\widehat{\mathfrak{g}})$, we get an injective map $[(\pi_1\pi_2)_{\mathbb{A}}] \to [(\pi_1)_{\mathbb{A}}] \otimes [(\pi_2)_{\mathbb{A}}]$. It was shown in [27, Lemma 2.20, Proposition 3.21] that tensoring with $\otimes_{\mathbb{A}}\mathbb{C}$ and pulling back by the automorphism of $\mathfrak{g}[t]$ induced by $t \to t-1$, gives rise to a map of $\mathfrak{g}[t]$-modules $[\pi_{\mathbb{C}}] \to [(\pi_1)_{\mathbb{C}}] \otimes [(\pi_2)_{\mathbb{C}}]$. Here are the main steps in proving the theorem. Retain the notation of the theorem.

We prove that the choice of parameters in Definition 4.1 along with Theorem 2.2 implies that there exists a surjective map of $\mathfrak{g}[t]$-modules $D(\lambda_1, \lambda_2) \to [\pi_{\mathbb{C}}] \to 0$.

Next note that one can define canonically, elements $\pi_1, \pi_2 \in \mathcal{P}_{\mathbb{Z}}^+(1)$ with $\mathrm{wt}\,\pi_s = \lambda_s$, $s = 1, 2$ and $\pi = \pi_1\pi_2$. Moreover, using the results in [9, 15], one has the following isomorphisms of graded $\mathfrak{g}[t]$-modules,

$$[(\pi_1)_{\mathbb{C}}] \cong_{\mathfrak{g}[t]} D(1, \lambda_1), \qquad [(\pi_2)_{\mathbb{C}}] \cong_{\mathfrak{g}[t]} D(1, \lambda_2).$$

The results in [6] show that there is an injective map of $\mathbb{U}_q(\widehat{\mathfrak{g}})$-modules $[\pi] \to [\pi_1] \otimes [\pi_2]$. As discussed earlier, one can use [27, Lemma 2.20, Proposition 3.21] to show that the image of the induced map

$$[\pi_{\mathbb{C}}] \to [(\pi_1)_{\mathbb{C}}] \otimes [(\pi_2)_{\mathbb{C}}] \cong D(1, \lambda_1) \otimes D(1, \lambda_2)$$

is $D(\lambda_1, \lambda_2)$. Hence we have a composition of surjective maps

$$D(\lambda_1, \lambda_2) \twoheadrightarrow [\pi_{\mathbb{C}}] \twoheadrightarrow D(\lambda_1, \lambda_2)$$

and Theorem 4.1 is proved.

4.4 The Connection with the Category $\mathcal{C}_\xi$

We discuss the relationship of our work with that of [19, 20, 31] and restrict ourselves to D_n. Let $\xi : \{1, 2, \cdots, n\} \to \mathbb{Z}$ and define the bipartite quiver on the Dynkin diagram of D_n given by

$$\xi(i) = \xi(i+1) \pm 1, \;\; \xi(i) = \xi(i+2), \;\; \xi(n-1) = \xi(n).$$

Let $\mathcal{C}_\xi$ be the full subcategory of finite-dimensional representations of $\mathbb{U}_q(\widehat{\mathfrak{g}})$ defined as follows: an object of $\mathcal{C}_\xi$ has all its Jordan–Holder components of the form $[\boldsymbol{\pi}]$, where $\boldsymbol{\pi} \in \mathcal{P}^+$ is a product of terms of the form $\boldsymbol{\omega}_{i,a}$, $a \in \{q^{\xi(i)\pm 1}\}$, $1 \le i \le n$. The following was proved in [19, 20] for D_4 and in [31] for D_n.

Theorem 4.2 *The category $\mathcal{C}_\xi$ is closed under taking tensor products. The Grothendieck ring of $\mathcal{C}_\xi$ is a monoidal categorification of a cluster algebra of type D_n with n frozen variables.* □

Recall that the cluster variables in a cluster algebra of type D_n are indexed by elements of $R^+ \cup \{-\alpha_i : 1 \le i \le n\}$. The cluster variable associated with a root $\alpha_{i,j}$ of D_n corresponds to a representation $[\boldsymbol{\pi}]$ where $\boldsymbol{\pi} \in \mathcal{P}^+_{\mathbb{Z}}(1)$ is given by $\boldsymbol{\pi} = \boldsymbol{\omega}_{i,a_i}\boldsymbol{\omega}_{i+1,a_{i+1}} \cdots \boldsymbol{\omega}_{j,a_j}$ and the $a_k s$ are determined by requiring that $\boldsymbol{\pi} \in \mathcal{P}^+_{\mathbb{Z}}(1)$ and $a_i = q^{\xi(i)\pm 1}$ if $\xi(i) = \xi(i-1) \pm 1$. Theorem 4.1 gives a character formula for the prime objects corresponding to these cluster variables. More generally if we relax the condition that ξ define a bipartite quiver, then the elements of $\mathcal{P}^+_{\mathbb{Z}}(1)$ should correspond to cluster variables in a suitable monoidal categorification. This was done in detail for A_n in [3] and it was non-trivial to identify the cluster variable corresponding to $[\boldsymbol{\pi}]$ when ξ is not bipartite or monotonic.

For D_n, even in the bipartite case, it seems to be much more difficult to give the character of the prime representations corresponding to roots of the form $\beta_{i,j}$. However, preliminary calculations suggest that the graded limits of these will also admit a flag where the successive quotients are Demazure modules, but possibly of level bigger than two. We hope to return to these problems elsewhere.

References

1. Rekha Biswal, Vyjayanthi Chari, Peri Shereen and Jeffrey Wand, *Macdonald Polynomials and level two Demazure modules for affine* $\mathfrak{sl}_{n+1}$, (2019) arXiv:1910.05848.
2. M, Brito and F. Pereira *Graded Limits of Simple Tensor Product of Kirillov–Reshetikhin Modules for* $U_q(\tilde{\mathfrak{sl}}_{n+1})$, Comm. in Algebra, 44, **10**, (2016), 4504–4518.
3. M. Brito, and V. Chari, *Tensor products and q-characters of HL-modules and monoidal categorifications*, Journal de l'École Polytechnique — Mathématiques, **6**, (2019), 581–619.
4. M. Brito, V. Chari, and A. Moura,*Demazure modules of level two and prime representations of quantum affine* sl_{n+1},J. Inst. Math. Jussieu, 31 pages, 2015.
5. V. Chari, *On the fermionic formula and the Kirillov-Reshetikhin conjecture*, Int. Math. Res. Notices **12** (2001), 629–654.
6. V. Chari, *Braid group actions and tensor products*, Int. Math. Res. Notices (2002), 357–382.
7. V. Chari and S. Loktev, *Weyl, Demazure and fusion modules for the current algebra of* sl_{r+1}, Adv. Math. **207** (2006), no.2, 928–960
8. V. Chari and A. Moura, *The restricted Kirillov–Reshetikhin modules for the current and twisted current algebras,* Comm. Math. Phys. **266** (2006), no. 2, 431–454.
9. V. Chari, A. Pressley, *Weyl Modules for Classical and Quantum Affine Algebras* Represent Theory, **5** (2001), 191–223.
10. V. Chari and R. Venkatesh, *Demazure modules, fusion products and Q-systems*,Comm. Math. Phy. **333** (2), (2015), 799–830, no. 1, 191–216.
11. P. Di Francesco and R. Kedem, *Proof of the combinatorial Kirillov-Reshetikhin conjecture*, Int. Math. Res. Not. IMRN **7** (2008).
12. E. Frenkel and E. Mukhin, *Combinatorics of q-Characters of Finite-Dimensional Representations of Quantum Affine Algebras*, Comm. Math. Phy. **216**, no. 1, (2001), pp 23–57.
13. E. Frenkel and E. Mukhin, *The Hopf algebra* $RepU_q(gl_\infty)$, Selecta Math. (N.S.) **8**, no. 4, (2002), 537–635.
14. E. Frenkel and N. Reshetikhin, *The q-Characters of Representations of Quantum Affine Algebras and Deformations of W-Algebras*, Recent Developments in Quantum Affine Algebras and related topics, Cont. Math., vol. 248, (1999), 163–205.
15. G. Fourier, P. Littelmann, *Weyl modules, Demazure modules, KR-modules, crystals, fusion products and limit constructions*, Adv. Math., 211 (2)(2007), pp. 566–593.
16. G. Hatayama, A. Kuniba, M. Okado, T. Takagi and Z. Tsuboi, *Paths, crystals and fermionic formulae*, MathPhys odyssey, 2001, Prog. Math. Phys., 23, Birkhauser Boston, Boston, MA, (2002), 205–272,
17. G. Hatayama, A. Kuniba, M. Okado, T. Takagi and Y. Yamada, *Remarks on fermionic formula, in Recent developments in quantum affine algebras and related topics*, (Raleigh, NC, 1998), Contemp. Math., 248, Amer. Math. Soc., Providence, RI (1999) 243–291.
18. D. Hernandez, *The Kirillov-Reshetikhin conjecture and solutions of T-systems*, J. Reine Angew. Math. **596**, (2006), 63–87/
19. D. Hernandez and B. Leclerc, *Cluster algebras and quantum affine algebras*, Duke Math. J. **154** (2010), 265–341, DOI 10.1215/00127094-2010-040.
20. D. Hernandez and B. Leclerc, *Monoidal categorifications of cluster algebras of type A and D*, Symmetries, Integrable Systems and Representations, Springer Proceedings in Mathematics & Statistics 40 (2013), 175–193.
    Boston, Boston, MA, 2006, 131–169.
21. R. Kedem, *A pentagon of identities, graded tensor products, and the Kirillov–Reshetikhin conjecture*, New trends in quantum integrable systems, World Sci. Publ. (2011), 173–193.
22. A.N. Kirillov and N. Reshetikhin, *Representations of Yangians and multiplicities of the inclusion of the irreducible components of the tensor product of representations of simple Lie algebras*, J. Soviet Math. **52**, no. 3, 3156–3164 (1990); translated from Zap. Nauchn. Sem. Leningrad. Otdel. Mat. Inst. Steklov. (LOMI) 160, Anal. Teor. Chisel i Teor. Funktsii. 8, 211–221, 301 (1987).

23. S. Kumar, *Proof of the Parthasarathy-Ranga Rao-Varadarajan conjecture,* Invent. Math. **93** (1988), 117–130.
24. C. Lenart, S. Naito, D. Sagaki, A. Schilling, and M. Shimozono, *A uniform model for Kirillov-Reshetikhin crystals I*: Lifting the parabolic quantum Bruhat graph, Int. Math. Res. Not., no. 7, (201, 1848–1901.
25. J.R. Li and K. Naoi, *Graded limits of minimal affinizations over the quantum loop algebra of type* G_2, Algebras and Representation Theory 19 **4** (2016), 957–973.
26. O. Mathieu, *Construction d'un groupe de Kac-Moody et applications*, Compositio Math. **69** (1989), no. 1, 37–60.
27. A. Moura, *Restricted limits of minimal affinizations*, Pacific J. Math. **244** (2010), 359–397.
28. A. Moura and F. Pereira, *Graded limits of minimal affinizations and beyond: the multiplicity free case for type* E_6, Algebra and Discrete Mathematics **12** (2011), 69–115.
29. H. Nakajima, *Quiver Varieties and t-Analogs of q-Characters of Quantum Affine Algebras*, Ann. of Math. **160** (2004), 1057–1097.
30. H. Nakajima, *t-analogs of q-characters of Kirillov-Reshetikhin modules of quantum affine algebras*, Represent. Theory **7**, (2003) 259–274.
31. H. Nakajima,Quiver varieties and cluster algebras, Kyoto J. Math.**51** (2011), 71–126.
32. K. Naoi, *Fusion products of Kirillov-Reshetikhin modules and the* $X = M$ *conjecture*, Adv. Math. **231** (2012), 1546–1571.
33. K. Naoi, *Demazure modules and graded limits of minimal affinizations*, Represent. Theory **17** (2013), 524–556.
34. K. Naoi, *Graded limits of minimal affinizations in type D*, SIGMA **10** (2014), 047, 20 pages.
35. K. Naoi, *Defining relations of fusion products and Schur positivity*, Toyama Mathematical Journal, **37** (2015), 87–106.
36. B. Ravinder, *Generalized Demazure modules and fusion products*, J. Algebra, **476** (2017), 186–215.
37. M. Okado and A. Schilling, *Existence of Kirillov-Reshetikhin crystals for nonexceptional types*. Represent. Theory **12** (2008), 186–207.

Cylindric Rhombic Tableaux and the Two-Species ASEP on a Ring

Sylvie Corteel, Olya Mandelshtam, and Lauren Williams

Abstract The asymmetric simple exclusion process (ASEP) is a model of particles hopping on a one-dimensional lattice of n sites. It was introduced around 1970 (Macdonald et al., Biopolymers, 6, 1968; Spitzer, Adv Math, 5:246–290, 1970), and since then has been extensively studied by researchers in statistical mechanics, probability, and combinatorics. Recently the ASEP on a lattice with open boundaries has been linked to Koornwinder polynomials (Corteel and Williams, to appear in Selecta Math, 2015; Cantini, Ann Henri Poincaré, 18(4):1121–1151, 2017), and the ASEP on a ring has been linked to Macdonald polynomials (Cantini et al., J Phys A, 48(38):384001, 25, 2015). In this article we study the two-species asymmetric simple exclusion process (ASEP) on a ring, in which two kinds of particles ("heavy" and "light"), as well as "holes," can hop both clockwise and counterclockwise (at rates 1 or t depending on the particle types) on a ring of n sites. We introduce some new tableaux on a cylinder called *cylindric rhombic tableaux* (CRT) and use them to give a formula for the stationary distribution of the two-species ASEP—each probability is expressed as a sum over all CRT of a fixed type. When λ is a partition in $\{0, 1, 2\}^n$, we then give a formula for the nonsymmetric Macdonald polynomial E_λ and the symmetric Macdonald polynomial P_λ by refining our tableaux formulas for the stationary distribution.

Keywords Asymmetric exclusion process · Macdonald polynomials

S. Corteel
Department of Mathematics, University of California Berkeley, Berkeley, CA, USA
e-mail: corteel@berkeley.edu

O. Mandelshtam (✉)
Department of Combinations and Optimization, University of Waterloo, Waterloo, CA, Canada
e-mail: omandels@uwaterloo.ca

L. Williams
Department of Mathematics, Harvard University, Cambridge, MA, USA
e-mail: williams@math.harvard.edu

A. Alekseev et al. (eds.), *Representation Theory, Mathematical Physics, and Integrable Systems*, Progress in Mathematics 340,
https://doi.org/10.1007/978-3-030-78148-4_5

1 Introduction

Introduced around 1970 [MGP68, Spi70], the asymmetric simple exclusion process (ASEP) is a model of interacting particles hopping on a one-dimensional lattice of n sites. It has been extensively studied by researchers in statistical mechanics [DEHP93, USW04], probability [Lig05, Lig75, FM07, BC14], and combinatorics [DS05, Ang06, BE04, CW07, CW11, CMW17]. Recently the ASEP on a lattice with open boundaries has been linked to Koornwinder polynomials [CW15, Can17], and the ASEP on a ring has been linked to Macdonald polynomials [CdGW15]. In particular, it was shown in [CdGW15] that when $q = 1$ and $x_i = 1$ for all i, the Macdonald polynomial P_λ is the *partition function* for the multispecies ASEP on a ring.

In this article we study the two-species asymmetric simple exclusion process (ASEP) on a ring, in which two kinds of particles ("heavy" and "light") hop on a lattice of n sites arranged in a ring. Two adjacent particles, or a particle and a hole, can switch places at a rate t or 1, depending on their relative weights. We introduce some new tableaux on a cylinder called *cylindric rhombic tableaux* (CRT) and use them to give a formula for the stationary distribution of the ASEP—each probability is expressed as a sum over the weights of all CRT of a fixed type, where the weight of each CRT is a series. When λ is a partition in $\{0, 1, 2\}^n$, we then give a formula for the nonsymmetric Macdonald polynomial E_λ and the symmetric Macdonald polynomial P_λ by refining our tableaux formulas for the stationary distribution.

When $t = 0$, the asymmetric simple exclusion process is called the *totally asymmetric simple exclusion process* or TASEP. Ferrari and Martin [FM07] studied the multispecies (k-species) TASEP on a ring and gave combinatorial formulas for the stationary distribution in terms of *multiline queues*; they viewed the k-TASEP on a ring as a projection of a Markov process on multiline queues, which can be viewed as a coupled system of k single species TASEPs. This work was recently generalized by Martin [Mar18] to the case of ASEP (i.e. t is general). Matrix product formulas were found for the probabilities of the TASEP using probabilistic methods in [EFM09] and generalized to the ASEP case in [PEM09a] with an explicit construction in [AAMP12]. From the statistical mechanics side, other formulas for the k-TASEP were found by interpreting the Ferrari-Martin process as a combinatorial R matrix in [KMO15]. The inhomogeneous multispecies TASEP was also studied in [AL14], with a graphical construction that generalized the Ferrari-Martin algorithm for the 2-TASEP, and a general conjecture for the k-TASEP which was proved using a generalized Matrix ansatz in [AM13].

Multiline queues have been used to study many aspects of the ASEP [AL14, AL18]; in this case we give a bijection between CRT and multiline queues, which is related to recent work by the second author [Man17]. Also note that Haglund-Haiman-Loehr have a tableaux formula for both symmetric [HHL05a] and nonsymmetric Macdonald polynomials [HHL05b] using *nonattacking fillings*; we explain the relation between nonattacking fillings and multiline queues in [CMW18] (they are in bijection when the partition has distinct parts; but in general there are more nonattacking fillings than multiline queues).

Remark 1.1 In some sense the results of this paper are subsumed by the results of [CMW18], in that the latter has combinatorial formulas that work for Macdonald polynomials associated with arbitrary partitions (not just $\lambda \in \{0, 1, 2\}^n$). However, since these cylindric rhombic tableaux are significantly different than multiline queues, and our methods of proof use the Matrix Ansatz rather than the Hecke algebra, we thought that this paper might be of independent interest.

2 The ASEP on a Ring

We now define the two-species asymmetric simple exclusion process (ASEP) on a ring.

Definition 2.1 Let k, r, and ℓ be nonnegative integers which sum to n, and let t be a constant such that $0 \leq t \leq 1$. Let States(k, r, ℓ) be the set of all words of length n in $\{0, 1, 2\}^n$ consisting of k 0's, r 1's, and ℓ 2's. We consider indices modulo n; i.e. if $\mu = \mu_1 \dots \mu_n \in \{0, 1, 2\}^n$, then $\mu_{n+1} = \mu_1$. The *two-species asymmetric simple exclusion process* $ASEP(k, r, \ell)$ on a ring is the Markov chain on States(k, r, ℓ) with transition probabilities $P_{\mu,\nu}$ between states $\mu, \nu \in$ States(k, r, ℓ):

- If $\mu = AijB$ and $\nu = AjiB$, where A and B are words in $\{0, 1, 2\}^*$ and $i > j$ are letters in $\{0, 1, 2\}$, then $P_{\mu,\nu} = \frac{t}{n}$ and $P_{\nu,\mu} = \frac{1}{n}$.
- Otherwise, $P_{\mu,\nu} = 0$ for $\nu \neq \mu$ and $P_{\mu,\mu} = 1 - \sum_{\mu \neq \nu} P_{\mu,\nu}$.

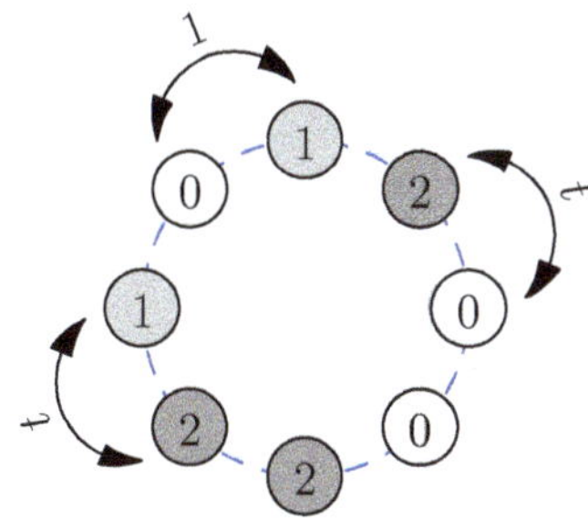

Fig. 1 The two-species ASEP on a lattice with 8 sites. There are three holes (0's), two light particles (1's), and three heavy particles (2's), so we refer to this Markov chain as $frm{-}eASEP(3, 2, 3)$

We think of the 1's and 2's as representing two types of particles ("light" and "heavy") which can occupy the sites; each 0 denotes an empty site (Fig. 1).

The following Matrix Ansatz [DJLS93] (see also [PEM09b]) is a useful tool for computing these probabilities in terms of the trace of a certain matrix product.

Theorem 2.2 (Matrix Ansatz, [DJLS93, Section 8]) *Suppose that A_0, A_1, and A_2 are matrices (typically infinite) that satisfy the following relations:*

$$A_0A_2 = tA_2A_0 + (1-t)(A_0 + A_2), \qquad A_0A_1 = tA_1A_0 + (1-t)A_1,$$
$$A_1A_2 = tA_2A_1 + (1-t)A_1. \tag{1}$$

Given $\mu \in \mathrm{States}(k, r, \ell)$, *we let* $\mathrm{Mat}(\mu)$ *denote the product of matrices obtained from* μ *by substituting* A_0 *for* 0, A_1 *for* 1, *and* A_2 *for* 2. *Then in the* $frm{-}eASEP(k, r, \ell)$, *the steady state probability* $\Pr(\mu)$ *of state* μ *is given by*

$$\Pr(\mu) = \frac{1}{Z_{k,r,\ell}} \operatorname{tr}(\mathrm{Mat}(\mu)),$$

where $Z_{k,r,\ell}$ *is the* partition function *defined by* $[x^k y^r z^\ell] \operatorname{tr}((xA_0 + yA_1 + zA_2)^{k+r+\ell})$.

3 Probabilities for the Two-Species ASEP Using Cylindric Rhombic Tableaux

In this section we define some new combinatorial objects that we call *cylindric rhombic tableaux* (or CRT), and then in Theorem 3.17 we use them to give combinatorial formulas for the steady state probabilities of the ASEP. The proof of our formulas uses the Matrix Ansatz. Our combinatorial objects will be fillings of certain diagrams composed of squares and rhombi. The squares have two horizontal and two vertical edges, while the rhombi have two vertical edges as well as two diagonal edges (of slope 1), see Fig. 2. The fact that states of the $ASEP$ are words in $\{0, 1, 2\}^*$ is related to the fact that there are three types of lines making up the sides of a square or rhombus: vertical, diagonal, and horizontal.

Definition 3.1 A *(generalized) row* in a CRT is a connected strip of squares and rhombi which are adjacent along their vertical edges, see the left diagram in Fig. 2. A *square column* is a connected strip of squares, which are adjacent along their horizontal edges; and a *rhombic column* is a connected strip of rhombi, which are adjacent along their diagonal edges.

Definition 3.2 Given $\mu \in \{0, 1, 2\}^*$, we define $\mu|_{12}$ to be the subword of μ consisting of 1's and 2's. An *μ-strip* is a generalized row composed of adjacent squares and rhombi which is obtained by reading $\mu|_{12}$ and appending a square for each 2 and a rhombus for each 1 to the left of the row; see the left diagram in Fig. 2.

Definition 3.3 Given $\mu \in \{0, 1, 2\}^*$, we define the *μ-path* $P(\mu)$ to be the lattice path consisting of south, southwest, and west steps obtained by reading μ and mapping a 0 to a south step, a 1 to a southwest step, and a 2 to a west step; see the bold path at the right of Fig. 2.

Definition 3.4 Let $\mu \in \mathrm{States}(k, r, \ell)$. Define the *$\mu$-diagram* $\mathcal{H}(\mu)$ to be the shape consisting of k μ-strips stacked on top of each other, together with the path $P(\mu)$ superimposed onto the shape so that it connects the northeast and southwest corners.

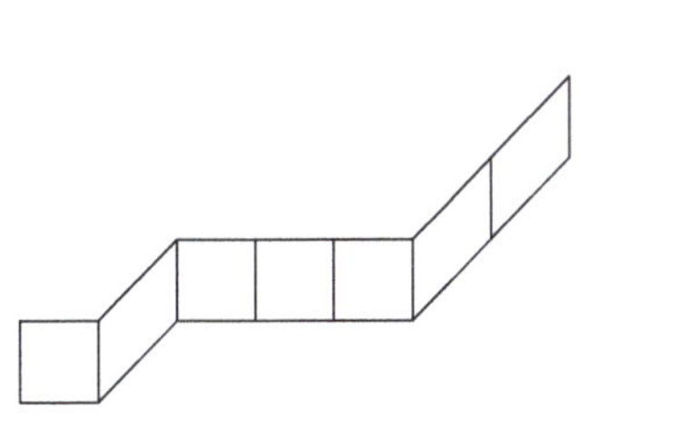

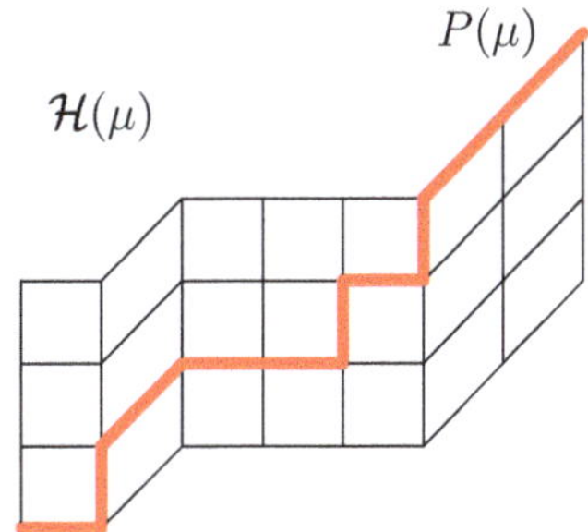

Fig. 2 For $\mu = 1102022102$, the μ-strip is shown at the left, while $\mathcal{H}(\mu)$ is shown at the right, with the path $P(\mu)$ superimposed in bold

(If $k = 0$ then $\mathcal{H}(\mu)$ is defined to be just the path $P(\mu)$.) See the diagram at the right in Fig. 2. We identify the two vertical edges on either end of each row; in this way we view the shape on a cylinder. Thus the rightmost tile is adjacent to the leftmost tile in each row.

Note that for $\mu \in \text{States}(k, r, \ell)$, $\mathcal{H}(\mu)$ has k rows, r rhombic columns, and ℓ square columns. For example, in Fig. 2, $\mathcal{H}(\mu)$ has 3 rows, 3 rhombic columns, and 4 square columns. For $m \in [k]$, let $\text{row}(m)$ denote the m'th row, numbered from bottom to top.

Definition 3.5 (Cylindric Rhombic Tableau and Arrow Ordering) Choose a word $\mu \in \text{States}(k, r, \ell)$. A *cylindric rhombic tableau* (CRT) T of type μ is a placement of up-arrows into the square tiles of the diagram $\mathcal{H}(\mu)$ so that there is *at most* one up-arrow in each column. (We allow columns to be empty.) We denote the set of cylindric rhombic tableaux of type μ by $\text{CRT}(\mu)$.

An *arrow ordering* of T is a labeling of the arrows in each row by the numbers $1, \ldots, i$, where i is the number of arrows in that row. Let $\text{arr}(T)$ denote the total number of arrows in T. We let σ^i denote the labeling of the arrows in $\text{row}(i)$, and let $\{\sigma^i\} = (\sigma^1, \ldots, \sigma^k)$.

For an example, see Fig. 3.

Definition 3.6 We say an arrow in tile s is *pointing at* a tile s' if they are in the same column and s is below s' when we read from bottom to top. We call a square tile *free* if the tile is empty and there is no arrow pointing to it. (Note that freeness does not depend on the path $P(\mu)$.)

We will define the *weight* of each cylindric rhombic tableau. To do so, we need to introduce a few combinatorial statistics.

Definition 3.7 Given a subset I of a finite sequence U where $|U| = m$, we let $\text{Sym}_{I,U}$ denote the set of total orders on I, which we also call *partial permutations*. We write the elements of $\text{Sym}_{I,U}$ as strings of length m, with a $*$ denoting elements not in I.

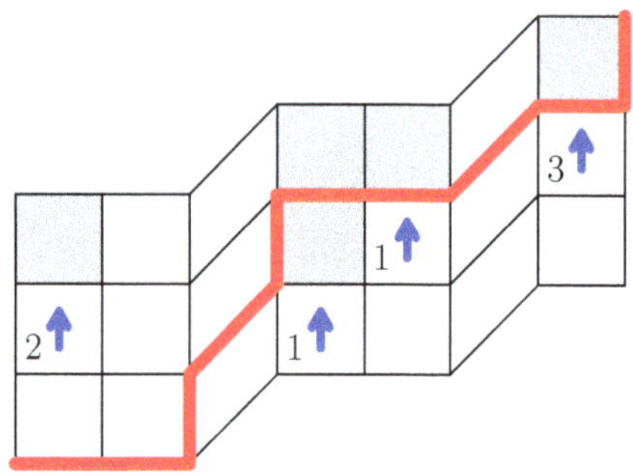

Fig. 3 A cylindric rhombic tableau T of type 0212201022 with a chosen arrow ordering σ. The square tiles that are not free are grey

For example, if $U = \{1, 2, \ldots, 6\}$ and $I = \{1, 2, 4\}$, then there are $|I|!$ total orders on I, which we denote by

$$\mathrm{Sym}_{I,U} = \{1\,2 * 3 * *,\ 1\,3 * 2 * *,\ 2\,1 * 3 * *,\ 2\,3 * 1 * *,\ 3\,1 * 2 * *,\ 3\,2 * 1 * *\}.$$

Definition 3.8 (Disorder) Let $\widetilde{\mathrm{Sym}}_{I,U}$ denote the set of all sequences that can be obtained from the elements of $\mathrm{Sym}_{I,U}$ be inserting a 0 in an arbitrary position. Given $\tilde{\sigma} \in \widetilde{\mathrm{Sym}}_{I,U}$, we define its *disorder* $\mathrm{dis}(\tilde{\sigma})$ inductively as follows:

Reading the entries of $\tilde{\sigma}$ from left to right starting from the 0, we let $\mathrm{dis}_1(\tilde{\sigma})$ equal the number of $*$'s or numbers bigger than 1 we encounter before we reach the 1. We then let $\mathrm{dis}_2(\tilde{\sigma})$ be the number of $*$'s or numbers bigger than 2 we encounter if we travel from the 1 to the 2 from left to right, wrapping around to the beginning of $\tilde{\sigma}$ if necessary. Similarly, $\mathrm{dis}_i(\tilde{\sigma})$ is the number of $*$'s or numbers bigger than i we encounter if we travel from the $i-1$ to the i from left to right, wrapping around if necessary. Finally we define the *disorder* to be $\mathrm{dis}(\tilde{\sigma}) = \mathrm{dis}_1(\tilde{\sigma}) + \mathrm{dis}_2(\tilde{\sigma}) + \cdots + \mathrm{dis}_{|I|}(\tilde{\sigma})$.

If $\tilde{\sigma} = 0\ 2\ 1\ *\ 3\ *\ *$, then $\mathrm{dis}_1(\tilde{\sigma}) = 1$, $\mathrm{dis}_2(\tilde{\sigma}) = 4$, $\mathrm{dis}_3(\tilde{\sigma}) = 1$, and $\mathrm{dis}(\tilde{\sigma}) = 6$.

Remark 3.9 In a recent paper [KM17], a statistic very similar to disorder, called *betrayal*, was introduced on certain colored words in a formula for modified symmetric Macdonald polynomials $\tilde{H}_\lambda$. It would be interesting to understand the connection between the statistics on these different objects.

Definition 3.10 (From an Arrow Ordering to a Partial Permutation) Given a cylindric rhombic tableau T and an arrow ordering $\{\sigma^i\}$, we associate a partial permutation to each row of T as follows. We fix row(i) and read its elements from **right to left**, skipping over non-free square tiles, but recording free square tiles and rhombic tiles by a $*$, and arrows by their label. We also record the vertical line in $P(\mu)$ by a 0. We denote this partial permutation by $\tilde{\sigma}^i$.

For example, the rows of the tableau in Fig. 3 would give rise to the sequences $\tilde{\sigma}^1 = * * * 1 * 0 * *$, $\tilde{\sigma}^2 = 3 * 1\,0 * * 2$, and $\tilde{\sigma}^3 = 0 * * *$ (which are read from left to right).

We now define the *disorder* for arrow orderings of cylindric rhombic tableaux.

Definition 3.11 (Disorder of a CRT with an Arrow Ordering) Given a cylindric rhombic tableau T with k rows and an arrow ordering $\{\sigma^i\} = \{\sigma^1, \ldots, \sigma^k\}$, we define the *disorder* of $(T, \{\sigma^i\})$ to be

$$\mathrm{dis}(T, \{\sigma^i\}) = \sum_{i=1}^{k} \mathrm{dis}(\tilde{\sigma}^i).$$

Example 3.12 Using $(T, \{\sigma^i\})$ from Fig. 3, we compute $\mathrm{dis}(* * * 1 * 0 * *) = 5$, $\mathrm{dis}(3 * 1\, 0 * * 2) = 5 + 2 + 0 = 7$, and $\mathrm{dis}(0 * * *) = 0$, so $\mathrm{dis}(T, \{\sigma^i\}) = 12$.

We let $[i] = [i]_t$ denote the t-analogue of the positive integer i, that is, $[i] = \frac{1-t^i}{1-t} = 1 + t + \ldots + t^{i-1}$. We also let $[i]! = [1][2]\ldots[i]$.

Definition 3.13 The *t-weight* $\mathrm{wt}_t(T)$ of a cylindric rhombic tableau T of type $\mu \in \mathrm{States}(k, r, \ell)$ is computed as follows.

Given an arrow ordering $\{\sigma^i\}$ of the arrows in T, we define

$$\mathrm{wt}_t(T, \{\sigma^i\}) = t^{\mathrm{dis}(T, \{\sigma^i\})}.$$

We then define the *t-weight* of T to be

$$\mathrm{wt}_t(T) = \frac{[r + \ell - \mathrm{arr}(T)]!}{[r + \ell]!} \sum_{\{\sigma^i\}} \mathrm{wt}_t(T, \{\sigma^i\}),$$

where $\{\sigma^i\}$ varies over all possible arrow orderings of T.

Example 3.14 Continuing Example 3.12, with Fig. 3, we have $r = 2$, $\ell = 5$, and $\mathrm{arr}(T) = 4$. Thus $\frac{[r+\ell-\mathrm{arr}(T)]!}{[r+\ell]!} = \frac{1}{[7][6][5][4]}$.

To compute $\mathrm{wt}_t(T)$, we need to consider all possible arrow orderings. Note that:

- There is only one arrow ordering of row(1) and of row(3), so the weight contributed to $\mathrm{wt}_t(T)$ by the possible arrow orderings of row(1) and row(3) is just t^5.
- If we represent the arrows versus rhombic/free tiles in row(2) by x's and $*$'s, respectively, then the content of row(2) can be encoded by the sequence $x * x\, 0 * * x$. We have $\mathrm{dis}(1 * 2\, 0 * * 3) = 6$, $\mathrm{dis}(1 * 3\, 0 * * 2) = 8$, $\mathrm{dis}(2 * 1\, 0 * * 3) = 11$, $\mathrm{dis}(2 * 3\, 0 * * 1) = 3$, $\mathrm{dis}(3 * 1\, 0 * * 2) = 7$, and $\mathrm{dis}(3 * 2\, 0 * * 1) = 6$. Thus the weight contributed by the possible arrow labeling of row(2) (only one of which is shown in Fig. 3) is $t^3+2t^6+t^7+t^8+t^{11}$.

Letting I_i denote the positions of the arrows in row(i) and U_i denote the positions of the arrows and free tiles in row(i), we can write the total weight of this tableau for all possible arrow orderings $\{\sigma^i\} = (\sigma^1, \ldots, \sigma^5)$ as

$$\mathrm{wt}_t(T) = \frac{1}{[7][6][5][4]} \prod_{m=1}^{3} \sum_{\sigma^m \in \mathrm{Sym}_{I_m,U_m}} t^{\mathrm{dis}(\tilde{\sigma}^m)} = \frac{t^5(t^3 + 2t^6 + t^7 + t^8 + t^{11})}{[7][6][5][4]}.$$

Remark 3.15 It is interesting to note that if $I = U$, disorder on $\mathrm{Sym}_{I,U} = \mathrm{Sym}_{|I|}$ is a Mahonian statistic, i.e. it has the same distribution as *inversions*.

Definition 3.16 (Combinatorial Partition Function) Given $\mu \in \mathrm{States}(k, r, \ell)$, we define

$$\mathrm{Tab}_t(\mu) := \sum_{T \in \mathrm{CRT}(\mu)} \mathrm{wt}_t(T).$$

We also define the *combinatorial partition function* of the cylindric rhombic tableaux to be

$$\mathcal{Z}_{k,r,\ell}(t) = \sum_{\mu \in \mathrm{States}(k,r,\ell)} \mathrm{Tab}_t(\mu).$$

We are finally ready to state the first main result of this paper.

Theorem 3.17 *Consider the two-species asymmetric simple exclusion process $ASEP(k, r, \ell)$. Then the steady state probability of being in state μ, where $\mu \in$ $\mathrm{States}(k, r, \ell)$, is*

$$\Pr(\mu) = \frac{\mathrm{Tab}_t(\mu)}{\mathcal{Z}_{k,r,\ell}(t)},$$

where $\mathrm{Tab}_t(\mu)$ *and* $\mathcal{Z}_{k,r,\ell}(t)$ *are as in Definition 3.16.*

Example 3.18 To compute the steady state probability Pr(221100) of the state 221100 of the $ASEP(2, 2, 2)$, we need to sum the weights of all cylindric rhombic tableaux of type 221100, see Fig. 4. We then find that

$$\Pr(221100) = \frac{(t+1)(t^4 + t^3 + 6t^2 + t + 6)}{[4][3]Z_{2,2,2}}.$$

Example 3.19 To compute the steady state probability Pr(201021) of the state 201021 of the $ASEP(2, 2, 2)$, we need to sum the weights of all cylindric rhombic tableaux of type 201021, see Fig. 5. We then find that

$$\Pr(201021) = \frac{(t+1)(t^2 + t + 1)(2t^2 + t + 2)}{[4][3]\mathcal{Z}_{2,2,2}(t)}.$$

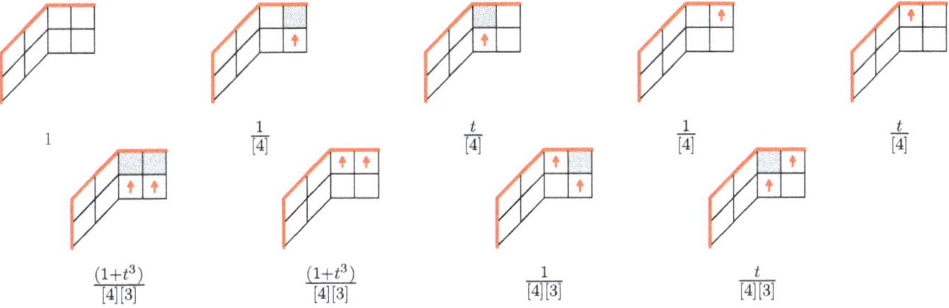

Fig. 4 The cylindric rhombic tableaux of type 221100 and their weights

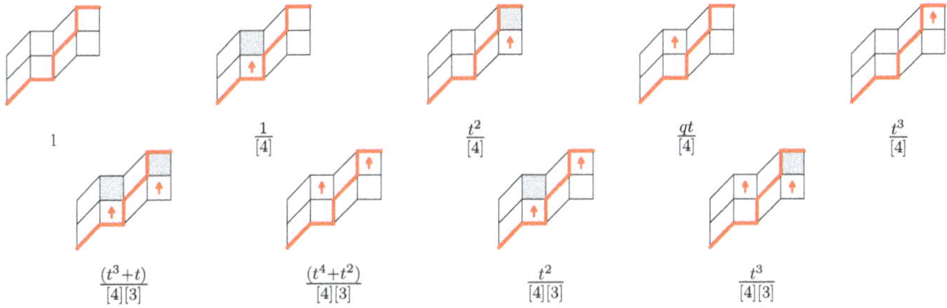

Fig. 5 The cylindric rhombic tableaux of type 201021 and their weights

Remark 3.20 For a given state μ of the $ASEP(k, r, \ell)$, we can multiply $\mathrm{Tab}_t(\mu)$ by the scalar $[r + \ell]!/[r]!$ to obtain polynomials in t with positive coefficients. The resulting polynomials display a "particle-hole symmetry," see Problems 6.11 and 6.12.

4 Formulas for Macdonald Polynomials Using Cylindric Rhombic Tableaux

Symmetric Macdonald polynomials [Mac95] are a family of multivariable orthogonal polynomials indexed by partitions, whose coefficients depend on two parameters q and t. In recent works [CdGW15, CdGW], Cantini, de Gier, and Wheeler gave a link between the multispecies exclusion process on a ring and Macdonald polynomials. In this section we will give a combinatorial formula for Macdonald polynomials in a special case; the proof of our formula uses some results from [CdGW15].

Let $F = \mathbb{Q}(q, t)$ be the field of rational functions in q and t, and let m_λ denote the monomial symmetric polynomial indexed by the partition λ. The Macdonald polynomials are defined as follows.

Definition 4.1 Let $\langle \cdot, \cdot \rangle$ denote the Macdonald inner product on power sum symmetric functions [Mac95, Chapter VI, Equation (1.5)], where $<$ denotes the dominance order on partitions [Mac95, Chapter I, Section 1]. The *Macdonald polynomial* $P_\lambda(x_1, \ldots, x_n; q, t)$ is the unique homogeneous symmetric polynomial in $x_1, \ldots, x_n$ with coefficients in F which satisfies

$$\langle P_\lambda, P_\mu \rangle = 0, \text{ for } \lambda \neq \mu,$$

$$P_\lambda(x_1, \ldots, x_n; q, t) = m_\lambda(x_1, \ldots, x_n) + \sum_{\mu < \lambda} c_{\lambda,\mu}(q, t) m_\mu(x_1, \ldots, x_n),$$

i.e. the coefficients $c_{\lambda,\mu}(q, t)$ of the lower degree terms are determined by the orthogonality conditions.

The *nonsymmetric Macdonald polynomials* E_μ, which are indexed by compositions, were later defined by Opdam [Opd95] and Cherednik [Che95b, Che95a] as joint eigenfunctions of a family of commuting operators in the double affine Hecke algebra, with P_λ can be expressed as a linear combination of the E_μ for μ ranging over all permutations of λ. For more details, see [Mac95].

In this section we enhance our weight function on tableaux, to include an additional parameter q and variables $x_1, \ldots, x_n$ (where $n = k + r + \ell$). We then give our second main result, which is a formula for certain Macdonald polynomials in terms of cylindric rhombic tableaux. In particular, we will give a formula for the nonsymmetric Macdonald polynomial E_λ and a formula for the symmetric Macdonald polynomial P_λ, where λ is any partition in $\{0, 1, 2\}^*$. Note that Haglund, Haiman, and Loehr have given combinatorial formulas for both the nonsymmetric Macdonald polynomials and the symmetric Macdonald polynomials in terms of *nonattacking fillings of composition diagrams* [HHL05a, HHL05b]; it would be interesting to understand how our formulas relate to theirs.

Definition 4.2 We refer to the left and right border of a cylindric tableau (which are identified) as its *vertical boundary*. Given a cylindric rhombic tableau T with path $P(\mu)$ and arrow ordering $\{\sigma^i\}$, for each row row(i), we define $\text{cyc}(T, \sigma^i)$ to be the number of times we cross the vertical boundary if we start at the vertical line in $P(\mu)$ in row i and then travel from right to left (wrapping around if necessary) to the arrow labeled 1, then the arrow labeled 2, and so on. We define the *cycling* of $(T, \{\sigma^i\})$ to be

$$\text{cyc}(T, \{\sigma^i\}) = \sum_i \text{cyc}(T, \sigma^i).$$

Recall that a *recoil* of a (partial) permutation σ is a pair $(j + 1, j)$ such that $\sigma^{-1}(j + 1) < \sigma^{-1}(j)$. In other words, it is a pair of values $(j + 1, j)$ where $j + 1$ appears to the left of j in σ. For example, the partial permutation $3 * 10 * * 2$ has two recoils, $(1, 0)$ and $(3, 2)$. Note that $\text{cyc}(T, \sigma)$ for a given row's arrow ordering $\{\sigma^i\}$ is equal to the number of recoils of $\tilde{\sigma}$ (see Definition 3.10). The cycling statistic defined above will contribute to the power of q associated with each tableau and arrow ordering.

Now given a CRT with path $P(\mu)$, let us number the steps of $P(\mu)$ from northeast to southwest using the numbers $1, 2, \ldots, n$, where $n = |\mu|$; see Fig. 6. This allows us to give every row and column of $\mathcal{H}(\mu)$ a unique integer label, and we will subsequently refer to row i and column j using this labeling.

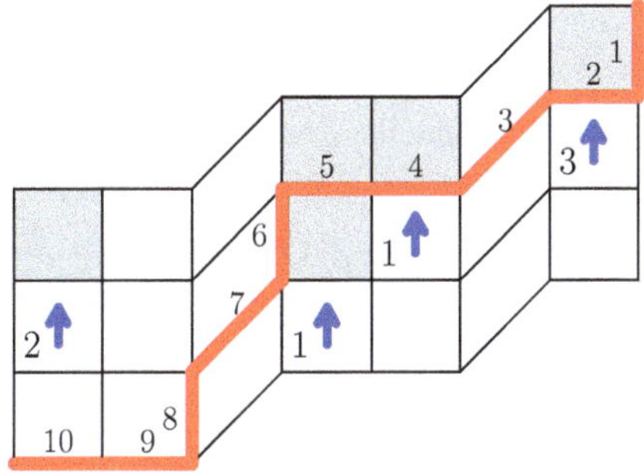

Fig. 6 A cylindric rhombic tableau of type $\mu = 0212201022$, with the steps of $P(\mu)$ labeled from 1 to 10

Definition 4.3 The *x-weight* $\mathrm{wt}_x(T, \{\sigma^i\})$ of a cylindric rhombic tableau T of type $\mu \in \mathrm{States}(k, r, \ell)$ with an arrow ordering $\{\sigma^i\}$ is computed as follows.

For each arrow a in T, if its label given by the arrow ordering is the maximum among all arrows in its row, then we set $\mathrm{wt}(a) = x_i$, where i is the row label of the square containing a. Otherwise, we set $\mathrm{wt}(a) = x_j$, where j is the column label of the square containing a.

For each column c of squares in T, if c contains no arrows, we set $\mathrm{wt}(c) = x_j$, where j is the column label of c. Otherwise we set $\mathrm{wt}(c) = 1$.

We also define $\mathrm{wt}_{12}(\mu) := \prod_{k \in \mathrm{Pos}_{12}(\mu)} x_k$, where $\mathrm{Pos}_{12}(\mu) = \{i \mid \mu_i = 1 \text{ or } 2\}$.

Finally we define the *x-weight* $\mathrm{wt}_x(T, \{\sigma^i\})$ of $(T, \{\sigma^i\})$ to be

$$\mathrm{wt}_x(T, \{\sigma^i\}) = \mathrm{wt}_{\mathrm{Pos}_{12}}(\mu) \prod_a \mathrm{wt}(a) \prod_c \mathrm{wt}(c),$$

where the products are over all arrows a and columns c of squares of T.

Remark 4.4 It follows from the above definition that given a CRT T of type $\mu \in \mathrm{States}(k, r, \ell)$, the x-weight of T is a monomial in $x_1 \ldots x_n$ of degree $r + 2\ell$.

Example 4.5 Figure 6 shows a cylindric rhombic tableau T of type 0212201022 together with an arrow ordering $\{\sigma^i\}$. In this example we have $\prod_a \mathrm{wt}(a) = x_4 x_6 x_8 x_{10}$, $\prod_c \mathrm{wt}(c) = x_9$, $\mathrm{wt}_{\mathrm{Pos}_{12}}(\mu) = x_2 x_3 x_4 x_5 x_7 x_9 x_{10}$. Therefore

$$\mathrm{wt}_x(T, \{\sigma^i\}) = x_2 x_3 x_4^2 x_5 x_6 x_7 x_8 x_9^2 x_{10}^2.$$

Given a positive integer i, let $[i]_{qt} = \frac{1-qt^i}{1-t}$, and let $[i]_{qt}! = [i]_{qt}[i-1]_{qt}\dots[1]_{qt}$. Note that when $q = 1$, $[i]_{qt}$ recovers the quantity $[i] = [i]_t$ we defined earlier. Finally we are ready to define the qtx-weight of a cylindric rhombic tableau.

Definition 4.6 Let T be a cylindric rhombic tableau of type $\mu \in \mathrm{States}(k, r, \ell)$, and and let $\{\sigma^i\}$ be an arrow ordering of its arrows. The *qtx-weight* $\mathrm{wt}_{qtx}(T, \{\sigma^i\})$ is defined to be

$$\mathrm{wt}_{qtx}(T, \{\sigma^i\}) = t^{\mathrm{dis}(T,\{\sigma^i\})} q^{\mathrm{cyc}(T,\{\sigma^i\})}\, \mathrm{wt}_x(T, \{\sigma^i\}). \tag{2}$$

We then define the *qtx-weight* of T to be

$$\mathrm{wt}_{qtx}(T) = \frac{[r+\ell-\mathrm{arr}(T)]_{qt}!}{[r+\ell]_{qt}!} \sum_{\{\sigma^i\}} \mathrm{wt}_{qtx}(T, \{\sigma^i\}),$$

where $\{\sigma^i\}$ varies over all possible arrow orderings of T.

Definition 4.7 Given $\mu \in \mathrm{States}(k, r, \ell)$, we define

$$\mathrm{Tab}_{qtx}(\mu) := \sum_{T \in \mathrm{CRT}(\mu)} \mathrm{wt}_{qtx}(T).$$

Example 4.8 Figure 7 shows the cylindric rhombic tableaux of type 201021. The sum of the weights of all the tableaux is $\mathrm{Tab}_{qtx}(201021) = x_1^2x_3x_5^2x_6 + \frac{(x_1+qt^2x_5)x_1x_3x_4x_5x_6}{[4]_{qt}} + \frac{(tx_1+qt^3x_5)x_1x_2x_3x_5x_6}{[4]_{qt}} + \frac{q(t^3x_1+tx_5)x_1x_3x_4x_5x_6}{[4]_{qt}[3]_{qt}} + \frac{q(t^4x_1+t^2x_5)x_1x_2x_3x_5x_6}{[4]_{qt}[3]_{qt}} + \frac{q(t^2+t^3)x_1x_2x_3x_4x_5x_6}{[4]_{qt}[3]_{qt}}$.

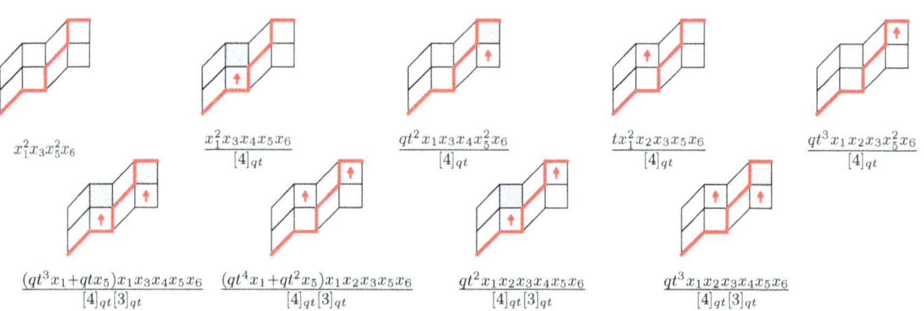

Fig. 7 All cylindric rhombic tableaux of type 201021, along with their weights

The second main result of this paper is the following.

Theorem 4.9 *For any partition $\lambda = (\lambda_1, \dots, \lambda_n)$ of the form $2^\ell 1^r 0^k$, we have that the nonsymmetric Macdonald polynomial E_λ is given by*

$$E_\lambda(x_1,\ldots,x_n;q,t) = \mathrm{Tab}_{qtx}(2^\ell 1^r 0^k). \tag{3}$$

Moreover the symmetric Macdonald polynomial $\mathcal{P}_\lambda$ is given by

$$\mathcal{P}_\lambda(x_1,\ldots,x_n;q,t) = \sum_\mu \mathrm{Tab}_{qtx}(\mu), \tag{4}$$

where the sum runs through all distinct permutations μ of λ.

Example 4.10 Using SageMath [The18], we find that the nonsymmetric Macdonald polynomial $E_{221100} = E_{221100}(x_1,\ldots,x_6;q,t)$ equals

$$E_{221100} = x_1^2x_2^2x_3x_4 + \frac{q(x_1+x_2)(x_5+x_6)x_1x_2x_3x_4}{[3]_{qt}} + \frac{q^2(1+t)x_1x_2x_3x_4x_5x_6}{[3]_{qt}[4]_{qt}}.$$

This agrees with the sum of the weights of the tableaux of type $\mu = 221100$, see Fig. 8.

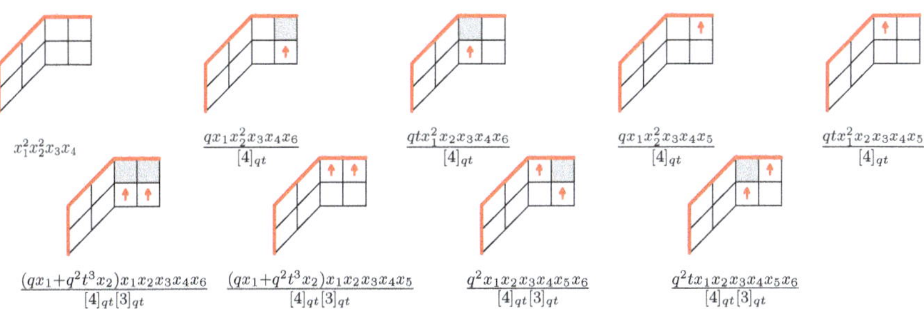

Fig. 8 All cylindric rhombic tableaux of type 221100, along with their weights

5 The Matrix Ansatz and the Results of Cantini-deGier-Wheeler

In order to prove Theorems 3.17 and 4.9, we need to introduce some matrices from [CdGW15], which can be used to compute certain Macdonald polynomials.

Definition 5.1 ([CdGW15, (53)]) We define semi-infinite matrices $A_0(x)$, $A_1(x)$, $A_2(x)$, and S, whose rows and columns are indexed by $\mathbb{Z}_{\geq 0}$.

Let $A_0(x) = (A_0(x)_{i,j})$ be defined by

$$A_0(x)_{i,j} = \begin{cases} 1 \text{ if } i = j \\ x \text{ if } i = j-1 \\ 0 \text{ otherwise.} \end{cases}$$

Let $A_2(x) = (A_2(x)_{i,j})$ be defined by

$$A_2(x)_{i,j} = \begin{cases} x^2 & \text{if } i = j \\ x(1-t^i) & \text{if } i = j+1 \\ 0 & \text{otherwise.} \end{cases}$$

Let $A_1(x) = (A_1(x)_{i,j})$ be a **diagonal** matrix defined by

$$A_1(x)_{i,i} = xt^i.$$

Let $S = (S_{i,j})$ be a **diagonal** matrix defined by

$$S_{i,i} = q^i.$$

Cantini, deGier, and Wheeler [CdGW15] proved that Macdonald polynomials can be computed in terms of the matrices above as follows. (We restrict to the setting where compositions have parts equal to 0, 1, or 2.)

Theorem 5.2 ([CdGW15, (16),(24), Lemma 3]) *Given a composition $\mu = (\mu_1, \dots, \mu_n) \in \{0, 1, 2\}^n$, let λ be the partition obtained from μ by sorting its parts, and set*

$$\Omega_\lambda(q,t) = \prod_{1 \le i < j \le s} (1 - q^{j-i} t^{\lambda'_i - \lambda'_j}), \tag{5}$$

where s is the largest part of λ. We define

$$f_\mu(x_1, \dots, x_n; q, t) = \Omega_\lambda \operatorname{tr}[A_{\mu_1}(x_1) \dots A_{\mu_n}(x_n) S]. \tag{6}$$

For any partition $\lambda = (\lambda_1, \dots, \lambda_n) \in \{0, 1, 2\}^*$, the nonsymmetric Macdonald polynomial E_λ is given by[1]

$$E_\lambda(x_1, \dots, x_n; q, t) = f_\lambda(x_1, \dots, x_n; q, t). \tag{7}$$

Moreover the symmetric Macdonald polynomial $\mathcal{P}_\lambda$ is given by

[1]Note that (7) does not hold if we replace λ with an arbitrary composition. Instead the two families of polynomials (the E's and the f's) are related via a triangular change of basis, see [CdGW15, (23)].

$$\mathcal{P}_\lambda(x_1,\ldots,x_n;q,t) = \sum_{\mu} f_\mu(x_1,\ldots,x_n;q,t), \tag{8}$$

where the sum runs through all distinct permutations μ of λ.[2]

6 The Proofs of Theorems 3.17 and 4.9

In this section we prove our main results. We start by sketching an outline of the proofs.

1. We show that the matrices from Definition 5.1 satisfy certain relations generalizing those of the Matrix Ansatz Eq. (1), see Lemma 6.1.
2. We use the relations from Lemma 6.1 to prove that traces of matrix products $\mathrm{Mat}(\mu)$ in A_0, A_1, A_2 satisfy a certain recurrence, see Theorem 6.5. This recurrence allows us to reduce the computation of traces of matrix products in A_0, A_1, A_2, to the computation of traces of matrix products in A_1 and A_2.
3. We show that the weight generating functions for tableaux $\mathrm{Tab}_{qtx}(\mu)$ satisfy an analogous recurrence, see Theorem 6.10.
4. We verify that the base cases (i.e. corresponding to words in 1's and 2's) agree up to the scalar factor $(1-qt^r)$ where $\mu \in \mathrm{States}(k,r,\ell)$, see Lemmas 6.2 and 6.9. It follows that $\mathrm{Tab}_{qtx}(\mu) = (1-qt^r)\,\mathrm{tr}(\mathrm{Mat}(\mu))$.
5. Since Lemma 6.1 generalizes the relations of Eq. (1), Item 6 and Theorem 2.2 imply that Theorem 3.17 holds.
6. Using Item 6, it follows that $\mathrm{Tab}_{qtx}(\mu)$ agrees with the quantity $f_\mu(x_1,\ldots,x_n)$ from Eq. (6), up to normalization.
7. To verify that Theorem 4.9 is true (i.e. we are getting the actual Macdonald polynomials E_λ and P_λ as opposed to scalar multiples of them), we can check the coefficient of x_λ in $\mathrm{Tab}_{qtx}(\lambda)$ when $\lambda = 2^\ell 1^r 0^k$. There is a unique CRT of type $2^\ell 1^r 0^k$ with x-weight equal to x_λ; this is the CRT with no arrows, so its weight is just x_λ. Similarly, one can check that the coefficient of x_λ in E_λ is also 1, for instance, by using the formula of Haglund-Haiman-Loehr [HHL08] and verifying that there is a unique nonattacking filling with x-weight x_λ.

[2] Note that [CdGW15] uses some unusual conventions for Macdonald polynomials. In particular, the polynomial computed in [CdGW15, page 10] and [CdGW, Section 4] is what we (and SageMath and [HHL08]) would refer to as $E_{(2,2,1,1,0,0)}(x_6,x_5,\ldots,x_1;q,t)$, rather than $E_{(0,0,1,1,2,2)}(x_1,x_2\ldots,x_6;q,t)$. We have stated Theorem 5.2 so as to be consistent with our conventions (and those of SageMath and [HHL08]), so it looks slightly different than the version given in [CdGW15].

6.1 Relations Among the Matrices from Definition 5.1

The following lemma gives some relations among the matrices. Note that (9) and (12) below are special cases of [CdGW15, (25) and (27)]. Meanwhile (10) and (11) appear somewhat related to [CdGW15, (26)] but are not equivalent to it.

Lemma 6.1

$$A_0(x)A_0(y) = A_0(y)A_0(x) \tag{9}$$

$$A_0(x)A_2(y) = tA_2(y)A_0(x) + (1-t)A_2(y) + xy(1-t)A_0(y) \tag{10}$$

$$A_0(x)A_1(y) = tA_1(y)A_0(x) + (1-t)A_1(y) \tag{11}$$

$$A_0(x)S = SA_0(qx) \tag{12}$$

Proof The proof is a series of simple calculations. It suffices to prove (10) and (11). To prove (10), note that

$$(A_0(x)A_2(y))_{i,j} = \begin{cases} y^2 + xy(1-t^{i+1}) & \text{if } i = j \\ xy^2 & \text{if } i = j-1 \\ y(1-t^i) & \text{if } i = j+1 \\ 0 & \text{otherwise} \end{cases}$$

and also

$$(tA_2(y)A_0(x) + (1-t)A_2(y) + xy(1-t)A_0(y))_{i,j} =$$

$$\begin{cases} t(y^2 + xy(1-t^i)) + y^2(1-t) + xy(1-t) & \text{if } i = j \\ txy^2 + xy(1-t)y & \text{if } i = j-1 \\ yt(1-t^i) + (1-t)y(1-t^i) & \text{if } i = j+1 \\ 0 & \text{otherwise.} \end{cases}$$

To prove (11), note that

$$(A_0(x)A_1(y))_{i,j} = \begin{cases} yt^i & \text{if } i = j \\ xt^{i+1} & \text{if } i = j-1 \\ 0 & \text{otherwise} \end{cases}$$

and also

$$(tA_1(y)A_0(x) + (1-t)A_1(y))_{i,j} = \begin{cases} yt^{i+1} + (1-t)yt^i & \text{if } i = j \\ xt^{i+1} & \text{if } i = j-1 \\ 0 & \text{otherwise.} \end{cases}$$

□

6.2 *The Recurrence for Matrix Products*

In this section we give a recurrence for traces of certain matrix products. We start by verifying a base case.

Lemma 6.2 *Let $\mu \in \{1,2\}^n$ be a composition with n parts which has 2's precisely in positions $(h_1, \dots h_\ell)$, and let $W_1 \dots W_n$ be the corresponding matrix product, with A_2's in positions $h_1, \dots, h_\ell$ and A_1's elsewhere. Let $r = n - \ell$. Then*

$$\mathrm{tr}(W_1(x_1)\dots W_n(x_n)S) = \frac{x_1 \dots x_n \prod_{j=1}^{\ell} x_{h_j}}{1 - qt^r}.$$

Proof One can easily check that for $n \geq 1$

$$(W_1(x_1)\dots W_n(x_n)S)_{i,i} = \begin{cases} x_1^2(W_2(x_2)\dots W_n(x_n)S)_{i,i} & \text{if } W_1 = A_2 \\ x_1 t^i (W_2(x_2)\dots W_n(x_n)S)_{i,i} & \text{if } W_1 = A_1 \end{cases}$$

and $S_{i,i} = q^i$. Therefore

$$\begin{aligned}\mathrm{tr}(W_1(x_1)\dots W_n(x_n)S) &= \sum_i (W_1(x_1)\dots W_n(x_n))_{i,i} \\ &= \sum_i q^i t^{ri} x_1 \dots x_n \prod_{j=1}^{\ell} x_{h_j}.\end{aligned}$$

□

To state the recurrence, we need some notation.

Definition 6.3 Given $\mu \in \{0,1,2\}^*$ a word of length n in States(k,r,ℓ), we let Mat(μ) denote the product of matrices obtained from μ by substituting a $A_0(x_i)$ (respectively, $A_1(x_i)$, $A_2(x_i)$) for each 0, 1, or 2 in the ith position of μ, and followed by S. For example, if $\mu = 012201$, then Mat$(\mu) = A_0(x_1)A_1(x_2)A_2(x_3)A_2(x_4)A_0(x_5)A_1(x_6)S$.

For $J \subseteq [n]$ and μ a word of length n, we let Mat$(\mu)|_J$ be the subword of Mat(μ) obtained by restricting to positions J. For example, if $\mu = 012201$, then Mat$(\mu)|_{\{2,4,5\}} = A_1(x_2)A_2(x_4)A_0(x_5)S$.

Definition 6.4 Given $\mu \in \{0,1,2\}^*$, let $\mathrm{Pos}_2(\mu) = \{i \mid \mu_i = 2\}$ and let $\mathrm{Pos}_{12}(\mu) = \{i \mid \mu_i = 1 \text{ or } 2\}$. Given a partial permutation $\sigma \in \mathrm{Sym}_{I,\mathrm{Pos}_{12}(\mu)}$, and a choice of $d \in [n]$ such that $\mu_d = 0$, we define $\tilde{\sigma}$ to be the sequence obtained from σ by inserting a 0 into σ in the position that represents the relative position of μ_d in $\mathrm{Pos}_{12}(\mu)$. For example, set $\mu = 0121021$ and $d = 5$. Then $\mathrm{Pos}_{12}(\mu) = \{2,3,4,6,7\}$. If we choose $I = \{3,6\} \subset \mathrm{Pos}_2(\mu)$, and $\sigma = *2*1* \in \mathrm{Sym}_{I,\mathrm{Pos}_{12}(\mu)}$, then $\tilde{\sigma} = *2*01*$. If $I = \emptyset$, we define $\tilde{\sigma}^{-1}(0) = d$.

Given $I = \{i_1, i_2, \ldots, i_m\} \subset [n]$, we let x_I denote $x_{i_1} \ldots x_{i_m}$.

Theorem 6.5 *Consider the matrices $A_0(x)$, $A_1(x)$, $A_2(x)$ from Sect. 5, and let $\mu \in$ States(k, r, ℓ) with $n = k + r + \ell$, where $r \geq 1$. Suppose $t < 1$. Let $d \in [n]$ be such that $\mu_d = 0$. Then we have that* tr(Mat(μ)) *is equal to*

$$\sum_{I \subseteq \mathrm{Pos}_2(\mu)} \frac{[r+\ell-|I|]_{qt}!}{[r+\ell]_{qt}!}(x_I)^2 x_d \,\mathrm{tr}(\mathrm{Mat}(\mu)|_{[n]\setminus I\cup\{d\}}) \sum_{\sigma \in \mathrm{Sym}_{I,\mathrm{Pos}_{12}(\mu)}} \frac{t^{\mathrm{dis}(\tilde{\sigma})} q^{\mathrm{rec}(\tilde{\sigma})}}{x_{\tilde{\sigma}^{-1}(|I|)}}. \tag{13}$$

Note that by Definition 3.10 we have $x_{\tilde{\sigma}^{-1}(0)} = x_d$, and so $I = \emptyset$ gives the term $\mathrm{tr}(\mathrm{Mat}(\mu)|_{[n]\setminus\{d\}})$ in the above.

Example 6.6 If $\mu = 0212$ (so that $k = 1$, $r = 1$, $\ell = 2$), and $d = 1$, then Theorem 6.5 says that

$$\begin{aligned}
\mathrm{tr}(A_0(x_1)A_2(x_2)A_1(x_3)A_2(x_4)S) &= \mathrm{tr}(A_2(x_2)A_1(x_3)A_2(x_4)S) \\
&+ [r+\ell]_{qt}(\mathrm{tr}(A_1(x_3)A_2(x_4)S)x_1x_2t^{\mathrm{dis}(1**)} + \mathrm{tr}(A_2(x_2)A_1(x_3)S)x_1x_4t^{\mathrm{dis}(**1)}) \\
&+ [r+\ell]_{qt}[r+\ell-1]_{qt}\,\mathrm{tr}(A_1(x_3)S)(x_1x_2t^{\mathrm{dis}(1*2)} + x_1x_4qt^{\mathrm{dis}(2*1)}).
\end{aligned}$$

Proof In the expression Mat(μ), we replace the A_0 in position d by a $\tilde{A}_0$ so that we can keep track of this "marked" A_0. Without loss of generality, using the fact that $\mathrm{tr}(M_1 \ldots M_n) = \mathrm{tr}(M_2 \ldots M_n M_1)$, we can assume that $d = 1$.

Using (1), we will apply the operations below (and only these ones) to tr(Mat(μ)) until $\tilde{A}_0$ is annihilated in every term on the right-hand side.

$$\tilde{A}_0(x)A_2(y) = tA_2(y)\tilde{A}_0(x) + xy(1-t)\tilde{A}_0(y) + (1-t)A_2(y) \tag{14}$$

$$\tilde{A}_0(x)A_1(y) = tA_1(y)\tilde{A}_0(x) + (1-t)A_1(y) \tag{15}$$

$$\tilde{A}_0(x)A_0(y) = A_0(y)\tilde{A}_0(x) \tag{16}$$

$$\tilde{A}_0(x)S = S\tilde{A}_0(qx). \tag{17}$$

More specifically, we think of (14) as giving us the choice of either moving the $\tilde{A}_0$ to the right past an A_2 (picking up a factor of t), or annihilating an A_2 or annihilating the $\tilde{A}_0$ (in each case picking up a factor of $(1-t)$). Similarly, (15) gives us the choice of moving the $\tilde{A}_0$ to the right past an A_1 picking up a factor of t, or annihilating the $\tilde{A}_0$ and picking up a factor of $(1-t)$. (16) allows us to move the $\tilde{A}_0$ to the right past a A_0, and, if we have moved the $\tilde{A}_0$ to the end of the word, (17) allows us to move it back to the beginning.

After applying (14) through (17) as long as possible, we will be left with terms obtained from tr(Mat(μ)) by:

- deleting some subset $I \subseteq \mathrm{Pos}_2(\mu)$ of the A_2's, having chosen a certain order $\sigma \in \mathrm{Sym}_{I,\mathrm{Pos}_{12}(\mu)}$ in which to delete them

- either deleting or not deleting the $\tilde{A}_0$; in the latter case, that means that we wind up commuting the $\tilde{A}_0$ past all the remaining A_1 and A_2 letters of $\mu|_{[n]\setminus I}$ infinitely many times.

We obtain that $\mathrm{tr}(\mathrm{Mat}(\mu))$ is equal to the following:

$$\sum_{I \subseteq \mathrm{Pos}_2(\mu)} (x_I)^2 x_d$$

$$\cdot \Big[\sum_{\sigma \in \mathrm{Sym}_{I,\mathrm{Pos}_{12}(\mu)}} \frac{1}{x_{\tilde{\sigma}^{-1}(|I|)}} \prod_{i=0}^{|I|-1} (1-t)(1+qt^{r+\ell-i}+q^2t^{2(r+\ell-i)}+\cdots)t^{\mathrm{dis}_{i+1}(\tilde{\sigma})}q^{\mathrm{cyc}_{i+1}(\tilde{\sigma})} \Big]$$

$$\cdot \Big[\sum_{m=1}^{r+\ell-|I|} (1-t)(1+t^{r+\ell-|I|}+t^{2(r+\ell-|I|)}+\cdots)t^{m-1}\,\mathrm{tr}(\mathrm{Mat}(\mu)|_{[n]\setminus I\cup\{d\}})$$

$$+ \lim_{j\to\infty} t^{j(r+\ell-|I|)}\,\mathrm{tr}(\mathrm{Mat}(\mu)|_{[n]\setminus I})(x_d \to q^j x_d) \Big]. \tag{18}$$

We use the notation $\delta_{(*)}$ to represent the Kronecker delta, which returns 1 if $(*)$ is true and 0 otherwise, and $\mathrm{cyc}_{i+1}(\tilde{\sigma}) = \delta_{(\tilde{\sigma}^{-1}(i+1)<\tilde{\sigma}^{-1}(i))}$.

Let us explain the factor in (18): this is the factor we pick up in deleting the chosen (possibly empty) set $I \subseteq \mathrm{Pos}_2(\mu)$ of A_2's. If $I = \emptyset$, we simply get 1. Otherwise, suppose we delete $|I| = s > 0$ A_2's, in the order and positions specified by $\sigma \in \mathrm{Sym}_{I,\mathrm{Pos}_{12}(\mu)}$. Let $u_i = \tilde{\sigma}^{-1}(i)$ be the label of the i'th A_2 to be deleted. We start by deleting the A_2 with label u_1: to do so, we first commute the $\tilde{A}_0$ past all letters of the word μ a total of j_1 times (where $j_1 \geq 0$), thus picking up a factor of $t^{j_1(r+\ell)}$ with $\tilde{A}_0(x)$ becoming $\tilde{A}_0(q^{j_1}x)$; we then apply some number $m < r+\ell$ of commutations to bring the $\tilde{A}_0$ adjacent to this A_2. Note that $m = \mathrm{dis}_1(\tilde{\sigma})$, and so we pick up a factor of $t^{\mathrm{dis}_1(\tilde{\sigma})}$ with $\tilde{A}_0(x)$ becoming $\tilde{A}_0(q^{\mathrm{cyc}_1(\tilde{\sigma})}x)$. We then delete the A_2 with label u_1, picking up a factor of $x_{u_1}x_d q^{j_1}q^{\mathrm{cyc}_1(\tilde{\sigma})}(1-t)$. Similarly, to delete the A_2 with label u_2, we commute the $\tilde{A}_0$ past all remaining letters of the word μ a total of $j_2 \geq 0$ times, picking up $t^{j_2(r+\ell-1)}$ and with $\tilde{A}_0(x)$ becoming $\tilde{A}_0(q^{j_2}x)$, then apply $\mathrm{dis}_2(\tilde{\sigma})$ commutations to move the $\tilde{A}_0$ from position u_1 to u_2 with $\tilde{A}_0(x)$ becoming $\tilde{A}_0(q^{\mathrm{cyc}_2(\tilde{\sigma})}x)$. We then delete that A_2, picking up a factor of $x_{u_2}x_{u_1}q^{j_2}q^{\mathrm{cyc}_2(\tilde{\sigma})}(1-t)$. We continue in this fashion until the last A_2, which has label u_s: when this is deleted, we pick up a factor of $x_{u_s}x_{u_{s-1}}q^{j_s}q^{\mathrm{cyc}_s(\tilde{\sigma})}(1-t)$. Thus the overall contribution of the x's is $\frac{(x_I)^2 x_d}{x_{\tilde{\sigma}^{-1}(|I|)}}$, which is how we obtain the factor in (18).

After annihilating the chosen A_2's, we then either delete the $\tilde{A}_0$, or we do not. If we do delete the $\tilde{A}_0$, we obtain the sum in the first line of (18). Again we possibly cycle the $\tilde{A}_0$ through all the remaining $r+\ell-|I|$ A_1 and A_2 letters of μ j times, then commute it past m more letters, where $0 \leq m \leq r+\ell-|I|-1$. Note that here, even though the $\tilde{A}_0(x)$ does become $\tilde{A}_0(q^j x)$, there are no further components arising from the term $xy(1-t)\tilde{A}_0(y)$ in (15); thus the variable that the $\tilde{A}_0$ carries never enters into the equation, and so no q's are collected in the final expression.

If we do not ultimately delete the $\tilde{A}_0$, then we necessarily cycle the $\tilde{A}_0$ around the remaining letters of μ indefinitely, resulting in the term $\lim_{j\to\infty} t^{j(r+\ell-s)}\,\mathrm{tr}(\mathrm{Mat}(\mu)|_{[n]\setminus I})(x_d \to q^j x_d)$, where the notation $x_d \to q^j x_d$ means that we substitute $q^j x_d$ for x_d in $\tilde{A}_0$; this is the second line of (18).

Now we can simplify the terms within the sums obtained above. Since $t < 1$, the terms involving the limit go to 0. In the top line of (18), we have that $\sum_{m=1}^{r+\ell-s}(1-t)(1+t^{r+\ell-s}+\ldots)t^{m-1}$ is equal to 1. To simplify the bottom line of (18), we recall that $\sum_{i=0}^{|I|-1}\mathrm{dis}_{i+1}(\tilde{\sigma}) = \mathrm{dis}(\tilde{\sigma})$ and that by definition $\sum_{i=0}^{|I|-1}\mathrm{cyc}_{i+1}(\tilde{\sigma}) = \mathrm{rec}(\tilde{\sigma})$.

We thus obtain the desired identity (13). □

Observe that at $q = x_1 = \cdots = x_n = 1$, the recurrence (13) becomes

$$\mathrm{tr}(\mathrm{Mat}(\mu)) = \sum_{I\subseteq \mathrm{Pos}_2(\mu)} \frac{[r+\ell-|I|]!}{[r+\ell]!}\,\mathrm{tr}(\mathrm{Mat}(\mu)|_{[n]\setminus I\cup\{d\}}) \sum_{\sigma\in \mathrm{Sym}_{I,\mathrm{Pos}_{12}(\mu)}} t^{\mathrm{dis}(\tilde{\sigma})}. \tag{19}$$

Remark 6.7 The case $k = 0$ is trivial, since in this case the stationary distribution of the ASEP is uniform. From Lemma 6.2 we obtain $\mathrm{tr}(\mathrm{Mat}(\mu)) = \frac{1}{1-qt^n} x_{\mathrm{Pos}_{12}(\mu)} x_{\mathrm{Pos}_2(\mu)}$. We also get the uniform distribution at $t = 1$.

Remark 6.8 We will not actually need the full generality of Theorem 6.5; it is enough to know Theorem 6.5 in the case that $d \in [n]$ is maximal such that $\mu_d = 0$.

6.3 The Recurrence for Weight Generating Functions of Tableaux

We again start by giving a base case.

Lemma 6.9 *Let $\mu \in$ States$(0, r, \ell)$ with $r + \ell = n$, i.e. it is a composition with parts equal to* 1 *or* 2. *Let $h_1, \ldots, h_\ell$ be the positions of the* 2*'s. Then we have*

$$\mathrm{Tab}_{qtx}(\mu) = x_1 \ldots x_n \prod_{j=1}^{\ell} x_{h_j}.$$

Proof Lemma 6.9 follows directly from the definitions and, in particular, Definition 4.6. □

In what follows, if $\mu \in$ States(k, r, ℓ) with $k+r+\ell = n$, and if $J \subset [n]$, then we let $\mathrm{Tab}_{qtx}(\mu|_J)$ be the weight generating function for cylindric rhombic tableaux of type $\mu|_J$ obtained if we label the path $P(\mu)$ in each tableau using the numbers J. (So that the x-weight of each tableau is a monomial in x_j's for $j \in J$.)

Theorem 6.10 *Let $\mu \in \text{States}(k, r, \ell)$ with $n = k + r + \ell$. Let $d \in [n]$ be maximal such that $\mu_d = 0$. Then $\text{Tab}_{qtx}(\mu)$ equals*

$$x_d \sum_{s=0}^{\ell} \frac{[r+\ell-s]!}{[r+\ell]!} \sum_{\substack{I \subseteq \text{Pos}_2(\mu) \\ |I|=s}} x_I^2 \, \text{Tab}_{qtx}(\mu|_{[n]\setminus I \cup \{d\}}) \sum_{\sigma^1 \in \text{Sym}_{I,\text{Pos}_{12}(\mu)}} \frac{t^{\text{dis}(\tilde{\sigma}^1)} q^{\text{rec}(\tilde{\sigma}^1)}}{x_{(\tilde{\sigma}^1)^{-1}(s)}}.$$

Proof We prove that Definition 4.6 satisfies the recurrence using a bijective proof. Since $d \in [n]$ is maximal such that $\mu_d = 0$, removing μ_d from $\mu_1 \dots \mu_n$ corresponds to deleting the bottom row row(1) from any $T \in \text{CRT}(\mu)$.

Choose some $T \in \text{CRT}(\mu)$ with k rows. Suppose that row(1) contains s_1 up-arrows in columns with positions corresponding to $I \subseteq \text{Pos}_2(\mu)$. Define $\hat{T}$ to be the tableau with $k-1$ rows obtained by removing row(1) as well as the s_1 columns corresponding to I, and then gluing together the remaining boxes in the obvious way.

If $s_1 = 0$, $\hat{T}$ is simply the same tableau with row labeled d removed, whose weight is $\frac{x_d}{x_{(\tilde{\sigma}^1)^{-1}(0)}} \text{Tab}_{qtx}(\mu|_{[n]\setminus\{d\}})$, which is simply $\text{Tab}_{qtx}(\mu|_{[n]\setminus\{d\}})$.

When $s_1 \geq 1$, clearly $\hat{T} \in \text{CRT}(\mu|_{[n]\setminus I \cup \{d\}})$, and in fact the set of $T \in \text{CRT}(\mu)$ with s_1 up-arrows in locations I in row(1) maps bijectively to the set $\text{CRT}(\mu|_{[n]\setminus I \cup \{d\}})$. Moreover if we choose an arrow ordering $\{\sigma^i\} = \{\sigma^1, \dots, \sigma^k\}$ for T, then this induces an arrow ordering $\{\hat{\sigma}^i\} = \{\sigma^2, \dots, \sigma^k\}$ for $\hat{T}$, with the property that $\text{dis}(\hat{\sigma}^i) = \text{dis}(\sigma^{i+1})$ and $\text{cyc}(\hat{\sigma}^i) = \text{cyc}(\sigma^{i+1})$ for $i = 1, \dots, k-1$. Therefore $\sum_T \text{wt}_{qtx}(T)$, where the sum is over $T \in \text{CRT}(\mu)$ with s_1 arrows in locations I of row(1), is equal to $x_d \, \text{Tab}_{qtx}(\mu|_{[n]\setminus I \cup \{d\}})$ times the contribution of weights from all possible orderings $\sigma^1 \in \text{Sym}_{I,\text{Pos}_{12}(\mu)}$. By Definition 4.6, the possible choices of arrow orderings contribute

$$\sum_{\sigma^1 \in \text{Sym}_{I,\text{Pos}_{12}(\mu)}} t^{\text{dis}(\tilde{\sigma}^1)} q^{\text{rec}(\tilde{\sigma}^1)} \frac{x_I^2}{x_{\tilde{\sigma}^{-1}(s_1)}}$$

to the weight. □

6.4 Symmetries of the Tableaux

The following statements can be proved using Theorem 3.17 and the symmetries of the Markov chain $ASEP(k, r, \ell)$. However, it is not obvious how to give a combinatorial proof using the tableaux.

Problem 6.11 For any $\mu \in \text{States}(k, r, \ell)$, define $\mu^{\odot} \in \text{States}(\ell, r, k)$ to be the *complement* of μ, which is the word obtained by replacing each 0 by a 2 and vice versa. For example, for $\mu = 2210$, $\mu^{\odot} = 0012$. Give a combinatorial proof that

$$\text{Tab}_t(\mu) = t^{\ell(\ell+2r-1)/2}\,\text{Tab}_{1/t}(\mu^{\odot}).$$

Note that $\frac{\ell(\ell+2r-1)}{2}$ is the degree of $\text{Tab}_t(\mu)$.

Problem 6.12 For any $\mu = \mu_1 \dots \mu_n \in \text{States}(k, r, \ell)$, define $\mu^T := \mu_n^{\odot} \dots \mu_1^{\odot}$ to be the "particle-hole symmetry" word. For example, for $\mu = 2210$, we have $\mu^T = 2100$. Give a combinatorial proof that

$$\text{Tab}_t(\mu)[r+k]_t! = \text{Tab}_t(\mu^T)[r+\ell]_t!$$

We next compute $\text{Tab}_t(\mu)$ when $t = 1$.

Corollary 6.13 *Let $\mu \in \text{States}(k, r, \ell)$ with $k + r + \ell = n$. Then*

$$\text{Tab}_t(\mu)(t=1) = \binom{n}{\ell}\frac{\ell!r!}{(r+\ell)!}.$$

Proof For $\mu \in \text{States}(k, r, \ell)$, the *$\mu$-diagram* $\mathcal{H}(\mu)$ has k rows. Each row in $\mathcal{H}(\mu)$ contains ℓ square tiles, each of which is either empty or contains an arrow. Since we are setting $t = 1$, we do not need to compute the disorder of any arrow placements, we simply need to determine how many arrow placements and arrow orderings there are.

Suppose we are selecting an arrow placement for $\mathcal{H}(\mu)$. We first choose the total number s of arrows to place in the square tiles, where $0 \leq s \leq \ell$; there are $\binom{\ell}{s}$ choices for the s columns that will contain these arrows. Let $s_1 + \dots + s_k = s$ be a composition representing the number of arrows placed in the rows $\text{row}(1), \dots, \text{row}(k)$. Once $s_1, \dots, s_k$ are chosen (in $\binom{s+k-1}{k-1}$ ways), there are $\binom{s}{s_1,\dots,s_k}$ ways to select which arrows go in which rows, and $s_i!$ possible orderings of the arrows in $\text{row}(i)$, for each $i \in \{1, \dots, k\}$. Finally, given that $\text{arr}(T) = s$, we have that the factor $\frac{[r+\ell-\text{arr}(T)]!}{[r+\ell]!}$ in Definition 3.13 is equal to $\frac{(r+\ell-s)!}{(r+\ell)!}$. Thus we obtain

$$\begin{aligned}
\text{Tab}_t(\mu)(t=1) &= \sum_{0\leq s\leq \ell}\ \sum_{s_1+\dots+s_k=s}\binom{\ell}{s}\binom{s}{s_1,\dots,s_k}s_1!\dots s_k!\frac{(r+\ell-s)!}{(r+\ell)!}\\
&= \sum_{0\leq s\leq \ell}\binom{s+k-1}{k-1}\binom{\ell}{s}\frac{s!(r+\ell-s)!}{(r+\ell)!}\\
&= \frac{r!\ell!}{(r+\ell)!}\sum_{0\leq s\leq \ell}\binom{s+k-1}{k-1}\binom{r+\ell-s}{r}\\
&= \frac{r!\ell!}{(r+\ell)!}\binom{k+r+\ell}{\ell} = \frac{r!\ell!}{(r+\ell)!}\binom{n}{\ell}.
\end{aligned}$$

□

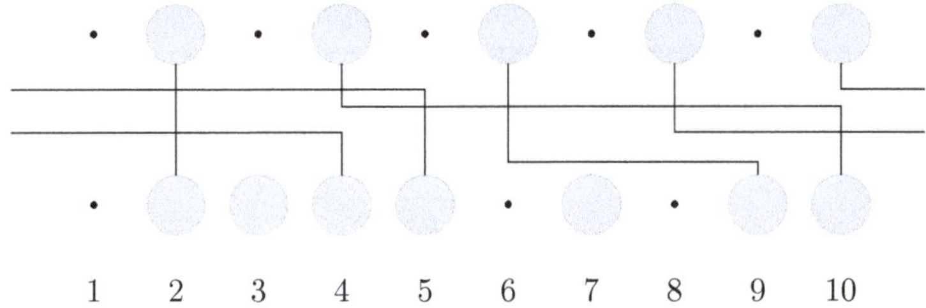

Fig. 9 An example of a two-line queue of type 0212201022

7 A Bijection from Cylindric Rhombic Tableaux to Two-Line Queues

In this section we present a bijection between cylindric rhombic tableaux and two-line queues which are equivalent to the multiline queues of Martin [Mar18].

Definition 7.1 A two-line queue of size n is a two rowed array Q on a cylinder where the entries can be { ,•} where means that there is a ball at the site and • means the site is empty. There exists a partial matching between the balls in the top row and the bottom row such that:

- All the balls of the top row are matched
- A ball in the bottom row is allowed to not be matched only if there is no ball in the same column in the top row.

For each matching of a top row ball from column i to a bottom row ball in column j, we draw an edge from left to right, wrapping around if necessary. See Fig. 9. In this queue, the top row balls in columns 2, 3, 6, 8, 10 are matched with the bottom row balls in columns 2, 10, 9, 4, 5, respectively.

The *type* of a two-line queue is a word in $\{0, 1, 2\}^*$ which is read off the bottom row from left to right: an empty site is read as a 0, an unmatched ball is read as a 1, and a matched ball is read as a 2. The type of the queue in Fig. 9 is 0212201022.

To each queue, we associate a weight in $x_1, \ldots, x_n, q, t$. Each ball in column i has weight x_i. We also give a weight to the edges that connect balls in different columns. We explore the queue with a simple algorithm. We call a ball *restricted* if it has another ball besides itself in its column.

1. At initialization, all bottom row balls are considered *free*, and all balls are unmatched.
2. Let i be the column containing the rightmost unrestricted top row ball that has not yet been matched. If there are no remaining unmatched unrestricted top row balls, we are done.
3. To compute the weight of a matching from the top row ball in column i, let free be the number of free bottom row balls remaining at this point. Suppose the ball in column i is matched to the bottom row ball in column j. Then skipped is the

number of free bottom row balls that are skipped over to get from column i to column j while moving to the right, wrapping around if necessary. The weight of that matching is

$$\frac{q^{\delta(i>j)} t^{\text{skipped}}}{[\text{free}]_{qt}},$$

where δ denotes the Kronecker delta. In other words,

- if $i < j$, the weight of that matching is $t^{\text{skipped}}/[\text{free}]_{qt}$ where skipped is the number of free bottom row balls in columns u such that $i \leq u < j$.
- if $i > j$, the weight of that matching is $qt^{\text{skipped}}/[\text{free}]_{qt}$ where skipped is the number of free bottom row balls in columns u such that $u \geq i$ or $u < j$.

The bottom row ball in column j that has been matched is now no longer free.

4. If there is a top row ball in column j, continue to Step 3, setting $i = j$. Otherwise, go to Step 2.

Now the *weight of the two-line queue* is the product of the weight of the edges times the weight of the balls.

For example, let us compute the weight of the queue in Fig. 9. We start with the top row ball in column 8. It is matched with the bottom row ball in column 4, by cycling around and skipping 4 free balls out of a total of 7 free balls. The weight of this edge is thus $qt^4/[7]_{qt}$.

Since there is a top row ball in column 4, that is the next one we match. This ball is matched with the bottom row ball in column 10, by skipping 3 free balls out of a total of 6 remaining free balls. The weight of this edge is thus $t^3/[6]_{qt}$.

We continue with the top row ball in column 10. It is matched with the bottom row ball in column 5 by cycling around and skipping 2 free balls out of a total of 5 remaining free balls. The weight of this edge is thus $qt^2/[5]_{qt}$.

The next ball to be matched is the rightmost unrestricted unmatched top row ball, which is in column 6: this one is matched to the bottom row ball in column 9 by skipping 1 free ball out of a total of 4 remaining free balls. The weight of this edge is $t/[4]_{qt}$.

There are no remaining unmatched unrestricted top row balls, so therefore the weight of this two-line queue is

$$\frac{q^2 t^{10} x_2^2 x_3 x_4^2 x_5 x_6 x_7 x_8 x_9 x_{10}^2}{[7]_{qt}[6]_{qt}[5]_{qt}[4]_{qt}}.$$

Remark 7.2 There is some recent work [AGS18] that considers the usual multiline queues (at $t = 0$) with $\{x_i\}$ weights as we have defined them here; in that paper, the authors call them *multiline queues with spectral parameters*.

We will exhibit a construction that will prove:

Theorem 7.3 *There exists a bijection between CRTs of type* $\mu \in \text{States}(k, r, \ell)$ *and two-line queues of type* μ*. This bijection is weight preserving.*

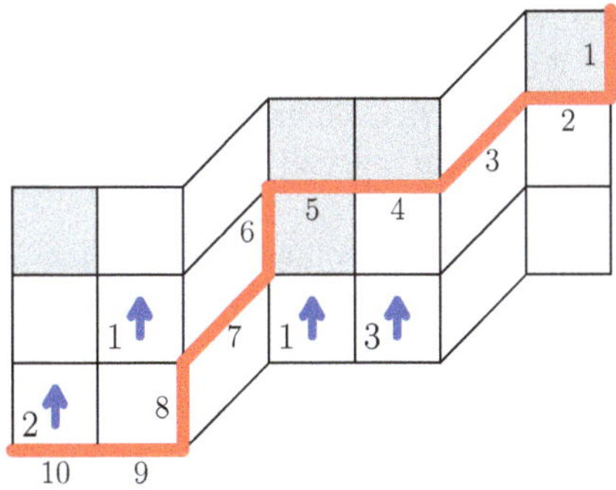

Fig. 10 A CRT of type $\mu = 0212201022$ that corresponds to the two-line queue of Figure 9

Proof We present the bijection. Given a CRT T of type $\mu \in \text{States}(k, r, \ell)$, we label the rows and columns of T by the label of the corresponding edge in $P(\mu)$. We build a queue Q from T. We first fill the two rows of the queue with the following rules. For each i, the site in column i of the bottom row is empty if and only if $\mu_i = 0$; otherwise it contains a ball. The site in column i of the top row is empty if and only if one of the following occurs:

- $\mu_i = 1$,
- $\mu_i = 0$ and the row i of T is empty, or
- $\mu_i = 2$ and the column i of T contains an arrow which has the largest label in its row.

We now explain how to match the balls in Q.

- A restricted top row ball in column i is matched to the bottom row ball in column i if and only if the column i of T is empty.
- An unrestricted top row ball in column i is matched to the bottom row ball in column j if and only if there exists in T an arrow labeled 1 in row i and column j.
- A restricted top row ball in column i is matched to the bottom row ball in column j where $i \neq j$ if and only if there exists in T an arrow labeled with some $k > 1$ in column j, and in the same row there is an arrow labeled $k - 1$ in column i.

It is a simple exercise to check that the weight of the Q is equal to the weight of T, and the construction is bijective. □

Example of the Bijection We start with the CRT T in Fig. 10, where we have labeled the edges of $P(\mu)$ from 1 to 10 to correspond with the labels of the columns of the two-line queue Q. We first fill the bottom row of Q by putting balls in all sites except for those in columns 1, 6, and 8, since those correspond to the vertical edge labels in T.

Now we fill the top row of Q. We put an empty site in column 1, as row 1 of T is empty. The edges 3 and 6 or $P(\mu)$ are diagonal, so we put an empty site in columns 3 and 6 of Q. Finally the columns 5 and 9 of T contain an arrow with the largest label in its corresponding row, and therefore we put an empty site in columns 5 and 9 of Q. The rest of the sites are filled with balls.

We proceed to match balls between the two rows of Q, starting with the empty columns of T. Column 2 of T is empty, so the top row ball in column 2 is matched to the bottom row ball in column 2. Now we look at the non-empty rows of T from bottom to top.

In row 8 and column 4 of T, there is an arrow labeled 1: therefore the top row ball in column 8 or Q is matched with the bottom row ball in column 4.

In row 8 and column 10 of T, there is an arrow labeled 2: therefore the top row ball in column 4 of Q is matched to the bottom row ball in column 10.

In row 8 and column 5 of T, there is an arrow labeled 3: therefore the top row ball in column 10 of Q is matched to the bottom row ball in column 5.

We now look at row 6 of T. In column 9, there is an arrow labeled 1: therefore the top row ball in column 6 of Q is matched to the bottom row ball in column 9.

Having recorded all the arrows in T, we get the two-line queue of Fig. 9.

Acknowledgments SC was partially funded by the "Combinatoire à Paris" projet Emergences 2013–2017 and by "ALEA Sorbonne" projet IDEX USPC. LW was partially supported by NSF grant DMS-1600447.

References

[AAMP12] Chikashi Arita, Arvind Ayyer, Kirone Mallick, and Sylvain Prolhac. Generalized matrix ansatz in the multispecies exclusion process—the partially asymmetric case. *J. Phys. A*, 45(19):195001, 16, 2012.

[AGS18] Erik Aas, Darij Grinberg, and Travis Scrimshaw. Multiline queues with spectral parameters. 2018. arXiv:1810.08157.

[AL14] Arvind Ayyer and Svante Linusson. An inhomogeneous multispecies TASEP on a ring. *Adv. in Appl. Math.*, 57:21–43, 2014.

[AL18] Erik Aas and Svante Linusson. Continuous multi-line queues and TASEP. *Ann. Inst. Henri Poincaré D*, 5(1):127–152, 2018.

[AM13] Chikashi Arita and Kirone Mallick. Matrix product solution of an inhomogeneous multi-species TASEP. *J. Phys. A*, 46(8):085002, 11, 2013.

[Ang06] Omer Angel. The stationary measure of a 2-type totally asymmetric exclusion process. *J. Combin. Theory Ser. A*, 113(4):625–635, 2006.

[BC14] Alexei Borodin and Ivan Corwin. Macdonald processes. *Probab. Theory Related Fields*, 158(1–2):225–400, 2014.

[BE04] R. Brak and J. W. Essam. Asymmetric exclusion model and weighted lattice paths. *J. Phys. A*, 37(14):4183–4217, 2004.

[Can17] Luigi Cantini. Asymmetric simple exclusion process with open boundaries and Koornwinder polynomials. *Ann. Henri Poincaré*, 18(4):1121–1151, 2017.

[CdGW] Luigi Cantini, Jan de Gier, and Michael Wheeler. Matrix product and sum rule for Macdonald polynomials. FPSAC abstract.

[CdGW15] Luigi Cantini, Jan de Gier, and Michael Wheeler. Matrix product formula for Macdonald polynomials. *J. Phys. A*, 48(38):384001, 25, 2015.

[Che95a] Ivan Cherednik. Double affine Hecke algebras and Macdonald's conjectures. *Ann. of Math. (2)*, 141(1):191–216, 1995.

[Che95b] Ivan Cherednik. Nonsymmetric Macdonald polynomials. *Internat. Math. Res. Notices*, (10):483–515, 1995.

[CMW17] Sylvie Corteel, Olya Mandelshtam, and Lauren Williams. Combinatorics of the two-species ASEP and Koornwinder moments. *Adv. Math.*, 321:160–204, 2017.

[CMW18] Sylvie Corteel, Olya Mandelshtam, and Lauren Williams. From multiline queues to Macdonald polynomials via the exclusion process. 2018.

[CW07] Sylvie Corteel and Lauren K. Williams. Tableaux combinatorics for the asymmetric exclusion process. *Adv. in Appl. Math.*, 39(3):293–310, 2007.

[CW11] Sylvie Corteel and Lauren K. Williams. Tableaux combinatorics for the asymmetric exclusion process and Askey-Wilson polynomials. *Duke Math. J.*, 159(3):385–415, 2011.

[CW15] Sylvie Corteel and Lauren Williams. Macdonald-Koornwinder moments and the two-species exclusion process. *Selecta Math.*, 24:2275–2317, 2018.

[DEHP93] B. Derrida, M. R. Evans, V. Hakim, and V. Pasquier. Exact solution of a 1D asymmetric exclusion model using a matrix formulation. *J. Phys. A*, 26(7):1493–1517, 1993.

[DJLS93] B. Derrida, S. A. Janowsky, J. L. Lebowitz, and E. R. Speer. Exact solution of the totally asymmetric simple exclusion process: shock profiles. *J. Statist. Phys.*, 73(5–6):813–842, 1993.

[DS05] Enrica Duchi and Gilles Schaeffer. A combinatorial approach to jumping particles. *J. Combin. Theory Ser. A*, 110(1):1–29, 2005.

[EFM09] Martin R. Evans, Pablo A. Ferrari, and Kirone Mallick. Matrix representation of the stationary measure for the multispecies TASEP. *J. Stat. Phys.*, 135(2):217–239, 2009.

[FM07] Pablo A. Ferrari and James B. Martin. Stationary distributions of multi-type totally asymmetric exclusion processes. *Ann. Probab.*, 35(3):807–832, 2007.

[HHL05a] J. Haglund, M. Haiman, and N. Loehr. A combinatorial formula for Macdonald polynomials. *J. Amer. Math. Soc.*, 18(3):735–761, 2005.

[HHL05b] J. Haglund, M. Haiman, and N. Loehr. Combinatorial theory of Macdonald polynomials. I. Proof of Haglund's formula. *Proc. Natl. Acad. Sci. USA*, 102(8):2690–2696, 2005.

[HHL08] J. Haglund, M. Haiman, and N. Loehr. A combinatorial formula for nonsymmetric Macdonald polynomials. *Amer. J. Math.*, 130(2):359–383, 2008.

[KM17] Ryan Kaliszewski and Jennifer Morse. Colorful combinatorics and Macdonald polynomials. 2017. arXiv:1710.00801.

[KMO15] Atsuo Kuniba, Shouya Maruyama, and Masato Okado. Multispecies TASEP and combinatorial *R*. *J. Phys. A*, 48(34):34FT02, 19, 2015.

[Lig75] Thomas M. Liggett. Ergodic theorems for the asymmetric simple exclusion process. *Trans. Amer. Math. Soc.*, 213:237–261, 1975.

[Lig05] Thomas M. Liggett. *Interacting particle systems*. Classics in Mathematics. Springer-Verlag, Berlin, 2005. Reprint of the 1985 original.

[Mac95] I. G. Macdonald. *Symmetric functions and Hall polynomials*. Oxford Mathematical Monographs. The Clarendon Press, Oxford University Press, New York, second edition, 1995. With contributions by A. Zelevinsky, Oxford Science Publications.

[Man17] Olya Mandelshtam. Toric tableaux and the inhomogeneous two-species TASEP on a ring. *Adv. in Appl. Math.*, 133:101958, 2020.

[Mar18] James B. Martin. Stationary distributions of the multi-type ASEPs. 2018. arXiv:1810.10650.

[MGP68] J Macdonald, J Gibbs, and A Pipkin. Kinetics of biopolymerization on nucleic acid templates. *Biopolymers*, 6, 1968.

[Opd95] Eric M. Opdam. Harmonic analysis for certain representations of graded Hecke algebras. *Acta Math.*, 175(1):75–121, 1995.

[PEM09a] S. Prolhac, M. R. Evans, and K. Mallick. The matrix product solution of the multispecies partially asymmetric exclusion process. *J. Phys. A*, 42(16):165004, 25, 2009.

[PEM09b] S. Prolhac, M. R. Evans, and K. Mallick. The matrix product solution of the multispecies partially asymmetric exclusion process. *J. Phys. A*, 42(16):165004, 25, 2009.

[Spi70] Frank Spitzer. Interaction of Markov processes. *Advances in Math.*, 5:246–290 (1970), 1970.

[The18] The Sage Developers. *SageMath, the Sage Mathematics Software System (Version v8.2)*, 2018. http://www.sagemath.org.

[USW04] Masaru Uchiyama, Tomohiro Sasamoto, and Miki Wadati. Asymmetric simple exclusion process with open boundaries and Askey-Wilson polynomials. *J. Phys. A*, 37(18):4985–5002, 2004.

Macdonald Operators and Quantum Q-Systems for Classical Types

Philippe Di Francesco and Rinat Kedem

To Nicolai Reshetikhin on his 60th birthday

Abstract We propose solutions of the quantum Q-systems of types B_N, C_N, D_N in terms of q-difference operators, generalizing our previous construction for the Q-system of type A. The difference operators are interpreted as q-Whittaker limits of discrete time evolutions of Macdonald-van Diejen type operators. We conjecture that these new operators act as raising and lowering operators for q-Whittaker functions, which are special cases of graded characters of fusion products of KR-modules.

1 Introduction

The characters of tensor products of KR-modules of Yangians, quantum affine algebras, or affine algebras have fermionic formulas, generalizing Bethe's original counting formula of the Bethe eigenstates of the Heisenberg spin chain [Bet31]. The fermionic formulas, in the case of KR-modules of Yangians of the classical Lie algebras $\mathfrak{g} = ABCD$, were conjectured by Kirillov and Reshetikin [KR87] and further generalized in [HKO$^+$99]. They were proved for the case of any simple Lie algebra $\mathfrak{g}$ in [AK07, DFK08]. In the course of the proof, it becomes clear that there is

P. Di Francesco (✉)
Department of Mathematics, University of Illinois MC-382, Urbana, IL, USA

Institut de Physique Théorique du Commissariat à l'Energie Atomique, Unité de Recherche associée du CNRS, Gif sur Yvette Cedex, France
e-mail: philippe@illinois.edu; philippe.di-francesco@cea.fr

R. Kedem
Department of Mathematics, University of Illinois MC-382, Urbana, IL, USA
e-mail: rinat@illinois.edu

A. Alekseev et al. (eds.), *Representation Theory, Mathematical Physics, and Integrable Systems*, Progress in Mathematics 340,
https://doi.org/10.1007/978-3-030-78148-4_6

a close connection between solutions of recursion relations known as the Q-systems and the fermionic formulas.

In [KR87, HKO+99], q-analogs of the fermionic characters of tensor products of KR-modules, or graded characters, are also presented. There are several interpretations of the grading of these tensor products, which gives rise to these q-analogues, or graded characters: (1) as linearized energy function for the corresponding generalized inhomogeneous Heisenberg spin chain, (2) as a charge function for the crystal limit of the corresponding quantum affine algebra modules, when it exists, or (3) as the natural grading of the underlying affine algebra [FL99]. The latter definition was used in [AK07, DFK14, Lin19] to prove the q-graded fermionic character formulas. It turns out that there is a close connection between the graded character formulas and q-deformed versions of Q-systems known as the quantum Q-systems.

Quantum Q-systems are defined by quantizing [BZ05] the cluster algebraic [FZ02] structure of the classical Q-systems [Ked08, DFK09]. They are recursion relations for non-commuting variables $\{\mathcal{Q}_{a,k}\}$, with a running over the Dynkin labels of $\mathfrak{g}$ and $k \in \mathbb{Z}$.

Graded characters are Weyl-symmetric functions with coefficients in $\mathbb{Z}_+[q]$. The relation with the quantum Q-system can be schematically described as follows (details can be found in [DFK14]). One can construct a linear functional ϕ from the ordered product of (opposite, with parameter q^{-1}) quantum Q-system solutions $\mathcal{Q}^*_{a,k}$ to the graded characters of the corresponding tensor product:[1]

$$\chi_{\mathbf{n}}(q^{-1}; \mathbf{x}) = \phi\left(\prod_{a,k}^{\rightarrow} (\mathcal{Q}^*_{a,k})^{n_{a,k}}\right), \tag{1.1}$$

where $\chi_{\mathbf{n}}(q; \mathbf{x})$ is the graded character of the tensor product (or fusion product in the sense of [FL99]) of KR-modules $\otimes_{a,k} \mathrm{KR}_{k\omega_a}^{\otimes n_{a,k}}$, $\mathbf{n} = \{n_{a,k}\}$ denoting the collection of tensor powers, and $\mathbf{x} = (x_1, x_2, \ldots, x_N)$. Here, $\mathrm{KR}_{k\omega_a}$ is a $\mathfrak{g}$-module, the restriction of the KR-module of the (quantum) affine algebra module, which has highest weight $k\omega_a$, $k \in \mathbb{Z}_+$ and ω_a being one of the fundamental weights of $\mathfrak{g}$. The arrow on top of the product sign refers to a specific ordering of the terms [DFK14, Lin19], in which long and short roots play different roles.

Tensoring the tensor product above by an extra factor $\mathrm{KR}_{k'\omega_b}$ corresponds to the insertion of the factor $\mathcal{Q}^*_{b,k'}$ on the right in the product inside the functional, if k' is sufficiently large, so that the product remains ordered.

In the case of $\mathfrak{g} = \mathfrak{sl}_N$ we introduced [DFK18, DFK17] a set of q-difference operators $\mathcal{D}_{b,k'}$, acting on the space of symmetric functions of $\mathbf{x}$ to the right, and representing the insertion on the right of the factor $\mathcal{Q}^*_{b,k'}$ in the linear functional of

[1]The functional uses only "half-space" solutions $\mathcal{Q}^*_{a,k}$ of the quantum Q-system for $k \in \mathbb{Z}_+$, with a special prescription which amounts to evaluate $\mathcal{Q}^*_{a,0}$ to 1.

Eq. (1.1). The following diagram explains the action by difference operators on the space of symmetric functions:

$$\begin{array}{ccc}
\overrightarrow{\prod_{a,k}}(\mathcal{Q}^*_{a,k})^{n_{a,k}} & \xrightarrow{\mathcal{Q}^*_{b,k'}} & \left(\overrightarrow{\prod_{a,k}}(\mathcal{Q}^*_{a,k})^{n_{a,k}}\right)\mathcal{Q}^*_{b,k'} \\
\downarrow \phi & & \downarrow \phi \\
\chi_{\mathbf{n}}(q^{-1};\mathbf{x}) & \xrightarrow{\mathcal{D}_{b,k'}} & \chi_{\mathbf{n}+\epsilon_{b,k'}}(q^{-1};\mathbf{x})
\end{array}$$

The result is an explicit expression for graded characters as the iterated action of q-difference operators $\mathcal{D}_{a,k}$ on the constant 1. The difference operators satisfy the quantum Q-system. We refer to them as the functional representation of the quantum Q-system.

In [DFK18], it was observed that the difference operators $\mathcal{D}_{a,k}$ for $k = 0$ are the $t \to \infty$ limit of the Macdonald difference operators of type A, of which Macdonald polynomials form a set of common eigenfunctions. Thus, in [DFK19], we identified the t-deformation of the A type quantum Q-system as the spherical Double Affine Algebra Hecke (sDAHA) of type A. This is the algebra which underlies Macdonald theory [Mac95, Che05].

In the $t \to \infty$ limit, in which Macdonald polynomials tend to (dual) q-Whittaker functions with parameter q^{-1}, we further identified in [DFK18] the operator representation $\mathcal{D}_{a,k}$ of the quantum Q-system when $k = 1$ (resp. $k = -1$) as the $t \to \infty$ limits of the Kirillov-Noumi [KN99] raising (resp. lowering) operators. These operators, acting on a Macdonald polynomial indexed by some Young diagram, have the effect of adding (resp. subtracting) a column of a boxes to the Young diagram. Equivalently, using the correspondence of the Young diagram λ with a dominant $\mathfrak{gl}_N$-weight, it corresponds to adding or subtracting the fundamental weight ω_a.

As a consequence, in the case where all KR-modules in the tensor product are fundamental modules ($n_{a,k} = 0$ unless $k = 1$), the graded characters are identified as limits of Macdonald polynomials as $t \to \infty$ or $t = 0$ upon changing $q \to q^{-1}$, and can therefore be identified with specialized q-Whittaker functions. See also [LNS+17].

We remark that the above difference operators can be compared to the so-called minuscule monopole operators representing the Coulomb branches of 4D $N = 4$ quiver gauge theories [BFN16] in the particular case of the "Jordan quiver" with a single node, and when the equivariant parameter t is taken to infinity (with the result of imposing that the representation N is trivial).

It is natural to look for the generalization of the functional representation of the quantum Q-systems, corresponding to the affine algebras of types BCD. These are described in terms of the root systems of the finite BCD type. Motivated by the results in type A, we expect the functional representation of the other type quantum Q-systems to involve the $t \to \infty$ limits of the corresponding BCD type generalized Macdonald operators. In this paper, we present, without proof, a set of such difference operators. Our main Conjecture 3 is that these operators satisfy the relevant quantum Q-systems relations.

The construction of the BCD type difference operators is best understood by thinking of the quantum Q-systems as evolution equations in the discrete time variable $k \in \mathbb{Z}$ for the elements $\mathcal{Q}_{a,k}$. In type A, from the relation to sDAHA, we noted in [DFK19] that the discrete time evolution $k \to k+1$ is given by the adjoint action of a generator of the $SL_2(\mathbb{Z})$ symmetry of DAHA. The latter is also expressed as the adjoint action of the "Gaussian" function γ^{-1} of the variables $\mathbf{x}$, where

$$\gamma \equiv \gamma(\mathbf{x}) = e^{\sum_{i=1}^{N} \frac{\mathrm{Log}(x_i)}{2\mathrm{Log}(q)}}, \tag{1.2}$$

and such that $\mathcal{D}_{a,k+1} \propto \gamma^{-1}\mathcal{D}_{a,k}\gamma$.

The aim of the present paper is to present constructions of difference operator solutions to the quantum Q-systems of types BCD, such that they coincide, at $k = 0$, with the $t \to \infty$ limit of suitable Macdonald operators, and which have raising/lowering properties at $k = \pm 1$. To this end, we first identify the $t \to \infty$ limits of suitable Macdonald-type operators in types BCD by use of works of Macdonald and van Diejen [Mac01, vD95, vDE11]. Next, we construct their time evolution, by a suitable Gaussian conjugation. Our main result is the Conjectures 3 and 5.1 stating that (1) these operators obey a renormalized version of the quantum Q-systems in types BCD and (2) the operators at times $k = \pm 1$ act as raising/lowering operators on the corresponding q-Whittaker functions.

2 Q-Systems and Quantum Q-Systems

2.1 *Weights and Roots*

Let $\mathfrak{g}$ be a Lie algebra of classical type, $\mathfrak{g} \in \{A_{N-1}, D_N, B_N, C_N\}$. For each of these, we list in Table 1 the standard data of fundamental weights ω_a, the simple roots α_a, and the conditions on the non-increasing sequence $\lambda = (\lambda_1 \geq \lambda_2 \geq \cdots \geq \lambda_N)$ corresponding to dominant weights $\sum_a n_a\omega_a = \sum_i \lambda_i e_i$ ($n_a \in \mathbb{Z}_+$). In Table 1, the set $\{e_a\}_{a=1}^N$ is the standard basis of $\mathbb{R}^N$, whereas $\{\hat{e}_a = e_a - \epsilon/N\}_{a=1}^{N-1}$ where ϵ is the sum over all the basis elements.

We also denote by t_a the integers $\frac{2}{||\alpha_a||^2}$, so that $t_a = 1$ for long roots and $t_a = 2$ for the short roots in types BC.

Table 1 Root and weight data for the classical Lie algebras

Algebra	Fundamental weights ω_a	Simple roots α_a	λ
A_{N-1}	$\omega_a = \sum_{i=1}^{a} \hat{e}_i,\ a \in [1, N-1]$	$e_a - e_{a+1},\ a < N$	$\lambda_a \in \mathbb{Z}_+$
B_N	$\omega_a = \begin{cases} \sum_{i=1}^{a} e_i, & a < N; \\ \frac{1}{2}\sum_{i=1}^{N} e_i, & a = N. \end{cases}$	$e_a - e_{a+1},\ a < N$ e_N	$\lambda_a \in \mathbb{Z}_+$ for all a or $\lambda_a \in \mathbb{Z}_+ + \frac{1}{2}$ for all a
C_N	$\omega_a = \sum_{i=1}^{a} e_i,\ a \in [1, N]$	$e_a - e_{a+1},\ a < N$ $2e_N$	$\lambda_a \in \mathbb{Z}_+$
D_N	$\omega_a = \sum_{i=1}^{a} e_i,\ a < N-1$ $\omega_{N-1} = \frac{1}{2}(\omega_{N-2} + e_{N-1} - e_N)$ $\omega_N = \frac{1}{2}(\omega_{N-2} + e_{N-1} + e_N)$	$e_a - e_{a+1},\ a < N$ $e_a + e_{a+1},\ a < N$	$\lambda_a \in \mathbb{Z}$ for all a or $\lambda_a \in \mathbb{Z} + \frac{1}{2}$for all a, $\lambda_{N-1} \geq \lvert\lambda_N\rvert \geq 0$.

2.2 *The Classical Q-Systems for Untwisted Affine ABCD*

The Q-systems are recursion relations for the variables $\{Q_{a,k}\}$ where a is a label in the Dynkin diagram, and k is any integer. We list the Q-systems associated with types $ABCD$ [KR87, KNS94]. The boundary condition $Q_{0,k} = 0$ is assumed in all cases.

$$
\begin{aligned}
\mathfrak{g} = A_{N-1}:\ & Q_{a,k+1}Q_{a,k-1} = Q_{a,k}^2 - Q_{a+1,k}Q_{a-1,k}, && (a \in [1, N-1]), \\
& Q_{N,k} = 1. && (2.1) \\
\mathfrak{g} = B_N:\ & Q_{a,k+1}Q_{a,k-1} = Q_{a,k}^2 - Q_{a+1,k}Q_{a-1,k}, && (a \in [1, N-2]), \\
& Q_{N-1,k+1}\, Q_{N-1,k-1} = (Q_{N-1,k})^2 - Q_{N,2k}\, Q_{N-2,k}, && (2.2) \\
& Q_{N,2k+1}\, Q_{N,2k-1} = (Q_{N,2k})^2 - (Q_{N-1,k})^2, && \\
& Q_{N,2k+2}\, Q_{N,2k} = (Q_{N,2k+1})^2 - Q_{N-1,k+1}\, Q_{N-1,k}. && \\
\mathfrak{g} = C_N:\ & Q_{a,k+1}Q_{a,k-1} = Q_{a,k}^2 - Q_{a+1,k}Q_{a-1,k}, && (a \in [1, N-2]), \\
& Q_{N-1,2k+1}Q_{N-1,2k-1} = Q_{N-1,2k}^2 - Q_{N-2,2k}Q_{N,k}^2, && (2.3) \\
& Q_{N-1,2k+2}Q_{N-1,2k} = Q_{N-1,2k+1}^2 - Q_{N-2,2k+1}Q_{N,k+1}Q_{N,k}. && \\
& Q_{N,k+1}\, Q_{N,k-1} = Q_{N,k}^2 - Q_{N-1,2k}, && \\
\mathfrak{g} = D_N:\ & Q_{a,k+1}Q_{a,k-1} = Q_{a,k}^2 - Q_{a+1,k}Q_{a-1,k}, && (a \in [1, N-3]), \\
& Q_{N-2,k+1}Q_{N-2,k-1} = Q_{N-2,k}^2 - Q_{N,k}Q_{N-1,k}Q_{N-3,k}, && (2.4)
\end{aligned}
$$

$$Q_{a,k+1}Q_{a,k-1} = Q_{a,k}^2 - Q_{N-2,k} \qquad (a \in \{N-1, N\}).$$

These recursion relations (for $k \geq 1$) were originally observed [KR87] to be relations satisfied by characters of finite-dimensional irreducible Yangian modules.

Each set of the Q-systems is associated with a cluster algebra:

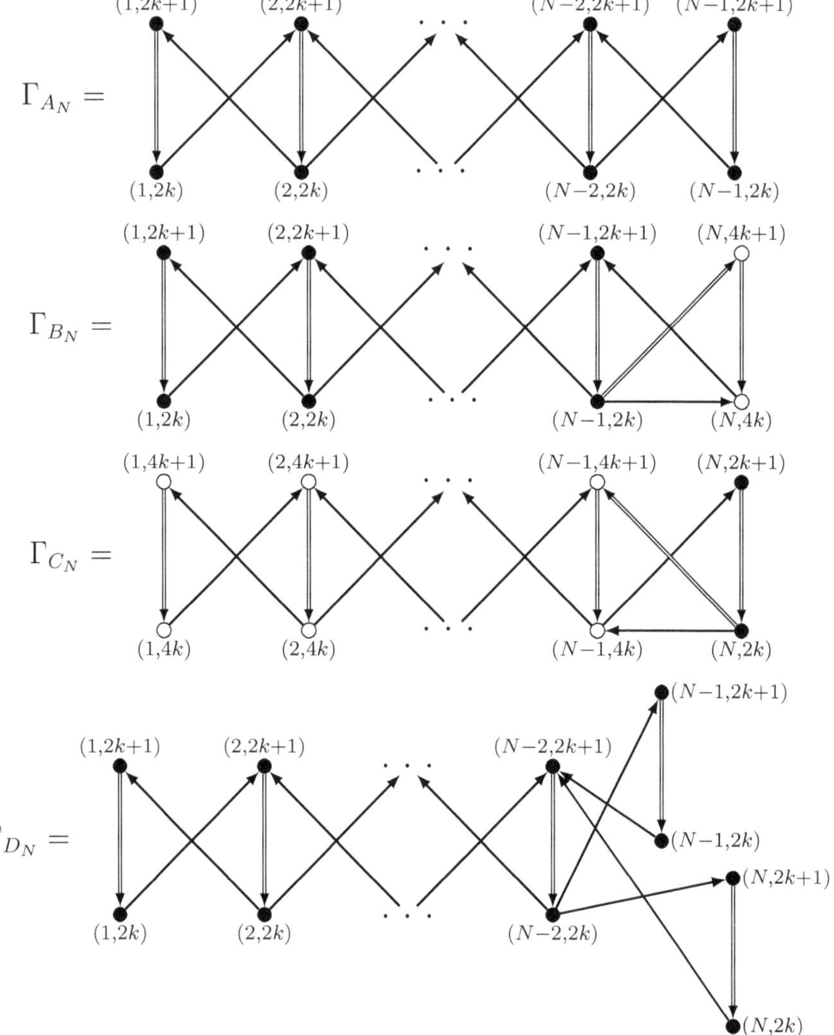

Fig. 1 The quivers for the A_{N-1}, D_N, B_N, C_N Q-system cluster algebras. We have indicated a generic Q-system cluster along the bipartite belt: each vertex labelled (a, k) corresponds to a cluster variable $Q_{a,k}$. Nodes corresponding to short roots are denoted by empty circles

Theorem 2.1 ([Ked08, DFK09]) *For each algebra $\mathfrak{g}$, the variables $\{Q_{a,k} : a \in [1,r], k \in \mathbb{Z}\}$, up to a simple rescaling which eliminates the minus sign on the right hand side, are cluster variables in a corresponding cluster algebra. Each of the Q-system relations is an exchange relation in the cluster algebra.*

The cluster algebras are defined via a $2r \times 2r$ skew-symmetric exchange matrix B, (r being the rank of $\mathfrak{g}$), or quiver Γ as in Fig. 1, which depends only on the Cartan matrix C of $\mathfrak{g}$:

$$B = \left(\begin{array}{c|c} C^T - C & -C^T \\ \hline C & 0 \end{array}\right). \tag{2.5}$$

This exchange matrix, together with the initial cluster variables $\mathbf{X} = (Q_{a,0}; Q_{a,1})_{a=1}^r$, defines the cluster algebra. The cluster variables $\{Q_{a,k}\}$ are obtained from a generalized bipartite evolution of the initial cluster $(\mathbf{X}, B)$ [DFK09]. The subset of mutations on the (generalized) bipartite belt which generates all the cluster variables corresponding to the Q-system algebra was given in [DFK09, Theorem 3.6].

2.3 Quantum Q-Systems

One of the advantages of formulating the Q-system relations in terms of cluster algebra mutations is that there is a canonical quantization of the cluster algebra, using the canonical Poisson structure [GSV10] and its quantization [BZ05].

The quantum cluster algebra attached to a non-degenerate, skew-symmetric matrix B is the non-commutative algebra generated by the cluster variables $\mathbf{X} = (X_i)$ at an initial cluster and their inverses, with exchange matrix B, as well as the cluster variables at all mutation equivalent clusters. Within the cluster $(\mathbf{X}, B)$ the cluster variables q-commute according to a skew-symmetric Λ proportional to the inverse of B:

$$X_i X_j = q^{\Lambda_{ij}} X_j X_i.$$

The exchange relations are given by normal-ordering of the classical mutations:

$$X_i' =: \prod_{j:B_{ij}>0} X_j^{B_{ij}} X_i^{-1} : + : \prod_{j:B_{ij}<0} X_j^{-B_{ij}} X_i^{-1} : .$$

Here, given a monomial $\prod X_i^{b_i}$ such that $X_i X_j = q^{a_{i,j}} X_j X_i$, the normal ordered product is

$$: X_1^{b_1} \cdots X_\ell^{b_\ell} := q^{-\frac{1}{2}\sum_{i<j} a_{i,j} b_i b_j} X_1^{b_1} \cdots X_\ell^{b_\ell}.$$

The exchange matrices (2.5) corresponding to the Q-system cluster algebras are skew-symmetric and invertible. We use the associated quantum cluster algebra to define the quantum Q-systems as the quantized exchange relations corresponding to the exchange relations appearing in the classical Q-systems.

Let Λ be a $2r \times 2r$ skew-symmetric matrix, with $\Lambda^T B = -D$, a diagonal integer matrix with negative integer entries. Then

$$\Lambda = \left(\begin{array}{c|c} 0 & \lambda \\ \hline -\lambda^T & \lambda^T - \lambda \end{array} \right), \tag{2.6}$$

where λ is proportional to the inverse Cartan matrix. We use the following normalizations:

$$A_{N-1} : \lambda_{ab} = C^{-1}_{ab} = \min(a,b) - \tfrac{ab}{N}$$

$$B_N : \lambda_{ab} = 2C^{-1}_{ab} = \begin{cases} 2\min(a,b), & 1 \le a \le N,\ 1 \le b \le N-1; \\ \min(a,b), & 1 \le a \le N = b, \end{cases}$$

$$C_N : \lambda_{ab} = 2C^{-1}_{ab} = \begin{cases} 2\min(a,b), & 1 \le a \le N-1,\ 1 \le b \le N; \\ \min(a,b), & 1 \le b \le N = a, \end{cases}$$

$$D_N : \lambda_{ab} = 2C^{-1}_{ab} = \begin{cases} 2\min(a,b), & 1 \le a,b \le N-2, \\ b, & 1 \le b \le N-2,\ a \in \{N-1,N\}, \\ a, & 1 \le a \le N-2,\ b \in \{N-1,N\}, \\ \frac{1}{2}N, & a = b \in \{N-1,N\}, \\ \frac{1}{2}(N-2), & a \ne b \in \{N-1,N\}. \end{cases} \tag{2.7}$$

This choice of normalization results, in the cases B_N, C_N, D_N, in the value $\lambda_{a,b} = 2\min(a,b)$ whenever a, b correspond to the part of the Dynkin diagram that forms an A-type chain, that is, $a, b \in [1, N-1]$ for types BC and $a, b \in [1, N-2]$ for type D.

The quantum Q-systems are recursion relations for the non-commuting variables $\mathcal{Q}_{a,k}$. These take the form

$$\begin{aligned} A_{N-1} : \ & \mathcal{Q}_{a,k}\,\mathcal{Q}_{b,p} = q^{\lambda_{a,b}(p-k)}\,\mathcal{Q}_{b,p}\,\mathcal{Q}_{a,k}, \\ & q^{\lambda_{a,a}}\,\mathcal{Q}_{a,k+1}\,\mathcal{Q}_{a,k-1} = \mathcal{Q}_{a,k}^2 - q^{\frac{1}{2}}\,\mathcal{Q}_{a+1,k}\,\mathcal{Q}_{a-1,k}, \quad (a \in [1,N]), \\ & \mathcal{Q}_{0,k} = 1, \qquad \mathcal{Q}_{N+1,k} = 0. \end{aligned} \tag{2.8}$$

$$\begin{aligned} B_N : \ & \mathcal{Q}_{a,k}\,\mathcal{Q}_{b,p} = q^{p\lambda_{a,b} - k\lambda_{b,a}}\,\mathcal{Q}_{b,p}\,\mathcal{Q}_{a,k}, \\ & q^{2a}\,\mathcal{Q}_{a,k+1}\,\mathcal{Q}_{a,k-1} = \mathcal{Q}_{a,k}^2 - q\,\mathcal{Q}_{a+1,k}\,\mathcal{Q}_{a-1,k}\,\mathcal{Q}_{\alpha-1,k}, \quad (\alpha \in [1, N-2]), \\ & q^{2N-2}\,\mathcal{Q}_{N-1,k+1}\,\mathcal{Q}_{N-1,k-1} = (\mathcal{Q}_{N-1,k})^2 - q\,\mathcal{Q}_{N,2k}\,\mathcal{Q}_{N-2,k}, \end{aligned} \tag{2.9}$$

$$
\begin{aligned}
& q^N \mathcal{Q}_{N,2k+1}\, \mathcal{Q}_{N,2k-1} = (\mathcal{Q}_{N,2k})^2 - q(\mathcal{Q}_{N-1,k})^2, \\
& q^N \mathcal{Q}_{N,2k+2}\, \mathcal{Q}_{N,2k} = (\mathcal{Q}_{N,2k+1})^2 - q^N \mathcal{Q}_{N-1,k+1}\, \mathcal{Q}_{N-1,n}, \\
& \mathcal{Q}_{0,k} = 1. \\
C_N : \; & \mathcal{Q}_{a,k}\, \mathcal{Q}_{b,p} = q^{p\lambda_{a,b} - k\lambda_{b,a}}\, \mathcal{Q}_{b,p}\, \mathcal{Q}_{a,k}, \\
& q^{2a} \mathcal{Q}_{a,k+1}\, \mathcal{Q}_{a,k-1} = \mathcal{Q}_{a,k}^2 - q\, \mathcal{Q}_{a+1,k}\, \mathcal{Q}_{a-1,k}, \qquad (\alpha \in [1, N-2]), \\
& q^{2N-2} \mathcal{Q}_{N-1,2k+1} \mathcal{Q}_{N-1,2k-1} = \mathcal{Q}_{N-1,2k}^2 - q \mathcal{Q}_{N-2,2k} \mathcal{Q}_{N,k}^2, \qquad (2.10) \\
& q^{2N-2} \mathcal{Q}_{N-1,2k+2} \mathcal{Q}_{N-1,2k} = \mathcal{Q}_{N-1,2k+1}^2 - q^{1+\frac{N}{2}}\, \mathcal{Q}_{N-2,2k+1} \mathcal{Q}_{N,k+1} \mathcal{Q}_{N,k}, \\
& q^N \mathcal{Q}_{N,k+1}\, \mathcal{Q}_{N,k-1} = \mathcal{Q}_{N,k}^2 - q \mathcal{Q}_{N-1,2k}, \\
& \mathcal{Q}_{0,k} = 1. \\
D_N : \; & \mathcal{Q}_{a,k}\, \mathcal{Q}_{b,p} = q^{\lambda_{a,b}(p-k)}\, \mathcal{Q}_{b,p}\, \mathcal{Q}_{a,k}, \\
& q^{2a} \mathcal{Q}_{a,k+1}\, \mathcal{Q}_{a,k-1} = \mathcal{Q}_{a,k}^2 - q\, \mathcal{Q}_{a+1,k}\, \mathcal{Q}_{a-1,k} \qquad (a \in [1, N-3]), \\
& q^{2(N-2)} \mathcal{Q}_{N-2,k+1}\, \mathcal{Q}_{N-2,k-1} = \mathcal{Q}_{N-2,k}^2 - q\, \mathcal{Q}_{N,k}\, \mathcal{Q}_{N-1,k}\, \mathcal{Q}_{N-3,k}, \qquad (2.11) \\
& q^{\frac{N}{2}} \mathcal{Q}_{N-1,k+1}\, \mathcal{Q}_{N-1,k-1} = \mathcal{Q}_{N-1,k}^2 - q\, \mathcal{Q}_{N-2,k}, \\
& q^{\frac{N}{2}} \mathcal{Q}_{N,k+1}\, \mathcal{Q}_{N,k-1} = \mathcal{Q}_{N,k}^2 - q\, \mathcal{Q}_{N-2,k}, \\
& \mathcal{Q}_{0,k} = 1.
\end{aligned}
$$

In each case, the q-commutation relation, i.e. the first equation in each set, holds only for variables within the same cluster, hence the restriction on possible values of the second index. For example, in the case A_{N-1}, the restriction is $|p-k| \leq |a-b| + 1$. In general, the clusters consisting of Q-system solutions only are parameterized by generalized Motzkin paths [Lin19].

Remark 2.2 The general boundary condition in (2.8) the quantum Q-system of type A_{N-1} is $\mathcal{Q}_{N+1,k} = 0$, as opposed to the condition $Q_{N,k} = 1$ for the classical system (2.1), which is compatible but more restrictive. This means that there is an additional set of variables in the center of the quantum cluster algebra which satisfy $\mathcal{Q}_{N,k+1}\mathcal{Q}_{N,k-1} = \mathcal{Q}_{N,k}^2$. This is consistent with the extension of the definition (2.7) of the matrix $\lambda_{a,b}$ in type A to an $N \times N$ matrix using the same formula, so that $\lambda_{a,N} = 0$. The most general solution subject to the boundary conditions has $\mathcal{Q}_{N,k} = \mathcal{Q}_{N,1}^k \mathcal{Q}_{N,0}^{k-1}$, in terms of the two central elements $\mathcal{Q}_{N,1}$ and $\mathcal{Q}_{N,0}$. This more general system can be embedded into a cluster algebra with coefficients.

3 The A_{N-1} Case Solution: Generalized Macdonald Difference Operators and Quantum Determinants

3.1 *Renormalized Quantum Q-System*

With the choice of boundary condition for the type A quantum Q-system as in Remark 2.2, the quantum Q-system is homogeneous with respect to the grading $\deg(\mathcal{Q}_{a,k}) := ak$ ($a \in [1, N]$). We adjoin an invertible degree operator $\Delta^{1/N}$ to the algebra, such that

$$\Delta\, \mathcal{Q}_{a,k} = q^{ak}\, \mathcal{Q}_{a,k}\, \Delta, \qquad a \in [1, N], k \in \mathbb{Z}.$$

Using $\lambda_{a,a} + \frac{a}{N} = a$ and $\lambda_{a+1,a+1} + \lambda_{a-1,a-1} - 2\lambda_{a,a} = -\frac{2}{N}$, the renormalized variables

$$\widetilde{\mathcal{Q}}_{a,k} = q^{-\frac{1}{2}(k+\frac{N}{2})\lambda_{a,a}}\, \mathcal{Q}_{a,k}\, \Delta^{\frac{a}{N}}, \qquad a \in [1, N], k \in \mathbb{Z}, \tag{3.1}$$

satisfy the *renormalized quantum Q-system*:

$$\begin{aligned} &\widetilde{\mathcal{Q}}_{a,k}\, \widetilde{\mathcal{Q}}_{b,k'} = q^{(k'-k)\min(a,b)}\, \widetilde{\mathcal{Q}}_{b,k'}\, \widetilde{\mathcal{Q}}_{a,k}, \quad |k-k'| \le |a-b|+1, \\ &q^a\, \widetilde{\mathcal{Q}}_{a,k+1}\, \widetilde{\mathcal{Q}}_{a,k-1} = \widetilde{\mathcal{Q}}_{a,k}^2 - \widetilde{\mathcal{Q}}_{a+1,k}\, \widetilde{\mathcal{Q}}_{a-1,k}, \qquad (a \in [1, N]), \\ &\widetilde{\mathcal{Q}}_{0,k} = 1, \qquad \widetilde{\mathcal{Q}}_{N+1,k} = 0. \end{aligned} \tag{3.2}$$

The choice $\mathcal{Q}_{N,0} = 1$ implies $\widetilde{\mathcal{Q}}_{N,0} = \Delta$. Defining $A = \widetilde{\mathcal{Q}}_{N,1}\widetilde{\mathcal{Q}}_{N,0}^{-1}$, a homogeneous element of degree N, we have $\widetilde{\mathcal{Q}}_{N,k} = A^k\, \Delta$. The quiver with coefficients corresponding to this cluster algebra is illustrated in Fig. 2.

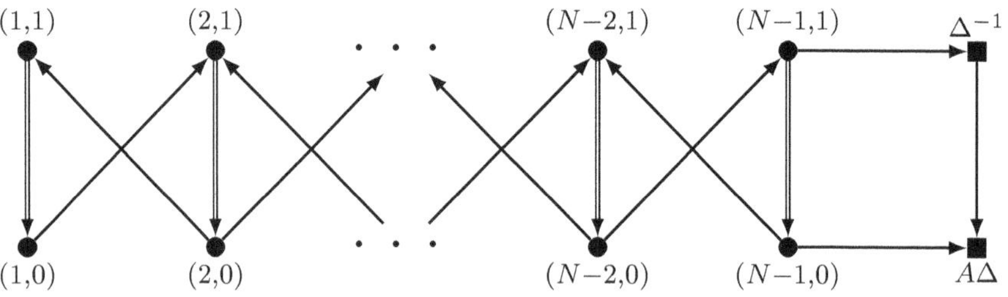

Fig. 2 The type A quantum Q-system quiver corresponding to the initial seed $\{\mathcal{Q}_{a,0}, \mathcal{Q}_{a,1}\}$. Square nodes denote coefficients

3.2 The Quantum Determinant

The exchange relations (3.2) define a quantum determinant: The variables $\widetilde{\mathcal{Q}}_{a,k}$ with $a > 1$ are polynomials in the variables $\{\widetilde{\mathcal{Q}}_{1,k'} : |k' - k| \leq a - 1\}$. Below, we use the notation $\widetilde{\mathcal{Q}}_k := \widetilde{\mathcal{Q}}_{1,k}$. The quantum determinant is best defined in terms of generating functions.

Definition 3.1 Given a set of integers $k_1, \ldots, k_a \in \mathbb{Z}$, define the Hankel matrix

$$(\widetilde{\mathcal{Q}}_{k_i+i-j})_{1\leq i,j\leq a}. \tag{3.3}$$

The quantum determinant of this matrix, denoted by $|\widetilde{\mathcal{Q}}[k_1, k_2, \ldots, k_a]|_q$, is given by the coefficients of the generating function:

$$\sum_{k_1,\ldots,k_a\in\mathbb{Z}} u_1^{k_1} \cdots u_a^{k_a} \, |\widetilde{\mathcal{Q}}[k_1, \ldots, k_a]|_q = \prod_{1\leq i<j\leq a} \left(1 - q\frac{u_j}{u_i}\right) \widetilde{\mathcal{Q}}(u_1)\widetilde{\mathcal{Q}}(u_2)\cdots \widetilde{\mathcal{Q}}(u_\alpha),$$

where

$$\widetilde{\mathcal{Q}}(u) := \sum_{k\in\mathbb{Z}} u^k \, \widetilde{\mathcal{Q}}_k.$$

The quantum determinant $|\widetilde{\mathcal{Q}}[k_1, \ldots, k_a]|_q$ is a homogeneous polynomial of degree a in the $\widetilde{\mathcal{Q}}_k$s.

Theorem 3.2 *[DFK17] The solutions $\widetilde{\mathcal{Q}}_{a,k}$ of the system* (3.2) *with $a \geq 1$ and $k \in \mathbb{Z}$ are the quantum determinants*

$$\widetilde{\mathcal{Q}}_{a,k} = |\widetilde{\mathcal{Q}}[\underbrace{k, k, \ldots, k}_{a \text{ times}}]|_q.$$

3.3 The Functional Representation of the Quantum Q-System

We recall the functional representation of the renormalized quantum Q-system (3.2) $\rho(\widetilde{\mathcal{Q}}_{a,k}) = M_{a,k}$ [DFK18], which act on the space of symmetric functions of N variables $x_1, x_2, \ldots, x_N$.

Theorem 3.3 ([DFK18, DFK17]) *Let Γ_i be the q-shift operator acting on the space of functions in N variables, defined by $\Gamma_i f(x_1, \ldots, x_i, \ldots, x_N) = f(x_1, \ldots, qx_i, \ldots, x_N)$. The q-difference operators*

$$M_{a,k} = \sum_{\substack{I\subset[1,N]\\|I|=a}} \Big(\prod_{i\in I} x_i\Big)^k \prod_{\substack{i\in I\\j\notin I}} \frac{x_i}{x_i - x_j} \prod_{i\in I} \Gamma_i, \quad a \in [1, N], k \in \mathbb{Z} \tag{3.4}$$

satisfy the quantum Q-system relations (3.2).

In this representation,

$$A = x_1 x_2 \cdots x_N, \qquad \Delta = \Gamma_1 \Gamma_2 \cdots \Gamma_N. \tag{3.5}$$

Remark 3.4 For $\mathfrak{sl}_N$-characters, the products A and Δ are taken to be equal to 1. The more general boundary condition here corresponds to $\mathfrak{gl}_N$-characters.

3.4 The Spherical Double Affine Hecke Algebra

The reader will have recognized that the difference operators $M_{a,0}$ in (3.4) are the limit $t \to \infty$ of the (renormalized) Macdonald difference operators in type A. This is the main observation which led to the results of [DFK19], where it is shown that the spherical DAHA [Che05] of type A_{N-1} is the natural t-deformation of the quantum Q-system.

An important observation in [DFK19] is that the evolution in the discrete time variable k is induced by the adjoint action of one of the generators of the $SL_2(\mathbb{Z})$ symmetry of the DAHA.

Theorem 3.5 ([DFK19]) *The discrete time evolution of the operators $M_{a,k}$ is induced by the adjoint action of the Gaussian γ^{-1}, with γ as in* (1.2)*:*

$$M_{a,k} = q^{-ak/2}\, \gamma^{-k}\, M_{a,0}\, \gamma^k. \tag{3.6}$$

The difference operators $\{M_{a,k}\}$, together with the elementary symmetric functions of $\mathbf{x}$, are the image in the functional representation of the generators of the spherical DAHA.

The double affine Hecke algebra and corresponding Macdonald operators are defined for other Lie algebras. In the following sections, we will use this as the inspiration to give conjectures for the functional representations of the quantum Q-systems for the other classical types by following the inverse reasoning: Starting from the appropriate choice of Macdonald difference operators, act with the adjoint action of γ to obtain the discrete time-evolved operators, imitating the contents of Theorem 3.5. The resulting q-difference operators, in the limit $t \to \infty$, are (conjecturally) solutions of renormalized quantum Q-systems for the other classical types.

3.5 Raising and Lowering Operators

The dual q-Whittaker functions are defined as

$$\Pi_\lambda \equiv \Pi_\lambda(q^{-1}; \mathbf{x}) = \lim_{t \to \infty} P_\lambda(q, t; \mathbf{x}),$$

where the P_λ are the Macdonald polynomials for type A_{N-1} [Mac95]. Here, the partition λ corresponds to a dominant integral weight of $\mathfrak{gl}_N$, $\lambda = \sum_{i=1}^N \lambda_i e_i$ in the standard basis of $\mathbb{R}^N$, with $\lambda_1 \geq \lambda_2 \geq \cdots \geq \lambda_N \geq 0$. The polynomials P_λ are common monic eigenfunctions of the Macdonald operators D_a

$$D_a = \sum_{\substack{I\subset[1,N]\\ |I|=a}} \prod_{\substack{i\in I\\ j\notin I}} \frac{tx_i - x_j}{x_i - x_j} \prod_{i\in I} \Gamma_i, \quad a \in [1, N], \tag{3.7}$$

with

$$D_a P_\lambda = t^{-\frac{a(a-1)}{2}} e_a(q^{\lambda_1} t^{N-1}, q^{\lambda_2} t^{N-2}, \ldots, q^{\lambda_N}) P_\lambda,$$

where e_a are the elementary symmetric functions in N variables. Using $\lim_{t\to\infty} t^{-a(N-a)} D_a = M_{a,0}$, this means that the q-Whittaker functions are eigenfunctions of the difference operators $M_{a,0}$:

$$M_{a,0}\, \Pi_\lambda = q^{(\lambda,\omega_a)}\, \Pi_\lambda, \quad a \in [1, N], \tag{3.8}$$

where $\omega_a = e_1 + e_2 + \cdots + e_a$ are the fundamental weights of $\mathfrak{gl}_N$.

In [DFK18] we showed that the operators $M_{a;\pm 1}$ are the limit $t \to \infty$ of the raising and lowering operators for Macdonald polynomials constructed by Kirillov and Noumi [KN99]. However, the commutation relations in (3.2) provide an easier proof, which we present here.

Theorem 3.6 *The operators $M_{a,\pm 1}$ act on the polynomials Π_λ as follows:*

$$M_{a,1}\, \Pi_\lambda = q^{(\lambda,\omega_a)}\, \Pi_{\lambda+\omega_a}, \tag{3.9}$$

$$M_{a,-1}\, \Pi_\lambda = q^{(\lambda,\omega_a)} \left(1 - q^{-(\lambda,\alpha_a)}\right) \Pi_{\lambda-\omega_a}, \tag{3.10}$$

where $\alpha_a = e_a - e_{a+1}$ are the simple roots of $\mathfrak{sl}_N$.

Note that if λ is a partition, $\lambda + \omega_a$ is also a partition, and if $\lambda - \omega_a$ is not a partition, the scalar factor $(1 - q^{(\lambda,\alpha_a)})$ vanishes.

Proof Up to scalar multiple, the polynomials Π_λ are uniquely determined by the collection of eigenvalues $q^{(\omega_a,\lambda)}$ corresponding to the diagonal action of the operators $M_{a,0}$, $a = 1, 2, \ldots, N$. The commutation relations in (3.2) are

$$M_{a,0}\, M_{b,\pm 1} = q^{\pm \min(a,b)}\, M_{b,0\pm 1}\, M_{a,0}.$$

Applying this to Π_λ and using (3.8),

$$M_{a,0}\, M_{b,\pm 1} \Pi_\lambda = q^{(\lambda,\omega_a)\pm \min(a,b)}\, M_{b,\pm 1} \Pi_\lambda = q^{(\lambda\pm\omega_b,\omega_a)} M_{b,\pm 1} \Pi_\lambda,$$

since $\min(a,b) = (\omega_b, \omega_a)$. That is, $M_{b,\pm 1}\Pi_\lambda$ is an eigenfunction of $M_{a,0}$ with eigenvalue $q^{(\lambda\pm\omega_b,\omega_a)}$. Therefore there exist scalars $c^{\pm}_{\lambda,b}$ such that $M_{b,\pm 1}\Pi_\lambda = c^{\pm}_{\lambda,b}\,\Pi_{\lambda\pm\omega_b}$.

Recall that the Macdonald polynomials P_λ, and therefore Π_λ, have a triangular decomposition with respect to the monomial symmetric functions with leading term m_λ. The following analysis provides the scalar factor $c^{+}_{\lambda,b}$. When $|x_1| \gg |x_2| \gg \cdots \gg |x_N|$, $\Pi_\lambda = x_1^{\lambda_1}\cdots x_N^{\lambda_N} +$ lower order and $M_{b,1} = x_1 x_2 \ldots x_b\,\Gamma_1\Gamma_2\cdots\Gamma_b +$ lower order. Thus,

$$
\begin{aligned}
M_{b,1}\,\Pi_\lambda &= q^{\lambda_1+\cdots+\lambda_b} x_1^{\lambda_1+1}\cdots x_b^{\lambda_b+1} x_{b+1}^{\lambda_{b+1}}\cdots x_N^{\lambda_N} + \text{lower order} \\
&= q^{(\lambda,\omega_b)}\Pi_{\lambda+\omega_b} + \text{lower order},
\end{aligned}
$$

and thus $c^{+}_{\lambda,b} = q^{(\lambda,\omega_b)}$.

Applying the exchange relation in (3.2) with $k=0$ to Π_λ, we get

$$
q^a\, M_{a,1}\, M_{a,-1}\,\Pi_\lambda = q^a\, c^{+}_{\lambda-\omega_a,a}\, c^{-}_{\lambda,a}\Pi_\lambda = (q^{2(\lambda,\omega_a)} - q^{(\lambda,\omega_{a+1}+\omega_{a-1})})\Pi_\lambda,
$$

which shows that $c^{-}_{\lambda,a} = q^{(\lambda,\omega_a)} - q^{(\lambda,\omega_{a+1}+\omega_{a-1}-\omega_a)}$ and the Theorem follows. □

3.6 Graded Characters in Terms of Difference Operators

The difference operators (3.4) can be used to efficiently generate graded characters (1.1). Theorem 3.3 is a necessary condition for the following theorem:

Theorem 3.7 *([DFK18], Corollary 18): Starting with the trivial character* $\chi_0 = 1$, *the difference operators* (3.4) *act consecutively to generate the character of the graded tensor product of KR-modules* $\otimes_{a,i}\mathrm{KR}_{i\omega_a}^{\otimes n_{a,i}}$ *as follows:*

$$
\chi_{\mathbf{n}}(q^{-1},\mathbf{x}) = q^{-\frac{1}{2}Q(\mathbf{n})}\prod_{a=1}^{N-1}(M_{a,k})^{n_{a,k}}\prod_{a=1}^{N-1}(M_{a,k-1})^{n_{a,k-1}}\cdots\prod_{a=1}^{N-1}(M_{a,1})^{n_{a,1}}\,1, \tag{3.11}
$$

where

$$
Q(\mathbf{n}) = \sum_{a,b=1}^{N-1}\sum_{i,j\geq 1} n_{a,i}\min(i,j)\min(a,b)\,n_{b,j} - \sum_{a=1}^{N-1}\sum_{i\geq 1} i\,a\,n_{a,i}.
$$

Alternatively, when $k \geq \max\{j : n_{a,j} > 0\}$,

$$D_{a,k}\chi_{\mathbf{n}}(q^{-1}; \mathbf{x}) = q^{(\omega_a, \sum_{b,j} j n_{b,j}\omega_b)} \chi_{\mathbf{n}+\epsilon_{a,k}}(q^{-1}; \mathbf{x}),$$

where we write $\mathbf{n} = \sum_{b,j} n_{b,j}\epsilon_{b,j}$.

4 The Quantum Q-System Conjectures for Types BCD

In order to formulate the conjectures for the functional representation of the quantum Q-systems of types D_N, B_N, C_N, we start with the results of Macdonald and van Diejen. There are N algebraically independent commuting Hamiltonians in the sDAHA of those types. Our goal is to find a set that will be the seed, in the q-Whittaker limit, of the Q-system solutions.

4.1 *Macdonald and van Diejen Operators*

In [Mac01], Macdonald constructed certain difference operators for types D_N, B_N, C_N, corresponding to minuscule coweights. For these types, there are, respectively, 3, 1, and 1 minuscule coweights, indexed by some of the extremal nodes of the Dynkin diagrams. The corresponding Macdonald operators are

$$D_N: \quad \mathcal{E}_1^{(D_N)} = \sum_{\epsilon=\pm 1} \sum_{i=1}^{N} \prod_{j\neq i} \frac{1-tx_i^{\epsilon}x_j}{1-x_i^{\epsilon}x_j} \frac{tx_i^{\epsilon}-x_j}{x_i^{\epsilon}-x_j} \Gamma_i^{2\epsilon}, \tag{4.1}$$

$$\mathcal{E}_{N-1}^{(D_N)} = \sum_{\substack{\epsilon_1,\ldots,\epsilon_N=\pm 1 \\ \epsilon_1\epsilon_2\cdots\epsilon_N=-1}} \prod_{1\leq i<j\leq N} \frac{1-tx_i^{\epsilon_i}x_j^{\epsilon_j}}{1-x_i^{\epsilon_i}x_j^{\epsilon_j}} \prod_{i=1}^{N} \Gamma_i^{\epsilon_i}, \tag{4.2}$$

$$\mathcal{E}_{N}^{(D_N)} = \sum_{\substack{\epsilon_1,\ldots,\epsilon_N=\pm 1 \\ \epsilon_1\epsilon_2\cdots\epsilon_N=1}} \prod_{1\leq i<j\leq N} \frac{1-tx_i^{\epsilon_i}x_j^{\epsilon_j}}{1-x_i^{\epsilon_i}x_j^{\epsilon_j}} \prod_{i=1}^{N} \Gamma_i^{\epsilon_i}. \tag{4.3}$$

$$B_N: \quad \mathcal{E}_1^{(B_N)} = \sum_{\epsilon=\pm 1} \sum_{i=1}^{N} \frac{1-tx_i^{\epsilon}}{1-x_i^{\epsilon}} \prod_{j\neq i} \frac{1-tx_i^{\epsilon}x_j}{1-x_i^{\epsilon}x_j} \frac{tx_i^{\epsilon}-x_j}{x_i^{\epsilon}-x_j} \Gamma_i^{2\epsilon}. \tag{4.4}$$

$$C_N: \quad \mathcal{E}_N^{(C_N)} = \sum_{\epsilon_1,\ldots,\epsilon_N=\pm 1} \prod_{i=1}^{N} \frac{1-tx_i^{2\epsilon_i}}{1-x_i^{2\epsilon_i}} \prod_{1\leq i<j\leq N} \frac{1-tx_i^{\epsilon_i}x_j^{\epsilon_j}}{1-x_i^{\epsilon_i}x_j^{\epsilon_j}} \prod_{i=1}^{N} \Gamma_i^{\epsilon_i}. \tag{4.5}$$

The Macdonald operator for C_N is of "order" N, namely acts by shifts of N variables in each term, as opposed to the operators $\mathcal{E}_1^{(D_N)}$ and $\mathcal{E}_1^{(B_N)}$, which are linear combinations of shifts of a single variable. A first order difference operator for C_N, $\mathcal{E}_1^{(C_N)}$, can be obtained using the commuting operators constructed in [vD95, vDE11]. We choose the following first order C_N operator, which is a particular linear combination of the identity and the first order van Diejen operator:

$$\mathcal{E}_1^{(C_N)} = (1+t^{N+1})\frac{1-t^N}{1-t} + \sum_{\epsilon=\pm 1}\sum_{i=1}^{N} \frac{1-tx_i^{2\epsilon}}{1-x_i^{2\epsilon}}\frac{1-tq^2x_i^{2\epsilon}}{1-q^2x_i^{2\epsilon}} \times \prod_{j\neq i}\frac{1-tx_i^{\epsilon}x_j}{1-x_i^{\epsilon}x_j}\frac{tx_i^{\epsilon}-x_j}{x_i^{\epsilon}-x_j}(\Gamma_i^{2\epsilon}-1) \tag{4.6}$$

This particular choice ensures that the corresponding eigenfunctions are the Macdonald polynomials $P_\lambda(\mathbf{x})$ of type C_N with eigenvalues

$$t^N\,\hat{e}_1(\{t^{N+1-i}q^{2\lambda_i}\}_{i=1}^N), \qquad \text{where } \hat{e}_1(z_1,z_2,\ldots,z_N) = \sum_{i=1}^{N}(z_i+z_i^{-1}).$$

In particular, for $P_0(\mathbf{x}) = 1$, we have $\mathcal{E}_1^{(C_N)}1 = t^N\,\hat{e}_1(\{t,t^2,\ldots,t^N\}) = (1+t^{N+1})\frac{1-t^N}{1-t}$.

4.2 The q-Whittaker Limit

Each of the operators $\mathcal{E}_i^{(\mathfrak{g})}$ in Eqs. (4.1)–(4.6) have suitable limits as $t\to\infty$, denoted by

$$M_{i,0}^{(\mathfrak{g})} := \lim_{t\to\infty} t^{-a_i^{(\mathfrak{g})}}\mathcal{E}_i^{(\mathfrak{g})},$$

where:

$$a_1^{(D_N)} = 2(N-1), \quad a_{N-1}^{(D_N)} = a_N^{(D_N)} = \frac{N(N-1)}{2},$$
$$a_1^{(B_N)} = 2N-1,$$
$$a_1^{(C_N)} = 2N, \quad a_N^{(C_N)} = \frac{N(N+1)}{2}.$$

It is also necessary to define the additional operator in type C_N:

$$M_{1,1}^{(C_N)} := \sum_{\epsilon=\pm 1} \sum_{i=1}^{N} \frac{x_i^{2\epsilon}}{x_i^{2\epsilon}-1} \frac{q^2 x_i^{2\epsilon}}{q^2 x_i^{2\epsilon}-1} \prod_{j\neq i} \frac{x_i^{\epsilon} x_j}{x_i^{\epsilon} x_j - 1} \frac{x_i^{\epsilon}}{x_i^{\epsilon} - x_j} x_i^{-\epsilon} (x_i^{2\epsilon} \Gamma_i^{2\epsilon} - q^{-2}). \tag{4.7}$$

4.3 Discrete Time Evolution by Adjoint Action of the Gaussian

By analogy with the A_{N-1} case (see Theorem 3.5), define the Gaussian function for types BCD:

$$\gamma := e^{\sum_{i=1}^{N} \frac{(\mathrm{Log}(x_i))^2}{4\mathrm{Log}(q)}}. \tag{4.8}$$

Note the slight modification $q \to q^2$ compared to the A type Gaussian (1.2).

The adjoint action of Gaussian function on the difference operators of Sect. 4.2 induces a discrete time evolution:

Definition 4.1 Let $k \in \mathbb{Z}$. The difference operators $M_{a,k}^{(\mathfrak{g})}$ are defined as follows:

$$\begin{aligned}
D_N: \quad & M_{1,k}^{(D_N)} := q^{-k}\,\gamma^{-k}\, M_{1,0}^{(D_N)}\,\gamma^k \\
& M_{N-1,k}^{(D_N)} := q^{-kN/4}\,\gamma^{-k}\, M_{N-1,0}^{(D_N)}\,\gamma^k \\
& M_{N,k}^{(D_N)} := q^{-kN/4}\,\gamma^{-k}\, M_{N,0}^{(D_N)}\,\gamma^k \\
B_N: \quad & M_{1,k}^{(B_N)} := q^{-k}\,\gamma^{-k}\, M_{1,0}^{(B_N)}\,\gamma^k \\
C_N: \quad & M_{1,2k}^{(C_N)} := q^{-2k}\,\gamma^{-2k}\, M_{1,0}^{(C_N)}\,\gamma^{2k} \\
& M_{1,2k+1}^{(C_N)} := q^{-2k}\,\gamma^{-2k}\, M_{1,1}^{(C_N)}\,\gamma^{2k} \\
& M_{N,k}^{(C_N)} := q^{-kN/2}\,\gamma^{-2k}\, M_{N,0}^{(C_N)}\,\gamma^{2k}.
\end{aligned}$$

Note that in type C, the definition for the time evolution of $M_{1,k}^{(C_N)}$ splits into two separate evolutions for even and odd k, involving the operators $M_{1,0}^{(C_N)}$ for even k and $M_{1,1}^{(C_N)}$ of (4.7) for odd k. This is due to the fact that α_1 is a short root in type C. The same parity phenomenon is expected for the short root α_N of B_N.

Using the simple relation $\gamma^{-1}\Gamma_i^2\gamma = qx_i\Gamma_i^2$ for γ as in (4.8), we see that Definition 4.1 results immediately in the following explicit expressions:

$$\text{Type } D_N: \quad M_{1,k}^{(D_N)} = \sum_{\epsilon=\pm1}\sum_{i=1}^{N}\prod_{j\neq i}\frac{x_i^\epsilon x_j}{x_i^\epsilon x_j - 1}\frac{x_i^\epsilon}{x_i^\epsilon - x_j}x_i^{k\epsilon}\Gamma_i^{2\epsilon}$$

$$M_{N-1,k}^{(D_N)} = \sum_{\substack{\epsilon_1,\ldots,\epsilon_N=\pm1\\ \epsilon_1\epsilon_2\cdots\epsilon_N=-1}}\prod_{1\le i<j\le N}\frac{x_i^{\epsilon_i}x_j^{\epsilon_j}}{x_i^{\epsilon_i}x_j^{\epsilon_j}-1}\prod_{i=1}^{N}x_i^{\frac{k\epsilon_i}{2}}\prod_{i=1}^{N}\Gamma_i^{\epsilon_i} \tag{4.9}$$

$$M_{N,k}^{(D_N)} = \sum_{\substack{\epsilon_1,\ldots,\epsilon_N=\pm1\\ \epsilon_1\epsilon_2\cdots\epsilon_N=1}}\prod_{1\le i<j\le N}\frac{x_i^{\epsilon_i}x_j^{\epsilon_j}}{x_i^{\epsilon_i}x_j^{\epsilon_j}-1}\prod_{i=1}^{N}x_i^{\frac{k\epsilon_i}{2}}\prod_{i=1}^{N}\Gamma_i^{\epsilon_i} \tag{4.10}$$

$$\text{Type } B_N: \quad M_{1,k}^{(B_N)} = \sum_{\epsilon=\pm1}\sum_{i=1}^{N}\frac{x_i^\epsilon}{x_i^\epsilon-1}\prod_{j\neq i}\frac{x_i^\epsilon x_j}{x_i^\epsilon x_j - 1}\frac{x_i^\epsilon}{x_i^\epsilon - x_j}x_i^{k\epsilon}\Gamma_i^{2\epsilon}$$

$$\text{Type } C_N: \quad M_{1,2k}^{(C_N)} = q^{-2k} + \sum_{\epsilon=\pm1}\sum_{i=1}^{N}\frac{x_i^{2\epsilon}}{x_i^{2\epsilon}-1}\frac{q^2x_i^{2\epsilon}}{q^2x_i^{2\epsilon}-1}$$
$$\times\prod_{j\neq i}\frac{x_i^\epsilon x_j}{x_i^\epsilon x_j - 1}\frac{x_i^\epsilon}{x_i^\epsilon - x_j}(x_i^{2k\epsilon}\Gamma_i^{2\epsilon} - q^{-2k})$$

$$M_{1,2k-1}^{(C_N)} = \sum_{\epsilon=\pm1}\sum_{i=1}^{N}\frac{x_i^{2\epsilon}}{x_i^{2\epsilon}-1}\frac{q^2x_i^{2\epsilon}}{q^2x_i^{2\epsilon}-1}\prod_{j\neq i}\frac{x_i^\epsilon x_j}{x_i^\epsilon x_j - 1}\frac{x_i^\epsilon}{x_i^\epsilon - x_j}x_i^{-\epsilon}(x_i^{2k\epsilon}\Gamma_i^{2\epsilon} - q^{-2k})$$

$$M_{N,k}^{(C_N)} = \sum_{\epsilon_1,\ldots,\epsilon_N=\pm1}\prod_{i=1}^{N}\frac{x_i^{2\epsilon_i}}{x_i^{2\epsilon_i}-1}\prod_{1\le i<j\le N}\frac{x_i^{\epsilon_i}x_j^{\epsilon_j}}{x_i^{\epsilon_i}x_j^{\epsilon_j}-1}\prod_{i=1}^{N}x_i^{k\epsilon_i}\prod_{i=1}^{N}\Gamma_i^{\epsilon_i}. \tag{4.11}$$

These expressions motivate the main conjectures of this paper.

4.4 Other Macdonald Operators via Quantum Determinants

The list of operators defined in the previous section can be completed to include $M_{a,k}^{(\mathfrak{g})}$ for the Dynkin labels a which are not in the list above, by using quantum determinants.

Comparing the commutation relations and mutation relations of the quantum BCD Q-systems involving only the Dynkin labels $\{1,\ldots,N-1\}$ (types BC) and $\{1,\ldots,N-2\}$ (type D), we see that the relations are the same as in type A of (3.2), expressed in the variable $M_{a,k}$, up to the change of parameter $q \mapsto q^2$ (which results in a change in the first factor $q^a \mapsto q^{2a}$, and the commutation relations $q^{\min(a,b)} \mapsto q^{2\min(a,b)}$). This observation leads to the following definitions:

Definition 4.2 For types BCD, define the difference operators $M_{a,k}^{(\mathfrak{g})}$ by the following quantum determinants:

$$M_{a,k}^{(\mathfrak{g})} := |\mathbf{M}^{(\mathfrak{g})}[\underbrace{k, k, \ldots, k}_{a \text{ times}}]|_{q^2} = M_{k,k,\ldots,k}^{(\mathfrak{g})} \qquad (a = 1, 2, \ldots, n_{\mathfrak{g}}), \tag{4.12}$$

where $n_{D_N} = N - 2$, and $n_{B_N} = n_{C_N} = N - 1$. Here the matrix $\mathbf{M}^{(\mathfrak{g})}(\{\mathbf{a}\})$ is obtained by replacing $M_{1,k}$ with $M_{1,k}^{(\mathfrak{g})}$ in the defining expression (3.3), and the quantum determinant has parameter q^2 instead of q in the Vandermonde factor.

By homogeneity of the quantum determinant as a polynomial in the variables $M_{1,k}^{(\mathfrak{g})}$, Definition (4.12) is compatible with the discrete time evolution relations:

$$\begin{aligned}
M_{a,k}^{(D_N)} &= q^{-ka}\,\gamma^{-k}\,M_{a,0}^{(D_N)}\,\gamma^k, \qquad (a = 1, 2, \ldots, N-2);\\
M_{a,k}^{(B_N)} &= q^{-ka}\,\gamma^{-k}\,M_{a,0}^{(B_N)}\,\gamma^k, \qquad (a = 1, 2, \ldots, N-1);\\
M_{a,2k+\eta}^{(C_N)} &= q^{-2ka}\,\gamma^{-2k}\,M_{a,\eta}^{(C_N)}\,\gamma^{2k}, \qquad (a = 1, 2, \ldots, N-1, \eta = 0, 1),
\end{aligned}$$

with the Gaussian function γ as in (4.8).

4.5 *The Quantum Q-System Conjectures*

The main conjecture of this paper is the following:

Conjecture 3 *For $G = D_N, B_N, C_N$, the Macdonald operators $M_{a,n}^{(\mathfrak{g})}$ obey the following renormalized versions (we omit the superscript $\mathfrak{g}$ for simplicity) of the quantum Q-systems (2.12–2.10):*

Type D_N :

$$\begin{aligned}
M_{a,k}\,M_{b,p} &= q^{\Lambda_{a,b}(p-k)}\,M_{b,p}\,M_{a,k}\\
q^{2a}\,M_{a,k+1}M_{a,k-1} &= M_{a,k}^2 - M_{a+1,k}M_{a-1,k} \qquad (a \in [1, N-3])\\
q^{2(N-2)}\,M_{N-2,k+1}M_{N-2,k-1} &= M_{N-2,k}^2\\
&\quad -q^{-\frac{(N-2)k}{2}}\,M_{N,k}M_{N-1,k}M_{N-3,k} \qquad (4.13)\\
q^{\frac{N}{2}}\,M_{N-1,k+1}M_{N-1,k-1} &= M_{N-1,k}^2 - q^{\frac{(N-4)k}{2}}\,M_{N-2,k}\\
q^{\frac{N}{2}}\,M_{N,k+1}M_{N,k-1} &= M_{N,k}^2 - q^{\frac{(N-4)k}{2}}\,M_{N-2,k}
\end{aligned}$$

$$
\begin{aligned}
\textit{Type } B_N: \qquad M_{a,k}\, M_{b,p} &= q^{\Lambda_{a,b}p-\Lambda_{b,a}k}\, M_{b,p}\, M_{a,k} \\
q^{2a}\, M_{a,k+1}\, M_{a,k-1} &= (M_{a,k})^2 - M_{a+1,k}\, M_{a-1,k} \quad (a \in [1, N-2]) \\
q^{2N-2}\, M_{N-1,k+1}\, M_{N-1,k-1} &= (M_{N-1,k})^2 - M_{N,2k}\, M_{N-2,k} \qquad (4.14) \\
q^{N}\, M_{N,2k+1}\, M_{N,2k-1} &= (M_{N,2k})^2 - q^{-2k}(M_{N-1,k})^2 \\
q^{N}\, M_{N,2k+2}\, M_{N,2k} &= (M_{N,2k+1})^2 \\
&\quad - q^{N-1-(2k+1)} M_{N-1,k+1}\, M_{N-1,k} \qquad (4.15)
\end{aligned}
$$

$$
\begin{aligned}
\textit{Type } C_N: \qquad M_{a,k}\, M_{b,p} &= q^{\Lambda_{a,b}p-\Lambda_{b,a}k}\, M_{b,p}\, M_{a,k} \\
q^{2a} M_{a,k+1}\, M_{a,k-1} &= M_{a,k}^2 - M_{a+1,k} M_{a-1,k} \qquad (a \in [1, N-2]) \\
q^{2N-2} M_{N-1,2k+1} M_{N-1,2k-1} &= M_{N-1,2k}^2 - q^{-Nk}\, M_{N-2,2k} M_{N,k}^2 \\
q^{2N-2} M_{N-1,2k+2} M_{N-1,2k} &= M_{N-1,2k+1}^2 - q^{-Nk}\, M_{N-2,2k+1} M_{N,k+1} M_{N,k} \\
q^{N} M_{N,k+1}\, M_{N,k-1} &= M_{N,k}^2 - q^{(N-2)k}\, M_{N-1,2k} \qquad (4.16)
\end{aligned}
$$

In the case of B_N we have not given an *a priori* definition of the order N operators $M_{N,k}$. They can be constructed as a suitable linear combination of the limits at $t \to \infty$ of van Diejen operators [vD95, vDE11], compatible with the quantum Q-system in type B. These operators can be defined also by taking the quantum Q-system as the defining set of equations. First, define $M_{N,2k} = |M(\{k, k, \dots, k\})|_{q^2}$ (k repeated N times), compatible with Eq. (4.14), in which $M_{N,2k}$ plays the role of the quantum determinant of size N. Then Eq. (4.15) gives information about $M_{N,2k+1}$:

$$
(M_{N,2k+1})^2 = q^N\, M_{N,2k+2}\, M_{N,2k} + q^{N-1-(2k+1)} M_{N-1,k+1}\, M_{N-1,k}.
$$

This can be used to determine the relevant difference operators. The fact that all the other equations of the system are satisfied is a highly non-trivial check.

These conjectures have been checked numerically up to $N = 6$. We illustrate them in cases of small rank in Appendix A.

Lemma 4.4 *The difference operators $M_{a,k}$ are a functional representation of the quantum Q-system (2.12–2.10) up to the following rescaling:*

$$
\begin{aligned}
D_N: \qquad M_{a,k} &= q^{\frac{a(a+1)}{2}-a(N+k)}\, \mathcal{Q}_{a,k} \qquad (\alpha \in [1, N-2]) \\
M_{N-1,k} &= q^{-\frac{N(N-1)}{4}-\frac{Nk}{4}}\, \mathcal{Q}_{N-1,k} \\
M_{N,k} &= q^{-\frac{N(N-1)}{4}-\frac{Nk}{4}}\, \mathcal{Q}_{N,k} \qquad (4.17) \\
B_N: \qquad M_{a,k} &= q^{\frac{a^2}{2}-a(N+k)}\, \mathcal{Q}_{a,k} \qquad (a \in [1, N-1]) \\
M_{N,k} &= q^{-\frac{N^2}{2}-\frac{Nk}{2}}\, \mathcal{Q}_{N,k} \qquad (4.18)
\end{aligned}
$$

$$C_N: \qquad M_{a,k} = q^{\frac{a(a-1)}{2}-a(N+k)}\,\mathcal{Q}_{a,k} \qquad (a \in [1, N-1])$$
$$M_{N,k} = q^{-\frac{N(N+1)}{4}-\frac{Nk}{2}}\,\mathcal{Q}_{N,k} \tag{4.19}$$

Proof By straightforward inspection. □

5 The Raising/Lowering Operator Conjectures for Types BCD

In this section, we present conjectures that extend the result of Theorem 3.6 to types BCD. These involve the action of the difference operators $M_{a,k}$ on the dual q-Whittaker functions:

$$\Pi_\lambda^{(\mathfrak{g})}(q^{-1};\mathbf{x}) := \lim_{t\to\infty} P_\lambda^{(\mathfrak{g})}(q,t;\mathbf{x}).$$

The general idea is that while the operators $M_{a,0}^{(\mathfrak{g})}$ are all limits of Macdonald operators, for which the q-Whittaker functions $\Pi_\lambda^{(\mathfrak{g})}$ are common eigenfunctions, the operators $M_{a,\pm 1}^{(\mathfrak{g})}$ are simple raising and lowering operators on those eigenfunctions.

Conjecture 5.1 *The operators $M_{a,0}^{(\mathfrak{g})}$ and $M_{a,\pm 1}^{(\mathfrak{g})}$ have the following action on the q-Whittaker functions $\Pi_\lambda^{(\mathfrak{g})}$, valid for all $a \in [1, N]$:*

$$D_N, B_N: \qquad M_{a,0}\,\Pi_\lambda = q^{2t_a(\lambda,\omega_a)}\,\Pi_\lambda$$
$$M_{a,1}\,\Pi_\lambda = q^{2t_a(\lambda,\omega_a)}\,\Pi_{\lambda+\omega_a}$$
$$M_{a,-1}\,\Pi_\lambda = q^{2t_a(\lambda,\omega_a)}\left(1-q^{-2t_a(\alpha_a,\lambda)}\right)\Pi_{\lambda-\omega_a} \tag{5.1}$$
$$C_N: \qquad M_{a,0}\,\Pi_\lambda = q^{t_a(\omega_a,\lambda)}\,\Pi_\lambda$$
$$M_{a,1}\,\Pi_\lambda = q^{t_a(\omega_a,\lambda)}\,\Pi_{\lambda+\omega_a}$$
$$M_{a,-1}\,\Pi_\lambda = q^{t_a(\omega_a,\lambda)}\left(1-q^{-t_a(\alpha_a,\lambda)}\right)\Pi_{\lambda-\omega_a}, \tag{5.2}$$

where ω_a and α_a are the fundamental weights and simple roots of the corresponding algebras, $t_a = 2$ for the short roots in types BC, and $t_a = 1$ otherwise.

The eigenvalue equations (the first equation in each set) are a consequence of the Macdonald eigenvalue equations of the finite t case. As in type A, up to a scalar multiple, the raising equations (the second equation in each set) are a consequence of the quantum commutation relations $M_{a,0}^{(\mathfrak{g})} M_{b,1}^{(\mathfrak{g})} = q^{\lambda_{a,b}^{(\mathfrak{g})}} M_{b,1}^{(\mathfrak{g})} M_{a,0}^{(\mathfrak{g})}$. When applied to the q-Whittaker functions, this gives

$$M_{a,0}^{(\mathfrak{g})}\left(M_{b,1}^{(\mathfrak{g})}\,\Pi_{\lambda}^{(\mathfrak{g})}\right) = q^{\lambda_{a,b}^{(\mathfrak{g})}} M_{b,1}^{(\mathfrak{g})} M_{a,0}^{(\mathfrak{g})}\,\Pi_{\lambda}^{(\mathfrak{g})} = q^{\epsilon^{(\mathfrak{g})} t_a^{(\mathfrak{g})}(\omega_a^{(\mathfrak{g})},\lambda+\omega_b^{(\mathfrak{g})})} M_{b,1}^{(\mathfrak{g})}\,\Pi_{\lambda}^{(\mathfrak{g})},$$

where we have defined $\epsilon^{(\mathfrak{g})} = 2$ for $\mathfrak{g} = B_N, D_N$ and $\epsilon^{(\mathfrak{g})} = 1$ for $\mathfrak{g} = C_N$, and used the fact that $\lambda_{a,b}^{(\mathfrak{g})} = \epsilon^{(\mathfrak{g})} t_a^{(\mathfrak{g})}(\omega_a^{(\mathfrak{g})}, \omega_b^{(\mathfrak{g})})$. As the simultaneous eigenvalue property for the action of all $M_{a,0}$ determines the eigenvectors uniquely, up to a scalar multiple, we deduce that $M_{b,1}^{(\mathfrak{g})}\,\Pi_{\lambda}^{(\mathfrak{g})}$ must be proportional to $\Pi_{\lambda+\omega_b}^{(\mathfrak{g})}$. The conjecture concerns simply the explicit value of the proportionality factor $q^{\epsilon^{(\mathfrak{g})} t_a^{(\mathfrak{g})}(\omega_a^{(\mathfrak{g})},\lambda)}$, which we checked numerically up to $N = 6$.

Similarly, the action of the lowering operator (third equation in each set) follows from the eigenvalue and raising operator actions and the quantum Q-system equations (4.13–4.16) of Conjecture 3. This is readily seen by applying the exchange relations involving $M_{a,1}$ and $M_{a,-1}$ to Π_λ.

Examples of q-Whittaker functions and raising/lowering difference operators are given in Appendix A.

6 The Graded Character Conjectures for Types BCD

The main Conjecture 3 is a necessary ingredient in proving the following conjecture about the expression of the graded characters in terms of difference operators. Let $I_<$ be the subset labels of the short roots the Dynkin diagram in types BC. Recall that $t_a = 2$ for $a \in I_<$, and $t_a = 1$ otherwise.

Conjecture 6.1 *The graded characters of the tensor products of KR-modules in types BCD can be expressed using the iterated action of the difference operators in types BCD on the polynomial 1:*

$$\chi_{\mathbf{n}}^{(\mathfrak{g})}(q^{-1};\mathbf{x}) = q^{-\frac{1}{2}Q^{(\mathfrak{g})}(\mathbf{n})} \prod_{\ell=k}^{1}\left(\prod_{a\in I}(M_{a,t_a\ell}^{(\mathfrak{g})})^{n_{a,t_a\ell}} \prod_{a\in I_<}(M_{a,t_a\ell-1}^{(\mathfrak{g})})^{n_{a,t_a\ell-1}}\right)\cdot 1,$$

where k is chosen large enough to cover all the non-zero n's, and

$$Q^{(\mathfrak{g})}(\mathbf{n}) := \sum_{a,b=1}^{N}\sum_{i,j\geq 1} n_{a,i}\,\frac{\lambda_{a,b}^{(\mathfrak{g})}}{t_a^{(\mathfrak{g})}}\min(t_b^{(\mathfrak{g})} i, t_a^{(\mathfrak{g})} j)\, n_{b,j} - \sum_{a=1}^{N}\sum_{i\geq 1} i\,\lambda_{a,a}^{(\mathfrak{g})}\, n_{a,i}.$$

The ordering of the difference operators in Conjecture 6.1 is consistent with [DFK14, Lin19] and is determined by the order of mutations in the bipartite belt of the quantum cluster algebra.

In particular, Conjecture 6.1 allows one to interpret the so-called level 1 graded characters corresponding to only possibly non-zero $n_a = n_{a,1}$, as the limiting Macdonald polynomials (or dual q-Whittaker functions) Π_λ, with the correspondence:

$$\chi_{\{n_a\}}^{(\mathfrak{g})}(q^{-1}, \mathbf{x}) = q^{-\frac{1}{2}Q^{(\mathfrak{g})}(\{n_a\})} \prod_{a\in I_>} \left(M_{a,1}^{(\mathfrak{g})}\right)^{n_a} \prod_{a\in I_<} \left(M_{a,1}^{(\mathfrak{g})}\right)^{n_a} \cdot 1 = \Pi_\lambda^{(\mathfrak{g})}(q^{-1}; \mathbf{x}), \tag{6.1}$$

where $I_> = I \setminus I_<$ is the set of long root labels, $\lambda = \sum_a n_a \omega_a$, and the quadratic form $Q^{(\mathfrak{g})}$ is

$$\begin{aligned} Q^{(\mathfrak{g})}(\{n_a\}) &= \sum_{a,b=1}^{N} n_a \frac{\lambda_{a,b}^{(\mathfrak{g})}}{t_a^{(\mathfrak{g})}} \min(t_a^{(\mathfrak{g})}, t_b^{(\mathfrak{g})})\, n_b - \sum_{a=1}^{N} \lambda_{a,a}^{(\mathfrak{g})} n_a \\ &= \epsilon^{(\mathfrak{g})} \left\{ \sum_{a,b=1}^{N} n_a\, (\omega_a^{(\mathfrak{g})}, \omega_b^{(\mathfrak{g})}) \min(t_a^{(\mathfrak{g})}, t_b^{(\mathfrak{g})})\, n_b - \sum_{a=1}^{N} t_a^{(\mathfrak{g})} (\omega_a^{(\mathfrak{g})}, \omega_a^{(\mathfrak{g})})\, n_a \right\}. \end{aligned}$$

Alternatively, for a fixed $\mathbf{n}$, acting with $M_{a,k}^{(\mathfrak{g})}$ such that $t_a k \geq \max\{t_b j : n_{b,j} > 0\}$ on $\chi_{\mathbf{n}}$ gives

$$M_{a,k}^{(\mathfrak{g})}\, \chi_{\mathbf{n}}^{(\mathfrak{g})}(q^{-1}; \mathbf{x}) = q^{\epsilon^{(\mathfrak{g})} \sum_{j,b} (\omega_a^{(\mathfrak{g})}, \omega_b^{(\mathfrak{g})}) \min(t_b^{(\mathfrak{g})} k, t_a^{(\mathfrak{g})} j) n_{b,j}}\, \chi_{\mathbf{n}+\epsilon_{a,k}}^{(\mathfrak{g})}(q^{-1}; \mathbf{x}),$$

where, again, $\mathbf{n} = \sum n_{b,j} \epsilon_{b,j}$.

It is a non-trivial exercise to check the compatibility between (6.1) and the raising operator conditions on the second lines of (5.1) and (5.2). Indeed, we note that for a short root label a, we may only act with $M_{a,1}^{(\mathfrak{g})}$ on characters obtained themselves by short root raising operators say $\prod_{b\in I_<} (M_{b,1}^{(\mathfrak{g})})^{n_b} \cdot 1$:

$$\begin{aligned} M_{a,1}^{(\mathfrak{g})}\, \chi_{\{n_b\}}^{(\mathfrak{g})}(q^{-1}, \mathbf{x}) &= q^{-\frac{1}{2}Q^{(\mathfrak{g})}(\{n_b\})} \prod_{b\in I_<} (M_{b,1}^{(\mathfrak{g})})^{n_b+\delta_{a,b}} \cdot 1 \\ &= q^{\frac{1}{2}(Q^{(\mathfrak{g})}(\{n_b+\delta_{a,b}\}) - Q^{(\mathfrak{g})}(\{n_b\}))}\, \chi_{\{n_b+\delta_{a,b}\}}^{(\mathfrak{g})}(q^{-1}, \mathbf{x}). \end{aligned}$$

We compute

$$\begin{aligned} Q^{(\mathfrak{g})}(\{n_b+\delta_{a,b}\}) - Q^{(\mathfrak{g})}(\{n_b\}) &= \epsilon^{(\mathfrak{g})} \sum_{b\in I_<} n_b \min(t_a^{(\mathfrak{g})}, t_b^{(\mathfrak{g})})(\omega_a^{(\mathfrak{g})}, \omega_b^{(\mathfrak{g})}) \\ &= \epsilon^{(\mathfrak{g})}\, t_a^{(\mathfrak{g})} (\omega_a^{(\mathfrak{g})}, \sum_b n_b \omega_b^{(\mathfrak{g})}) = \epsilon^{(\mathfrak{g})}\, t_a^{(\mathfrak{g})} (\omega_a^{(\mathfrak{g})}, \lambda) \end{aligned}$$

by use of $t_a^{(\mathfrak{g})} = t_b^{(\mathfrak{g})} = 2$. This is in agreement with (5.1) and (5.2), upon writing $\lambda = \sum n_b \omega_b$. For a long root label a, we have similarly:

$$Q^{(\mathfrak{g})}(\{n_b + \delta_{a,b}\}) - Q^{(\mathfrak{g})}(\{n_b\}) = \epsilon^{(\mathfrak{g})} \sum_{b \in I} n_b \min(t_a^{(\mathfrak{g})}, t_b^{(\mathfrak{g})})(\omega_a^{(\mathfrak{g})}, \omega_b^{(\mathfrak{g})})$$
$$= \epsilon^{(\mathfrak{g})} (\omega_a^{(\mathfrak{g})}, \sum_b n_b \omega_b^{(\mathfrak{g})}) = \epsilon^{(\mathfrak{g})} t_a^{(\mathfrak{g})} (\omega_a, \lambda)$$

by use of $t_a^{(\mathfrak{g})} = 1$. This is again in agreement with (5.1) and (5.2).

This special case is also in agreement with [LNS$^+$17].

7 Conclusion

7.1 *Summary: Macdonald Operators and Quantum Cluster Algebra*

In this paper we have presented two main conjectures about the difference operators $M_{a,k}^{(\mathfrak{g})}$ corresponding to root systems of types BCD.

The first conjecture states that the difference operators $\{M_{a,k}^{(\mathfrak{g})}\}$ satisfy renormalized quantum Q-systems of types BCD. The commuting difference operators $\{M_{a,0}^{(\mathfrak{g})}\}$ can be obtained as the limits $t \to \infty$ of appropriate Macdonald operators for the relevant Lie root systems. The operators $M_{a,k}^{(\mathfrak{g})}$ are their $SL_2(\mathbb{Z})$ "discrete time evolution."

The second conjecture is that the difference operators $M_{a,\pm 1}^{(\mathfrak{g})}$ at times $k = \pm 1$ act as raising and lowering operators on q-Whittaker functions $\Pi_\lambda^{(\mathfrak{g})}$.

The quantum Q-systems are mutations in the corresponding quantum cluster algebras. From this point of view, the sets $S_\pm := \{M_{a,0}^{(\mathfrak{g})}, M_{a,\pm 1}^{(\mathfrak{g})}\}$ are two possible valid initial cluster seeds, and the conjectures state that, as in type A, these are formed of the Macdonald and raising (reps. lowering) operators, at $t \to \infty$. This raises a number of questions regarding cluster variables in general: from preliminary inspection, it appears that *all cluster variables* in the corresponding cluster algebra are difference operators as well, a quite surprising property which goes way beyond the usual Laurent property of quantum cluster algebra, which would only imply that all cluster variables are Laurent polynomials of those in an initial cluster. These other difference operators should tell us something new about Macdonald theory (including in type A).

Finally we note that the second conjecture, if true, gives a perturbative starting point for constructing generalizations of finite t, A-type raising operators of Kirillov and Noumi for other types. Indeed, we expect the finite t operators to have a finite $1/t$ expansion, whose coefficients are themselves difference operators.

Conjecture 5.1 provides the leading order term in this expansion. We will return to this question in a future publication.

7.2 Towards Proving the Conjectures

The second conjecture of this paper offers a possible strategy for proving both conjectures which goes as follows. The second conjecture indeed can be restated as follows: the Macdonald and raising operators $M_{a,0}^{(\mathfrak{g})}$ and $M_{a,1}^{(\mathfrak{g})}$ act on the basis Π_λ of Weyl-symmetric polynomials of $\mathbf{x}$ as very simple *dual* operators X_a, P_a:

$$M_{a,0}^{(\mathfrak{g})}\,\Pi_\lambda^{(\mathfrak{g})} = \Pi_\lambda^{(\mathfrak{g})}\,X_a, \quad M_{a,1}^{(\mathfrak{g})}\,\Pi_\lambda^{(\mathfrak{g})} = \Pi_\lambda^{(\mathfrak{g})}\,P_a,$$

where the operators X_a, P_a act to the left on the variables $\Lambda_a = q^{2\lambda_a}$. More precisely the operators X_a acts diagonally on the basis $\Pi_\lambda^{(\mathfrak{g})}$ with eigenvalues $q^{\epsilon_a^{(\mathfrak{g})}(\omega_a^{(\mathfrak{g})},\lambda)}$, whereas P_a includes a multiplicative shift of Λ variables. As a result, the operators X_a, P_a obey the simple (opposite) commutation relations: $X_a\,P_b = q^{-\lambda_{a,b}^{(\mathfrak{g})}}\,P_b\,X_a$.

Reversing the logic entirely, we may *start from the data* of operators X_a, P_a, and *define* the left action of the operators $M_{a,k}^{(\mathfrak{g})}$ via the (opposite) renormalized quantum Q-systems of this paper. To identify the left action and the right one, we need to construct the Gaussian operator in X_a, P_a variables, and the conjectures will follow from the anti-homomorphism mapping left and right actions. We will pursue this program elsewhere.

7.3 sDAHA, EHA and the t-Deformation of Quantum Q-Systems

In [DFK19], we investigated the natural t-deformation of the A type quantum Q-system provided by the type A sDAHA, also expressed as a quotient of the elliptic Hall algebra (EHA) [Sch12], or the quantum toroidal algebra of $\mathfrak{gl}_1$. There exist sDAHAs for all classical types [Che05], and this would be the natural candidate for a generalization of the A type results. We may use as starting points the (q,t)-Macdonald operators of Sect. 4.1, and their time evolution via the suitable $SL_2(\mathbb{Z})$ action, to derive current algebra relations that will generalize EHA or quantum toroidal algebra relations, ideally giving rise to new interesting algebras.

Acknowledgments RK and PDF acknowledge support by NFS grant DMS18-02044. PDF is partially supported by the Morris and Gertrude Fine endowment. RK thanks the Institut de Physique Théorique of CEA/Saclay for hospitality.

Appendix A: Examples

This appendix gathers a few inter-connected examples, which give non-trivial checks of the conjectures of this paper. We give here the examples of the B_2 and C_2 cases, that of the A_3 and D_3, and finally the D_4 case. We verify in particular that the symmetries of the Dynkin diagrams are also reflected on our difference operators.

A.1 Weyl-Invariant Schur Functions

Definition A.1 We define the Weyl-invariant Schur functions $s_\lambda^{(\mathfrak{g})}(\mathbf{x})$ for $\mathfrak{g} = A_{N-1}, B_N, C_N, D_N$ to be the characters of the irreducible representations of corresponding weight [Miz03]:[2]

$$s_\lambda^{(A_{N-1})}(\mathbf{x}) = \frac{\det_{1\le i,j\le N}\left(x_i^{N-j+\lambda_j}\right)}{\det_{1\le i,j\le N}\left(x_i^{N-j}\right)} \tag{A.1}$$

$$s_\lambda^{(B_N)}(\mathbf{x}) = \frac{\det_{1\le i,j\le N}\left(x_i^{N-j+\lambda_j+\frac{1}{2}} - \frac{1}{x_i^{N-j+\lambda_j+\frac{1}{2}}}\right)}{\det_{1\le i,j\le N}\left(x_i^{N-j+\frac{1}{2}} - \frac{1}{x_i^{N-j+\frac{1}{2}}}\right)} \tag{A.2}$$

$$s_\lambda^{(C_N)}(\mathbf{x}) = \frac{\det_{1\le i,j\le N}\left(x_i^{N-j+\lambda_j+1} - \frac{1}{x_i^{N-j+\lambda_j+1}}\right)}{\det_{1\le i,j\le N}\left(x_i^{N-j+1} - \frac{1}{x_i^{N-j+1}}\right)} \tag{A.3}$$

$$s_\lambda^{(D_N)}(\mathbf{x}) = \frac{\det_{1\le i,j\le N}\left(x_i^{N-j+\lambda_j} - \frac{1}{x_i^{N-j+\lambda_j}}\right)}{\det_{1\le i,j\le N}\left(x_i^{N-j} + \frac{1}{x_i^{N-j}}\right)} + \frac{\det_{1\le i,j\le N}\left(x_i^{N-j+\lambda_j} + \frac{1}{x_i^{N-j+\lambda_j}}\right)}{\det_{1\le i,j\le N}\left(x_i^{N-j} + \frac{1}{x_i^{N-j}}\right)}. \tag{A.4}$$

The Weyl-invariant Schur functions form a basis of the space of Weyl-invariant (Laurent) polynomials. In the following sections, we write the dual q-Whittaker functions Π_λ in this Schur basis. We drop the superscript $(\mathfrak{g})$ for simplicity.

[2] In [Miz03], the author restricts the definition to λ's that are actual partitions. Here we include all possible dominant weights.

Remark A.2 Note that Eq. (A.1) for $\mathfrak{gl}_N$ Schur functions gives the $\mathfrak{sl}_N$ Schur functions, upon restriction to $x_1x_2\cdots x_N = 1$ and then noting that $s_\lambda = s_{\lambda-\epsilon}$ where $\epsilon = e_1 + e_2 + \cdots + e_N$.

A.2 The B_2 Case

A.2.1 M Operators

$$M_{1,k} = \sum_{\epsilon=\pm 1} x_1^{k\epsilon} \frac{x_1^\epsilon}{x_1^\epsilon - 1} \frac{x_1^\epsilon x_2}{x_1^\epsilon x_2 - 1} \frac{x_1^\epsilon}{x_1^\epsilon - x_2} \Gamma_1^{2\epsilon} + x_2^{k\epsilon} \frac{x_2^\epsilon}{x_2^\epsilon - 1} \frac{x_2^\epsilon x_1}{x_2^\epsilon x_1 - 1} \frac{x_2^\epsilon}{x_2^\epsilon - x_1} \Gamma_2^{2\epsilon}$$

$$M_{2,2k} = q^{-2k} + \sum_{\epsilon_1,\epsilon_2=\pm 1} \frac{x_1^{\epsilon_1}}{x_1^{\epsilon_1} - 1} \frac{x_2^{\epsilon_2}}{x_2^{\epsilon_2} - 1} \frac{x_1^{\epsilon_1}x_2^{\epsilon_2}}{x_1^{\epsilon_1}x_2^{\epsilon_2} - 1} \frac{q^2x_1^{\epsilon_1}x_2^{\epsilon_2}}{q^2x_1^{\epsilon_1}x_2^{\epsilon_2} - 1} \times (x_1^{k\epsilon_1}x_2^{k\epsilon_2}\Gamma_1^{2\epsilon_1}\Gamma_2^{2\epsilon_2} - q^{-2k})$$

$$M_{2,2k-1} = \sum_{\epsilon_1,\epsilon_2=\pm 1} \frac{x_1^{\epsilon_1}}{x_1^{\epsilon_1} - 1} \frac{x_2^{\epsilon_2}}{x_2^{\epsilon_2} - 1} \frac{x_1^{\epsilon_1}x_2^{\epsilon_2}}{x_1^{\epsilon_1}x_2^{\epsilon_2} - 1} \frac{q^2x_1^{\epsilon_1}x_2^{\epsilon_2}}{q^2x_1^{\epsilon_1}x_2^{\epsilon_2} - 1} \times (x_1^{\epsilon_1}x_2^{\epsilon_2})^{-\frac{1}{2}} (x_1^{k\epsilon_1}x_2^{k\epsilon_2}\Gamma_1^{2\epsilon_1}\Gamma_2^{2\epsilon_2} - q^{-2k})$$

Here $M_{1,k}$ is given by the k-th iterate conjugation of (4.4) w.r.t. the Gaussian, and $M_{2,2k}$ by the quantum determinant $M_{1,k}^2 - q^2M_{1,k+1}M_{1,k-1}$.

A.2.2 Dual q-Whittaker Functions

$$\Pi_{0,0} = s_{0,0}$$

$$\Pi_{1,0} = s_{1,0}$$

$$\Pi_{2,0} = s_{2,0} + q^{-2}s_{1,1} + q^{-4}s_{0,0}$$

$$\Pi_{1,1} = s_{1,1} + q^{-2}s_{1,0} + q^{-2}s_{0,0}$$

$$\Pi_{3,0} = s_{3,0} + \frac{1+q^2}{q^4}s_{2,1} + q^{-6}s_{1,1} + \frac{1+q^2+q^4}{q^8}s_{1,0}$$

$$\Pi_{2,1} = s_{2,1} + q^{-2}s_{2,0} + \frac{1+q^2}{q^4}s_{1,1} + \frac{1+q^2}{q^4}s_{1,0} + q^{-6}s_{0,0}$$

$$\Pi_{\frac{1}{2},\frac{1}{2}} = s_{\frac{1}{2},\frac{1}{2}}$$

$$\Pi_{\frac{3}{2},\frac{1}{2}} = s_{\frac{3}{2},\frac{1}{2}} + q^{-2} s_{\frac{1}{2},\frac{1}{2}}$$

$$\Pi_{\frac{5}{2},\frac{1}{2}} = s_{\frac{5}{2},\frac{1}{2}} + q^{-2} s_{\frac{3}{2},\frac{3}{2}} + \frac{1+q^2}{q^4} s_{\frac{3}{2},\frac{1}{2}} + \frac{1+q^2}{q^6} s_{\frac{1}{2},\frac{1}{2}}$$

$$\Pi_{\frac{3}{2},\frac{3}{2}} = s_{\frac{3}{2},\frac{3}{2}} + \frac{1+q^2}{q^4} s_{\frac{3}{2},\frac{1}{2}} + \frac{1+q^2+q^4}{q^6} s_{\frac{1}{2},\frac{1}{2}}$$

A.3 The C_2 Case

A.3.1 The $B_2 \leftrightarrow C_2$ Symmetry

The B_2 case can be mapped onto the C_2 case, by interchanging the roles of the two fundamental weights. More precisely, let us denote by $\omega_a = \omega_a^{(B_2)}$ (resp. $\omega'_a = \omega_a^{(C_2)}$). The variables x_1, x_2 of the B_2 case can be thought of as $x_i = e^{e_i}$, with $e_1 = \omega_1$ and $e_2 = 2\omega_2 - \omega_1$, and similarly for C_2, where $x'_i = e^{e'_i}$, $e'_1 = \omega'_1$ and $e'_2 = \omega'_2 - \omega'_1$. The mapping $(\omega_1, \omega_2) \mapsto (\omega'_2, \omega'_1)$ sends

$$x_1 = e^{\omega_1} \mapsto e^{\omega'_2} = x'_1 x'_2, \quad x_2 = e^{2\omega_2 - \omega_1} \mapsto e^{2\omega'_1 - \omega'_2} = \frac{x'_1}{x'_2}. \tag{A.5}$$

Similarly, we have

$$\Gamma_1^2 \mapsto \Gamma'_1 \Gamma'_2, \qquad \Gamma_2^2 \mapsto \Gamma'_1 {\Gamma'_2}^{-1}. \tag{A.6}$$

The map (A.5–A.6) sends the operators $M_{a,k}^{(B_2)} \mapsto M_{3-a,k}^{(C_2)}$ for $a = 1, 2$ and $k \in \mathbb{Z}$, and the Macdonald polynomials $P_{\lambda_1,\lambda_2}^{(B_2)} \mapsto P_{\lambda_1+\lambda_2,\lambda_1-\lambda_2}^{(C_2)}$ and similarly for the q-Whittaker functions, using

$$\begin{aligned} \lambda = \lambda_1 e_1 + \lambda_2 e_2 &= (\lambda_1 - \lambda_2)\omega_1 + 2\lambda_2\omega_2 \mapsto (\lambda_1 - \lambda_2)\omega'_2 + 2\lambda_2\omega'_1 \\ &= (\lambda_1 + \lambda_2)e'_1 + (\lambda_1 - \lambda_2)e'_2. \end{aligned}$$

A.3.2 M Operators

From the definitions in type C,

$$\begin{aligned} M_{1,2k} = q^{-2k} + \sum_{\epsilon=\pm 1} \Bigg\{ &\frac{x_1^{2\epsilon}}{x_1^{2\epsilon}-1} \frac{q^2 x_1^{2\epsilon}}{q^2 x_1^{2\epsilon}-1} \frac{x_1^{\epsilon} x_2}{x_1^{\epsilon} x_2 - 1} \frac{x_1^{\epsilon}}{x_1^{\epsilon} - x_2} (x_1^{2k\epsilon} \Gamma_1^{2\epsilon} - q^{-2k}) \\ &+ \frac{x_2^{2\epsilon}}{x_2^{2\epsilon}-1} \frac{q^2 x_2^{2\epsilon}}{q^2 x_2^{2\epsilon}-1} \frac{x_2^{\epsilon} x_1}{x_2^{\epsilon} x_1 - 1} \frac{x_2^{\epsilon}}{x_2^{\epsilon} - x_1} (x_2^{2k\epsilon} \Gamma_2^{2\epsilon} - q^{-2k}) \Bigg\}, \end{aligned}$$

$$M_{1,2k-1} = \sum_{\epsilon=\pm1} \left\{ \frac{x_1^{2\epsilon}}{x_1^{2\epsilon}-1} \frac{q^2 x_1^{2\epsilon}}{q^2 x_1^{2\epsilon}-1} \frac{x_1^\epsilon x_2}{x_1^\epsilon x_2 - 1} \frac{x_1^\epsilon}{x_1^\epsilon - x_2} x_1^{-\epsilon}(x_1^{2k\epsilon}\Gamma_1^{2\epsilon} - q^{-2k}) \right.$$

$$\left. + \frac{x_2^{2\epsilon}}{x_2^{2\epsilon}-1} \frac{q^2 x_2^{2\epsilon}}{q^2 x_2^{2\epsilon}-1} \frac{x_2^\epsilon x_1}{x_2^\epsilon x_1 - 1} \frac{x_2^\epsilon}{x_2^\epsilon - x_1} x_2^{-\epsilon}(x_2^{2k\epsilon}\Gamma_2^{2\epsilon} - q^{-2k}) \right\},$$

$$M_{2,k} = \sum_{\epsilon_1,\epsilon_2=\pm1} \frac{x_1^{2\epsilon_1}}{x_1^{2\epsilon_1}-1} \frac{x_2^{2\epsilon_2}}{x_2^{2\epsilon_2}-1} \frac{x_1^{\epsilon_1}x_2^{\epsilon_2}}{x_1^{\epsilon_1}x_2^{\epsilon_2}-1} (x_1^{k\epsilon_1}x_2^{k\epsilon_2})\,\Gamma_1^{\epsilon_1}\Gamma_2^{\epsilon_2}.$$

The expected symmetry between $M_{1,k}^{(B_2)} \mapsto M_{2,k}^{(C_2)}$ and $M_{2,k}^{(B_2)} \mapsto M_{1,k}^{(C_2)}$ is easily checked.

A.3.3 Dual q-Whittaker Functions

$$\Pi_{0,0} = s_{0,0}$$

$$\Pi_{1,0} = s_{1,0}$$

$$\Pi_{2,0} = s_{2,0} + q^{-2}s_{1,1} + q^{-2}s_{0,0}$$

$$\Pi_{1,1} = s_{1,1}$$

$$\Pi_{3,0} = s_{3,0} + \frac{1+q^2}{q^4}s_{2,1} + \frac{1+q^2+q^4}{q^6}s_{1,0}$$

$$\Pi_{2,1} = s_{2,1} + q^{-2}s_{2,0} + q^{-2}s_{1,0}$$

$$\Pi_{4,0} = s_{4,0} + \frac{1+q^2+q^4}{q^6}s_{3,1} + \frac{1+q^4}{q^8}s_{2,2} + \frac{(1+q^2+q^4)(1+q^4)}{q^{10}}s_{2,0}$$

$$+ \frac{1+2q^2+q^4+q^6}{q^{10}}s_{1,1} + \frac{1+q^4+q^8}{q^{12}}s_{0,0}$$

$$\Pi_{3,1} = s_{3,1} + q^{-2}s_{2,2} + \frac{(1+q^2)}{q^4}s_{2,0} + \frac{1+q^2}{q^4}s_{1,1} + q^{-6}s_{0,0}$$

$$\Pi_{2,2} = s_{2,2} + q^{-2}s_{2,0} + q^{-4}s_{0,0}.$$

The expected symmetry relations between B_2 and C_2 q-Whittaker functions are easily checked, using the explicit expressions for the relevant Schur functions (A.2–A.3).

A.4 The $\mathfrak{sl}_4$ and D_3 Cases

A.4.1 The $\mathfrak{sl}_4 \leftrightarrow D_3$ Symmetry

Compared to the $\mathfrak{gl}_4$ case, the symmetric functions of the case $\mathfrak{sl}_4$ involve the extra condition that $x_1x_2x_3x_4 = 1$, and accordingly $\Gamma_1\Gamma_2\Gamma_3\Gamma_4 = 1$ (see Remark 3.4). This is implemented by imposing the extra condition $e_1 + e_2 + e_3 + e_4 = 0$ under which:

$$e_1 = \omega_1, \; e_2 = \omega_2 - \omega_1, \; e_3 = \omega_3 - \omega_2, \; e_4 = -\omega_3.$$

Primed variables are used for D_3:

$$e_1' = \omega_1', \; e_2' = \omega_2' + \omega_3' - \omega_1', \; e_3' = \omega_3' - \omega_2'.$$

We use the mapping

$$\omega_1 \mapsto \omega_3', \quad \omega_2 \mapsto \omega_1', \quad \omega_3 \mapsto \omega_1'.$$

This is equivalent to the changes of variables (using $x_i = e^{e_i}, x_i' = e^{e_i'}$):

$$x_1 \mapsto \sqrt{x_1'x_2'x_3'}, \quad x_2 \mapsto \sqrt{\frac{x_1'}{x_2'x_3'}}, \quad x_3 \mapsto \sqrt{\frac{x_2'}{x_1'x_3'}}, \quad x_4 \mapsto \sqrt{\frac{x_3'}{x_1'x_2'}}. \tag{A.7}$$

Moreover, to account for our choice of normalization we must also take $q \mapsto q^2$, which results in

$$\Gamma_1 \mapsto \Gamma_1'\Gamma_2'\Gamma_3', \quad \Gamma_2 \mapsto \frac{\Gamma_1'}{\Gamma_2'\Gamma_3'}, \quad \Gamma_3 \mapsto \frac{\Gamma_2'}{\Gamma_1'\Gamma_3'}, \quad \Gamma_4 \mapsto \frac{\Gamma_3'}{\Gamma_1'\Gamma_2'}. \tag{A.8}$$

The above transformations send the $\mathfrak{sl}_4$ Macdonald operators to the D_3 ones, namely $M_{1,k} \mapsto M_{3,k}'$, $M_{2,k} \mapsto M_{1,k}'$ and $M_{3,k} \mapsto M_{2,k}'$. The corresponding mapping of Macdonald polynomials is

$$P^{(\mathfrak{sl}_4)}_{\lambda_1,\lambda_2,\lambda_3,\lambda_4} \mapsto P^{(D_3)}_{\frac{\lambda_1+\lambda_2-\lambda_3-\lambda_4}{2},\frac{\lambda_1-\lambda_2+\lambda_3-\lambda_4}{2},\frac{\lambda_1-\lambda_2-\lambda_3+\lambda_4}{2}}. \tag{A.9}$$

A.4.2 M Operators

The $\mathfrak{sl}_4$ operators are

$$M_{1,k}^{(\mathfrak{sl}_4)} = \sum_{i=1}^{4} x_i^k \prod_{j\neq i} \frac{x_i}{x_i - x_j} \Gamma_i$$

$$M_{2,k}^{(\mathfrak{sl}_4)} = \sum_{1\leq i<j\leq 4}^{4} (x_i x_j)^k \prod_{m\neq i,j} \frac{x_m}{x_m - x_i} \frac{x_m}{x_m - x_j} \Gamma_i \Gamma_j$$

$$M_{3,k}^{(\mathfrak{sl}_4)} = \sum_{i=1}^{4} x_i^{-k} \prod_{j\neq i} \frac{x_j}{x_j - x_i} \Gamma_i^{-1},$$

where $x_1 x_2 x_3 x_4 = 1$ and $\Gamma_1 \Gamma_2 \Gamma_3 \Gamma_4 = 1$.

The D_3 operators are

$$M_{1,k}^{(D_3)} = \sum_{\epsilon=\pm 1} \sum_{i=1}^{3} x_i^k \prod_{j\neq i} \frac{x_i^{\epsilon} x_j}{x_i^{\epsilon} x_j - 1} \frac{x_i^{\epsilon}}{x_i^{\epsilon} - x_j} \Gamma_i^{2\epsilon}$$

$$M_{2,k}^{(D_3)} = \sum_{\substack{\epsilon_1,\epsilon_2,\epsilon_3=\pm 1 \\ \epsilon_1\epsilon_2\epsilon_3=-1}} \prod_{i=1}^{3} x_i^{k\epsilon_i} \prod_{1\leq i<j\leq 3} \frac{x_i^{\epsilon_i} x_j^{\epsilon_j}}{x_i^{\epsilon_i} x_j^{\epsilon_j} - 1} \prod_{i=1}^{3} \Gamma_i^{\epsilon_i}$$

$$M_{3,k}^{(D_3)} = \sum_{\substack{\epsilon_1,\epsilon_2,\epsilon_3=\pm 1 \\ \epsilon_1\epsilon_2\epsilon_3=1}} \prod_{i=1}^{3} x_i^{k\epsilon_i} \prod_{1\leq i<j\leq 3} \frac{x_i^{\epsilon_i} x_j^{\epsilon_j}}{x_i^{\epsilon_i} x_j^{\epsilon_j} - 1} \prod_{i=1}^{3} \Gamma_i^{\epsilon_i}.$$

It is straightforward to see that the change of variables (A.7) and (A.8) map the M operators as follows: $M_{1,k}^{(\mathfrak{sl}_4)} \mapsto M_{3,k}^{(D_3)}$, $M_{2,k}^{(\mathfrak{sl}_4)} \mapsto M_{1,k}^{(D_3)}$, and $M_{3,k}^{(\mathfrak{sl}_4)} \mapsto M_{2,k}^{(D_3)}$.

A.4.3 Dual q-Whittaker Functions

In terms of the Schur functions (A.1), the first few A_3 dual q-Whittaker functions are

$$\Pi_{0,0,0,0} = 1$$

$$\Pi_{1,0,0,0} = s_{1,0,0,0}$$

$$\Pi_{2,0,0,0} = s_{2,0,0,0} + q^{-1} s_{1,1,0,0}$$

$$\Pi_{1,1,0,0} = s_{1,1,0,0}$$

$$\Pi_{3,0,0,0} = s_{3,0,0,0} + \frac{1+q}{q^2} s_{2,1,0,0} + q^{-3} s_{1,1,1,0}$$
$$\Pi_{2,1,0,0} = s_{2,1,0,0} + q^{-1} s_{1,1,1,0}$$
$$\Pi_{1,1,1,0} = s_{1,1,1,0}$$
$$\Pi_{4,0,0,0} = s_{4,0,0,0} + \frac{1+q+q^2}{q^3} s_{3,1,0,0} + \frac{1+q^2}{q^4} s_{2,2,0,0} + \frac{1+q+q^2}{q^5} s_{2,1,1,0} + q^{-6} s_{1,1,1,1}$$
$$\Pi_{3,1,0,0} = s_{3,1,0,0} + q^{-1} s_{2,2,0,0} + \frac{1+q}{q^2} s_{2,1,1,0} + q^{-3} s_{1,1,1,1}$$
$$\Pi_{2,2,0,0} = s_{2,2,0,0} + q^{-1} s_{2,1,1,0} + q^{-2} s_{1,1,1,1}$$
$$\Pi_{2,1,1,0} = s_{2,1,1,0} + q^{-1} s_{1,1,1,1}$$
$$\Pi_{1,1,1,1} = s_{1,1,1,1}.$$

In terms of the Schur functions (A.4), the first few D_3 dual q-Whittaker functions are

$$\Pi_{0,0,0} = 1$$
$$\Pi_{1,0,0} = s_{1,0,0}$$
$$\Pi_{2,0,0} = s_{2,0,0} + q^{-2} s_{1,1,0} + q^{-4} s_{0,0,0}$$
$$\Pi_{1,1,0} = s_{1,1,0} + q^{-2} s_{0,0,0}$$
$$\Pi_{3,0,0} = s_{3,0,0} + \frac{1+q^2}{q^4} s_{2,1,0} + q^{-6} (s_{1,1,1} + s_{1,1,-1}) + \frac{1+q^2+q^4}{q^8} s_{1,0,0}$$
$$\Pi_{2,1,0} = s_{2,1,0} + q^{-2} (s_{1,1,1} + s_{1,1,-1}) + \frac{1+q^2}{q^4} s_{1,0,0}$$
$$\Pi_{1,1,\epsilon} = s_{1,1,\epsilon} + q^{-2} s_{1,0,0}$$
$$\Pi_{\frac{1}{2},\frac{1}{2},\frac{\epsilon}{2}} = s_{\frac{1}{2},\frac{1}{2},\frac{\epsilon}{2}}$$
$$\Pi_{\frac{3}{2},\frac{1}{2},\frac{\epsilon}{2}} = s_{\frac{3}{2},\frac{1}{2},\frac{\epsilon}{2}} + q^{-2} s_{\frac{1}{2},\frac{1}{2},-\frac{\epsilon}{2}}$$
$$\Pi_{\frac{5}{2},\frac{1}{2},\frac{\epsilon}{2}} = s_{\frac{5}{2},\frac{1}{2},\frac{\epsilon}{2}} + q^{-2} s_{\frac{3}{2},\frac{3}{2},\frac{\epsilon}{2}} + \frac{1+q^2}{q^4} s_{\frac{3}{2},\frac{1}{2},-\frac{\epsilon}{2}} + \frac{1+q^2}{q^6} s_{\frac{1}{2},\frac{1}{2},\frac{\epsilon}{2}}$$
$$\Pi_{\frac{3}{2},\frac{3}{2},\frac{\epsilon}{2}} = s_{\frac{3}{2},\frac{3}{2},\frac{\epsilon}{2}} + q^{-2} s_{\frac{3}{2},\frac{1}{2},-\frac{\epsilon}{2}} + \frac{1+q^2}{q^4} s_{\frac{1}{2},\frac{1}{2},\frac{\epsilon}{2}}$$
$$\Pi_{\frac{3}{2},\frac{3}{2},\frac{3\epsilon}{2}} = s_{\frac{3}{2},\frac{3}{2},\frac{3\epsilon}{2}} + \frac{1+q^2}{q^4} s_{\frac{3}{2},\frac{1}{2},\frac{\epsilon}{2}} + q^{-6} s_{\frac{1}{2},\frac{1}{2},-\frac{\epsilon}{2}}.$$

The relations (A.9) are easily checked for these polynomials. First we note that the mapping (A.9) extends to the Weyl-invariant Schur functions as well. We then simply have to check that the coefficients in the D_3 case match those in the $\mathfrak{sl}_4$ case up to $q \to q^2$. For instance, we have $\Pi^{(\mathfrak{sl}_4)}_{2,1,0,0} \mapsto \Pi^{(D_3)}_{\frac{3}{2},\frac{1}{2},\frac{1}{2}}$ as the coefficient of $s^{(\mathfrak{sl}_4)}_{1,1,1,0}$ (resp. $s^{(D_3)}_{\frac{1}{2},\frac{1}{2},-\frac{\epsilon}{2}}$) is q^{-1} (resp. q^{-2}). Similarly the coefficients in $\Pi^{(\mathfrak{sl}_4)}_{3,0,0,0}$ and $\Pi^{(D_3)}_{\frac{3}{2},\frac{3}{2},\frac{3}{2}}$ also agree up to $q \to q^2$.

A.5 The D_4 Case

A.5.1 Symmetries

There are two simple symmetries of the Dynkin diagram which induce symmetries of the Macdonald operators and polynomials:

- The $\mathbb{Z}_2$ automorphism of the Dynkin diagram under which $1 \to 1$, $2 \to 2$ and $3 \leftrightarrow 4$,
- The $\mathbb{Z}_3$ automorphism of the Dynkin diagram under which $1 \to 3 \to 4 \to 1$ and $2 \to 2$.

Recall the relations

$$e_1 = \omega_1,\ e_2 = \omega_2 - \omega_1,\ e_3 = \omega_4 + \omega_3 - \omega_2,\ e_4 = \omega_4 - \omega_3.$$

The $\mathbb{Z}_2$ symmetry under:

$$\omega_1 \mapsto \omega_1,\ \omega_2 \mapsto \omega_2,\ \omega_3 \mapsto \omega_4,\ \omega_4 \mapsto \omega_3$$

induces the transformations

$$x_1 \mapsto x_1,\ x_2 \mapsto x_2,\ x_3 \mapsto x_3,\ x_4 \mapsto \frac{1}{x_4}. \tag{A.10}$$

Similarly, we have

$$\Gamma_1 \mapsto \Gamma_1,\ \Gamma_2 \mapsto \Gamma_2,\ \Gamma_3 \mapsto \Gamma_3,\ \Gamma_4 \mapsto \Gamma_4^{-1}.$$

We will check that under these transformations, we have $M_{1,k} \mapsto M_{1,k}$, $M_{2,k} \mapsto M_{2,k}$, $M_{3,k} \mapsto M_{4,k}$ and $M_{4,k} \mapsto M_{3,k}$ for Macdonald operators, and

$$P_{\lambda_1,\lambda_2,\lambda_3,\lambda_4} \mapsto P_{\lambda_1,\lambda_2,\lambda_3,-\lambda_4}$$

for Macdonald polynomials as well as q-Whittaker functions.

The $\mathbb{Z}_3$ symmetry sends

$$\omega_1 \mapsto \omega_3,\ \omega_2 \mapsto \omega_2,\ \omega_3 \mapsto \omega_4,\ \omega_4 \mapsto \omega_1$$

therefore induces the transformations

$$x_1 \mapsto \sqrt{\frac{x_1x_2x_3}{x_4}},\ x_2 \mapsto \sqrt{\frac{x_1x_2x_4}{x_3}},\ x_3 \mapsto \sqrt{\frac{x_1x_3x_4}{x_2}},\ x_4 \mapsto \sqrt{\frac{x_1}{x_2x_3x_4}}. \tag{A.11}$$

Similarly, we have

$$\Gamma_1^2 \mapsto \frac{\Gamma_1\Gamma_2\Gamma_3}{\Gamma_4},\ \Gamma_2^2 \mapsto \frac{\Gamma_1\Gamma_2\Gamma_4}{\Gamma_3},\ \Gamma_3^2 \mapsto \frac{\Gamma_1\Gamma_3\Gamma_4}{\Gamma_2},\ \Gamma_4^2 \mapsto \frac{\Gamma_1}{\Gamma_2\Gamma_3\Gamma_4}.$$

We expect that under these transformations, we have $M_{1,k} \mapsto M_{3,k} \mapsto M_{4,k} \mapsto M_{1,k}$ as well as $M_{2,k} \mapsto M_{2,k}$ for Macdonald operators, and

$$P_{\lambda_1,\lambda_2,\lambda_3,\lambda_4} \mapsto P_{\frac{\lambda_1+\lambda_2+\lambda_3+\lambda_4}{2},\frac{\lambda_1+\lambda_2-\lambda_3-\lambda_4}{2},\frac{\lambda_1-\lambda_2+\lambda_3-\lambda_4}{2},\frac{-\lambda_1+\lambda_2+\lambda_3-\lambda_4}{2}}$$

for Macdonald polynomials as well as q-Whittaker functions.

The fact that our operators/polynomials obey these symmetry relations is a highly non-trivial check of our construction.

A.5.2 M Operators

$$M_{1,k} = \sum_{\epsilon=\pm 1}\sum_{i=1}^{4}\prod_{j\neq i}\frac{x_i^\epsilon x_j}{x_i^\epsilon x_j - 1}\frac{x_i^\epsilon}{x_i^\epsilon - x_j} x_i^{k\epsilon}\,\Gamma_i^{2\epsilon}$$

$$M_{2,k} = M_{1,k}^2 - q^2 M_{1,k+1}\,M_{1,k-1}$$

$$M_{3,k} = \sum_{\substack{\epsilon_1,\epsilon_2,\epsilon_3,\epsilon_4=\pm 1\\ \epsilon_1\epsilon_2\epsilon_3\epsilon_4=-1}}\ \prod_{1\le i<j\le 4}\frac{x_i^{\epsilon_i}x_j^{\epsilon_j}}{x_i^{\epsilon_i}x_j^{\epsilon_j}-1}\prod_{i=1}^{4}x_i^{\frac{k\epsilon_i}{2}}\prod_{i=1}^{4}\Gamma_i^{\epsilon_i}$$

$$M_{4,k} = \sum_{\substack{\epsilon_1,\epsilon_2,\epsilon_3,\epsilon_4=\pm 1\\ \epsilon_1\epsilon_2\epsilon_3\epsilon_4=1}}\ \prod_{1\le i<j\le 4}\frac{x_i^{\epsilon_i}x_j^{\epsilon_j}}{x_i^{\epsilon_i}x_j^{\epsilon_j}-1}\prod_{i=1}^{4}x_i^{\frac{k\epsilon_i}{2}}\prod_{i=1}^{N}\Gamma_i^{\epsilon_i}.$$

The expected symmetry relations under both $\mathbb{Z}_2$ and $\mathbb{Z}_3$ automorphisms are easily checked. For instance, in the $\mathbb{Z}_2$ case, we see immediately that $M_{3,k}$ and $M_{4,k}$ are interchanged under the transformation $(x_4, \Gamma_4) \to (x_4^{-1}, \Gamma_4^{-1})$ while all other x_i, Γ_i remain unchanged, as the transformation amounts to changing $\epsilon_4 \to -\epsilon_4$ in the summation.

In the general D_N case, the $\mathbb{Z}_2$ symmetry of the Dynkin diagram that interchanges the two end-nodes $N-1$ and N implies the symmetry under $x_N \mapsto 1/x_N$ while all other x's remain unchanged, together with $\Gamma_N \mapsto \Gamma_N^{-1}$. It is clear that under this transformation, we have $M_{N-1,k}^{(D_N)} \mapsto M_{N,k}^{(D_N)}$ and $M_{N,k}^{(D_N)} \mapsto M_{N-1,k}^{(D_N)}$ Indeed, Eqs. (4.9) and (4.10) are interchanged under the change of summation variable $\epsilon_N \to -\epsilon_N$, which amounts exactly to $x_N \to 1/x_N$ and $\Gamma_N \to \Gamma_N^{-1}$.

A.5.3 Dual q-Whittaker Functions

$$\Pi_{0,0,0,0} = s_{0,0,0,0}$$

$$\Pi_{1,0,0,0} = s_{1,0,0,0}$$

$$\Pi_{2,0,0,0} = s_{2,0,0,0} + q^{-2}s_{1,1,0,0} + q^{-4}s_{0,0,0,0}$$

$$\Pi_{1,1,0,0} = s_{1,1,0,0} + q^{-2}s_{0,0,0,0}$$

$$\Pi_{3,0,0,0} = s_{3,0,0,0} + \frac{1+q^2}{q^4}s_{2,1,0,0} + q^{-6}s_{1,1,1,0} + \frac{1+q^2+q^4}{q^8}s_{1,0,0,0}$$

$$\Pi_{2,1,0,0} = s_{2,1,0,0} + q^{-2}s_{1,1,1,0} + \frac{1+q^2}{q^4}s_{1,0,0,0}$$

$$\Pi_{1,1,1,0} = s_{1,1,1,0} + q^{-2}s_{1,0,0,0}$$

$$\begin{aligned}\Pi_{4,0,0,0} = s_{4,0,0,0} &+ \frac{1+q^2+q^4}{q^6}s_{3,1,0,0} + \frac{1+q^4}{q^8}s_{2,2,0,0} + \frac{1+q^2+q^4}{q^{10}}s_{2,1,1,0} \\ &+ q^{-12}(s_{1,1,1,1} + s_{1,1,1,-1}) + \frac{(1+q^4)(1+q^2+q^4)}{q^{12}}s_{2,0,0,0} \\ &+ \frac{(1+q^4)(1+q^2+q^4)}{q^{14}}s_{1,1,0,0} + \frac{1+q^4+q^8}{q^{16}}s_{0,0,0,0}\end{aligned}$$

$$\begin{aligned}\Pi_{3,1,0,0} = s_{3,1,0,0} &+ q^{-2}s_{2,2,0,0} + \frac{1+q^2}{q^4}s_{2,1,1,0} + q^{-6}(s_{1,1,1,1} + s_{1,1,1,-1}) \\ &+ \frac{1+q^2+q^4}{q^6}s_{2,0,0,0} + \frac{1+q^2+2q^4}{q^8}s_{1,1,0,0} + \frac{1+q^4}{q^{10}}s_{0,0,0,0}\end{aligned}$$

$$\begin{aligned}\Pi_{2,2,0,0} = s_{2,2,0,0} &+ q^{-2}s_{2,1,1,0} + q^{-4}(s_{1,1,1,1} + s_{1,1,1,-1}) \\ &+ q^{-4}s_{2,0,0,0} + \frac{1+q^2+q^4}{q^6}s_{1,1,0,0} + \frac{1+q^4}{q^8}s_{0,0,0,0}\end{aligned}$$

$$\begin{aligned}\Pi_{2,1,1,0} = s_{2,1,1,0} &+ q^{-2}(s_{1,1,1,1} + s_{1,1,1,-1}) + q^{-2}s_{2,0,0,0} \\ &+ \frac{1+q^2}{q^4}s_{1,1,0,0} + q^{-6}s_{0,0,0,0}\end{aligned}$$

$$\Pi_{1,1,1,\epsilon} = s_{1,1,1,\epsilon} + q^{-2} s_{1,1,0,0} + q^{-4} s_{0,0,0,0}$$

$$\Pi_{\frac{1}{2},\frac{1}{2},\frac{1}{2},\frac{\epsilon}{2}} = s_{\frac{1}{2},\frac{1}{2},\frac{1}{2},\frac{\epsilon}{2}}$$

$$\Pi_{\frac{3}{2},\frac{1}{2},\frac{1}{2},\frac{\epsilon}{2}} = s_{\frac{3}{2},\frac{1}{2},\frac{1}{2},\frac{\epsilon}{2}} + q^{-2} s_{\frac{1}{2},\frac{1}{2},\frac{1}{2},-\frac{\epsilon}{2}}$$

$$\Pi_{\frac{5}{2},\frac{1}{2},\frac{1}{2},\frac{\epsilon}{2}} = s_{\frac{5}{2},\frac{1}{2},\frac{1}{2},\frac{\epsilon}{2}} + q^{-2} s_{\frac{3}{2},\frac{3}{2},\frac{1}{2},\frac{\epsilon}{2}} + \frac{1+q^2}{q^4} s_{\frac{3}{2},\frac{1}{2},\frac{1}{2},-\frac{\epsilon}{2}} + \frac{1+q^2}{q^6} s_{\frac{1}{2},\frac{1}{2},\frac{1}{2},\frac{\epsilon}{2}}$$

$$\Pi_{\frac{3}{2},\frac{3}{2},\frac{1}{2},\frac{\epsilon}{2}} = s_{\frac{3}{2},\frac{3}{2},\frac{1}{2},\frac{\epsilon}{2}} + q^{-2} s_{\frac{3}{2},\frac{1}{2},\frac{1}{2},-\frac{\epsilon}{2}} + \frac{1+q^2}{q^4} s_{\frac{1}{2},\frac{1}{2},\frac{1}{2},\frac{\epsilon}{2}}$$

for $\epsilon = \pm 1$.

The expected symmetry relations under the $\mathbb{Z}_3$ Dynkin automorphism are easily checked using the explicit expressions for Weyl-invariant D-type Schur functions (A.4). As an illustration, the reader can check that $\Pi_{2,0,0,0} \mapsto \Pi_{1,1,1,-1}$, as a consequence of $s_{2,0,0,0} \mapsto s_{1,1,1,-1}$, $s_{1,1,0,0} \mapsto s_{1,1,0,0}$ and $s_{0,0,0,0} \mapsto s_{0,0,0,0}$ under the transformation (A.11). The $\mathbb{Z}_2$ symmetry is simply the covariance of Π under $\epsilon \to -\epsilon$.

References

[AK07] Eddy Ardonne and Rinat Kedem. Fusion products of Kirillov-Reshetikhin modules and fermionic multiplicity formulas. *J. Algebra*, 308(1):270–294, 2007.

[Bet31] Hans Bethe. Zur Theorie der Metalle. I. Eigenwerte und Eigenfunktionen der linearen Atomkette. *Zeitschrift für Physik 71*, 71:205–226, 1931.

[BFN16] Alexander Braverman, Michael Finkelberg, and Hiraku Nakajima. Coulomb branches of 3d n=4 quiver gauge theories and slices in the affine Grassmannian, 2016. Preprint arXiv:arXiv:1604.03625v5 [math.RT].

[BZ05] Arkady Berenstein and Andrei Zelevinsky. Quantum cluster algebras. *Adv. Math.*, 195(2):405–455, 2005.

[Che05] Ivan Cherednik. *Double affine Hecke algebras*, volume 319 of *London Mathematical Society Lecture Note Series*. Cambridge University Press, Cambridge, 2005.

[DFK08] Philippe Di Francesco and Rinat Kedem. Proof of the combinatorial Kirillov-Reshetikhin conjecture. *Int. Math. Res. Not. IMRN*, (7):Art. ID rnn006, 57, 2008.

[DFK09] Philippe Di Francesco and Rinat Kedem. Q-systems as cluster algebras. II. Cartan matrix of finite type and the polynomial property. *Lett. Math. Phys.*, 89(3):183–216, 2009.

[DFK14] Philippe Di Francesco and Rinat Kedem. Quantum cluster algebras and fusion products. *Int. Math. Res. Not. IMRN*, (10):2593–2642, 2014.

[DFK17] Philippe Di Francesco and Rinat Kedem. Quantum Q systems: from cluster algebras to quantum current algebras. *Lett. Math. Phys.*, 107(2):301–341, 2017.

[DFK18] P. Di Francesco and R. Kedem. Difference equations for graded characters from quantum cluster algebra. *Transform. Groups*, 23(2):391–424, 2018.

[DFK19] Philippe Di Francesco and Rinat Kedem. (q,t)-Deformed Q-Systems, DAHA and Quantum Toroidal Algebras via Generalized Macdonald Operators. *Comm. Math. Phys.*, 369(3):867–928, 2019.

[FL99] Boris Feigin and Sergey Loktev. On generalized Kostka polynomials and the quantum Verlinde rule. In *Differential topology, infinite-dimensional Lie algebras, and applications*, volume 194 of *Amer. Math. Soc. Transl. Ser. 2*, pages 61–79. Amer. Math. Soc., Providence, RI, 1999.

[FZ02] Sergey Fomin and Andrei Zelevinsky. Cluster algebras. I. Foundations. *J. Amer. Math. Soc.*, 15(2):497–529, 2002.

[GSV10] Michael Gekhtman, Michael Shapiro, and Alek Vainshtein. *Cluster algebras and Poisson geometry*, volume 167 of *Mathematical Surveys and Monographs*. American Mathematical Society, Providence, RI, 2010.

[HKO$^+$99] G. Hatayama, A. Kuniba, M. Okado, T. Takagi, and Y. Yamada. Remarks on fermionic formula. In *Recent developments in quantum affine algebras and related topics (Raleigh, NC, 1998)*, volume 248 of *Contemp. Math.*, pages 243–291. Amer. Math. Soc., Providence, RI, 1999.

[Ked08] Rinat Kedem. Q-systems as cluster algebras. *J. Phys. A*, 41(19):194011, 14, 2008.

[KN99] A. N. Kirillov and M. Noumi. q-difference raising operators for Macdonald polynomials and the integrality of transition coefficients. In *Algebraic methods and* q-*special functions (Montréal, QC, 1996)*, volume 22 of *CRM Proc. Lecture Notes*, pages 227–243. Amer. Math. Soc., Providence, RI, 1999.

[KNS94] Atsuo Kuniba, Tomoki Nakanishi, and Junji Suzuki. Functional relations in solvable lattice models. I. Functional relations and representation theory. *Internat. J. Modern Phys. A*, 9(30):5215–5266, 1994.

[KR87] A. N. Kirillov and N. Yu. Reshetikhin. Representations of Yangians and multiplicities of the inclusion of the irreducible components of the tensor product of representations of simple Lie algebras. *Zap. Nauchn. Sem. Leningrad. Otdel. Mat. Inst. Steklov. (LOMI)*, 160(Anal. Teor. Chisel i Teor. Funktsii. 8):211–221, 301, 1987.

[Lin19] Mingyan Simon Lin. Quantum q-systems and fermionic sums – the non-simply laced case, 2019. *Int. Math. Res. Not. IMRN*, 2:805–854, 2021.

[LNS$^+$17] Cristian Lenart, Satoshi Naito, Daisuke Sagaki, Anne Schilling, and Mark Shimozono. A uniform model for Kirillov-Reshetikhin crystals II. Alcove model, path model, and $P = X$. *Int. Math. Res. Not. IMRN*, (14):4259–4319, 2017.

[Mac01] I. G. Macdonald. Orthogonal polynomials associated with root systems. *Sém. Lothar. Combin.*, 45:Art. B45a, 40, 2000/01.

[Mac95] I. G. Macdonald. *Symmetric functions and Hall polynomials*. Oxford Mathematical Monographs. The Clarendon Press, Oxford University Press, New York, second edition, 1995. With contributions by A. Zelevinsky, Oxford Science Publications.

[Miz03] Hiroshi Mizukawa. Factorization of Schur functions of various types. *J. Algebra*, 269(1):215–226, 2003.

[Sch12] Olivier Schiffmann. Drinfeld realization of the elliptic Hall algebra. *J. Algebraic Combin.*, 35(2):237–262, 2012.

[vD95] J. F. van Diejen. Commuting difference operators with polynomial eigenfunctions. *Compositio Math.*, 95(2):183–233, 1995.

[vDE11] J. F. van Diejen and E. Emsiz. A generalized Macdonald operator. *Int. Math. Res. Not. IMRN*, (15):3560–3574, 2011.

The Meromorphic R-Matrix of the Yangian

Sachin Gautam, Valerio Toledano Laredo, and Curtis Wendlandt

To Kolya Reshetikhin, on his 60th birthday.

Abstract Let $\mathfrak{g}$ be a complex semisimple Lie algebra and $Y_\hbar(\mathfrak{g})$ its Yangian. Drinfeld proved that the universal R-matrix $\mathcal{R}(s)$ of $Y_\hbar(\mathfrak{g})$ gives rise to rational solutions of the QYBE on irreducible, finite-dimensional representations of $Y_\hbar(\mathfrak{g})$. This result was recently extended by Maulik–Okounkov to symmetric Kac–Moody algebras and representations arising from geometry. We show that rationality ceases to hold on arbitrary finite-dimensional representations, if one requires such solutions to be natural and compatible with tensor products. Equivalently, the tensor category of finite-dimensional representations of $Y_\hbar(\mathfrak{g})$ does not admit rational commutativity constraints. We construct instead two meromorphic commutativity constraints, which are related by a unitarity condition. Each possesses an asymptotic expansion in s which has the same formal properties as $\mathcal{R}(s)$, and therefore coincides with it by uniqueness. In particular, we give a constructive proof of the existence of $\mathcal{R}(s)$. Our construction relies on the Gauss decomposition $\mathcal{R}^+(s) \cdot \mathcal{R}^0(s) \cdot \mathcal{R}^-(s)$ of $\mathcal{R}(s)$. The divergent abelian term $\mathcal{R}^0$ was resummed on finite-dimensional representations by the first two authors in Gautam and Toledano Laredo (Publ Math Inst Hautes Études Sci 125:267–337, 2017). In the present paper, we construct $\mathcal{R}^\pm(s)$, prove

S. Gautam
Department of Mathematics, The Ohio State University, Columbus, OH, USA
e-mail: gautam.42@osu.edu

C. Wendlandt
Department of Mathematics and Statistics, University of Saskatchewan, Saskatoon, SK, Canada
e-mail: wendlandt@math.usask.ca

V. Toledano Laredo (✉)
Department of Mathematics, Northeastern University, Boston, MA, USA
e-mail: v.toledanolaredo@neu.edu

A. Alekseev et al. (eds.), *Representation Theory, Mathematical Physics, and Integrable Systems*, Progress in Mathematics 340,
https://doi.org/10.1007/978-3-030-78148-4_7

that they are rational on finite-dimensional representations, and that they intertwine the standard coproduct of $Y_\hbar(\mathfrak{g})$ and the deformed Drinfeld coproduct introduced in *loc. cit.*

1 Introduction

1.1

Let $\mathfrak{g}$ be a complex, semisimple Lie algebra with an invariant inner product $(\cdot,\cdot)$, and $Y_\hbar(\mathfrak{g})$ the corresponding Yangian, which is a Hopf algebra deforming the current algebra $U(\mathfrak{g}[z])$ introduced by Drinfeld [4]. We assume that $\hbar \in \mathbb{C}^\times$ is fixed throughout. Drinfeld proved that $Y_\hbar(\mathfrak{g})$ possesses a unique universal R-matrix. Specifically, let $\Delta : Y_\hbar(\mathfrak{g}) \to Y_\hbar(\mathfrak{g}) \otimes Y_\hbar(\mathfrak{g})$ be the coproduct of $Y_\hbar(\mathfrak{g})$, and $\tau_s : Y_\hbar(\mathfrak{g}) \to Y_\hbar(\mathfrak{g})$ the one–parameter group of automorphisms which quantizes the shift automorphism $z \mapsto z+s$ of $U(\mathfrak{g}[z])$. Note that, if s is considered as a variable, τ_s may be regarded as a homomorphism $Y_\hbar(\mathfrak{g}) \to Y_\hbar(\mathfrak{g})[s]$. Then, the following holds.

Theorem ([4, Thm. 3])

1. *There is a unique formal series*

$$\mathcal{R}(s) = 1 + \sum_{k=1}^{\infty} \mathcal{R}_k s^{-k} \in Y_\hbar(\mathfrak{g})^{\otimes 2}[\![s^{-1}]\!]$$

such that the following holds[1] *in* $Y_\hbar(\mathfrak{g})^{\otimes 2}[s; s^{-1}]\!]$:

$$\tau_s \otimes \mathbf{1} \circ \Delta^{\mathrm{op}}(a) = \mathcal{R}(s) \cdot \tau_s \otimes \mathbf{1} \circ \Delta(a) \cdot \mathcal{R}(s)^{-1} \tag{1.1}$$

for any $a \in Y_\hbar(\mathfrak{g})$, *and*

$$\Delta \otimes \mathbf{1}(\mathcal{R}(s)) = \mathcal{R}_{13}(s) \cdot \mathcal{R}_{23}(s) \tag{1.2}$$

$$\mathbf{1} \otimes \Delta(\mathcal{R}(s)) = \mathcal{R}_{13}(s) \cdot \mathcal{R}_{12}(s) \tag{1.3}$$

2. $\mathcal{R}$ *satisfies the following identities*

- *1–jet:* $\mathcal{R}(s) = 1 + \hbar s^{-1}\Omega_{\mathfrak{g}} + O(s^{-2})$
- *Unitarity:* $\mathcal{R}(s)^{-1} = \mathcal{R}_{21}(-s)$
- *Translation:* $\tau_a \otimes \tau_b(\mathcal{R}(s)) = \mathcal{R}(s+a-b)$

where $\Omega_{\mathfrak{g}} \in \mathfrak{g} \otimes \mathfrak{g}$ *is the Casimir tensor corresponding to* $(\cdot,\cdot)$.

[1] Our conventions differ slightly from those of [4], where the intertwining equation (1.1) is written as $\mathbf{1} \otimes \tau_s \circ \Delta^{\mathrm{op}}(a) = \mathcal{R}(s)^{-1} \cdot \mathbf{1} \otimes \tau_s \circ \Delta(a) \cdot \mathcal{R}(s)$. Thus, our $\mathcal{R}(s)$ is Drinfeld's $\mathcal{R}(-s)^{-1}$.

3. *$\mathcal{R}$ is a solution of the quantum Yang–Baxter equation (QYBE)*[2]

$$\mathcal{R}_{12}(s_1)\mathcal{R}_{13}(s_1+s_2)\mathcal{R}_{23}(s_2)=\mathcal{R}_{23}(s_2)\mathcal{R}_{13}(s_1+s_2)\mathcal{R}_{12}(s_1) \tag{1.4}$$

1.2

One of the main goals of this paper is to clarify the analytic nature of the formal power series $\mathcal{R}(s)$, and of the solutions of the QYBE obtained from it. Let V_1, V_2 be two finite-dimensional representations of $Y_\hbar(\mathfrak{g})$, and $\mathcal{R}_{V_1,V_2}(s) \in \mathrm{End}(V_1\otimes V_2)[[s^{-1}]]$ the corresponding evaluation of $\mathcal{R}(s)$. Drinfeld proved that $\mathcal{R}_{V_1,V_2}(s)$ has a zero radius of convergence in general [4, Examples 1,2], but nevertheless gives rise to a rational solution of the QYBE as follows.

Theorem ([4, Thm. 4]) *Assume that V_1 and V_2 are irreducible with highest weight vectors v_1, v_2, and let $\rho_{V_1,V_2}(s) \in 1+s^{-1}\mathbb{C}[[s^{-1}]]$ be the matrix element of $\mathcal{R}_{V_1,V_2}(s)$ given by*

$$\mathcal{R}_{V_1,V_2}(s)\, v_1\otimes v_2=\rho_{V_1,V_2}(s)\, v_1\otimes v_2$$

Then,

1. $\mathsf{R}_{V_1,V_2}(s)=\mathcal{R}_{V_1,V_2}(s)\cdot\rho_{V_1,V_2}(s)^{-1}$ *is a rational function of s.*
2. *If $V_1=V=V_2$,* (1.4) *together with the factorisation*

$$\mathcal{R}_{V_1,V_2}(s)=\mathsf{R}_{V_1,V_2}(s)\cdot\rho_{V_1,V_2}(s) \tag{1.5}$$

   imply that $\mathsf{R}_{V,V}(s)$ is a rational solution of the QYBE.

More recently, a geometric construction of R-matrices corresponding to the (extended) Yangian of a symmetric Kac–Moody algebra was given by Maulik–Okounkov [20], which provides in particular an alternative construction of rational solutions of the QYBE on the equivariant cohomology of Nakajima quiver varieties.

1.3

One of the byproducts of this paper is to extend the factorisation (1.5) to an arbitrary pair of (not necessarily irreducible) finite-dimensional representations. In this case,

[2]The QYBE may be viewed as an equation in $Y_\hbar(\mathfrak{g})^{\otimes 3}[s_1;s_1{}^{-1}]][[s_2^{-1}]]$ by expanding it as if $|s_2|\gg|s_1|$, that is setting $(s_1+s_2)^{-1}=\sum_{k\geq 0}(-1)^k s_1^k s_2^{-k-1}$, or as an equation in $Y_\hbar(\mathfrak{g})^{\otimes 3}[s_2;s_2{}^{-1}]][[s_1^{-1}]]$, by setting $(s_1+s_2)^{-1}=\sum_{k\geq 0}(-1)^k s_2^k s_1^{-k-1}$. The precise statement of (3) above is that (1.4) holds in either of these cases.

the divergent factor $\rho_{V_1,V_2}(s)$ takes values in $\mathrm{End}(V_1 \otimes V_2)[\![s^{-1}]\!]$, and intertwines the action of $Y_\hbar(\mathfrak{g})$ given by $\Delta_s = \tau_s \otimes \mathbf{1} \circ \Delta$, whereas the rational factor $\mathsf{R}_{V_1,V_2}(s)$ intertwines Δ_s and $\Delta_s^{\mathrm{op}} = \tau_s \otimes \mathbf{1} \circ \Delta^{\mathrm{op}}$. However, since $\rho_{V_1,V_2}(s)$ is not scalar–valued in general, it is not clear whether $\mathsf{R}_{V_1,V_2}(s)$ satisfies the QYBE when $V_1 = V_2$.

We prove in fact that, even for $\mathfrak{g} = \mathfrak{sl}_2$, no rational intertwiner $\mathsf{R}_{V_1,V_2}(s) \in \mathrm{End}(V_1 \otimes V_2)$ exists which is defined for any $V_1, V_2 \in \mathrm{Rep}_{\mathrm{fd}}(Y_\hbar(\mathfrak{g}))$, is natural in V_1 and V_2, and satisfies the cabling identities (1.2) and (1.3). Equivalently, the tensor category of finite-dimensional representations of $Y_\hbar(\mathfrak{g})$ does not admit rational commutativity constraints. In particular, this raises the question of whether one can consistently define rational solutions of the QYBE on all finite-dimensional representations of $Y_\hbar(\mathfrak{g})$.[3]

1.4

In the present paper, we propose an alternative solution to this issue, by constructing *meromorphic* commutativity constraints on $\mathrm{Rep}_{\mathrm{fd}}(Y_\hbar(\mathfrak{g}))$, and in particular consistent meromorphic solutions of the QYBE on all $V \in \mathrm{Rep}_{\mathrm{fd}}(Y_\hbar(\mathfrak{g}))$. Namely, we prove that the universal R-matrix of $Y_\hbar(\mathfrak{g})$, while generally divergent on a pair of finite-dimensional representations V_1, V_2, can be canonically *resummed*, in two distinct ways. This yields a *pair* of meromorphic functions

$$\mathcal{R}^{\uparrow}_{V_1,V_2}(s), \mathcal{R}^{\downarrow}_{V_1,V_2}(s) : \mathbb{C} \to \mathrm{End}(V_1 \otimes V_2)$$

which are natural with respect to V_1, V_2, satisfy the intertwining relation (1.1), the cabling identities (1.2) and (1.3), as well as the translation property. The function $\mathcal{R}^{\uparrow}_{V_1,V_2}(s)$ (resp. $\mathcal{R}^{\downarrow}_{V_1,V_2}(s)$) is asymptotic to $\mathcal{R}_{V_1,V_2}(s)$ as $s \to \infty$ with $\mathrm{Re}(s/\hbar) > 0$ (resp. $\mathrm{Re}(s/\hbar) < 0$), and is related to $\mathcal{R}^{\downarrow}_{V_1,V_2}(s)$ by the unitarity relation

$$\mathcal{R}^{\uparrow}_{V_1,V_2}(s)^{-1} = \mathcal{R}^{\downarrow}_{V_2,V_1}(-s)^{21}$$

The situation is somewhat analogous to the case of the quantum loop algebra $U_q(L\mathfrak{g})$. In that case, if $\mathscr{R} \in U_q(L\mathfrak{g})\widehat{\otimes} U_q(L\mathfrak{g})$ is the universal R-matrix, then

$$\mathscr{R}^{\infty}(z) = \tau_z \otimes \mathbf{1}(\mathscr{R}) \in U_q(L\mathfrak{g})^{\otimes 2}[\![z^{-1}]\!] \text{ and } \mathscr{R}^{0}(z) = \mathbf{1} \otimes \tau_z(\mathscr{R}) \in U_q(L\mathfrak{g})^{\otimes 2}[\![z]\!]$$

converge, near $z = \infty$ and $z = 0$ respectively, to meromorphic functions of $z \in \mathbb{C}^\times$ on the tensor product $\mathcal{V}_1 \otimes \mathcal{V}_2$ of any two finite-dimensional representations [7, 16],

[3]The Maulik–Okounkov construction mentioned in Sect. 1.2 provides a partial solution to this question, since an arbitrary representation of $Y_\hbar(\mathfrak{g})$ may not have a geometric realisation.

and are related by $\mathscr{R}^{\infty}_{\mathcal{V}_1,\mathcal{V}_2}(z)^{-1} = \mathscr{R}^{0}_{\mathcal{V}_2,\mathcal{V}_1}(z^{-1})^{21}$. In the case of $U_q(L\mathfrak{g})$, however, $\mathscr{R}^{\infty}(z)$ and $\mathscr{R}^{0}(z)$ are convergent as is, and do not need to be resummed.

1.5

Our approach does not rely on Drinfeld's cohomological construction of $\mathcal{R}(s)$ to carry out the resummation. It produces the functions $\mathcal{R}^{\uparrow/\downarrow}_{V_1,V_2}(s)$ through a direct, explicit construction, which shows in particular that they have an asymptotic expansion as $s \to \infty$. The fact that the latter coincides with $\mathcal{R}_{V_1,V_2}(s)$, and therefore *a posteriori* that $\mathcal{R}_{V_1,V_2}(s)$ can be resummed, follows from the fact that the asymptotic expansion can be lifted to $Y_\hbar(\mathfrak{g})^{\otimes 2}[\![s^{-1}]\!]$, and shown to have the properties which uniquely determine $\mathcal{R}(s)$ by Theorem 1.1. In particular, our construction yields an independent, and constructive proof of the existence of $\mathcal{R}(s)$.

1.6

Our construction can be motivated by the following considerations. The R-matrix of the Yangian is expected to arise as the canonical element in $DY_\hbar(\mathfrak{g})\widehat{\otimes}DY_\hbar(\mathfrak{g})$, where $DY_\hbar(\mathfrak{g}) \supset Y_\hbar(\mathfrak{g})$ is the double Yangian of $\mathfrak{g}$, which is a quantisation of the graded Drinfeld double

$$\left(\mathfrak{g}[z^{\pm 1}], \mathfrak{g}[z], z^{-1}\mathfrak{g}[z^{-1}]\right)$$

of $\mathfrak{g}[z]$. Although a detailed understanding of $DY_\hbar(\mathfrak{g})$ is still lacking at present (see, however, [17] and [24]), this suggests that, given a triangular decomposition $\mathfrak{g} = \mathfrak{n}_+ \oplus \mathfrak{h} \oplus \mathfrak{n}_-$ of $\mathfrak{g}$, $\mathcal{R}(s)$ should have a corresponding Gauss decomposition

$$\mathcal{R}(s) = \mathcal{R}^+(s) \cdot \mathcal{R}^0(s) \cdot \mathcal{R}^-(s) \tag{1.6}$$

where $\mathcal{R}^0(s)$ quantises the canonical element in $\mathfrak{h}[z]\widehat{\otimes}z^{-1}\mathfrak{h}[z^{-1}]$, and $\mathcal{R}^{\pm}(s)$ those in $\mathfrak{n}_{\pm}[z]\widehat{\otimes}z^{-1}\mathfrak{n}_{\mp}[z^{-1}]$ respectively. Moreover, the unitarity of $\mathcal{R}(s)$ suggests that

$$\mathcal{R}^0(s)^{-1} = \mathcal{R}^0(-s)^{21} \qquad \text{and} \qquad \mathcal{R}^+(s)^{-1} = \mathcal{R}^-(-s)^{21}$$

Accordingly, we construct each factor $\mathcal{R}^0(s)$, $\mathcal{R}^-(s)$, $\mathcal{R}^+(s) = (\mathcal{R}^-(-s)^{21})^{-1}$, and their resummation on finite-dimensional representations separately.

1.7

Khoroshkin–Tolstoy gave a heuristic formula for $\mathcal{R}^0$ [17], as the exponential of an infinite sum in the abelian subalgebra of $DY_\hbar(\mathfrak{g})$ which quantises $\mathfrak{h}[z, z^{-1}]$. In [12], the first two named authors gave a precise version of this formula, where the exponent takes values in the abelian subalgebra $Y_\hbar^0(\mathfrak{g})$ of $Y_\hbar(\mathfrak{g})$ which quantises $\mathfrak{h}[z]$. We showed moreover that this expression can be resummed on a tensor product $V_1 \otimes V_2$ of finite-dimensional representations in two different ways. This yields two meromorphic functions $\mathcal{R}^{0,\uparrow}_{V_1,V_2}(s)$, $\mathcal{R}^{0,\downarrow}_{V_1,V_2}(s)$, which have the same asymptotic expansion on $\pm\,\mathrm{Re}(s/\hbar) > 0$, and are related by $\mathcal{R}^{0,\uparrow}_{V_1,V_2}(s)^{-1} = \mathcal{R}^{0,\downarrow}_{V_2,V_1}(-s)^{21}$.

1.8

An important discovery of [12] is that these abelian R-matrices play a similar role to that of the full R-matrix of $Y_\hbar(\mathfrak{g})$, but with respect to the *deformed Drinfeld tensor product*. The latter was introduced in [12] by degenerating the Drinfeld tensor product of the quantum loop algebra introduced by Hernandez [15]. It gives rise to a family of actions of $Y_\hbar(\mathfrak{g})$ on the vector space $V_1 \otimes V_2$, which is denoted by $V_1 \underset{\mathrm{D},s}{\otimes} V_2$ and is a rational function of a parameter $s \in \mathbb{C}$. The tensor product $\underset{\mathrm{D},s}{\otimes}$ is associative, in that the identification of vector spaces

$$(V_1 \underset{\mathrm{D},s_1}{\otimes} V_2) \underset{\mathrm{D},s_2}{\otimes} V_3 = V_1 \underset{\mathrm{D},s_1+s_2}{\otimes} (V_2 \underset{\mathrm{D},s_2}{\otimes} V_3)$$

intertwines the action of $Y_\hbar(\mathfrak{g})$ for any $s_1, s_2 \in \mathbb{C}$, and endows $\mathrm{Rep}_{\mathrm{fd}}(Y_\hbar(\mathfrak{g}))$ with the structure of a meromorphic tensor category in the sense of [22, 23].

The endomorphisms $\mathcal{R}^{0,\uparrow/\downarrow}_{V_1,V_2}$ are meromorphic commutativity constraints with respect to $\underset{\mathrm{D},s}{\otimes}$, that is they satisfy the representation theoretic version of the identities (1) of Theorem 1.1. In the present paper, we complement the results of [12] by lifting $\underset{\mathrm{D},s}{\otimes}$ to a *deformed Drinfeld coproduct*

$$\underset{\mathrm{D},s}{\Delta} : Y_\hbar(\mathfrak{g}) \to \big(Y_\hbar(\mathfrak{g}) \otimes Y_\hbar(\mathfrak{g})\big)[s; s^{-1}]\!]$$

and the common asymptotic expansion of $\mathcal{R}^{0,\uparrow/\downarrow}_{V_1,V_2}(s)$ to an element

$$\mathcal{R}^0(s) \in \big(Y_\hbar(\mathfrak{g}) \otimes Y_\hbar(\mathfrak{g})\big)[\![s^{-1}]\!]$$

which satisfy the identities (1) of Theorem 1.1, with $\tau_s \otimes \mathbf{1} \circ \Delta$ replaced by $\underset{\mathrm{D},s}{\Delta}$, and $\mathcal{R}(s)$ by $\mathcal{R}^0(s)$.

1.9

The central ingredient of the present paper is the construction of $\mathcal{R}^{\pm}(s)$, which is based on the following. The fact that $\mathcal{R}(s)$ (resp. $\mathcal{R}^0(s)$) conjugates the standard coproduct $\Delta_s = \tau_s \otimes \mathbf{1} \circ \Delta$ (resp. the deformed Drinfeld coproduct $\underset{\mathrm{D},s}{\Delta}$) to its opposite, together with the Gauss decomposition (1.6), suggest that $\mathcal{R}^-(s)$ should conjugate the standard coproduct Δ_s to the deformed Drinfeld coproduct $\underset{\mathrm{D},s}{\Delta}$. This is consistent with the fact that an analogous statement holds for the quantum loop algebra, and the related construction of twists conjugating quantum coproducts corresponding to different polarisations of a Manin triple given in [6]. In this case, the standard (resp. Drinfeld) coproducts on $U_q(L\mathfrak{g})$ correspond, respectively, to the polarisations

$$\mathfrak{g}[z] \oplus z^{-1}\mathfrak{g}[z^{-1}] = \mathfrak{g}[z^{\pm 1}] = \left(\mathfrak{n}_-[z^{\pm 1}] \oplus \mathfrak{h}[z]\right) \oplus \left(z^{-1}\mathfrak{h}[z^{-1}] \oplus \mathfrak{n}_+[z^{\pm 1}]\right)$$

1.10

We prove that this intertwining property uniquely determines an element $\mathcal{R}^-(s)$, provided it is required to lie in $\left(Y_\hbar^-(\mathfrak{g}) \otimes Y_\hbar^+(\mathfrak{g})\right)[\![s^{-1}]\!]$ and have constant term 1, where $Y_\hbar^{\pm}(\mathfrak{g}) \subset Y_\hbar(\mathfrak{g})$ is the subalgebra deforming $U(\mathfrak{n}_{\pm}[z])$. We show in fact that, under this triangularity assumption, $\mathcal{R}^-(s)$ is uniquely determined by the requirement that it intertwines the standard and Drinfeld coproducts of the loop generators $t_{i,0}, t_{i,1}$ of $Y_\hbar(\mathfrak{g})$ which deform $\mathfrak{h} \oplus \mathfrak{h} \otimes z \subset \mathfrak{h}[z]$.

We then show that, for any $V_1, V_2 \in \mathrm{Rep}_{\mathrm{fd}}(Y_\hbar(\mathfrak{g}))$, $\mathcal{R}^-_{V_1,V_2}(s)$ is a rational function of s. Moreover, the following cocycle identity holds for any $V_1, V_2, V_3 \in \mathrm{Rep}_{\mathrm{fd}}(Y_\hbar(\mathfrak{g}))$

$$\mathcal{R}^-_{V_1 \underset{\mathrm{D},s_1}{\otimes} V_2, V_3}(s_2) \cdot \mathcal{R}^-_{V_1,V_2}(s_1) = \mathcal{R}^-_{V_1, V_2 \underset{\mathrm{D},s_2}{\otimes} V_3}(s_1+s_2) \cdot \mathcal{R}^-_{V_2,V_3}(s_2) \tag{1.7}$$

Together with the identities satisfied by $\mathcal{R}^0(s)$, this guarantees that the product $\mathcal{R}(s) = \mathcal{R}^+(s) \cdot \mathcal{R}^0(s) \cdot \mathcal{R}^-(s)$, where $\mathcal{R}^+(s) = (\mathcal{R}^-(-s)^{21})^{-1}$, satisfies the identities (1.1)–(1.3) on any pair of finite-dimensional representations. A separation of points argument then implies that $\mathcal{R}(s)$ satisfies Drinfeld's uniqueness criterion of the universal R-matrix of $Y_\hbar(\mathfrak{g})$, and in particular coincides with it.[4]

Finally, since $\mathcal{R}^-_{V_1,V_2}(s)$ is a rational function of s, the product

$$\mathcal{R}^{\uparrow/\downarrow}_{V_1,V_2}(s) = \mathcal{R}^+_{V_1,V_2}(s) \cdot \mathcal{R}^{0,\uparrow/\downarrow}_{V_1,V_2}(s) \cdot \mathcal{R}^-_{V_1,V_2}(s) \tag{1.8}$$

[4]The passage to finite-dimensional representations is dictated by the fact that the cocycle identity (1.7) does not appear to have a natural lift to $Y_\hbar(\mathfrak{g})$. Indeed, when lifted to $Y_\hbar(\mathfrak{g})$, the left–hand side lies in $Y_\hbar(\mathfrak{g})^{\otimes 3}[s_1; s_1{}^{-1}]\!][\![s_2^{-1}]\!]$, while the right–hand side lies in $Y_\hbar(\mathfrak{g})^{\otimes 3}[s_2; s_2{}^{-1}]\!][\![s_1^{-1}]\!]$.

is a resummation of $\mathcal{R}_{V_1,V_2}(s)$, as well a meromorphic commutativity constraint on $\mathrm{Rep}_{\mathrm{fd}}(Y_\hbar(\mathfrak{g}))$ with respect to the standard tensor product.

1.11

Our results may be rephrased as follows. As proved in [12], and mentioned above, finite-dimensional representations of $Y_\hbar(\mathfrak{g})$, together with the deformed Drinfeld tensor product $\underset{\mathrm{D},s}{\otimes}$ and one of the resummed abelian R-matrices $\mathcal{R}^{0,\uparrow/\downarrow}(s)$ is a meromorphic braided tensor category.

Similarly, $\mathrm{Rep}_{\mathrm{fd}}(Y_\hbar(\mathfrak{g}))$ endowed with the deformed standard tensor product $\otimes_s = \otimes \circ (\tau_s^* \otimes \mathbf{1})$ is a meromorphic (in fact, polynomial) tensor category. Our construction of the resummed R-matrices $\mathcal{R}^{\uparrow/\downarrow}(s)$ endows this category with a meromorphic braiding.[5] Moreover, the element $\mathcal{R}^-(s)$ is a rational braided tensor structure on the identity functor

$$(\mathrm{Rep}_{\mathrm{fd}}(Y_\hbar(\mathfrak{g})), \underset{\mathrm{D},s}{\otimes}, \mathcal{R}^{0,\uparrow/\downarrow}) \to (\mathrm{Rep}_{\mathrm{fd}}(Y_\hbar(\mathfrak{g})), \otimes_s, \mathcal{R}^{\uparrow/\downarrow})$$

That is, $\mathcal{R}^-(s)$ gives rise to a system of natural isomorphisms of $Y_\hbar(\mathfrak{g})$–modules $\mathcal{R}^-_{V_1,V_2}(s) : V_1 \otimes_s V_2 \to V_1 \underset{\mathrm{D},s}{\otimes} V_2$, which is compatible with the (trivial) associativity constraints and the meromorphic braidings, *i.e.*, such that the following diagrams commute for any $V_1, V_2, V_3 \in \mathrm{Rep}_{\mathrm{fd}}(Y_\hbar(\mathfrak{g}))$:

$$\begin{array}{ccc}
(V_1 \otimes_{s_1} V_2) \otimes_{s_2} V_3 & = & V_1 \otimes_{s_1+s_2} (V_2 \otimes_{s_2} V_3) \\
{\scriptstyle \mathcal{R}^-_{V_1,V_2}(s_1)\otimes 1_{V_3}} \Big\downarrow & & \Big\downarrow {\scriptstyle 1_{V_1}\otimes \mathcal{R}^-_{V_2,V_3}(s_2)} \\
(V_1 \underset{\mathrm{D},s_1}{\otimes} V_2) \otimes_{s_2} V_3 & & V_1 \otimes_{s_1+s_2} (V_2 \underset{\mathrm{D},s_2}{\otimes} V_3) \\
{\scriptstyle \mathcal{R}^-_{V_1 \underset{\mathrm{D},s_1}{\otimes} V_2, V_3}(s_2)} \Big\downarrow & & \Big\downarrow {\scriptstyle \mathcal{R}^-_{V_1, V_2 \underset{\mathrm{D},s_2}{\otimes} V_3}(s_1+s_2)} \\
(V_1 \underset{\mathrm{D},s_1}{\otimes} V_2) \underset{\mathrm{D},s_2}{\otimes} V_3 & = & V_1 \underset{\mathrm{D},s_1+s_2}{\otimes} (V_2 \underset{\mathrm{D},s_2}{\otimes} V_3)
\end{array}$$

[5] An analogous statement was proved by Kazhdan–Soibelman for the quantum loop algebra in [16]. As pointed out in 1.4, however, in the case of $U_q(L\mathfrak{g})$ no resummation of the universal R-matrix of $U_q(L\mathfrak{g})$ is needed.

as dictated by the cocycle equation (1.7), and

$$\begin{array}{ccc} V_1(s)\otimes V_2 & \xrightarrow{(1\,2)\circ\mathcal{R}^{\uparrow/\downarrow}_{V_1,V_2}(s)} & V_2\otimes V_1(s) \\ \Big\downarrow{\scriptstyle \mathcal{R}^{-}_{V_1,V_2}(s)} & & \Big\downarrow{\scriptstyle \mathcal{R}^{-}_{V_2,V_1}(-s)} \\ V_1(s)\underset{D,0}{\otimes} V_2 & \xrightarrow[(1\,2)\circ\mathcal{R}^{0,\uparrow/\downarrow}_{V_1,V_2}(s)]{} & V_2\underset{D,0}{\otimes} V_1(s) \end{array}$$

which follows from the Gauss decomposition (1.8), together with the fact that $(\mathcal{R}^-(-s)^{21})^{-1} = \mathcal{R}^+(s)$.

1.12 Outline of the Paper

We review the definition of $Y_\hbar(\mathfrak{g})$ in Sect. 2, and that of the standard and Drinfeld coproducts in Sect. 3. In Sect. 4, we prove the existence and uniqueness of $\mathcal{R}^-(s)$, and establish its various properties. In Sect. 5, we give an explicit expressions for $\mathcal{R}^-(s)$ when $\mathfrak{g} = \mathfrak{sl}_2$. In Sect. 6, we review the construction of $\mathcal{R}^{0,\uparrow/\downarrow}(s)$ given in [12]. We then explicitly lift its asymptotic expansion to $Y^0_\hbar(\mathfrak{g})^{\otimes 2}[\![s^{-1}]\!]$, and prove that it satisfies properties analogous to those of Drinfeld's R-matrix, but with respect to the deformed Drinfeld coproduct. We also prove that there is no rational commutativity constraint on $\mathrm{Rep}_{\mathrm{fd}}(Y_\hbar(\mathfrak{g}))$. Combined with the results of Sect. 4, we obtain the same assertions for the standard tensor product in Sect. 7. We give a proof of the uniqueness of the universal R-matrix of the Yangian in Appendix B, thus completing the proof that our construction gives rise to Drinfeld's R-matrix. In Sect. 8, we restate our results in the language of meromorphic tensor categories. In the final Sect. 9, we discuss the analogous case of the quantum loop algebra, and relate the two by means of the meromorphic tensor functor constructed in [12]. Appendix A contains a proof due to Drinfeld that finite-dimensional representations separate points of $Y_\hbar(\mathfrak{g})$.

2 The Yangian $Y_\hbar(\mathfrak{g})$

2.1

Let $\mathfrak{g}$ be a complex, semisimple Lie algebra and $(\cdot,\cdot)$ an invariant, symmetric, non-degenerate bilinear form on $\mathfrak{g}$. Let $\mathfrak{h}\subset\mathfrak{g}$ be a Cartan subalgebra of $\mathfrak{g}$, $\{\alpha_i\}_{i\in\mathbf{I}}\subset\mathfrak{h}^*$ a basis of simple roots of $\mathfrak{g}$ relative to $\mathfrak{h}$ and $a_{ij} = 2(\alpha_i,\alpha_j)/(\alpha_i,\alpha_i)$ the entries of the corresponding Cartan matrix $\mathbf{A}$. Let $\Phi_+ \subset \mathfrak{h}^*$ be the corresponding set of positive roots, and $\mathsf{Q} = \mathbb{Z}\Phi_+ = \bigoplus_{i\in\mathbf{I}} \mathbb{Z}\alpha_i \subset \mathfrak{h}^*$ the root lattice. We assume that $(\cdot,\cdot)$ is normalised so that the square length of short roots is 2. Set $d_i = (\alpha_i,\alpha_i)/2 \in \{1,2,3\}$, so that $d_i a_{ij} = d_j a_{ji}$ for any $i,j \in \mathbf{I}$. In addition, we set $h_i = \nu^{-1}(\alpha_i)/d_i$ and choose root vectors $x_i^\pm \in \mathfrak{g}_{\pm\alpha_i}$ such that $[x_i^+, x_i^-] = d_i h_i$, where $\nu : \mathfrak{h} \to \mathfrak{h}^*$ is the isomorphism determined by $(\cdot,\cdot)$.

2.2 *The Yangian $Y_\hbar(\mathfrak{g})$ [5]*

Let $\hbar \in \mathbb{C}$. The Yangian $Y_\hbar(\mathfrak{g})$ is the $\mathbb{C}$–algebra generated by elements $\{x_{i,r}^\pm, \xi_{i,r}\}_{i\in\mathbf{I}, r\in\mathbb{Z}_{\geq 0}}$, subject to the following relations.

(Y1) For any $i,j \in \mathbf{I}$, $r,s \in \mathbb{Z}_{\geq 0}$: $[\xi_{i,r}, \xi_{j,s}] = 0$.
(Y2) For $i,j \in \mathbf{I}$ and $s \in \mathbb{Z}_{\geq 0}$: $[\xi_{i,0}, x_{j,s}^\pm] = \pm d_i a_{ij} x_{j,s}^\pm$.
(Y3) For $i,j \in \mathbf{I}$ and $r,s \in \mathbb{Z}_{\geq 0}$:

$$[\xi_{i,r+1}, x_{j,s}^\pm] - [\xi_{i,r}, x_{j,s+1}^\pm] = \pm\hbar\frac{d_i a_{ij}}{2}(\xi_{i,r}x_{j,s}^\pm + x_{j,s}^\pm \xi_{i,r}).$$

(Y4) For $i,j \in \mathbf{I}$ and $r,s \in \mathbb{Z}_{\geq 0}$:

$$[x_{i,r+1}^\pm, x_{j,s}^\pm] - [x_{i,r}^\pm, x_{j,s+1}^\pm] = \pm\hbar\frac{d_i a_{ij}}{2}(x_{i,r}^\pm x_{j,s}^\pm + x_{j,s}^\pm x_{i,r}^\pm).$$

(Y5) For $i,j \in \mathbf{I}$ and $r,s \in \mathbb{Z}_{\geq 0}$: $[x_{i,r}^+, x_{j,s}^-] = \delta_{ij}\xi_{i,r+s}$.
(Y6) Let $i \neq j \in \mathbf{I}$ and set $m = 1 - a_{ij}$. For any $r_1, \cdots, r_m \in \mathbb{Z}_{\geq 0}$ and $s \in \mathbb{Z}_{\geq 0}$:

$$\sum_{\pi\in\mathfrak{S}_m} \left[x_{i,r_{\pi(1)}}^\pm, \left[x_{i,r_{\pi(2)}}^\pm, \left[\cdots, \left[x_{i,r_{\pi(m)}}^\pm, x_{j,s}^\pm\right]\cdots\right]\right]\right] = 0.$$

We denote by $Y_\hbar^0(\mathfrak{g})$ and $Y_\hbar^\pm(\mathfrak{g})$ the unital subalgebras of $Y_\hbar(\mathfrak{g})$ generated by $\{\xi_{i,r}\}_{i\in\mathbf{I}, r\in\mathbb{Z}_{\geq 0}}$ and $\{x_{i,r}^\pm\}_{i\in\mathbf{I}, r\in\mathbb{Z}_{\geq 0}}$, respectively.

2.3

Assume henceforth that $\hbar \neq 0$, and define $\xi_i(u), x_i^\pm(u) \in Y_\hbar(\mathfrak{g})[[u^{-1}]]$ by

$$\xi_i(u) = 1 + \hbar\sum_{r\geq 0}\xi_{i,r}u^{-r-1} \quad \text{and} \quad x_i^\pm(u) = \hbar\sum_{r\geq 0} x_{i,r}^\pm u^{-r-1}.$$

Proposition **([11, Prop.2.3])** *The relations (Y1)–(Y6) are respectively equivalent to the following identities in* $Y_\hbar(\mathfrak{g})[u, v; u^{-1}, v^{-1}]\!]$.

($\mathcal{Y}1$) *For any* $i, j \in \mathbf{I}$, $[\xi_i(u), \xi_j(v)] = 0$.

($\mathcal{Y}2$) *For any* $i, j \in \mathbf{I}$, $[\xi_{i,0}, x_j^{\pm}(u)] = \pm d_i a_{ij} x_j^{\pm}(u)$.

($\mathcal{Y}3$) *For any* $i, j \in \mathbf{I}$, *and* $a = \hbar d_i a_{ij}/2$:

$$(u - v \mp a)\xi_i(u)x_j^{\pm}(v) = (u - v \pm a)x_j^{\pm}(v)\xi_i(u) \mp 2a x_j^{\pm}(u \mp a)\xi_i(u).$$

($\mathcal{Y}4$) *For any* $i, j \in \mathbf{I}$, *and* $a = \hbar d_i a_{ij}/2$:

$$\begin{aligned}&(u - v \mp a)x_i^{\pm}(u)x_j^{\pm}(v)\\&\quad= (u - v \pm a)x_j^{\pm}(v)x_i^{\pm}(u) + \hbar\left([x_{i,0}^{\pm}, x_j^{\pm}(v)] - [x_i^{\pm}(u), x_{j,0}^{\pm}]\right).\end{aligned}$$

($\mathcal{Y}5$) *For any* $i, j \in \mathbf{I}$:

$$(u - v)[x_i^{+}(u), x_j^{-}(v)] = -\delta_{ij}\hbar\left(\xi_i(u) - \xi_i(v)\right).$$

($\mathcal{Y}6$) *For any* $i \neq j \in \mathbf{I}$, $m = 1 - a_{ij}$, $r_1, \cdots, r_m \in \mathbb{Z}_{\geq 0}$, *and* $s \in \mathbb{Z}_{\geq 0}$:

$$\sum_{\pi \in \mathfrak{S}_m}\left[x_i^{\pm}(u_{\pi_1}), \left[x_i^{\pm}(u_{\pi(2)}), \left[\cdots, \left[x_i^{\pm}(u_{\pi(m)}), x_j^{\pm}(v)\right]\cdots\right]\right]\right] = 0.$$

Remark When $\mathfrak{g} = \mathfrak{sl}_2$, we will write ξ_r, $x_r^{\pm}$, $\xi(u)$ and $x^{\pm}(u)$ in place of $\xi_{i,r}$, $x_{i,r}^{\pm}$, $\xi_i(u)$ and $x_i^{\pm}(u)$, respectively.

2.4 Alternative Generators of $Y_\hbar^0(\mathfrak{g})$

Let $\{t_{i,r}\}_{i\in\mathbf{I}, r\in\mathbb{Z}_{\geq 0}} \subset Y_\hbar^0(\mathfrak{g})$ be the generators defined by

$$t_i(u) = \hbar\sum_{r\geq 0} t_{i,r}u^{-r-1} := \log(\xi_i(u))$$

In particular, $t_{i,0} = \xi_{i,0}$ and

$$t_{i,1} = \xi_{i,1} - \frac{\hbar}{2}\xi_{i,0}^2 \tag{2.1}$$

The relations (Y2) and (Y3) of $Y_\hbar(\mathfrak{g})$ imply that for any $i, j \in \mathbf{I}$ and $r \in \mathbb{Z}_{\geq 0}$,

$$[t_{i,1}, x_{j,r}^{\pm}] = \pm d_i a_{ij} x_{j,r+1}^{\pm} \tag{2.2}$$

so that $t_{i,1}$ act as shift operators on the generators $x_{j,r}^{\pm}$.

The relation (2.2) also implies that $\{\xi_{i,0}, x_{i,0}^{\pm}, t_{i,1}\}_{i\in\mathbf{I}}$ generate $Y_\hbar(\mathfrak{g})$ as an algebra. We refer the reader to [18, Thm. 1.2] for a presentation of $Y_\hbar(\mathfrak{g})$ given in terms of these generators, and to [13, Thm. 2.13] for a refinement of this result.

2.5 Shift Automorphism

The group of translations of the complex plane acts on $Y_\hbar(\mathfrak{g})$ by

$$\tau_a(y_r) = \sum_{s=0}^{r} \binom{r}{s} a^{r-s} y_s$$

where $a \in \mathbb{C}$ and y is one of $\xi_i, t_i, x_i^{\pm}$. In terms of the generating series introduced in Sects. 2.3 and 2.4, we have

$$\tau_a(y(u)) = y(u-a)$$

Given a representation V of $Y_\hbar(\mathfrak{g})$ and $a \in \mathbb{C}$, set $V(a) = \tau_a^*(V)$.

2.6 PBW Theorem

Consider the loop filtration $\mathcal{F}_\bullet(Y_\hbar(\mathfrak{g}))$ on $Y_\hbar(\mathfrak{g})$ defined by $\deg(y_r) = r$ for each of the generators $y = \xi_i, x_i^{\pm}$. Note that $\deg(t_{i,r}) = r$. The Hopf algebra structure on $Y_\hbar(\mathfrak{g})$ preserves this filtration, and endows $\mathrm{gr}(Y_\hbar(\mathfrak{g}))$ with the structure of a graded Hopf algebra. The PBW Theorem for $Y_\hbar(\mathfrak{g})$ [19] (see also [8, Thm. B.6] and [14, Prop. 2.2]) is equivalent to the assertion that the assignments

$$x_i^{\pm}.z^r \mapsto \bar{x}_{i,r}^{\pm} \qquad \text{and} \qquad d_i h_i.z^r \mapsto \bar{\xi}_{i,r}$$

uniquely extend to an isomorphism of graded Hopf algebras

$$U(\mathfrak{g}[z]) \xrightarrow{\sim} \mathrm{gr}(Y_\hbar(\mathfrak{g})), \tag{2.3}$$

where, for any fixed $k \in \mathbb{Z}_{\geq 0}$ and element $y_k \in \mathcal{F}_k(Y_\hbar(\mathfrak{g}))$,

$$\bar{y}_k \in \mathrm{gr}_k(Y_\hbar(\mathfrak{g})) := \mathcal{F}_k(Y_\hbar(\mathfrak{g}))/\mathcal{F}_{k-1}(Y_\hbar(\mathfrak{g}))$$

is defined to be the image of y_k in the k-th graded component $\mathrm{gr}_k(Y_\hbar(\mathfrak{g}))$ of the associated graded algebra $\mathrm{gr}(Y_\hbar(\mathfrak{g}))$.

Henceforth, we shall freely make use of the above identification without further comment. Similarly, we will exploit the fact that it allows us to identify $U(\mathfrak{g}[z]) \otimes U(\mathfrak{g}[w])$ with $\mathrm{gr}(Y_\hbar(\mathfrak{g}) \otimes Y_\hbar(\mathfrak{g})) \cong \mathrm{gr}(Y_\hbar(\mathfrak{g})) \otimes \mathrm{gr}(Y_\hbar(\mathfrak{g}))$, the associated graded algebra of $Y_\hbar(\mathfrak{g}) \otimes Y_\hbar(\mathfrak{g})$ with respect to the tensor product filtration $\mathcal{F}_\bullet(Y_\hbar(\mathfrak{g}) \otimes Y_\hbar(\mathfrak{g}))$ induced by $\mathcal{F}_\bullet(Y_\hbar(\mathfrak{g}))$.

2.7 The Embedding $U(\mathfrak{g}) \subset Y_\hbar(\mathfrak{g})$

Since $\mathrm{gr}_0(Y_\hbar(\mathfrak{g})) = \mathcal{F}_0(Y_\hbar(\mathfrak{g})) \subset Y_\hbar(\mathfrak{g})$, the isomorphism (2.3) restricts to an embedding of $U(\mathfrak{g})$ into $Y_\hbar(\mathfrak{g})$, given by

$$x_i^\pm \mapsto x_{i,0}^\pm \qquad \text{and} \qquad d_i h_i \mapsto \xi_{i,0}$$

We shall henceforth identify $U(\mathfrak{g}) \subset Y_\hbar(\mathfrak{g})$, with the above embedding implicitly understood. Viewed as a module over $\mathfrak{h} \subset Y_\hbar(\mathfrak{g})$, we then have $Y_\hbar(\mathfrak{g}) = \bigoplus_{\beta \in \mathsf{Q}} Y_\hbar(\mathfrak{g})_\beta$

A second embedding $\mathrm{T} : \mathfrak{h} \to Y_\hbar(\mathfrak{g})$ is given by setting $\mathrm{T}(d_i h_i) = t_{i,1}$ for all $i \in \mathbf{I}$, where $t_{i,1}$ is defined by (2.1). The relation (2.2) then reads

$$[\mathrm{T}(h), x_{i,r}^\pm] = \pm\alpha_i(h) x_{i,r+1}^\pm \tag{2.4}$$

and implies in particular that, for any $h \in \alpha_i^\perp$ and $r \geq 0$,

$$[\mathrm{T}(h), x_{i,r}^\pm] = 0 \tag{2.5}$$

2.8 Formal Series Filtration

For any $k \in \mathbb{Z}$, set

$$\begin{aligned} \mathcal{F}_k\left(Y_\hbar(\mathfrak{g})^{\otimes 2}[s; s^{-1}]\!]\right) &= s^k \prod_{n \geq 0} \mathcal{F}_n(Y_\hbar(\mathfrak{g})^{\otimes 2}) s^{-n} \\ &= \left\{ \sum_{m \leq M} y_m s^m \in Y_\hbar(\mathfrak{g})^{\otimes 2}[s; s^{-1}]\!] \;\middle|\; \deg(y_m) \leq k - m \right\} \end{aligned} \tag{2.6}$$

For $k \leq 0$, we shall write this as $\mathcal{F}_k(Y_\hbar(\mathfrak{g})^{\otimes 2}[\![s^{-1}]\!])$, for obvious reasons.

The above spaces generate a $\mathbb{Z}$-filtered algebra

$$\bigcup_{k\in\mathbb{Z}} \mathcal{F}_k(Y_\hbar(\mathfrak{g})^{\otimes 2}[s;s^{-1}]\!]) \subset Y_\hbar(\mathfrak{g})^{\otimes 2}[s;s^{-1}]\!]$$

with associated graded algebra that can (and will) be identified with the $\mathbb{C}[s^{\pm 1}]$–submodule of $(U(\mathfrak{g}[z]) \otimes U(\mathfrak{g}[w]))[s;s^{-1}]\!]$ generated by

$$\prod_{n\geq 0} (U(\mathfrak{g}[z]) \otimes U(\mathfrak{g}[w]))_n s^{-n}$$

where $(U(\mathfrak{g}[z]) \otimes U(\mathfrak{g}[w]))_n$ is the n-th graded component of $U(\mathfrak{g}[z]) \otimes U(\mathfrak{g}[w])$.

If $\mathfrak{X}(s) \in \mathcal{F}_k(Y_\hbar(\mathfrak{g})^{\otimes 2}[s;s^{-1}]\!])$, we denote by

$$\begin{aligned}\overline{\mathfrak{X}(s)} &= \mathfrak{X}(s) \mod \mathcal{F}_{k-1}(Y_\hbar(\mathfrak{g})^{\otimes 2}[s;s^{-1}]\!]) \\ &\in s^k \prod_{n\geq 0} (U(\mathfrak{g}[z]) \otimes U(\mathfrak{g}[w]))_n s^{-n} \\ &\subset (U(\mathfrak{g}[z]) \otimes U(\mathfrak{g}[w]))[s;s^{-1}]\!]\end{aligned}$$

the image of $\mathfrak{X}(s)$ in the k–th graded component of the associated graded algebra. Note in passing that if $\mathfrak{X}(s) \in \mathcal{F}_k(Y_\hbar(\mathfrak{g})^{\otimes 2}[s;s^{-1}]\!])$, with $k < 0$, then $\exp(\mathfrak{X}(s)) - 1 \in \mathcal{F}_k(Y_\hbar(\mathfrak{g})^{\otimes 2}[s;s^{-1}]\!])$, and

$$\overline{\exp(\mathfrak{X}(s)) - 1} = \overline{\mathfrak{X}(s)}. \tag{2.7}$$

2.9 Rationality

The following rationality property is due to Beck–Kac [1] and Hernandez [15] for the analogous case of the quantum loop algebra, and to the first two authors for $Y_\hbar(\mathfrak{g})$. In the form below, the result appears in [11, Prop. 3.6].

Proposition *Let V be a $Y_\hbar(\mathfrak{g})$-module on which $\{\xi_{i,0}\}_{i\in\mathbf{I}}$ act semisimply with finite-dimensional weight spaces. Then, for every weight μ of V, the generating series*

$$\xi_i(u) \in \mathrm{End}(V_\mu)[\![u^{-1}]\!] \qquad \textit{and} \qquad x_i^\pm(u) \in \mathrm{Hom}(V_\mu, V_{\mu\pm\alpha_i})[\![u^{-1}]\!]$$

are the Laurent expansions at ∞ of rational functions of u. Specifically,

$$x_i^\pm(u) = \hbar u^{-1}\left(1 \mp \frac{\mathrm{ad}(t_{i,1})}{2d_i u}\right)^{-1} x_{i,0}^\pm$$

and

$$\xi_i(u) = 1 + [x_i^+(u), x_{i,0}^-]$$

If V is a finite-dimensional $Y_\hbar(\mathfrak{g})$–module, we define $\sigma(V) \subset \mathbb{C}$ to be the (finite) set of poles of the rational $\mathrm{End}(V)$–valued functions $\{\xi_i(u), x_i^\pm(u)\}_{i\in\mathbf{I}}$.

3 The Standard and Drinfeld Coproducts

We review the definition of the standard coproduct on $Y_\hbar(\mathfrak{g})$ following [13], and the deformed Drinfeld tensor product on its finite-dimensional representations introduced in [12]. We then lift the latter to a deformed Drinfeld coproduct $\underset{\mathrm{D},s}{\Delta}$: $Y_\hbar(\mathfrak{g}) \to Y_\hbar(\mathfrak{g})^{\otimes 2}[s; s^{-1}]\!]$.

3.1 *Standard Coproduct*

Set

$$\mathsf{r} = \sum_{\beta\in\Phi_+} x_{\beta,0}^- \otimes x_{\beta,0}^+ \tag{3.1}$$

where $x_{\beta,0}^\pm \in \mathfrak{g}_{\pm\beta} \subset Y_\hbar(\mathfrak{g})$ are root vectors such that $(x_{\beta,0}^-, x_{\beta,0}^+) = 1$. The coproduct $\Delta : Y_\hbar(\mathfrak{g}) \to Y_\hbar(\mathfrak{g}) \otimes Y_\hbar(\mathfrak{g})$ is defined by the following formulae

$$\Delta(\xi_{i,0}) = \xi_{i,0} \otimes 1 + 1 \otimes \xi_{i,0}$$

$$\Delta(x_{i,0}^\pm) = x_{i,0}^\pm \otimes 1 + 1 \otimes x_{i,0}^\pm$$

$$\begin{aligned}\Delta(t_{i,1}) &= t_{i,1} \otimes 1 \, + 1 \otimes t_{i,1} + \hbar\, \mathrm{ad}(\xi_{i,0} \otimes 1)\mathsf{r} \\ &= t_{i,1} \otimes 1 \, + 1 \otimes t_{i,1} - \hbar \sum_{\beta\in\Phi_+} (\beta, \alpha_i) x_{\beta,0}^- \otimes x_{\beta,0}^+\end{aligned}$$

We refer the reader to [13, §4.2] for a proof that Δ is an algebra homomorphism. It is immediate that Δ is coassociative (see [13, §4.5]).

3.2 *Deformed Drinfeld Tensor Product*

We review below the definition of the deformed Drinfeld tensor product introduced in [12, Section 4.4]. Let $V, W \in \mathrm{Rep}_{\mathrm{fd}}(Y_\hbar(\mathfrak{g}))$, and $\sigma(V), \sigma(W) \subset \mathbb{C}$ their sets of poles. Let $s \in \mathbb{C}$ be such that $\sigma(V)+s$ and $\sigma(W)$ are disjoint, and define an action of the generators of $Y_\hbar(\mathfrak{g})$ on $V \otimes W$ by

$$\underset{\mathrm{D},s}{\Delta}(\xi_i(u)) = \xi_i(u-s) \otimes \xi_i(u)$$

$$\underset{\mathrm{D},s}{\Delta}(x_i^+(u)) = x_i^+(u-s) \otimes 1 + \oint_{C_2} \frac{1}{u-v} \xi_i(v-s) \otimes x_i^+(v)dv$$

$$\underset{\mathrm{D},s}{\Delta}(x_i^-(u)) = \oint_{C_1} \frac{1}{u-v} x_i^-(v-s) \otimes \xi_i(v)dv + 1 \otimes x_i^-(u)$$

where

- C_1 encloses $\sigma(V)+s$ and none of the points in $\sigma(W)$.
- C_2 encloses $\sigma(W)$ and none of the points in $\sigma(V)+s$.
- The integral $\oint_{C_1}$ (resp. $\oint_{C_2}$) is understood to mean the holomorphic function of u it defines for u outside of C_1 (resp. C_2).

Note that in terms of the generators $t_{i,r}$ of $Y_\hbar^0(\mathfrak{g})$,

$$\underset{\mathrm{D},s}{\Delta}(t_i(u)) = t_i(u-s) \otimes 1 + 1 \otimes t_i(u)$$

Theorem ([12, Thm. 4.6])

1. *The formulae above define an action of $Y_\hbar(\mathfrak{g})$ on $V \otimes W$. The corresponding representation is denoted by $V \underset{\mathrm{D},s}{\otimes} W$.*
2. *The action of $Y_\hbar(\mathfrak{g})$ on $V \underset{\mathrm{D},s}{\otimes} W$ is a rational function of s, with poles contained in $\sigma(W)-\sigma(V)$.*
3. *The identification of vector spaces*

$$(V_1 \underset{\mathrm{D},s_1}{\otimes} V_2) \underset{\mathrm{D},s_2}{\otimes} V_3 = V_1 \underset{\mathrm{D},s_1+s_2}{\otimes} (V_2 \underset{\mathrm{D},s_2}{\otimes} V_3)$$

intertwines the action of $Y_\hbar(\mathfrak{g})$.

4. *If $V \cong \mathbb{C}$ is the trivial representation of $Y_\hbar(\mathfrak{g})$, then*

$$V \underset{\mathrm{D},s}{\otimes} W = W \qquad \text{and} \qquad W \underset{\mathrm{D},s}{\otimes} V = W(s)$$

5. *The following holds for any $s, t \in \mathbb{C}$,*

$$V(t) \underset{\mathrm{D},s}{\otimes} W(t) = (V \underset{\mathrm{D},s}{\otimes} W)(t) \qquad \text{and} \qquad V(t) \underset{\mathrm{D},s}{\otimes} W = V \underset{\mathrm{D},s+t}{\otimes} W$$

In particular, $V \underset{\mathrm{D},s}{\otimes} W(t) = (V \underset{\mathrm{D},s-t}{\otimes} W)(t)$.

6. *The following holds for any $s \in \mathbb{C}$,*

$$\sigma(V \underset{\mathrm{D},s}{\otimes} W) \subset (s + \sigma(V)) \cup \sigma(W)$$

3.3 Laurent Expansion of the Deformed Drinfeld Tensor Product

Proposition *The Laurent expansion at $s = \infty$ of the formulae of Sect. 3.2 is given by*

$$\underset{\mathrm{D},s}{\Delta}(t_{i,r}) = \tau_s(t_{i,r}) \otimes 1 + 1 \otimes t_{i,r}$$

and

$$\underset{\mathrm{D},s}{\Delta}(x_{i,r}^{+}) = \tau_s(x_{i,r}^{+}) \otimes 1 + 1 \otimes x_{i,r}^{+}$$

$$+ \hbar \sum_{N \geq 0} \left(\sum_{n=0}^{N} (-1)^{n+1} \binom{N}{n} \xi_{i,n} \otimes x_{i,r+N-n}^{+} \right) s^{-N-1}$$

$$\underset{\mathrm{D},s}{\Delta}(x_{i,r}^{-}) = \tau_s(x_{i,r}^{-}) \otimes 1 + 1 \otimes x_{i,r}^{-}$$

$$+ \hbar \sum_{N \geq 0} \left(\sum_{n=0}^{N} (-1)^{n+1} \binom{N}{n} x_{i,r+n}^{-} \otimes \xi_{i,N-n} \right) s^{-N-1}$$

Proof The expansion of $\underset{\mathrm{D},s}{\Delta}(t_{i,r})$ follows from $\underset{\mathrm{D},s}{\Delta}(t_i(u)) = \tau_s(t_i(u)) \otimes 1 + 1 \otimes t_i(u)$. Expanding in u^{-1} yields

$$\underset{\mathrm{D},s}{\Delta}(x_{i,r}^{+}) = \tau_s(x_{i,r}^{+}) \otimes 1 + \frac{1}{\hbar} \oint_{C_2} v^r \xi_i(v - s) \otimes x_i^{+}(v)\, dv$$

Expanding now $\xi_i(v - s)$ with respect to s^{-1} by using $(1 - x)^{-p-1} = \sum_{m\geq 0}\binom{p+m}{p} x^m$ yields:

$$\begin{aligned}\xi_i(v-s) &= 1 + \hbar\sum_{p\geq 0}\xi_{i,p}(v-s)^{-p-1}\\ &= 1 + \hbar\sum_{p\geq 0}\xi_{i,p}(-s)^{-p-1}\sum_{q\geq 0}\binom{p+q}{p} v^q s^{-q}\\ &= 1 + \hbar\sum_{m\geq 0} s^{-m-1}\sum_{\substack{p,q\geq 0\\ p+q=m}}(-1)^{p+1}\binom{m}{p} v^q \xi_{i,p}\end{aligned}$$

Substituting gives

$$\begin{aligned}\underset{\mathrm{D},s}{\Delta}(x^+_{i,r}) = \tau_s(x^+_{i,r})\otimes 1 + \frac{1}{\hbar}\oint_{C_2} v^r 1\otimes x_i^+(v)\,dv\\ + \sum_{m\geq 0} s^{-m-1}\sum_{\substack{p,q\geq 0\\ p+q=m}}(-1)^{p+1}\binom{m}{p}\oint_{C_2} v^{r+q}\xi_{i,p}\otimes x_i^+(v)\,dv\end{aligned}$$

which is the claimed result since, for any $a \geq 0$, $\oint_{C_2} v^a x_i^+(v)\,dv = \hbar x^+_{i,a}$. The expansion of $\underset{\mathrm{D},s}{\Delta}(x^-_{i,r})$ is obtained in the same way. □

3.4 Deformed Drinfeld Coproduct

We now lift the deformed Drinfeld tensor product to an algebra homomorphism

$$\underset{\mathrm{D},s}{\Delta} : Y_\hbar(\mathfrak{g}) \to \big(Y_\hbar(\mathfrak{g})\otimes Y_\hbar(\mathfrak{g})\big)[s; s^{-1}]].$$

Theorem *The Laurent expansions of Sect. 3.3 give rise to an algebra homomorphism*

$$\underset{\mathrm{D},s}{\Delta} : Y_\hbar(\mathfrak{g}) \to \big(Y_\hbar(\mathfrak{g})\otimes Y_\hbar(\mathfrak{g})\big)[s; s^{-1}]]$$

The deformed Drinfeld coproduct $\underset{\mathrm{D},s}{\Delta}$ has the following properties.

1. It is compatible with the counit ϵ, that is

$$\epsilon\otimes\mathbf{1}\circ\underset{\mathrm{D},s}{\Delta} = \mathbf{1} \qquad \textit{and} \qquad \mathbf{1}\otimes\epsilon\circ\underset{\mathrm{D},s}{\Delta} = \tau_s$$

2. *For every $x \in Y_\hbar(\mathfrak{g})$, the following holds in $(Y_\hbar(\mathfrak{g}) \otimes Y_\hbar(\mathfrak{g})[a])[s; s^{-1}]]$*

$$\tau_a \otimes \tau_a \circ \underset{\mathrm{D},s}{\Delta}(x) = \underset{\mathrm{D},s}{\Delta} \circ \tau_a(x) \qquad \text{and} \qquad \tau_a \otimes \mathbf{1} \circ \underset{\mathrm{D},s}{\Delta}(x) = \underset{\mathrm{D},s+a}{\Delta}(x)$$

3. *$\underset{\mathrm{D},s}{\Delta}$ is a filtered homomorphism, that is*

$$\underset{\mathrm{D},s}{\Delta}(\mathcal{F}_k(Y_\hbar(\mathfrak{g}))) \subset \mathcal{F}_k(Y_\hbar(\mathfrak{g})^{\otimes 2}[s; s^{-1}]])$$

for each $k \geq 0$, where $\mathcal{F}_\bullet(Y_\hbar(\mathfrak{g})^{\otimes 2}[s; s^{-1}]])$ is the filtration defined in (2.6).

Proof Using Theorem 4.1, and the fact that Δ_s is an algebra homomorphism, we conclude the same for $\underset{\mathrm{D},s}{\Delta}$. Properties (1) and (2) above are easy to verify directly from the definition. Property (3) follows immediately from the explicit formulas given in Proposition 3.3. □

Remark The coassociativity property of the Drinfeld tensor product $\underset{\mathrm{D},s}{\otimes}$ does not appear to have a natural lift to $\underset{\mathrm{D},s}{\Delta}$. The candidate identity

$$\underset{\mathrm{D},s_1}{\Delta} \otimes \mathbf{1} \circ \underset{\mathrm{D},s_2}{\Delta}(x) = \mathbf{1} \otimes \underset{\mathrm{D},s_2}{\Delta} \circ \underset{\mathrm{D},s_1+s_2}{\Delta}(x) \tag{3.2}$$

holds if x is one of the commuting generators $t_{i,r}$. However, if x is an arbitrary element of $Y_\hbar(\mathfrak{g})$, the left–hand side and right–hand side lie, respectively, in

$$Y_\hbar(\mathfrak{g})^{\otimes 3}[s_1; s_1^{-1}]][s_2; s_2^{-1}]] \qquad \text{and} \qquad Y_\hbar(\mathfrak{g})^{\otimes 3}[s_2; s_2^{-1}]][s_1; s_1^{-1}]]$$

and cannot therefore be directly compared. The coassociativity of $\underset{\mathrm{D},s}{\otimes}$ implies, however, that the evaluation of the left– and right–hand sides of (3.2) on a tensor product $V_1 \otimes V_2 \otimes V_3$ of finite-dimensional representations are, respectively, the expansions at $|s_2| \gg |s_1|$ and $|s_1| \gg |s_2|$ of the same rational function.

4 The Element $\mathcal{R}^-(s)$

4.1

Set $\mathsf{Q}_+ = \bigoplus_{i\in\mathbf{I}} \mathbb{Z}_{\geq 0}\, \alpha_i \subset \mathfrak{h}^*$. The following is one of the main result of this paper.

Theorem

1. *There is a unique element*

$$\mathcal{R}^-(s) \in (Y_\hbar(\mathfrak{g}) \otimes Y_\hbar(\mathfrak{g}))[[s^{-1}]]$$

which is

- *strictly lower triangular, that is* $\mathcal{R}^-(s) = \sum_{\beta,\gamma\in\mathsf{Q}_+} \mathcal{R}^-(s)_{\beta,\gamma}$, *with*

$$\mathcal{R}^-(s)_{\beta,\gamma} \in (Y_\hbar(\mathfrak{g})_{-\beta} \otimes Y_\hbar(\mathfrak{g})_\gamma)[\![s^{-1}]\!] \quad \text{and} \quad \mathcal{R}^-(s)_{0,0} = 1 \otimes 1$$

- *of weight zero*
- *such that, for any* $i \in \mathbf{I}$,

$$\mathcal{R}^-(s) \cdot \Delta_s\left(t_{i,1}\right) = \underset{\mathrm{D},s}{\Delta}\left(t_{i,1}\right) \cdot \mathcal{R}^-(s) \tag{4.1}$$

2. *The element* $\mathcal{R}^-(s)$ *lies in* $(Y_\hbar^-(\mathfrak{g}) \otimes Y_\hbar^+(\mathfrak{g}))[\![s^{-1}]\!]$, *and has the following additional properties.*

   - *For any* $i \in \mathbf{I}$,

$$\mathcal{R}^-(s) \cdot \Delta_s\left(x_{i,0}^{\pm}\right) = \underset{\mathrm{D},s}{\Delta}\left(x_{i,0}^{\pm}\right) \cdot \mathcal{R}^-(s) \tag{4.2}$$

   - *For any* $a, b \in \mathbb{C}$,

$$\tau_a \otimes \tau_b\left(\mathcal{R}^-(s)\right) = \mathcal{R}^-(s + a - b)$$

   - $\mathcal{R}^-(s) - 1 \in \mathcal{F}_{-1}(Y_\hbar(\mathfrak{g})^{\otimes 2}[\![s^{-1}]\!])$, *with semiclassical limit given by*

$$\overline{\mathcal{R}^-(s) - 1} = \frac{\hbar\mathsf{r}}{z + s - w} \in (U(\mathfrak{g}[z]) \otimes U(\mathfrak{g}[w]))[\![s^{-1}]\!]$$

   In particular, $\mathcal{R}^-(s) = 1 + \hbar\mathsf{r}/s + O(s^{-2})$.

3. *Let* $V_1, V_2 \in \mathrm{Rep}_{\mathrm{fd}}(Y_\hbar(\mathfrak{g}))$, *and* $\mathcal{R}^-_{V_1,V_2}(s) \in \mathrm{End}(V_1 \otimes V_2)[\![s^{-1}]\!]$ *the corresponding evaluation of* $\mathcal{R}^-(s)$. *Then,*

   - $\mathcal{R}^-_{V_1,V_2}(s)$ *is the Taylor expansion at* $s = \infty$ *of a rational function.*
   - *The following cocycle equation holds for any* $V_1, V_2, V_3 \in \mathrm{Rep}_{\mathrm{fd}}(Y_\hbar(\mathfrak{g}))$

$$\mathcal{R}^-_{V_1 \underset{\mathrm{D},s_1}{\otimes} V_2, V_3}(s_2) \cdot \mathcal{R}^-_{V_1,V_2}(s_1) = \mathcal{R}^-_{V_1, V_2 \underset{\mathrm{D},s_2}{\otimes} V_3}(s_1 + s_2) \cdot \mathcal{R}^-_{V_2,V_3}(s_2) \tag{4.3}$$

Remark Analogously to Remark 3.4, the cocycle equation (4.3) only holds as an identity of rational functions with values in $\mathrm{End}(V_1 \otimes V_2 \otimes V_3)$, and does not seem to possess a natural lift to $Y_\hbar(\mathfrak{g})$. Indeed, the candidate identity

$$\underset{\mathrm{D},s_1}{\Delta} \otimes \mathbf{1}(\mathcal{R}^-(s_2)) \cdot \mathcal{R}^-_{12}(s_1) = \mathbf{1} \otimes \underset{\mathrm{D},s_2}{\Delta}\left(\mathcal{R}^-(s_1 + s_2)\right) \cdot \mathcal{R}^-_{23}(s_2)$$

does not make sense since the left–hand side lies in $Y_\hbar(\mathfrak{g})^{\otimes 3}[s_1; s_1{}^{-1}]\![\![s_2^{-1}]\!]$, while the right–hand side lies in $Y_\hbar(\mathfrak{g})^{\otimes 3}[s_2; s_2{}^{-1}]\![\![s_1^{-1}]\!]$.

The rest of the section is devoted to the proof of Theorem 4.1.

4.2 Existence and Uniqueness of $\mathcal{R}^-(s)$

The triangularity and zero–weight assumptions are equivalent to the requirement that $\mathcal{R}^-(s)$ have the form

$$\mathcal{R}^-(s) = \sum_{\beta\in\mathsf{Q}_+} \mathcal{R}^-(s)_\beta \quad \text{with} \quad \mathcal{R}^-(s)_\beta \in (Y_\hbar(\mathfrak{g})_{-\beta} \otimes Y_\hbar(\mathfrak{g})_\beta)[\![s^{-1}]\!] \tag{4.4}$$

and $\mathcal{R}^-(s)_0 = 1 \otimes 1$. We shall construct $\mathcal{R}^-(s)_\beta$ recursively in β, and prove that the sum over β converges in the s–adic topology. In fact, define $\nu : \mathsf{Q}_+ \to \mathbb{Z}_{\geq 0}$ by

$$\nu(\beta) = \min\left\{k \in \mathbb{Z}_{\geq 0} | \beta = \alpha^{(1)} + \cdots + \alpha^{(k)}, \alpha^{(1)}, \ldots, \alpha^{(k)} \in \Phi_+\right\}$$

with $\nu(0) = 0$. Then, we shall prove that $\mathcal{R}^-(s)_\beta \in s^{-\nu(\beta)} Y_\hbar(\mathfrak{g})^{\otimes 2}[\![s^{-1}]\!]$.

Let r be given by (3.1) and, for any $h \in \mathfrak{h}$, set

$$\mathsf{r}(h) = \mathrm{ad}(h \otimes 1)(\mathsf{r}) = -\sum_{\gamma\in\Phi_+} \gamma(h)\, x^-_{\gamma,0} \otimes x^+_{\gamma,0}.$$

By Sects. 3.1 and 3.3,

$$\underset{\mathrm{D},s}{\Delta}(\mathrm{T}(h)) = \sum_{a=1}^{2} \mathrm{T}(h)^{(a)} + sh^{(1)}$$

$$\Delta_s(\mathrm{T}(h)) = \sum_{a=1}^{2} \mathrm{T}(h)^{(a)} + sh^{(1)} + \hbar\mathsf{r}(h),$$

where we use the standard notation: $X^{(1)} = X \otimes 1$ and $X^{(2)} = 1 \otimes X$. The intertwining equation (4.1) therefore reads as $A(h)\mathcal{R}^-(s) = 0$ where, for any $h \in \mathfrak{h}$,

$$\begin{aligned} A(h) &= \lambda(\underset{\mathrm{D},s}{\Delta}(\mathrm{T}(h))) - \rho\,(\Delta_s(\mathrm{T}(h))) \\ &= \mathrm{ad}\left(\mathrm{T}(h)^{(1)} + \mathrm{T}(h)^{(2)} + sh^{(1)}\right) - \hbar\rho(\mathsf{r}(h)) \end{aligned}$$

and λ, ρ denote left and right multiplication, respectively. In components, this reads:

$$(\mathcal{T}(h) - s\beta(h))\, \mathcal{R}^-(s)_\beta = -\hbar \sum_{\alpha\in\Phi_+} \alpha(h)\mathcal{R}^-(s)_{\beta-\alpha}\, x^-_{\alpha,0} \otimes x^+_{\alpha,0}$$

where $\mathcal{T}(h) = \mathrm{ad}(\mathrm{T}(h)^{(1)} + \mathrm{T}(h)^{(2)})$ and $\mathcal{R}^-(s)_\gamma = 0$ if $\gamma \notin \mathsf{Q}_+$.

If $h \in \mathfrak{h}$ is such that $\beta(h) \neq 0$ for any nonzero $\beta \in \mathsf{Q}_+$, this yields

$$\mathcal{R}^-(s)_\beta = \frac{\hbar}{s\beta(h)} \left(1 - \frac{\mathcal{T}(h)}{s\beta(h)}\right)^{-1} \sum_{\alpha\in\Phi_+} \alpha(h)\mathcal{R}^-(s)_{\beta-\alpha} x^-_{\alpha,0} \otimes x^+_{\alpha,0} \tag{4.5}$$

$$= \hbar \sum_{k\geq 0} \frac{\mathcal{T}(h)^k}{(s\beta(h))^{k+1}} \sum_{\alpha\in\Phi_+} \alpha(h)\mathcal{R}^-(s)_{\beta-\alpha} x^-_{\alpha,0} \otimes x^+_{\alpha,0} \tag{4.6}$$

This shows that $\mathcal{R}^-(s)$ is uniquely determined by $\mathcal{R}^-(s)_0$, and that it lies in $(Y_\hbar^-(\mathfrak{g}) \otimes Y_\hbar^+(\mathfrak{g}))[\![s^{-1}]\!]$ if $\mathcal{R}^-(s)_0$ does, since $Y_\hbar^-(\mathfrak{g}), Y_\hbar^+(\mathfrak{g})$ are invariant under $\mathrm{ad}\, \mathrm{T}(h)$.

Fix now $h \in \mathfrak{h} \setminus \bigcup_{\beta\in(\mathsf{Q}_+\setminus\{0\})} \mathrm{Ker}\, \beta$. The above equations can be used to define elements $\mathcal{R}^-(s)_\beta$ recursively on the height of β, starting from $\mathcal{R}^-(s)_0 = 1$, which lie in $s^{-\nu(\beta)} Y_\hbar(\mathfrak{g})^{\otimes 2}[\![s^{-1}]\!]$. The corresponding sum $\mathcal{R}^-(s) = \sum_{\beta\in\mathsf{Q}_+} \mathcal{R}^-(s)_\beta$ is therefore well–defined, and satisfies $A(h)\mathcal{R}^-(s) = 0$. We claim that it also satisfies $A(h')\mathcal{R}^-(s) = 0$ for any $h' \in \mathfrak{h}$. Note that

$$[A(h), A(h')] = \lambda[\underset{\mathrm{D},s}{\Delta}(\mathrm{T}(h)), \underset{\mathrm{D},s}{\Delta}(\mathrm{T}(h'))] - \rho[\Delta_s(\mathrm{T}(h)), \Delta_s(\mathrm{T}(h'))]$$

which vanishes because Δ_s is an algebra homomorphism, and $\underset{\mathrm{D},s}{\Delta}(\mathrm{T}(h)), \underset{\mathrm{D},s}{\Delta}(\mathrm{T}(h')) \in Y_\hbar^0(\mathfrak{g})^{\otimes 2}[s]$. Thus, $A(h')\mathcal{R}^-(s)$ satisfies

$$A(h)A(h')\mathcal{R}^-(s) = A(h')A(h)\mathcal{R}^-(s) = 0$$

Since $A(h')\mathcal{R}^-(s)$ is also triangular, with

$$\left(A(h')\mathcal{R}^-(s)\right)_0 = \mathrm{ad}\left(\mathrm{T}(h')^{(1)} + \mathrm{T}(h')^{(2)} + sh'^{(1)}\right)\mathcal{R}^-(s)_0 = 0$$

it follows by uniqueness that $A(h')\mathcal{R}^-(s) = 0$ as claimed.

4.3 Translation Invariance

The identity $\tau_a \otimes \tau_b \left(\mathcal{R}^-(s)\right) = \mathcal{R}^-(s + a - b)$ follows by uniqueness, since both sides are strictly lower triangular, intertwine

$$\tau_a \otimes \tau_b \circ \Delta_s(t_{i,m}) = \tau_{a-b} \otimes \mathbf{1} \circ \Delta_s(\tau_b(t_{i,m})) = \Delta_{s+a-b}(\tau_b(t_{i,m}))$$

and $\tau_a \otimes \tau_b \circ \underset{\mathrm{D},s}{\Delta}(t_{i,m}) = \underset{\mathrm{D},s+a-b}{\Delta}(\tau_b(t_{i,m}))$ for $i \in \mathbf{I}$ and $m = 0, 1$, and the span of $\{t_{i,m}\}_{i\in\mathbf{I},m=0,1}$ is invariant under τ_b.

4.4 Semiclassical Limit

Since $\mathcal{T}(h)$ is a filtered operator of degree 1, the recursion (4.6) shows that $\mathcal{R}^-(s)_\beta \in \mathcal{F}_{-\nu(\beta)}(Y_\hbar(\mathfrak{g})^{\otimes 2}[\![s^{-1}]\!])$ for any $\beta \in \mathsf{Q}_+$. In particular, $\mathcal{R}^-(s) - 1 \in \mathcal{F}_{-1}(Y_\hbar(\mathfrak{g})^{\otimes 2}[\![s^{-1}]\!])$ and, mod $\mathcal{F}_{-2}(Y_\hbar(\mathfrak{g})^{\otimes 2}[\![s^{-1}]\!])$,

$$\mathcal{R}^-(s) - 1 = \sum_{\beta:\nu(\beta)=1} \mathcal{R}^-(s)_\beta = \frac{\hbar}{s} \sum_{\alpha\in\Phi_+} \sum_{k\geq 0} \frac{\mathcal{T}(h)^k}{(s\alpha(h))^k} x^-_{\alpha,0} \otimes x^+_{\alpha,0}$$

whose image in $\mathcal{F}_{-1}(Y_\hbar(\mathfrak{g})^{\otimes 2}[\![s^{-1}]\!])/\mathcal{F}_{-2}(Y_\hbar(\mathfrak{g})^{\otimes 2}[\![s^{-1}]\!])$ is $\hbar\mathsf{r}/(z + s - w)$.

4.5 Rationality

The argument given in Sect. 4.2 can be carried out in $\mathrm{End}(V_1 \otimes V_2)$ rather than $Y_\hbar(\mathfrak{g})^{\otimes 2}$, and shows the existence and uniqueness of an element

$$\mathcal{R}^-_{V_1,V_2}(s) = \sum_{\beta\in\mathsf{Q}_+} \mathcal{R}^-_{V_1,V_2}(s)_\beta \in \mathrm{End}(V_1 \otimes V_2)[\![s^{-1}]\!]$$

with

$$[h \otimes 1, \mathcal{R}^-_{V_1,V_2}(s)_\beta] = -\beta(h)\, \mathcal{R}^-_{V_1,V_2}(s)_\beta = -[1 \otimes h, \mathcal{R}^-_{V_1,V_2}(s)_\beta]$$

for any $h \in \mathfrak{h}$, $\mathcal{R}^-_{V_1,V_2}(s)_0 = 1$, and

$$\mathcal{R}^-_{V_1,V_2}(s) \cdot \Delta_s\left(t_{i,1}\right) = \underset{\mathrm{D},s}{\Delta}\left(t_{i,1}\right) \cdot \mathcal{R}^-_{V_1,V_2}(s)$$

The recursive construction of $\mathcal{R}^-_{V_1,V_2}(s)$ given by (4.5) shows that each $\mathcal{R}^-_{V_1,V_2}(s)_\beta$ is a rational function of s, regular at $s = \infty$. Since only finitely many

$\mathcal{R}^{-}_{V_1,V_2}(s)_\beta$ are non-zero by finite-dimensionality, $\mathcal{R}^{-}_{V_1,V_2}(s)$ is therefore rational. It follows by uniqueness that the Taylor expansion of $\mathcal{R}^{-}_{V_1,V_2}(s)$ is the evaluation of $\mathcal{R}^{-}(s)$ on $V_1 \otimes V_2$.

4.6 Cocycle Equation

Let $V_1, V_2, V_3 \in \mathrm{Rep}_{\mathrm{fd}}(Y_\hbar(\mathfrak{g}))$. We obtain below an alternative version of the cocycle equation, namely

$$\mathcal{R}^{-}_{V_1,V_2}(s_1) \cdot \mathcal{R}^{-}_{V_1 \underset{s_1}{\otimes} V_2, V_3}(s_2) = \mathcal{R}^{-}_{V_2,V_3}(s_2) \cdot \mathcal{R}^{-}_{V_1, V_2 \underset{s_2}{\otimes} V_3}(s_1+s_2) \tag{4.7}$$

Note that (4.7) and (4.3) are equivalent, provided the intertwining equation (4.2) is established. The latter will be proved in Sect. 4.8, by relying in part on (4.7).[6]

The intertwining property of $\mathcal{R}^{-}(s)$ implies that

$$\begin{aligned}
&\mathcal{R}^{-}_{V_1,V_2}(s_1) \cdot \mathcal{R}^{-}_{V_1 \underset{s_1}{\otimes} V_2, V_3}(s_2) \cdot \pi_{(V_1 \underset{s_1}{\otimes} V_2) \underset{s_2}{\otimes} V_3}(t_{i,1}) \\
&\qquad = \mathcal{R}^{-}_{V_1,V_2}(s_1) \cdot \pi_{(V_1 \underset{s_1}{\otimes} V_2) \underset{\mathrm{D},s_2}{\otimes} V_3}(t_{i,1}) \cdot \mathcal{R}^{-}_{V_1 \underset{\mathrm{D},s_1}{\otimes} V_2, V_3}(s_2) \\
&\qquad\qquad = \pi_{(V_1 \underset{\mathrm{D},s_1}{\otimes} V_2) \underset{\mathrm{D},s_2}{\otimes} V_3}(t_{i,1}) \cdot \mathcal{R}^{-}_{V_1,V_2}(s_1) \cdot \mathcal{R}^{-}_{V_1 \underset{s_1}{\otimes} V_2, V_3}(s_2),
\end{aligned}$$

where the second equality stems from the fact that

$$\pi_{(V_1 \underset{s_1}{\otimes} V_2) \underset{\mathrm{D},s_2}{\otimes} V_3}(t_{i,1}) = \pi_{(V_1 \underset{s_1}{\otimes} V_2)}(\tau_{s_2}(t_{i,1})) + \pi_{V_3}(t_{i,1})$$

Similarly,

$$\begin{aligned}
&\mathcal{R}^{-}_{V_2,V_3}(s_2) \cdot \mathcal{R}^{-}_{V_1, V_2 \underset{s_2}{\otimes} V_3}(s_1+s_2) \cdot \pi_{V_1 \underset{s_1+s_2}{\otimes} (V \underset{s_2}{\otimes} V_3)}(t_{i,1}) \\
&\qquad = \mathcal{R}^{-}_{V_2,V_3}(s_2) \cdot \pi_{V_1 \underset{\mathrm{D},s_1+s_2}{\otimes} (V \underset{s_2}{\otimes} V_3)}(t_{i,1}) \cdot \mathcal{R}^{-}_{V_1, V_2 \underset{s_2}{\otimes} V_3}(s_1+s_2) \\
&\qquad\qquad = \pi_{V_1 \underset{\mathrm{D},s_1+s_2}{\otimes} (V \underset{\mathrm{D},s_2}{\otimes} V_3)}(t_{i,1}) \cdot \mathcal{R}^{-}_{V_2,V_3}(s_2) \cdot \mathcal{R}^{-}_{V_1, V_2 \underset{s_2}{\otimes} V_3}(s_1+s_2)
\end{aligned}$$

[6] The cocycle equation (4.3) arises from the definition of a tensor structure on a functor $F : \mathcal{C} \to \mathcal{D}$ as a system of natural isomorphisms $J_{U,V} : F(U) \otimes_{\mathcal{D}} F(V) \to F(U \otimes_{\mathcal{C}} V)$, by interpreting $\mathcal{R}^{-}(s)$ as a tensor structure on the identity functor $(\mathrm{Rep}_{\mathrm{fd}}(Y_\hbar(\mathfrak{g})), \underset{\mathrm{D},s}{\otimes}) \to (\mathrm{Rep}_{\mathrm{fd}}(Y_\hbar(\mathfrak{g})), \underset{s}{\otimes})$. Similarly, (4.7) arises by adopting the opposite convention of a tensor structure as a system of isomorphisms $K_{U,V} : F(U \otimes_{\mathcal{C}} V) \to F(U) \otimes_{\mathcal{D}} F(V)$, and interpreting $\mathcal{R}^{-}(s)$ as a tensor structure on $(\mathrm{Rep}_{\mathrm{fd}}(Y_\hbar(\mathfrak{g})), \underset{s}{\otimes}) \to (\mathrm{Rep}_{\mathrm{fd}}(Y_\hbar(\mathfrak{g})), \underset{\mathrm{D},s}{\otimes})$.

Since $\underset{s}{\otimes}$, $\underset{\mathrm{D},s}{\otimes}$ are coassociative, both sides of the cocycle equation (4.7) are therefore solutions of

$$X \cdot \Delta_{s_1} \otimes \mathbf{1} \circ \Delta_{s_2}(t_{i,1}) = \underset{\mathrm{D},s_1}{\Delta} \otimes \mathbf{1} \circ \underset{\mathrm{D},s_2}{\Delta}(t_{i,1}) \cdot X \tag{4.8}$$

By Sects. 3.1 and 3.3,

$$\underset{\mathrm{D},s_1}{\Delta} \otimes \mathbf{1} \circ \underset{\mathrm{D},s_2}{\Delta}(\mathrm{T}(h)) = \sum_{a=1}^{3} \mathrm{T}(h)^{(a)} + (s_1 + s_2)\, h^{(1)} + s_2\, h^{(2)}$$

$$\Delta_{s_1} \otimes \mathbf{1} \circ \Delta_{s_2}(\mathrm{T}(h)) = \sum_{a=1}^{3} \mathrm{T}(h)^{(a)} + (s_1 + s_2)\, h^{(1)} + s_2\, h^{(2)} + \hbar\, (\mathsf{r}(h)_{13} + \mathsf{r}(h)_{23})$$

The conclusion now follows by noting that, analogously to Sect. 4.2, (4.8) admits at most one rational solution $X(s_1, s_2)$ with values in $\mathrm{End}(V_1 \otimes V_2 \otimes V_3)$, provided it is strictly lower triangular, that is of the form $X = \sum_{\beta,\gamma \in \mathsf{Q}_+} X_{\beta,\gamma}$, where

$$X_{\beta,\gamma} \in \mathrm{End}(V_1)[-\beta] \otimes \mathrm{End}(V_2)[\beta - \gamma] \otimes \mathrm{End}(V_3)[\gamma]$$

and $X_{0,0} = \mathbf{1}$.

4.7 *Rank 1 Reduction*

Consider the intertwining identity

$$\mathcal{R}^-(s) \cdot \Delta_s(x_{i,0}^{\pm}) = \underset{\mathrm{D},s}{\Delta}(x_{i,0}^{\pm}) \cdot \mathcal{R}^-(s) \tag{4.9}$$

We claim that it holds for any $\mathfrak{g}$ and $i \in \mathbf{I}$ if, and only if it holds for $\mathfrak{g} = \mathfrak{sl}_2$. Set $\mathsf{Q}^{(i)} = \mathsf{Q}/\mathbb{Z}\alpha_i \subset \mathfrak{h}^*/\mathbb{C}\alpha_i = (\alpha_i^{\perp})^*$, and let $\mathsf{Q}_+^{(i)}$ be the image of Q_+ in $\mathsf{Q}^{(i)}$. Both sides of (4.9) are lower triangular and of weight zero with respect to the adjoint action of $\alpha_i^{\perp} \subset \mathfrak{h}$ *i.e.*, lie in $\bigoplus_{\beta \in \mathsf{Q}_+^{(i)}} (Y_\hbar(\mathfrak{g})_{-\beta}^{(i)} \otimes Y_\hbar(\mathfrak{g})_{\beta}^{(i)})[[s^{-1}]]$ where, for $\gamma \in \mathsf{Q}^{(i)}$,

$$Y_\hbar(\mathfrak{g})_{\gamma}^{(i)} = \{x \in Y_\hbar(\mathfrak{g}) \mid [h, x] = \gamma(h)x,\ h \in \alpha_i^{\perp}\}$$

Since both intertwine $\Delta_s(\mathrm{T}(h))$ and $\underset{\mathrm{D},s}{\Delta}(\mathrm{T}(h))$ for $h \in \alpha_i^\perp$, it follows by uniqueness that they are equal if, and only if, their projections onto $(Y_\hbar(\mathfrak{g})_0^{(i)} \otimes Y_\hbar(\mathfrak{g})_0^{(i)})[\![s^{-1}]\!]$ coincide.

Let $\pi_i \in \mathrm{End}(Y_\hbar(\mathfrak{g}))$ be the projection onto $Y_\hbar(\mathfrak{g})_0^{(i)} = \bigoplus_{m\in\mathbb{Z}} Y_\hbar(\mathfrak{g})_{m\alpha_i}$. Then,

$$\pi_i \otimes \pi_i(\mathcal{R}^-(s) \cdot \Delta_s(x_{i,0}^\pm)) = \pi_i \otimes \pi_i(\mathcal{R}^-(s)) \cdot \Delta_s(x_{i,0}^\pm)$$

$$\pi_i \otimes \pi_i(\underset{\mathrm{D},s}{\Delta}(x_{i,0}^\pm) \cdot \mathcal{R}^-(s)) = \underset{\mathrm{D},s}{\Delta}(x_{i,0}^\pm) \cdot \pi_i \otimes \pi_i(\mathcal{R}^-(s))$$

Let $\varphi_i : Y_{d_i\hbar}(\mathfrak{sl}_2) \to Y_\hbar(\mathfrak{g})$ be the algebra homomorphism given by $x_r^\pm \mapsto d_i^{-1/2} x_{i,r}^\pm$ and $\xi_r \mapsto d_i^{-1}\xi_{i,r}$. Then,

$$\Delta_s(x_{i,0}^\pm) = \sqrt{d_i} \cdot \varphi_i^{\otimes 2} \circ \Delta_s(x_0^\pm) \qquad \text{and} \qquad \underset{\mathrm{D},s}{\Delta}(x_{i,0}^\pm) = \sqrt{d_i} \cdot \varphi_i^{\otimes 2} \circ \underset{\mathrm{D},s}{\Delta}(x_0^\pm).$$

We claim that $\pi_i \otimes \pi_i(\mathcal{R}^-(s)) = \varphi_i \otimes \varphi_i(\mathcal{R}^-_{(\mathfrak{sl}_2)}(s))$ so that, by the foregoing, (4.9) holds for any $\mathfrak{g}$ and $i \in \mathbf{I}$ if, and only if it holds for $\mathfrak{g} = \mathfrak{sl}_2$. The claim follows by uniqueness, since both $\pi_i \otimes \pi_i(\mathcal{R}^-(s))$ and $\varphi_i \otimes \varphi_i(\mathcal{R}^-_{(\mathfrak{sl}_2)}(s))$ are strictly lower triangular and of weight zero with respect to the action of $\xi_{i,0}$, and both intertwine

$$\varphi_i^{\otimes 2} \circ \Delta_s(t_i) = d_i^{-1} \pi_i^{\otimes 2} \circ \Delta_s(t_{i,1})$$

and

$$\varphi_i^{\otimes 2} \circ \underset{\mathrm{D},s}{\Delta}(t_i) = d_i^{-1} \underset{\mathrm{D},s}{\Delta}(t_{i,1}) = d_i^{-1} \pi_i^{\otimes 2} \circ \underset{\mathrm{D},s}{\Delta}(t_{i,1}).$$

4.8 Rank 1 Intertwining Relations

Assume $\mathfrak{g} = \mathfrak{sl}_2$, and consider the identity

$$\mathcal{R}^-(s) \cdot \Delta_s(x_0^-) = \underset{\mathrm{D},s}{\Delta}(x_0^-) \cdot \mathcal{R}^-(s) \tag{4.10}$$

The latter can be proved by a lengthy, direct calculation.[7] We give below an alternative proof, which relies on the cocycle identity (4.3) satisfied by $\mathcal{R}^-(s)$ to reduce it to the case when $\mathcal{R}^-(s)$ is acting on $\mathbb{C}^2 \otimes V$, where V is an arbitrary finite-dimensional representation.

By Appendix A, it is sufficient to prove that (4.10) holds on the tensor product $V_1 \otimes V_2$, where V_1 (resp. V_2) is chosen from a collection $\mathcal{V}_1$ (resp. $\mathcal{V}_2$) of finite-

[7]The calculation is carried out in §5 of the earlier version of this paper, arXiv:1907.03525 v1.

dimensional representations of $Y_\hbar(\mathfrak{g})$ which is stable under tensor product, contains the trivial representation, and a representation whose restriction to $\mathfrak{g}$ is faithful. We choose $\mathcal{V}_2$ to consist of all finite-dimensional representations of $Y_\hbar(\mathfrak{g})$, while $\mathcal{V}_1$ consists of arbitrary tensor products $\mathbb{C}^2(a_1) \otimes \cdots \otimes \mathbb{C}^2(a_m)$ of evaluation representations.

For any $V_1, V_2, V_3 \in \mathrm{Rep}_{\mathrm{fd}}(Y_\hbar(\mathfrak{g}))$, the cocycle identity (4.7) implies that (4.10) holds on $(V_1(s_1) \otimes V_2) \otimes V_3$ if it holds on $V_1 \otimes V_2$, $V_2 \otimes V_3$, and $V_1 \otimes (V_2(s_2) \otimes V_3)$. Indeed,

$$
\begin{aligned}
&\mathcal{R}^-_{V_1 \underset{s_1}{\otimes} V_2, V_3}(s_2) \cdot \pi_{(V_1 \underset{s_1}{\otimes} V_2) \underset{s_2}{\otimes} V_3}(x_0^-) \\
&= \mathcal{R}^-_{V_1,V_2}(s_1)^{-1} \cdot \mathcal{R}^-_{V_2,V_3}(s_2) \cdot \mathcal{R}^-_{V_1, V_2 \underset{s_2}{\otimes} V_3}(s_1+s_2) \cdot \pi_{V_1 \underset{s_1+s_2}{\otimes} (V_2 \underset{s_2}{\otimes} V_3)}(x_0^-) \\
&= \mathcal{R}^-_{V_1,V_2}(s_1)^{-1} \cdot \mathcal{R}^-_{V_2,V_3}(s_2) \cdot \pi_{V_1 \underset{\mathrm{D},s_1+s_2}{\otimes} (V_2 \underset{s_2}{\otimes} V_3)}(x_0^-) \cdot \mathcal{R}^-_{V_1, V_2 \underset{s_2}{\otimes} V_3}(s_1+s_2) \\
&= \mathcal{R}^-_{V_1,V_2}(s_1)^{-1} \cdot \pi_{V_1 \underset{\mathrm{D},s_1+s_2}{\otimes} (V_2 \underset{\mathrm{D},s_2}{\otimes} V_3)}(x_0^-) \cdot \mathcal{R}^-_{V_2,V_3}(s_2) \cdot \mathcal{R}^-_{V_1, V_2 \underset{s_2}{\otimes} V_3}(s_1+s_2) \\
&= \mathcal{R}^-_{V_1,V_2}(s_1)^{-1} \cdot \pi_{(V_1 \underset{\mathrm{D},s_1}{\otimes} V_2) \underset{\mathrm{D},s_2}{\otimes} V_3}(x_0^-) \cdot \mathcal{R}^-_{V_2,V_3}(s_2) \cdot \mathcal{R}^-_{V_1, V_2 \underset{s_2}{\otimes} V_3}(s_1+s_2) \\
&= \pi_{(V_1 \underset{s_1}{\otimes} V_2) \underset{\mathrm{D},s_2}{\otimes} V_3}(x_0^-) \cdot \mathcal{R}^-_{V_1,V_2}(s_1)^{-1} \cdot \mathcal{R}^-_{V_2,V_3}(s_2) \cdot \mathcal{R}^-_{V_1, V_2 \underset{s_2}{\otimes} V_3}(s_1+s_2) \\
&= \pi_{(V_1 \underset{s_1}{\otimes} V_2) \underset{\mathrm{D},s_2}{\otimes} V_3}(x_0^-) \cdot \mathcal{R}^-_{V_1 \underset{s_1}{\otimes} V_2, V_3}(s_2)
\end{aligned}
$$

It is therefore sufficient to check (4.10) on $\mathbb{C}^2(a) \otimes V$, where V is an arbitrary finite-dimensional representation of $Y_\hbar(\mathfrak{g})$, and $a \in \mathbb{C}$. Moreover, by using the translation invariance of $\mathcal{R}^-(s)$, $\underset{s}{\otimes}$, and $\underset{\mathrm{D},s}{\otimes}$, it is sufficient to consider the case $a = 0$.

Let now $x^\pm, \xi$ be the standard nilpotent and semisimple generators of $\mathfrak{sl}_2$ acting on $\mathbb{C}^2$. Then, the following define an action of $Y_\hbar(\mathfrak{sl}_2)$ on $\mathbb{C}^2$:

$$
x^\pm(u) = \hbar \frac{x^\pm}{u} \qquad \text{and} \qquad \xi(u) = 1 + \hbar \frac{\xi}{u}
$$

By Sects. 3.1 and 3.2, $\pi_{\mathbb{C}^2 \underset{s}{\otimes} V}(x_0^-) = x^- \otimes 1 + 1 \otimes x_0^-$, and

$$
\pi_{\mathbb{C}^2 \underset{\mathrm{D},s}{\otimes} V}(x_0^-) = \oint_{C_1} \frac{x^-}{v-s} \otimes \xi_i(v)\, dv + 1 \otimes x_0^- = x^- \otimes \xi(s) + 1 \otimes x_0^-
$$

where C_1 encloses $\sigma(\mathbb{C}^2) + s = \{s\}$ and none of the points in $\sigma(V)$.

On the other hand, since $x^-(u)x^-(v) = 0$ on $\mathbb{C}^2$, Sect. 5.4 and Theorem 5.5 yield

$$\mathcal{R}^-_{\mathbb{C}^2,V}(s) = 1 - \oint_{C_2} \frac{x^-}{u-s} \otimes x^+(u)\, du = 1 + \oint_{C_1} \frac{x^-}{u-s} \otimes x^+(u)\, du$$
$$= 1 + x^- \otimes x^+(s)$$

where C_2 encloses $\sigma(V)$ and none of the points in $\sigma(\mathbb{C}^2) + s = s$, and the second equality follows because the residue of the integrand at infinity is 0. Therefore,

$$\mathcal{R}^-_{\mathbb{C}^2,V}(s) \cdot \pi_{\mathbb{C}^2 \underset{s}{\otimes} V}(x_0^-) = x^- \otimes 1 + 1 \otimes x_0^- + x^- \otimes x^+(s)x_0^-$$
$$\pi_{\mathbb{C}^2 \underset{\mathrm{D},s}{\otimes} V}(x_0^-) \cdot \mathcal{R}^-_{\mathbb{C}^2,V}(s) = x^- \otimes \xi(s) + 1 \otimes x_0^- + x^- \otimes x_0^- x^+(s)$$

so that

$$\mathcal{R}^-_{\mathbb{C}^2,V}(s) \cdot \pi_{\mathbb{C}^2 \underset{s}{\otimes} V}(x_0^-) - \pi_{\mathbb{C}^2 \underset{\mathrm{D},s}{\otimes} V}(x_0^-) \cdot \mathcal{R}^-_{\mathbb{C}^2,V}(s) = x^- \otimes \left(1 + [x^+(s), x_0^-] - \xi(s)\right)$$

which is equal to zero since $[x^+(s), x_0^-] = \xi(s) - 1$.

The identity $\mathcal{R}^-(s) \cdot \Delta_s(x_0^+) = \underset{\mathrm{D},s}{\Delta}(x_0^+) \cdot \mathcal{R}^-(s)$ is proved in a similar way, by taking $\mathcal{V}_1$ to consist of all finite-dimensional representations, $\mathcal{V}_2$ of tensor products $\mathbb{C}^2(a_1) \otimes \cdots \otimes \mathbb{C}^2(a_m)$, and using the cocycle identity to reduce this to a check on $V \otimes \mathbb{C}^2$, for an arbitrary $V \in \mathrm{Rep}_{\mathrm{fd}}(Y_\hbar(\mathfrak{g}))$.

5 The Element $\mathcal{R}^-(s)$ for $\mathfrak{sl}_2$

In this section, we give an explicit formula for the element $\mathcal{R}^-(s)$ when $\mathfrak{g} = \mathfrak{sl}_2$.

5.1

It will be convenient to consider the generating series

$$\mathcal{R}^-(s, z) = \sum_{n \geq 0} \mathcal{R}^-_{n\alpha}(s) z^n \in (Y_\hbar(\mathfrak{sl}_2) \otimes Y_\hbar(\mathfrak{sl}_2))[z][\![s^{-1}]\!]$$

where α is the positive root of $\mathfrak{g}$. Let $G = PSL_2(\mathbb{C})$ be the complex Lie group of adjoint type corresponding to $\mathfrak{g}$, and $H \subset G$ its maximal torus with Lie algebra $\mathfrak{h}$. We identify H with $\mathbb{C}^\times$ via the character corresponding to $-\alpha$.

In particular, $\mathcal{R}^-(s,z) = \mathrm{Ad}(z^{(1)})\mathcal{R}^-(s)$. Moreover, $\mathcal{R}^-(s,z)$ satisfies

$$\mathcal{R}^-(s,z)\cdot\Delta_s^z(t_1) = \underset{\mathrm{D},s}{\Delta^z}(t_1)\cdot\mathcal{R}^-(s,z)$$

where $\Delta_s^z = \mathrm{Ad}(z^{(1)})\circ\Delta_s^z$ and $\underset{\mathrm{D},s}{\Delta} = \mathrm{Ad}(z^{(1)})\circ\underset{\mathrm{D},s}{\Delta}$, so that

$$\underset{\mathrm{D},s}{\Delta^z}(t_1) = \sum_{a=1}^{2} t_1^{(a)} + sh^{(1)} \qquad \text{and} \qquad \Delta_s^z(t_1) = \sum_{a=1}^{2} t_1^{(a)} + sh^{(1)} + \hbar z\mathsf{r}$$

5.2

The intertwining equation for $\mathcal{R}^-(s,z)$ may be written as the following ODE, together with the initial condition $\mathcal{R}^-(s,0) = 1\otimes 1$

$$\left(sz\partial_z - \frac{1}{2}\,\mathrm{ad}(t_1\otimes 1 + 1\otimes t_1)\right)\mathcal{R}^-(s,z) = \mathcal{R}^-(s,z)\cdot\hbar z\mathsf{r} \tag{5.1}$$

where $\mathsf{r} = x_0^-\otimes x_0^+$.

Lemma *Set $\omega(s,z) = \mathcal{R}^-(s,z)^{-1}\cdot z\partial_z\mathcal{R}^-(s,z)$. Then, $\omega(s,z)$ satisfies*

$$(sz\partial_z - \mathcal{T} + \hbar z\,\mathrm{ad}(\mathsf{r}))\,\omega(s,z) = \hbar\mathsf{r}z \tag{5.2}$$

where $\mathcal{T} = \dfrac{1}{2}\,\mathrm{ad}(t_1\otimes 1 + 1\otimes t_1)$.

Proof Since $(sz\partial_z - \mathcal{T})$ is a derivation, we have

$$\begin{aligned}(sz\partial_z - \mathcal{T})\,\omega(s,z) &= -\hbar z\mathsf{r}\omega(s,z) + \mathcal{R}^-(s,z)^{-1}z\partial_z(\mathcal{R}^-(s,z)\hbar z\mathsf{r})\\ &= -\hbar z[\mathsf{r},\omega(s,z)] + \hbar z\mathsf{r}\end{aligned}$$

□

5.3 Formula for $\omega(s,z)$

Given a vector space $\mathcal{H}$, let

$$\oint - du : \mathcal{H}[u;u^{-1}]\!] \to \mathcal{H}$$

be the formal residue, given by taking the coefficient of u^{-1}.

Proposition *The series* $\omega(s,z)$ *is given by*

$$\omega(s,z) = \sum_{k\geq 1} z^k \frac{(-1)^k}{k\hbar} \oint x^-(u-s)^k \otimes x^+(u)^k \, du \tag{5.3}$$

Proof Write $\omega(s,z) = \sum_{k\geq 1} \omega_k(s) z^k$. For $k = 1$, (5.2) yields

$$\begin{aligned}\omega_1(s) &= (s - \mathcal{T})^{-1} \hbar \mathsf{r} = \hbar \sum_{m\geq 0} s^{-m-1} \mathcal{T}^m \mathsf{r} \\ &= \hbar \sum_{m\geq 0} s^{-m-1} \sum_{n=0}^{m} (-1)^n \binom{m}{n} x_n^- \otimes x_{m-n}^+\end{aligned}$$

where the last equality follows from $[t_1/2, x_k^\pm] = \pm x_{k+1}^\pm$. On the other hand,

$$\begin{aligned}x^-(u-s) &= \hbar \sum_{n\geq 0} x_n^- (u-s)^{-n-1} \\ &= -\hbar \sum_{n\geq 0} (-1)^n x_n^- s^{-n-1} \sum_{a\geq 0} \binom{n+a}{n} u^a s^{-a} \\ &= -\hbar \sum_{m\geq 0} s^{-m-1} \sum_{n=0}^{m} (-1)^n \binom{m}{n} x_n^- u^{m-n}\end{aligned}$$

which shows that ω_1 is given by (5.3).

Let now $k \geq 2$, and set $I_k(v,s) = \dfrac{(-1)^k}{k\hbar} x^-(v-s)^k \otimes x^+(v)^k$. We claim that

$$(sk - \mathcal{T}) I_k(v,s) = -[\hbar \mathsf{r}, I_{k-1}(v,s)]$$

which proves in particular that (5.3) satisfies (5.2). We shall need the commutation relation $[x_0^\pm, x^\pm(u)] = \mp x^\pm(u)^2$. The latter follows from relation ($\mathcal{Y}$4) of Proposition 2.3, namely

$$(u-v\mp\hbar)x^\pm(u)x^\pm(v) - (u-v\pm\hbar)x^\pm(u)x^\pm(v) = \hbar([x_0^\pm, x^\pm(v)] - [x^\pm(u), x_0^\pm]),$$

by taking $u = v$. Thus, for every $k \geq 1$, $\mathrm{ad}(x_0^\pm) \cdot x^\pm(u)^k = \mp k x^\pm(u)^{k+1}$.

Using this, and $\left[\frac{t_1}{2}, x^{\pm}(u)\right] = \pm(ux^{\pm}(u) - \hbar x_0^{\pm})$, yields

$$\left[\frac{t_1}{2}, x^{\pm}(u)^k\right] = \pm\sum_{j=1}^{k} x^{\pm}(u)^{j-1}(ux^{\pm}(u) - \hbar x_0^{\pm})x^{\pm}(u)^{k-j}$$

$$= \pm kux^{\pm}(u)^k \mp \hbar \sum_{j=1}^{k} x^{\pm}(u)^{k-1} x_0^{\pm} x^{\pm}(u)^{k-j}$$

$$= \pm kux^{\pm}(u)^k \mp \hbar k x^{\pm}(u)^{k-1} x_0^{\pm} + \hbar \frac{k(k-1)}{2} x^{\pm}(u)^k$$

Thus,

$$(sk - \mathcal{T})\, x^{-}(v-s)^k \otimes x^{+}(v)^k = -\hbar k(k-1) x^{-}(v-s)^k \otimes x^{+}(v)^k$$
$$- \hbar k x^{-}(v-s)^{k-1} x_0^{-} \otimes x^{+}(v)^k + \hbar k x^{-}(v-s)^k \otimes x^{+}(v)^{k-1} x_0^{+}.$$

Together with $[A \otimes B, C \otimes D] = [A, C] \otimes [B, D] + [A, C] \otimes DB + CA \otimes [B, D]$, this yields

$$[x_0^{-} \otimes x_0^{+}, x^{-}(u)^k \otimes x^{+}(u)^k]$$
$$= [x_0^{-}, x^{-}(u)^k] \otimes [x_0^{+}, x^{+}(u)^k] + [x_0^{-}, x^{-}(u)^k] \otimes x^{+}(u)^k x_0^{+}$$
$$+ x^{-}(u)^k x_0^{-} \otimes [x_0^{+}, x^{+}(u)^k]$$
$$= -k^2 x^{-}(u)^{k+1} \otimes x^{+}(u)^{k+1} + k x^{-}(u)^{k+1} \otimes x^{+}(u)^k x_0^{+}$$
$$- k x^{-}(u)^k x_0^{-} \otimes x^{+}(u)^{k+1}$$

as claimed. □

5.4 Formula for $\omega_{V_1,V_2}(s, z)$

If V_1, V_2 are finite-dimensional representations of $Y_\hbar(\mathfrak{g})$, Theorem 4.1 (3) implies that $\omega_{V_1,V_2}(s, z) = \pi_{V_1} \otimes \pi_{V_2}(\omega(s, z))$ is a rational function of s. It follows from the lemma below that

$$\omega_{V_1,V_2}(s, z) = \sum_{k \geq 1} \frac{(-1)^k z^k}{k\hbar} \oint_{C_2} x^{-}(u-s)^k \otimes x^{+}(u)^k \, du$$

where C_2 encloses $\sigma(V_2)$ and none of the points in $\sigma(V_1)+s$. Note that in this case the sum over k is finite since $x^{\pm}(u)$ are nilpotent on V_1, V_2.

Lemma *Let A be a finite-dimensional dimensional algebra over $\mathbb{C}$, and $f, g : \mathbb{C} \to A$ rational functions which are regular at ∞. Consider the integral*

$$I(s) = \oint_{C_2} f(u-s)g(u)\,du$$

where $s \in \mathbb{C}$, and C_2 is a contour enclosing all poles of $g(u)$ and none of those of $f(u-s)$. Then, the Taylor expansion $\widehat{I}(s)$ of $I(s)$ at $s=\infty$ is equal to

$$\widehat{I}(s) = \oint \widehat{f}(u-s)\widehat{g}(u)\,du$$

where $\oint - du$ is the formal residue defined in Sect. 5.3, $\widehat{f}, \widehat{g}$ are the Taylor series of f, g at ∞, and $\widehat{f}(u-s)$ is expanded in $A[u][\![s^{-1}]\!]$.

Proof $\widehat{I}(s)$ is equal to $\oint_{C_2} \widehat{f}(u-s)g(u)du$. Since $\widehat{f}(u-s) \in A[u][\![s^{-1}]\!]$, it suffices to prove that $\oint_{C_2} p(u)g(u)du = \oint p(u)\widehat{g}(u)du$ for any $p \in A[u]$, which follows by deformation of contour. □

5.5 Formula for $\mathcal{R}^-(s)$

Integrating $z\partial_z\mathcal{R}^-(s,z) = \mathcal{R}^-(s,z)\cdot\omega(s,z)$ and setting $z=1$ yields the following corollary of Proposition 5.3.

Theorem *The element $\mathcal{R}^-(s) \in Y_\hbar(\mathfrak{sl}_2)^{\otimes 2}[\![s^{-1}]\!]$ is given by*

$$\mathcal{R}^-(s) = 1^{\otimes 2} + \sum_{n\geq 1} \sum_{\substack{k_1,\ldots,k_r\in\mathbb{Z}_{\geq 1}\\ k_1+\cdots+k_r=n}} \frac{1}{k_1(k_1+k_2)\cdots(k_1+\cdots+k_r)} \omega(s)_{k_1}\cdots\omega(s)_{k_r}$$

where

$$\omega(s)_k = \frac{(-1)^k}{k\hbar} \oint x^-(u-s)^k \otimes x^+(u)^k\,du$$

Remark It is an interesting problem to give an explicit formula for $\mathcal{R}^-(s)$ for $\mathfrak{g} \not\cong \mathfrak{sl}_2$.

6 The Universal and the Meromorphic Abelian R-Matrices of $Y_\hbar(\mathfrak{g})$

In this section, we review the construction of the meromorphic abelian R-matrix of $Y_\hbar(\mathfrak{g})$ given in [12]. We then show that it gives rise to rational intertwiners $V_1(s) \underset{\mathrm{D},0}{\otimes} V_2 \to V_2 \underset{\mathrm{D},0}{\otimes} V_1(s)$ for any finite-dimensional representations V_1, V_2. We also prove that these cannot be chosen to be both natural and compatible with the Drinfeld tensor product. Finally, we lift the meromorphic abelian R-matrix to obtain a universal abelian R-matrix for the deformed Drinfeld coproduct.

6.1 The Endomorphism $\mathcal{A}_{V_1,V_2}(s)$ [12]

Let $V \in \mathrm{Rep}_{\mathrm{fd}}(Y_\hbar(\mathfrak{g}))$. For any $i \in \mathbf{I}$, let $\sigma_V(\xi_i)$ be the set of poles of $\xi_i(u)^{\pm 1}$ acting on V, and

$$\mathsf{X}_V(\xi_i) = \bigcup_{a \in \sigma_V(\xi_i)} [0, a]$$

the union of straight line segments joining 0 to points in $\sigma_V(\xi_i)$. Then, the generating series $t_i(v) = \hbar \sum_{r \geq 0} t_{i,r} v^{-r-1}$ introduced in Sect. 2.4 converges to a holomorphic function $t_i(v) : \mathbb{C} \setminus \mathsf{X}_V(\xi_i) \to \mathrm{End}_{\mathbb{C}}(V)$, which is uniquely determined by $\exp(t_i(v)) = \xi_i(v)$ and $t_i(\infty) = 0$ [12, §5.4].

Let $V_1, V_2 \in \mathrm{Rep}_{\mathrm{fd}}(Y_\hbar(\mathfrak{g}))$, $s \in \mathbb{C}$, and define $\mathcal{A}_{V_1,V_2}(s) \in \mathrm{End}_{\mathbb{C}}(V_1 \otimes V_2)$ by

$$\mathcal{A}_{V_1,V_2}(s) = \exp\left(- \sum_{\substack{i,j \in \mathbf{I} \\ r \in \mathbb{Z}}} c_{ij}^{(r)} \oint_{C_1} \frac{dt_i(v)}{dv} \otimes t_j\left(v + s + \frac{(\ell + r)\hbar}{2}\right) dv \right)$$

where

- C_1 is a contour enclosing $\sigma_{V_1}(\xi_i)$.
- $\ell = mh^\vee$, with $h^\vee$ the dual Coxeter number of $\mathfrak{g}$ and $m = \frac{(\theta,\theta)}{2}$ for $\theta \in \Phi_+$ the highest root.

- The non-negative integers $c_{ij}^{(r)}$ are the entries of the following matrix [12, Appendix A][8]

$$\left(c_{ij}(q) = \sum_{r \in \mathbb{Z}} c_{ij}^{(r)} q^r\right) = [\ell]_q \cdot \left([d_i a_{ij}]_q\right)^{-1}$$

- s is large enough so that $t_j(v+s+\hbar(\ell+r)/2)$ is an analytic function of v within C_1, for every $j \in \mathbf{I}$ and $r \in \mathbb{Z}$ such that $c_{ij}^{(r)} \neq 0$ for some $i \in \mathbf{I}$.

Then, $\mathcal{A}_{V_1,V_2}(s)$ is a rational function of s, regular at ∞, with an expansion of the form $\mathbf{1} - \frac{\ell\hbar}{s^2}\Omega_{\mathfrak{h}} + O(s^{-3})$, where $\Omega_{\mathfrak{h}} \in \mathfrak{h} \otimes \mathfrak{h}$ is the Cartan part of the Casimir tensor [12, Thm. 5.5]. Moreover, $[\mathcal{A}_{V_1,V_2}(s), \mathcal{A}_{V_1,V_2}(s')] = 0$ for any $s, s' \in \mathbb{C}$.

6.2 The Meromorphic Abelian R-Matrix of $Y_\hbar(\mathfrak{g})$ [12]

Consider the additive difference equation determined by $\mathcal{A}_{V_1,V_2}(s)$

$$\mathcal{R}^0_{V_1,V_2}(s+\ell\hbar) = \mathcal{A}_{V_1,V_2}(s) \cdot \mathcal{R}^0_{V_1,V_2}(s) \tag{6.2}$$

It admits two meromorphic solutions $\mathcal{R}^{0,\uparrow/\downarrow}_{V_1,V_2}(s)$, which are uniquely determined by the requirement that $\mathcal{R}^{0,\uparrow}_{V_1,V_2}(s)$ (resp. $\mathcal{R}^{0,\downarrow}_{V_1,V_2}(s)$) is holomorphic and invertible for $\mathrm{Re}(s/\ell\hbar) \gg 0$ (resp. $\mathrm{Re}(s/\ell\hbar) \ll 0$), and possesses an asymptotic expansion of the form $1 + O(s^{-1})$ as $s \to \infty$ in any halfplane of the form $\mathrm{Re}(s/\ell\hbar) > m$ (resp. $\mathrm{Re}(s/\ell\hbar) < m$). These solutions are explicitly given by the infinite products

$$\mathcal{R}^{0,\uparrow}_{V_1,V_2}(s) = \overrightarrow{\prod_{n\geq 0}} \mathcal{A}_{V_1,V_2}(s+n\ell\hbar)^{-1} \quad \text{and} \quad \mathcal{R}^{0,\downarrow}_{V_1,V_2}(s) = \overrightarrow{\prod_{n\geq 1}} \mathcal{A}_{V_1,V_2}(s-n\ell\hbar)$$

The product defining $\mathcal{R}^{0,\uparrow}_{V_1,V_2}(s)$ (resp. $\mathcal{R}^{0,\downarrow}_{V_1,V_2}(s)$) converges uniformly on compact subsets of the complement of $\mathcal{Z} - \ell\hbar\mathbb{Z}_{\geq 0}$ (resp. $\mathcal{P} + \ell\hbar\mathbb{Z}_{>0}$), where $\mathcal{Z}$ (resp. $\mathcal{P}$) is the set of poles of $\mathcal{A}(s)^{-1}$ (resp. $\mathcal{A}(s)$).

Then, the following holds [12, Thm 5.9].

[8] It was proved in [12, Appendix A] that $c_{ij}(q) \in \mathbb{Z}_{\geq 0}[q, q^{-1}]$. It is clear from the definition that $c_{ij}(q) = c_{ji}(q) = c_{ij}(q^{-1})$, and the matrix identity can be expanded as

$$\sum_{k \in \mathbf{I}} c_{ik}(q)[d_k a_{kj}]_q = \delta_{ij}[\ell]_q \quad \forall i, j \in \mathbf{I} \tag{6.1}$$

Theorem *Fix $\varepsilon \in \{\uparrow, \downarrow\}$. Then, the following holds for any $V_1, V_2, V_3 \in \mathrm{Rep}_{\mathrm{fd}}(Y_\hbar(\mathfrak{g}))$*

1. *The map*

$$(1\,2) \circ \mathcal{R}^{0,\varepsilon}_{V_1,V_2}(s) : V_1(s) \underset{\mathrm{D},0}{\otimes} V_2 \to V_2 \underset{\mathrm{D},0}{\otimes} V_1(s)$$

   is a morphism of $Y_\hbar(\mathfrak{g})$–modules, which is natural in V_1 and V_2.

2. *The following cabling identities hold:*

$$\mathcal{R}^{0,\varepsilon}_{V_1 \underset{\mathrm{D},s_1}{\otimes} V_2, V_3}(s_2) = \mathcal{R}^{0,\varepsilon}_{V_1,V_3}(s_1+s_2) \cdot \mathcal{R}^{0,\varepsilon}_{V_2,V_3}(s_2)$$

$$\mathcal{R}^{0,\varepsilon}_{V_1, V_2 \underset{\mathrm{D},s_2}{\otimes} V_3}(s_1+s_2) = \mathcal{R}^{0,\varepsilon}_{V_1,V_3}(s_1+s_2) \cdot \mathcal{R}^{0,\varepsilon}_{V_1,V_2}(s_1)$$

3. *For any $a, b \in \mathbb{C}$,*

$$\mathcal{R}^{0,\varepsilon}_{V_1(a),V_2(b)}(s) = \mathcal{R}^{0,\varepsilon}_{V_1,V_2}(s+a-b)$$

4. *The following unitary condition holds:*

$$(1\,2) \circ \mathcal{R}^{0,\uparrow}_{V_1,V_2}(-s) \circ (1\,2)^{-1} = \mathcal{R}^{0,\downarrow}_{V_2,V_1}(s)^{-1}$$

5. *$\mathcal{R}^{0,\uparrow/\downarrow}_{V_1,V_2}(s)$ have the same asymptotic expansion, as $s \to \infty$ in any halfplane of the form $\pm\mathrm{Re}(s/\ell\hbar) > m$, which is of the form*

$$\mathcal{R}^{0,\uparrow/\downarrow}_{V_1,V_2}(s) \sim 1 + \hbar\Omega_{\mathfrak{h}} s^{-1} + O(s^{-2})$$

6.3 Existence of a Rational Intertwiner

Let $V_1, V_2 \in \mathrm{Rep}_{\mathrm{fd}}(Y_\hbar(\mathfrak{g}))$.

Theorem *There is a rational map $\mathsf{R}^0_{V_1,V_2} : \mathbb{C} \to \mathrm{Aut}_{\mathbb{C}}(V_1 \otimes V_2)$, which is normalised by $\mathsf{R}^0_{V_1,V_2}(\infty) = \mathbf{1}$ and such that*

$$(1\,2) \circ \mathsf{R}^0_{V_1,V_2}(s) : V_1(s) \underset{\mathrm{D},0}{\otimes} V_2 \to V_2 \underset{\mathrm{D},0}{\otimes} V_1(s)$$

intertwines the action of $Y_\hbar(\mathfrak{g})$. In particular, $V_1(s) \underset{\mathrm{D},0}{\otimes} V_2$ and $V_2 \underset{\mathrm{D},0}{\otimes} V_1(s)$ are isomorphic as $Y_\hbar(\mathfrak{g})$–modules for all but finitely many values of s.

Proof The map $\mathsf{R}^0_{V_1,V_2}(s)$ will be obtained from the meromorphic intertwiners $\mathcal{R}^{0,\uparrow/\downarrow}_{V_1,V_2}(s)$ as follows. Consider the difference equation (6.2) satisfied by $\mathcal{R}^{0,\uparrow/\downarrow}_{V_1,V_2}(s)$. Its monodromy is given by

$$\eta^0_{V_1,V_2}(s) = \mathcal{R}^{0,\uparrow}_{V_1,V_2}(s)^{-1} \cdot \mathcal{R}^{0,\downarrow}_{V_1,V_2}(s)$$

By construction, $\eta^0_{V_1,V_2}$ is an $\ell\hbar$–periodic function of s, and in fact a rational function of $z = \exp\left(\frac{2\pi\iota s}{\ell\hbar}\right)$ which takes the value $\mathbf{1}$ at $z = 0, \infty$ [11, §4.8]. Moreover, by Theorem 6.2 (1), $\eta^0_{V_1,V_2}(s)$ commutes with the action of $Y_\hbar(\mathfrak{g})$ on $V_1 \underset{\mathrm{D},s}{\otimes} V_2$. In fact, the periodicity of $\eta^0_{V_1,V_2}$ implies that $\eta^0_{V_1,V_2}(s)$ commutes with the action of $Y_\hbar(\mathfrak{g})$ on $V_1 \underset{\mathrm{D},s'}{\otimes} V_2$ for any s' of the form $s + \ell\hbar m$, $m \in \mathbb{Z}$. Since that action is rational in s', it follows that $\eta^0_{V_1,V_2}$ takes values in the subalgebra $\mathcal{Z} \subset \mathrm{End}_{\mathbb{C}}(V_1 \otimes V_2)$ which consists of elements commuting with the action of $Y_\hbar(\mathfrak{g})$ on $V_1 \underset{\mathrm{D},s'}{\otimes} V_2$ for any $s' \in \mathbb{C}$.

Consider now the following factorisation problem. Find two meromorphic functions $\mathfrak{X}^{0,\uparrow/\downarrow}_{V_1,V_2}(s) : \mathbb{C} \to \mathcal{Z}$ such that

1. $\mathfrak{X}^{0,\uparrow/\downarrow}_{V_1,V_2}(s)$ is holomorphic and invertible for $\mathrm{Re}(\pm s/\hbar) \gg 0$.
2. $\mathfrak{X}^{0,\uparrow/\downarrow}_{V_1,V_2}(s)$ possesses an asymptotic expansion of the form $1 + O(s^{-1})$, valid in any half–plane of the form $\mathrm{Re}(s/\ell\hbar) \gtrless m$.
3. $\eta^0_{V_1,V_2}(s) = \mathfrak{X}^{0,\uparrow}_{V_1,V_2}(s)^{-1} \cdot \mathfrak{X}^{0,\downarrow}_{V_1,V_2}(s)$.

Since $[\eta^0_{V_1,V_2}(s), \eta^0_{V_1,V_2}(s')] = 0$ for any s, s', such a factorisation exists, and can be obtained explicitly, once a choice of representatives of poles of $\eta^0(s)$ modulo translations by $\ell\hbar\mathbb{Z}$ is made [11, §4.14]. We remark that the consistency equation, required upon such a choice in [11, §4.14], is vacuous in our case, since $\mathcal{A}_{V_1,V_2}(s) = 1 + O(s^{-2})$.

Summarising, $\eta^0_{V_1,V_2}(s)$ admits two factorisations in $\mathrm{End}_{\mathbb{C}}(V_1 \otimes V_2)$, namely

$$\mathfrak{X}^{0,\uparrow}_{V_1,V_2}(s)^{-1} \cdot \mathfrak{X}^{0,\downarrow}_{V_1,V_2}(s) = \eta^0_{V_1,V_2}(s) = \mathcal{R}^{0,\uparrow}_{V_1,V_2}(s)^{-1} \cdot \mathcal{R}^{0,\downarrow}_{V_1,V_2}(s)$$

Set now

$$\mathsf{R}^0_{V_1,V_2}(s) = \mathcal{R}^{0,\uparrow}_{V_1,V_2}(s) \cdot \mathfrak{X}^{0,\uparrow}_{V_1,V_2}(s)^{-1} = \mathcal{R}^{0,\downarrow}_{V_1,V_2}(s) \cdot \mathfrak{X}^{0,\downarrow}_{V_1,V_2}(s)^{-1}$$

Then, $(1\,2) \circ \mathsf{R}^0_{V_1,V_2}(s) : V_1(s) \underset{\mathrm{D},0}{\otimes} V_2 \to V_2 \underset{\mathrm{D},0}{\otimes} V_1(s)$ intertwines the action of $Y_\hbar(\mathfrak{g})$, and $\mathsf{R}^0_{V_1,V_2}$ is a rational function of s equal to $\mathbf{1}$ at $s = \infty$ [11, §4.11]. □

6.4 Non-existence of Rational Commutativity Constraints

Theorem 6.3 raises the question of whether the rational factor $\mathsf{R}^0_{V_1,V_2}(s)$ may be chosen consistently for any pair of representations V_1, V_2 so as to satisfy the cabling identities of Theorem 6.2. The following shows that not to be the case.

Theorem *There is no function* $\mathsf{R}^0_{V_1,V_2} : \mathbb{C} \to \mathrm{Aut}_{\mathbb{C}}(V_1 \otimes V_2)$ *which is rational, defined for any* $V_1, V_2 \in \mathrm{Rep}_{\mathrm{fd}}(Y_\hbar(\mathfrak{g}))$*, and such that the following conditions hold.*

1. $(1\,2) \circ \mathsf{R}^0_{V_1,V_2}(s) : V_1(s) \underset{\mathrm{D},0}{\otimes} V_2 \to V_2 \underset{\mathrm{D},0}{\otimes} V_1(s)$ *is* $Y_\hbar(\mathfrak{g})$*–linear, and natural in* V_1 *and* V_2.
2. *For any* $V_1, V_2, V_3 \in \mathrm{Rep}_{\mathrm{fd}}(Y_\hbar(\mathfrak{g}))$

$$\mathsf{R}^0_{V_1 \underset{\mathrm{D},s_1}{\otimes} V_2, V_3}(s_2) = \mathsf{R}^0_{V_1,V_3}(s_1+s_2) \cdot \mathsf{R}^0_{V_2,V_3}(s_2) \tag{6.3}$$

$$\mathsf{R}^0_{V_1, V_2 \underset{\mathrm{D},s_2}{\otimes} V_3}(s_1+s_2) = \mathsf{R}^0_{V_1,V_3}(s_1+s_2) \cdot \mathsf{R}^0_{V_1,V_2}(s_2) \tag{6.4}$$

Proof Note first if V_1 is the trivial one–dimensional representation $\mathbf{1}$ of $Y_\hbar(\mathfrak{g})$, the first cabling identity (6.3) and part (4) of Theorem 3.2 imply that $\mathsf{R}^0_{\mathbf{1},V_3}(s) = \mathrm{Id}_{V_3}$. Setting $V_2 = \mathbf{1}$ in (6.3) then yields $\mathsf{R}^0_{V_1(a),V_3}(s) = \mathsf{R}^0_{V_1,V_3}(s+a)$. Similarly, upon setting $V_2 = \mathbf{1}$ first, and then $V_3 = \mathbf{1}$, the second cabling identity (6.4) implies that $\mathsf{R}^0_{V_1,\mathbf{1}}(s) = \mathrm{Id}_V$, and that $\mathsf{R}^0_{V_1,V_2(b)}(s) = \mathsf{R}^0_{V_1,V_2}(s-b)$.

We now take $\mathfrak{g} = \mathfrak{sl}_2$ and proceed by contradiction, assuming the existence of a rational $\mathsf{R}^0_{V_1,V_2}(s)$ with the stated properties. We will use the following facts about $Y_\hbar(\mathfrak{sl}_2)$. There is a two–dimensional representation $\mathbb{C}^2$ of $Y_\hbar(\mathfrak{sl}_2)$ (see Sect. 4.8). Explicitly, in a fixed basis v_+, v_-, the action of $Y_\hbar(\mathfrak{sl}_2)$ is given by the following 2×2 matrices.

$$\xi(u) = \mathbf{1} + \frac{\hbar}{u}\begin{pmatrix} 1 & 0 \\ 0 & -1 \end{pmatrix} \qquad \text{and} \qquad x^+(u) = \frac{\hbar}{u}\begin{pmatrix} 0 & 1 \\ 0 & 0 \end{pmatrix} = x^-(u)^T$$

Further, there is a $Y_\hbar(\mathfrak{sl}_2)$–linear map $\mathbf{1} \to \mathbb{C}^2 \underset{\mathrm{D},\hbar}{\otimes} \mathbb{C}^2$ given by $1 \mapsto v_+ \otimes v_-$.

Given $V \in \mathrm{Rep}_{\mathrm{fd}}(Y_\hbar(\mathfrak{g}))$, we consider $\mathsf{R}^0_{\mathbb{C}^2,V}(s) \in \mathrm{End}_{\mathbb{C}}(\mathbb{C}^2 \otimes V)$ as a 2×2–matrix with entries in $\mathrm{End}(V)$ given by

$$\mathsf{R}^0_{\mathbb{C}^2,V}(s) = \begin{pmatrix} \alpha(s) & \beta(s) \\ \gamma(s) & \delta(s) \end{pmatrix}$$

The intertwining property of $\mathsf{R}^0_{\mathbb{C}^2,V}(s)$ reads

$$\mathsf{R}^0_{\mathbb{C}^2,V}(s) \cdot \pi_{\mathbb{C}^2(s) \underset{\mathrm{D},0}{\otimes} V}(x) = (\pi_{V \underset{\mathrm{D},0}{\otimes} \mathbb{C}^2(s)}(x))^{21} \cdot \mathsf{R}^0_{\mathbb{C}^2,V}(s) \tag{6.5}$$

for any $x \in Y_\hbar(\mathfrak{g})$. For $x = \xi_0$,

$$\pi_{\mathbb{C}^2(s) \underset{\mathrm{D},0}{\otimes} V}(\xi_0) = (\pi_{V \underset{\mathrm{D},0}{\otimes} \mathbb{C}^2(s)}(\xi_0))^{21} = \xi_0 \otimes 1 + 1 \otimes \xi_0 = \begin{pmatrix} 1+\xi_0 & 0 \\ 0 & -1+\xi_0 \end{pmatrix}$$

the intertwining relation (6.5) yields

$$[\xi_0, \alpha] = 0 \qquad [\xi_0, \delta] = 0 \qquad [\xi_0, \beta] = -2\beta \qquad [\xi_0, \gamma] = 2\gamma$$

For $x = t_1$, $\pi_{V_1 \underset{\mathrm{D},0}{\otimes} V_2}(t_1) = \pi_{V_1}(t_1) + \pi_{V_2}(t_1)$, so that

$$\begin{aligned} \pi_{\mathbb{C}^2(s) \underset{\mathrm{D},0}{\otimes} V}(t_1) = (\pi_{V \underset{\mathrm{D},0}{\otimes} \mathbb{C}^2(s)}(t_1))^{21} &= (t_1 + s\xi_0) \otimes 1 + 1 \otimes t_1 \\ &= (\xi_1 - \frac{\hbar}{2}\xi_0^2 + s\xi_0) \otimes 1 + 1 \otimes t_1 \end{aligned}$$

Since ξ_1 acts by 0 on $\mathbb{C}^2$, $\pi_{\mathbb{C}^2(s) \underset{\mathrm{D},0}{\otimes} V}(t_1)$ acts by the matrix

$$\begin{pmatrix} -\hbar/2 + s + t_1 & 0 \\ 0 & \hbar/2 - s + t_1 \end{pmatrix}$$

The relation (6.5) then implies that $[t_1, \alpha] = 0$, $[t_1, \delta] = 0$, $[t_1, \beta] = -2s\beta$ and $[t_1, \gamma] = 2s\gamma$. The last two relations imply in turn that $\beta = \gamma = 0$ since $\mathrm{ad}(t_1) \pm s$ is invertible on $\mathrm{End}(V)$ for all but finitely many values of s.

For $x = x_0^-$, Sect. 3.2 implies that the $-+$ coefficient of $\pi_{V \underset{\mathrm{D},0}{\otimes} \mathbb{C}^2(s)}(x_0^-)$ is $\mathbf{1}_V$, while that of $\pi_{\mathbb{C}^2(s) \underset{\mathrm{D},0}{\otimes} V}(x_0^-)$ is equal to $\oint_{C_1} \frac{\xi(v)}{v-s} = \xi(s)$, where C_1 encloses $\sigma(\mathbb{C}^2(s)) = \{s\}$ and none of the points in $\sigma(V)$. Taking $-+$ coefficients in (6.5) then yields the relation $\alpha(s) = \delta(s) \cdot \xi(s)$. Similarly, taking the $+-$ coefficients in the intertwining relation (6.5) with $x = x_0^+$ yields $\alpha = \xi \cdot \delta$. Summarising, (6.5) implies that

$$\mathsf{R}^0_{\mathbb{C}^2, V}(s) = \begin{pmatrix} \alpha(s) & 0 \\ 0 & \delta(s) \end{pmatrix}$$

where $\alpha, \delta \in \mathrm{End}(V)(s)$ commute with ξ_0, t_1, and satisfy $\alpha = \delta \cdot \xi = \xi \cdot \delta$.

We now use the cabling relation (6.3)

$$\mathsf{R}^0_{\mathbb{C}^2_{(1)} \underset{\mathrm{D},\hbar}{\otimes} \mathbb{C}^2_{(2)}, V}(s) = \mathsf{R}^0_{\mathbb{C}^2_{(1)}, V}(s+\hbar) \cdot \mathsf{R}^0_{\mathbb{C}^2_{(2)}, V}(s)$$

in $\mathrm{End}_{\mathbb{C}}(\mathbb{C}^2 \otimes \mathbb{C}^2 \otimes V)$, where the subscripts are added to emphasize the order of the tensors. The right–hand side applied to $v_+ \otimes v_- \otimes v$ yields $\alpha(s+\hbar) \cdot \delta(s)\, v$. On the other hand, by naturality,

$$\mathsf{R}^0_{\mathbb{C}^2_{(1)}\underset{\mathrm{D},\hbar}{\otimes}\mathbb{C}^2_{(2)},V}(s)\, v_+\otimes v_-\otimes v = \mathsf{R}^0_{\mathbb{C},V}(s)1\otimes v = v$$

Combining these relations yields $\alpha(s+\hbar)\cdot\alpha(s) = \xi(s)$. Taking now $V=\mathbb{C}^2$, this equation implies that the coefficient $a(s)$ of v_+ in $\alpha(s)v_+$ satisfies the additive difference equation

$$\frac{a(s+2\hbar)}{a(s)} = \frac{c(s+\hbar)}{c(s)} = \frac{s(s+2\hbar)}{(s+\hbar)^2}$$

where $c(s) = (s+\hbar)/s$ is the matrix coefficient of $\xi(s)$ corresponding to v_+. This equation has a unique solution $\varphi(s)$ which is holomorphic and non-zero for $\mathrm{Re}(s/2\hbar)\gg 0$ and is asymptotic to $1+O(s^{-1})$ in that domain (see, e.g., [11, §4]). Clearly,

$$\varphi(s) = \frac{\Gamma\left(\frac{s}{2\hbar}\right)\cdot\Gamma\left(\frac{s+2\hbar}{2\hbar}\right)}{\Gamma\left(\frac{s+\hbar}{2\hbar}\right)^2}$$

which is not a rational function. □

6.5 Taylor Expansion of $\mathcal{A}_{V_1,V_2}(s)$

Let $V_1, V_2\in\mathrm{Rep}_{\mathrm{fd}}(Y_\hbar(\mathfrak{g}))$. We determine in Sect. 6.6 the asymptotic expansion of $\mathcal{R}^{0,\uparrow/\downarrow}_{V_1,V_2}(s)$ as $s\to\infty$. As a preliminary step, we compute the Taylor series of the endomorphism $\mathcal{A}_{V_1,V_2}(s)$ introduced in Sect. 6.2. Let

$$\mathcal{L}_{V_1,V_2}(s) = -\sum_{\substack{i,j\in\mathbf{I}\\ r\in\mathbb{Z}}} c_{ij}^{(r)}\oint_{C_1}\frac{dt_i(v)}{dv}\otimes t_j\left(v+s+\frac{(\ell+r)\hbar}{2}\right)dv$$

be the logarithm of $\mathcal{A}_{V_1,V_2}$.

Lemma *The Taylor series of $\mathcal{L}_{V_1,V_2}$ at $s=\infty$ is given by the action on $V_1\otimes V_2$ of the element $\mathcal{L}(s)\in s^{-2}(Y^0_\hbar(\mathfrak{g})\otimes Y^0_\hbar(\mathfrak{g}))[\![s^{-1}]\!]$ defined by*

$$\mathcal{L}(s) = -\hbar^2\sum_{\substack{i,j\in\mathbf{I}\\ r\in\mathbb{Z}}} c_{ij}^{(r)}\, T_{\frac{\ell+r}{2}}\sum_{m\geq n\geq 0}(-1)^n(m+1)!s^{-m-2}\,\frac{t_{i,n}}{n!}\otimes\frac{t_{j,m-n}}{(m-n)!}$$

where $T_m f(s) = f(s+m\hbar)$.

Proof Given that

$$
\begin{aligned}
t_j(v+s) &= \hbar \sum_{m\geq 0} t_{j,m}(v+s)^{-m-1} \\
&= \hbar \sum_{m\geq 0} s^{-m-1} t_{j,m} \sum_{n\geq 0} (-1)^n \binom{m+n}{n} v^n s^{-n} \\
&= \hbar \sum_{m\geq 0} s^{-m-1} \sum_{n=0}^{m} (-1)^n \binom{m}{n} t_{j,n-m} v^n
\end{aligned}
$$

and that $t_i(v)' = -\hbar \sum_{a\geq 0}(a+1)t_{i,a}v^{-a-2}$, the expansion of the summand of $\mathcal{L}_{V_1,V_2}$ corresponding to a triple i, j, r is equal to

$$
\hbar^2 c_{ij}^{(r)} T_{\frac{\ell+r}{2}} \sum_{m\geq 0} s^{-m-1} \sum_{n=1}^{m} (-1)^n \binom{m}{n} n t_{i,n-1} \otimes t_{j,n-m}
$$

□

Remark Since $(m+1)!s^{-m-2} = (-1)^m \partial_s^m s^{-2}$, the above reads

$$
\mathcal{L}(s) = \sum_{\substack{i,j\in\mathbf{I}\\ r\in\mathbb{Z}}} c_{ij}^{(r)}\, T_{\frac{\ell+r}{2}}\, B_i(\partial_s) \otimes B_j(-\partial_s)\,(-s^{-2}) \tag{6.6}
$$

where $B_k(z) = \hbar \sum_{n\geq 0} \frac{t_{k,n}}{n!} z^n$ is the inverse Borel transform of $t_k(v)$.

6.6 Asymptotic Expansion of $\mathcal{R}^{0,\uparrow/\downarrow}_{V_1,V_2}(s)$

Let[9] $g(x) \in x^{-1}\mathbb{C}[[x^{-1}]]$ be the unique solution of the difference equation $g(x+1) = g(x) - x^{-2}$.

Define $\mathcal{R}^0(s) \in 1 + s^{-1}(Y_\hbar^0(\mathfrak{g}) \otimes Y_\hbar^0(\mathfrak{g}))[[s^{-1}]]$ by

$$
\log(\mathcal{R}^0(s)) = \frac{1}{\ell^2\hbar^2} \sum_{\substack{i,j\in\mathbf{I}\\ r\in\mathbb{Z}}} c_{ij}^{(r)} T_{\frac{\ell+r}{2}}\, B_i(\partial_s) \otimes B_j(-\partial_s)\, g\left(\frac{s}{\ell\hbar}\right) \tag{6.7}
$$

Let $V_1, V_2 \in \mathrm{Rep}_{\mathrm{fd}}(Y_\hbar(\mathfrak{g}))$. By Theorem 6.2, $\mathcal{R}^{0,\uparrow/\downarrow}_{V_1,V_2}(s)$ has an asymptotic expansion as $s \to \infty$ in any halfplane $\mathrm{Re}(s/\ell\hbar) \gtrless m$, which is of the form $1 + O(s^{-1})$.

[9] $g(x)$ is equal to $\sum_{k\geq 0} B_k x^{-k-1}$, and is the asymptotic expansion at $x = \infty$ of the trigamma function $\Psi_1(x) = \frac{d^2}{dx^2} \ln \Gamma(x) = \sum_{n\geq 0}(x+n)^{-2}$.

Proposition *The asymptotic expansion of* $\mathcal{R}^{0,\uparrow/\downarrow}_{V_1,V_2}(s)$ *as* $s\to\infty$ *is given by*

$$\mathcal{R}^0_{V_1,V_2}(s)=\pi_{V_1}\otimes\pi_{V_2}(\mathcal{R}^0(s))$$

Proof By definition of $\log(\mathcal{R}^0(s))$ and (6.6), we have

$$\begin{aligned}(T_\ell-1)&\log(\mathcal{R}^0(s))\\ &=\frac{1}{\ell^2\hbar^2}\sum_{\substack{i,j\in\mathbf{I}\\ r\in\mathbb{Z}}}c_{ij}^{(r)}T_{\frac{\ell+r}{2}}\,B_i(\partial_s)\otimes B_j(-\partial_s)\left(g\left(\frac{s}{\ell\hbar}+1\right)-g\left(\frac{s}{\ell\hbar}\right)\right)\\ &=\sum_{\substack{i,j\in\mathbf{I}\\ r\in\mathbb{Z}}}c_{ij}^{(r)}T_{\frac{\ell+r}{2}}\,B_i(\partial_s)\otimes B_j(-\partial_s)\left(-\frac{1}{s^2}\right)\\ &=\mathcal{L}(s)\end{aligned}$$

Thus, $\mathcal{R}^0_{V_1,V_2}(s)$ is the unique formal solution of

$$\mathcal{R}^0_{V_1,V_2}(s+\ell\hbar)=\mathcal{A}_{V_1,V_2}(s)\cdot\mathcal{R}^0_{V_1,V_2}(s)$$

and therefore equals the asymptotic expansion of $\mathcal{R}^{0,\uparrow/\downarrow}_{V_1,V_2}(s)$. □

6.7 Properties of $\mathcal{R}^0(s)$

The following is the universal analogue of Theorem 6.2.

Theorem $\mathcal{R}^0(s)\in(Y^0_\hbar(\mathfrak{g})\otimes Y^0_\hbar(\mathfrak{g}))[\![s^{-1}]\!]$ *has the following properties.*

1. *For every* $x\in Y_\hbar(\mathfrak{g})$, *the following holds in* $Y_\hbar(\mathfrak{g})^{\otimes 2}[s;s^{-1}]\!]$

$$\mathcal{R}^0(s)\cdot\underset{\mathrm{D},s}{\Delta}(x)=\underset{\mathrm{D},-s}{\Delta}{}^{(21)}(\tau_s(x))\cdot\mathcal{R}^0(s)$$

2. *The cabling identities*

$$\underset{\mathrm{D},s_1}{\Delta}\otimes\mathbf{1}\left(\mathcal{R}^0(s_2)\right)=\mathcal{R}^0_{13}(s_1+s_2)\cdot\mathcal{R}^0_{23}(s_2)$$

$$\mathbf{1}\otimes\underset{\mathrm{D},s_2}{\Delta}\left(\mathcal{R}^0(s_1+s_2)\right)=\mathcal{R}^0_{13}(s_1+s_2)\cdot\mathcal{R}^0_{12}(s_1)$$

hold in $Y^0_\hbar(\mathfrak{g})^{\otimes 3}[s_1][\![s_2^{-1}]\!]$ *and* $Y^0_\hbar(\mathfrak{g})^{\otimes 3}[s_2][\![s_1^{-1}]\!]$ *respectively.*

3. $\mathcal{R}^0(s)$ *is unitary*

$$\mathcal{R}^0(s)^{-1} = \mathcal{R}^0_{21}(-s)$$

4. *For any* $a, b \in \mathbb{C}$

$$(\tau_a \otimes \tau_b)(\mathcal{R}^0(s)) = \mathcal{R}^0(s + a - b)$$

5. $\mathcal{R}^0(s) - 1 \in \mathcal{F}_{-1}(Y_\hbar(\mathfrak{g})^{\otimes 2}[\![s^{-1}]\!])$, *with semiclassical limit given by*

$$\overline{\mathcal{R}^0(s) - 1} = \frac{\hbar\Omega_{\mathfrak{h}}}{z + s - w} \in (U(\mathfrak{g}[z]) \otimes U(\mathfrak{g}[w]))[\![s^{-1}]\!]$$

Statements (1)–(4) follow from that fact that $\mathcal{R}^0_{V_1,V_2}(s)$ is the asymptotic expansion of $\mathcal{R}^{0,\uparrow/\downarrow}_{V_1,V_2}(s)$ and Theorem 6.2, since finite-dimensional representations separate points in $Y_\hbar(\mathfrak{g})$ (Proposition A.1). For completeness, we give a direct proof of Theorem 6.7 below, which does not rely on this fact.

6.8 *Direct Proof of Theorem 6.7*

Proof of (2) Let $P^0 \subset Y^0_\hbar(\mathfrak{g})$ be the $\mathbb{C}$–linear span of $\{t_{i,r}\}_{i\in\mathbf{I},r\in\mathbb{Z}_{\geq 0}}$, so that $\log(\mathcal{R}^0(s)) \in (P^0 \otimes P^0)[\![s^{-1}]\!]$. The cabling identities follow from the fact that each $t_{i,r}$ is primitive with respect to the Drinfeld coproduct, that is satisfies

$$\underset{\mathrm{D},s}{\Delta}(t_{i,r}) = \tau_s(t_{i,r}) \otimes 1 + 1 \otimes t_{i,r}$$

and the fact that $Y^0_\hbar(\mathfrak{g})$ is a commutative subalgebra.

Proof of (3) Write

$$\begin{aligned}\log(\mathcal{R}^0(s)) &= \frac{1}{\ell^2\hbar^2} \sum_{\substack{i,j\in\mathbf{I}\\ r\in\mathbb{Z}}} c^{(r)}_{ij}\, T_{\frac{r}{2}}\, B_i(\partial_s) \otimes B_j(-\partial_s)\, g\left(\frac{s}{\ell\hbar} + \frac{1}{2}\right)\\ &= \frac{1}{\ell^2\hbar^2} \sum_{i,j\in\mathbf{I}} c_{ij}(e^{\frac{\hbar}{2}\partial_s})\, B_i(\partial_s) \otimes B_j(-\partial_s)\, g\left(\frac{s}{\ell\hbar} + \frac{1}{2}\right)\\ &= \frac{1}{\ell^2\hbar^2}\, \Omega(\partial_s)\, g\left(\frac{s}{\ell\hbar} + \frac{1}{2}\right)\end{aligned}$$

where $\Omega(z) = \sum_{i,j\in\mathbf{I}} c_{ij}(z) B_i(z) \otimes B_j(-z)$. The unitary condition follows from

$$\Omega^{21}(z) = \Omega(-z) \qquad \text{and} \qquad g\left(\frac{1}{2}+x\right) = -g\left(\frac{1}{2}-x\right)$$

The first identity holds because $c_{ij}(q) = c_{ij}(q^{-1}) = c_{ji}(q)$, and the second because $g(x) = -g(1-x)$, since both sides are solutions of the same difference equation.

Proof of (4) Since $\tau_a B_i(z) = e^{az} B_i(z)$,

$$\begin{aligned}\tau_a \otimes \tau_b \log(\mathcal{R}^0(s)) &= \frac{1}{\ell^2\hbar^2} \sum_{\substack{i,j\in\mathbf{I}\\ r\in\mathbb{Z}}} c_{ij}^{(r)} T_{\frac{\ell+r}{2}}\, B_i(\partial_s) \otimes B_j(-\partial_s) e^{(a-b)\partial_s}\, g\left(\frac{s}{\ell\hbar}\right)\\ &= \frac{1}{\ell^2\hbar^2} \sum_{\substack{i,j\in\mathbf{I}\\ r\in\mathbb{Z}}} c_{ij}^{(r)} T_{\frac{\ell+r}{2}}\, B_i(\partial_s) \otimes B_j(-\partial_s)\, g\left(\frac{s+a-b}{\ell\hbar}\right)\\ &= \log(\mathcal{R}^0(s+a-b))\end{aligned}$$

Proof of (5) By (2.7), it suffices to prove that $\log(\mathcal{R}^0(s))$ is an element of $\mathcal{F}_{-1}(Y_\hbar(\mathfrak{g})^{\otimes 2}[\![s^{-1}]\!])$, and that

$$\overline{\log(\mathcal{R}^0(s))} = \hbar \frac{\Omega_{\mathfrak{h}}}{z+s-w}$$

Note first that

$$t_{i,n} \otimes t_{j,m}\, (-1)^m \partial_s^m g\left(\frac{s}{\ell\hbar}\right) = t_{i,n} \otimes t_{j,m} \left(\frac{\ell\hbar \cdot m!}{s^{m+1}} + O(s^{-m-2})\right)$$

lies in $\mathcal{F}_{-1}(Y_\hbar(\mathfrak{g})^{\otimes 2}[\![s^{-1}]\!])$, and has symbol $\ell\hbar \cdot m!\, d_i h_i z^n \otimes d_j h_j w^m\, s^{-m-1}$. Since the shifts T_x preserve the filtration $\mathcal{F}_\bullet(Y_\hbar(\mathfrak{g})^{\otimes 2}[\![s^{-1}]\!])$ and act as the identity on its associated graded space, it follows that $\log(\mathcal{R}^0(s)) \in \mathcal{F}_{-1}(Y_\hbar(\mathfrak{g})^{\otimes 2}[\![s^{-1}]\!])$, and that

$$\begin{aligned}\overline{\log(\mathcal{R}^0(s))} &= \frac{\hbar}{\ell} \sum_{i,j\in\mathbf{I}} c_{ij}(1) \sum_{m\geq n\geq 0} (-1)^n \binom{m}{n} d_i h_i z^n \otimes d_j h_j w^{m-n}\, s^{-m-1}\\ &= \hbar \frac{\Omega_{\mathfrak{h}}}{z+s-w}\end{aligned}$$

where the last equality follows from the fact that

$$\sum_{j \in \mathbf{I}} c_{ij}(1) d_j h_j = \ell \sum_{j \in \mathbf{I}} (\mathbf{B}^{-1})_{ij} d_j h_j = \ell \varpi_i^\vee$$

with $\mathbf{B} = (d_i a_{ij})$, and $\varpi_i^\vee \in \mathfrak{h}$ the fundamental coweights.

Proof of (1) The intertwining relation is obvious for $x \in Y_\hbar^0(\mathfrak{g})$, since the latter is the commutative algebra generated by the elements $t_{i,r}$, which satisfy

$$\underset{\mathrm{D},s}{\Delta}(t_{i,r}) = \tau_s(t_{i,r}) \otimes 1 + 1 \otimes t_{i,r} = \underset{\mathrm{D},-s}{\Delta}^{(21)}(\tau_s t_{i,r})$$

Thus, it suffices to prove that, for any $k \in \mathbf{I}$

$$\mathcal{R}^0(s) \underset{\mathrm{D},s}{\Delta}(x_{k,0}^\pm) = \underset{\mathrm{D},-s}{\Delta}^{(21)}(x_{k,0}^\pm) \mathcal{R}^0(s)$$

We verify this identity for the $+$ case only. By Proposition 3.3, $\underset{\mathrm{D},s}{\Delta}(x_{k,0}^+)$ is equal to $x_{k,0}^+ \otimes 1 + 1 \otimes x_{k,0}^+ + \mathfrak{X}_k(s)$, where

$$\mathfrak{X}_k(s) = \hbar \sum_{N \geq 0} s^{-N-1} \sum_{n=0}^{N} (-1)^{n+1} \binom{N}{n} \xi_{k,n} \otimes x_{k,N-n}^+$$

We therefore have to prove that

$$\begin{aligned} \mathrm{Ad}(\mathcal{R}^0(s)) \cdot \left(x_{k,0}^+ \otimes 1 + 1 \otimes x_{k,0}^+\right) = x_{k,0}^+ \otimes 1 + 1 \otimes x_{k,0}^+ + \mathfrak{X}_k^{(21)}(-s) \\ - \mathrm{Ad}(\mathcal{R}^0(s)) \cdot \mathfrak{X}_k(s) \end{aligned}$$

We claim that $\mathrm{Ad}(\mathcal{R}^0(s)) \cdot (x_{k,0}^+ \otimes 1) = x_{k,0}^+ \otimes 1 + \mathfrak{X}_k^{(21)}(-s)$. Given this, the unitary condition (3) then implies that $\mathrm{Ad}(\mathcal{R}^0(s))^{-1} \cdot (1 \otimes x_{k,0}^+) = 1 \otimes x_{k,0}^+ + \mathfrak{X}_k(s)$ which, combined with the claim yields the required intertwining equation for $x_{k,0}^+$.

To prove the claim, we rely on the following commutation relation, which was obtained in [10, §2.9]

$$[B_i(z), x_{k,n}^+] = \frac{e^{\frac{d_i a_{ik} \hbar}{2} z} - e^{-\frac{d_i a_{ik} \hbar}{2} z}}{z} \cdot \left(\sum_{p \geq 0} x_{k,n+p}^+ \frac{z^p}{p!} \right)$$

Combining with the definition of $\Omega(z)$ given above, we can carry out the following computation, for each $k \in \mathbf{I}$, $n \in \mathbb{Z}_{\geq 0}$, and $y \in Y_\hbar^0(\mathfrak{g})$.

$$[\Omega(z), x_{k,n}^+ \otimes y] = \frac{1}{\ell^2\hbar^2}\sum_{j\in\mathbf{I}}\left(\sum_{i\in\mathbf{I}} c_{ij}\left(e^{\frac{\hbar}{2}z}\right)\left(\frac{e^{\frac{d_ia_{ik}\hbar}{2}z} - e^{-\frac{d_ia_{ik}\hbar}{2}z}}{z}\right)\right)$$

$$\cdot\left(\sum_{p\geq 0} x_{k,n+p}^+ \frac{(-z)^p}{p!}\right)\otimes B_j(z)y$$

$$= \frac{1}{\ell^2\hbar^2}\frac{e^{\frac{\ell\hbar}{2}z} - e^{-\frac{\ell\hbar}{2}z}}{z}\cdot\left(\sum_{p\geq 0} x_{k,n+p}^+ \frac{(-z)^p}{p!}\right)\otimes B_k(z)y$$

Note that we used the Eq. (6.1) satisfied by $(c_{ij}(q))$ above.

This calculation, combined with

$$\frac{e^{\frac{\ell\hbar}{2}\partial_s} - e^{-\frac{\ell\hbar}{2}\partial_s}}{\partial_s}\cdot g\left(\frac{s}{\ell\hbar}+\frac{1}{2}\right) = \frac{\ell^2\hbar^2}{s}$$

yields the commutation relation

$$[\log(\mathcal{R}^0(s)), x_{k,n}^+ \otimes y] = \sum_{p\geq 0}(-1)^p x_{k,n+p}^+ \otimes \left(\hbar\sum_{r\geq 0}\binom{r+p}{r} t_{k,r} s^{-r-p-1}\right) y$$

The claim now follows from

$$\mathrm{Ad}(\mathcal{R}^0(s)) = \exp(\mathrm{ad}(\log(\mathcal{R}^0(s))))$$

where both sides are acting on $V_k^+ \otimes Y_\hbar^0(\mathfrak{g})$, where V_k^+ is the $\mathbb{C}$–linear span of $\{x_{k,n}\}_{n\geq 0}$.

7 The Universal and the Meromorphic R-Matrices of $Y_\hbar(\mathfrak{g})$

In this section, we construct the meromorphic and universal R-matrices of $Y_\hbar(\mathfrak{g})$.

7.1 *The Meromorphic R-Matrix*

Given $V_1, V_2 \in \mathrm{Rep}_{\mathrm{fd}}(Y_\hbar(\mathfrak{g}))$ and $\varepsilon \in \{\uparrow, \downarrow\}$, define $\mathcal{R}_{V_1,V_2}^\varepsilon : \mathbb{C} \to \mathrm{End}(V_1 \otimes V_2)$ by

$$\mathcal{R}_{V_1,V_2}^\varepsilon(s) = \mathcal{R}_{V_1,V_2}^+(s)\cdot\mathcal{R}_{V_1,V_2}^{0,\varepsilon}(s)\cdot\mathcal{R}_{V_1,V_2}^-(s),$$

where $\mathcal{R}_{V_1,V_2}^+(s) = (1\,2)\circ\mathcal{R}_{V_2,V_1}^-(-s)^{-1}\circ(1\,2)$.

Theorem *The meromorphic function* $\mathcal{R}^{\varepsilon}_{V_1,V_2}(s)$ *has the following properties.*

1. *The map*

$$(1\,2)\circ\mathcal{R}^{\varepsilon}_{V_1,V_2}(s):V_1(s)\otimes V_2\to V_2\otimes V_1(s)$$

is a morphism of $Y_\hbar(\mathfrak{g})$*–modules, which is natural in* V_1, V_2.

2. *For any* $V_1, V_2, V_3\in \mathrm{Rep}_{\mathrm{fd}}(Y_\hbar(\mathfrak{g}))$,

$$\mathcal{R}^{\varepsilon}_{V_1\underset{s_1}{\otimes}V_2,V_3}(s_2)=\mathcal{R}^{\varepsilon}_{V_1,V_3}(s_1+s_2)\cdot\mathcal{R}^{\varepsilon}_{V_2,V_3}(s_2)$$

$$\mathcal{R}^{\varepsilon}_{V_1,V_2\underset{s_2}{\otimes}V_3}(s_1+s_2)=\mathcal{R}^{\varepsilon}_{V_1,V_3}(s_1+s_2)\cdot\mathcal{R}^{\varepsilon}_{V_1,V_2}(s_1).$$

In particular, the QYBE holds on $V_1\otimes V_2\otimes V_3$*:*

$$\mathcal{R}^{\varepsilon}_{V_1,V_2}(s_1)\mathcal{R}^{\varepsilon}_{V_1,V_3}(s_1+s_2)\mathcal{R}^{\varepsilon}_{V_2,V_3}(s_2)=\mathcal{R}^{\varepsilon}_{V_2,V_3}(s_2)\mathcal{R}^{\varepsilon}_{V_1,V_3}(s_1+s_2)\mathcal{R}^{\varepsilon}_{V_1,V_2}(s_1).$$

3. *For any* $a,b\in\mathbb{C}$,

$$\mathcal{R}^{\varepsilon}_{V_1(a),V_2(b)}(s)=\mathcal{R}^{\varepsilon}_{V_1,V_2}(s+a-b).$$

4. $\mathcal{R}^{\uparrow}_{V_1,V_2}(s)$ *and* $\mathcal{R}^{\downarrow}_{V_2,V_1}(s)$ *are related by the unitarity relation:*

$$(1\,2)\circ\mathcal{R}^{\uparrow}_{V_1,V_2}(-s)\circ(1\,2)=\mathcal{R}^{\downarrow}_{V_2,V_1}(s)^{-1}.$$

5. $\mathcal{R}^{\uparrow/\downarrow}_{V_1,V_2}(s)$ *have the same asymptotic expansion, which is of the form*

$$\mathcal{R}^{\uparrow/\downarrow}_{V_1,V_2}(s)\sim 1+\hbar\Omega_{\mathfrak{g}}s^{-1}+O(s^{-2})$$

as $s\to\infty$ *in any halfplane of the form* $\mathrm{Re}(s/\hbar)\gtrless m$.

Proof

(1) By definition,

$$(1\,2)\circ\mathcal{R}^{\varepsilon}_{V_1,V_2}(s)=\mathcal{R}^{-}_{V_2,V_1}(-s)^{-1}\cdot\left((1\,2)\circ\mathcal{R}^{0,\varepsilon}_{V_1,V_2}(s)\right)\cdot\mathcal{R}^{-}_{V_1,V_2}(s).$$

The result therefore follows from the fact that $\mathcal{R}^{-}_{V_1,V_2}(s)$ is a morphism of $Y_\hbar(\mathfrak{g})$–modules $V_1\underset{s}{\otimes}V_2\to V_1\underset{\mathrm{D},s}{\otimes}V_2$ (Theorem 4.1), and Theorem 6.2 (1).

(2) We will prove the following equivalent version of the first cabling identity

$$(1\,2\,3)\circ\mathcal{R}^{\varepsilon}_{V_1\underset{s_1}{\otimes}V_2,V_3}(s_2)=(1\,2)\left(\mathcal{R}^{\varepsilon}_{V_1,V_3}(s_1+s_2)\otimes\mathbf{1}\right)(2\,3)\left(\mathbf{1}\otimes\mathcal{R}^{\varepsilon}_{V_2,V_3}(s_2)\right)\tag{7.1}$$

By definition, the left–hand side is equal to

$$
\begin{aligned}
&\mathcal{R}^{-}_{V_3, V_1 \underset{s_1}{\otimes} V_2}(-s_2)^{-1} \cdot \left((1\,2\,3) \circ \mathcal{R}^{0,\varepsilon}_{V_1 \underset{s_1}{\otimes} V_2, V_3}(s_2) \right) \cdot \mathcal{R}^{-}_{V_1 \underset{s_1}{\otimes} V_2, V_3}(s_2) \\
=&\mathcal{R}^{-}_{V_3, V_1 \underset{s_1}{\otimes} V_2}(-s_2)^{-1} \cdot \left(\mathbf{1} \otimes \mathcal{R}^{-}_{V_1, V_2}(s_1)^{-1} \right) \cdot \left((1\,2\,3) \circ \mathcal{R}^{0,\varepsilon}_{V_1 \underset{\mathrm{D}, s_1}{\otimes} V_2, V_3}(s_2) \right) \cdot \\
&\quad \left(\mathcal{R}^{-}_{V_1, V_2}(s_1) \otimes \mathbf{1} \right) \cdot \mathcal{R}^{-}_{V_1 \underset{s_1}{\otimes} V_2, V_3}(s_2) \\
=&\mathcal{R}^{-}_{V_3, V_1 \underset{s_1}{\otimes} V_2}(-s_2)^{-1} \cdot \left(\mathbf{1} \otimes \mathcal{R}^{-}_{V_1, V_2}(s_1)^{-1} \right) \cdot (1\,2) \left(\mathcal{R}^{0,\varepsilon}_{V_1, V_3}(s_1 + s_2) \otimes \mathbf{1} \right) \cdot \\
&\quad (2\,3) \left(\mathbf{1} \otimes \mathcal{R}^{0,\varepsilon}_{V_2, V_3}(s_2) \right) \cdot \left(\mathcal{R}^{-}_{V_1, V_2}(s_1) \otimes \mathbf{1} \right) \cdot \mathcal{R}^{-}_{V_1 \underset{s_1}{\otimes} V_2, V_3}(s_2)
\end{aligned}
$$

In the first equality, we used Theorem 4.1 in order to change $V_1 \underset{s_1}{\otimes} V_2$ to $V_1 \underset{\mathrm{D}, s_1}{\otimes} V_2$, while the second equality follows from the cabling identity satisfied by $\mathcal{R}^{0,\varepsilon}(s)$ (Theorem 6.2 (2)).

Note that we have the following identity, which follows from the cocycle equation (4.7) after renaming variables

$$
\begin{aligned}
&\left(\mathcal{R}^{-}_{V_1, V_3}(s_1 + s_2) \otimes \mathbf{1} \right) \cdot \mathcal{R}^{-}_{V_1 \underset{s_1+s_2}{\otimes} V_3, V_2}(-s_2) = \\
&\left(\mathbf{1} \otimes \mathcal{R}^{-}_{V_3, V_2}(-s_2) \right) \cdot \mathcal{R}^{-}_{V_1, V_3 \underset{-s_2}{\otimes} V_2}(s_1)
\end{aligned}
$$

Inserting this operator and its inverse in the last line of the computation above allows us to write the left–hand side of (7.1) as $A(s_1, s_2) \cdot B(s_1, s_2)$, where

$$
\begin{aligned}
A(s_1, s_2) = &\mathcal{R}^{-}_{V_3, V_1 \underset{s_1}{\otimes} V_2}(-s_2)^{-1} \cdot \left(\mathbf{1} \otimes \mathcal{R}^{-}_{V_1, V_2}(s_1)^{-1} \right) \cdot (1\,2) \cdot \\
&\quad \left(\mathcal{R}^{0,\varepsilon}_{V_1, V_3}(s_1 + s_2) \otimes \mathbf{1} \right) \cdot \left(\mathcal{R}^{-}_{V_1, V_3}(s_1 + s_2) \otimes \mathbf{1} \right) \cdot \\
&\quad \mathcal{R}^{-}_{V_1 \underset{s_1+s_2}{\otimes} V_3, V_2}(-s_2)
\end{aligned}
$$

$$
\begin{aligned}
B(s_1, s_2) = &\left(\left(\mathbf{1} \otimes \mathcal{R}^{-}_{V_3, V_2}(-s_2) \right) \cdot \mathcal{R}^{-}_{V_1, V_3 \underset{-s_2}{\otimes} V_2}(s_1) \right)^{-1} \cdot \\
&\quad (2\,3) \left(\mathbf{1} \otimes \mathcal{R}^{0,\varepsilon}_{V_2, V_3}(s_2) \right) \cdot \left(\mathcal{R}^{-}_{V_1, V_2}(s_1) \otimes \mathbf{1} \right) \cdot \mathcal{R}^{-}_{V_1 \underset{s_1}{\otimes} V_2, V_3}(s_2)
\end{aligned}
$$

Thus, in order to prove (7.1), it is enough to show that

$$A(s_1,s_2)=(1\ 2)\circ\mathcal{R}^{\varepsilon}_{V_1,V_3}(s_1+s_2)\otimes\mathbf{1}\text{ and }B(s_1,s_2)=(2\ 3)\circ\mathbf{1}\otimes\mathcal{R}^{\varepsilon}_{V_2,V_3}(s_2)$$

We verify the latter below, the proof of the former, being entirely analogous, is omitted.

Using the cocycle equation (4.7) again, we have:

$$\begin{aligned}
B(s_1,s_2)&=\mathcal{R}^{-}_{V_1,V_3\underset{-s_2}{\otimes}V_2}(s_1)^{-1}\cdot\left(\mathbf{1}\otimes\mathcal{R}^{-}_{V_3,V_2}(-s_2)^{-1}\right)\cdot\\
&\qquad\left((2\ 3)\circ\mathbf{1}\otimes\mathcal{R}^{0,\varepsilon}_{V_2,V_3}(s_2)\right)\cdot\left(\mathbf{1}\otimes\mathcal{R}^{-}_{V_2,V_3}(s_2)\right)\cdot\mathcal{R}^{-}_{V_1,V_2\underset{s_2}{\otimes}V_3}(s_1+s_2)\\
&=\mathcal{R}^{-}_{V_1,V_3\underset{-s_2}{\otimes}V_2}(s_1)^{-1}\cdot\left((2\ 3)\circ\mathbf{1}\otimes\mathcal{R}^{\varepsilon}_{V_2,V_3}(s_2)\right)\cdot\mathcal{R}^{-}_{V_1,V_2\underset{s_2}{\otimes}V_3}(s_1+s_2)\\
&=\mathcal{R}^{-}_{V_1,V_3\underset{-s_2}{\otimes}V_2}(s_1)^{-1}\cdot\mathcal{R}^{-}_{V_1,V_3\underset{-s_2}{\otimes}V_2}(s_1)\cdot\left((2\ 3)\circ\mathbf{1}\otimes\mathcal{R}^{\varepsilon}_{V_2,V_3}(s_2)\right)\\
&=(2\ 3)\circ\mathbf{1}\otimes\mathcal{R}^{\varepsilon}_{V_2,V_3}(s_2)
\end{aligned}$$

In the second line, we have used the definition of $\mathcal{R}^{\varepsilon}(s)$, and in the third, Part (1) of this theorem.

(3) follows from Theorem 4.1 (2), and Theorem 6.2 (3).

(4) By definition,

$$\begin{aligned}
&(1\ 2)\circ\mathcal{R}^{\uparrow}_{V_1,V_2}(-s)\circ(1\ 2)\\
&\quad=\mathcal{R}^{-}_{V_2,V_1}(s)^{-1}\cdot\left((1\ 2)\circ\mathcal{R}^{0,\uparrow}_{V_1,V_2}(-s)\circ(1\ 2)\right)\cdot\mathcal{R}^{+}_{V_2,V_1}(s)^{-1}\\
&\quad=\mathcal{R}^{-}_{V_2,V_1}(s)^{-1}\cdot\left(\mathcal{R}^{0,\downarrow}_{V_2,V_1}(s)^{-1}\right)\cdot\mathcal{R}^{+}_{V_2,V_1}(s)^{-1}\\
&\quad=\mathcal{R}^{\downarrow}_{V_2,V_1}(s)^{-1}
\end{aligned}$$

where the second equality uses Theorem 6.2 (4).

(5) follows from Theorem 6.2 (5) and the Taylor series expansion of $\mathcal{R}^{-}_{V_1,V_2}(s)=1+\hbar\mathrm{r}/s+O(s^{-2})$ given in Theorem 4.1 (3).

□

7.2 Existence of a Rational Intertwiner

The following extends to an arbitrary pair of representations $V_1,V_2\in\mathrm{Rep}_{\mathrm{fd}}(Y_\hbar(\mathfrak{g}))$ a result due to Drinfeld, which is valid when V_1,V_2 are irreducible [4, Thm. 4], and

Maulik–Okounkov, which is valid when $\mathfrak{g}$ is simply–laced, and V_1, V_2 arise from geometry [20].

Theorem *There is a rational map* $\mathsf{R}_{V_1,V_2}(s) : \mathbb{C} \to \mathrm{End}_{\mathbb{C}}(V_1 \otimes V_2)$*, which is normalised by* $\mathsf{R}_{V_1,V_2}(\infty) = \mathbf{1}$ *and such that*

$$(1\,2) \circ \mathsf{R}_{V_1,V_2}(s) : V_1(s) \otimes V_2 \to V_2 \otimes V_1(s)$$

intertwines the action of $Y_\hbar(\mathfrak{g})$*. In particular,* $V_1(s) \otimes V_2$ *and* $V_2 \otimes V_1(s)$ *are isomorphic as* $Y_\hbar(\mathfrak{g})$*–modules for all but finitely many values of* s*.*

Proof Let $\mathsf{R}^0_{V_1,V_2}(s)$ be the rational operator such that $(1\,2) \circ \mathsf{R}^0_{V_1,V_2}(s)$ intertwines the action on $V_1(s) \underset{\mathrm{D},0}{\otimes} V_2$ and $V_2 \underset{\mathrm{D},0}{\otimes} V_1(s)$, as obtained in Theorem 6.3. Then,

$$\mathsf{R}_{V_1,V_2}(s) = \mathcal{R}^+_{V_1,V_2}(s) \cdot \mathsf{R}^0_{V_1,V_2}(s) \cdot \mathcal{R}^-_{V_1,V_2}(s)$$

yields the required map. □

7.3 Non Existence of Rational Commutativity Constraints

Theorem *There is no function* $\mathsf{R}_{V_1,V_2} : \mathbb{C} \to \mathrm{Aut}_{\mathbb{C}}(V_1 \otimes V_2)$ *which is rational, defined for any* $V_1, V_2 \in \mathrm{Rep}_{\mathrm{fd}}(Y_\hbar(\mathfrak{g}))$*, and such that the following holds*

1. $(1\,2) \circ \mathsf{R}_{V_1,V_2}(s) : V_1(s) \otimes V_2 \to V_2 \otimes V_1(s)$ *intertwines the action of* $Y_\hbar(\mathfrak{g})$*, and is natural in* V_1 *and* V_2*.*
2. *For any* $V_1, V_2, V_3 \in \mathrm{Rep}_{\mathrm{fd}}(Y_\hbar(\mathfrak{g}))$*,*

$$\mathsf{R}_{V_1 \underset{s_1}{\otimes} V_2, V_3}(s_2) = \mathsf{R}_{V_1,V_3}(s_1 + s_2) \cdot \mathsf{R}_{V_2,V_3}(s_2)$$

$$\mathsf{R}_{V_1, V_2 \underset{s_2}{\otimes} V_3}(s_1 + s_2) = \mathsf{R}_{V_1,V_3}(s_1 + s_2) \cdot \mathsf{R}_{V_1,V_2}(s_2)$$

Proof Assume that such a rational $\mathsf{R}_{V_1,V_2}(s)$ exists. Set

$$\mathsf{R}^0_{V_1,V_2}(s) = \mathcal{R}^+_{V_1,V_2}(s)^{-1} \cdot \mathsf{R}_{V_1,V_2}(s) \cdot \mathcal{R}^-_{V_1,V_2}(s)^{-1}$$

Then, the rational operator $\mathsf{R}^0_{V_1,V_2}(s)$ contradicts Theorem 6.4, and the claim follows. □

7.4 The Universal R-Matrix

We now turn our attention to obtaining a formal, universal analogue of Theorem 7.1. Consider the formal power series

$$\mathcal{R}(s) = \mathcal{R}^+(s) \cdot \mathcal{R}^0(s) \cdot \mathcal{R}^-(s) \in (Y_\hbar(\mathfrak{g}) \otimes Y_\hbar(\mathfrak{g}))[\![s^{-1}]\!]$$

where $\mathcal{R}^+(s) = \mathcal{R}_{21}^-(-s)^{-1}$. This series admits an expansion

$$\mathcal{R}(s) = 1 + \sum_{k=1}^{\infty} \mathcal{R}_k s^{-k}, \qquad \mathcal{R}_k \in \mathcal{F}_{k-1}(Y_\hbar(\mathfrak{g}) \otimes Y_\hbar(\mathfrak{g})).$$

Theorem *The formal power series $\mathcal{R}(s)$ has the following properties.*

1. *For every $x \in Y_\hbar(\mathfrak{g})$, the following holds in $Y_\hbar(\mathfrak{g})^{\otimes 2}[s; s^{-1}]\!]$*

$$\tau_s \otimes \mathbf{1} \circ \Delta^{\mathrm{op}}(x) = \mathcal{R}(s) \cdot \tau_s \otimes \mathbf{1} \circ \Delta(x) \cdot \mathcal{R}(s)^{-1}$$

2. *$\mathcal{R}(s)$ satisfies the cabling identities*

$$\Delta \otimes \mathbf{1}(\mathcal{R}(s)) = \mathcal{R}_{13}(s)\mathcal{R}_{23}(s)$$
$$\mathbf{1} \otimes \Delta(\mathcal{R}(s)) = \mathcal{R}_{13}(s)\mathcal{R}_{12}(s)$$

3. *$\mathcal{R}(s)$ is unitary*

$$\mathcal{R}(s)^{-1} = \mathcal{R}_{21}(-s).$$

4. *For any $a, b \in \mathbb{C}$, we have*

$$(\tau_a \otimes \tau_b)\mathcal{R}(s) = \mathcal{R}(s + a - b)$$

5. *$\mathcal{R}(s) - 1 \in \mathcal{F}_{-1}(Y_\hbar(\mathfrak{g})^{\otimes 2}[\![s^{-1}]\!])$, with semiclassical limit given by*

$$\overline{\mathcal{R}(s) - 1} = \frac{\hbar \Omega_{\mathfrak{g}}}{s + z - w} \in (U(\mathfrak{g}[z]) \otimes U(\mathfrak{g}[w]))[\![s^{-1}]\!]$$

   In particular, $\mathcal{R}(s) = 1 + \hbar s^{-1}\Omega_{\mathfrak{g}} + O(s^{-2})$.

6. *For any $V_1, V_2 \in \mathrm{Rep}_{\mathrm{fd}}(Y_\hbar(\mathfrak{g}))$ and $\varepsilon \in \{\uparrow, \downarrow\}$, we have*

$$\mathcal{R}^{\varepsilon}_{V_1,V_2}(s) \sim \mathcal{R}_{V_1,V_2}(s)$$

   as $s \to \infty$ in any halfplane of the form $\mathrm{Re}(s/\hbar) \gtrless m$. Here $\mathcal{R}_{V_1,V_2}(s) = \pi_{V_1} \otimes \pi_{V_2}(\mathcal{R}(s))$. That is, $\mathcal{R}_{V_1,V_2}(s)$ is equal to the asymptotic expansion of $\mathcal{R}^{\varepsilon}_{V_1,V_2}(s)$ from (5) of Theorem 7.1.

Proof Parts (1) and (3)–(6) are deduced directly from the definition of $\mathcal{R}(s)$ using the properties of $\mathcal{R}^-(s)$ and $\mathcal{R}^0(s)$ established in Theorems 4.1 and 6.7, respectively.

We note, however, that the cabling identities (2) do not follow directly from Theorems 4.1 and 6.7, as we do not have access to a formal, universal version of the cocycle equation (4.3) of Theorem 4.1. To remedy this, we shall instead make use of the fact that $\mathrm{Rep}_{\mathrm{fd}}(Y_\hbar(\mathfrak{g}))$ is sufficiently large to distinguish elements of $Y_\hbar(\mathfrak{g})$. More precisely, by Proposition A.1 (with $n = 3$), it is enough to prove that the following identities hold on $V_1 \otimes V_2 \otimes V_3$, for every $V_1, V_2, V_3 \in \mathrm{Rep}_{\mathrm{fd}}(Y_\hbar(\mathfrak{g}))$:

$$\begin{aligned}\mathcal{R}_{V_1\otimes V_2,V_3}(s) &= \mathcal{R}_{V_1,V_3}(s)\mathcal{R}_{V_2,V_3}(s)\\ \mathcal{R}_{V_1,V_2\otimes V_3}(s) &= \mathcal{R}_{V_1,V_3}(s)\mathcal{R}_{V_1,V_2}(s)\end{aligned} \tag{7.2}$$

Fix $\varepsilon \in \{\uparrow, \downarrow\}$ and consider the first equality. By (6), the left-hand side (resp. right-hand side) is equal to the uniquely determined asymptotic expansion of $\mathcal{R}^{\varepsilon}_{V_1\otimes V_2,V_3}(s)$ (resp. $\mathcal{R}^{\varepsilon}_{V_1,V_3}(s)\mathcal{R}^{\varepsilon}_{V_2,V_3}(s)$) as $s \to \infty$ in any halfplane of the form $\mathrm{Re}(s/\hbar) \gtrless m$. Moreover, by the first equality in (2) of Theorem 7.1, taken with $s_1 = 0$ and $s_2 = s$, we have

$$\mathcal{R}^{\varepsilon}_{V_1\otimes V_2,V_3}(s) = \mathcal{R}^{\varepsilon}_{V_1,V_3}(s)\mathcal{R}^{\varepsilon}_{V_2,V_3}(s).$$

Thus, we can conclude that the first cabling identity of (7.2) necessarily holds. An identical argument establishes the second identity. □

As an immediate consequence of the above theorem and the uniqueness assertion of Theorem 1.1 (see also Appendix B), we obtain the following corollary.

Corollary $\mathcal{R}(s)$ *is equal to Drinfeld's universal R-matrix.*

In particular, Theorem 7.4 provides an independent, and constructive proof of the existence of Drinfeld's universal R-matrix.

8 Meromorphic Tensor Structures

In this section, we reinterpret our results in the language of meromorphic tensor categories. We refer to [22, 23] for a more abstract and general treatment of meromorphic tensor categories. We caution the reader, however, that the framework developed in [22, 23] does not include examples where the tensor product depends non-trivially on a parameter, as is the case for the deformed Drinfeld tensor product. The setup of [22, 23] is also more general than needed for our purposes, in that it is adapted to *pseudo–tensor categories*, where the tensor product need not be defined for all pairs of representations, or be representable.

8.1 Drinfeld Tensor Product

Proposition

1. *The category* $(\mathrm{Rep}_{\mathrm{fd}}(Y_\hbar(\mathfrak{g})), \underset{\mathrm{D},s}{\otimes})$ *is a meromorphic (in fact, rational) tensor category over* $(\mathbb{C}, +)$.
2. *Each of the resummed abelian R-matrices* $\mathcal{R}^{0,\uparrow/\downarrow}(s)$ *is a meromorphic braiding on* $(\mathrm{Rep}_{\mathrm{fd}}(Y_\hbar(\mathfrak{g})), \underset{\mathrm{D},s}{\otimes})$.
3. $(\mathrm{Rep}_{\mathrm{fd}}(Y_\hbar(\mathfrak{g})), \underset{\mathrm{D},s}{\otimes})$ *does not admit a rational braiding.*

Proof

(1) $\mathrm{Rep}_{\mathrm{fd}}(Y_\hbar(\mathfrak{g}))$ admits an action of the additive group $(\mathbb{C}, +)$ given by $V \mapsto V(s)$. As recalled in Sect. 3.2, for every $V, W \in \mathrm{Rep}_{\mathrm{fd}}(Y_\hbar(\mathfrak{g}))$, there is a rational action of $Y_\hbar(\mathfrak{g})$ on $V \otimes W$ given by the deformed Drinfeld tensor product. The properties (1)–(5) of Theorem 3.2 mean exactly that $(\mathrm{Rep}_{\mathrm{fd}}(Y_\hbar(\mathfrak{g})), \underset{\mathrm{D},s}{\otimes})$ is a rational tensor category over $(\mathbb{C}, +)$.
(2) is the content of Theorem 6.2 (1)–(3).
(3) is a rephrasing of Theorem 6.4.

□

8.2 Standard Tensor Product

Proposition

1. *The category* $(\mathrm{Rep}_{\mathrm{fd}}(Y_\hbar(\mathfrak{g})), \underset{s}{\otimes})$ *is a meromorphic (in fact, polynomial) tensor category over* $(\mathbb{C}, +)$.
2. *Each of the resummed R-matrices* $\mathcal{R}^{\uparrow/\downarrow}(s)$ *is a meromorphic braiding on* $(\mathrm{Rep}_{\mathrm{fd}}(Y_\hbar(\mathfrak{g})), \underset{s}{\otimes})$.
3. $(\mathrm{Rep}_{\mathrm{fd}}(Y_\hbar(\mathfrak{g})), \underset{s}{\otimes})$ *does not admit a rational braiding.*

Proof

(1) This is a consequence of the fact that $V_1 \underset{s}{\otimes} V_2$ arises from the algebra homomorphism $\Delta_s : Y_\hbar(\mathfrak{g}) \to (Y_\hbar(\mathfrak{g}) \otimes Y_\hbar(\mathfrak{g}))[s]$. This tensor product satisfies the properties analogous to (3)–(5) of Theorem 3.2.
(2) is the content of (1)–(3) of Theorem 7.1.
(3) is a rephrasing of Theorem 7.3.

□

8.3 Meromorphic Tensor Structures

Proposition *$\mathcal{R}^-(s)$ is a rational braided tensor structure on the identity functor*

$$\left(\mathrm{Rep}_{\mathrm{fd}}(Y_\hbar(\mathfrak{g})), \underset{\mathrm{D},s}{\otimes}, \mathcal{R}^{0,\uparrow/\downarrow}(s)\right) \longrightarrow \left(\mathrm{Rep}_{\mathrm{fd}}(Y_\hbar(\mathfrak{g})), \underset{s}{\otimes}, \mathcal{R}^{\uparrow/\downarrow}(s)\right)$$

Proof By definition of a tensor structure on a functor, the statement means that, for every $V_1, V_2 \in \mathrm{Rep}_{\mathrm{fd}}(Y_\hbar(\mathfrak{g}))$, there is a rational $\mathrm{End}(V_1 \otimes V_2)$–valued function of s, $\mathcal{R}^-_{V_1,V_2}(s)$, which satisfies (1)–(3) of Theorem 4.1. Namely,

$$\mathcal{R}^-_{V_1,V_2}(s) : V_1 \underset{s}{\otimes} V_2 \to V_1 \underset{\mathrm{D},s}{\otimes} V_2$$

is a $Y_\hbar(\mathfrak{g})$–intertwiner such that $\mathcal{R}^-_{V_1(a),V_2(b)}(s) = \mathcal{R}^-_{V_1,V_2}(s + a - b)$, and the following diagram commutes, for every $V_1, V_2, V_3 \in \mathrm{Rep}_{\mathrm{fd}}(Y_\hbar(\mathfrak{g}))$

$$\begin{array}{ccc}
(V_1 \otimes_{s_1} V_2) \otimes_{s_2} V_3 & = & V_1 \otimes_{s_1+s_2} (V_2 \otimes_{s_2} V_3) \\
\Big\downarrow {\scriptstyle \mathcal{R}^-_{V_1,V_2}(s_1) \otimes 1_{V_3}} & & \Big\downarrow {\scriptstyle 1_{V_1} \otimes \mathcal{R}^-_{V_2,V_3}(s_1)} \\
(V_1 \underset{\mathrm{D},s_1}{\otimes} V_2) \otimes_{s_2} V_3 & & V_1 \otimes_{s_1+s_2} (V_2 \underset{\mathrm{D},s_2}{\otimes} V_3) \\
\Big\downarrow {\scriptstyle \mathcal{R}^-_{V_1 \underset{\mathrm{D},s_1}{\otimes} V_2, V_3}(s_2)} & & \Big\downarrow {\scriptstyle \mathcal{R}^-_{V_1, V_2 \underset{\mathrm{D},s_2}{\otimes} V_3}(s_1+s_2)} \\
(V_1 \underset{\mathrm{D},s_1}{\otimes} V_2) \underset{\mathrm{D},s_2}{\otimes} V_3 & = & V_1 \underset{\mathrm{D},s_1+s_2}{\otimes} (V_2 \underset{\mathrm{D},s_2}{\otimes} V_3)
\end{array}$$

Lastly, it is claimed in (3) that $\mathcal{R}^-(s)$ is compatible with the braidings on the two categories, that is satisfies

$$\begin{array}{ccc}
V_1(s) \otimes V_2 & \xrightarrow{(1\,2)\circ \mathcal{R}^{\uparrow/\downarrow}_{V_1,V_2}(s)} & V_2 \otimes V_1(s) \\
\Big\downarrow {\scriptstyle \mathcal{R}^-_{V_1,V_2}(s)} & & \Big\downarrow {\scriptstyle \mathcal{R}^-_{V_2,V_1}(-s)} \\
V_1(s) \underset{\mathrm{D},0}{\otimes} V_2 & \xrightarrow[(1\,2)\circ \mathcal{R}^{0,\uparrow/\downarrow}_{V_1,V_2}(s)]{} & V_2 \underset{\mathrm{D},0}{\otimes} V_1(s)
\end{array}$$

The commutativity of the diagram follows from the fact that, by definition

$$\mathcal{R}^{\uparrow/\downarrow}_{V_1,V_2}(s) = \mathcal{R}^{+}_{V_1,V_2}(s) \cdot \mathcal{R}^{0,\uparrow/\downarrow}_{V_1,V_2}(s) \cdot \mathcal{R}^{-}_{V_1,V_2}(s)$$

where $\mathcal{R}^{+}_{V_1,V_2}(s) = (1\,2) \circ \mathcal{R}^{-}_{V_2,V_1}(-s)^{-1} \circ (1\,2)$. □

9 Relation to the Quantum Loop Algebra $U_q(L\mathfrak{g})$

In this section, we review the construction of the tensor functor between finite-dimensional representations of $Y_\hbar(\mathfrak{g})$ and the quantum loop algebra $U_q(L\mathfrak{g})$ obtained in [11, 12]. We then disprove a conjecture stated in [12], and relate the meromorphic R-matrices of $Y_\hbar(\mathfrak{g})$ and $U_q(L\mathfrak{g})$.

9.1 *The Functor Γ [11]*

Set $q = \exp(\pi\iota\hbar)$ and assume that $\hbar \in \mathbb{C} \setminus \mathbb{Q}$, so that q is not a root of unity. Let $U_q(L\mathfrak{g})$ be the quantum loop algebra of $\mathfrak{g}$. We refer to [2, Ch. 12] and references therein for the definition and basic properties of $U_q(L\mathfrak{g})$.

A finite-dimensional representation $V \in \mathrm{Rep}_{\mathrm{fd}}(Y_\hbar(\mathfrak{g}))$ is said to be *non-congruent* if, for every $a, b \in \sigma(V)$, $a - b \notin \mathbb{Z}_{\neq 0}$. The full subcategory of such representations is denoted by $\mathrm{Rep}^{\mathrm{NC}}_{\mathrm{fd}}(Y_\hbar(\mathfrak{g})) \subset \mathrm{Rep}_{\mathrm{fd}}(Y_\hbar(\mathfrak{g}))$.

In [11], an exact, essentially surjective and faithful functor

$$\Gamma : \mathrm{Rep}^{\mathrm{NC}}_{\mathrm{fd}}(Y_\hbar(\mathfrak{g})) \to \mathrm{Rep}_{\mathrm{fd}}(U_q(L\mathfrak{g}))$$

is defined. Γ is such that $\Gamma(V) = V$ as vector spaces for any $V \in \mathrm{Rep}^{\mathrm{NC}}_{\mathrm{fd}}(Y_\hbar(\mathfrak{g}))$, and restricts to an isomorphism of categories on an explicit subcategory of $\mathrm{Rep}^{\mathrm{NC}}_{\mathrm{fd}}(Y_\hbar(\mathfrak{g}))$ determined by a choice of a branch of log.

Given $V \in \mathrm{Rep}^{\mathrm{NC}}_{\mathrm{fd}}(Y_\hbar(\mathfrak{g}))$, and $i \in \mathbf{I}$, consider the additive difference equation

$$\phi_i(u+1) = \xi_i(u)\phi_i(u)$$

determined by the action of the commuting current $\xi_i(u)$ of $Y_\hbar(\mathfrak{g})$ on V. The action of the commuting current $\psi_i(z)$ of $U_q(L\mathfrak{g})$ on V is given by the monodromy of this equation, that is by

$$\psi_i(z) = \lim_{n\to\infty} \xi_i(u+n)\cdots\xi_i(u-n)\Big|_{z=e^{2\pi\iota u}}$$

The action of the raising and lowering operators of $U_q(L\mathfrak{g})$ on V requires the non-congruence hypothesis. It is not relevant for our current discussion, and we refer to [11, §5] for details.

9.2 Meromorphic Tensor Structure on Γ [12]

Let $V_1, V_2 \in \mathrm{Rep}_{\mathrm{fd}}(Y_\hbar(\mathfrak{g}))$, and consider the abelian qKZ equation determined by $\mathcal{R}^{0,\uparrow/\downarrow}_{V_1,V_2}(s)$, that is the difference equation

$$\Phi(s+1) = \mathcal{R}^{0,\uparrow/\downarrow}_{V_1,V_2}(s) \cdot \Phi(s) \tag{9.1}$$

Let $\mathcal{J}^{\uparrow/\downarrow}_{V_1,V_2}(s) : \mathbb{C} \to \mathrm{End}_{\mathbb{C}}(V_1 \otimes V_2)$ be the left canonical solution of (9.1), which is uniquely determined by the requirement that it be holomorphic and invertible for $\mathrm{Re}(s) \ll 0$, and possess an asymptotic expansion of the form $(-s)^{\hbar\Omega}\left(1 + O(s^{-1})\right)$ as $s \to \infty$ in any halfplane of the form $\mathrm{Re}(s) < m$.

The deformed Drinfeld tensor product $\underset{\mathrm{D},\zeta}{\otimes}$ on finite-dimensional representations of $U_q(L\mathfrak{g})$ was introduced by D. Hernandez in [15], and further studied in [12].

Theorem ([12, Thm. 7.3]) *$\mathcal{J}^{\uparrow/\downarrow}(s)$ is a meromorphic tensor structure on the functor Γ, with respect to the deformed Drinfeld tensor products*

$$\left(\Gamma, \mathcal{J}^{\uparrow/\downarrow}(s)\right) : \left(\mathrm{Rep}^{\mathrm{NC}}_{\mathrm{fd}}(Y_\hbar(\mathfrak{g})), \underset{\mathrm{D},s}{\otimes}\right) \longrightarrow \left(\mathrm{Rep}_{\mathrm{fd}}(U_q(L\mathfrak{g})), \underset{\mathrm{D},\zeta}{\otimes}\right)$$

where $\zeta = \exp(2\pi\iota s)$.[10]

9.3 Tensor Structure with Respect to the Standard Coproducts

The Drinfeld coproduct of $U_q(L\mathfrak{g})$ is known to be conjugated to the standard coproduct by the lower triangular part $\mathscr{R}^-(\zeta)$ of the universal R-matrix of $U_q(L\mathfrak{g})$ (see, for example, [6]). Thus, for any $V_1, V_2 \in \mathrm{Rep}^{\mathrm{NC}}_{\mathrm{fd}}(Y_\hbar(\mathfrak{g}))$, we have the following isomorphisms of $U_q(L\mathfrak{g})$–modules

[10] In [12], the *right* canonical solution of the equations $\phi(s+1) = \mathcal{R}^{0,\uparrow/\downarrow}_{V_1,V_2}(s) \cdot \phi(s)$ is shown to give rise to a tensor structure on Γ. A similar computation shows that the left solution yields a tensor structure on a variant of Γ, which we denote Γ for simplicity.

$$\begin{array}{ccc}
\Gamma(V_1) \underset{\zeta}{\otimes} \Gamma(V_2) & \xrightarrow{J^{\uparrow/\downarrow}_{V_1,V_2}(s)} & \Gamma\left(V_1 \underset{s}{\otimes} V_2\right) \\
{\scriptstyle \mathscr{R}^{-}_{\Gamma(V_1),\Gamma(V_2)}(\zeta)}\big\downarrow & & \big\downarrow{\scriptstyle \mathcal{R}^{-}_{V_1,V_2}(s)} \\
\Gamma(V_1) \underset{\mathrm{D},\zeta}{\otimes} \Gamma(V_2) & \xrightarrow[\mathcal{J}^{\uparrow/\downarrow}_{V_1,V_2}(s)]{} & \Gamma\left(V_1 \underset{\mathrm{D},s}{\otimes} V_2\right)
\end{array}$$

where $\mathcal{V}_1 \underset{\zeta}{\otimes} \mathcal{V}_2 = \tau_\zeta^* \mathcal{V}_1 \otimes \mathcal{V}_2$ for any $\mathcal{V}_1, \mathcal{V}_2 \in \mathrm{Rep}_{\mathrm{fd}}(U_q(L\mathfrak{g}))$, and $J^{\uparrow/\downarrow}_{V_1,V_2}(s)$ is defined as the composition

$$J^{\uparrow/\downarrow}_{V_1,V_2}(s) = \mathcal{R}^{-}_{V_1,V_2}(s)^{-1} \cdot \mathcal{J}^{\uparrow/\downarrow}_{V_1,V_2}(s) \cdot \mathscr{R}^{-}_{\Gamma(V_1),\Gamma(V_2)}(\zeta) \tag{9.2}$$

Theorem 9.2 and Proposition 8.3 therefore imply the following

Corollary *$J^{\uparrow/\downarrow}_{V_1,V_2}(s)$ is a meromorphic tensor structure on the functor Γ, with respect to the standard tensor products*

$$\left(\Gamma, J^{\uparrow/\downarrow}(s)\right) : \left(\mathrm{Rep}^{\mathrm{NC}}_{\mathrm{fd}}(Y_\hbar(\mathfrak{g})), \underset{s}{\otimes}\right) \longrightarrow \left(\mathrm{Rep}_{\mathrm{fd}}(U_q(L\mathfrak{g})), \underset{\zeta}{\otimes}\right)$$

9.4 Non Regularity of $J^{\uparrow/\downarrow}_{V_1,V_2}(s)$

Since the tensor products $\underset{s}{\otimes}$ and $\underset{\zeta}{\otimes}$ are polynomial, the first two authors conjectured in [12, §2.13] that $J^{\uparrow/\downarrow}_{V_1,V_2}(s)$ is regular and invertible at $s = 0$. If so, $J^{\uparrow/\downarrow}_{V_1,V_2}(0)$ would give rise to a (non-meromorphic) tensor structure on the functor Γ with respect to the standard (unshifted) tensor products on $\mathrm{Rep}^{\mathrm{NC}}_{\mathrm{fd}}(Y_\hbar(\mathfrak{g}))$ and $\mathrm{Rep}_{\mathrm{fd}}(U_q(L\mathfrak{g}))$. The following shows that this is not the case.

Proposition *The meromorphic tensor structure $J^{\uparrow/\downarrow}(s)$ is either singular or not invertible at $s = 0$.*

Proof Since $\Gamma(V(a)) = \Gamma(V)(e^{2\pi\iota a})$ and each of the factors in the definition (9.2) of $J^{\uparrow/\downarrow}$ is compatible with shifts, we have

$$J^{\uparrow/\downarrow}_{V_1(a),V_2(b)}(s) = J^{\uparrow/\downarrow}_{V_1,V_2}(s + a - b)$$

Thus, if $J^{\uparrow/\downarrow}_{V_1,V_2}(s)$ were regular and invertible at $s = 0$ for every V_1, V_2, then for a fixed V_1, V_2 it would be holomorphic and invertible for all s. This cannot be true, as the following argument shows.

Assume that V_1, V_2 are irreducible, with highest weight vectors v_1, v_2 of weights $\lambda_1, \lambda_2 \in \mathfrak{h}^*$, and that $(\lambda_1, \lambda_2) \neq 0$. Let $W \subset V_1 \otimes V_2$ be the subspace spanned by $v_1 \otimes v_2$. The triangularity of $\mathcal{R}^-(s)$ and $\mathscr{R}^-(\zeta)$ implies that

$$J^{\uparrow}_{V_1,V_2}(s)\Big|_W = \mathcal{J}^{\uparrow}_{V_1,V_2}(s)\Big|_W$$

If the restriction of $\mathcal{J}^{\uparrow}_{V_1,V_2}(s)$ to W were holomorphic and invertible for every $s \in \mathbb{C}$, the same would be true for $\mathcal{R}^{0,\uparrow}_{V_1,V_2}(s)$, since by Sect. 9.1

$$\mathcal{J}^{\uparrow}_{V_1,V_2}(s) = \mathcal{R}^{0,\uparrow}_{V_1,V_2}(s+1) \cdot \mathcal{J}^{\uparrow}_{V_1,V_2}(s+1).$$

In turn, $\mathcal{A}_{V_1,V_2}(s)\big|_W$ would also be holomorphic and invertible, since (see Eq. (6.2) above)

$$\mathcal{R}^{0,\uparrow}_{V_1,V_2}(s+\ell\hbar) = \mathcal{A}_{V_1,V_2}(s)\mathcal{R}^{0,\uparrow}_{V_1,V_2}(s).$$

Since $\mathcal{A}_{V_1,V_2}(s)$ is a rational function of s such that $\mathcal{A}_{V_1,V_2}(\infty) = \mathbf{1}$, this implies that $\mathcal{A}_{V_1,V_2}(s)\big|_W \equiv \mathbf{1}$. The expansion $\mathcal{A}(s) = 1 - \frac{\ell\hbar}{s^2}\Omega_{\mathfrak{h}} + O(s^{-3})$, then yields $\Omega_{\mathfrak{h}}\big|_W = 0$, which contradicts the fact that $\Omega_{\mathfrak{h}} v_1 \otimes v_2 = (\lambda_1, \lambda_2) v_1 \otimes v_2$. □

Remark The above result leaves open the question of whether there exists a tensor functor between the (non meromorphic) tensor categories

$$\left(\mathrm{Rep}_{\mathrm{fd}}(Y_\hbar(\mathfrak{g})), \otimes\right) \to \left(\mathrm{Rep}_{\mathrm{fd}}(U_q(L\mathfrak{g})), \otimes\right)$$

9.5 *The Meromorphic Abelian R-Matrix of* $U_q(L\mathfrak{g})$

Assume now that $|q| \neq 1$. We review below the analogue of Theorem 6.2 for $U_q(L\mathfrak{g})$ obtained in [12], based on the results of [3].

Let $\mathcal{V}_1, \mathcal{V}_2$ be two finite-dimensional representations of $U_q(L\mathfrak{g})$. In [12, §8], a rational function $\mathscr{A}_{\mathcal{V}_1,\mathcal{V}_2}(\zeta) : \mathbb{P}^1 \to \mathrm{End}_{\mathbb{C}}(\mathcal{V}_1 \otimes \mathcal{V}_2)$ is defined via a contour integral formula involving the commuting generators of $U_q(L\mathfrak{g})$, which is analogous to the one given in Sect. 6.1. $\mathscr{A}_{\mathcal{V}_1,\mathcal{V}_2}(\zeta)$ is regular at $\zeta = 0, \infty$, and such that

$$\mathscr{A}_{\mathcal{V}_1,\mathcal{V}_2}(0) = \mathbf{1} = \mathscr{A}_{\mathcal{V}_1,\mathcal{V}_2}(\infty)$$

Moreover, $[\mathscr{A}_{\mathcal{V}_1,\mathcal{V}_2}(\zeta), \mathscr{A}_{\mathcal{V}_1,\mathcal{V}_2}(\zeta')] = 0$ for any ζ, ζ'.

Consider the (regular) difference equation with step $q^{2\ell}$ determined by $\mathscr{A}_{\mathcal{V}_1,\mathcal{V}_2}(\zeta)$

$$\overline{\mathscr{R}}(q^{2\ell}\zeta) = \mathscr{A}_{\mathcal{V}_1,\mathcal{V}_2}(\zeta) \cdot \overline{\mathscr{R}}(\zeta) \tag{9.3}$$

This equation admits two meromorphic solutions $\overline{\mathscr{R}}^{\uparrow}(\zeta), \overline{\mathscr{R}}^{\downarrow}(\zeta)$, which are uniquely determined by the requirement that $\overline{\mathscr{R}}^{\uparrow/\downarrow}(\zeta)$ be holomorphic near $z = q^{\pm\infty}$ and such that $\overline{\mathscr{R}}^{\uparrow/\downarrow}(q^{\pm\infty}) = 1$. These are explicitly given by

$$\overline{\mathscr{R}}^{\uparrow}(\zeta) = \overrightarrow{\prod_{n\geq 0}} \mathscr{A}_{\mathcal{V}_1,\mathcal{V}_2}(q^{2\ell n}\zeta)^{-1} \qquad \text{and} \qquad \overline{\mathscr{R}}^{\downarrow}(\zeta) = \overrightarrow{\prod_{n\geq 1}} \mathscr{A}_{\mathcal{V}_1,\mathcal{V}_2}(q^{-2\ell n}\zeta)$$

Now set

$$\mathscr{R}^{0,\uparrow/\downarrow}_{\mathcal{V}_1,\mathcal{V}_2}(\zeta) = \begin{cases} q^{\mp\Omega_{\mathfrak{h}}} \cdot \overline{\mathscr{R}}^{\uparrow/\downarrow}(\zeta) \text{ if } |q| < 1 \\ q^{\pm\Omega_{\mathfrak{h}}} \cdot \overline{\mathscr{R}}^{\uparrow/\downarrow}(\zeta) \text{ if } |q| > 1 \end{cases}$$

Theorem *[12, §8] The category* $\left(\mathrm{Rep}_{\mathrm{fd}}(U_q(L\mathfrak{g})), \underset{\mathrm{D},\zeta}{\otimes}, \mathscr{R}^{0,\uparrow/\downarrow}(\zeta)\right)$ *is a meromorphic braided tensor category.*

Remark Let $\mathscr{R}^0$ be the abelian part of the universal R-matrix of $U_q(L\mathfrak{g})$, and set

$$\mathscr{R}^0(\zeta) = \tau_\zeta \otimes \mathbf{1}(\mathscr{R}^0) \in U_q(L\mathfrak{g})^{\otimes 2}[[\zeta]]$$

It is easy to see that $\mathscr{R}^0_{\mathcal{V}_1,\mathcal{V}_2}(\zeta)$ satisfies the difference equation (9.3) [12, §8]. It follows by uniqueness that $\mathscr{R}^0_{\mathcal{V}_1,\mathcal{V}_2}(\zeta)$ is the Taylor expansion at $\zeta = 0$ of $\mathscr{R}^{0,\uparrow}_{\mathcal{V}_1,\mathcal{V}_2}(\zeta)$ if $|q| < 1$, and of $\mathscr{R}^{0,\downarrow}_{\mathcal{V}_1,\mathcal{V}_2}(\zeta)$ if $|q| > 1$. Similarly, if

$$\mathscr{R}^{0,\vee}(\zeta) = \mathbf{1} \otimes \tau_\zeta(\mathscr{R}^0) \in U_q(L\mathfrak{g})^{\otimes 2}[[\zeta^{-1}]]$$

then $\mathscr{R}^{0,\vee}_{\mathcal{V}_1,\mathcal{V}_2}(\zeta)$ is the Taylor expansion at $\zeta = \infty$ of $\mathscr{R}^{0,\downarrow}_{\mathcal{V}_1,\mathcal{V}_2}(\zeta)$ if $|q| < 1$, and of $\mathscr{R}^{0,\uparrow}_{\mathcal{V}_1,\mathcal{V}_2}(\zeta)$ if $|q| > 1$.

9.6 Abelian qDrinfeld–Kohno Theorem

Let now $V_1, V_2 \in \mathrm{Rep}_{\mathrm{fd}}(Y_\hbar(\mathfrak{g}))$, and consider the abelian qKZ equations (9.1) determined by $\mathcal{R}^{0,\uparrow/\downarrow}_{V_1,V_2}(s)$.

Theorem ([12, Thm. 9.3]) *The monodromy of the abelian qKZ Eqs.* (9.1) *is equal to* $\mathscr{R}^{0,\uparrow/\downarrow}_{\Gamma(V_1),\Gamma(V_2)}(\zeta)$. *In other words, the following holds*

$$\mathscr{R}^{0,\uparrow/\downarrow}_{\Gamma(V_1),\Gamma(V_2)}(\zeta) = \lim_{n\to\infty} \prod_{k=-n}^{n} \mathcal{R}^{0,\uparrow/\downarrow}_{V_1,V_2}(s+k)\Bigg|_{\zeta=\exp(2\pi\iota s)} \tag{9.4}$$

In [12, §9.6], this assertion is strengthened to include the abelian qKZ equations with values in $V_1 \otimes \cdots \otimes V_n$, where $V_i \in \mathrm{Rep}^{\mathrm{NC}}_{\mathrm{fd}}(Y_\hbar(\mathfrak{g}))$ and $n \geq 3$. Thus, Theorem 9.6 is an analogue of the Drinfeld–Kohno theorem for the abelian qKZ equations determined by $\mathcal{R}^{0,\uparrow/\downarrow}(s)$.

As is the case for the Drinfeld–Kohno theorem, Theorems 9.2 and 9.6 can be understood as defining a meromorphic *braided* tensor functor

$$\left(\mathrm{Rep}^{\mathrm{NC}}_{\mathrm{fd}}(Y_\hbar(\mathfrak{g})), \underset{\mathrm{D},s}{\otimes}, \mathcal{R}^{0,\uparrow/\downarrow}(s)\right) \longrightarrow \left(\mathrm{Rep}_{\mathrm{fd}}(U_q(L\mathfrak{g})), \underset{\mathrm{D},\zeta}{\otimes}, \mathscr{R}^{0,\uparrow/\downarrow}(\zeta)\right)$$

Unlike its non-meromorphic counterpart, this notion involves an ordered *pair* $(\mathcal{K}, \check{\mathcal{K}})$ of meromorphic tensor structures on the functor Γ, such that the following holds [12, Rem. 9.3][11]

$$\mathscr{R}^{0,\uparrow/\downarrow}_{\Gamma(V_1),\Gamma(V_2)}(\zeta) = \check{\mathcal{K}}_{V_2,V_1}(-s)_{21}^{-1} \cdot \mathcal{R}^{0,\uparrow/\downarrow}_{V_1,V_2}(s) \cdot \mathcal{K}_{V_1,V_2}(s) \tag{9.5}$$

Comparing (9.5) with (9.4), and using the definition of $\mathcal{J}_{V_1,V_2}$ given in Sect. 9.1, we see that one can take $\mathcal{K}_{V_1,V_2} = \mathcal{J}^{\uparrow/\downarrow}_{V_1,V_2}$, which is a regularisation of the product

$$\mathcal{R}^{0,\uparrow/\downarrow}_{V_1,V_2}(s-1) \cdot \mathcal{R}^{0,\uparrow/\downarrow}_{V_1,V_2}(s-2)\cdots$$

and $\check{\mathcal{K}}_{V_1,V_2} = \mathcal{J}^{\downarrow/\uparrow}_{V_1,V_2}$, which is a regularisation of

$$\mathcal{R}^{0,\downarrow/\uparrow}_{V_2,V_1}(s-1) \cdot \mathcal{R}^{0,\downarrow/\uparrow}_{V_1,V_2}(s-2)\cdots = \mathcal{R}^{0,\uparrow/\downarrow}_{V_2,V_1}(-s+1)_{21}^{-1} \cdot \mathcal{R}^{0,\uparrow/\downarrow}_{V_2,V_1}(-s+2)_{21}^{-1}\cdots$$

where we used the unitarity relation $\mathcal{R}^{0,\downarrow/\uparrow}_{V_1,V_2}(s) = \mathcal{R}^{0,\uparrow/\downarrow}_{V_2,V_1}(s)_{21}^{-1}$.

9.7 Meromorphic Braided Tensor Equivalence for Standard Coproducts

Let $\mathscr{R}$ be the universal R-matrix of $U_q(L\mathfrak{g})$, $\mathscr{R} = \mathscr{R}^+ \cdot \mathscr{R}^0 \cdot \mathscr{R}^-$ its Gauss decomposition, set $\mathscr{R}(\zeta) = \tau_\zeta \otimes \mathbf{1}(\mathscr{R})$, and $\mathscr{R}^\pm(\zeta) = \tau_\zeta \otimes \mathbf{1}(\mathscr{R}^\pm)$. Then, if

[11] For the meromorphic braided tensor structures discussed in Sect. 8.3, $\check{\mathcal{K}} = \mathcal{K}$.

$\mathcal{V}_1, \mathcal{V}_2 \in \mathrm{Rep}_{\mathrm{fd}}(U_q(L\mathfrak{g}))$, $\mathscr{R}^{\pm}_{\mathcal{V}_1,\mathcal{V}_2}(\zeta)$ are rational functions of ζ such that

$$\mathscr{R}^{+}_{\mathcal{V}_1,\mathcal{V}_2}(\zeta) = \mathscr{R}^{-}_{\mathcal{V}_2,\mathcal{V}_1}(\zeta^{-1})_{21}^{-1}$$

Define the meromorphic R-matrix of $U_q(L\mathfrak{g})$ by

$$\mathscr{R}^{\uparrow/\downarrow}_{\mathcal{V}_1,\mathcal{V}_2}(\zeta) = \mathscr{R}^{+}_{\mathcal{V}_1,\mathcal{V}_2}(\zeta) \cdot \mathscr{R}^{0,\uparrow/\downarrow}_{\mathcal{V}_1,\mathcal{V}_2}(\zeta) \cdot \mathscr{R}^{-}_{\mathcal{V}_1,\mathcal{V}_2}(\zeta) \tag{9.6}$$

By Remark 9.5, $\mathscr{R}_{\mathcal{V}_1,\mathcal{V}_2}(\zeta)$ is the Taylor expansion at $\zeta = 0$ of $\mathscr{R}^{\uparrow}_{\mathcal{V}_1,\mathcal{V}_2}(\zeta)$ if $|q| < 1$, and of $\mathscr{R}^{\downarrow}_{\mathcal{V}_1,\mathcal{V}_2}(\zeta)$ if $|q| > 1$.

Proposition *The pair* $\left(J^{\uparrow/\downarrow}_{V_1,V_2}(s), J^{\downarrow/\uparrow}_{V_1,V_2}(s)\right)$ *is a meromorphic braided tensor structure on the functor* Γ *with respect to the standard tensor products and meromorphic R-matrices*

$$\left(\mathrm{Rep}^{\mathrm{NC}}_{\mathrm{fd}}(Y_\hbar(\mathfrak{g})), \underset{s}{\otimes}, \mathcal{R}^{\uparrow/\downarrow}(s)\right) \longrightarrow \left(\mathrm{Rep}_{\mathrm{fd}}(U_q(L\mathfrak{g})), \underset{\zeta}{\otimes}, \mathscr{R}^{\uparrow/\downarrow}(\zeta)\right)$$

Proof By Corollary 9.3, and Sect. 9.6, we only need to check the compatibility of $\left(J^{\uparrow/\downarrow}_{V_1,V_2}(s), J^{\downarrow/\uparrow}_{V_1,V_2}(s)\right)$ with the meromorphic braidings, that is the relation

$$\mathscr{R}^{\uparrow/\downarrow}_{\Gamma(V_1),\Gamma(V_2)}(\zeta) = J^{\downarrow/\uparrow}_{V_2,V_1}(-s)_{21}^{-1} \cdot \mathcal{R}^{\uparrow/\downarrow}_{V_1,V_2}(s) \cdot J^{\uparrow/\downarrow}_{V_1,V_2}(s) \tag{9.7}$$

The Gauss decomposition (9.6) yields

$$\begin{aligned}
&\mathscr{R}^{\uparrow/\downarrow}_{\Gamma}(V_1), \Gamma(V_2)(\zeta)\\
&\qquad = \mathscr{R}^{+}_{\Gamma(V_1),\Gamma(V_2)}(\zeta) \cdot \mathscr{R}^{0,\uparrow/\downarrow}_{\Gamma(V_1),\Gamma(V_2)}(\zeta) \cdot \mathscr{R}^{-}_{\Gamma(V_1),\Gamma(V_2)}(\zeta)\\
&\qquad = \mathscr{R}^{+}_{\Gamma(V_1),\Gamma(V_2)}(\zeta) \cdot \mathcal{J}^{\downarrow/\uparrow}_{V_2,V_1}(-s)_{21}^{-1} \cdot \mathcal{R}^{0,\uparrow/\downarrow}_{V_1,V_2}(s) \cdot \mathcal{J}^{\uparrow/\downarrow}_{V_1,V_2}(s)\\
&\qquad\quad \cdot \mathscr{R}^{-}_{\Gamma(V_1),\Gamma(V_2)}(\zeta)\\
&\qquad = \mathscr{R}^{+}_{\Gamma(V_1),\Gamma(V_2)}(\zeta) \cdot \mathcal{J}^{\downarrow/\uparrow}_{V_2,V_1}(-s)_{21}^{-1} \cdot \mathcal{R}^{+}_{V_1,V_2}(s)^{-1} \cdot \mathcal{R}^{\uparrow/\downarrow}_{V_1,V_2}(s)\\
&\qquad\quad \cdot \mathcal{R}^{-}_{V_1,V_2}(s)^{-1} \cdot \mathcal{J}^{\uparrow/\downarrow}_{V_1,V_2}(s) \cdot \mathscr{R}^{-}_{\Gamma(V_1),\Gamma(V_2)}(\zeta)\\
&\qquad = J^{\downarrow/\uparrow}_{V_2,V_1}(-s)_{21}^{-1} \cdot \mathcal{R}^{\uparrow/\downarrow}_{V_1,V_2}(s) \cdot J^{\uparrow/\downarrow}_{V_1,V_2}(s)
\end{aligned}$$

where the second equality follows from the twist relation (9.5), the third from the definition of $\mathcal{R}^{\uparrow/\downarrow}_{V_1,V_2}(s)$ given in Sect. 7.1, and the fourth from the definition (9.2) of $J^{\uparrow/\downarrow}_{V_1,V_2}(s)$, together with the unitarity relations $\mathscr{R}^+_{\mathcal{V}_1,\mathcal{V}_2}(\zeta) = \mathscr{R}^-_{\mathcal{V}_2,\mathcal{V}_1}(\zeta^{-1})^{-1}_{21}$ and $\mathcal{R}^+_{V_1,V_2}(s) = \mathcal{R}^-_{V_2,V_1}(-s)^{-1}_{21}$. □

Remark Much like (9.4), the twist relation (9.7) can be regarded as a monodromy relation. Indeed, both

$$J^{\uparrow/\downarrow}_{V_1,V_2}(s) = \mathcal{R}^-_{V_1,V_2}(s)^{-1} \cdot \mathcal{J}^{\uparrow/\downarrow}_{V_1,V_2}(s) \cdot \mathscr{R}^-_{\Gamma(V_1),\Gamma(V_2)}(\zeta)$$

and

$$\begin{aligned}
\left(J^{\downarrow/\uparrow}_{V_2,V_1}(-s)^{-1}_{21} \cdot \mathcal{R}^{\uparrow/\downarrow}_{V_1,V_2}(s)\right)^{-1} &= \mathcal{R}^-_{V_1,V_2}(s)^{-1} \cdot \mathcal{R}^{0,\uparrow/\downarrow}_{V_1,V_2}(s)^{-1} \cdot \mathcal{J}^{\downarrow/\uparrow}_{V_2,V_1}(-s)_{21} \\
&\quad \cdot \mathscr{R}^-_{\Gamma(V_2),\Gamma(V_1)}(\zeta^{-1})_{21} \\
&= \mathcal{R}^-_{V_1,V_2}(s)^{-1} \cdot \mathcal{R}^{0,\downarrow/\uparrow}_{V_2,V_1}(-s)_{21} \cdot \mathcal{J}^{\downarrow/\uparrow}_{V_2,V_1}(-s)_{21} \\
&\quad \cdot \mathscr{R}^-_{\Gamma(V_2),\Gamma(V_1)}(\zeta^{-1})_{21}
\end{aligned}$$

are solutions of the difference equation

$$\Phi(s+1) = \left(\mathcal{R}^-(s+1)^{-1}_{V_1,V_2} \cdot \mathcal{R}^{0,\uparrow/\downarrow}_{V_1,V_2}(s) \cdot \mathcal{R}^-(s)_{V_1,V_2}\right) \cdot \Phi(s)$$

though, due to the presence of the factors $\mathscr{R}^-_{\Gamma(V_1),\Gamma(V_2)}(\zeta)$, $\mathscr{R}^-_{\Gamma(V_2),\Gamma(V_1)}(\zeta^{-1})_{21}$, neither is a canonical left or right solution. Unlike (9.4), however, the twist relation (9.7) should not be considered as a difference version of the (non-Abelian) Drinfeld–Kohno Theorem on $n = 2$ points for the following reasons.

1. As just pointed out, the difference equations underlying (9.7) are not the qKZ equations determined by $\mathcal{R}^{\uparrow/\downarrow}_{V_1,V_2}(s)$, but (a gauge transformation of) the abelian qKZ equations determined by $\mathcal{R}^{0,\uparrow/\downarrow}_{V_1,V_2}(s)$.
2. The qKZ equations of Frenkel–Reshetikhin [9] include a *dynamical parameter* $\lambda \in \mathfrak{h}$, and are given by

$$\Phi(s+1) = e^{\lambda} \otimes \mathbf{1} \cdot \mathcal{R}^{\uparrow/\downarrow}_{V_1,V_2}(s) \cdot \Phi(s)$$

Since the form of the asymptotics of solutions of this equation depends on the regularity of e^{λ}, its monodromy is a meromorphic function of λ, which is conjectured to be equivalent to $\mathscr{R}^{\uparrow/\downarrow}_{\Gamma(V_1)\Gamma(V_2)}(\zeta)$, via a λ-dependent gauge transformation.

Acknowledgments We would like to thank Pavel Etingof for several helpful discussions about qKZ equations and R-matrices. We are also grateful to Maria Angelica Cueto for helping us with the combinatorial aspects of the paper. The first author was supported through the Simons foundation collaboration grant 526947. The second author was supported through the NSF grant DMS–1802412. The third author was supported by an NSERC CGS D graduate award and an NSERC PDF postdoctoral fellowship.

A Separation of Points

A.1

Let $\mathcal{V}$ be a collection of finite-dimensional representations of $Y_\hbar(\mathfrak{g})$ such that

1. $\mathbb{C} \in \mathcal{V}$, and $V_1 \otimes V_2 \in \mathcal{V}$ for all $V_1, V_2 \in \mathcal{V}$,
2. There is a $V \in \mathcal{V}$ such that $\mathrm{Ker}(\pi_V|_{\mathfrak{g}}) = \{0\}$.

The goal of this section is to prove that the elements of $\mathcal{V}$ separate points in $Y_\hbar(\mathfrak{g})$. More generally, the following holds.

Proposition *Let $\mathcal{V}_1, \ldots, \mathcal{V}_n$ be collections of finite-dimensional representations of $Y_\hbar(\mathfrak{g})$ satisfying the conditions (1) and (2) above. Then, the ideal $\mathcal{J}_n \subset Y_\hbar(\mathfrak{g})^{\otimes n}$ defined by*

$$\mathcal{J}_n = \bigcap_{V_i \in \mathcal{V}_i} \mathrm{Ker}(\pi_{V_1} \otimes \cdots \otimes \pi_{V_n})$$

is the zero ideal.

Remark The analogous statement for $U(\mathfrak{g}[z])$ fails. Indeed, take $n = 1$ and let $V_{\mathfrak{g}}$ be any faithful, finite-dimensional $\mathfrak{g}$-module. Set $V = \mathrm{ev}^*(V_{\mathfrak{g}})$, where ev is the evaluation morphism

$$\mathrm{ev} : \mathfrak{g}[z] \to \mathfrak{g}, \quad f(z) \to f(0).$$

Then, V is a $U(\mathfrak{g}[z])$-module satisfying (2), and $\mathcal{V} = \{V^{\otimes n}\}_{n \in \mathbb{Z}_{\geq 0}}$ satisfies (1)–(2). However,

$$z\mathfrak{g}[z] \subset \bigcap_{k \geq 0} \mathrm{Ker}(\pi_{V^{\otimes k}})$$

The proof of the proposition is given in Sects. A.2 and A.5. In Sect. A.2, we reduce the task to proving that $\mathcal{J}_1 = \{0\}$. The reduction step is elementary, but is included for the sake of completeness. That the ideal $\mathcal{J}_1 \subset Y_\hbar(\mathfrak{g})$ vanishes is an unpublished result of V. Drinfeld's, whose proof in the $\hbar$–formal setting has been reproduced in [10, Prop. 8.8]. After recalling relevant background material on co-Poisson Hopf algebras and Lie bialgebras in Sects. A.3 and A.4, we explain in Sect. A.5 how to modify the argument given in [10] to deduce that $\mathcal{J}_1 = \{0\}$.

A.2 Reduction Step

Let $\mathcal{H}_1$ and $\mathcal{H}_2$ be associative algebras over $\mathbb{C}$. Assume in addition that, for each $i \in \{1, 2\}$, $\mathcal{H}_i$ is equipped with a family of representations $\mathcal{V}_i$ such that

$$\mathcal{J}_{\mathcal{V}_i} := \bigcap_{V \in \mathcal{V}_i} \mathrm{Ker}(\pi_V) = \{0\}$$

The following general result, coupled with a simple induction on n, implies that Proposition A.1 holds, provided $\mathcal{J}_1 = \{0\}$.

Lemma *We have*

$$\bigcap_{V_i \in \mathcal{V}_i} \mathrm{Ker}(\pi_{V_1} \otimes \pi_{V_2}) = \{0\}$$

Proof The lemma follows easily from the the identity

$$\mathrm{Ker}(\pi_{V_1} \otimes \pi_{V_2}) = (\mathbf{1} \otimes \pi_{V_2})^{-1}\big(\mathrm{Ker}(\pi_{V_1}) \otimes \mathrm{End}(V_2)\big),$$

the assumption $\mathcal{J}_{\mathcal{V}_i} = \{0\}$, and the fact that, for any vector spaces M, N and collection of subspaces $\{M_\lambda\}_{\lambda \in \Lambda} \subset M$, we have the following equality in $M \otimes N$:

$$\bigcap_{\lambda \in \Lambda} (M_\lambda \otimes N) = \left(\bigcap_{\lambda \in \Lambda} M_\lambda\right) \otimes N$$

□

A.3

We now pause to collect pertinent facts from the theories of co-Poisson Hopf algebras and Lie bialgebras. Fix a Lie algebra $\mathfrak{a}$ over $\mathbb{C}$, and recall that a co-Poisson algebra structure on $U(\mathfrak{a})$ is given by the additional data of a Poisson cobracket δ, i.e. a linear map

$$\delta : U(\mathfrak{a}) \to U(\mathfrak{a}) \wedge U(\mathfrak{a}) \subset U(\mathfrak{a})^{\otimes 2}$$

satisfying the co-Jacobi and co-Leibniz identities:

$$(\mathbf{1} + (1\,2\,3) + (1\,3\,2)) \circ \delta \otimes \mathbf{1} \circ \delta = 0$$
$$\mathbf{1} \otimes \Delta \circ \delta = \delta \otimes \mathbf{1} \circ \Delta + (1\,2) \circ \mathbf{1} \otimes \delta \circ \Delta$$

If in addition δ satisfies the compatibility condition

$$\delta(xy) = \delta(x)\Delta(y) + \Delta(x)\delta(y)$$

then $(U(\mathfrak{a}), \delta)$ is said to be a co-Poisson Hopf algebra. In this case, the cobracket δ is uniquely determined by its restriction $\delta|_{\mathfrak{a}}$ to $\mathfrak{a}$, which can be shown to define a Lie bialgebra structure on $\mathfrak{a}$. Conversely, every Lie bialgebra cocommutator on $\mathfrak{a}$ uniquely extends to a co-Poisson Hopf cobracket on $U(\mathfrak{a})$: see [2, Prop. 6.2.3].

A perhaps less well-known correspondence, which we shall exploit, takes place at the level of ideals. Recall that a co-Poisson bialgebra ideal of $(U(\mathfrak{a}), \delta)$ is a two-sided ideal J of $U(\mathfrak{a})$ satisfying the coideal and co-Poisson conditions

$$\Delta(\mathrm{J}) \subset \mathrm{J} \otimes U(\mathfrak{a}) + U(\mathfrak{a}) \otimes \mathrm{J}, \quad \epsilon(\mathrm{J}) = 0$$
$$\delta(\mathrm{J}) \subset \mathrm{J} \otimes U(\mathfrak{a}) + U(\mathfrak{a}) \otimes \mathrm{J}$$

where ϵ is the counit for $U(\mathfrak{a})$. Similarly, a Lie bialgebra ideal of $(\mathfrak{a}, \delta|_{\mathfrak{a}})$ is a Lie algebra ideal $\mathfrak{l}$ of $\mathfrak{a}$ satisfying

$$\delta(\mathfrak{l}) \subset \mathfrak{l} \otimes \mathfrak{a} + \mathfrak{a} \otimes \mathfrak{l}.$$

Lemma *Fix a co-Poisson Hopf algebra structure on $U(\mathfrak{a})$. Then the assignment*

$$\mathrm{J} \subset U(\mathfrak{a}) \mapsto \mathrm{J} \cap \mathfrak{a} \subset \mathfrak{a} \tag{A.1}$$

determines a bijective correspondence between co-Poisson bialgebra ideals J of $U(\mathfrak{a})$ and Lie bialgebra ideals $\mathfrak{l} \subset \mathfrak{a}$, with inverse

$$\mathfrak{l} \subset \mathfrak{a} \mapsto \langle \mathfrak{l} \rangle \subset U(\mathfrak{a}) \tag{A.2}$$

where $\langle \mathfrak{l} \rangle$ is the two-sided ideal generated by $\mathfrak{l}$.

When the underlying cobracket is taken to be trivial (that is, $\delta \equiv 0$), this reduces to the more familiar assertion (see [21, Prop. 4.8], for example) that (A.1) determines a bijective correspondence between bialgebra ideals of $U(\mathfrak{a})$ and Lie algebra ideals in $\mathfrak{a}$, with inverse (A.2). The lemma follows readily from this special case by a straightforward verification that (A.1) and (A.2) preserve any additional co-Poisson structure.

A.4

We now narrow our focus to $\mathfrak{a} = \mathfrak{g}[z]$. Let us begin by recalling how one passes from $Y_\hbar(\mathfrak{g})$ to the standard Lie bialgebra structure on $\mathfrak{g}[z]$, given by (A.4) below. Since $Y_\hbar(\mathfrak{g})$ is a filtered Hopf deformation of the cocommutative Hopf algebra $U(\mathfrak{g}[z])$ (see Sect. 2.6), the linear map

$$\Delta_{\mathrm{Alt}} := \Delta - \Delta^{\mathrm{op}} : Y_\hbar(\mathfrak{g}) \to Y_\hbar(\mathfrak{g}) \otimes Y_\hbar(\mathfrak{g})$$

is a filtered linear map of degree -1. That is, it satisfies

$$\Delta_{\mathrm{Alt}}(\mathcal{F}_k(Y_\hbar(\mathfrak{g}))) \subset \mathcal{F}_{k-1}(Y_\hbar(\mathfrak{g})^{\otimes 2}) \quad \forall\, k \in \mathbb{Z}_{\geq 0}$$

where $\mathcal{F}_{-n}(Y_\hbar(\mathfrak{g})^{\otimes 2}) = \{0\}$ for $n > 0$. We may therefore view it as a filtered map

$$(Y_\hbar(\mathfrak{g}), \mathcal{F}_\bullet(Y_\hbar(\mathfrak{g}))) \to (Y_\hbar(\mathfrak{g})^{\otimes 2}, \mathrm{F}_\bullet(Y_\hbar(\mathfrak{g})^{\otimes 2}))$$

where $\mathrm{F}_\bullet(Y_\hbar(\mathfrak{g})^{\otimes 2})$ is the vector space filtration on $Y_\hbar(\mathfrak{g})^{\otimes 2}$ defined by $\mathrm{F}_k(Y_\hbar(\mathfrak{g})^{\otimes 2}) := \mathcal{F}_{k-1}(Y_\hbar(\mathfrak{g})^{\otimes 2})$ for all $k \in \mathbb{Z}_{\geq 0}$. By definition, the semiclassical limit of Δ is the associated graded map

$$\delta := \mathrm{gr}\left(\frac{\Delta_{\mathrm{Alt}}}{\hbar}\right) : U(\mathfrak{g}[z]) \to \mathrm{gr}_{\mathrm{F}}(Y_\hbar(\mathfrak{g}) \otimes Y_\hbar(\mathfrak{g})) \cong U(\mathfrak{g}[z]) \otimes U(\mathfrak{g}[w]) \qquad \text{(A.3)}$$

It is a Poisson cobracket which endows $U(\mathfrak{g}[z])$ with the structure of a co-Poisson Hopf algebra. Moreover, this co-Poisson structure induces the standard Lie bialgebra structure on $\mathfrak{g}[z]$, with cocommutator

$$\delta|_{\mathfrak{g}[z]} : \mathfrak{g}[z] \to \mathfrak{g}[z] \otimes \mathfrak{g}[w] \cong (\mathfrak{g} \otimes \mathfrak{g})[z, w]$$

given on $f(z) \in \mathfrak{g}[z]$ by the formula

$$\delta(f(z)) = \left[f(z) \otimes 1 + 1 \otimes f(w), \frac{\Omega_{\mathfrak{g}}}{z - w} \right] \in (\mathfrak{g} \otimes \mathfrak{g})[z, w] \qquad \text{(A.4)}$$

We may summarize the above discussion concisely by saying that $Y_\hbar(\mathfrak{g})$ is a *filtered quantization* of the Lie bialgebra $\mathfrak{g}[z]$, equipped with the above cocommutator.

The last ingredient we shall need is the following lemma, which is immediately obtained from Corollary 8.9 of [10] with the help of Lemma A.3.

Lemma *Let* $\epsilon_U : U(\mathfrak{g}[z]) \to \mathbb{C}$ *be the counit.*

1. *If* $\mathfrak{l} \subset \mathfrak{g}[z]$ *is a Lie bialgebra ideal, then* $\mathfrak{l} = \{0\}$ *or* $\mathfrak{l} = \mathfrak{g}[z]$.
2. *If* $\mathrm{J} \subset U(\mathfrak{g}[z])$ *is a co-Poisson bialgebra ideal, then* $\mathrm{J} = \{0\}$ *or* $\mathrm{J} = \mathrm{Ker}(\epsilon_U)$.

A.5 Proof That $\mathcal{J}$ Vanishes

Consider now the filtration $\mathcal{F}_\bullet(\mathcal{J})$ on $\mathcal{J}$ given by $\mathcal{F}_k(\mathcal{J}) = \mathcal{F}_k(Y_\hbar(\mathfrak{g})) \cap \mathcal{J}$ for all $k \in \mathbb{Z}_{\geq 0}$, and the associated graded ideal

$$\mathrm{gr}(\mathcal{J}) = \bigoplus_{k \geq 0} \mathcal{F}_k(\mathcal{J})/\mathcal{F}_{k-1}(\mathcal{J}) \subset \mathrm{gr}(Y_\hbar(\mathfrak{g})) = U(\mathfrak{g}[z])$$

If $x \in \mathcal{J}$ is nonzero and $k \in \mathbb{Z}_{\geq 0}$ is minimal such that $x \in \mathcal{F}_k(\mathcal{J})$, then the image of x in $\mathrm{gr}_k(\mathcal{J})$ is a nonzero element. Hence, our task is reduced to proving that $\mathrm{gr}(\mathcal{J}) = \{0\}$.

Next, note that the condition (1) imposed on $\mathcal{V}$ guarantees that $\mathcal{J}$ is a bialgebra ideal in $Y_\hbar(\mathfrak{g})$. As Δ and ϵ are filtered, it follows that $\mathrm{gr}(\mathcal{J})$ is necessarily a co-Poisson bialgebra ideal of $U(\mathfrak{g}[z])$. Using Lemma A.4, we deduce that $\mathrm{gr}(\mathcal{J}) = \{0\}$ or $\mathrm{gr}(\mathcal{J}) = \mathrm{Ker}(\epsilon_U)$. If $\mathrm{gr}(\mathcal{J}) = \mathrm{Ker}(\epsilon_U)$, then $\mathfrak{g} \subset \mathrm{gr}(\mathcal{J})$, and thus

$$\mathfrak{g} \subset \mathrm{gr}_0(\mathcal{J}) = \mathcal{F}_0(\mathcal{J}) \subset \mathcal{J}$$

This contradicts the assumption (2) on $\mathcal{V}$, which guarantees that $\mathcal{J}$ intersects $\mathfrak{g}$ trivially. Hence, we may conclude that $\mathrm{gr}(\mathcal{J})$, and thus $\mathcal{J}$ itself, vanishes.

B Uniqueness of the Universal R-Matrix

The aim of this section is to give a proof of the uniqueness part of Drinfeld's theorem (Theorem 1.1). Namely, we assume that we are given two formal series

$$\mathcal{R}^{(1)}(s), \mathcal{R}^{(2)}(s) \in 1 + s^{-1}(Y_\hbar(\mathfrak{g}) \otimes Y_\hbar(\mathfrak{g}))[\![s^{-1}]\!]$$

satisfying the hypotheses of Theorem 1.1. That is, for $i = 1, 2$,

$$\Delta \otimes \mathbf{1}(\mathcal{R}^{(i)}(s)) = \mathcal{R}^{(i)}_{13}(s)\mathcal{R}^{(i)}_{23}(s)$$
$$\mathbf{1} \otimes \Delta(\mathcal{R}^{(i)}(s)) = \mathcal{R}^{(i)}_{13}(s)\mathcal{R}^{(i)}_{12}(s)$$

and, for every $a \in Y_\hbar(\mathfrak{g})$:

$$\tau_s \otimes \mathbf{1} \circ \Delta^{\mathrm{op}}(a) = \mathcal{R}^{(i)}(s) \cdot \tau_s \otimes \mathbf{1} \circ \Delta(a) \cdot \mathcal{R}^{(i)}(s)^{-1}.$$

Our main tool will be the following lemma.

B.1

Lemma *The Lie algebra of primitive elements*

$$\mathrm{Prim}^{\Delta}(Y_\hbar(\mathfrak{g})) = \{y \in Y_\hbar(\mathfrak{g}) \,:\, \Delta(y) = y \otimes 1 + 1 \otimes y\}$$

is equal to $\mathfrak{g}$.

Proof We shall again exploit the fact, recalled in Sect. A.4, that the Yangian $Y_\hbar(\mathfrak{g})$ provides a filtered quantization of the Lie bialgebra structure $(\mathfrak{g}[z], \delta)$ on $\mathfrak{g}[z]$ given by (A.4). By (A.3), this means that, for each $x \in \mathfrak{g}$ and $k \geq 0$, we have

$$\hbar \cdot \delta(x.z^k) = \Delta(y) - \Delta^{\mathrm{op}}(y) \mod \mathcal{F}_{k-2}(Y_\hbar(\mathfrak{g}) \otimes Y_\hbar(\mathfrak{g})) \tag{A.1}$$

for any $y \in \mathcal{F}_k(Y_\hbar(\mathfrak{g}))$ whose image $\bar{y} \in \mathrm{gr}_k(Y_\hbar(\mathfrak{g})) \subset U(\mathfrak{g}[z])$ coincides with $x.z^k$.

Now let $y \in Y_\hbar(\mathfrak{g})$ be an arbitrary nonzero primitive element. Assume that $k \geq 0$ is such that $y \in \mathcal{F}_k(Y_\hbar(\mathfrak{g})) \setminus \mathcal{F}_{k-1}(Y_\hbar(\mathfrak{g}))$. As Δ is filtered with $\mathrm{gr}(\Delta)$ recovering the standard coproduct on $U(\mathfrak{g}[z])$ (see Sect. 2.6), we can conclude that the image $\bar{y}$ of y in $\mathrm{gr}_k(Y_\hbar(\mathfrak{g})) \subset U(\mathfrak{g}[z])$ is a nonzero, primitive degree k element, and thus of the form $\bar{y} = x.z^k$ for some $x \in \mathfrak{g}$.

Using that $\Delta(y) = \Delta^{\mathrm{op}}(y)$, we deduce from (A.1) that $\delta(x.z^k) = 0$. On the other hand, by definition of δ, we have

$$0 = \delta(x.z^k) = \frac{z^k - w^k}{z - w}[x \otimes 1, \Omega_{\mathfrak{g}}]$$

Hence, $k = 0$ and $y \in \mathfrak{g} \subset Y_\hbar(\mathfrak{g})$. □

B.2

Now let $n \geq 1$ and $X \in Y_\hbar(\mathfrak{g}) \otimes Y_\hbar(\mathfrak{g})$ be such that $\mathcal{R}^{(1)}(s) - \mathcal{R}^{(2)}(s) = s^{-n}X + O(s^{-n-1})$. We will prove that $X = 0$.

Comparing the coefficients of s^{-n} on both sides of the cabling identities, we obtain the following

$$\Delta \otimes \mathbf{1}(X) = X_{13} + X_{23}$$

$$\mathbf{1} \otimes \Delta(X) = X_{13} + X_{12}$$

In other words, $X \in \mathrm{Prim}^{\Delta}(Y_\hbar(\mathfrak{g}))^{\otimes 2}$, that is, $X \in \mathfrak{g} \otimes \mathfrak{g} \subset Y_\hbar(\mathfrak{g}) \otimes Y_\hbar(\mathfrak{g})$, by the lemma above.

By the intertwining equation for $a \in \mathfrak{g}$, we conclude that $X \in (\mathfrak{g} \otimes \mathfrak{g})^{\mathfrak{g}}$. Hence X is a scalar multiple of the Casimir tensor: $X = c\Omega_{\mathfrak{g}}$, for some $c \in \mathbb{C}$.

Let us now consider the intertwining equation for $a = \mathrm{T}(h)$:

$$\mathrm{ad}(\mathrm{T}(h) \otimes 1 + 1 \otimes \mathrm{T}(h) + sh \otimes 1) \cdot \mathcal{R}^{(i)}(s) = \hbar \mathsf{r}^{21}(h)\mathcal{R}^{(i)}(s) + \hbar \mathcal{R}^{(i)}(s)\mathsf{r}(h)$$

Take the difference of the two equations, for $i = 1, 2$, and compare the coefficient of s^{-n+1}, to get $c[h \otimes 1, \Omega_{\mathfrak{g}}] = 0$, for every $h \in \mathfrak{h}$. But that means $c = 0$ and hence $X = 0$, which is what we wanted to show.

References

1. J. Beck and V. G. Kac, *Finite-dimensional representations of quantum affine algebras at roots of unity*, J. Amer. Math. Soc. **9** (1996), no. 2, 391–423. MR 1317228
2. V. Chari and A. Pressley, *A guide to quantum groups*, Cambridge University Press, Cambridge, 1994. MR 1300632
3. I. Damiani, *La R-matrice pour les algèbres quantiques de type affine non tordu*, Ann. Sci. École Norm. Sup. (4) **31** (1998), no. 4, 493–523. MR 1634087
4. V. G. Drinfeld, *Hopf algebras and the quantum Yang-Baxter equation*, Dokl. Akad. Nauk SSSR **283** (1985), no. 5, 1060–1064. MR 802128
5. V. G. Drinfeld, *A new realization of Yangians and of quantum affine algebras*, Dokl. Akad. Nauk SSSR **296** (1987), no. 1, 13–17. MR 914215
6. B. Enriquez, S. Khoroshkin, and S. Pakuliak, *Weight functions and Drinfeld currents*, Comm. Math. Phys. **276** (2007), no. 3, 691–725. MR 2350435
7. P. I. Etingof and A. A. Moura, *On the quantum Kazhdan-Lusztig functor*, Math. Res. Lett. **9** (2002), no. 4, 449–463. MR 1928865
8. Michael Finkelberg and Alexander Tsymbaliuk, *Shifted quantum affine algebras: integral forms in type A*, Arnold Math. J. **5** (2019), no. 2–3, 197–283. MR 4031357
9. I. B. Frenkel and N. Yu. Reshetikhin, *Quantum affine algebras and holonomic difference equations*, Comm. Math. Phys. **146** (1992), no. 1, 1–60. MR 1163666
10. S. Gautam and V. Toledano Laredo, *Yangians and quantum loop algebras*, Selecta Math. (N.S.) **19** (2013), no. 2, 271–336. MR 3090231
11. S. Gautam and V. Toledano Laredo, *Yangians, quantum loop algebras, and abelian difference equations*, J. Amer. Math. Soc. **29** (2016), no. 3, 775–824. MR 3486172
12. S. Gautam and V. Toledano Laredo, *Meromorphic tensor equivalence for Yangians and quantum loop algebras*, Publ. Math. Inst. Hautes Études Sci. **125** (2017), 267–337. MR 3668651
13. N. Guay, H. Nakajima, and C. Wendlandt, *Coproduct for Yangians of affine Kac-Moody algebras*, Adv. Math. **338** (2018), 865–911. MR 3861718
14. N. Guay, V. Regelskis, and C. Wendlandt, *Equivalences between three presentations of orthogonal and symplectic Yangians*, Lett. Math. Phys. **109** (2019), no. 2, 327–379. MR 3917347
15. D. Hernandez, *Drinfeld coproduct, quantum fusion tensor category and applications*, Proc. Lond. Math. Soc. (3) **95** (2007), no. 3, 567–608. MR 2368277
16. D. Kazhdan and Y.S. Soibelman, *Representations of quantum affine algebras*, Selecta Math. (N.S.) **1** (1995), no. 3, 537–595. MR 1366624
17. S. M. Khoroshkin and V. N. Tolstoy, *Yangian double*, Lett. Math. Phys. **36** (1996), no. 4, 373–402. MR 1384643
18. S. Z. Levendorskiĭ, *On generators and defining relations of Yangians*, J. Geom. Phys. **12** (1993), no. 1, 1–11. MR 1226802

19. S. Z. Levendorskiĭ, *On PBW bases for Yangians*, Lett. Math. Phys. **27** (1993), no. 1, 37–42. MR 1212024
20. D. Maulik and A. Okounkov, *Quantum groups and quantum cohomology*, Astérisque (2019), no. 408, ix+209. MR 3951025
21. D. Passman, *Elementary bialgebra properties of group rings and enveloping rings: an introduction to Hopf algebras*, Comm. Algebra **42** (2014), no. 5, 2222–2253. MR 3169701
22. Y. Soibelman, *Meromorphic tensor categories, quantum affine and chiral algebras. I*, Recent developments in quantum affine algebras and related topics (Raleigh, NC, 1998), Contemp. Math., vol. 248, Amer. Math. Soc., Providence, RI, 1999, pp. 437–451. MR 1745272
23. Y. S. Soibelman, *The meromorphic braided category arising in quantum affine algebras*, Internat. Math. Res. Notices (1999), no. 19, 1067–1079. MR 1725484
24. C. Wendlandt, *The formal shift operator on the Yangian double*, Int. Math. Res. Not. IMRN (2021), 59. https://doi.org/10.1093/imrn/rnab026

On Spectral Cover Equations in Simpson Integrable Systems

Anton A. Gerasimov and Samson L. Shatashvili

Abstract Following Simpson we consider the integrable system structure on the moduli spaces of Higgs bundles on a compact Kähler manifold X. We propose a description of the corresponding spectral cover of X as the fiberwise projective dual to a hypersurface in the projectivization $\mathbb{P}(\mathcal{T}_X \oplus \mathcal{O}_X)$ of the tangent bundle $\mathcal{T}_X$ to X. The defining equation of the hypersurface dual to the Simpson spectral cover is explicitly constructed in terms of the Higgs fields.

1 Introduction

The notion of a spectral curve, or, more precisely, a family of spectral curves plays an important role in the theory of integrable systems. Thus the phase space of an integrable system is interpreted as the total space of a family of Lagrangian abelian varieties identified with the Jacobian varieties of the corresponding spectral curves. This provides effective means to explicitly solve the integrable system via geometry of divisors on spectral curves. The standard construction of the families of spectral

A. A. Gerasimov (✉)
Laboratory for Quantum Field Theory and Information, Institute for Information Transmission Problems, RAS, Moscow, Russia
e-mail: anton.a.gerasimov@gmail.com

S. L. Shatashvili
School of Mathematics, Trinity College Dublin, Dublin, Ireland

Hamilton Mathematics Institute, TCD, Dublin 2, Ireland

Chair Israel Gelfand, Institut des Hautes Etudes Scientifiques, Bures-sur-Yvette, France

Simons Center for Geometry and Physics, Stony Brook, NY, USA

Laboratory for Quantum Field Theory and Information, Institute for Information Transmission Problems, RAS, Moscow, Russia
e-mail: Samson@maths.tcd.ie

A. Alekseev et al. (eds.), *Representation Theory, Mathematical Physics, and Integrable Systems*, Progress in Mathematics 340,
https://doi.org/10.1007/978-3-030-78148-4_8

curves is via characteristic polynomials of the Lax operators depending on the spectral parameter.

In the original constructions the spectral parameter is a coordinate function on an underlying elliptic curve or its degeneration, and the spectral curve is a finite cover of the underlying curve realized as a subvariety of its cotangent bundle. More general construction was proposed by Hitchin [H]. The phase space of the integrable system is identified with the moduli space of Higgs bundles, a certain degeneration of the moduli space of complex G-bundles over an arbitrary algebraic curve Σ. The corresponding spectral curve is given by a finite cover of Σ realized as a hypersurface in the cotangent bundle $T^*\Sigma$, defined by the characteristic polynomial of the Higgs field on Σ.

One of the important applications of the spectral curve construction is to the method of the quantum separation of variables [Sk1], [Sk2]. In this approach the equation defining spectral curve as a hypersurface in $T^*\Sigma$ transforms into the defining equation for corresponding quantum integrable system eigenfunctions in the separated variables. The Hitchin integrable systems, both classical and quantum, are much studied and have found many applications in representation theory, quantum field theory and string theory.

A higher-dimensional generalization of the integrable systems was proposed by Simpson [S1]. The phase space of the corresponding integrable system is the moduli space of Higgs bundles on a Kähelr manifold X and the role of the spectral curve is played by a spectral cover of X realized as a subvariety of the total space of some vector bundle over X. A particularly interesting choice of the vector bundle is given by the cotangent bundle T^*X to X. By analogy with the Hitchin description of spectral curves, it is possible to provide a set of polynomial equations describing the spectral cover of X as a subvariety of T^*X. However, in the higher-dimensional case the corresponding set of equations turns out to be highly overdetermined. This might potentially be a problem in the constructive approach to finding explicit solutions of the corresponding integrable system.

In this note we point out a curious property of the Simpson spectral cover. It is projectively dual to a hypersurface defined by a single explicit equation. More precisely—consider the projective compactification $\mathbb{P}(\mathcal{T}_X^* \oplus \mathcal{O}_X)$ of the holomorphic cotangent bundle $\mathcal{T}_X^*$. We propose a description of the Simpson spectral cover in $\mathbb{P}(\mathcal{T}_X^* \oplus \mathcal{O}_X)$ as a fiberwise projective dual to a hypersurface in the projective compactification $\mathbb{P}(\mathcal{T}_X \oplus \mathcal{O})$ of the tangent bundle $\mathcal{T}_X$. This hypersurface is defined by a single fiberwise polynomial function on $\mathbb{P}(\mathcal{T}_X \oplus \mathcal{O}_X)$ (see Eq. (4.8)). We prove this statement over the open subset $U \subset X$ where the Higgs field is diagonalizable (see Theorem 4.1).

In the case of X being an algebraic curve this equation essentially coincides with the Hitchin equation of the spectral curve. The formulation in terms of the tangent bundle instead of the cotangent bundle bears a resemblance with the Legendre transform between the Hamiltonian and the Lagrangian formulations of classical mechanics. One might hope that the proposed description of the Simpson spectral cover via projective duality will be useful in various applications of the theory of integrable systems associated with higher-dimensional algebraic varieties. As this is one of our primary motivations to study Simpson's spectral cover we include rather detailed discussion of the corresponding integrable systems.

The theory of the Simpson integrable systems is an exciting area of research expanding the traditional horizons of the theory of integrable systems. For instance, the Simpson integrable systems, similar to the Hitchin systems, allow a non-commutative deformation akin to the Knizhnik-Zamolodchikov connection leading to quantum versions of the spectral covers. The corresponding higher-dimensional analog of the quantum spectral curve shall be an important part of the higher-dimensional holomorphic/chiral quantum field theory, which is an interesting direction to pursue. Note in this regard that the phase spaces of the Simpson integrable systems appear naturally in the description of D-brane string backgrounds and thus via dualities appear in various related problems. One might also expect applications of the Simpson integrable systems to some kind of higher-dimensional version of the geometric Langlands correspondence.

2 Hitchin Integrable Systems

We start by briefly recalling the construction of the integrable systems associated with the holomorphic vector bundles over algebraic curves due to Hitchin [H].

Let E be a rank n smooth complex vector bundles over a Riemann surface Σ. Let $\Omega^{p,q}_{\Sigma}$ be complex vector bundles of differential forms of type (p,q) and $\Omega^{p,q}_{\Sigma}(E) = \Omega^{p,q}_{\Sigma} \otimes E$. To define a holomorphic vector bundle $\mathcal{E}$ associated with E it is enough to pick a differential operator $\overline{\partial}_{\mathcal{E}} : \Omega^{0,0}_{\Sigma}(E) \to \Omega^{0,1}_{\Sigma}(E)$ such that $\overline{\partial}_{\mathcal{E}}(fs) = (\overline{\partial} f)s + f\overline{\partial}_{\mathcal{E}}s$, for $s \in \Omega^{0,0}_{\Sigma}(E)$ and $f \in C^{\infty}(\Sigma)$. Then the holomorphic sections of $\mathcal{E}$ satisfy the equation $\overline{\partial}_{\mathcal{E}}s = 0$. The space of complex structures on the C^{∞}-bundle E is an infinite-dimensional affine space modeled on $\Omega^{0,1}_{\Sigma}(\mathrm{End}(\mathcal{E}))$. The group $\mathcal{G}$ of automorphisms of the vector bundle E acts on the operator $\overline{\partial}_{\mathcal{E}}$ via conjugation $\overline{\partial}_{\mathcal{E}} \to g^{-1}\overline{\partial}_{\mathcal{E}}g$. The moduli space $\mathcal{N}$ of stable complex structures on a complex C^{∞} vector bundle E is obtained by taking the quotient of the space $\mathcal{A}^s$ of stable complex structures on E by the action of the automorphism group $\mathcal{G}$. Recall that the holomorphic vector bundle $\mathcal{E}$ is stable iff for every proper subbundle $\mathcal{F} \subset \mathcal{E}$ one has $\frac{\deg(\mathcal{E})}{\mathrm{rank}(\mathcal{E})} > \frac{\deg(\mathcal{F})}{\mathrm{rank}(\mathcal{F})}$. In general the quotient space $\mathcal{N} = \mathcal{A}^s/\mathcal{G}$ is a smooth non-compact space, but if the rank n and the first Chern class $c_1(\mathcal{E})$ are mutually prime then the space $\mathcal{N}$ is compact.

The cotangent bundle $T^*\mathcal{N}$ to the moduli space $\mathcal{N}$ can be succinctly described via Hamiltonian reduction of the space of Higgs bundles with respect to the action of the group of automorphisms $\mathcal{G}$. For a rank n complex vector bundle E let us consider the cotangent space $T^*\mathcal{A}^s$ naturally identified with the space of pairs $(\overline{\partial}_{\mathcal{E}}, \Phi)$ of complex structures on E defined by $\overline{\partial}_{\mathcal{E}}$ and $\Phi \in \Omega^{1,0}_{\Sigma} \otimes \mathrm{End}(\mathcal{E})$. This space has a canonical holomorphic symplectic structure of the cotangent bundle. The Hamiltonian reduction over the zero set of moment map with respect to the action of the automorphism group $\mathcal{G}$ provides a realization of the cotangent bundle $T^*\mathcal{N}$. Points of $T^*\mathcal{N}$ parametrize Higgs bundles, i.e. pairs $(\overline{\partial}_{\mathcal{E}}, \Phi)$ satisfying the equation

$$\overline{\partial}_{\mathcal{E}}\Phi = 0. \tag{2.1}$$

There exists a more relaxed stability condition for Higgs bundles (wherein Φ-stable subbundles are considered) leading to the Hitchin moduli space $\mathcal{M}$ of stable Higgs bundles such that $T^*\mathcal{N} \subset \mathcal{M}$ [H].

The total space of the cotangent bundle $T^*\mathcal{N}$ is naturally a holomorphic symplectic manifold with the holomorphic symplectic structure descending from the $\mathcal{G}$-invariant holomorphic symplectic structure

$$\Omega = \int_{\Sigma} \operatorname{Tr} \delta\bar{A} \wedge \delta\Phi, \tag{2.2}$$

on $T^*\mathcal{A}^s$. Here $\bar{A} \in \Omega^{0,1}_{\Sigma}(\operatorname{End}(\mathcal{E}))$ are affine coordinates on the space of complex structures on $\mathcal{E}$, i.e. for two complex structures we have $\overline{\partial}^{(2)}_{\mathcal{E}} = \overline{\partial}^{(1)}_{\mathcal{E}} + \bar{A}^{(2)} - \bar{A}^{(1)}$ and thus the two-form (2.2) is well-defined. This holomorphic symplectic structure admits an extension to the moduli space $\mathcal{M}$ of stable Higgs bundles.

According to Hitchin [H], the symplectic space $\mathcal{M}$ is an algebraically completely integrable system. The corresponding set of mutually commuting algebraic functions is provided by the Hitchin map

$$\mathcal{M} \longrightarrow \oplus_{i=1}^{n} H^0(\Sigma, \operatorname{Sym}^i(\Omega^{1,0}_{\Sigma})), \tag{2.3}$$

sending the pair $(\overline{\partial}_{\mathcal{E}}, \Phi)$ to the vector $(\operatorname{Tr}\Phi, \operatorname{Tr}\Phi^2, \cdots, \operatorname{Tr}\Phi^n)$. To construct the explicit basis of the algebraic Hamiltonian functions we fix a basis $\{\mu_{i,j}\}$, $i = 1, \ldots, n$, $j = 1, \ldots, n_i$, $n_i = \dim H^1(\Sigma, \operatorname{Sym}^{i-1}\mathcal{T}_{\Sigma})$ in the space $\oplus_{i=1}^{n} H^1(\Sigma, \operatorname{Sym}^{i-1}\mathcal{T}_{\Sigma})$ where $\mathcal{T}_{\Sigma}$ is the holomorphic tangent bundle to Σ. Then the ring of integrable Hamiltonians is generated by the elements

$$H_{i,j} = \int_{\Sigma} \mu_{i,j} \operatorname{Tr}\Phi^i, \qquad i = 1, \ldots, n, \, , \qquad j = 1, \ldots, n_i. \tag{2.4}$$

Mutual commutativity of the Hamiltonians (2.4) follows from their invariance under the action of the automorphism group $\mathcal{G}$ and their obvious mutual commutativity considered as functions on the space $T^*\mathcal{A}^s$.

The generic fibers of the Hitchin map (2.3) are compact Lagrangian abelian varieties identified with the Jacobi varieties of the algebraic curves parametrized by the target space of the map (2.4). This family of curves, known as a family of spectral curves of the integrable system, allows the following construction. One constructs a coherent sheaf $\hat{\mathcal{E}}$ on the total space of $T^*\Sigma$ of the cotangent bundle to Σ so that the spectral curve is given by the support of $\hat{\mathcal{E}}$. To define a coherent sheaf on $T^*\Sigma$ is the same as to define a sheaf of $\pi_*\mathcal{O}_{T^*\Sigma}$-modules on Σ. Here the direct image $\pi_*\mathcal{O}_{T^*\Sigma}$ is the sheaf of polynomial algebras in one variable over $\mathcal{O}_{\Sigma}$ with the stalk over a point $z \in \Sigma$ given by the algebra of polynomial functions over the fiber $T_z^*\Sigma$. Using the Higgs field Φ we can supply $\mathcal{E}$ with the structure of a $\pi_*\mathcal{O}_{T^*\Sigma}$-module. For this it is enough to allow the generator of the polynomial algebra to act as multiplication by the matrix $\Phi_z(z)$ where we locally have $\Phi = \Phi_z(z)dz$. Now the spectral curve associated with $(\overline{\partial}_{\mathcal{E}}, \Phi)$ is given by the support of the coherent sheaf $\hat{\mathcal{E}}$ on $T^*\Sigma$. Let $U \subset \Sigma$ be an open subset over which the Higgs field can be diagonalized with non-coincident eigenvalues. Then the construction above realizes an open part of the spectral curve as a non-ramified n-sheet cover of U. The spectral curve then provides the n-sheet cover of Σ but, in general, ramified over some points. It is clear that the spectral curve, constructed this way, can be conveniently defined as the space of solutions of the characteristic equation of the Higgs field

$$\det(w - \Phi(z)) = 0, \tag{2.5}$$

where w is a linear coordinate on the fibers of the line bundle $T^*\Sigma$. The coefficients of the spectral cover equation

$$\det(w - \Phi(z)) = \textstyle\sum_{m=0}^{n}(-1)^m w^{n-m} \mathrm{Tr}_{\wedge^m}\Phi \tag{2.6}$$

provide another set of generators of the ring of integrable Hamiltonians

$$\tilde{H}_{m,j} = \textstyle\int_{\Sigma} \mu_{m,j} \mathrm{Tr}_{\wedge^m}\Phi. \tag{2.7}$$

Here we denote by $\mathrm{Tr}_{\wedge^m}\Phi$ the trace of Φ in the m-th fundamental representation of GL_n realized in the space of m-th exterior powers of the standard representation $\mathbb{C}^n$.

At the end of this short review of the Hitchin construction let us note that the explicit description (2.5) of the spectral curve via the characteristic equation is widely used in the theory of integrable systems and its applications.

3 Simpson Integrable Systems

A generalization of the Hitchin integrable systems for a higher-dimensional base was introduced by Simpson [S1] using the notion of Higgs bundles in arbitrary dimensions. In the following we consider a particular case of Simpson's construction based on Higgs pairs associated with the cotangent bundle of the underlying manifold.

Let X be a d-dimensional Kähler manifold X with the Kähelr form ω. Let E be a rank n complex vector bundle over X. The integrable complex structure on E can be defined via a differential operator $\overline{\partial}_{\mathcal{E}} : \Omega_X^{0,0}(E) \to \Omega_X^{0,1}(E)$ such that $\overline{\partial}_{\mathcal{E}}(fs) = (\overline{\partial} f)s + f\overline{\partial}_{\mathcal{E}}s$, for $s \in \Omega_X^{0,0}(\mathcal{E})$ and $f \in C^{\infty}(X)$ and such that the integrability condition $\overline{\partial}_{\mathcal{E}}^2 = 0$ holds. Let $\mathcal{E}$ be the corresponding holomorphic vector bundle over X. The moduli space of stable integrable complex structures modulo the action of the automorphism group $\mathcal{G}$ of the vector bundle is a non-singular and in general non-compact Kähelr manifold.

The generalized Higgs pair $(\overline{\partial}_{\mathcal{E}}, \Phi)$ is a pair of: 1. an operator $\overline{\partial}_{\mathcal{E}}$ defining an integrable complex structure on E, and 2. a holomorphic section Φ of $\Omega_X^{1,0} \otimes \mathrm{End}(\mathcal{E})$ which squares to zero. Thus we have the following defining equations for the Higgs pair $(\overline{\partial}_{\bar{\mathcal{E}}}, \Phi)$

$$\overline{\partial}_{\mathcal{E}}^2 = 0, \qquad \overline{\partial}_{\mathcal{E}} \Phi = 0, \qquad \Phi \wedge \Phi = 0. \tag{3.1}$$

Note that the last equation can be written in components as the commutativity conditions

$$[\Phi_i, \Phi_j] = 0, \qquad \Phi = \sum_{i=1}^{d} \Phi_i(z) dz^i. \tag{3.2}$$

In the following we consider the case of $\mathcal{E}$ having trivial rational Chern classes. In this case there is a one-to-one correspondence between isomorphism classes of the Higgs bundles defined above and the rank n complex local systems on X [S1].

Let $\mathcal{M}$ be the moduli space of stable Higgs bundles considered modulo action of the automorphism group $\mathcal{G}$. The stability condition on Higgs bundles is given by the standard stability condition for vector bundles with the exception that only Φ-stable subbundles are considered [S1]. The space $\mathcal{M}$ is a non-singular holomorphic symplectic manifold with the holomorphic symplectic structure obtained by the reduction from the holomorphic symplectic structure

$$\Omega = \int_X \omega^{d-1} \, \mathrm{Tr} \delta \bar{A} \wedge \delta \Phi, \tag{3.3}$$

on the space of pairs $(\overline{\partial}_{\mathcal{E}}, \Phi)$ consisting of not necessary integrable complex structures $\overline{\partial}_{\mathcal{E}}$ and $\Phi \in \Omega_X^{1,0} \otimes \mathrm{End}(\mathcal{E})$. Here $\bar{A}$ is the affine coordinate on the space of not necessary integrable complex structures. The reason for the existence of the holomorphic symplectic structure (and moreover the hyperkähelr structure) is the non-linear version of the Hodge decomposition on the first non-abelian complex cohomology of X (see [S2] and the references therein). The precise construction is analogous to the construction of the holomorphic symplectic structure (and more generally hyperkähler structure) on the moduli space of rank n complex local systems on X (see [F] and references in [S2]); moreover, the holomorphic symplectic structure on the moduli space of Higgs bundles can be obtained from the holomorphic structure on the moduli space of complex local systems by degeneration using the identification of two moduli spaces [S1, S2]. For a different approach see also [DM].

The moduli space $\mathcal{M}$ is an algebraically completely integrable system. The analog of the Hitchin map (2.3)

$$\mathcal{M} \longrightarrow \oplus_{i=1}^{n} H^0(X, \mathrm{Sym}^i(\Omega_X^{1,0})) \tag{3.4}$$

sends the Higgs field Φ to the array $(\mathrm{Tr}\Phi, \mathrm{Tr}\Phi^2, \cdots, \mathrm{Tr}\Phi^n)$ of symmetric GL_n-invariant polynomial functions. Precisely, let us contract the matrix valued one form Φ with a vector field v on X and consider the traces $\mathrm{Tr}(\iota_v\Phi)^i$ of its power. This gives a degree i symmetric function of v and thus an element of $H^0(X, \mathrm{Sym}^i(\Omega_X^{1,0}))$. The explicit set of Hamiltonians can be described using a basis $\{\mu_{i,j}\}$, $i = 1, \dots, n$, $j = 1, \dots, n_i$ in each vector space $H^d(X, \mathrm{Sym}^{i-1}\mathcal{T}_X \otimes \Omega_X^{d,0})$ where $\mathcal{T}_X$ is the tangent bundle to X (this is similar to Hitchin case, see Sect. 2). The generators of the ring of integrable Hamiltonians are then given by

$$H_{i,j} = \int_X \mu_{i,j}\mathrm{Tr}\Phi^i. \tag{3.5}$$

The fibers of the projection (3.4) are Lagrangian abelian varieties that are Lagrangian with respect to the symplectic structure on $\mathcal{M}$. Simpson in [S1] defined an analog of the Hitchin spectral curve given by a covering space $\tilde{X} \rightarrow X$. In good cases these spectral covers are non-singular, and the abelian varieties in the fibration (3.4) are identified with the moduli spaces of line bundles on $\tilde{X}$.

These spectral covers can be represented as subvarieties of the cotangent space T^*X. Precisely they can be realized via supports of coherent sheaves $\hat{\mathcal{E}}$ on the total space of T^*X associated with the holomorphic vector bundle $\mathcal{E}$ corresponding to the holomorphic structure on E defined by $\overline{\partial}_{\mathcal{E}}$. This construction straightforwardly generalizes the corresponding construction of the spectral curve for the Hitchin integrable system described in the previous Section. To provide a description of $\hat{\mathcal{E}}$ we shall construct a sheaf of $\pi_*\mathcal{O}_{T^*X}$-modules over X. Note that $\pi_*\mathcal{O}_{T^*X}$ is a sheaf of polynomial algebras of d-variables over $\mathcal{O}_X$. Using the Higgs field Φ we supply $\mathcal{E}$ with the structure of a $\pi_*\mathcal{O}_{T^*X}$-module by allowing the generators of the polynomial algebra act as multiplications by the matrices $\Phi_k(z)$ where we locally have $\Phi = \sum_{k=1}^d \Phi_k(z)dz^k$. Now the spectral cover $\tilde{X}$ associated with $(\overline{\partial}_{\mathcal{E}}, \Phi)$ is given by the support of the coherent sheaf $\hat{\mathcal{E}}$ on T^*X. An open part of the spectral cover is realized as a non-ramified n-sheet cover of an open subset of X over which at least one component Φ_k of the Higgs field can be diagonalized with non-coincident eigenvalues. Let us give an explicit local construction of the spectral covering locally over X. Let $(z^1, \dots, z^d)$ be local coordinates over a small open subset U of X and $(w_1, \dots, w_d)$ be local linear coordinates in the fibers of the cotangent bundle T^*U. Suppose that over U the Higgs field can be chosen in the diagonal form

$$\Phi = \sum_{k=1}^d \sum_{a=1}^n E_{aa}\Phi_k^a\, dz^k, \tag{3.6}$$

where E_{ab} are elementary $(n \times n)$-matrices $(E_{ab})_{\alpha\beta} = \delta_{a\alpha}\delta_{b\beta}$. Then the support of $\hat{\mathcal{E}}$ is locally a collection of n copies of U_0 defined by the union of n embeddings of U into T^*U

$$(z^1, \dots, z^d) \longrightarrow (z^1, \dots, z^d, \Phi_1^a(z), \dots \Phi_d^a(z)), \qquad a = 1, \dots, n. \tag{3.7}$$

This provides the structure of a degree n non-ramified covering of U.

The above construction of the spectral cover can be easily reformulated in terms of a set of defining characteristic equations (see, e.g., [KOP]) generalizing the spectral curve description via characteristic equation (2.5) for the Hitchin integrable systems. Let v be a section of the holomorphic tangent bundle $\mathcal{T}_X$. Then by construction the linear operator $(\iota_v\Phi - \iota_v w)$ has non-trivial kernel when $(z, w) \in \tilde{X}$ and thus $\det(\iota_v\Phi - \iota_v w) = 0$ on $\tilde{X}$. This holds for any vector field v. Thus points of the spectral cover shall satisfy the following equation

$$\det(\Phi - w) = 0, \tag{3.8}$$

as an element of the n-th symmetric power of the cotangent bundle. This is a direct analog of (2.5).

The set of equations is overdetermined as the spectral cover is a codimension d subspace in T^*X while the condition (3.8) provides $\dim(S^n\mathbb{C}^d) = \binom{n+d-1}{n}$-equations on T^*X. This renders the description (3.8) of the Simpson spectral cover less explicit than the analogous equation (2.5) of the Hitchin spectral cover. In the following we demonstrate that there is a convenient description of the Simpson spectral cover by a single equation but in another space. This construction uses fiberwise projective duality.

4 Spectral Cover via Projective Duality

Let us start by recalling the standard projective duality relation between linear subspaces in complex projective spaces. Let $\mathbb{P}(V)$, $\dim V = d + 1$ be a projective space and $\mathbb{P}(V^*)$ be its dual. We fix dual homogeneous coordinates $(\xi^0, \ldots, \xi^d)$ and $(\eta_0, \ldots, \eta_d)$ in V and V^* correspondingly. A linear k-dimensional subspace $W_k \subset \mathbb{P}(V)$ can be represented by a $(k + 1)$-dimensional subspace of V and thus be described by a set of $d - k$ linear homogeneous equations

$$\textstyle\sum_{j=0}^{d} A^i_j \xi^j = 0, \qquad i = 1, \ldots, d - k\,. \tag{4.1}$$

For instance, the hyperplane in $\mathbb{P}(V)$ is a space of solutions of a linear homogeneous equation

$$\textstyle\sum_{j=0}^{d} \alpha_j \xi^j = 0. \tag{4.2}$$

Linear subspaces generically intersect according to their dimensions, i.e. a linear k-dimensional subspace W_k in $\mathbb{P}(V)$ intersects a linear m-dimensional subspace W_m over a linear subspace of dimension $k + m - d$ for $k + m \geq d$.

The projective duality interchanges dimension k linear space in $\mathbb{P}(V)$, $V = \mathbb{C}^{d+1}$, and dimension $(d - k - 1)$ linear subspaces in $\mathbb{P}(V^*)$. Let us parametrize points of the k-dimensional space $W_k \subset \mathbb{P}(V)$ by a $(k+1)$-tuple $(s^0, \dots, s^k) \in \mathbb{C}^{k+1}$ of variables as follows

$$\xi^i = \sum_{j=0}^{k} A^i_j s^j, \qquad i = 0, \dots, d. \tag{4.3}$$

The corresponding dual $(d-k-1)$-linear subspace in $\mathbb{P}(V^*)$ is defined by the set of linear homogeneous equations

$$\sum_{i=0}^{d} \eta_i A^i_j = 0, \qquad j = 0, \dots, k. \tag{4.4}$$

In particular to the hyperplane H in $\mathbb{P}^d(V)$, defined by the linear equation (4.2), corresponds the dual point in $\mathbb{P}(V^*)$ given by the line in V^*

$$L_H = \{(\lambda\alpha_0, \dots, \lambda\alpha_d) | \lambda \in \mathbb{C}^*\}. \tag{4.5}$$

This line can be also characterized as a solution of a system of linear homogeneous equations

$$\sum_{j=0}^{d} \beta^j_i \eta_j = 0, \qquad i = 1, \dots, d, \tag{4.6}$$

and (4.2) and (4.6) are related by the consistency condition

$$\sum_{j=0}^{d} \beta^j_i \alpha_j = 0, \qquad i = 1, \dots, d. \tag{4.7}$$

The projective duality respects incidence relations between various linear subspaces. For instance, two points belong to the same hyperplane in $\mathbb{P}(V)$ are dual to the two hyperplanes in $\mathbb{P}(V^*)$ intersecting at the point dual to the hyperplane in $\mathbb{P}(V)$.

This classical projective duality picture allows various generalizations (see, e.g., [GKZ, Chapter 1]). Let $Z \in \mathbb{P}(V)$ be a closed irreducible algebraic subvariety. A hyperplane $H \subset \mathbb{P}(V)$ is said to be tangential to Z if there exists a smooth point $z \in Z$ such that $z \in H$ and the tangent space to H at z contains the tangent space to Z at z. Denote by $Z^* \subset \mathbb{P}(V^*)$ the closure of the set of all hyperplanes tangent to Z. The variety Z^* is called the projective dual to Z. More generally given a non-linear bundle of projective spaces over X, e.g., the projectivization $\mathbb{P}(\mathcal{E}_X \oplus \mathcal{O}_X)$ of a vector bundle $\mathcal{E}_X$ over the base X one can consider the bundle of the dual projective spaces $\mathbb{P}(\mathcal{E}^*_X \oplus \mathcal{O}_X)$. In this case the classical projective duality holds fiberwise.

The main point of this note is the following statement.

Theorem 4.1 *Let X be a compact complex manifold of dimension d, and consider the projectivization $\mathbb{P}(\mathcal{T}_X \oplus \mathcal{O}_X)$ of the tangent bundle $\mathcal{T}_X$. Let $\mathcal{E}$ be a holomorphic vector bundle on X of rank n, equipped with a Higgs pair $(\overline{\partial}_{\mathcal{E}}, \Phi)$, and let $\tilde{X}$ be the associated Simpson spectral cover. Let $U \subset X$ be a subset with coordinates $z = (z^1, \dots, z^d)$ and let $(\xi^0, \dots, \xi^d)$ be the corresponding homogeneous fiber* *coordinates on the projective bundle* $\mathbb{P}(\mathcal{T}_U \oplus \mathcal{O}_U)$ *over* U. *Consider the hypersurface* $\widehat{U}$ *in* $\mathbb{P}(\mathcal{T}_U \oplus \mathcal{O}_U)$ *defined by the equation*

$$\det(\textstyle\sum_{j=1}^{d} \xi^j \Phi_j(z) + \xi^0) = 0, \tag{4.8}$$

where Φ_j *are given by the local expression* $\Phi = \sum_{j=1}^{d} \Phi_j(z) dz^j$ *over* U. *Then the fiberwise projective dual to the hypersurface* $\widehat{U}$ *is the projective compactification of the Simpson spectral cover* $\tilde{X}|_U$ *over the locus* $U \subset X$.

Proof By the assumption the Higgs field can be diagonalized on the subset $U \subset X$

$$\Phi = \textstyle\sum_{a=1}^{n} \sum_{j=1}^{d} E_{aa} \Phi_j^a(z) dz^j. \tag{4.9}$$

Here E_{ab} are elementary $(n \times n)$-matrices $y(E_{ab})_{\alpha\beta} = \delta_{a\alpha}\delta_{b\beta}$. Due to (3.2) this is equivalent to the condition that at least one of the Higgs field components is diagonalizable. Then the corresponding hypersurface (4.8) in $\mathbb{P}(\mathcal{T}_U \oplus \mathcal{O}_U)$ is given by the equation

$$\textstyle\prod_{a=1}^{n} (\sum_{j=1}^{d} \Phi_j^a(z) \xi^j + \xi^0) = 0, \tag{4.10}$$

and thus is a family of unions of n hyperplanes

$$\textstyle\sum_{j=1}^{d} \Phi_j^a(z) \xi^j + \xi^0 = 0, \qquad a = 1, \ldots, n, \tag{4.11}$$

in the projective fibers of $\mathbb{P}(\mathcal{T}_U \oplus \mathcal{O}_U)$ parametrized by points of U. Applying the projective duality correspondence (4.2), (4.5) between hyperplanes and points we obtain a family of n points $a = 1, \ldots, n$

$$(\eta_0, \eta_1, \ldots, \eta_d) = (\lambda, \lambda \Phi_1^a(z), \ldots, \lambda \Phi_d^a(z)), \qquad \lambda \in \mathbb{C}^*, \tag{4.12}$$

in the fibers of the dual projective family $\mathbb{P}(\mathcal{T}_U^* \oplus \mathcal{O}_U)$. This indeed defines the spectral cover of $U \subset X$ as a subset of $\mathbb{P}(\mathcal{T}_U^* \oplus \mathcal{O}_U)$. □ □

We expect that the assumption that Φ *be diagonalizable over* U *is unnecessary, and that this description can be compactified by adding components where the Higgs field is not diagonalizable.* This then would provide a concise projective dual description (4.8) of the Simpson spectral cover.

Let us stress that the proposed construction (4.8) is trivially connected with the Hitchin spectral cover equation (2.5). Following (4.8) the spectral cover $\widetilde{\Sigma}$ of the algebraic curve Σ associated with the Higgs pair $(\bar{\partial}_{\mathcal{E}}, \Phi)$ is defined by the fiberwise homogeneous equation in $\mathbb{P}(\mathcal{T}_\Sigma \oplus \mathcal{O}_\Sigma)$

$$\det(\xi^1 \Phi_z(z) + \xi^0) = 0, \qquad \Phi = \Phi_z(z) dz, \tag{4.13}$$

and let us consider the diagonalized Higgs field (4.9) over a subspace $U \subset \Sigma$. The dimension of the projective fiber is equal to one and thus the projective duality maps points in the fibers of the projective bundle $\mathbb{P}(\mathcal{T}_U \oplus \mathcal{O}_U)$ to the points of the projective fibers of $\mathbb{P}(\mathcal{T}_U^* \oplus \mathcal{O}_U)$ according to the incidence relation (4.7). The incidence relation (4.6) is easily solved so that the equation for the dual curve in $\mathbb{P}(T_U^* \oplus \mathcal{O}_U)$ is given by

$$\det(\eta_1 - \eta_0 \Phi_z(z)) = 0, \tag{4.14}$$

which is indeed the projective compactification of the Hitchin curve

$$\det(w - \Phi_z(z)) = 0. \tag{4.15}$$

The Hitchin description (2.5) is recovered in the chart $\eta_0 = 1$, $\eta_1 = w$.

Acknowledgments We are grateful to L. Katzarkov, A. Rosly and L. Takhtajan for enjoyable discussions.

References

[DM] D. Donagi, E. Markman, *Spectral covers, algebraically completely integrable, Hamiltonian systems, and moduli of bundles*, Springer L.N.M 1620 (1996), 1–119.

[F] A. Fujiki, *Hyper-Kähler structure on the moduli space of flat bundles*, Prospects in complex geometry (Katata and Kyoto, 1989), Springer L.N.M. 1468 (1991), 1–83.

[GKZ] I. M. Gelfand, M. M. Kapranov, and A.V. Zelevinsky, *Discriminants, Resultants, and Multidimensional Determinants*, Birkhäuser Boston, 1994, 523 pp.

[H] N. Hitchin, *Stable bundles and integrable systems*, Duke Math. J. 54 (1987), no. 1, 91–114.

[KOP] L. Katzarkov, D. Orlov, T. Pantev *Notes on Higgs bundles and D-branes*, see webpage of CIMPA-CIMAT-SWAGP Workshop *Moduli Spaces and Mathematical Physics*, 21.01.2013-02.02.2013;

[S1] C. Simpson, *Moduli of representations of the fundamental group of a smooth projective variety.*, I: Publ. Math. I.H.E.S. 79 (1994), 47–129; II: Publ. Math. I.H.E.S. 80 (1994), 5–79.

[S2] C. Simpson, *Non-abelian Hodge theory*. Proceedings, ICM-90, Kyoto Springer, Tokyo (1991), 198–230.

[Sk1] E.K. Sklyanin, *Separation of Variables in the Gaudin model*, Zapiski Nauchnykh Seminarov LOMI 164 (1987), 151–169, (in Russian); J. Sov. Math. 47 (1989), 2473–2488, (English transl.).

[Sk2] E.K. Sklyanin, *Separation of variables. New trends* in Quantum field theory, integrable models and beyond (Kyoto, 1994). Progr. Theoret. Phys. Suppl. 118 (1995), 35–60.

Peter-Weyl Bases, Preferred Deformations, and Schur-Weyl Duality

Anthony Giaquinto, Alex Gilman, and Peter Tingley

Dedicated to Kolya Reshetikhin on the occasion of his 60th birthday

Abstract We discuss the deformed function algebra $\mathcal{O}_\hbar(G)$ of a simply connected reductive algebraic group G over $\mathbb{C}$ using a basis consisting of matrix elements of finite dimensional representations. This leads to a preferred deformation, meaning one where the structure constants of comultiplication are unchanged. The structure constants of multiplication are controlled by quantum $3j$ symbols. We then discuss connections earlier work on preferred deformations that involved Schur-Weyl duality.

1 Introduction

Let G be a connected reductive algebraic group over $\mathbb{C}$, and let $\mathfrak{g}$ be its Lie algebra. Associated with this data are two Hopf algebras, the commutative function algebra $\mathcal{O}(G)$ and the cocommutative universal enveloping algebra $U(\mathfrak{g})$. During the 1980s, various non-commutative and non-cocommutative quantizations of these Hopf algebras were introduced. The first example, now known as $U_\hbar(\mathfrak{sl}_2)$, was discovered by Kulish and Reshetikhin in [KR81] in relation to the quantum inverse scattering method. Later, Drinfeld and Jimbo independently introduced the well studied quantized universal enveloping algebra $U_\hbar(\mathfrak{g})$, see [Dr87] and [Ji85]. On

A. Giaquinto (✉) · P. Tingley
Department of Mathematics and Statistics, Loyola University, Chicago, IL, USA
e-mail: agiaqui@luc.edu; ptingley@luc.edu

A. Gilman
School of Physics and Astronomy, University of Minnesota, Minneapolis, MN, USA
e-mail: agilman@physics.umn.edu

A. Alekseev et al. (eds.), *Representation Theory, Mathematical Physics, and Integrable Systems*, Progress in Mathematics 340,
https://doi.org/10.1007/978-3-030-78148-4_9

the $\mathcal{O}(G)$ side the first was the quantum matrix bialgebra $\mathcal{O}_\hbar(M_2)$ introduced by Faddeev and Takhtajan in [FT86], constructed using the monodromy matrix for the quantum Lax operator of the Liouville model. This approach was fully developed in the landmark work [FRT90] of Faddeev, Reshetikhin, and Takhtajan in which a quantum Yang-Baxter R-matrix is used to deform the defining relations of the classical series of coordinate algebras $\mathcal{O}(G)$.

We should also mention a few other early approaches. In [W87a, W87b], Woronowicz developed the theory of compact quantum groups in the C^*-algebra framework by introducing the quantization $SU_\mu(2)$ in which the parameter μ is a positive real number. Another early work on the quantized algebra of functions on $SU(2)$ is the paper [SV88] by Vaksman and Soibelman [SV88]. Matrix coefficients of finite dimensional representations play a key role in this theory. Another approach due to Manin [Man87] constructs quantum coendomorphism bialgebras as universal objects coacting on a pair of quantum linear spaces.

Since $U(\mathfrak{g})$ is rigid as an algebra and $\mathcal{O}(G)$ is rigid as a coalgebra, the fact that $U_\hbar(\mathfrak{g})$ and $\mathcal{O}_\hbar(G)$ are formal deformations implies that their finite dimensional representations and corepresentations correspond exactly to those for $U(\mathfrak{g})$ and $\mathcal{O}(G)$, respectively. In particular, these categories are semi-simple/cosemisimple. With this in mind, a dual approach may be taken by first studying the monoidal categories of corepresentations of $\mathcal{O}_\hbar(G)$ and $\mathcal{O}(G)$. Once enough is known about these categories, one can follow the generalized Tannaka-Krein theory to reconstruct the Hopf algebras, see [JS90]. The resulting $\mathcal{O}_\hbar(G)$ must necessarily be isomorphic to $\mathcal{O}(G)[[\hbar]]$ as a coalgebra.

Focusing more sharply on our main point, it is a natural question to find a so-called preferred presentations of $U_\hbar(\mathfrak{g})$ and $\mathcal{O}_\hbar(G)$, where the algebra structure is completely unchanged for $U_\hbar(\mathfrak{g})$ and the coalgebra structure is completely unchanged for $\mathcal{O}_\hbar(G)$. With the usual generators and relations descriptions this seems to be hard since, for the natural bases, all structures are varying.

The purpose of this note is to discuss a preferred presentation for $\mathcal{O}_\hbar(G)$. The starting point is to view $\mathcal{O}_\hbar(G)$ as the restricted dual Hopf algebra of $U_\hbar(\mathfrak{g})$. The preferred presentation is achieved from a Peter-Weyl basis of $\mathcal{O}_\hbar(G)$—a basis consisting of matrix elements of finite dimensional representations. The structure constants for the preferred presentation involve the quantum $3j$-symbols from physics, which encode the decomposition of a tensor product of irreducible representations into irreducibles. These coefficients have numerous applications and in the rank one case have been extensively studied, see [KK89, KR88, V89].

We finish by describing how this relates to Schur-Weyl duality in type A, and hence to some earlier work by Gerstenhaber, Giaquinto and Schack [GGS92, Gia92] on preferred deformations. There the quantum matrix bialgebra $\mathcal{O}_\hbar(M_n)$ is viewed as the elements of the tensor algebra $T(M_n^*)$ which are fixed by the action of a certain quantum symmetric group. This is a subgroup of the cactus group studied in e.g. [KT09], and as discussed there is related to using Drinfeld's unitarized R-matrix from [Dr90] in place of the usual R-matrix. If V is the vector representation, then the image of this group in $\text{End}(V^{\otimes n})$ generates the usual action of the Hecke algebra.

In [GGS92, Gia92] the decomposition of tensor space $V^{\otimes n}$ into quantum symmetric elements is obtained with the aid of the Woronowicz quantization of $U(\mathfrak{sl}_n)$, which acts as skew derivations of $V^{\otimes n}$ associated with certain automorphisms. These automorphisms coincide with the exponentials $\exp(\hbar H_i)$ of the standard Cartan generators H_i of $\mathfrak{sl}_n$. Thus the images of the Woronowicz quantization and $U_\hbar(\mathfrak{sl}_n)$ in $\mathrm{End}(V^{\otimes n})[[\hbar]]$ coincide and so the Schur-Weyl decompositions of $V^{\otimes n}$ are the same for either of these two quantizations. The use of the Woronowicz quantization was motivated by the fact that its finite dimensional representations correspond exactly to those of $U_\hbar(\mathfrak{sl}_n)$ or $U(\mathfrak{sl}_n)$, and one does not have to exclude the non-type-1 representation that appear for the rational form $U_q(\mathfrak{sl}_n)$. The disadvantage is that the Woronowicz quantization does not give a bialgebra structure (see [GGS92, p26]).

We do not carefully address the preferred presentation of $U_\hbar(\mathfrak{g})$ in this note. The dual Peter-Weyl basis does give a preferred presentation, but of a certain completion of $U_\hbar(\mathfrak{g})$. Finding a preferred presentation for $U_\hbar(\mathfrak{g})$ itself seems more difficult. An algebra isomorphism from $U_\hbar(\mathfrak{sl}_2)$ to $U(\mathfrak{sl}_2)$ was given in [CP94, Proposition 6.4.6], giving the preferred presentation in that case. Only recently in [AG17] was an explicit trivialization of $U_\hbar(\mathfrak{sl}_n)$ given by Appel and Gautam. This isomorphism is induced by a map between the quantum loop algebra of $\mathfrak{sl}_n$ and a completion of the Yangian.

This note is organized as follows. In Sect. 2 we discuss background on deformation theory. In Sect. 3 we construct a preferred deformation using a Peter-Weyl basis. In Sect. 4 we restrict to type A and reformulate the construction using Schur-Weyl duality, then discuss how this relates to some older work.

2 Deformations

2.1 *Formal Deformations*

Let B be a bialgebra over $\mathbb{C}$. A $\mathbb{C}[[\hbar]]$-bialgebra $B_\hbar$ is a *formal deformation* of B if it is a topologically free $\mathbb{C}[[\hbar]]$-module together with an isomorphism $B_\hbar/hB_\hbar \simeq B$. If $B_\hbar$ is a formal deformation of B, we can choose an identification of $B_\hbar$ with $B[[\hbar]]$ as a $\mathbb{C}[[\hbar]]$-module. The multiplication μ and comultiplication Δ of $B_\hbar$ then necessarily have the form

$$\mu(a,b) = \mu_0(a,b) + \hbar\mu_1(a,b) + \hbar^2\mu_2(a,b) + \cdots$$
$$\Delta(a) = \Delta_0(a) + \hbar\Delta_1(a) + \hbar^2\Delta_2(a) + \cdots$$

where $\mu_i : B \otimes B \to B$ and $\Delta_i : B \to B \otimes B$ are linear maps and μ_0 and Δ_0 are the undeformed multiplication and comultiplication of B.

2.2 Equivalence and Preferred Presentations

Deformations $B_\hbar$ and $B'_\hbar$ are equivalent if there is a $\mathbb{C}[[\hbar]]$-bialgebra isomorphism $\phi : B_\hbar \to B'_\hbar$ which reduces to the identity modulo $\hbar$. It is known that *every* deformation of the universal enveloping algebra $U(\mathfrak{g})$ is equivalent to one in which $\mu = \mu_0$. That is, it is a trivial deformation of the algebra structure, and so the representation theory of $U_\hbar(\mathfrak{g})$ and $U(\mathfrak{g})$ is identical. Dually, every deformation of $\mathcal{O}(G)$ is equivalent to one in which $\Delta = \Delta_0$, so the co-representation theory is unchanged. A *preferred presentation* of a deformation of $U(\mathfrak{g})$ or $\mathcal{O}(G)$ is one with unchanged multiplication or comultiplication.

A natural question is to find preferred presentations of $U_\hbar(\mathfrak{g})$ and $\mathcal{O}_\hbar(G)$. To do so requires an identification of their underlying vector spaces with $U(\mathfrak{g})[[\hbar]]$ and $\mathcal{O}(G)[[\hbar]]$ as $\mathbb{C}[[\hbar]]$-modules. This can be accomplished, for example, by finding bases of $U_\hbar(\mathfrak{g})$ and $\mathcal{O}_\hbar(G)$ which reduce to bases of $U(\mathfrak{g})$ and $\mathcal{O}(G)$ modulo $\hbar$. However, most choices of bases do not provide preferred presentations: both multiplication and comultiplication depend on $\hbar$. This is true, in particular, for the various PBW type bases in the literature.

We now arrive at an interesting juncture: Once $\mathcal{O}_\hbar(G)$ is shown to be a formal deformation, we know its irreducible representations are the same as those for $\mathcal{O}(G)$, just tensored with $\mathbb{C}[[\hbar]]$. We can then consider a Peter-Weyl type basis of $\mathcal{O}_\hbar(G)$, meaning a basis consisting of matrix elements of irreducible representations. We shall see that this provides the sought after preferred presentation of $\mathcal{O}_\hbar(G)$.

2.3 Standard Deformation of $U(\mathfrak{g})$

Consider the standard Chevalley generators E_i, F_i, H_i of $U(\mathfrak{g})$. The deformation $U_\hbar(\mathfrak{g})$ is usually defined to be the algebra with these same generators, but deformed relations and a deformed coproduct. The main structure we will need here is the coproduct, so we state that explicitly:

$$\Delta E_i = E_i \otimes e^{-\hbar H_i} + 1 \otimes E_i$$

$$\Delta F_i = F_i \otimes 1 + e^{\hbar H_i} \otimes F_i$$

$$\Delta H_i = H_i \otimes 1 + 1 \otimes H_i$$

The rest of the structure can be found in many places, see e.g. [CP94]. The relations for multiplication are also deformed. For instance, in $\mathfrak{sl}_2$ the undeformed relation $EF - FE = 2H$ becomes

$$EF - FE = \frac{e^{\hbar H} - e^{-\hbar H}}{e^{\hbar} - e^{-\hbar}}$$

So the deformation is certainly not preferred with respect to any PBW type basis.

Remark In the general reductive case we should actually include more H generators, using the notion of root datum. See [Lus93, §2.2]. This does not cause significant issues.

2.4 Standard Deformation of $\mathcal{O}(M_n)$

Here we give the deformed relations for $\mathcal{O}(M_n)$, as constructed by Faddeev-Reshetikhin-Takhtajan [FRT90]. We focus on the case $n = 2$, and for simplicity of presentation define $q = e^{\hbar}$. Consider the coordinate functions

$$X = \begin{bmatrix} a & b \\ c & d \end{bmatrix} = \begin{bmatrix} e_{11}^* & e_{12}^* \\ e_{21}^* & e_{22}^* \end{bmatrix}$$

on the space of 2×2 matrices. The FRT formalism starts with a solution R to the quantum Yang-Baxter equation ($R_{12}R_{13}R_{23} = R_{23}R_{13}R_{12}$) and imposes the relations $RX_1X_2 = X_2X_1R$ where $X_1 = X \otimes I$ and $X_2 = I \otimes X$. In coordinates,

$$X_1X_2 = \begin{bmatrix} a^2 & ab & ba & b^2 \\ ac & ad & bc & bd \\ ca & cb & da & db \\ c^2 & cd & dc & d^2 \end{bmatrix} \qquad X_2X_1 = \begin{bmatrix} a^2 & ba & ab & b^2 \\ ca & da & cb & db \\ ac & bc & ad & bd \\ c^2 & dc & cd & d^2 \end{bmatrix}$$

and

$$R = \begin{bmatrix} q & 0 & 0 & 0 \\ 0 & 1 & 0 & 0 \\ 0 & q - q^{-1} & 1 & 0 \\ 0 & 0 & 0 & q \end{bmatrix}$$

This produces the relations

$$\begin{gathered} ab = qba, \quad ac = qca, \quad bd = qdb, \quad cd = qdc, \\ bc = cb, \quad ad - da = (q - q^{-1})bc \end{gathered} \tag{1}$$

The coproduct is defined on generators by

$$\Delta\left(\begin{bmatrix} a & b \\ c & d \end{bmatrix}\right) = \begin{bmatrix} a & b \\ c & d \end{bmatrix} \otimes \begin{bmatrix} a & b \\ c & d \end{bmatrix} = \begin{bmatrix} a \otimes a + b \otimes c & a \otimes b + b \otimes d \\ c \otimes a + d \otimes c & c \otimes b + d \otimes d \end{bmatrix}$$

and is extended multiplicatively to monomials of higher degree. For example,

$$\Delta(a^2) = a^2 \otimes a^2 + ab \otimes ac + ba \otimes ca + b^2 \otimes c^2 = a^2 \otimes a^2 + (1+q^{-2}) ab \otimes bc + b^2 \otimes c^2$$

This is dependent on q, so the deformation is not preferred, at least when using the PBW basis $\{a^i\, b^j\, c^k\, d^l \quad | \quad i, j, k, l \in \mathbb{Z}_{\geq 0}\}$ to identify $\mathcal{O}_\hbar(M_n)$ with $\mathcal{O}(M_n)[[\hbar]]$.

3 Peter-Weyl Bases and Preferred Deformations

Another approach to deforming $\mathcal{O}(G)$ is by duality: one simply defines $\mathcal{O}_\hbar(G)$ as the restricted dual of $U_\hbar(\mathfrak{g})$. In this setting the Peter-Weyl basis arises naturally. It is with this basis that we get a preferred presentation of $\mathcal{O}_\hbar(G)$.

3.1 The Peter-Weyl Basis

Since $\mathcal{O}_\hbar(G)$ is cosemisimple, the restricted dual definition implies

$$\mathcal{O}_\hbar(G) \simeq \oplus_\lambda \mathrm{End}(V_\lambda)^*$$

where λ runs over the dominant integral weights of G, the V_λ are the corresponding representations of $U_\hbar(\mathfrak{g})$, and the isomorphism is as coalgebras over $k[[\hbar]]$. See e.g. [KS97, Chapter 11]. This becomes an isomorphism of Hopf algebras if one defines multiplication on $\oplus_\lambda \mathrm{End}(V_\lambda)$ as the dual of the coproduct for $U_\hbar(\mathfrak{g})$.

For any λ, $\mathrm{End}(V_\lambda)$ is naturally identified with $V_\lambda \otimes V_\lambda^*$. Taking duals, we identify $\mathrm{End}(V_\lambda)^*$ with $V_\lambda^* \otimes V_\lambda$. This gives a natural way to choose a basis for $\mathcal{O}_\hbar(G)$: pick dual bases B_λ and B_λ^* for each pair V_λ, V_λ^*. Then

$$\bigsqcup_\lambda \{Y^* \otimes X : X, Y \in B_\lambda\}$$

is a basis for $\mathcal{O}_\hbar(G)$, which we call a **Peter-Weyl basis**. The pairing of $U(\mathfrak{g})$ with $\mathcal{O}(G)$ is given by, for $Y^* \otimes X \in \mathcal{O}(G)$ and $u \in U(\mathfrak{g})$,

$$\langle Y^* \otimes X, u \rangle = Y^* u(X)$$

Remark One often reverses order of factors when taking duals of tensor products. We have not done so, in part to match conventions in [GGS92].

3.2 Comultiplication

We are identifying $\prod_\lambda \mathrm{End} V_\lambda$ with a completion of $U_\hbar(\mathfrak{g})$, and $\mathcal{O}_\hbar(G)$ with the dual Hopf algebra to this. Thus, in both $\mathcal{O}(G)$ and $\mathcal{O}_\hbar(G)$, comultiplication is the dual of multiplication in $\prod_\lambda \mathrm{End} V_\lambda$. In coordinates, for $X, Y \in B_\lambda$,

$$\Delta(Y^* \otimes X) = \sum_{Z \in B_\lambda} (Y^* \otimes Z) \otimes (Z^* \otimes X) \tag{2}$$

3.3 Multiplication (Abstract)

Multiplication is the dual to comultiplication in $U_\hbar(\mathfrak{g})$. In coordinates this means, for $X_1, Y_1 \in B_\lambda$, $X_2, Y_2 \in B_\mu$, and any $u \in U(\mathfrak{g})$,

$$((Y_1^* \otimes X_1)(Y_2^* \otimes X_2))(u) = (Y_1^* \otimes Y_2^*)\Delta u(X_1 \otimes X_2) \tag{3}$$

To be explicit we need to express this in terms of the Peter-Weyl basis. The resulting structure constants are closely related to the famous 3j symbols from physics.

3.4 3*j* Symbols

These are often studied just for SL(2), but we need the following more general notion. For each triple λ, μ, ν, choose a basis $\{\phi_1, \cdots, \phi_{c^\nu_{\lambda,\mu}}\}$ for the space of embeddings of $V_\nu \hookrightarrow V_\lambda \otimes V_\mu$. For $X_1 \in B_\lambda$, $X_2 \in B_\mu$, $X_3 \in B_\nu$, write

$$X_1 \otimes X_2 = \sum_\nu \sum_{1 \le k \le c^\nu_{\lambda,\mu}} \sum_{X_3 \in B_\nu} \begin{pmatrix} \lambda & \mu & \nu \\ X_1 & X_2 & X_3 \end{pmatrix}_k \phi_k(X_3)$$

The constants $\begin{pmatrix} \lambda & \mu & \nu \\ X_1 & X_2 & X_3 \end{pmatrix}_k$ are called the $3j$ symbols.

Taking duals gives a basis $\{\phi_1^*, \cdots, \phi^*_{c^\nu_{\lambda,\mu}}\}$ of the space of surjections $V_\lambda^* \otimes V_\mu^* \to V_\nu^*$. We then get dual $3j$ symbols defined by

$$\phi_k^*(Y_1^* \otimes Y_2^*) = \sum_{Y_3^* \in B_\nu^*} \overline{\begin{pmatrix} \lambda & \mu & \nu \\ Y_1^* & Y_2^* & Y_3^* \end{pmatrix}_k} Y_3^*$$

This can be done just as easily for representations of $U(\mathfrak{g})$ or $U_\hbar(\mathfrak{g})$.

Remark In the SL(2) case, there is a unique (up to signs) orthonormal weight basis, so a chosen Peter-Weyl basis. The spaces of embeddings $V_\nu \hookrightarrow V_\lambda \otimes V_\mu$ are 1 dimensional, and the inner product can be used to normalize the embedding, fixing the $3j$ symbols. As mentioned earlier, these have been calculated extensively. Using orthonormal bases also implies that the $3j$ symbols and dual $3j$ symbols coincide exactly.

3.5 Structure Constants for Multiplication

It is now immediate from definitions that the structure constants for multiplication in the Peter-Weyl basis are given by, for $X_1, Y_1 \in B_\lambda$ and $X_2, Y_2 \in B_\mu$,

$$(Y_1^* \otimes X_1)(Y_2^* \otimes X_2) = \sum_{\substack{\nu \in P_+ \\ X_3, Y_3 \in B_\nu}} \left[\sum_{1 \le k \le c_{\lambda,\mu}^\nu} \overline{\begin{pmatrix} \lambda & \mu & \nu \\ Y_1^* & Y_2^* & Y_3^* \end{pmatrix}_k} \begin{pmatrix} \lambda & \mu & \nu \\ X_1 & X_2 & X_3 \end{pmatrix}_k \right] Y_3^* \otimes X_3 \tag{4}$$

Multiplication as defined in (3) does not depend on the basis $\{\phi_1, \ldots, \phi_k\}$, so

$$\sum_{1 \le k \le c_{\lambda,\mu}^\nu} \overline{\begin{pmatrix} \lambda & \mu & \nu \\ Y_1^* & Y_2^* & Y_3^* \end{pmatrix}_k} \begin{pmatrix} \lambda & \mu & \nu \\ X_1 & X_2 & X_3 \end{pmatrix}_k$$

must be independent of the choice of basis $\{\phi_1, \ldots, \phi_k\}$ as well.

3.6 Preferred Presentation of $\mathcal{O}_\hbar(G)$

The set of irreducible representations V_λ of $U(\mathfrak{g})$ and $U_\hbar(\mathfrak{g})$ correspond exactly, and the comultiplication from Sect. 3.2 does not reference $U(\mathfrak{g})$ at all, so is unchanged under deformation. The multiplication from Sect. 3.3 does change when we move to $U_\hbar(\mathfrak{g})$, since its definition uses the coproduct of $U(\mathfrak{g})$, which is deformed in $U_\hbar(\mathfrak{g})$. But the presentation in Sect. 3.5 is still valid. The only difference is that the spaces of embeddings $V_\nu \hookrightarrow V_\lambda \otimes V_\mu$ change.

In order to see $\mathcal{O}_\hbar(\mathfrak{g})$ as a preferred deformation of $\mathcal{O}(\mathfrak{g})$, one must simply choose

- A basis for each $U_\hbar(\mathfrak{g})$ module V_λ which specializes to a basis at $h = 0$,
- A basis for each space of $U_\hbar(\mathfrak{g})$-homomorphisms $V_\nu \hookrightarrow V_\lambda \otimes V_\mu$ which also specializes to a basis at $\hbar = 0$. This leads to a definition of quantum 3j symbols.

Then the construction above gives a deformation where the structure constants of comultiplication are manifestly identical, and the structure constants for multiplication are given by (4), but with the 3j symbols replaces by their deformed counterparts.

Remark We have relied on the fact that we already have a (non-preferred) deformation of $U(\mathfrak{g})$ to construct our preferred presentation of $\mathcal{O}_\hbar(G)$. One might try to use this approach to construct a deformation from scratch, by simply deforming the spaces of embeddings $V_\nu \hookrightarrow V_\lambda \otimes V_\mu$. However, this deformation is not arbitrary: one needs to ensure that $\mathcal{O}(G)$ remains a Hopf algebra. Directly ensuring this seems difficult.

3.7 *Preferred Deformation of* $U(\mathfrak{g})$

This method also gives a preferred presentation of a completion of $U_\hbar(\mathfrak{g})$ by working with the topological basis $\{f^\lambda_{b,c} = b \otimes c^* \in \operatorname{End} V_\lambda\}$. The operations are the duals those of $\mathcal{O}_\hbar(G)$. However, $U_\hbar(\mathfrak{g})$ is a proper subalgebra of $\prod_\lambda \operatorname{End}(V_\lambda)$, and the preferred presentation does not restrict in any nice way. It is also unclear how to relate the Peter-Weyl type bases with the Chevalley generators. So this approach is not really satisfactory.

3.8 *Non-simply connected Groups and Matrix Algebras*

The condition that G be simply connected is not really needed. A non-simply connected reductive group G' is always the quotient of a corresponding simply connected one, and the category of finite dimensional representations of G' is a sub-tensor-category of the category of finite dimension representations of G. The irreducible representations of that category are parameterized by λ in the positive part of some sub-lattice P' of the weight lattice of G. The whole story then goes through by realizing $\mathcal{O}(G')$ as $\oplus \operatorname{End}(V_\lambda)^*$, where now one restricts to $\lambda \in P'_+$.

In type A one can also consider $\mathcal{O}(\mathrm{GL}_k)$, and again the story goes through without significant changes, only now there are more representations than for $\mathcal{O}(\mathrm{SL}_k)$, since any irreducible representation can be tensored with any integer power of the determinant representation. In Sect. 4 we will actually work with $\mathcal{O}(M_k)$, the function algebra on the algebra of all $k \times k$ matrices. This is isomorphic to $\oplus_\lambda \operatorname{End} V^*_\lambda$, where now the λ index the polynomial representations of GL_k. These λ's are naturally indexed by partitions with at most k parts.

4 Relation to Schur-Weyl Duality

For the case of GL_k, or SL_k, [Gia92, GGS92] studied another approach to finding a preferred deformation. Their approach most naturally realizes $\mathcal{O}(M_k)$, the function algebra on the algebraic monoid of $k \times k$ matrices. We now discuss how their results naturally arise in our framework.

4.1 General Categorical Discussion

Identifying $\mathcal{O}(G)$ with $\oplus_\lambda \mathrm{End}(V_\lambda)^*$ as we have is not necessarily the most natural thing to do, since it requires choosing a representation in each isomorphism class of simples. In fact, any $f \in \mathrm{End}(V)^*$, for any representation V, gives a function of G. However, different elements of $\mathrm{End}(V)^*$ can give identical functions on G, so $\mathcal{O}(G)$ should be identified with a quotient of $\oplus_V \mathrm{End}(V)^*$. This is a badly infinite sum, but, ignoring that for now, multiplication is simple: given two elements of $\mathcal{O}(G)$, $f \in \mathrm{End}(V)^*$, $g \in \mathrm{End}(W)^*$,

$$fg = f \otimes g \in \mathrm{End}(V \otimes W)^* \simeq \mathrm{End}(V)^* \otimes \mathrm{End}(W)^*$$

We work with $\oplus_\lambda (\mathrm{End} V_\lambda)^*$ essentially because every element of $\mathcal{O}(G)$ appears exactly once. This avoids the need to take a quotient, but makes multiplication more difficult to describe.

4.2 Using $V^{\otimes n}$, Undeformed

Now we will restrict to considering $\mathcal{O}(M_k)$. Then there is another natural space which encodes every function exactly once:

$$\oplus_n ((\mathrm{End} V^{\otimes n})^{S_n})^*$$

where V is the vector representation. To see this, recall that, by Schur-Weyl duality,

$$V^{\otimes n} \simeq \oplus_\lambda V_\lambda \boxtimes W_\lambda$$

where λ ranges over all partitions of n with at most k rows, and V_λ, W_λ are the irreducible representations of M_k and S_n, respectively. Then, by Schur's lemma,

$$\oplus_n (\mathrm{End} V^{\otimes n})^{S_n} \simeq \oplus_\lambda \mathrm{End} V_\lambda$$

We need to understand this identification explicitly. Fix λ and choose any $w, w^* \in W_\lambda, W_\lambda^*$ with $w^*(w) = 1$. Then, for any $c^* \otimes b \in \mathrm{End}(V_\lambda)^*$, the element $(c^* \otimes w^*) \otimes (b \otimes w) \in \mathrm{End}(V^{\otimes n})^*$ gives the same function on $U_q(\mathfrak{gl}_n)$ as $c^* \otimes b$, but it is not S_n invariant. To fix that, consider the Young symmetrizer $P = \frac{1}{n!} \sum_{\sigma \in S_n} \sigma$.

Then

$$P((c^* \otimes w^*) \otimes (b \otimes w))$$

is clearly S_n invariant, and gives the same function on $U_q(\mathfrak{gl}_n)$. Explicitly,

$$P((b^* \otimes w^*) \otimes (c \otimes w)) = \frac{1}{n!} \sum_{\sigma \in S_n} (b^* \otimes \sigma w^*) \otimes (c \otimes \sigma w)$$

Since there is only one S_n invariant element corresponding to a given function, this is independent of the choice of w, w^*.

Multiplication is then given by, for $f \in (\mathrm{End} V^{\otimes n})^{S_n}$ and $g \in \mathrm{End}(V^{\otimes m})^{S_m}$,

$$fg = P_{n+m}(f \otimes g)$$

Comultiplication would normally be given by, for any dual bases C and C^* of $V^{\otimes n}$,

$$\Delta(Y^* \otimes X) = \sum_{Z \in C} (Y^* \otimes Z) \otimes (Z^* \otimes X)$$

However, while this is in $\mathrm{End}(V^{\otimes n}) \otimes \mathrm{End}(V^{\otimes n})$, and corresponds to the correct element of $\mathcal{O}(M_k) \otimes \mathcal{O}(M_k)$, it is not in $(\mathrm{End} V^{\otimes n})^{S_n} \otimes (\mathrm{End} V^{\otimes n})^{S_n}$. To fix this, apply the symmetrizer P to each of the two factors to get something in the right space which gives the same function on $U_\hbar(\mathrm{gl}_n)$. The correct definition becomes

$$\Delta(Y^* \otimes X) = (P \otimes P) \sum_{Z \in C} (Y^* \otimes Z) \otimes (Z^* \otimes X)$$

4.3 Using $V^{\otimes n}$, Deformed

By quantum Schur-Weyl duality $V^{\otimes n} \simeq \oplus_\lambda V_\lambda \boxtimes W_\lambda$, where V is now the vector representation of $U_\hbar(\mathfrak{gl}_n)$, the V_λ are the polynomial representation of $U_\hbar(\mathfrak{gl}_n)$, and the W_λ are the irreducible representations of the Hecke algebra H_n corresponding to partitions with at most k rows. Then, by Schur's lemma,

$$\oplus_n((\mathrm{End}(V^{\otimes n})^{H_n})^* \simeq \oplus_\lambda \mathrm{End}(V_\lambda)^* = \mathcal{O}(M_n)$$

where the H_n superscript means H_n equivariant functions. The space on the left can be written as $\oplus((V^*)^{\otimes n} \otimes V^{\otimes n})^{H_n}$, where the H_n still means equivariant.

The key to understanding the operations in Sect. 4.2 was to understand the projection

$$(V^*)^n \otimes V^n \to ((V^*)^n \otimes V^n)^{S_n}$$

where S_n acts simultaneously on the two factors. This was defined as the Young symmetrizer P acting simultaneously on both factors, but to quantize we need a different characterization, since it is not clear how to have H_n act on $(V^*)^n \otimes V^n$.

The crucial thing is that P acts on $\oplus_n(\mathrm{End} V^n)^*$ as a projection so that, for any $\phi \in (V^*)^n \otimes V^n$, ϕ and $P(\phi)$ define the same function on $U(\mathfrak{gl}_k)$. In this form, there is no problem giving the deformed definition.

Definition 4.1 $\pi : \oplus_n(V^*)^{\otimes n} \otimes V^{\otimes n} \to \oplus_n((V^*)^{\otimes n} \otimes V^{\otimes n})^{H_n}$ is the unique projection such that, for any $\phi \in (V^*)^{\otimes n} \otimes V^{\otimes n}$, ϕ and $\pi(\phi)$ define the same function on $U_\hbar(\mathfrak{g})$.

This induces a Hopf algebra structure on $\oplus((V^*)^{\otimes n} \otimes V^{\otimes n})^{H_n}$ because the subset of $\oplus((V^*)^{\otimes n} \otimes V^{\otimes n})$ consisting of elements that define the zero function on $U_\hbar(\mathfrak{g})$ is a Hopf ideal. Multiplication and comultiplication on $\oplus((V^*)^{\otimes n} \otimes V^{\otimes n})^{H_n}$ are given by

$$\Delta(Y^* \otimes X) = (\pi \otimes \pi) \sum_{Z \in C} (Y^* \otimes Z) \otimes (Z^* \otimes X) \tag{5}$$

$$fg = \pi(f \otimes g) \tag{6}$$

Remark It would be nice to have a more explicit formula for π. In the case $n = 2$ such a formula is known. As we shall see in Sect. 4.5, the H_2-equivariant endomorphisms of $V^* \otimes V^*$ are determined by an involution Q. It follows that $\pi = \frac{1+Q}{2}$, see [Gia92, GGS92].

In general one might try to replace P with the q-symmetrizer from [Gyo86]. This does give a natural analogue of P acting on $V^{\otimes n}$, but we would need it to act on $(V^*)^{\otimes n} \otimes V^{\otimes n}$. If the T_i are the generators of the Hecke algebra, the appropriate action on $(V^*)^n$ should replace T_i with T_i^{-1}, and these satisfy a different set of Hecke-algebra relations. So the Hecke algebra does not even naturally act on $(V^*)^{\otimes n} \otimes V^{\otimes n}$. In fact no symmetrizer that acts by simultaneous permutations in the $(V^*)^{\otimes n}$ and $V^{\otimes n}$ can work, as then the relation $ad - da = (q - q^{-1})bc$ in (1) would not be possible.

4.4 Preferred Presentation

We can now construct a preferred presentation of $\mathcal{O}_\hbar(M_n)$.

- Fix the Schur-Weyl duality isomorphism $V^{\otimes n} \simeq \oplus_\lambda V_\lambda \boxtimes W_\lambda$. Of course one then gets a corresponding isomorphism $(V^{\otimes n})^* \simeq \oplus_\lambda V_\lambda^* \boxtimes W_\lambda^*$.
- Fix bases B_λ for each V_λ, and P_λ for each W_λ, and their dual bases B_λ^* for each V_λ^*, and P_λ^* for each W_λ^*, in such a way that all specialize at $h = 0$.
- Then $\bigcup_\lambda \left\{ X^\lambda_{b,c^*} := \frac{1}{\dim W_\lambda} \sum_{p \in P_\lambda} (c^* \boxtimes p^*) \otimes (b \boxtimes p) : b \in B_\lambda, c^* \in B_\lambda^* \right\}$ is a basis for $\mathcal{O}_\hbar(M_n)$. As a function on $U_\hbar(\mathfrak{g})$, the element X^λ_{b,c^*} agrees with $c^* \otimes b \in (\mathrm{End} V_\lambda)^*$.
- Comultiplication and multiplication are given by (5) and (6).

Since X^λ_{b,c^*} agrees with $c^* \otimes b \in (\mathrm{End} V_\lambda)^*$, the structure constants of multiplication and comultiplication in this basis must agree with (2) and (4). It is an interesting exercise to directly obtain these formulae from (5) and (6).

4.5 Comparing with Previous Work

We now compare the current approach with the “method of quantum symmetry” from [Gia92, GGS92]. The starting point there is to view $\mathcal{O}(M_k)$ as the symmetric algebra $SX = \bigoplus_{n \geq 0} (X^{\otimes n})^{S_n}$, where $X = V^* \otimes V$. To quantize, S_n is replaced by a “quantum symmetric group” qS_n with generators $\tau_1, \ldots, \tau_{n-1}$ and relations $\tau_i^2 = \mathrm{Id}$ and $\tau_i \tau_j = \tau_j \tau_i$ if $|i - j| > 1$. Note that if the braid relations $\tau_i \tau_{i+1} \tau_i = \tau_{i+1} \tau_i \tau_{i+1}$ are added then we have the Artin presentation of S_n. As mentioned in the introduction, qS_n is a subgroup of the cactus group.

To describe the qS_n-action on $X^{\otimes n}$ we first deform the flip operator $\sigma : V^* \otimes V^* \to V^* \otimes V^*$ where $\sigma(\alpha \otimes \beta) = \beta \otimes \alpha$. Let $r = \sum_{i<j} e_{ij} \wedge e_{ji} = \frac{1}{2}(e_{ij} \otimes e_{ji} - e_{ji} \otimes e_{ij})$. This is the standard unitary solution to the modified classical Yang-Baxter equation associated with $\mathcal{O}_\hbar(M_k)$. Define an involution of $V^* \otimes V^*$ by $Q = (\exp(-\hbar r))\sigma(\exp \hbar r)$. With this there is an action of qS_n on $(V^*)^{\otimes n}$ where τ_i acts as Q in factors i and $i+1$ and the identity elsewhere. Taking duals there is a corresponding action on $V^{\otimes d}$ and hence qS_n acts diagonally on $(V^*)^{\otimes n} \otimes V^{\otimes n} = X^{\otimes n}$.

One of the main results of [Gia92, GGS92] is that the set of invariant elements of the tensor algebra TX is a bialgebra which is isomorphic to $\mathcal{O}_\hbar(M_k)$. Moreover, the comultiplication in $\bigoplus_{n \geq 0} (X^{\otimes n})^{qS_n}$ is independent of $\hbar$ and coincides with the usual comultiplication in $\mathcal{O}(M_k) = \bigoplus_{n \geq 0} (X^{\otimes n})^{S_n}$. Thus this construction yields the desired preferred presentation of $\mathcal{O}_\hbar(M_k)$.

This essentially coincides with our construction. Using the notation of [GGS92, §10], $(k\langle M(n)\rangle^*, \otimes)$ is naturally the tensor algebra of $V^* \otimes V$, which we identify with $\oplus_n (\mathrm{End} V^{\otimes n})^*$, and think of as functions on $U_\hbar(\mathfrak{g})$. The space $\mathrm{sk}_q\langle M(n)^*\rangle$ is generated by the images of the operators $\frac{1}{2}(Id - \tau_i)$ acting on $(V^* \otimes V)^{\otimes n}$, and these images are easily seen to define the zero function on $U_\hbar(\mathfrak{g})$. So, the quotient in the top line of the diagram in [GGS92, Theorem 10.8] is by a set of elements which are all the zero function on $U_\hbar(\mathfrak{g})$, and by comparing dimensions it agrees with our π. Thus the comultiplication given in [Gia92, GGS92] coincides exactly with (2) and (5), and the multiplication is described using the projection formula (6).

The expression for the multiplicative structure constants in terms of $3j$ symbols is largely new to this paper, although the multiplication formulas for quantum linear spaces given in [Gia92, GGS92] can easily be expressed in the $3j$ symbol notation, and this in turn gives some of the structure constants for $\mathcal{O}(M_n)$. So this idea really dates to those papers as well.

4.6 Deriving the R-Matrix Relations in $\mathcal{O}_q(M_2)$

We now derive the last two relations in the FRT construction of $\mathcal{O}_q(M_2)$ (see Sect. 2.4) in our language (the others are simpler). One could also see that the constructions agree by directly showing that the $\frac{1}{2}(\mathrm{Id} - \tau)$ action on $X \otimes X$ gives the FRT relations.

The variables a, b, c, d in our language are

$$a = e_1 \otimes e_1^*, \quad b = e_1 \otimes e_2^*, \quad c = e_2 \otimes e_1^*, \quad d = e_2 \otimes e_2^*$$

As a representation of $U_\hbar(\mathfrak{gl}_2)$, $V \otimes V \simeq W \oplus T$, where W is a three dimensional representation and T is one dimensional. These have basis

$$W : \{e_1 \otimes e_1, e_2 \otimes e_1 + q e_1 \otimes e_2, e_2 \otimes e_2\}, \qquad T : \{e_2 \otimes e_1 - q^{-1} e_1 \otimes e_2\}$$

Let $s = e_2 \otimes e_1 + q e_1 \otimes e_2$, $t = e_2 \otimes e_1 - q^{-1} e_1 \otimes e_2$. Then $\{s, t\}$ spans the 0 weight space of $V \otimes V$. Let $\{s^*, t^*\}$ be the dual basis of this weight space. Then

$$e_1 \otimes e_2 = \frac{s - t}{q + q^{-1}}, \quad e_2 \otimes e_1 = \frac{q^{-1} s + q t}{q + q^{-1}}$$
$$e_1^* \otimes e_2^* = q s^* - q^{-1} t^*, \quad e_2^* \otimes e_1^* = s^* + t^*$$

The Hecke algebra is the algebra of operators commuting with the action of $U_\hbar(\mathfrak{gl}_2)$, so it is spanned by the projections onto W and T. Thus $s^* \otimes s$ and $t^* \otimes t$ are both H_2 equivariant. Both $s^* \otimes t$ and $t^* \otimes s$ are zero as functions on $U_\hbar(\mathfrak{gl}_2)$ by Schur's lemma, so these are both killed by π. Thus

$$\begin{aligned}
ad &= \pi((e_1^* \otimes e_2^*) \otimes (e_1 \otimes e_2)) \\
&= \pi\left((qs^* - q^{-1}t^*) \otimes \frac{s-t}{q+q^{-1}}\right) \\
&= \frac{q}{q+q^{-1}} s^* \otimes s + \frac{q^{-1}}{q+q^{-1}} t^* \otimes t
\end{aligned}$$

$$\begin{aligned}
da &= \pi((e_2^* \otimes e_1^*) \otimes (e_2 \otimes e_1)) \\
&= \pi\left((s^* + t^*) \otimes \frac{q^{-1}s + qt}{q+q^{-1}}\right) \\
&= \frac{q^{-1}}{q+q^{-1}} s^* \otimes s + \frac{q}{q+q^{-1}} t^* \otimes t
\end{aligned}$$

$$\begin{aligned}
bc &= \pi((e_2^* \otimes e_1^*) \otimes (e_1 \otimes e_2)) \\
&= \pi\left((s^* + t^*) \otimes \frac{s-t}{q+q^{-1}}\right) \\
&= \frac{1}{q+q^{-1}} s^* \otimes s - \frac{1}{q+q^{-1}} t^* \otimes t
\end{aligned}$$

$$\begin{aligned}
cb &= \pi((e_1^* \otimes e_2^*) \otimes (e_2 \otimes e_1)) \\
&= \pi\left((qs^* - q^{-1}t^*) \otimes \frac{q^{-1}s + qt}{q+q^{-1}}\right) \\
&= \frac{1}{q+q^{-1}} s^* \otimes s - \frac{1}{q+q^{-1}} t^* \otimes t
\end{aligned}$$

Now the relation $bc = cb$ is obvious, and $ad - da = (q - q^{-1})bc$ is a simple calculation.

References

[AG17] Andrea Appel and Sachin Gautam. An explicit isomorphism between quantum and classical $\mathfrak{sl}_n$. To appear in Transformation Groups. arXiv:1712.03601v2

[CP94] V. Chari and A. Pressley. *A Guide to Quantum Groups*, Cambridge University Press, 1994.

[Dr87] Drinfel'd, V. G. Quantum groups. Proceedings of the International Congress of Mathematicians, Vol. 1, 2 (Berkeley, Calif., 1986), 798–820, Amer. Math. Soc., Providence, RI, 1987.

[Dr90] V. G. Drinfel'd. Quasi-Hopf algebras, *Leningrad Math. J.* **1** (1990), no. 6, 1419–1457.

[FRT90] N. Yu. Reshetikhin, L. A. Takhtadzhyan, L. D. Faddeev, "Quantization of Lie groups and Lie algebras", Algebra i Analiz, 1:1 (1989), 178–206; Leningrad Math. J., 1:1 (1990), 193–225

[FT86] Ludwig D. Faddeev and Leon A. Takhtajan. Liouville model on the lattice, Lect.Notes Phys. 246 (1986) 166–179.

[GGS92] Murray Gerstenhaber, Anthony Giaquinto and Samuel D. Schack. Quantum symmetry. In: Kulish P.P. (eds) Quantum Groups. Lecture Notes in Mathematics, vol 1510 (1992). Springer, Berlin, Heidelberg.

[Gia92] Anthony Giaquinto. Quantization of tensor representations and deformation of matrix bialgebras. *J. Pure Appl. Algebra* 79 (1992), no. 2, 169–190.

[Gyo86] Akihiko Gyoja. A q-analogue of Young symmetrizer. Osaka Journal of Mathematics. 23(4), 1986, 841–852.

[Ji85] https://link.springer.com/article/10.1007%2FBF00704588

[JS90] André Joyal and Ross Street. An introduction to Tannaka duality and quantum groups, in Part II of Category Theory, Proceedings, Como 1990, eds. A. Carboni, M. C. Pedicchio and G. Rosolini, Lectures Notes in Mathematics 1488, Springer, Berlin, 1991, 411–492.

[KK89] Hendrik Tjerk Koelink and Tom H. Koornwinder. The Clebsch-Gordan coefficients for the quantum group $S_\mu(2)$ and q-Hahn polynomials, Indagationes Mathematicae (Proceedings). Vol. 92. No. 4. North-Holland, 1989.

[KT09] Joel Kamnitzer and Peter Tingley. The crystal commutor and Drinfeld's unitarized R-matrix. *J. Algebraic Combin.* 29 Issue 3 (2009), 315–335.

[KS97] Klimyk, Anatoli; Schmüdgen, Konrad. Quantum groups and their representations. Texts and Monographs in Physics. Springer-Verlag, Berlin, 1997.

[KR81] P.P. Kulish and N.Y. Reshetikhin. Quantum linear problem for the sine-Gordon equation and higher representations, J Math Sci (1983) 23: 2435. (Translation of (Zap. Nauchnykh Semin. POMI 101 (1981) 101–110.)

[KR88] A. N. Kirillov and N. Yu. Reshetikhin. Representations of the algebra $U_q(\mathfrak{sl}(2))$, q-orthogonal polynomials, and invariants of links, in: Adv. Series Math. Phys., 7, World Scientific (1989), 285–339.

[Lus93] Lusztig, George. *Introduction to quantum groups.* Volume 110 of Progress in Mathematics. Birkhaüser Boston Inc., Boston, MA, 1993.

[Man87] Yu. I. Manin. Some remarks on Koszul algebras and quantum groups Annales de l'institut Fourier, tome 37, no 4 (1987), 191–205.

[SV88] Y. S. Soibelman and L. L. Vaksman, Algebra of functions on the quantum group SU(2), Funktsional. Anal, i Prilozhen. 22 (1988), no. 3, 1–14; English transl., Functional Anal. Appl. 22 (1988), 170–181.

[V89] L. L. Vaksman, q-analogues of Clebsch-Gordan coefficients, and the algebra of functions on the quantum group $SU(2)$. (English. Russian original), Sov. Math., Dokl. 39, No. 3, 467–470 (1989; Zbl 0693.33006); translation from Dokl. Akad. Nauk SSSR 306, No. 2, 269–271 (1989).

[W87a] S. Woronowicz, Twisted SU(2) group. An example of a noncommutative differential calculus, Publ. Res. Inst. Math. Sci. 23 (1987), 117–181.

[W87b] S.L. Woronowicz. Compact Matrix Pseudogroups. Commun. Math. Phys. 111 (1987), 613–665.

Quantum Periodicity and Kirillov–Reshetikhin Modules

David Hernandez

To Nicolai Reshetikhin on his 60th birthday

Abstract We give a proof of the periodicity of quantum T-systems of type $A_n \times A_\ell$ with certain spiral boundary conditions. Our proof is based on the categorification of the T-system in terms of the representation theory of quantum affine algebras, more precisely on relations between classes of Kirillov–Reshetikhin modules and of evaluation modules.

1 Introduction

The Q-system was introduced by Kirillov and Reshetikhin [KR] as a system of relations between characters of certain simple finite-dimensional representations of the quantum affine algebra $\mathcal{U}_q(\hat{sl}_{n+1})$, now called Kirillov–Reshetikhin modules

$$Q_{a,b}^2 = Q_{a-1,b}Q_{a+1,b} + Q_{a,b+1}Q_{a,b-1},$$

where $1 \leq a \leq n$ and b is a non-negative integer. Inspired by this work, the T-system was written in [KNS] as a refined version of the Q-system depending on a spectral parameter u:

$$T_{a,b}(u-1)T_{a,b}(u+1) = T_{a-1,b}(u)T_{a+1,b}(u) + T_{a,b+1}(u)T_{a,b-1}(u).$$

It was conjectured that it is satisfied by the classes of the Kirillov–Reshetikhin modules.

D. Hernandez (✉)
Université de Paris et Sorbonne Université, CNRS, IMJ-PRG, Paris, France
e-mail: david.hernandez@u-paris.fr

A. Alekseev et al. (eds.), *Representation Theory, Mathematical Physics, and Integrable Systems*, Progress in Mathematics 340,
https://doi.org/10.1007/978-3-030-78148-4_10

In a fundamental work [FR], Frenkel and Reshetikhin introduced a character theory for finite-dimensional representations of quantum affine algebras, called the q-characters. Then, the T-system is satisfied by q-characters of Kirillov–Reshetikhin modules in all types [N2, H2] and so by their classes as conjectured above.

In another direction, Zamolodchikov initiated in [Z] a long series of work on the periodicity of solutions of T-systems with certain boundary conditions, which culminated in the work of Keller [Kel2] with a very general uniform proof of the periodicity of T-systems associated with a pair (Δ, Δ') of Dynkin diagrams (see [IIKNS, Kel1] for reviews and references).

In this note, we propose a simple proof of the periodicity (and half-periodicity) of T-systems of type $A_n \times A_\ell$ and of its quantum version (in the sense of [N2, HL2]), with certain spiral boundary conditions (more general than the unit condition usually considered). We follow the approach in [HL1, Section 12.1] where the proof of the commutative periodicity in type $A_n \times A_1$ is obtained with formulas for solutions in terms of q-characters. Indeed, we find solutions in terms of certain evaluation representations, containing Kirillov-Reshetikhin modules, but not only. More precisely, our solutions are given in terms of their q, t-characters as defined by Nakajima [N1].

The quantum periodicity (and half-periodicity) established in this note should also follow from the analog results in the commutative case (with unit boundary condition) mentioned above and from results in [BZ, CKLP] (indeed the approach in [Kel2] is based on the study of the periodicity of a sequence of mutations in a certain cluster algebra). Our direct method gives an explicit solution in terms of q, t-characters.

The chapter is organized as follows. In Sect. 2, we state the main periodicity and quantum periodicity results and we give several examples. In Sect. 3, we give the necessary reminders on the representation theory of quantum affine algebras. In Sect. 4, we recall how the T-system appears in the Grothendieck ring of the category of representations, and we prove that it also has other incarnations. We conclude in Sect. 5 with the proof of the quantum periodicity.

2 Periodicity and Quantum Periodicity

In this section, we state the periodicity and quantum periodicity of T-systems that we establish in this note.

2.1 Periodicity

Let us first state the commutative $A_n \times A_\ell$-periodicity. Let

$$I = \{1, \cdots, n\} \text{ and } J = \{1, \cdots, \ell\}.$$

We work on the lattice

$$\Lambda = \{(a, b, u) \in I \times J \times \mathbb{Z} | a + b + u \in 2\mathbb{Z}\}.$$

Let us consider a family of commuting variables $(T_{a,b}(u))_{(a,b,u)\in\Lambda}$ satisfying the T-system (sometimes called octahedron relation):

$$T_{a,b}(u-1)T_{a,b}(u+1) = T_{a-1,b}(u)T_{a+1,b}(u) + T_{a,b+1}(u)T_{a,b-1}(u),$$

for any $(a, b, u+1) \in \Lambda$.

So that the system is well defined, we have to fix the boundary conditions, that is, the values of

$$T_{0,b}(u)\ ,\ T_{a,\ell+1}(u)\ ,\ T_{n+1,b}(u) \text{ and } T_{a,0}(u).$$

The first choice is to set all values to 1; this is called the unit boundary condition (see [IIKNS]). The system is already non-trivial with such a choice. We will also consider the following boundary condition. Let $(\mathcal{F}_r)_{r\in I}$ be formal variables that we call coefficients. We also set $\mathcal{F}_0 = \mathcal{F}_{n+1} = 1$. We set the following boundary conditions (for u modulo $2(\ell + n + 2)$):

$$T_{n+1,m}(u) = \begin{cases} \mathcal{F}_{\frac{u+n+m+3}{2}} & \text{for } -n-3 \leq u+m \leq n-1 \\ 1 & \text{for } 0 \leq u+m-n+1 \leq 2\ell+2. \end{cases}$$

$$T_{0,m}(u) = \begin{cases} \mathcal{F}_{\frac{u-m+2}{2}} & \text{for } -2 \leq u-m \leq 2n, \\ 1 & \text{for } 0 \leq u-m-2n \leq 2(\ell+1). \end{cases}$$

$$T_{k,\ell+1}(u) = \begin{cases} \mathcal{F}_{\frac{u-k-\ell+1}{2}} & \text{for } 0 \leq u-\ell-k+1 \leq 2n+2, \\ 1 & \text{for } \ell+1 \leq u-2n-k \leq 3(\ell+1). \end{cases}$$

$$T_{k,0}(u) = \begin{cases} \mathcal{F}_{\frac{u+k+2}{2}} & \text{for } -2 \leq u+k \leq 2n, \\ 1 & \text{for } 0 \leq u-2n+k \leq 2(\ell+1). \end{cases}$$

This is a particular case of the spiral boundary condition (see [IIKNS]).

Remark 2.1 For $(a, b, u) \in \Lambda$, an induction on $u \geq a+b-2$ shows that $T_{a,b}(u)$ is a rational fraction in the

$$X_{k,m} = T_{k,m}(k+m-2) \text{ for } (k,m) \in I \times J$$

and in the coefficients

$$X_{k,0} = T_{k,0}(k-2) = \mathcal{F}_k.$$

We have the following periodicity.

Theorem 2.2 *For any $(a, b, u) \in \Lambda$, we have the half-periodicity property:*

$$T_{a,b}(u) = T_{n+1-a,\ell+1-b}(u+n+\ell+2).$$

It implies that $T_{a,b}(u)$ is $2(n+\ell+2)$-periodic in u.

Example 2.3 Let $n = \ell = 1$. The non-trivial boundary conditions are

$$T_{0,1}(1) = T_{1,2}(3) = T_{2,1}(5) = T_{1,0}(7) = \mathcal{F}_1.$$

Set $X = X_{1,1} = T_{1,1}(0)$.
The values of $T_{1,1}(u)$ for $u = 0, 2, 4, 6, 8$ are, respectively,

$$X\,,\ \frac{\mathcal{F}_1+1}{X}\,,\ X\,,\ \frac{\mathcal{F}_1+1}{X}\,,\ X.$$

Example 2.4 Let $n = 1$ and $\ell = 2$. The non-trivial boundary conditions are

$$T_{0,1}(1) = T_{0,2}(2) = T_{1,3}(4) = T_{2,2}(6) = T_{2,1}(7) = T_{1,0}(9) = \mathcal{F}_1.$$

Set $X_1 = X_{1,1} = T_{1,1}(0)$ and $X_2 = X_{1,2} = T_{1,2}(1)$.
The values of $T_{1,1}(t)$ for $t = 0, 2, 4, 6, 8, 10$ are

$$X_1\,,\ \frac{\mathcal{F}_1+X_2}{X_1}\,,\ \frac{X_1+1}{X_2}\,,\ X_2\,,\ \frac{\mathcal{F}_1X_1+\mathcal{F}_1+X_2}{X_1X_2}\,,\ X_1.$$

The values of $T_{1,2}(u)$ for $u = 1, 3, 5, 7, 9, 11$ are, respectively,

$$X_2\,,\ \frac{\mathcal{F}_1X_1+\mathcal{F}_1+X_2}{X_1X_2}\,,\ X_1\,,\ \frac{\mathcal{F}_1+X_2}{X_1}\,,\ \frac{X_1+1}{X_2}\,,\ X_2.$$

Example 2.5 Let $n = 2$ and $\ell = 1$. The non-trivial boundary conditions are

$$T_{0,1}(1) = T_{1,2}(3) = T_{2,2}(4) = T_{3,1}(6) = T_{2,0}(8) = T_{1,0}(9) = \mathcal{F}_1.$$

$$T_{1,0}(1) = T_{0,1}(3) = T_{1,2}(5) = T_{2,2}(6) = T_{3,1}(8) = T_{2,0}(11) = \mathcal{F}_2.$$

Set $X_1 = X_{1,1} = T_{1,1}(0)$ and $X_2 = X_{2,1} = T_{2,1}(1)$.

The values of $T_{1,1}(t)$ for $t = 0, 2, 4, 6, 8, 10$ are

$$X_1\,,\ \frac{\mathcal{F}_1X_2+\mathcal{F}_2}{X_1}\,,\ \frac{X_1+\mathcal{F}_2}{X_2}\,,\ X_2\,,\ \frac{X_1+\mathcal{F}_2+\mathcal{F}_1X_2}{X_1X_2}\,,\ X_1.$$

The values of $T_{2,1}(u)$ for $u = 1, 3, 5, 7, 9, 11$ are, respectively,

$$X_2\,,\ \frac{X_1+\mathcal{F}_2+\mathcal{F}_1X_2}{X_1X_2}\,,\ X_1\,,\ \frac{\mathcal{F}_2+\mathcal{F}_1X_2}{X_1}\,,\ \frac{X_1+\mathcal{F}_2}{X_2}\,,\ X_2.$$

2.2 *Quantum Periodicity*

Let us now state the quantum version of the $A_n \times A_\ell$-periodicity (see also [KN] and [DFK] for $n = 1$).

We work now with quasi-commuting variables $(X_{a,b})_{(a,b)\in I\times J}$:

$$X_{a,b} * X_{c,d} = t^{\gamma(a,b;c,d)-\gamma(c,d;a,b)}X_{c,d} * X_{a,b}.$$

To define the power of t, we use the inverse $\tilde{C}(z)$ of the quantized Cartan matrix

$$C(z) = ((z+z^{-1})\delta_{i,j} - \delta_{i+1,j} - \delta_{i-1,j})_{i,j\in I}.$$

For $p \in \mathbb{Z}$ and $a, c \in I$, we denote by $\tilde{C}_{a,c}(p)$ the coefficient of z^p in the expansion in z of $\tilde{C}_{a,c}(p)$. We set

$$\gamma(a,b;c,d)=\tilde{C}_{a,c}(2\ell-2b+c-a+1)+\tilde{C}_{a,c}(2\ell-2b+c-a-1)+\cdots+\tilde{C}_{a,c}(2d-2b+c-a+1).$$

The relation is also extended to $b = 0$ or $d = 0$, so that we get the quasi-commutation rule with the coefficients. In particular, for $r < r'$, we have

$$\mathcal{F}_r * \mathcal{F}_{r'} = t^{\tilde{C}_{r,r'}(2\ell+r'-r+1)+\tilde{C}_{r,r'}(2\ell+r'-r-1)+\cdots+\tilde{C}_{r,r'}(2\ell+r-r'+3)}\mathcal{F}_{r'} * \mathcal{F}_r \tag{1}$$

and $\mathcal{F}_r * \mathcal{F}_{r+1} = t^{\tilde{C}_{r,r+1}(2\ell+2)}\mathcal{F}_{r+1} * \mathcal{F}_r$. This is derived from $\tilde{C}_{i,j}(k) = 0$ if $k \leq |j-i|$ (which can be observed, for example, in the formula in [GTL, Appendix A.3]).

The quasi-commuting variables $(X_{a,b})_{(a,b)\in I\times J}$ with the $\mathcal{F}_t$ generate a quantum torus $\mathcal{T}_t$ over $\mathbb{Z}[t^{\pm 1/2}]$. We denote its fraction field by K_t. It has an antimultiplicative bar involution satisfying $\bar{t} = t^{-1}$ and so that the $X_{a,b}$, $\mathcal{F}_t$ are bar-invariant.

For each product m of various $X_{a,b}^{\pm 1}$, $\mathcal{F}_t^{\pm 1}$, there is a unique $\alpha \in \mathbb{Z}$ so that $t^{\alpha/2}m$ is bar-invariant. This is called a commutative monomial. The commutative monomials form a $\mathbb{Z}[t^{\pm 1/2}]$-basis of $\mathcal{T}_t$.

Example 2.6 For $n = \ell = 1$, we have

$$X_1 * \mathcal{F}_1 = t^2 \mathcal{F}_1 * X_1.$$

For $n = 1$ and $\ell = 2$, we have

$$X_1 * X_2 = t^{-2} X_2 * X_1 \, , X_1 * \mathcal{F}_1 = \mathcal{F}_1 * X_1 \, , X_2 * \mathcal{F}_1 = \mathcal{F}_1 * X_2.$$

For $n = 2$ and $\ell = 1$, we have

$$X_1 * X_2 = t X_2 * X_1 \, , X_1 * \mathcal{F}_1 = t \mathcal{F}_1 * X_1 \, , X_1 * \mathcal{F}_2 = \mathcal{F}_2 * X_1,$$

$$X_2 * \mathcal{F}_2 = t \mathcal{F}_2 * X_2 \, , X_2 * \mathcal{F}_1 = t \mathcal{F}_1 * X_2 \, , \mathcal{F}_1 * \mathcal{F}_2 = t^{-1} \mathcal{F}_2 * \mathcal{F}_1.$$

We fix the same boundary conditions as for the commutative setting above.

Theorem 2.7 *Consider a family of bar-invariant $T_{a,b}(u) \in K_t$ satisfying:*

$$T_{a,b}(u-1) * T_{a,b}(u+1) \in t^{\mathbb{Z}/2} T_{a-1,b}(u) * T_{a+1,b}(u) + t^{\mathbb{Z}/2} T_{a,b+1}(u) * T_{a,b-1}(u)$$

for $(a, b, u+1) \in \Lambda$. We assume the same initial conditions as in Remark 2.1 and the same spiral boundary conditions as in the classical setting. Then, for $(a, b, u) \in \Lambda$,

$$T_{a,b}(u) = T_{n+1-a,\ell+1-b}(u + n + \ell + 2).$$

It implies that $T_{a,b}(u)$ is $2(n + \ell + 2)$-periodic in u.

The classical periodicity in Theorem 2.2 follows directly from this theorem. We propose a simple proof based on the representation theory of quantum affine algebras and on their Kirillov–Reshetikhin modules.

Example 2.8 Let us study Examples 2.3, 2.4, and 2.5 above. In these examples, let us just replace each Laurent monomial in the $X_{a,b}$ by the corresponding commutative monomial in the quantum torus. We get bar-invariant elements in $\mathcal{T}_t$ and we keep the notation $T_{a,b}(u)$. Let us verify that they satisfy the quantum T-system.

Let $n = \ell = 1$. We get

$$T_{1,1}(0) * T_{1,1}(2) = t\mathcal{F}_1 + 1, \quad T_{1,1}(2) * T_{1,1}(4) = t^{-1}\mathcal{F}_1 + 1.$$

Let $n = 1$ and $\ell = 2$.

$$T_{1,1}(0) * T_{1,1}(2) = \mathcal{F}_1 + t^{-1} T_{1,2}(1), \quad T_{1,2}(1) * T_{1,2}(3) = \mathcal{F}_1 + t^{-1} T_{1,1} \quad (2),$$

$$T_{1,1}(2) * T_{1,1}(4) = 1 + t^{-1} T_{1,2}(3), \quad T_{1,2}(3) * T_{1,2}(5) = 1 + t^{-1} \mathcal{F}_1 * T_{1,1} \quad (4),$$

$$T_{1,1}(4) * T_{1,1}(6) = 1 + t^{-1} T_{1,2}(5), \quad T_{1,2}(5) * T_{1,2}(7) = \mathcal{F}_1 + t^{-1} T_{1,1} \quad (6),$$

$$T_{1,1}(6) * T_{1,1}(8) = \mathcal{F}_1 + t^{-1}T_{1,2}(7), \quad T_{1,2}(7) * T_{1,2}(9) = 1 + t^{-1}T_{1,1}(8),$$

$$T_{1,1}(8) * T_{1,1}(10) = 1 + t^{-1}T_{1,2}(9) * \mathcal{F}_1, \quad T_{1,2}(9) * T_{1,2}(11) = 1 + t^{-1}T_{1,1}(10).$$

Let $n = 2$ and $\ell = 1$.

$$T_{1,1}(0) * T_{1,1}(2) = t^{\frac{1}{2}}T_{2,1}(1) * \mathcal{F}_1 + \mathcal{F}_2, \quad T_{2,1}(1) * T_{2,1}(3) = tT_{1,1}(2) + 1,$$

$$T_{1,1}(2) * T_{1,1}(4) = t^{\frac{3}{2}}T_{2,1}(3) * \mathcal{F}_2 + \mathcal{F}_1, \quad T_{2,1}(3) * T_{2,1}(5) = t^{\frac{1}{2}}T_{1,1}(4) + t^{-\frac{1}{2}}\mathcal{F}_1,$$

$$T_{1,1}(4) * T_{1,1}(6) = t^{\frac{1}{2}}T_{2,1}(5) + t^{-\frac{1}{2}}\mathcal{F}_2, \quad T_{2,1}(5) * T_{2,1}(7) = t^{\frac{3}{2}}\mathcal{F}_1 * T_{1,1}(6) + \mathcal{F}_2,$$

$$T_{1,1}(6) * T_{1,1}(8) = tT_{2,1}(7) + 1, \quad T_{2,1}(7) * T_{2,1}(9) = t^{\frac{1}{2}}\mathcal{F}_2 * T_{1,1}(8) + \mathcal{F}_1,$$

$$T_{1,1}(8) * T_{1,1}(10) = t^{\frac{1}{2}}T_{2,1}(9) + t^{-\frac{1}{2}}\mathcal{F}_1, \quad T_{2,1}(9) * T_{2,1}(11) = t^{\frac{1}{2}}T_{1,1}(10) + t^{-\frac{1}{2}}\mathcal{F}_2.$$

3 Finite-Dimensional Representations of Quantum Affine Algebras

We recall the main definitions and properties of finite-dimensional representations of the quantum affine algebra associated with sl_{n+1}.

3.1 Quantum Affine Algebras

All vector spaces, algebras, and tensor products are defined over $\mathbb{C}$.

Let $C = (C_{i,j})_{0 \le i,j \le n}$ be the Cartan matrix of type $A_n^{(1)}$, that is,

$$C_{i,j} = 2\delta_{i,j} - \delta_{i,j+1} - \delta_{i+1,j},$$

where $n+1$ is identified with 0. Fix $q \in \mathbb{C}^*$, which is not a root of unity.

The *quantum affine algebra* $\mathcal{U}_q(\mathfrak{g})$ is defined by generators $k_i^{\pm 1}$, $x_i^{\pm}$ $(0 \le i \le n)$ and relations

$$k_i k_j = k_j k_i \,,\, k_i x_j^{\pm} = q^{\pm C_{i,j}} x_j^{\pm} k_i \,,\, [x_i^+, x_j^-] = \delta_{i,j} \frac{k_i - k_i^{-1}}{q - q^{-1}},$$

$$\sum_{p=0 \cdots 1-C_{i,j}} (-1)^p (x_i^{\pm})^{(1-C_{i,j}-p)} x_j^{\pm} (x_i^{\pm})^{(p)} = 0 \text{ (for } i \neq j),$$

where we denote $\left(x_i^{\pm}\right)^{(p)} = \left(x_i^{\pm}\right)^p / [p]_q$ for $0 \le p \le 2$, where $[p]_q = (q^p - q^{-p})(q - q^{-1})^{-1}$.

It is a Hopf algebra with a coproduct $\Delta : \mathcal{U}_q(\mathfrak{g}) \to \mathcal{U}_q(\mathfrak{g}) \otimes \mathcal{U}_q(\mathfrak{g})$ defined for $0 \leq i \leq n$ by

$$\Delta(k_i) = k_i \otimes k_i \ , \ \Delta(x_i^+) = x_i^+ \otimes 1 + k_i \otimes x_i^+ \ , \ \Delta(x_i^-) = x_i^- \otimes k_i^{-1} + 1 \otimes x_i^- .$$

Let $\overline{\mathfrak{g}} = sl_{n+1}$ be the finite-dimensional simple Lie algebra of Cartan matrix $(C_{i,j})_{i,j\in I}$. We denote, respectively, by ω_i, α_i, and $\alpha_i^\vee$ $(i \in I)$ the fundamental weights, the simple roots, and the simple coroots of $\overline{\mathfrak{g}}$. We use the standard partial ordering $\leq$ on the weight lattice P of $\overline{\mathfrak{g}}$.

The algebra $\mathcal{U}_q(\mathfrak{g})$ has another set of generators, the *Drinfeld generators*, denoted by

$$x_{i,m}^{\pm} \ , k_i^{\pm 1} \ , h_{i,r} \ , c^{\pm 1/2} \text{ for } i \in I, \ m \in \mathbb{Z}, \ r \in \mathbb{Z} \setminus \{0\}.$$

We have $x_i^{\pm} = x_{i,0}^{\pm}$ for $i \in I$. A complete set of relations for the Drinfeld generators was obtained in [B, D]. In particular, the multiplication defines a surjective linear morphism

$$\mathcal{U}_q^-(\mathfrak{g}) \otimes \mathcal{U}_q(\mathfrak{h}) \otimes \mathcal{U}_q^+(\mathfrak{g}) \to \mathcal{U}_q(\mathfrak{g}), \tag{2}$$

where $\mathcal{U}_q^{\pm}(\mathfrak{g})$ is the subalgebra generated by the $x_{i,m}^{\pm}$ $(i \in I, m \in \mathbb{Z})$ and $\mathcal{U}_q(\mathfrak{h})$ is the subalgebra generated by the $k_i^{\pm 1}$, $h_{i,r}$, and $c^{\pm 1/2}$ $(i \in I, r \in \mathbb{Z} \setminus \{0\})$.

3.2 Finite-Dimensional Representations

We refer to [CH] for generalities on the category $\mathcal{C}$ of finite-dimensional representations of $\mathcal{U}_q(\mathfrak{g})$. For $i \in I$, the action of k_i on any object of $\mathcal{C}$ is diagonalizable with eigenvalues in $\pm q^{\mathbb{Z}}$. Without loss of generality, we can assume that $\mathcal{C}$ is the category of *type 1* finite-dimensional representations (see [CP2]), i.e., we assume that for any object of $\mathcal{C}$, the eigenvalues of k_i are in $q^{\mathbb{Z}}$ for $i \in I$. The simple objects of $\mathcal{C}$ are parametrized by n-tuples of polynomials $(P_i(u))_{i\in I}$ satisfying $P_i(0) = 1$ (they are called the *Drinfeld polynomials*) [CP1, CP2].

In type A, there is a family of evaluation morphisms $ev_a : \mathcal{U}_q(\mathfrak{g}) \to \mathcal{U}_q(\overline{\mathfrak{g}})$ parametrized by $a \in \mathbb{C}^*$. Hence, for V a simple finite-dimensional representation of $\mathcal{U}_q(\overline{\mathfrak{g}})$, by pullback, we get an evaluation representation $(V)_a$. If the highest weight of V is a multiple of a fundamental weight, then V is a Kirillov–Reshetikhin module. In the particular case of a fundamental weight, we get the fundamental representations $V_i(a) = (V(\omega_i))_a$ of the Drinfeld polynomials $(1, \cdots, 1, 1 - za, 1, \cdots, 1)$ with a non-trivial polynomial in position i. Their classes generate the Grothendieck ring $K_0(\mathcal{C})$ of the category $\mathcal{C}$, which is a polynomial ring in the variables $[V_i(a)]$ as proved in [FR]. In general, simple finite-dimensional representations are not evaluation modules.

For $\omega \in P$, the *weight space* V_ω of an object V in $\mathcal{C}$ is the set of *weight vectors* of weight ω, i.e., of vectors $v \in V$ satisfying $k_i v = q^{(\omega(\alpha_i^\vee))} v$ for any $i \in I$.

The elements $c^{\pm 1/2}$ act by identity on any object V of $\mathcal{C}$, and so the action of the $h_{i,r}$ commutes. Since the $h_{i,r}$, $i \in I$, $r \in \mathbb{Z} \setminus \{0\}$, also commute with the k_i, $i \in I$, every object in $\mathcal{C}$ can be decomposed as a direct sum of generalized eigenspaces of the $h_{i,r}$ and k_i. More precisely, by the Frenkel–Reshetikhin theory of q-characters [FR], the eigenvalues of the $h_{i,r}$ and k_i can be *encoded* by *monomials* m in formal variables $Y_{i,a}^{\pm 1}$ ($i \in I, a \in \mathbb{C}^*$). Let $\mathcal{M}$ be the set of such monomials (also called *l-weights*). Given $m \in \mathcal{M}$ and an object V in $\mathcal{C}$, let V_m be the subspace of V of common pseudo-eigenvectors of the $h_{i,r}$, k_i with pseudo-eigenvalues associated with m (also called *l-weight space*). Thus,

$$V = \bigoplus_{m \in \mathcal{M}} V_m.$$

If $v \in V_m$, then v is a weight vector of weight

$$\omega(m) = \sum_{i \in I, a \in \mathbb{C}^*} u_{i,a}(m) \omega_i \in P,$$

where we denote $m = \prod_{i \in I, a \in \mathbb{C}^*} Y_{i,a}^{u_{i,a}(m)}$. For $v \in V_m$, we set $\omega(v) = \omega(m)$.

The *q-character morphism* is an injective ring morphism

$$\chi_q : \mathrm{Rep}(\mathcal{U}_q(\mathfrak{g})) \to \mathcal{Y} = \mathbb{Z}\left[Y_{i,a}^{\pm 1}\right]_{i \in I, a \in \mathbb{C}^*},$$

$$\chi_q(V) = \sum_{m \in \mathcal{M}} \dim(V_m) m.$$

If $V_m \neq \{0\}$, we say that m is an *l-weight of* V.

A monomial $m \in \mathcal{M}$ is said to be *dominant* if $u_{i,a}(m) \geq 0$ for any $i \in I, a \in \mathbb{C}^*$. For V a simple object in $\mathcal{C}$, let $M(V)$ be the *highest weight monomial* of $\chi_q(V)$, that is, so that $\omega(M(V))$ is maximal for the partial ordering on P. $M(V)$ is dominant and characterizes the isomorphism class of V (it is equivalent to the data of the Drinfeld polynomials). Hence, to a dominant monomial M is associated a simple representation $L(M)$. For $i \in I$ and $a \in \mathbb{C}^*$, we have, for example, the fundamental representation $V_i(a) = L\left(Y_{i,a}\right)$. The simple modules of the highest weight monomial

$$X_{i,\alpha}^{\beta} = Y_{i,q^\alpha} Y_{i,q^{\alpha+2}} \cdots Y_{i,q^{\alpha+2(\beta-1)}}$$

for some $i \in I, \alpha \in \mathbb{Z}$, and $\beta \geq 1$ are Kirillov–Reshetikhin modules. We will also use the notation $X_{i,\alpha}^{\beta} = 1$ for $\beta \leq 0$.

Example 3.1 The q-character of the fundamental representation $L(Y_a)$ of $\mathcal{U}_q(\hat{sl}_2)$ is

$$\chi_q(L(Y_a)) = Y_a + Y_{aq^2}^{-1}.$$

The q-characters of evaluation modules, including Kirillov–Reshetikhin modules and fundamental modules, are known explicitly (see references in the introduction of [H3]). The formulas involve the monomials $A_{i,a}$ defined in [FR] for $i \in I$, $a \in \mathbb{C}^*$ by

$$A_{i,a} = Y_{i,aq^{-r_i}} Y_{i,aq^{r_i}} \times \prod_{\{j \in I | C_{i,j} = -1\}} Y_{j,a}^{-1}.$$

3.3 *Quantum Grothendieck Ring*

The Grothendieck ring $K_0(\mathcal{C})$ has a t-deformation called the quantum Grothendieck ring $K_t(\mathcal{C})$ as constructed in [VV, N1, H1] (we use the version of [H1, HL2]). It is a $\mathbb{Z}[t^{\pm 1/2}]$-subalgebra of a quantum torus $\mathcal{Y}_t$, and simple objects $L(m)$ have corresponding classes $[L(m)]_t \in K_t(\mathcal{C})$. A quantum version of a result in [FM] gives the following [N1]:

$$[L(m)]_t \in m * (1 + \mathbb{Z}[t^{\pm 1/2}, A_{i,c}^{-1}]_{i \in I, c \in \mathbb{C}^*}). \quad (3)$$

In other words, m is maximal for the Nakajima partial ordering on monomials, that is, $M \preceq M'$ if $M'M^{-1}$ is a product of variables $A_{i,c}$.

If a simple module V is thin, that is, if its ℓ-weight spaces are of dimension 1, then $[V]_t$ is a sum of commutative monomials (defined as in Sect. 2.2) and can be identified with its q-character (see [HL2, Corollary 5.3]). In type A, all simple evaluation modules are thin.

4 Relations in the Grothendieck Ring

We recall how the T-system originally occurs in the representation theory of quantum affine algebras, and we also establish another incarnation of the T-system in the Grothendieck ring $K_0(\mathcal{C})$ (which we call horizontal T-system).

We denote $I = \{1, \cdots, n\}$ and $J = \{1, \cdots, \ell\}$ as above.

4.1 Original T-Systems

For $1 \leq i \leq n$ and $0 \leq m \leq p \leq \ell$, consider the Kirillov–Reshetikhin module

$$\beta(m, p)^i = L(X_{i,i+2m}^{p-m+1}).$$

We extend the notation to $i = 0$ and $i = n + 1$ by setting $\beta(m, p)^i = 1$ in these cases.

For $i \in I$ and $0 \leq m \leq p < \ell$, we have the T-system in $K_0(\mathcal{C})$:

$$\beta(m, p)^i \beta(m+1, p+1)^i = \beta(m, p+1)^i \beta(m+1, p)^i + \beta(m+1, p+1)^{i-1} \beta(m, p)^{i+1}.$$

See the list of references in the introduction of [H3]. It can be deformed into the quantum T-system in $K_t(\mathcal{C})$ (see [N2, HL2]):

$$\beta(m, p)^i * \beta(m+1, p+1)^i = t^\lambda \beta(m, p+1)^i * \beta(m+1, p)^i + t^\mu \beta(m+1, p+1)^{i-1} * \beta(m, p)^{i+1}, \tag{4}$$

for some $\lambda, \mu \in \mathbb{Z}/2$, which depend on m, p, i (they can be explicitly computed, but this is not relevant for the following).

4.2 Horizontal T-Systems

The T-system has another incarnation in $K_0(\mathcal{C})$.

For $0 \leq i \leq j \leq n + 1$ and $0 \leq m \leq \ell + 1$, consider the evaluation module

$$\alpha(i, j)^m = L(M_{[i,j]}^m) \text{ where } M_{[i,j]}^m = X_{i,i}^m X_{j,j+2m}^{\ell+1-m}.$$

Some of these representations are Kirillov–Reshetikhin modules:

$$\alpha(i, n+1)^{m+1} = \beta(0, m)^i \text{ and } \alpha(0, i)^m = \beta(m, \ell)^i \text{ for } 0 \leq m \leq \ell \text{ and } 0 \leq i \leq n+1. \tag{5}$$

For $0 \leq i \leq n + 1$ and $j \geq 0$, we will denote

$$F_i = L(X_{i,i}^{\ell+1}) = \alpha(i, i)^m = \beta(0, \ell)^i = \alpha(i, i+j)^{\ell+1} = \alpha(i-j, i)^0. \tag{6}$$

Note that $F_0 = F_{n+1} = 1$.

Theorem 4.1 *For $0 \leq i < j \leq n$ and $m \in J$, there are $\lambda, \lambda' \in \mathbb{Z}/2$, so that*

$$\alpha(i, j)^m * \alpha(i+1, j+1)^m = t^\lambda \alpha(i, j+1)^m * \alpha(i+1, j)^m + t^{\lambda'} \alpha(i, j)^{m+1} * \alpha(i+1, j+1)^{m-1}.$$

Remark 4.2

(i) This relation is "orthogonal" to the original quantum T-system in the sense that the spectral parameter is replaced by the vertex of the Dynkin diagram.
(ii) At $t = 1$, it can be shown that the relation comes from a non-split exact sequence obtained by a normalized R-matrix, as for the original T-system (see [N2, H2]). In fact, it can be checked that the two tensor products associated with the right-hand terms correspond to simple modules. Using [C, Theorem 4], the proof is analogous to the one for the T-system.
(iii) At $t = 1$, this relation can be seen as an extended T-systems in [MY].
(iv) For the limit values of i, j, the relation involves both the α and the β-modules and so connects the two families. The specialization at $t = 1$ reads

$$\beta(m,\ell)^j\alpha(1,j+1)^m = \beta(m,\ell)^{j+1}\alpha(1,j)^m + \beta(m+1,\ell)^j\alpha(1,j+1)^{m-1} (\text{ for } i = 0)$$

$$\alpha(i,n)^m\beta(0,m-1)^{i+1} = \beta(0,m-1)^i\alpha(i+1,n)^m + \alpha(i,n)^{m+1}\beta(0,m-2)^{i+1} (\text{ for } j = 0).$$

Proof By Frenkel and Reshetikhin [FR] and Frenkel and Mukhin [FM], a q-character is determined uniquely by the multiplicity of its dominant monomials. We will use the notation $A_{i,\lambda}$ instead of A_{i,q^λ} for $i \in I$, $\lambda \in \mathbb{Z}$. First, we prove that $\alpha(i,j)^m * \alpha(i+1,j+1)^m$ has $2m+1$ dominant monomials:

$$M_1 = M_{[i,j]}^m M_{[i+1,j+1]}^m \, , \, M_2 = M_1 A_{i+1,i+2m}^{-1} A_{i+2,i+1+2m}^{-1} \cdots A_{j,j+2m-1}^{-1},$$

$$M_{2r} = M_2 \prod_{2\leq p\leq r} (A_{i,i+2m-2p+3}A_{i+1,i+2m-2p+2})^{-1} \, , \, M_{2r+1} = M_{2r}A_{i,i+2m-2r+1}^{-1},$$

where $1 \leq r \leq m$. It is clear that these monomials occur. Indeed, M_1 is the product of the highest monomials. For $2 \leq r \leq 2m+1$, we decompose

$$M_r = (M_{[i+1,j+1]}^m(M_2M_1^{-1})) \times (M_{[i,j]}^m(M_rM_2^{-1})).$$

Now, consider $M \neq M_1$ a dominant monomial that occurs. We factorize

$$M = M_1 M' M'',$$

where M' (respectively, M'') is a monomial of $(M_{[i,j]}^m)^{-1}\alpha(i,j)^m$ (respectively, of $(M_{[i+1,j+1]}^m)^{-1}\alpha(i+1,j+1)^m$). As M is dominant, we have $M'M'' \in \mathbb{Z}[A_{k,r}^{-1}]_{k\in I, r\leq j+2\ell}$. Then,

$$M' \in \mathbb{Z}[A_{k,r}^{-1}]_{k\leq j-1, r\in\mathbb{Z}}$$

and that there is $R \geq 0$ such that

$$M'' \in (A_{j,j+2m-1} A_{j,j+2m-3} \cdots A_{j,j+2m-1-2R})^{-1} \mathbb{Z}[A_{k,r}^{-1}]_{k \leq j-1, r \in \mathbb{Z}}.$$

The monomial $\tilde{M} = M(X^{\ell+1-m}_{j+1,j+1+2m} X^{\ell+1-m}_{j,j+2m})^{-1}$ is a monomial of $\chi_q(L(X^m_{i,i} X^m_{i+1,i+1}))$ which has a unique dominant monomial. Hence, $\tilde{M} Y_{j,j+2m}$ is dominant. So,

$$\tilde{M}' = \tilde{M}(A_{j,j+2m-1} \cdots A^{-1}_{i+1,i+2m})$$

is a monomial of $\chi_q(L(X^m_{i,i} X^m_{i+1,i+1}))$. If $\tilde{M}' = X^m_{i,i} X^m_{i+1,i+1}$, then $M = M_2$. Otherwise, $\tilde{M} A_{i,i+2m-1}$ is a monomial of $\chi_q(L(X^m_{i,i} X^m_{i+1,i+1}))$. We continue by induction, and so M is one of the M_r.

This also implies that each M_r occurs with multiplicity which is a power of t.

Similarly, we get that $\alpha(i, j+1)^m * \alpha(i+1, j)^m$ has $m+1$ dominant monomials that are the M_{2r+1} for $0 \leq r \leq m$. We also get that $\alpha(i, j)^{m+1} * \alpha(i+1, j+1)^{m-1}$ has m dominant monomials that are the M_{2r} for $1 \leq r \leq m$.

To conclude, we have to check that the powers of t match: this can be done using positivity in the quantum Grothendieck ring as in [HL2, Section 5.10] or directly as in [HO, section 9]. □

5 Proof of Periodicity

In this section, we finish the proof of the quantum periodicity.

It suffices to identify the $T_{a,b}(t)$ with variables satisfying the T-system, the half-periodicity, and such that the variables corresponding to the $X_{a,b}$ are algebraically independent. We will identify the $T_{a,b}(t)$ with certain q,t-characters of minimal affinizations, that is, elements of the quantum torus $\mathcal{Y}_t$.

For $0 \leq k \leq n+1$, $0 \leq m \leq \ell+1$ and $u \in \mathbb{Z}$ so that $k+m+u \in 2\mathbb{Z}$, we set

$$T_{k,m}(u) = \begin{cases} \alpha(\frac{u+2-k-m}{2}, \frac{u+2+k-m}{2})^m & \text{for } 0 \leq u+2-k-m \leq 2(n+1-k), \\ \beta(\frac{u-2n+k-m}{2}, \frac{u-2n-2+k+m}{2})^{n+1-k} & \text{for } m \leq u-2n+k \leq 2\ell-m+2, \\ \alpha(\frac{u-2n-2\ell+k+m-2}{2}, \frac{u-k-2\ell+m}{2})^{\ell+1-m} & \text{for } 0 \leq u-2n-2\ell-2+m+k \leq 2k, \\ \beta(\frac{u-2-2n-k+m-2\ell}{2}, \frac{u-2n-2-k-m}{2})^k & \text{for } -m \leq u-2\ell-2n-k-2 \leq m. \end{cases}$$

This defines $T_{k,m}(u)$ for $0 \leq u-m-k+2 \leq 2n+2\ell+4$, and we extend the definition for any u by $2(n+\ell+2)$-periodicity.

Remark 5.1

(i) The formulas in all cases are compatible thanks to relations (5).
(ii) Identifying the class F_r defined in (6) with $\mathcal{F}_r$, we recover boundary conditions of Sect. 2.

The $X_{k,m}$ quasi-commute, with the same rules as in Sect. 2. The relations in (4) and Theorem 4.1 imply that the $T_{a,b}(u)$ satisfy the quantum T-system for a distinguished choice of the powers of t (let us call it the distinguished powers).

The $(X_{k,m})_{(k,m)\in I\times(J\cup\{0\})}$ form a family of algebraically independent variables. We may argue as in [HL4]. Let us explain this point for completeness: all the representations we consider belong to the monoidal category $\mathcal{C}_\ell^o$ of representations whose classes belong to the subring of the Grothendieck ring $K_0(\mathcal{C})$ generated by the classes of fundamental representations $[L(Y_{k,k+2m})]$ for $(k,m)\in I\times(J\cup\{0\})$. Then, there is an injective ring morphism

$$\chi_q^T : K_0(\mathcal{C}_\ell^o) \to \mathcal{Y}$$

called truncated q-character morphism [HL3]: it is defined so that for $L(m)$ a simple module in $\mathcal{C}_\ell^o$, $\chi_q^T(L(m))$ is obtained from $\chi_q(L(m))$ by removing the monomials m' so that in the factorization of $m'm^{-1}$, a factor of the form $A_{k,k+2\ell+1}^{-1}$, $k\in I$, occurs. Now, by Hernandez [H2], the $\chi_q^T(X_{k,m}) = X_{k,k+2m}^{\ell+1-m}$ are just monomials that are clearly algebraically independent.

As by construction we have $T_{a,b}(u) = T_{n+1-a,\ell+1-b}(u+n+\ell+2)$, we get the result for the quantum T-system with the distinguished powers of t.

To conclude, it suffices to check that the powers of t correspond automatically to the distinguished choice. We consider a solution, and we prove by induction on $u\geq a+b-2$ that the $T_{a,b}(u)$ correspond to the q,t-characters and that the powers of t are given by the distinguished choice. As discussed above, the $X_{a,b}$ are algebraically independent, so we can identify the $T_{a,b}(u)$ for $u=a+b-2$, with the corresponding q,t-characters. In general, we have a relation

$$T_{a,b}(U+1)*T_{a,b}(U-1) = t^\alpha T_{a-1,b}(U)*T_{a+1,b}(U)+t^\beta T_{a,b+1}(U)*T_{a,b-1}(U),$$

for some $\alpha,\beta\in\mathbb{Z}/2$. For $u\leq U$, we have $T_{a,b}(u) = M_{a,b}(u)\chi_{a,b}(u)$, where $M_{a,b}(u)$ is a monomial in the quantum torus and $\chi_{a,b}(u)$ is a polynomial in the $A_{i,c}^{-1}$ with coefficients in $\mathbb{Z}[t^{\pm 1}]$ and with constant term 1 (see (3)). Then, $(\chi_{a,b}(u))^{-1}$ is a formal power series in the $A_{i,c}^{-1}$. Each term of the sum

$$T_{a,b}(U+1)=t^\alpha T_{a-1,b}(U)*T_{a+1,b}(U)*(T_{a,b}(U-1))^{-1}+t^\beta T_{a,b+1}(U)*T_{a,b-1}(U)*(T_{a,b}(U-1))^{-1}$$

is a monomial multiplied by such a formal power series. The highest monomial is

$$t^\alpha M_{a-1,b}(U)*M_{a+1,b}(U)*(M_{a,b}(U-1))^{-1},$$

which only appears in the first term. As $T_{a,b}(U+1)$ is bar-invariant, it imposes that α is the power of the distinguished choice. Then, one may consider

$$T_{a,b}(U+1)-t^\alpha T_{a-1,b}(U)*T_{a+1,b}(U)*(T_{a,b}(U-1))^{-1}.$$

The same arguments identify β with the distinguished choice. Hence, $T_{a,b}(U+1)$ satisfies the equation as the corresponding q,t-character and so is equal to it.

Example 5.2 Let us study Examples 2.3, 2.4, and 2.5 above. Let $n=\ell=1$. We get

$$L(Y_3)*L(Y_1)=tL(Y_1Y_3)+1, \quad L(Y_1)*L(Y_3)=t^{-1}L(Y_1Y_3)+1.$$

Let $n=1$ and $\ell=2$.

$$L(Y_3Y_5)*L(Y_1)=L(Y_1Y_3Y_5)+t^{-1}L(Y_5), \quad L(Y_5)*L(Y_1Y_3)=L(Y_1Y_3Y_5)+t^{-1}L(Y_1),$$

$$L(Y_1)*L(Y_3)=1+t^{-1}L(Y_1Y_3), \quad L(Y_1Y_3)*L(Y_3Y_5)=1+t^{-1}L(Y_1Y_3Y_5)*L(Y_3),$$

$$L(Y_3)*L(Y_5)=1+t^{-1}L(Y_3Y_5), \quad L(Y_3Y_5)*L(Y_1)=L(Y_1Y_3Y_5)+t^{-1}L(Y_5),$$

$$L(Y_5)*L(Y_1Y_3)=L(Y_1Y_3Y_5)+t^{-1}L(Y_1), \quad L(Y_1)*L(Y_3)=1+t^{-1}L(Y_1Y_3),$$

$$L(Y_1Y_3)*L(Y_3Y_5)=1+t^{-1}L(Y_3)*L(Y_1Y_3Y_5), \quad L(Y_3)*L(Y_5)=1+t^{-1}L(Y_3Y_5).$$

Let $n=2$ and $\ell=1$.

$$L(Y_{1,3})*L(Y_{1,1}Y_{2,4})=t^{\frac{1}{2}}L(Y_{2,4})*L(Y_{1,1}Y_{1,3})+L(Y_{2,2}Y_{2,4}),$$

$$L(Y_{2,4})*L(Y_{1,1})=tL(Y_{1,1}Y_{2,4})+1,$$

$$L(Y_{1,1}Y_{2,4})*L(Y_{2,2})=t^{\frac{3}{2}}L(Y_{1,1})*L(Y_{2,2}Y_{2,4})+L(Y_{1,1}Y_{1,3}),$$

$$L(Y_{1,1})*L(Y_{1,3})=t^{\frac{1}{2}}L(Y_{2,2})+t^{-\frac{1}{2}}L(Y_{1,1}Y_{1,3}),$$

$$L(Y_{2,2})*L(Y_{2,4})=t^{\frac{1}{2}}L(Y_{1,3})+t^{-\frac{1}{2}}L(Y_{2,2}Y_{2,4}),$$

$$L(Y_{1,3})*L(Y_{1,1}Y_{2,4})=t^{\frac{3}{2}}L(Y_{1,1}Y_{1,3})*L(Y_{2,4})+L(Y_{2,2}Y_{2,4}),$$

$$L(Y_{2,4})*L(Y_{1,1})=tL(Y_{1,1}Y_{2,4})+1,$$

$$L(Y_{1,1}Y_{2,4})*L(Y_{2,2})=t^{\frac{1}{2}}L(Y_{2,2}Y_{2,4})*L(Y_{1,1})+L(Y_{1,1}Y_{1,3}),$$

$$L(Y_{1,1})*L(Y_{1,3})=t^{\frac{1}{2}}L(Y_{2,2})+t^{-\frac{1}{2}}L(Y_{1,1}Y_{1,3}),$$

$$L(Y_{2,2})*L(Y_{2,4})=t^{\frac{1}{2}}L(Y_{1,3})+t^{-\frac{1}{2}}L(Y_{2,2}Y_{2,4}).$$

Acknowledgments The author is very grateful to Bernard Leclerc for many discussions over the years and to Bernhard Keller for useful remarks and explanations about [Kel2] and its consequences. The author would like to thank Laura Fedele for her careful reading and for pointing typos in a former version of this work. The author is supported by the European Research Council under the European Union's Framework Programme H2020 with ERC Grant Agreement number 647353 Qaffine.

References

[B] **J. Beck**, *Braid group action and quantum affine algebras*, Comm. Math. Phys. **165** (1994), no. 3, 555–568

[BZ] **A. Berenstein and A. Zelevinsky**, *Quantum cluster algebras* Adv. Math. **195** (2005), no. 2, 405–455.

[C] **V. Chari**, *Braid group actions and tensor products*, Int. Math. Res. Not. **2003**, no. 7, 357–382

[CH] **V. Chari and D. Hernandez**, *Beyond Kirillov-Reshetikhin modules*, in Quantum affine algebras, extended affine Lie algebras, and their applications, Contemp. Math., **506**, 49–81, 2010.

[CKLP] **G. Cerulli Irelli, B. Keller, D. Labardini-Fragoso, P-G. Plamondon**, *Linear independence of cluster monomials for skew-symmetric cluster algebras*, Compos. Math. **149** (2013), no. 10, 1753–1764.

[CP1] **V. Chari and A. Pressley**, *Quantum affine algebras*, Comm. Math. Phys. **142** (1991), no. 2, 261–283

[CP2] **V. Chari and A. Pressley**, *A Guide to Quantum Groups*, Cambridge University Press, Cambridge (1994)

[D] **I. Damiani**, *From the Drinfeld realization to the Drinfeld-Jimbo presentation of affine quantum algebras : Injectivity*, Publ. Res. Inst. Math. Sci. **51** (2015), 131–171.

[DFK] **P. Di Francesco and R. Kedem**, *The solution of the quantum A_1 T-system for arbitrary boundary*, Comm. Math. Phys. **313** (2012), no. 2, 329–350.

[FM] **E. Frenkel and E. Mukhin**, *Combinatorics of q-Characters of Finite-Dimensional Representations of Quantum Affine Algebras*, Comm. Math. Phys., vol **216** (2001), no. 1, 23–57.

[FR] **E. Frenkel and N. Reshetikhin**, *The q-Characters of Representations of Quantum Affine Algebras and Deformations of W-Algebras*, Recent Developments in Quantum Affine Algebras and related topics, Cont. Math., vol. **248** (1999), 163–205

[GTL] **S. Gautam and V. Toledano Laredo**, *Meromorphic tensor equivalence for Yangians and quantum loop algebras*, Publ. Math. Inst. Hautes Études Sci. **125** (2017), 267–337.

[H1] **D. Hernandez**, *Algebraic approach to q,t-characters*, Adv. Math. **187**, 1–52 (2004).

[H2] **D. Hernandez**, *The Kirillov-Reshetikhin conjecture and solutions of T-systems*, J. Reine Angew. Math. **596** (2006), 63–87.

[H3] **D. Hernandez**, *On minimal affinizations of representations of quantum groups*, Comm. Math. Phys. **277** (2007), no. 1, 221–259

[HL1] **D. Hernandez and B. Leclerc**, *Cluster algebras and quantum affine algebras*, Duke Math. J. **164** (2015), no. 12, 2407–2460.

[HL2] **D. Hernandez and B. Leclerc**, *Quantum Grothendieck rings and derived Hall algebras*, J. Reine Angew. Math. **701** (2015), 77–126.

[HL3] **D. Hernandez and B. Leclerc**, *Monoidal categorifications of cluster algebras of type A and D*, in Symmetries, integrable systems and representations, Proc. Math. Stat. **40** (2013), 175–193.

[HL4] **D. Hernandez and B. Leclerc**, *A cluster algebra approach to q-characters of Kirillov-Reshetikhin modules*, J. Eur. Math. Soc. **18** (2016), no. 5, 1113–1159.

[HO] **D. Hernandez and H. Oya**, *Quantum Grothendieck ring isomorphisms, cluster algebras and Kazhdan-Lusztig algorithm*, Adv. Math. **347** (2019), 192–272.

[IIKNS] **R. Inoue, O. Iyama, A. Kuniba, T. Nakanishi and J. Suzuki**, *Periodicities of T-systems and Y-systems*, Nagoya Math. J. **197** (2010), 59–174.

[Kel1] **B. Keller**, *Algèbres amassées et applications (d'après Fomin-Zelevinsky,...)*, Séminaire Bourbaki. Vol. 2009/2010. Exposés 1012–1026. Astérisque No. 339 (2011), Exp. No. 1014, vii, 63–90.

[Kel2] **B. Keller**, *The periodicity conjecture for pairs of Dynkin diagrams*, Ann. of Math. (2) **177** (2013), no. 1, 111–170.

[KN] **R. Kashaev and T. Nakanishi**, *Classical and Quantum Dilogarithm Identities*, SIGMA **7** (2011), 102.

[KNS] **A. Kuniba, T. Nakanishi and J. Suzuki**, *Functional relations in solvable lattice models. I. Functional relations and representation theory*, Internat. J. Modern Phys. A 9 (1994), no. 30, 5215–5266.

[KR] **A. Kirillov and N. Reshetikhin**, *Representations of Yangians and multiplicities of the inclusion of the irreducible components of the tensor product of representations of simple Lie algebras*, J. Soviet Math. 52, no. 3, 3156–3164 (1990); translated from Zap. Nauchn. Sem. Leningrad. Otdel. Mat. Inst. Steklov. (LOMI) 160, Anal. Teor. Chisel i Teor. Funktsii. 8, 211–221, 301 (1987).

[MY] **E. Mukhin and C. Young**, *Extended T-systems*, Selecta Math. (N.S.) **18** (2012), no. 3, 591–631.

[N1] **H. Nakajima**, *Quiver varieties and t-analogs of q-characters of quantum affine algebras*, Ann. of Math. (2) **160**, no. 3, 1057–1097 (2004).

[N2] **H. Nakajima**, *t-analogs of q-characters of Kirillov-Reshetikhin modules of quantum affine algebras*, Represent. Theory **7** (2003), 259–274.

[VV] **M. Varagnolo and E. Vasserot**, *Perverse sheaves and quantum Grothendieck rings*, Studies in memory of Issai Schur (Chevaleret/Rehovot, 2000), Progr. Math. **210**, Birkhäuser Boston, Boston, MA, 345–365 (2003).

[Z] **A. B. Zamolodchikov**, *On the thermodynamic Bethe ansatz equations for reflectionless ADE scattering theories*, Phys. Lett. B 253 (1991), no. 3–4, 391–394.

A Note on the E-Polynomials of a Stratification of the Hilbert Scheme of Points

Yi-Ning Hsiao and Andras Szenes

Abstract The stratification associated with the number of generators of the ideals of the punctual Hilbert scheme of points on the affine plane has been studied since the 1970s. In this paper, we present an elegant formula for the E-polynomials of these strata.

1 Introduction

The study of the topology of the Hilbert scheme of points on the affine plane has brought a wealth of results in several branches of mathematics: in geometric representation theory, theory of symmetric polynomials, singularities, symplectic geometry, enumerative geometry, etc. [Ch1, Gö1, Ha, Ia, Gr, Le, KST, Na2, Na1, OS]. In this article, we study the strata of the Hilbert scheme of points on the complex plane associated with the number of generators of the elements of this space considered as ideals.

Let $X = \mathbb{C}^2$ be the affine plane and denote by $H^{[n]}$ the Hilbert scheme of n points on X; this is the space parametrizing colength-n ideals of the polynomial ring $R = \mathbb{C}[x, y]$:

$$H^{[n]} = \{I \trianglelefteq R \mid \dim_{\mathbb{C}} R/I = n\}.$$

The space $H^{[n]}$ is naturally endowed with the structure of a smooth $2n$-dimensional complex variety [Fo], and a universal sequence of R-modules:

$$I \to R \to R/I$$

over $I \in H^{[n]}$.

Y.-N. Hsiao · A. Szenes (✉)
Section de mathématiques, Université de Genève, Geneva, Switzerland
e-mail: Yi-Ning.Hsiao@unige.ch; Andras.Szenes@unige.ch

A. Alekseev et al. (eds.), *Representation Theory, Mathematical Physics, and Integrable Systems*, Progress in Mathematics 340,
https://doi.org/10.1007/978-3-030-78148-4_11

Counted with multiplicities, the support of an ideal $I \in H^{[n]}$ is n points in the plane, i.e., an element of $S^n\mathbb{C}^2$, the nth symmetric product of $\mathbb{C}^2$. We will represent the elements of $S^n\mathbb{C}^2$ as formal sums of points in $\mathbb{C}^2$, and the resulting morphism, the Abel-Jacobi map, will be denoted by $\mathrm{AJ} : H^{[n]} \to S^n\mathbb{C}^2$.

The *Briançon variety* or *the punctual Hilbert scheme* $B^{[n]}$ is the subvariety of $H^{[n]}$, which collects all ideals supported at $(0, 0)$:

$$B^{[n]} = \{I \trianglelefteq R \mid \dim_{\mathbb{C}} R/I = n,\ \mathrm{AJ}(I) = n \cdot (0, 0)\}.$$

The variety $B^{[n]}$ is an $(n-1)$-dimensional, compact, and irreducible variety, which is smooth for $n = 1, 2$ only; it is a deformation retract of $H^{[n]}$ [Br].

Given an ideal $I \in H^{[n]}$, there are distinct points $p_1, \ldots, p_k \in X$ such that I may be expressed as the intersection of ideals

$$I = I_{p_1} \cap \cdots \cap I_{p_k}, \tag{1}$$

where I_{p_i} is supported at $\{p_i\}$ for $i = 1, \ldots, k$. We will write I_0 for I_p when $p = (0, 0)$. Let

$$\sigma(I) = \dim_{\mathbb{C}} (R/I_0)$$

be the multiplicity of I at $(0, 0)$; with this notation then I_0 is an element of $B^{[\sigma(I)]}$.

Denote by $\mathfrak{m} := \langle x, y\rangle$ the maximal ideal of R at $(0, 0)$, and introduce another $\mathbb{N}$-valued function on $H^{[n]}$: the minimal number of generators of I_0:

$$\mu(I) = \dim_{\mathbb{C}} (I_0/\mathfrak{m}I_0).$$

Indeed, by Nakayama's Lemma, every basis of $I_0/\mathfrak{m}I_0 \simeq \mathbb{C}^{\mu(I)}$ lifts to a minimal set of generators of I_0 considered as an R-module. This function is a classical invariant studied by A. Iarrobino in [Ia] (See §2).

An important set of examples of ideals in $H^{[n]}$ is the set $\Pi(n)$ of *monomial ideals*, i.e., ideals generated by monomials in x and y. For such an ideal I, the monomials that are not contained in I form a basis of the n-dimensional space R/I. The exponents

$$\left\{(p, q) \in \mathbb{N}^2 \,\middle|\, x^p y^q \notin I\right\}$$

then associate a finite subset of $\mathbb{N}^2$ to I, and this defines a one-to-one correspondence between monomial ideals in $H^{[n]}$ and Young diagrams of size $n \in \mathbb{N}$, which, in turn, are in bijection with partitions of n. Note that given a diagram representing a monomial ideal I, the value of $\mu(I)$ is the number of "concave corners" of the diagram (Figs. 1 and 2):

In the present paper, we study the geometric and topological invariants of strata associated with the integer invariants μ and σ.

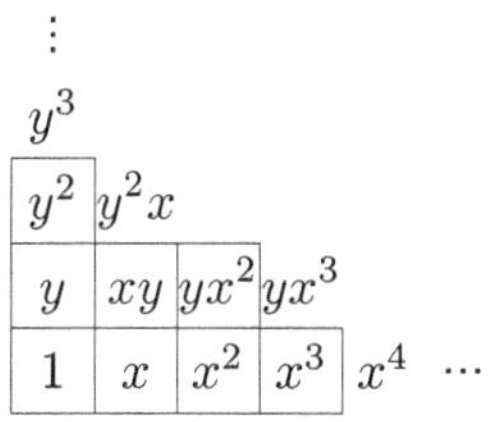

$\vdots$

y^3

$y^2 \quad y^2x$

$y \quad xy \quad yx^2 \quad yx^3$

$1 \quad x \quad x^2 \quad x^3 \quad x^4 \quad \cdots$

Fig. 1 The diagram corresponding to the ideal $\langle y^3, y^2x, yx^3, x^3\rangle$

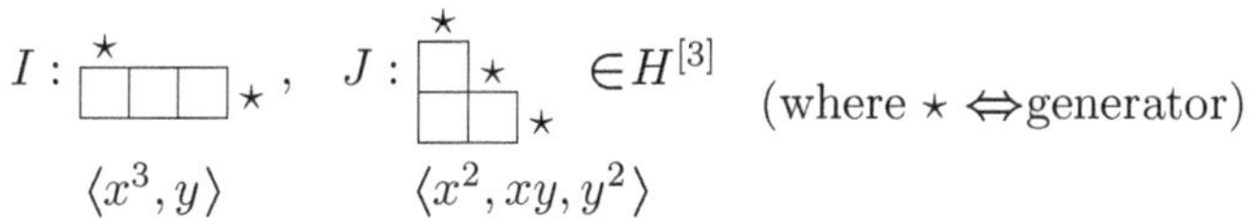

$I: \star \ \star, \quad J: \star \ \star \ \star \quad \in H^{[3]} \quad$ (where $\star \Leftrightarrow$generator)

$\langle x^3, y\rangle \qquad \langle x^2, xy, y^2\rangle$

Fig. 2 The monomial ideals I, J have $\mu(I) = 2$ and $\mu(J) = 3$, respectively

We will focus on the *E-polynomial* (or *the Hodge-Deligne polynomial*), whose motivic properties facilitate its calculation [De1, De2, HR]. We present the definitions and relevant properties of the E-polynomial in Appendix 1.

Notation If the mixed Hodge numbers $h^{p,q;j}(Z)$ of an algebraic variety Z vanish unless $p = q$, then we say that the mixed Hodge structure of Z is *pure*. In this case, the E-polynomial $E(Z; u, v)$ may be written as a polynomial in the degree-2 variable $t = uv$, and we can introduce the simplified notation $E(Z; t)$:

$$E(Z;t) := E\left(Z; \sqrt{t}, \sqrt{t}\right) = \sum_{p,q,j} h_c^{p,q;j}(Z)\, u^p v^q s^j \,|_{u,v=\sqrt{t} \text{ and } s=-1} .$$

The calculations of the E-polynomials of Hilbert schemes of points on smooth surfaces goes back to the works of L. Göttsche and J. Cheah [Gö2, Ch1, Ch2]. A version of their result is the following.

Theorem 1 *The generating function of the E-polynomials of a smooth surface S has the form*

$$\sum_{n=0}^{\infty} E\left(S^{[n]}; u, v\right) s^n = \prod_{d=1}^{\infty} \prod_{p,q} \left(\frac{1}{1 - u^{p+d-1} v^{q+d-1} s^d}\right)^{e_{p,q}(S)}, \tag{2}$$

where $e_{p,g} := \sum_k (-1)^k h_c^{p,q,k}(Z)$.

In particular, for the case of $\mathbb{C}^2$, we have (cf. [ES])

$$\sum_{n=0}^{\infty} E\left(H^{[n]}; t\right) q^n = \prod_{d=1}^{\infty} \frac{1}{1 - t^{d+1} q^d}. \tag{3}$$

In this article, we present a refinement of these formulas by calculating the E-polynomials of the strata

$$B_m^{[n]} := \left\{ I \in B^{[n]} \,\middle|\, \mu(I) = m \right\}, \quad H_m^{[n]} := \left\{ I \in H^{[n]} \,\middle|\, \mu(I) = m \right\}$$

associated with the invariant $\mu(I)$ of ideals introduced above.

Theorem 2 *The generating function of the E-polynomials of the strata $B_m^{[n]}$ of the Briançon variety is given by*

$$\sum_{n=0}^{\infty} E\left(B_m^{[n]}; t\right) q^n = \prod_{i=1}^{m-1} \frac{1}{1-t^{i+1}} \cdot \sum_{a=1}^{m} \left((-1)^{a+1} t^{\binom{a}{2}+m-1} \begin{bmatrix} m \\ a \end{bmatrix}_t \prod_{k=0}^{\infty} \frac{1-q^k t^{k-a}}{1-q^k t^{k-1}} \right), \tag{4}$$

$$\textit{where} \begin{bmatrix} m \\ a \end{bmatrix}_t = \begin{cases} \displaystyle\prod_{i=0}^{a-1} \frac{1-t^{m-i}}{1-t^{i+1}} & \textit{if } 1 \le a \le m \\ 1 & \textit{if } a = 0 \\ 0 & \textit{if } a > m. \end{cases}$$

Remark 3 Note that in the summand indexed by $a = 1$ on the right-hand side of (4), the product cancels, and hence this term does not depend on q; in particular, it does not contribute to $E\left(B_m^{[n]}; t\right)$ for $n > 0$.

Theorem 4 *The E-polynomial of the refined stratum $H_m^{[n]}$ has the generating function*

$$\sum_{n=0}^{\infty} E\left(H_m^{[n]}; t\right) q^n = \prod_{i=1}^{m-1} \frac{1}{1-t^{i+1}} \cdot \sum_{a=1}^{m} \left((-1)^{a} t^{\binom{a}{2}+m} \begin{bmatrix} m \\ a \end{bmatrix}_t \prod_{k=0}^{\infty} \frac{1-q^k t^{k-a}}{1-q^k t^{k+1}} \right). \tag{5}$$

Example 5 Here are some examples of the generating function from Theorem 2:

- **The case** $m = 2$:

$$\begin{aligned} \sum_{n=0}^{\infty} E\left(B_2^{[n]}; t\right) q^n &= \frac{t}{1-t} + \frac{t^2}{t^2-1} \cdot \prod_{k=0}^{\infty} \frac{1-t^{k-2}q^k}{1-t^{k-1}q^k} \\ &= \frac{t}{1-t}\left(1 - \prod_{k=1}^{\infty} \frac{1-t^{k-2}q^k}{1-t^{k-1}q^k}\right) \\ &= \frac{t}{1-t} + \prod_{k=1}^{\infty}\left(1-t^{k-2}q^k\right) \cdot \prod_{k=0}^{\infty} \frac{1}{1-t^{k-1}q^k}. \end{aligned}$$

- **The case** $m = 3$:

$$\sum_{n=0}^{\infty} E\left(B_3^{[n]}; t\right) q^n = \frac{t^2}{(1-t)\left(1-t^2\right)} - \frac{t^3}{(1-t)\left(1-t^2\right)} \cdot \prod_{k=0}^{\infty} \frac{1-t^{k-2}q^k}{1-t^{k-1}q^k}$$
$$+ \frac{t^5}{\left(1-t^2\right)\left(1-t^3\right)} \cdot \prod_{k=0}^{\infty} \frac{1-t^{k-3}q^k}{1-t^{k-1}q^k}$$
$$= \frac{t^2}{\left(1-t^2\right)(1-t)} - \frac{t^2}{(1-t)^2} \cdot \prod_{k=1}^{\infty} \frac{1-t^{k-2}q^k}{1-t^{k-1}q^k}$$
$$+ \frac{t^3}{(1-t^2)(1-t)} \cdot \prod_{k=1}^{\infty} \frac{1-t^{k-3}q^k}{1-t^{k-1}q^k}.$$

- **The case** $m = 4$:

$$\sum_{n=0}^{\infty} E\left(B_4^{[n]}; t\right) q^n = \frac{t^3}{(1-t)\left(1-t^2\right)\left(1-t^3\right)}$$
$$- \frac{t^4}{(1-t)\left(1-t^2\right)^2} \cdot \prod_{d=0}^{\infty} \frac{1-t^{d-2}q^d}{1-t^{d-1}q^d}$$
$$+ \frac{t^6}{(1-t)\left(1-t^2\right)\left(1-t^3\right)} \cdot \prod_{d=0}^{\infty} \frac{1-t^{d-3}q^d}{1-t^{d-1}q^d}$$
$$- \frac{t^9}{\left(1-t^2\right)\left(1-t^3\right)\left(1-t^4\right)} \cdot \prod_{d=0}^{\infty} \frac{1-t^{d-4}q^d}{1-t^{d-1}q^d}.$$

A key role in our calculations is played by the *refined incidence varieties* inside $H^{[n]} \times H^{[n+r]}$:

$$H^{[n,n+r]} := \left\{(I, J) \in H^{[n]} \times H^{[n+r]} \,\middle|\, I \supset J \supseteq \mathfrak{m}I\right\}.$$

Note that for $(I, J) \in H^{[n,n+r]}$, I/J is supported at $(0, 0)$.

These spaces were introduced in [NY] by H. Nakajima and K. Yoshioka, where they appeared as examples of the rank 1 case of the moduli spaces of stable perverse coherent sheaves on the blow-up of $\mathbb{P}^2$ at a point.

The key idea suggested to us by A. Oblomkov is that the fibers of the projection $\pi : H^{[n,n+r]} \to H^{[n]}$ over $H_m^{[n]} = \{I \mid \mu(I) = m\}$ are Grassmannians:

$$\begin{array}{ccc} \mathrm{Gr}_r\left(\mathbb{C}^m\right) \to \pi^{-1}\left(H_m^{[n]}\right) \subseteq H^{[n,n+r]} \\ \downarrow \pi \\ H_m^{[n]} \subseteq H^{[n]} \end{array} \tag{6}$$

This idea appears in [OS] and [ORS] in the context of a conjecture relating the HOMFLY polynomial of the link of a plane curve singularity C to the E-polynomials of the Hilbert schemes of points supported on C. We leverage this idea to obtain our formulas.

The paper is organized as follows. In Sect. 1, we introduce the refined Hilbert scheme and state some geometric and topological properties that will be needed later. In Sect. 2, we prove our main result: Theorem 2. Then in Sect. 3, as an application of Theorem 2, we compute the generating function of $E\left(H_m^{[n]}; t\right)$. In Sect. 4, we present a formula for the Euler characteristics of $B_m^{[n]}$. In Appendix 1, we present the definitions and relevant properties of the E-polynomial. We list a table of examples of $E\left(B_m^{[n]}; t\right)$ in Appendix 5.

2 The Refined Hilbert Schemes

2.1 Definition and Basic Properties

Definition 6 Let $n, r \in \mathbb{N}$, $r \geq 1$. The refined Hilbert scheme $H^{[n,n+r]} \subset H^{[n]} \times H^{[n,n+r]}$ is defined as

$$H^{[n,n+r]} = \{(I, J) \in H^{[n]} \times H^{[n,n+r]} \,|\, I \supset J \supseteq \mathfrak{m}I\}. \tag{1}$$

Here, the condition $I \supset J \supseteq \mathfrak{m}I$ is equivalent to the requirement that the quotient space I/J lies in the kernel of the multiplication by x and y, in other words, it carries a trivial R-module structure. It follows that for an element $(I, J) \in H^{[n,n+r]}$, the quotient I/J is supported at $(0, 0)$.

Example 7 If $(I, J) \in H^{[1,3]}$, then $I/J \simeq \mathbb{C}^2$ is a subspace of $I/\mathfrak{m}I \simeq \mathbb{C}^{\mu(I)}$. Then $\mu(I)$ must be at least 2, and the only ideal $I \in H^{[1]}$ of codimension 1 with $\mu(I) \geq 2$ is the maximal ideal $\langle x, y\rangle = \mathfrak{m}$. Moreover, since $\dim_{\mathbb{C}} \mathfrak{m}/\mathfrak{m}^2 = 2$, the only possible $J \in H^{[3]}$ containing $\mathfrak{m}I$ is $\mathfrak{m}^2$. Thus $H^{[1,3]}$ consists of the single point $\left(\mathfrak{m}, \mathfrak{m}^2\right)$.

The refined Hilbert schemes first appeared in the work of H. Nakajima and K. Yoshioka in [NY] in the form of a special case of a moduli space of coherent sheaves on the blow-up of $\mathbb{P}^2$ at a point that they study. We will not recall the general definition of this moduli space here; for our purposes it is sufficient to observe that in Sections 5.3 and 5.4 of [NY] they explicitly identify the rank-1 case of this space, $\widehat{M}^0(c)$, with our refined Hilbert schemes as defined above.

We will use the following key result of this work (cf. [NY, Corollary 3.7]):

Theorem 8 (H. Nakajima, K. Yoshioka) *The refined Hilbert scheme $H^{[n,n+r]}$ is smooth and of complex dimension $2n - r(r-1)$.*

The algebraic torus $T \simeq (\mathbb{C}^*)^2$ acts on $\mathbb{C}^2$ by

$$t \cdot (x, y) \mapsto (t_1 x, t_2 y)$$

for $(t_1, t_2) \in T$, $(x, y) \in \mathbb{C}^2$. This induces an action on the refined Hilbert schemes $H^{[n,n+r]}$. The set of fixed points of this action $\left(H^{[n,n+r]}\right)^T$ is parametrized by pairs of monomial ideals $(I, J) \in H^{[n]} \times H^{[n+r]}$ such that $I/J \simeq \mathbb{C}^r$ as a trivial R-module. To give a description of such T-fixed point using Young diagrams, we call the boxes of a Young diagram Δ that have no other boxes above and to the right of them, the *elbows* of Δ (cf. Fig. 3). Let $(I, J) \in (H^{[n,n+r]})^T$ and Δ_I, Δ_J be the corresponding Young diagrams. Since $I \supset J$, Δ_I is a subdiagram of Δ_J. Moreover, the quotient I/J corresponds to a subset $S_{I/J}$ of Δ_J of r elbows and $\Delta_I = \Delta_J \setminus S_{I/J}$. Therefore, we may represent a T-fixed point (I, J) by a pair $(\Delta_J, S_{I/J})$ of a Young diagram Δ_J with $n + r$ boxes, and a subset $S_{I/J}$ of r marked elbows of Δ_J and we denote the set of these pairs $(\Delta_J, S_{I/J})$ by $\Pi(n, r)$.

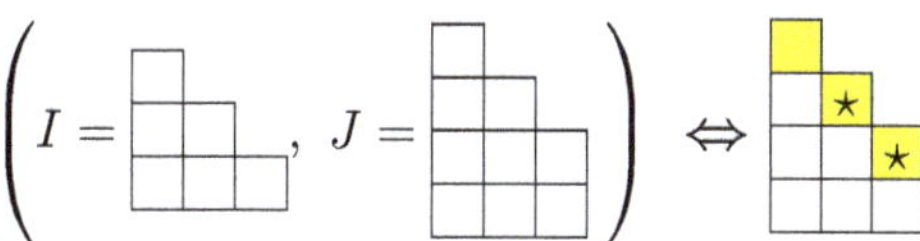

Fig. 3 A fixed point of $H^{[6,9]}$

As we will see, the topology of $H^{[n,n+r]}$ is closely related to the function μ on $H^{[n]}$.

Proposition 9 *If $I \in H^{[n]}$ has $\mu(I) = k$, then $n \geq \frac{k(k-1)}{2} = \binom{k}{2}$. Moreover, the equality holds if and only if $I = \mathfrak{m}^{k-1}$.*

Proof We look for the maximal value of $\mu(I)$ for $I \in H^{[n]}$. We recall that if $I_\lambda \in (H^{[n]})^T$ and Δ_λ is the corresponding Young diagram with n boxes, then $\mu(I_\lambda)$ is equal to the number of "elbows" of Δ_λ.

We claim that

$$\max\left\{\mu(I) \,\middle|\, I \in H^{[n]}\right\} = \max\left\{\mu(I) \,\middle|\, I \in \left(H^{[n]}\right)^T\right\}. \tag{2}$$

To see this, we consider the action of the one-parameter subgroup $\phi : \mathbb{C}^* \to T$, $t \mapsto (t, t^N)$ on $H^{[n]}$: $t \cdot f(x, y) = f(t^{-1}x, t^{-N}y)$, with $N \in \mathbb{N}$ large.

For each $I \in H^{[n]}$, the limit $t \to 0$ of the action $\phi(t) \cdot (x, y) = (t^{-1}x, t^{-N}y)$ is a T-fixed point I_λ [Na1]. Let $U_\lambda := \left\{ I \in H^{[n]} \,\middle|\, \lim_{t\to 0}\phi(t) \cdot I = I_\lambda \right\}$. We claim that $\mu(I_\lambda) \geq \mu(I)$ for all $I \in U_\lambda$. Since every $I \in H^{[n]}$ supported at the origin must contain the ideal $\mathfrak{m}^n$, without loss of generality, we can consider the image of I in the quotient space $R/\mathfrak{m}^n$. Taking the limit $t \to 0$ of this $\mathbb{C}^*$-action induces a total ordering "$y \succ x$" in $R/\mathfrak{m}^n$ and the limit $\lim_{t\to 0}\phi(t) \cdot I = I_\lambda$ is, in fact, the initial ideal of I with respect to this ordering $\succ$. Now, if $I \in U_\lambda$ and G_I is a reduced Gröbner basis of I, then by definition, $\mu(I_\lambda) = |G_I|$.

One can compute a reduced Gröbner basis can be from any generating set of I by the Buchberger Algorithm:

(1) Start with a generating set B, calculate S-polynomials of $g, h \in B$

$$S(g, h) := \operatorname{lcm}(\mathrm{LM}(g), \mathrm{LM}(h)) \left(\frac{g}{\mathrm{LT}(g)} - \frac{f}{\mathrm{LT}(f)} \right),$$

where $\mathrm{LM}(f)$ is the leading monomial of f and $\mathrm{LM}(f)$ is the leading term of f for $f \in \mathbb{C}[x, y]$.

(2) Reduce $S(g, h)$ (relative to B) until the result is not further reducible and add it to B if it is non-zero.

(3) Repeat previous steps with new $B' = B \cup \{\text{non-zero } S(g, h)\}$ until all possible pairs $g, h \in B$ have vanishing S-polynomial.

The resulting generating set (with new added polynomials) is the reduced Göbner basis of I.

It follows from this algorithm that the number of elements of the final reduced Göbner basis can only be greater than the original generating set.

Since I is the intersection of ideals with I_0 and $\mu(I) = \mu(I_0)$, we have $\mu(I) \leq |G_I| = \mu(I_\lambda)$ by the Buchberger Algorithm construction of the reduced Göbner basis. We conclude that the function $\mu(I)$, $I \in H^{[n]}$ reaches its maximum value in $\left(H^{[n]}\right)^T$.

Now, by Eq. (2), the question of finding the maximal value of $\mu(I)$ for $I \in H^{[n]}$ reduces to a combinatorial question for Young diagrams. In this context, $\max\left\{\mu(I) \,\middle|\, I \in H^{[n]}\right\}$ is equal to the maximal number of elbows that a Young diagram with n boxes can have. It follows that if I is an ideal with $\dim_{\mathbb{C}} R/I = \binom{k}{2}$ and $\mu(I) = k$, then it can only be $\mathfrak{m}^{k-1} = \langle y^{k-1}, y^{k-2}x, \ldots, yx^{k-2}, x^{k-1} \rangle$, which is the monomial ideal corresponding to the partition $\lambda = k - 1 \geq k - 2 \geq \cdots \geq 1$. □

An immediate consequence of Proposition 9 is the following corollary:

Corollary 10 *We have* $\sigma(I) \geq \binom{r}{2}$, *and* $H^{[n,n+r]}$ *consists of a single point* $\left(\mathfrak{m}^{r-1}, \mathfrak{m}^r\right)$ *if* $n = \binom{r}{2}$.

In particular, this shows that the refined Hilbert scheme $H^{[n,n+r]}$ is empty if $n < \binom{r}{2}$.

2.2 *The Calculation of the E-Polynomial*

According to a theorem of Bialynicki-Birula [Bi1, Bi2], a $\mathbb{C}^*$-action on a smooth projective variety with isolated fixed points induces a decomposition of M into affine spaces $M = \bigsqcup_{p \in M^T} \mathbb{A}^{\alpha(p)}$, where $\alpha(p)$ is the number of positive weights of T_pM. In this case, the compactly supported Poincaré polynomial of M is $\sum_{p \in M^T} t^{\alpha(p)}$ and it agrees with the E-polynomial of M. In our notation, we can write

$$E\,(M; t) = P\left(M; \sqrt{t}^{-1}\right) t^{\dim_{\mathbb{C}} M} = \sum_{p \in M^T} t^{\alpha(p)}. \tag{3}$$

The Hilbert scheme of n points on the plane is not projective, and the punctual Hilbert scheme $B^{[n]}$ is not smooth, but for these varieties the affine Bialynicki-Birula holds nevertheless. The reason is that the Bialynicki-Birula decompositions corresponding to appropriately chosen $\mathbb{C}*$-actions are compatible with $H^{[n]}$ and $B^{[n]}$, respectively, and this, in particular, implies the purity of the Hodge structures of these spaces. The details of the relevant arguments can be found in [ES, HR2, Na1]. The resulting decompositions are

$$H^{[n]} = \bigsqcup_{p \in \Pi(n)} \mathbb{A}^{\alpha(p)} \text{ and } B^{[n]} = \bigsqcup_{p \in \Pi(n)} \mathbb{A}^{2n-\alpha(p)},$$

where $\alpha(p)$ is the number of positive weights of $T_pH^{[n]}$. The same arguments are equally valid for $H^{[n,n+r]}$ and $B^{[n,n+r]}$, and this gives us the following formulas:

Proposition 11 *We have*

$$E\left(H^{[n,n+r]}; t\right) = \sum_{p \in \Pi(n,r)} t^{\alpha(p)} \text{ and } E\left(B^{[n,n+r]}; t\right) = \sum_{p \in \Pi(n,r)} t^{2n-r(r-1)-\alpha(p)},$$

where $\alpha(p)$ is the number of positive weights of $T_pH^{[n,n+r]}$.

Notation

- For a pair of "elbows" $\boxed{\star}$ and $\boxed{\blacktriangle}$ of a Young diagram represented by coordinates $(a, b), (c, d) \in \mathbb{Z}^2_{\geq 1}$ (assuming that $a < c, d < b$), there is the box $\boxed{\bullet} = (a, d)$ in the diagram.
  Given a pair $\left(\Delta_J, S_{I/J}\right)$, we denote by

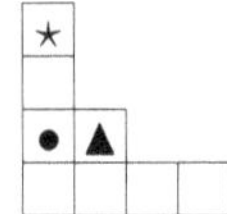

An example for ⋆ $= (1,4)$, ▲ $= (2,2)$ and • $= (1,2)$.

$$Q_{I,J} = \left\{(a,d) \in \Delta_J \,\middle|\, \exists (a,b), (c,d) \in S_{I/J} \text{ with } a < c, d < b\right\}$$

the set of boxes obtained this way.

- For a box $\square$ of a Young diagram, we define $a(\square)$ to be the number of the boxes above $\square$ and $l(\square)$ to be the number of the boxes to the right of $\square$.
- Denote by $T_i : (t_1, t_2) \to t_i, i = 1, 2$ the one-dimensional representations of the torus group corresponding to the two coordinates.

To find the E-polynomial of $H^{[n,n+r]}$, we apply the character formula of Proposition 5.2. in [NY] for the tangent space of a T-fixed point of the moduli space $\widehat{M}^0(c)$, which is identified with $H^{[n,n+r]}$ in Sections 5.3. and 5.4 of [NY].

Theorem 12 (H. Nakajima, K. Yoshioka) *Let (I, J) be a T-fixed point of $H^{[n,n+r]}$ with corresponding marked Young diagram $\left(\Delta_J, S_{I/J}\right)$, where $\Delta_J = \Delta_I \cup S_{I/J}$. Then the character of the tangent space $T_{(I,J)}H^{[n,n+r]}$ as a T-module is given by*

$$\sum_{\square} \left(T_1^{-l_{\Delta_J}(\square)} T_2^{a_{\Delta_I}(\square)+1} + T_1^{l_{\Delta_I}(\square)+1} T_2^{-a_{\Delta_J}(\square)}\right), \tag{4}$$

where the summation runs over all empty boxes $\square$ of $\Delta_J \setminus \left(S_{I/J} \bigcup Q_{I,J}\right)$ (Fig. 4).

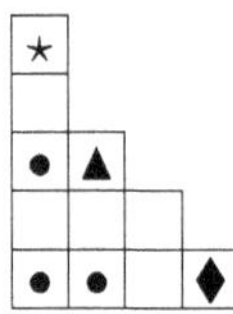

Fig. 4 The summation runs over all empty boxes "$\square$"

Corollary 13 *The generating function of the E-polynomial of $H^{[n,n+r]}$ has the form*

$$\sum_{n=\binom{r}{2}}^{\infty} E\left(H^{[n,n+r]}; t\right) q^n = q^{\binom{r}{2}} \left(\prod_{d=1}^{\infty} \frac{1}{1 - t^{(d+1)} q^d}\right) \left(\prod_{d=1}^{r} \frac{1}{1 - t^d q^d}\right). \tag{5}$$

Proof By Proposition 11, this is equivalent to finding the generating function of the compactly supported Poincaré polynomials of $H^{[n,n+r]}$. We follow the argument of [NY, Corollary 5.3 and 5.4], where a formula for the Poincaré polynomial of $H^{[n,n+r]}$ is derived. After adjusting the signs in this formula in order to pass to compactly supported cohomology, we obtain expression (5). □

3 Proof of Theorem 2

We consider the refined Hilbert scheme strata

$$H_m^{[n]} := \left\{ I \in H^{[n]} \,\middle|\, \mu(I) = m \right\},$$

$$H_m^{[n]}(s) := \left\{ I \in H^{[n]} \,\middle|\, \mu(I) = m, \sigma(I) = s \right\}$$

and which induce stratifications of $H^{[n]}$.

If $J \in H^{[n,n+r]}$ satisfies the condition $I \supset J \supseteq \mathfrak{m}I$, then J is fully determined by its image in $I/\mathfrak{m}I$. parametrized by $Gr(r, I/\mathfrak{m}I)$, the Grassmannian of r-dimensional subspaces of $I/\mathfrak{m}I \simeq \mathbb{C}^m$. Then the fibers of the map $\pi : H^{[n,n+r]} \to H^{[n]}$ over an ideal $I \in H_m^{[n]}$ are isomorphic to $Gr(r, \mathbb{C}^m)$. Vector bundles are Zariski locally trivial ([SGA, XI/Theorem 5]), hence this Grassmannian bundle $\pi^{-1}(H_m^{[n]})$ is a locally trivial fibration with fiber $Gr(r, \mathbb{C}^m)$.

Since Grassmannians $Gr(r, \mathbb{C}^m)$ are projective and smooth, their E-polynomials and Poincaré polynomials are equal:

$$E\left(Gr(r, \mathbb{C}^m); t\right) = P\left(Gr(r, \mathbb{C}^m); \sqrt{t}\right) = \begin{bmatrix} m \\ r \end{bmatrix}_t,$$

where $\begin{bmatrix} m \\ r \end{bmatrix}_t := \prod_{i=1}^r \frac{1-t^{m-i+1}}{1-t^i}$.

We can thus apply property (6) to the fibration $H^{[n,n+r]} \to H^{[n]}$ and obtain the following equality:

$$E\left(H^{[n,n+r]}; t\right) = \sum_{m=1}^{\mu_n^{\max}} E\left(H_m^{[n]}; t\right) E\left(Gr_r(\mathbb{C}^m); t\right) = \sum_{m=1}^{\mu_n^{\max}} E\left(H_m^{[n]}; t\right) \begin{bmatrix} m \\ r \end{bmatrix}_t, \tag{1}$$

where $\mu_n^{\max} := \max\left\{\mu(I) \,|\, I \in H^{[n]}\right\} = \max\left\{\mu(I) \,\middle|\, I \in \left(H^{[n]}\right)^T\right\}$.

Furthermore, presenting an ideal $I \in H^{[n]}$ as an intersection of ideals $I_0 \cap I'$, where I_0 is the part supported on $(0, 0)$ and I' is the part supported on $Y_0 := \mathbb{C}^2 \backslash \{(0, 0)\}$, we obtain the decomposition

$$H_m^{[n]} \simeq \bigsqcup_{s=0}^{n} \left(B_m^{[s]} \times Y_0^{[n-s]} \right). \tag{2}$$

Using the motivic properties of the E-polynomial, we can conclude

$$E\left(H_m^{[n]}; t\right) = \sum_{s=0}^{n} E\left(Y_0^{[n-s]}; t\right) \cdot E\left(B_m^{[s]}; t\right). \tag{3}$$

Our goal is to find the E-polynomials of $H_m^{[k]}$ and $B_m^{[k]}$ using Eqs. (1) and (3). We will consider all $H_m^{[k]}$ and $B_m^{[k]}$, $m, k \in \mathbb{N}$, at the same time and we define the following infinite matrices:
$\mathcal{X} := \left(E\left(H_i^{[j]}; t\right)\right)_{i \geq 1, j \geq 0}$, $\mathcal{B} := \left(E\left(B_i^{[j]}; t\right)\right)_{i \geq 1, j \geq 0}$, $\mathcal{R} := \left(E\left(H^{[j, j+i]}; t\right)\right)_{i \geq 1, j \geq 0}$,

$$\mathcal{G} := \left(E\left(Gr_i(\mathbb{C}^j); t\right)\right)_{i, j \geq 1} = \left(\left[{j \atop i}\right]_t\right)_{i, j \geq 1} \text{ and } \mathcal{A} := \left(E\left(Y_0^{[j-i]}; t\right)\right)_{i, j \geq 1}.$$

Proposition 14 *The matrices $\mathcal{X}$, $\mathcal{B}$, $\mathcal{R}$, $\mathcal{G}$ and $\mathcal{A}$ satisfy*

$$\mathcal{G}\mathcal{X} = \mathcal{R} \text{ and } \mathcal{B}\mathcal{A} = \mathcal{X}.$$

Proof A direct calculation of matrix products gives

$$\begin{aligned} \mathcal{G}\mathcal{X} &= \left(\textstyle\sum_{k=1}^{\infty} \mathcal{G}_{ik} \mathcal{X}_{kj}\right)_{i \geq 1, j \geq 0} \\ &= \left(\textstyle\sum_{k=1}^{\infty} E\left(Gr_i(\mathbb{C}^k); t\right) E\left(H_k^{[j]}; t\right)\right)_{i \geq 1, j \geq 0} \\ \text{(by Eq. (1))} &= \left(E\left(H^{[j, j+i]}; t\right)\right)_{i \geq 1, j \geq 0} = \mathcal{R}. \end{aligned}$$

Similarly, we have the product $\mathcal{B}\mathcal{A}$

$$\begin{aligned} \mathcal{B}\mathcal{A} &= \left(\textstyle\sum_{k=0}^{\infty} \mathcal{B}_{ik} \mathcal{A}_{kj}\right)_{i \geq 1, j \geq 1} \\ &= \left(\textstyle\sum_{k=0}^{\infty} E\left(B_i^{[k]}; t\right) E\left(Y_0^{[j-k]}; t\right)\right)_{i \geq 1, j \geq 1} \\ \text{(by Eq. (3))} &= \left(E\left(H_i^{[j]}; t\right)\right)_{i \geq 1, j \geq 0} = \mathcal{X}. \end{aligned}$$

□

By definition, $\mathcal{G}$ and $\mathcal{A}$ are upper triangular matrices with 1s on the diagonal, so they are invertible with upper triangular inverses. Then the matrices $\mathcal{X}$ and $\mathcal{B}$ may be expressed as products of matrices

$$\begin{cases} \mathcal{X} &= \mathcal{G}^{-1}\mathcal{R} \\ \mathcal{B} &= \mathcal{X}\mathcal{A}^{-1} = \mathcal{G}^{-1}\mathcal{R}\mathcal{A}^{-1}. \end{cases} \tag{4}$$

Thus to compute the E-polynomials $E\left(H_m^{[n]};t\right) = \mathcal{X}_{m,n}$ and $E\left(B_m^{[n]};t\right) = \mathcal{B}_{m,n}$, it is sufficient to find the inverse matrices of $\mathcal{G}$ and $\mathcal{A}$.

Proposition 15 *The matrix* $\mathcal{G} = \left(\left[\begin{smallmatrix} j \\ i \end{smallmatrix}\right]_t\right)_{i,j\geq 1}$ *has the inverse*

$$\mathcal{G}^{-1} = \left((-1)^{j-i}t^{\binom{j-i}{2}}\left[\begin{smallmatrix} j \\ i \end{smallmatrix}\right]_t\right)_{i,j\geq 1}.$$

Proof Denoting by δ_{ij} the Kronecker delta-function, we have

$$(\mathcal{G}^{-1}\mathcal{G})_{ij} = \sum_{k=1}^{\infty}\mathcal{G}_{ik}^{-1}\mathcal{G}_{kj} = \sum_{k=1}^{j}\mathcal{G}_{ik}^{-1}\begin{bmatrix} j \\ k \end{bmatrix}_t = \delta_{ij}. \tag{5}$$

We apply the following orthogonality relation for the q-binomial coefficients ([Co, p.118-p.119]): For every $0 \leq i \leq j$, one has

$$\delta_{ij} = \sum_{k=i}^{j}(-1)^{k-i}q^{\binom{k-i}{2}}\begin{bmatrix} j \\ k \end{bmatrix}_q\begin{bmatrix} k \\ i \end{bmatrix}_q \overset{\left(\left[\begin{smallmatrix} k \\ i \end{smallmatrix}\right]_q=0 \text{ if } k<i\right)}{=\!=\!=} \sum_{k=1}^{j}(-1)^{k-i}q^{\binom{k-i}{2}}\begin{bmatrix} j \\ k \end{bmatrix}_q\begin{bmatrix} k \\ i \end{bmatrix}_q,$$

and we thus

$$\sum_{k=1}^{j}\mathcal{G}_{ik}^{-1}\begin{bmatrix} j \\ k \end{bmatrix}_t = \delta_{ij} = \sum_{k=1}^{j}(-1)^{k-i}t^{\binom{k-i}{2}}\begin{bmatrix} j \\ k \end{bmatrix}_t\begin{bmatrix} k \\ i \end{bmatrix}_t. \tag{6}$$

By comparing the coefficients of $\left[\begin{smallmatrix} j \\ k \end{smallmatrix}\right]_t$ in (6), we obtain

$$\mathcal{G}_{ik}^{-1} = (-1)^{k-i}t^{\binom{k-i}{2}}\begin{bmatrix} k \\ i \end{bmatrix}_t.$$

□

Our next step is to calculate the inverse of $\mathcal{A}$. To this end, we first need the generating function of E-polynomials of the Hilbert scheme of points on the punctured plane Y_0. By Theorem 1, this generating function is

$$\sum_{n=0}^{\infty} E\left(Y_0^{[n]}; t\right) q^n = \prod_{d=1}^{\infty} \frac{1 - t^{d-1} q^d}{1 - t^{d+1} q^d}. \tag{7}$$

As an exercise in our motivic calculus, we present below a direct proof of (7) using the knowledge of $E\left(H^{[n,n+r]}; t\right)$ and $E(B^{[n]}; t)$.

Proof of (7) First, we observe that from the decomposition in (2): $H^{[n]} \simeq \bigsqcup_{s=0}^{n} Y_0^{[s]} \times B^{[n-s]}$, we have a bijective morphism

$$\bigsqcup_{s=0}^{n} Y_0^{[s]} \times B^{[n-s]} \to H^{[n]}$$

by sending a pair of subschemes in $Y_0^{[s]} \times B^{[n-s]}$ to the union of the two.

We recall that the formula (3) for the generating function of the E-polynomials of $H^{[n]}$ has the form

$$\sum_{n=0}^{\infty} E\left(H^{[n]}; t\right) q^n = \prod_{d=1}^{\infty} \frac{1}{1 - t^{d+1} q^d}. \tag{8}$$

Applying the motivicity of the E-polynomial to this decomposition (2), we obtain the equality

$$\sum_{n=0}^{\infty} E\left(H^{[n]}; t\right) q^n = \sum_{n=0}^{\infty} \left(\sum_{s=0}^{n} E\left(Y_0^{[s]}; t\right) E\left(B^{[n-s]}; t\right) \right) q^n. \tag{9}$$

After the change of variable $k = n - s$, the right-hand side of the Eq. (9) has the form of a doubly infinite summation

$$\begin{aligned}
&\sum_{n=0}^{\infty} \left(\sum_{s=0}^{n} E\left(Y_0^{[s]}; t\right) E\left(B^{[n-s]}; t\right) \right) q^n \\
&\overset{(k=n-s)}{=\!=} \sum_{k=0}^{\infty} \left(\sum_{s=0}^{\infty} E\left(Y_0^{[s]}; t\right) E\left(B^{[k]}; t\right) \right) q^{s+k} \\
&= \sum_{k=0}^{\infty} \sum_{s=0}^{\infty} E\left(Y_0^{[s]}; t\right) q^s E\left(B^{[k]}; t\right) q^k \\
&\overset{(s,k \text{ are independent})}{=\!=} \left(\sum_{k=0}^{\infty} E\left(B^{[k]}; t\right) q^k \right) \left(\sum_{s=0}^{\infty} E\left(Y_0^{[s]}; t\right) q^s \right).
\end{aligned} \tag{10}$$

As we pointed out in the beginning of the section, the Briançon variety $B^{[k]}$ admits a cell decomposition as well and thus $E\left(B^{[k]};t\right) = P\left(B^{[k]};\sqrt{t}^{-1}\right)t^{\dim_{\mathbb{C}} B^{[k]}}$. We replace pieces $E\left(B^{[k]};t\right)$ by $P\left(B^{[k]};\sqrt{t}^{-1}\right)t^{\dim_{\mathbb{C}} B^{[k]}}$ in Eq. (10):

$$\left(\sum_{k=0}^{\infty} P\left(B^{[k]};\sqrt{t}^{-1}\right)t^{\dim_{\mathbb{C}} B^{[k]}}q^k\right)\left(\sum_{s=0}^{\infty} E\left(Y_0^{[s]};t\right)q^s\right),$$

which is equal to

$$\left(\sum_{k=0}^{\infty} P\left(H^{[k]};\sqrt{t}\right)q^k\right)\left(\sum_{s=0}^{\infty} E\left(Y_0^{[s]};t\right)q^s\right), \tag{11}$$

since $H^{[n]}$ and $B^{[n]}$ are homotopic. Recall that $P\left(H^{[n]};\sqrt{t}\right)$ has the generating function

$$\sum_{n=0}^{\infty} P\left(H^{[n]};\sqrt{t}\right)q^n = \prod_{d=1}^{\infty}\frac{1}{1-t^{d-1}q^d}.$$

Thus we have

$$\sum_{n=0}^{\infty} E\left(H^{[n]};t\right)q^n = \prod_{d=1}^{\infty}\frac{1}{1-t^{d+1}q^d} = \left(\prod_{d=1}^{\infty}\frac{1}{1-t^{d-1}q^d}\right)\left(\sum_{s=0}^{\infty} E\left(Y_0^{[s]};t\right)q^s\right).$$

Therefore

$$\sum_{n=0}^{\infty} E\left(Y_0^{[n]};t\right)q^n = \prod_{d=1}^{\infty}\frac{1-t^{d-1}q^d}{1-t^{d+1}q^d}.$$

□

Example 16 We list some E-polynomials $E\left(Y_0^{[n]};t\right)$ for $0 \leq n \leq 8$:

n	$E\left(Y_0^{[n]};t\right)$
0	1
1	t^2-1
2	$t^4+t^3-t^2-t$
3	$t^6+t^5-2t^3-t^2+t$
4	$t^8+t^7+t^6-t^5-3t^4+t^2$
5	$t^{10}+t^9+t^8-3t^6-3t^5+t^4+2t^3$
6	$t^{12}+t^{11}+t^{10}+t^9-t^8-4t^7-3t^6+3t^5+2t^4-t^3$
7	$t^{14}+t^{13}+t^{12}+t^{11}-3t^9-6t^8-t^7+5t^6+2t^5-t^4$
8	$t^{16}+t^{15}+t^{14}+t^{13}+t^{12}-t^{11}-5t^{10}-6t^9+t^8+7t^7+t^6-2t^5$

Before stating the result about the matrix $\mathcal{A}^{-1}$, we define the *dual E-polynomial* $\check{E}$ of a complex variety Z as

$$\check{E}(Z;t) := H(Z;\sqrt{t},\sqrt{t},-1). \tag{12}$$

When Z is a connected and smooth, we can compute $\check{E}(Z;t)$ from $E(Z;t)$ by Poincaré duality (2):

$$\check{E}(Z;t) = E\left(Z;t^{-1}\right)t^{\dim_{\mathbb{C}} Z}.$$

The Hilbert scheme $Y_0^{[n]}$ is a complex $4n$-dimensional smooth variety, and hence we have

$$\check{E}\left(Y_0^{[n]};t\right) = t^{\dim_{\mathbb{C}} Y_0^{[n]}} E\left(Y_0^{[n]};t^{-1}\right) = t^{2n} E\left(Y_0^{[n]};t^{-1}\right).$$

Then from the definition of the generating function of $\check{E}$, we have

$$\begin{aligned}\sum_{n=0}^{\infty}\check{E}\left(Y_0^{[n]};t\right)q^n = \sum_{n=0}^{\infty}E\left(Y_0^{[n]};t^{-1}\right)t^{2n}q^n &\overset{(\tilde{q}=t^2q)}{=\!=\!=} \prod_{d=1}^{\infty}\frac{1-t^{1-d}\tilde{q}^d}{1-t^{-d-1}\tilde{q}^d}\\ &= \prod_{d=1}^{\infty}\frac{1-t^{1-d}(t^2q)^d}{1-t^{-d-1}(t^2q)^d},\end{aligned}$$

and we finally obtain

$$\sum_{n=0}^{\infty}\check{E}\left(Y_0^{[n]};t\right)q^n = \prod_{d=1}^{\infty}\frac{1-t^{d+1}q^d}{1-t^{d-1}q^d}. \tag{13}$$

We are now ready to calculate $\mathcal{A}^{-1}$.

Proposition 17 *The matrix $\mathcal{A}$ has the inverse*

$$\mathcal{A}^{-1} = \left(\mathcal{A}_{ij}^{-1}\right)_{\cdot,\cdot} = \left(\check{E}\left(Y_0^{[j-i]}\right)\right)_{\cdot,\cdot}$$

More explicitly,

$$\mathcal{A}^{-1} = \begin{pmatrix} 1 & \check{E}\left(Y_0^{[1]}\right) & \check{E}\left(Y_0^{[2]}\right) & \check{E}\left(Y_0^{[3]}\right) & & \check{E}\left(Y_0^{[k]}\right) & \\ 0 & 1 & \check{E}\left(Y_0^{[1]}\right) & \check{E}\left(Y_0^{[2]}\right) & \dots & \check{E}\left(Y_0^{[k-1]}\right) & \dots \\ \vdots & 0 & 1 & \check{E}\left(Y_0^{[1]}\right) & & \vdots & \\ & & 0 & 1 & & \check{E}\left(Y_0^{[k-j]}\right) & \\ & & \vdots & 0 & & \vdots & \\ & & & & & 1 & \end{pmatrix}, \tag{14}$$

where $\check{E}\left(Y_0^{[j-i]}\right)=t^{2(j-i)}E\left(Y_0^{[j-i]};t^{-1}\right)$ *is the dual E-polynomial of* $Y_0^{[j-i]}$.

Proof Denote by $\mathcal{C}$ be the infinite matrix in (14). The (i,j)-th entry of the matrix $\mathcal{A}\cdot\mathcal{C}$ is given by the sum

$$\begin{aligned}(\mathcal{AC})_{ij}&=\sum_{k=i}^{j}E\left(Y_0^{[k-i]};t\right)\check{E}\left(Y_0^{[j-k]};t\right)\\&=\sum_{l=0}^{j-i}E\left(Y_0^{[l]};t\right)\check{E}\left(Y_0^{[j-i-l]};t\right).\end{aligned}$$

Note that Proposition 7 and (13) yield the product of the generating functions

$$\begin{aligned}&\sum_{n=0}^{\infty}\left(\sum_{i=0}^{n}E\left(Y_0^{[i]};t\right)\check{E}\left(Y_0^{[n-i]};t\right)\right)q^n\\&\quad=\left(\sum_{n=0}^{\infty}E\left(Y_0^{[n]};t\right)q^n\right)\left(\sum_{n=0}^{\infty}\check{E}\left(Y_0^{[n]};t\right)q^n\right)=1.\end{aligned}$$

Therefore, $(\mathcal{A}\cdot\mathcal{C})_{ij}=\delta_{ij}$ and this implies $\mathcal{C}=\mathcal{A}^{-1}$. □

We are now ready to prove Theorem 2.

Proof of Theorem 2 The E-polynomial $E\left(B_m^{[n]};t\right)$ is equal to the (m,n)-th entry of the matrix product $\mathcal{B}=\mathcal{X}\mathcal{A}^{-1}=\mathcal{G}^{-1}\mathcal{R}\mathcal{A}^{-1}$ which is given by the sum

$$\begin{aligned}\sum_k\sum_j\mathcal{G}_{mk}^{-1}\mathcal{R}_{kj}\mathcal{A}_{jn}^{-1}&=\sum_{k=m}^{\mu_n^{\max}}\sum_{j=\frac{k(k-1)}{2}}^{n}(-1)^{k-m}t^{\binom{k-m}{2}}\begin{bmatrix}k\\m\end{bmatrix}_t\\&\quad\times E\left(H^{[j,j+k]};t\right)\check{E}\left(Y_0^{[n-j]};t\right).\end{aligned}$$

Substituting this into the generating function, we obtain

$$\begin{aligned}&\sum_{n=0}^{\infty}E\left(B_m^{[n]};t\right)q^n\\&=\sum_{n=0}^{\infty}\left(\sum_{k=m}^{\mu_n^{\max}}\sum_{j=\frac{k(k-1)}{2}}^{n}(-1)^{k-m}t^{\binom{k-m}{2}}\begin{bmatrix}k\\m\end{bmatrix}_t E\left(H^{[j,j+k]};t\right)\check{E}\left(Y_0^{[n-j]};t\right)\right)q^n.\end{aligned}$$

Here we can let the indices k and j run from 0 to ∞ without changing the infinite sum since the space $H^{[j,j+k]} = \emptyset$ if $j \leq \frac{k(k-1)}{2}$ by Proposition 9, and $\begin{bmatrix} k \\ m \end{bmatrix}_t = 0$ if $k \leq m$. Recall that the generating functions of $E\left(H^{[n,n+r]}; t\right)$ and $\breve{E}\left(Y_0^{[n]}; t\right)$ are given by $q^{\binom{r}{2}}\left(\prod_{d=1}^{\infty} \frac{1}{1-t^{(d+1)}q^d}\right)\left(\prod_{d=1}^{r} \frac{1}{1-t^d q^d}\right)$ and $\prod_{d=1}^{\infty} \frac{1-t^{d+1}q^d}{1-t^{d-1}q^d}$, respectively. Then the generating function

$$\sum_{n=0}^{\infty} E\left(B_m^{[n]}; t\right) q^n$$

$$= \sum_{n=0}^{\infty} \left(\sum_{k=0}^{\infty} \sum_{j=0}^{\infty} (-1)^{k-m} t^{\binom{k-m}{2}} \begin{bmatrix} k \\ m \end{bmatrix}_t E\left(H^{[j,j+k]}; t\right) \breve{E}\left(Y_0^{[n-j]}; t\right) \right) q^n$$

is equal to the product

$$\prod_{d=1}^{\infty} \frac{1}{1-t^{d+1}q^d} \cdot \prod_{d=1}^{\infty} \frac{1-t^{d+1}q^d}{1-t^{d-1}q^d} \cdot \sum_{k=0}^{\infty} \left(q^{\binom{k}{2}} (-1)^{k-m} t^{\binom{k-m}{2}} \begin{bmatrix} k \\ m \end{bmatrix}_t \prod_{d=1}^{k} \frac{1}{1-t^d q^d} \right)$$

$$= \prod_{d=1}^{\infty} \frac{1}{1-t^{d-1}q^d} \cdot \sum_{k=0}^{\infty} \left((tq)^{\binom{k}{2}} (-1)^{k-m} t^{-km+\binom{m+1}{2}} \begin{bmatrix} k \\ m \end{bmatrix}_t \frac{1}{(tq)_k} \right), \tag{15}$$

where $(tq)_k := \prod_{d=1}^{k} \left(1 - t^d q^d\right)$. To continue the proof, we need the following lemma.

Lemma 18 *We have*

$$\sum_{i=0}^{m} (-1)^{m+i} t^{km - \binom{m}{2} + \binom{i}{2} - ik} \begin{bmatrix} m \\ i \end{bmatrix}_t = \prod_{i=0}^{m-1} \left(1 - t^{k-i}\right). \tag{16}$$

Proof of Lemma 18 Recall the Gauss's binomial formula ([KC, p.29]):

$$\prod_{k=0}^{n-1} (1 + aq^k) = \sum_{k=0}^{n} q^{\binom{k}{2}} \begin{bmatrix} n \\ k \end{bmatrix}_q a^k.$$

Applying this formula with $a = -t^{-k}, q = t$, we obtain

$$\begin{aligned}\prod_{i=0}^{m-1}\left(1-t^{k-i}\right) &= (-1)^m t^{mk-\binom{m}{2}} \prod_{i=0}^{m-1}\left(1+(-t^{-k})t^i\right)\\ &= (-1)^m t^{mk-\binom{m}{2}} \sum_{i=0}^{m} t^{\binom{i}{2}} \begin{bmatrix} m \\ i \end{bmatrix}_t (-t^{-k})^i\\ &= \sum_{i=0}^{m} (-1)^{m+i} t^{mk-\binom{m}{2}+\binom{i}{2}-ik} \begin{bmatrix} m \\ i \end{bmatrix}_t .\end{aligned}$$

□

We continue the calculation of the generating function of $E\left(B_m^{[n]}; t\right)$. We write $\begin{bmatrix} k \\ m \end{bmatrix}_t = \prod_{i=0}^{m-1} \frac{1-t^{k-i}}{1-t^{i+1}}$, and apply Lemma 18 to the product $\prod_{i=0}^{m-1}\left(1-t^{k-i}\right)$. Substituting the result into the generating function, we obtain

$$\begin{aligned}&\sum_{n=0}^{\infty} E\left(B_m^{[n]}; t\right) q^n\\ &= \prod_{d=1}^{\infty} \frac{1}{1-t^{d-1}q^d} \cdot \sum_{k=0}^{\infty}\left((tq)^{\binom{k}{2}}(-1)^{k-m} t^{-km+\binom{m+1}{2}} \begin{bmatrix} k \\ m \end{bmatrix}_t \frac{1}{(tq)_k}\right)\\ &= \prod_{d=1}^{\infty} \frac{1}{1-t^{d-1}q^d} \times (-1)^m\\ &\quad \times \sum_{k=0}^{\infty} \frac{\frac{(-1)^k (tq)^{\binom{k}{2}}}{(tq)_k} t^{-km+\binom{m+1}{2}} \sum_{a=0}^{m} (-1)^{m+a} t^{km-\binom{m}{2}+\binom{a}{2}-ak} \begin{bmatrix} m \\ a \end{bmatrix}_t}{\prod_{i=0}^{m-1}\left(1-t^{i+1}\right)}\\ &= \prod_{d=1}^{\infty} \frac{1}{1-t^{d-1}q^d} \cdot \prod_{i=0}^{m-1} \frac{1}{1-t^{i+1}} \cdot \sum_{a=0}^{m}\\ &\quad \times \left((-1)^a t^{m+\binom{a}{2}} \begin{bmatrix} m \\ a \end{bmatrix}_t \sum_{k=0}^{\infty} \frac{(-1)^k (tq)^{\binom{k}{2}} t^{-ak}}{(tq)_k}\right).\end{aligned}$$

Apply now the Euler identity (cf., e.g., [Co])

$$(z)_\infty = \sum_{n=0}^{\infty} \frac{(-1)^n z^n q^{\binom{n}{2}}}{(q)_n} = \prod_{n=0}^{\infty} \left(1 - zq^n\right) \tag{17}$$

to the infinite sum $\sum_{k=0}^{\infty} \frac{(-1)^k (tq)^{\binom{k}{2}} t^{-ak}}{(tq)_k}$ with the change of variables $\begin{cases} q \mapsto tq, \\ z \mapsto t^{-a} \end{cases}$.

We obtain

$$\begin{aligned}
\sum_{n=0}^{\infty} E\left(B_m^{[n]}; t\right) q^n &= \prod_{d=1}^{\infty} \frac{1}{1 - t^{d-1} q^d} \cdot \prod_{i=0}^{m-1} \frac{1}{1 - t^{i+1}} \cdot \sum_{a=0}^{m} \\
&\quad \times \left((-1)^a t^{m+\binom{a}{2}} \begin{bmatrix} m \\ a \end{bmatrix}_t \prod_{k=0}^{\infty} (1 - t^{-a} (tq)^k) \right) \\
&= \prod_{i=0}^{m-1} \frac{1}{1 - t^{i+1}} \cdot \sum_{a=0}^{m} \\
&\quad \times \left((-1)^a t^{m+\binom{a}{2}} \begin{bmatrix} m \\ a \end{bmatrix}_t \left(1 - t^{0-1}\right) \prod_{d=0}^{\infty} \frac{1 - t^{d-a} q^d}{1 - t^{d-1} q^d} \right) \\
&= \prod_{i=1}^{m-1} \frac{1}{1 - t^{i+1}} \cdot \sum_{a=0}^{m} \\
&\quad \times \left((-1)^{a+1} t^{m-1+\binom{a}{2}} \begin{bmatrix} m \\ a \end{bmatrix}_t \prod_{d=0}^{\infty} \frac{1 - t^{d-a} q^d}{1 - t^{d-1} q^d} \right).
\end{aligned}$$

Note that the infinite product $\prod_{d=0}^{\infty} \frac{1 - t^{d-a} q^d}{1 - t^{d-1} q^d} = 0$ when $a = 0$, and thus we arrive at our final result

$$\sum_{n=0}^{\infty} E\left(B_m^{[n]}; t\right) q^n = \prod_{i=1}^{m-1} \frac{1}{1 - t^{i+1}} \cdot \sum_{a=1}^{m} \left((-1)^{a+1} t^{m-1+\binom{a}{2}} \begin{bmatrix} m \\ a \end{bmatrix}_t \prod_{d=0}^{\infty} \frac{1 - t^{d-a} q^d}{1 - t^{d-1} q^d} \right).$$

□

4 The E-Polynomial of the Refined Strata $H_m^{[n]}$

Using Theorem 2, we can calculate the E-polynomial of the refined strata $H_m^{[n]}$ as well.

Theorem 19 *The E-polynomial of the refined stratum $H_m^{[n]}$ has the generating function*

$$\sum_{n=0}^{\infty} E\left(H_m^{[n]};t\right) q^n = \prod_{i=1}^{m-1} \frac{1}{1-t^{i+1}} \cdot \sum_{a=1}^{m} \left((-1)^a t^{\binom{a}{2}+m} \begin{bmatrix} m \\ a \end{bmatrix}_t \prod_{k=0}^{\infty} \frac{1-q^k t^{k-a}}{1-q^k t^{k+1}} \right). \tag{1}$$

Remark 20 One can also obtain this formula directly from the entries of matrix products $\mathcal{X} = \mathcal{G}^{-1}\mathcal{R}$ in Eqs. (4) with a similar argument as in the Proof of Theorem 2.

Proof From relation (3), the E-polynomial of $H_m^{[n]}$ is a sum

$$E\left(H_m^{[n]};t\right) = \sum_{s=0}^{n} E\left((Y_0)^{[n-s]};t\right) E\left(B_m^{[s]};t\right).$$

Then the generating function of the E-polynomial $E\left(H_m^{[n]};t\right)$

$$\begin{aligned}\sum_{n=0}^{\infty} E\left(H_m^{[n]};t\right) q^n &= \sum_{n=0}^{\infty} \left(\sum_{j=0}^{n} E\left((Y_0)^{[n-j]};t\right) E\left(B_m^{[j]};t\right) \right) q^n \\ &= \left(\sum_{n=0}^{\infty} E\left(Y_0^{[n]};t\right) q^n \right) \left(\sum_{n=0}^{\infty} E\left(B_m^{[n]};t\right) q^n \right).\end{aligned}$$

Applying the formula of the generating functions in (7) and (2), we obtain

$$\begin{aligned}\sum_{n=0}^{\infty} E\left(H_m^{[n]};t\right) q^n &= \prod_{d=1}^{\infty} \frac{1-t^{d-1}q^d}{1-t^{d+1}q^d} \cdot \prod_{d=1}^{m-1} \frac{1}{1-t^{i+1}} \cdot \sum_{a=1}^{m} \\ &\times \left((-1)^{a+1} t^{m-1+\binom{a}{2}} \begin{bmatrix} m \\ a \end{bmatrix}_t \cdot \prod_{d=0}^{\infty} \frac{1-t^{d-a}q^d}{1-t^{d-1}q^d} \right)\end{aligned}$$

$$= \prod_{d=1}^{m-1} \frac{1}{1-t^{i+1}} \cdot \sum_{a=1}^{m}$$

$$\times \left((-1)^{a+1} t^{m+\binom{a}{2}} \begin{bmatrix} m \\ a \end{bmatrix}_t \cdot \frac{t^{-1}(1-t)}{1-t^{-1}} \cdot \prod_{d=0}^{\infty} \frac{1-t^{d-a}q^d}{1-t^{d+1}q^d} \right)$$

$$= \prod_{d=1}^{m-1} \frac{1}{1-t^{i+1}} \cdot \sum_{a=1}^{m} \left((-1)^{a} t^{m+\binom{a}{2}} \begin{bmatrix} m \\ a \end{bmatrix}_t \prod_{d=0}^{\infty} \frac{1-t^{d-a}q^d}{1-t^{d+1}q^d} \right).$$

□

Example 21 We list some of the first examples of $E\left(H_m^{[n]}; t\right)$:

$[t]$ (2)

n	$m=1$	$m=2$	$m=3$
1	t^2-1	1	0
2	$t^4+t^3-t^2-t$	t^2+t	0
3	$t^6+t^5-2t^3-t^2+t$	$t^4+2t^3+t^2-t-1$	1
4	$t^8+t^7+t^6-t^5-3t^4+t^2$	$t^6+2t^5+3t^4-2t^2-t$	t^2+t

It is instructive to compare these results with a geometric description of $H_m^{[n]}$ for the first few values of n.

- n=1: The stratum $H_1^{[1]} = Y_0$ consists of all the points of $\mathbb{C}^2$ except $(0,0)$. The other stratum $H_2^{[1]}$ is the single point $(0,0)$. Their E-polynomials are $E(Y_0) = t^2-1$ and $E(\mathrm{pt}) = 1$, respectively.
- n=2: The stratum $H_1^{[2]}$ of $m=1$ is the Hilbert scheme of two points on the punctured plane: $Y_0^{[2]}$. As for $m=2$, $H_2^{[2]} \simeq (Y_0 \times (0,0)) \bigcup B^{[2]}$.
- n=3: Similarly to the previous case, we have $H_1^{[3]} = Y_0^{[3]}$. The stratum of $m=2$ is the union $H_2^{[2]} \simeq \left(Y_0^{[2]} \times (0,0)\right) \cup \left(Y_0 \times B^{[2]}\right) \cup B_2^{[3]}$. The last stratum $H_3^{[3]}$ is the point represented by the maximal ideal $\langle x^2, xy, y^2 \rangle$.

5 Euler Characteristics of the Refined Strata

The specialization of $E\left(B_m^{[n]}; t\right)$ at $t=1$ is the topological Euler characteristic of $B_m^{[n]}$

$$\chi(B_m^{[n]}) = E\left(B_m^{[n]}; 1\right) = \chi\left(H_m^{[n]}\right).$$

Table 1 Examples of $\chi\left(B_m^{[n]}\right)$

n	$m=2$	$m=3$	$m=4$	$m=5$
0	0	0	0	0
1	1	0	0	0
2	2	0	0	0
3	2	1	0	0
4	3	2	0	0
5	2	5	0	0
6	4	6	1	0
7	2	11	2	0

It is not obvious how to obtain an explicit formula for the generating function of $\chi\left(B_m^{[n]}\right)$ from formula (4) in Theorem 2 because of the singularities at $t=1$. However, if we consider the specialization of Eq. (15) in the Proof of Theorem 2,

$$\sum_{n=0}^{\infty} E\left(B_m^{[n]};t\right)q^n = \prod_{d=1}^{\infty}\frac{1}{1-t^{d-1}q^d}\cdot\sum_{k=0}^{\infty}$$
$$\times\left((tq)^{\binom{k}{2}}(-1)^{k-m}t^{-km+\binom{m+1}{2}}\begin{bmatrix}k\\m\end{bmatrix}_t\frac{1}{(tq)_k}\right),$$

then we can pass to the limit at $t=1$, and we obtain the following formula:

$$\sum_{n=0}^{\infty}\chi\left(B_m^{[n]}\right)q^n = \prod_{d=1}^{\infty}\frac{1}{1-q^d}\cdot\sum_{k=0}^{\infty}\left(\frac{(-1)^{k-m}q^{\binom{k}{2}}}{(q)_k}\binom{k}{m}\right). \tag{1}$$

Some initial values are given in Table 1.

Appendix 1: Properties of the E-Polynomial

Let Z be a complex algebraic variety. P. Deligne established the existence of two filtrations on the j-th cohomology group of Z: the weight filtration W_*

$$0=W_{-1}\subseteq W_0\subseteq\cdots\subseteq W_{2j}=H^j(Z)$$

and the Hodge filtration F^*

$$H^j(Z)=F^0\supseteq F^1\supseteq\cdots\supseteq F^m\supseteq F^{2j}=0$$

such that, for each l, the filtration induced by F on the graded piece $\mathrm{gr}_l W := W_l/W_{l-1}$ endows $\mathrm{gr}_l W$ with a pure Hodge structure of weight l. One can define a mixed Hodge structure on the compactly supported cohomology $H_c^*(Z)$ as well [Fu].

We define the compactly supported mixed Hodge polynomial of Z to be the generating function of the compactly supported mixed Hodge numbers $h_c^{p,q;j}(Z) := \dim_{\mathbb{C}} \mathrm{gr}_p F\left(\mathrm{gr}_{p+q} W\left(H_c^j(Z)\right)\right)$:

$$H_c(Z; u, v, s) := \sum_{p,q,j} h_c^{p,q;j}(Z) u^p v^q s^j.$$

For a connected smooth variety Z of complex dimension d, the cohomological pairing

$$H^k(Z) \times H_c^{2d-k}(Z) \to H^{2d}(Z) \tag{2}$$

is a morphism of mixed Hodge structures. Then Poincaré duality implies the equality of mixed Hodge numbers:

$$h_c^{p,q;j}(Z) = h^{d-p,d-q;2d-j}(Z), \tag{3}$$

where $h^{p,q;j} := \dim_{\mathbb{C}} \mathrm{gr}_p F\left(\mathrm{gr}_{p+q} W(H^j(Z))\right)$ are the mixed Hodge numbers of Z. Equivalently, if $H(Z; u, v, s) := \sum_{p,q,j} h^{p,q;j}(Z) u^p v^q s^j$ is the mixed Hodge polynomial of Z, then we have

$$H_c(Z; u, v, s) = u^d v^d s^{2d} H(Z; u^{-1}, v^{-1}, s^{-1}). \tag{4}$$

The specialization $H_c(Z; u, v, s)|_{u=1,v=1}$ is the compactly supported Poincaré polynomial of Z. Moreover, if Z is a connected smooth variety, then by (3), we can express the compactly supported Poincaré polynomial in terms of $P(Z; s) := \sum_j \dim_{\mathbb{C}} H^j(Z) s^j$, the Poincaré polynomial of Z as follows:

$$H_c(Z; 1, 1, s) = P\left(Z; s^{-1}\right) s^{2d}. \tag{5}$$

The *E-polynomial* or *the Hodge-Deligne polynomial* of Z is defined to be the compactly supported mixed Hodge polynomial specialized at $s = -1$

$$E(Z; u, v) := \sum_{p,g,j} h_c^{p,q;j}(Z)(-1)^j u^p v^q.$$

Properties of the E-Polynomial

- **Additivity:** If $X \subset Z$ is a Zariski-closed subvariety, then

$$E(Z; u, v) = E(X; u, v) + E(Z \setminus X; u, v).$$

- **Factorization on fibrations:** If $f : Z \to B$ is a Zariski locally trivial fibration with fiber F, then

$$E(Z) = E(B) \cdot E(F). \tag{6}$$

In particular,

$$E(B \times F) = E(B) \cdot E(F).$$

- The specialization at $u = v = 1$ gives the topological Euler characteristic

$$E(Z; 1, 1) = \sum_{p,g,j} (-1)^j h_c^{p,q;j}(Z) = \chi(Z).$$

When Z is smooth and projective, the specialization

$$E(Z; -u, -v) = \sum_{p,q} \dim H^q(Z, \Omega^p) u^p v^q = \sum_{p,q} h^{p,q} u^p v^q$$

agrees with the Hodge polynomial of Z, and

$$E(Z; -y, 1) = \sum_{p,g,j} (-1)^j h^{p,q;j}(Z) y^p = \chi_y(Z)$$

is the Hirzebruch χ_y-genus of Z.

Example 22 Using the above properties, it is easy to verify that $E(\mathbb{C}; u, v) = uv$, and thus $E(\mathbb{C}^n; u, v) = (uv)^n$ and $E(\mathbb{C}^n \setminus 0; u, v) = (uv)^n - 1$. In particular, the E-polynomial of $\mathbb{C}^*$ is $uv - 1$.

Appendix 2: Tables of E-Polynomials of the Refined Strata

In Table 2, we list the first few instances of our main result, Theorem 2.

Some remarks on the geometric description of the strata $B_m^{[n]}$ for small values of n:

- n=1; $B^{[1]} = B_2^{[1]}$ is a single point, the origin $\{(0, 0)\}$. Thus $E\left(B_2^{[1]}; t\right) = E(\text{pt}; t) = 1$.
- n=2; $B^{[2]} = B_2^{[2]}$ is isomorphic to $\mathbb{P}^1$. We have thus $E\left(B_2^{[2]}; t\right) = E\left(\mathbb{P}^; t\right) = 1 + t$.
- n=3; $B^{[3]} = B_3^{[3]} \cup B_2^{[3]}$ has two strata: $B_3^{[3]}$ is a point given by the maximal ideal $\langle x^2, xy, y^2 \rangle$, and $B_2^{[3]}$, the curvilinear locus, is isomorphic to an affine bundle

Table 2 E-polynomials of the refined strata of the punctual Hilbert schemes $E\left(B_m^{[n]};t\right)$

n	$E\left(B_2^{[n]};t\right)$	$E\left(B_3^{[n]};t\right)$	$E\left(B_4^{[n]};t\right)$	$E\left(B_5^{[n]};t\right)$
1	1	0	0	0
2	$t+1$	0	0	0
3	t^2+t	1	0	0
4	t^3+2t^2	$t+1$	0	0
5	t^4+2t^3-t	$2t^2+2t+1$	0	0
6	$t^5+3t^4+t^3-t^2$	$2t^3+3t^2+t$	1	0
7	$t^6+3t^5+t^4-2t^3-t^2$	$3t^4+5t^3+3t^2$	$t+1$	0
8	$t^7+4t^6+2t^5-2t^4-t^3$	$3t^5+7t^4+4t^3-t$	$2t^2+2t+1$	0
9	$t^8+4t^7+3t^6-3t^5-2t^4$	$4t^6+9t^5+7t^4-2t^2-t$	$3t^3+4t^2+2t+1$	0
10	$t^9+5t^8+4t^7-3t^6-3t^5$	$4t^7+12t^6+10t^5+t^4-3t^3-2t^2$	$4t^4+6t^3+4t^2+t$	1
11	$t^{10}+5t^9+5t^8-4t^7-5t^6$	$5t^8+15t^7+15t^6+2t^5-5t^4-4t^3-t^2$	$5t^5+10t^4+7t^3+3t^2$	$t+1$
12	$t^{11}+6t^{10}+7t^9-3t^8-6t^7+t^5$	$5t^9+18t^8+19t^7+4t^6-8t^5-7t^4-2t^3$	$7t^6+14t^5+12t^4+5t^3-t$	$2t^2+2t+1$
13	$t^{12}+6t^{11}+8t^{10}-4t^9-9t^8-t^7+t^6$	$6t^{10}+22t^9+27t^8+7t^7-10t^6-11t^5-4t^4$	$8t^7+20t^6+18t^5+9t^4-2t^2-t$	$3t^3+4t^2+2t+1$
14	$t^{13}+7t^{12}+10t^{11}-3t^{10}-11t^9-2t^8+2t^7$	$6t^{11}+26t^{10}+34t^9+12t^8-13t^7-16t^6-6t^5+t^3$	$10t^8+26t^7+27t^6+13t^5-5t^3-3t^2-t$	$5t^4+7t^3+5t^2+2t+1$

over $\mathbb{P}^1$ [Br]. Then by the factorization property of the E-polynomial, we have $E\left(B_2^{[3]};t\right)=t(1+t)=t+t^2$.

We observe that the stratum of $B_k^{\left[\binom{k}{2}\right]}$ consists of a single point, the power of the maximal ideal $\mathfrak{m}^k$, e.g.,

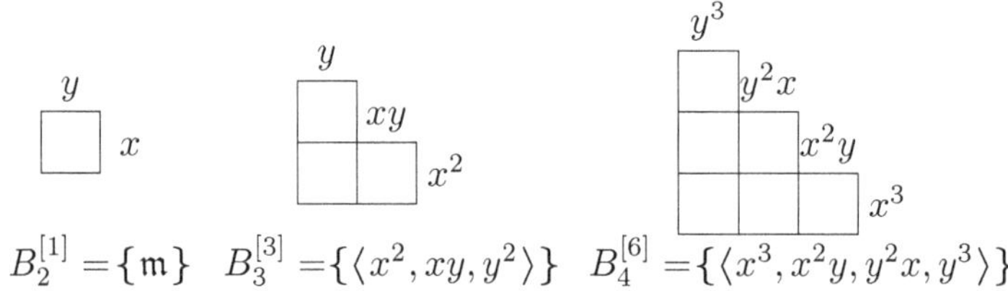

$$B_2^{[1]}=\{\mathfrak{m}\}\quad B_3^{[3]}=\{\langle x^2,xy,y^2\rangle\}\quad B_4^{[6]}=\{\langle x^3,x^2y,y^2x,y^3\rangle\}$$

and this is consistent with the fact that the corresponding entry of the table is the constant 1.

Acknowledgments We would like to thank Alexei Oblomkov for key suggestions, which started this project. We are also grateful to Tamas Hausel, Andrei Negut, and Rahul Pandharipande for useful discussions.

References

[Bi1] A. Białynicki-Birula, *Some theorems on actions of algebraic groups*, Ann. of Math. (2) **98** (1973), 480–497.

[Bi2] A. Białynicki-Birula, *Some properties of the decompositions of algebraic varieties determined by actions of a torus*, Bull. Acad. Polon. Sci. Sér. Sci. Math. Astronom. Phys. 24,(No.9):667–674, 1976.

[Br] J. Briançon, *Description de* $Hilb^n C\{x, y\}$, Invent. Math. **41** (1977), no.1, 45–89.

[Ch1] J. Cheah, *On the cohomology of Hilbert schemes of points*, J. Algebraic Geom. **5** (1996), no. 3, 479–511. MR1382733

[Ch2] J. Cheah, *The virtual Hodge polynomials of nested Hilbert schemes and related varieties.* Math. Z. **227** (1998), no. 3, 479–504.

[Co] L. Comtet, *Advanced Combinotorics*, revised and enlarged edition, D. Reidel Publishing Co., Dordrecht, 1974.

[De1] P. Deligne, *Théorie de Hodge. II* , Inst. Hautes Études Sci. Publ. Math. No. 40 (1971), 5–57.

[De2] P. Deligne, *Théorie de Hodge. III* , Inst. Hautes Études Sci. Publ. Math. No. 44 (1974), 5–77.

[ES] G. Ellingsrud, S.A. Strømme, *On the homology of the Hilbert scheme of points in the plane*, Invent. Math. **87** (1987), no. 2, 343–352.

[Fo] J. Fogarty, *Algebraic families on an algebraic surface*, Amer. J. Math **90** (1968), 511–521.

[Fu] W. Fulton, *Introduction to toric varieties*, Annals of Mathematics Studies, 131. The William H. Roever Lectures in Geometry. Princeton University Press, Princeton, NJ, 1993

[Gö1] L. Göttsche, *Hilbert schemes of points on surfaces*, Proceedings of the ICM, Beijing 2002, vol. 2, 483–494.

[Gö2] L. Göttsche, *Hilbert Schemes of Zero-Dimensional Subschemes of Smooth Varieties*, Lecture Notes in Mathematics, 1572. Springer-Verlag, Berlin, 1994. x+196 pp.

[Gr] I. Grojnowski, *Instantons and affine algebras. I. The Hilbert scheme and vertex operators*, Math. Res. Lett. 3 (1996), no. 2, 275–291.

[Ha] M. Haiman, *t,q-Catalan numbers and the Hilbert scheme*, Discrete Math. **193** (1998), no. 1–3, 201–224.

[HR] T. Hausel, F. Rodriguez-Villegas, *Mixed Hodge polynomials of character varieties*, Invent. Math. **174** (2008), no. 3, 555–624.

[HR2] T. Hausel, F. Rodriguez Villegas, *Cohomology of large semiprojective hyperkähler varieties*, Astérisque No. 370 (2015), 113–156.

[Ia] A. Iarrobino, *Punctual Hilbert schemes*, Mem. Amer. Math. Soc. **10** (1977), no. 188, viii+112 pp.

[KC] V. Kac, P. Cheung, *Quantum Calculus*, Universitext, Springer-Verlag, New York, 2002.

[KST] M. Kool, V. Shende, R. P. Thomast, *A short proof of the Göttsche conjecture*, Geom. Topol. **15** (2011), no. 1, 397–406.

[Le] M. Lehn, *Lectures on Hilbert schemes*, in *Algebraic structures and moduli spaces*, 1–30, CRM Proc. Lecture Notes, 38, Amer. Math. Soc., Providence, RI.

[Na1] H. Nakajima, *Lectures on Hilbert schemes of points on surfaces*, University Lecture Series, vol. 18, American Mathematical Society, Providence, RI, 1999.

[Na2] H. Nakajima *Heisenberg algebra and Hilbert schemes of points on projective surfaces*, Ann. of Math. (2) **145** (1997), no. 2, 379–388.

[NY] H. Nakajima, K. Yoshioka, *Perverse coherent sheaves on blow-up. II. wall-crossing and Betti numbers formula*, J. Algebraic Geom. **20** (2011), no. 1, 47–100.

[OS] A. Oblomkov, V. Shende, *The Hilbert scheme of a plane curve singularity and the HOMFLY polynomial of its link*, Duke Math. J. **161** (2012), no. 7, 1277–1303.

[ORS] A. Oblomkov and J. Rasmussen, V. Shende, *The Hilbert scheme of a plane curve singularity and the HOMFLY homology of its link*, Geom. Topol. **22** (2018), no. 2, 645–691.

[SGA] SGA1, arXiv:0206203.

Galois Action on VOA Gauge Anomalies

Theo Johnson-Freyd

For Kolya Reshetikhin on the occasion of his 60th birthday.

Abstract Assuming regularity of the fixed subalgebra, any action of a finite group G on a holomorphic VOA V determines a gauge anomaly $\alpha \in \mathrm{H}^3(G; \boldsymbol{\mu})$, where $\boldsymbol{\mu} \subset \mathbb{C}^\times$ is the group of roots of unity. We show that under Galois conjugation $V \mapsto {}^\gamma V$, the gauge anomaly transforms as $\alpha \mapsto \gamma^2(\alpha)$. This provides an a priori upper bound of 24 on the order of anomalies of actions preserving a $\mathbb{Q}$-structure, for example the Monster group $\mathbb{M}$ acting on its Moonshine VOA $V^\natural$. We speculate that each field $\mathbb{K}$ should have a "vertex Brauer group" isomorphic to $\mathrm{H}^3(\mathrm{Gal}(\bar{\mathbb{K}}/\mathbb{K}); \boldsymbol{\mu}^{\otimes 2})$. In order to motivate our constructions and speculations, we warm up with a discussion of the ordinary Brauer group, emphasizing the analogy between VOA gauging and quantum Hamiltonian reduction.

This chapter studies the behaviour of gauge (aka 't Hooft) anomalies under the action of Galois automorphisms. Our main theorem, stated and proved in Sect. 3, asserts that if V is a holomorphic VOA over $\mathbb{C}$ on which a finite group G acts with anomaly α, and if γ is a field automorphism of $\mathbb{C}$, then G acts on the Galois-

---

This project was started during the CIRM conference "Representation Theory, Mathematical Physics and Integrable Systems" celebrating the 60th birthday of Kolya Reshetikhin. It was from Kolya that I first learned about (quantum) Hamiltonian reduction, spin chains, conformal field theory, and modular tensor categories, and I thank him for his ongoing interest and inspiration. I also thank the anonymous referee for their thoughtful comments and corrections. Research at the Perimeter Institute for Theoretical Physics is supported by the Government of Canada through the Department of Innovation, Science and Economic Development Canada and by the Province of Ontario through the Ministry of Research, Innovation and Science.

T. Johnson-Freyd (✉)
Perimeter Institute for Theoretical Physics, Waterloo, ON, Canada
e-mail: theojf@pitp.ca

A. Alekseev et al. (eds.), *Representation Theory, Mathematical Physics, and Integrable Systems*, Progress in Mathematics 340,
https://doi.org/10.1007/978-3-030-78148-4_12

conjugate VOA ${}^{\gamma}V$ with anomaly $\gamma^2(\alpha)$, rather than, as might be naively expected, $\gamma(\alpha)$. One corollary is [JF19, Conjecture 1]: there is an a priori upper bound of 24 on the order of gauge anomalies of actions preserving a $\mathbb{Q}$-form of the VOA. In Sect. 4, we list some further questions and conjectures, including the possibility of a "vertex Brauer group".

These results are conditional on a widely believed regularity assumption, which asserts that the representation theory of the G-fixed sub-VOA V^G is finite semisimple. The regularity assumption is required in order for gauge anomalies to be well defined. We review the construction of gauge anomalies for VOAs in Sect. 2. As a warm-up, Sect. 1 discusses gauge anomalies in the case of actions on Azumaya algebras: anomalies live in $\mathrm{H}^2(G; \mathbb{K}^\times)$, with the standard Galois action; such anomalies lead quickly to the Galois-cohomological description of the Brauer group as $\mathrm{Br}(\mathbb{K}) \cong \mathrm{H}^2(\mathrm{Gal}(\bar{\mathbb{K}}/\mathbb{K}); \bar{\mathbb{K}}^\times)$.

1 Gauge Anomalies for Associative Algebras

The mathematics in this section is not new. Its goal is to present a familiar story in a way that will generalize to VOAs. A good source for the complete story is [GS06].

1.1 Azumaya Algebras

A $\mathbb{K}$-algebra A is *Azumaya* if it is finite-dimensional and simple, and its centre is exactly $\mathbb{K}$. Let A^{op} denote the opposite algebra. Then A is Azumaya if and only if its *enveloping algebra* $A \otimes A^{\mathrm{op}}$ is isomorphic to a matrix algebra. This is in turn equivalent to $A \otimes A^{\mathrm{op}}$ being Morita-trivial. Modules for $A \otimes A^{\mathrm{op}}$ are precisely A-A-bimodules, and so an Azumaya algebra is an algebra whose (monoidal) category of bimodules is equivalent to VECT.

1.2 Physical Interpretation

Azumaya algebras have the following interpretation in quantum physics. Consider a (one-dimensional) chain of atoms. To each atom, we associate a "local Hilbert space" given by the underlying vector space of A (so that a basis of A are the "spins" of the "spin chain"); if the atoms are indexed by $i \in \{\dots, 1, 2, \dots\}$, then we will write A_i for the copy of A at location i. The total Hilbert space is the tensor product $\bigotimes_i A_i$ of all the local Hilbert spaces. (That tensor product makes sense only for chains of finite length.)

We now place a Hamiltonian on this Hilbert space, which will be of "commuting projector" type. Specifically, consider the multiplication map $m : A \otimes A \to A$. It is a map of A-A-bimodules and, by our Azumaya condition, corresponds to a linear map $B \to \mathbb{K}$, where we have used the identification $\mathrm{BIMOD}(A) \simeq \mathrm{VECT}$ to identify ${}_A A_A \in \mathrm{BIMOD}(A)$ with $\mathbb{K} \in \mathrm{VECT}$ and to identify ${}_A A \otimes A_A \in \mathrm{BIMOD}(A)$ with some vector space $B \in \mathrm{VECT}$. (The vector space B will be noncanonically isomorphic to the underlying vector space of A.) The map $B \to \mathbb{K}$ is nonzero, hence a surjection, and so can be split: we can choose a bimodule map $\Delta : {}_A A_A \to {}_A A \otimes A_A$ such that $m \circ \Delta = \mathrm{id}_A$. Then $\Delta \circ m : A^{\otimes 2} \to A^{\otimes 2}$ is a projection (it squares to itself) onto a vector space isomorphic to A. Let id_i denote the identity on A_i, and for a set I of indices, let $\mathrm{id}_I = \bigotimes_{i \in I} \mathrm{id}_i$. Also write $(\Delta \circ m)_{i,i+1}$ for the map $\Delta \circ m$ acting on $A_i \otimes A_{i+1}$. The Hamiltonian we will use is

$$\sum_i \mathrm{id}_{<i} \otimes (\mathrm{id} - \Delta \circ m)_{i,i+1} \otimes \mathrm{id}_{>i+1}.$$

This Hamiltonian is gapped (i.e., it has a finite-dimensional lowest eigenspace, and its spectrum is discrete), and its vector space of ground states is isomorphic to A. Since the chain of atoms was 1-space-dimensional, together with the Hamiltonian we have a "gapped topological system of $(1+1)$-dimensional matter". It is expected that gapped physical systems admit low-energy effective descriptions in terms of TQFTs. For the spin chain model built from A, the low-energy TQFT is the one corresponding to A under the cobordism hypothesis as in [SP09].

The gapped topological phase (meaning the connected component of the space of all gapped systems under deformations that do not close the gap) constructed from a semisimple algebra A depends only on the Morita-equivalence class of A (see, e.g., [GJF19, §2.4]). To encode the algebra A itself, one can remember not just the phase of $(1 + 1)$-dimensional matter, but also a choice of boundary condition. Namely, the spin chain constructed above has an obvious "free" boundary condition, whose boundary observables form a copy of the associative algebra A. The bulk observables live in the centre $\mathcal{Z}(A)$, which is trivial by assumption. Thus, the bulk theory is almost trivial: it is "short-range entangled". The class of A in the Brauer group $\mathrm{Br}(\mathbb{K})$ can be considered a "gravitational anomaly" for the boundary system: the boundary system fails to be an absolute $(0 + 1)$-dimensional QFT only because of the Brauer class of A. Some discussion of absolute versus anomalous quantum systems, in the framework of TQFTs, is available in [FT14, JF20]; for a condensed matter discussion focussing on gauge anomalies, see [GJF19, §7].

1.3 *Construction of Gauge Anomalies*

Let A be Azumaya and $G \subset \mathrm{Aut}(A)$ a finite group of automorphisms of A. (One can be more general, but we will not.) We will associate to these data a *gauge anomaly* $\alpha \in \mathrm{H}^2(G; \mathbb{K}^\times)$, where G acts trivially on $\mathbb{K}^\times$.

There are many ways to define α. The quickest is to consider the exact sequence, valid for any algebra,

$$\{1\} \to \mathcal{Z}(A^\times) \to A^\times \to \mathrm{Aut}(A) \to \mathrm{Out}(A) \to \{1\}, \quad (1)$$

where the map $A^\times \to \mathrm{Aut}(A)$ picks out the inner automorphisms, and $\mathrm{Out}(A)$ denotes the group of outer automorphisms of A. Two simplifications occur when A is Azumaya. First, from the definition in terms of central simple algebras, we see that $\mathcal{Z}(A^\times) = \mathbb{K}^\times$. Second, $\mathrm{Out}(A)$ is trivial by the Skolem–Noether theorem. Let us quickly prove the second claim. Any automorphism $\phi : A \to A$ determines an invertible bimodule $A_\phi \in \mathrm{BIMOD}(A)$, which is A as a left A-module, but for which the right A-action is twisted by ϕ. But $\mathrm{BIMOD}(A) \simeq \mathrm{VECT}$, so there must be a bimodule isomorphism $A_\phi \overset{\sim}{\to} A$. This map must take the unit $1 \in A_\phi$ to some (necessarily invertible) element $f \in A$, and compatibility with the left A-action says that it takes $a \mapsto af$. But compatibility with the right A-action then says that $a\phi(b)f = afb$ for all a, and so ϕ is inner. All together, the sequence

$$\{1\} \to \mathbb{K}^\times \to A^\times \to \mathrm{Aut}(A) \to \{1\}$$

is seen to be exact and a central extension. The action $G \to \mathrm{Aut}(A)$ pulls back this central extension to an extension of G, classified by the anomaly $\alpha \in \mathrm{H}^2(G; \mathbb{K}^\times)$.

In terms of the physical description from Sect. 1.2, the anomaly α arises as follows. The action of G on A determines an action of G on the $(1+1)$-dimensional phase of matter built from A. Stack this G-equivariant phase of matter with a copy of the phase of matter built from A^{op} with trivial G-action. ("Stacking" means to consider the noninteracting tensor product of the two physical systems. Physically, you place them on separate sheets separated by an infinitely strong insulating material and then press the combined solar panel into a single layer.) Ignoring the G-symmetry, we find the trivial phase of matter, as it is built from the Morita-trivial algebra $A \otimes A^{\mathrm{op}}$. But the G-equivariance is nontrivial: we have built a "symmetry-protected trivial" phase. Bosonic SPT phases in $1+1$ dimensions are classified by ordinary group cohomology. (In the condensed matter sense, that claim is a widely believed assertion rather than a theorem; c.f. [GJF19, §5.1]. It is a theorem in the TQFT sense: it is equivalent to the abstract algebraic construction given above.)

There is a final construction of the anomaly α that is useful to describe, since it is the easiest to generalize to the VOA setting. (We will describe it assuming the action of G on A is faithful; it is not hard to extend the construction to the non-faithful case.) Consider the subalgebra $A^G \subset A$ fixed by the G-action. Provided G is finite and $\mathbb{K}$ has characteristic 0, this subalgebra remains semisimple. Its *commutant* is

$$B := \{b \in A \text{ s.t. } [b, a] = 0 \ \forall a \in A^G\}.$$

We claim that B is a twisted group algebra for G. To prove this, for each $g \in G$ set:

$$B_g := \{b \in A \text{ s.t. } ba = g(a)b \; \forall a \in A\}.$$

Then, B_g is one-dimensional by the Skolem–Noether theorem (it is at most one-dimensional since $\mathcal{Z}(A) \cong \mathbb{K}$, and it is at least one-dimensional since we have already shown that all automorphisms of A are inner) and $B_g \cap B_{g'} = \{0\}$ if $g \neq g'$ since the action is faithful. It is clear that $B_g B_{g'} = B_{gg'}$, and so $\bigoplus_{g\in G} B_g \subset A$ is a subalgebra. But the commutant of $\bigoplus_{g\in G} B_g$ in A is A^G, and so

$$B = \bigoplus_{g\in G} B_g$$

by the double commutant theorem.

Finally, being a twisted group algebra, B determines a cohomology class $\alpha \in \mathrm{H}^2(G; \mathbb{K}^\times)$, easily identified with the cohomology class that classified the central extension discussed above.

1.4 Gauging

Gauge anomalies arise as the obstruction to gauging symmetries. Consider first the case $A = \mathrm{Mat}_n(\mathbb{C})$, which corresponds via the bulk-boundary construction of Sect. 1.2 to a quantum mechanics model with Hilbert space $\mathbb{C}^n$. A gaugeable group of symmetries is a group G acting on $\mathbb{C}^n$; then the gauged theory has as its Hilbert space the fixed subspace $(\mathbb{C}^n)^G$. The anomaly must be trivialized in order to promote a group of automorphisms of A to a group of automorphisms of the Hilbert space.

For associative algebras, gauging is also called "quantum Hamiltonian reduction"; the choice of trivialization of the anomaly corresponds to the "quantum comoment map". Quantum Hamiltonian reduction can be written in various ways. Here is one. Suppose that G acts on A with trivialized anomaly, meaning that we have chosen a way to lift the homomorphism $G \to \mathrm{Aut}(A)$ to a homomorphism $G \to A^\times$. Then in particular the fixed subalgebra A^G and the untwisted group algebras $\mathbb{K}[G]$ are commutants in the Azumaya algebra A. This provides an isomorphism between their centres: $\mathcal{Z} = \mathcal{Z}(A^G) \cong \mathcal{Z}(\mathbb{K}[G])$, the algebra of class functions. The homomorphism $G \to \{1\}$ determines a homomorphism $\mathcal{Z}(\mathbb{K}[G]) \to \mathbb{K}$. Specifically, $\mathcal{Z}(\mathbb{K}[G])$ arises as the algebra of $\mathbb{K}$-valued global functions on the Deligne–Mumford stack G/G corresponding to the adjoint action, and $\mathbb{K} = \mathcal{O}(\mathrm{pt}/G)$, and the map $\mathcal{Z}(\mathbb{K}[G]) \to \mathbb{K}$ is restriction along the inclusion $\mathrm{pt}/G \to G/G$. The *gauging* of A by G is by definition the associative algebra:

$$A /\!\!/^{\beta} G := A^G \otimes_{\mathcal{Z}} \mathbb{K}. \tag{2}$$

The superscript emphasizes that the gauging depends on the choice of trivialization β of the anomaly α. The set of trivializations, if nonempty, is a torsor for $\mathrm{H}^1(G; \mathbb{K}^\times)$.

It can happen that $A /\!\!/^\beta G$ is the zero algebra. If it is not zero, then it is Azumaya. One way to see this is to realize $A /\!\!/^\beta G$ as the algebra of endomorphisms of the left A-module $A \otimes_G \mathbb{K}$ (where G acts on A by right multiplication via the homomorphism $\beta : G \to A^\times$). Indeed, such a description makes clear that if $A /\!\!/^\beta G \neq 0$, then A and $A /\!\!/^\beta G$ are Morita-equivalent (compare [JF16]).

1.5 Galois Action and the Cohomological Brauer Group

Given a $\mathbb{K}$-algebra A and a field automorphism γ of $\mathbb{K}$, there is a new $\mathbb{K}$-algebra ${}^\gamma A$ that is equal to A as a ring, but with $\mathbb{K}$-structure twisted by γ. If A is Azumaya, then so is ${}^\gamma A$, and if A has an action of G by algebra automorphisms, then so does ${}^\gamma A$. There are no surprises about how anomalies transform under twisting by γ: if G acts on A with gauge anomaly $\alpha \in \mathrm{H}^2(G; \mathbb{K}^\times)$, then it acts on ${}^\gamma A$ with anomaly $\gamma(\alpha)$. In other words, the anomaly α has coefficients not just in $\mathbb{K}^\times$-as-an-abelian-group, but in $\mathbb{K}^\times$-as-a-Galois-module.

This Galois-equivariance leads to a cohomological characterization of Azumaya algebras. Suppose that A and B are Azumaya algebras over $\mathbb{K}$ and that $\mathbb{K} \subset \mathbb{L}$ is a Galois extension with Galois group $\mathrm{Gal}(\mathbb{L}/\mathbb{K})$ such that $A \otimes_\mathbb{K} \mathbb{L} \cong B \otimes_\mathbb{K} \mathbb{L}$ as $\mathbb{L}$-algebras; let $f : A \otimes_\mathbb{K} \mathbb{L} \to B \otimes_\mathbb{K} \mathbb{L}$ denote a choice of isomorphism. Then $A \otimes_\mathbb{K} \mathbb{L}$ carries two different $\mathrm{Gal}(\mathbb{L}/\mathbb{K})$-actions, providing the different descent data: letting ρ denote the standard action of $\mathrm{Gal}(\mathbb{L}/\mathbb{K})$ on $\mathbb{K}$, the two actions are $\rho_A : \gamma \mapsto \mathrm{id}_A \otimes \rho(\gamma)$ and $\rho_B : \gamma \mapsto f^{-1} \circ (\mathrm{id}_B \otimes \rho(\gamma)) \circ f$. Neither $\rho_A(\gamma)$ nor $\rho_B(\gamma)$ is $\mathbb{L}$-linear, but the composition

$$\tilde{\kappa}(\gamma) := \rho_B(\gamma) \circ \rho_A(\gamma)^{-1}$$

is. This $\tilde{\kappa}$ is not a homomorphism from $\mathrm{Gal}(\mathbb{L}/\mathbb{K})$ to $\mathrm{Aut}(A \otimes_\mathbb{K} \mathbb{L})$, but it is a twisted cocycle—ρ_A determines an action of $\mathrm{Gal}(\mathbb{L}/\mathbb{K})$ on $\mathrm{Aut}(A \otimes_\mathbb{K} \mathbb{L})$—and so determines a twisted cohomology class

$$\kappa \in \mathrm{H}^1(\mathrm{Gal}(\mathbb{L}/\mathbb{K}); \mathrm{Aut}(A \otimes_\mathbb{K} \mathbb{L})).$$

It is a basic theorem that this class, together with A, determines the isomorphism type of B.

The earlier discussion of gauge anomalies amounts to the fact that there is a universal class $\alpha^{\mathrm{univ}} \in \mathrm{H}^2(\mathrm{Aut}(A \otimes_\mathbb{K} \mathbb{L}); \mathbb{L}^\times)$ classifying the extension $1 \to \mathbb{L}^\times \to (A \otimes_\mathbb{K} \mathbb{L})^\times \to \mathrm{Aut}(A \otimes_\mathbb{K} \mathbb{L}) \to 1$. Actually, our discussion did not quite imply this, because we worked only with finite groups G. But the image of the cocycle $\tilde{\kappa}$, together with its Galois-conjugates, is a finite subset of $\mathrm{Aut}(A \otimes_\mathbb{K} \mathbb{L})$ that is easily seen to be closed under composition, and so a subgroup, and so the earlier discussion does imply that α^{univ}, even if it did not exist on all of $\text{Aut}(A \otimes_{\mathbb{K}} \mathbb{L})$, does exist after being restricted to the image of κ.

The composition of $\alpha^{\text{univ}} \circ \kappa$ is a class in the twisted cohomology $\text{H}^2(\text{Gal}(\mathbb{L}/\mathbb{K}); \mathbb{L}^\times)$. Given the Morita-class of A, this twisted cohomology class determines the Morita-equivalence class of B.

Finally, it is a fundamental fact that for every Azumaya algebra B over $\mathbb{K}$, there is a Galois extension $\mathbb{K} \subset \mathbb{L}$ such that $B \otimes_{\mathbb{K}} \mathbb{L} \cong \text{Mat}_n(\mathbb{L})$ for some n. Fixing $A = \text{Mat}_n(\mathbb{K})$ and taking the union over extensions $\mathbb{L} \supset \mathbb{K}$, one discovers

$$\begin{aligned} \text{Br}(\mathbb{K}) &= \{\text{Azumaya algebras over } \mathbb{K} \text{ up to Morita-equivalence}\} \\ &\cong \text{H}^2(\text{Gal}(\bar{\mathbb{K}}/\mathbb{K}); (\bar{\mathbb{K}})^\times). \end{aligned}$$

The left-hand side is called the *Brauer group* of $\mathbb{K}$, and the right-hand side is called the *cohomological Brauer group*. (In positive characteristic, $\bar{\mathbb{K}}$ means the separable, rather than algebraic, closure.)

2 Gauge Anomalies for VOAs

In the remainder of this chapter, we will repeat the story from Sect. 1 with associative algebras (the observables for (0+1)-dimensional quantum field theories) replaced by vertex operator algebras ((1 + 1)-dimensional quantum field theories). The question of how to gauge an action of a finite group on a VOA was first studied in [DVVV89, DPR90]. The mathematical construction of gauge anomalies is due to [Kir02]; this section consists primarily of an interpretation of that paper.

2.1 Holomorphic VOAs

We will not review the definition of vertex operator algebra—an excellent reference is [LL04]—preferring instead to indicate the meanings of words used in VOA theory by analogizing with the associative algebra case. We first point out that the notion of VOA is purely algebraic: it makes sense over any field. We will work only over $\mathbb{C}$ and its sub-fields, which simplifies the story. (In positive characteristic, one must take care with divided powers. The introduction of vertex algebras in [Bor86] already included the positive-characteristic case.) In particular, if V is a VOA over $\mathbb{K}$, then so are its Galois-conjugates ${}^{\gamma}V$ for $\gamma \in \text{Gal}(\mathbb{K}/\mathbb{Q})$. In this chapter, VOAs are always simple, self-dual (i.e., admitting an invariant nondegenerate bilinear form), and $\mathbb{N}$-graded by the action of L_0.

Given a VOA V, there are various categories of "V-modules", with different regularity conditions placed on the module. We will focus on "admissible" modules (we will not give the definition), writing $\text{Rep}(V)$ for the category thereof. VOA modules correspond not to left modules for an associative algebra A but rather are analogous to A-A-bimodules: $\mathrm{REP}(V)$ is the VOA analog of $\mathrm{BIMOD}(A)$, and the Zhu algebra of V (defined so that the irreducible V-modules are in natural bijection with irreducible modules over the Zhu algebra; Zhu's construction in [Zhu96] is technical, but see [GG09] for a readable survey) is analogous to the enveloping algebra $A \otimes A^{\mathrm{op}}$.

The analog of semisimplicity for an algebra is called "regularity": a VOA V is *regular* if it is C_2-cofinite and the category $\mathrm{REP}(V)$ of admissible modules is semisimple with finitely many simples (a condition called "rationality", but this chapter avoids that term so as not to conflict with the question of definability over $\mathbb{Q}$). Suppose that we are working over $\mathbb{K} = \mathbb{C}$. Then, regularity implies that $\mathrm{REP}(V)$ is a modular tensor category [Hua08]. The braided monoidal structure is the VOA analog of the fact that $\mathrm{BIMOD}(A)$ is naturally a monoidal category.

The word analogous to "Azumaya" is "holomorphic": V is *holomorphic* if $\mathrm{REP}(V) \simeq \mathrm{VECT}$.

2.2 Physical Interpretation

It is believed that $(2+1)$-dimensional topological phases of matter are classified by MTCs together with some mild extra data about the central charge. (Namely, each MTC has a central charge c living in $\mathbb{Q}/8\mathbb{Z}$; the extra data is a lift to $\mathbb{Q}$. The root of unity $\alpha = \exp(2\pi ic/8)$ is sometimes called the *(gravitational) anomaly* of the MTC.) There is no proof of this belief, largely due to the lack of a definition of "$(2+1)$-dimensional topological phase of matter", but a TQFT version is proven in [BDSPV15], where it is confirmed that the Reshetikhin–Turaev construction [RT91] gives a precise classification of once-extended 3d TQFTs. If V is a regular VOA over $\mathbb{C}$, then the MTC $\mathrm{REP}(V)$ determines a $(2+1)$-dimensional topological phase of matter, and the VOA V arises as the algebra of observables for a distinguished $(1+1)$-dimensional chiral conformal boundary condition.

If V is holomorphic, then the bulk $(2+1)$-dimensional phase of matter is almost trivial: it is a short-range entangled (SRE) phase, encoding only the central charge of V. (The central charge corresponds to the "beyond group cohomology layer" in [GJF19, §5.2].) Consider a bulk-boundary system with V determining the boundary physics and with bulk physics this SRE phase. Since the bulk phase is SRE, any choice of local boundary coordinates identifies the boundary physics with a true quantum field theory, but the theory is not truly conformal: conformal transformations change the partition function of the theory in a simple way determined by c, and so c, or equivalently the bulks SRE phase, encodes the "gravitational anomaly" of the holomorphic CFT described by V.

2.3 Construction of Gauge Anomalies

Let V be holomorphic VOA defined over $\mathbb{C}$, and let $G \subset \mathrm{Aut}(V)$ be a finite group of VOA automorphisms of V. We will associate to these data a *gauge anomaly* $\alpha \in \mathrm{H}^3(G; \mathbb{C}^\times)$. We will do so by imitating the construction from Sect. 1.3. There is no VOA analog of the exact sequence (1). But we can consider the analog of the commutant subalgebras $A^G \subset A \supset B \cong \mathbb{K}^\alpha[G]$.

First, it makes perfect sense in VOA theory to consider the G-fixed subalgebra $V^G \subset V$. There is a version of commutant subalgebras that can be said entirely within the language of vertex algebras, called the "coset construction" [GKO85, FZ92]: the *coset* of a sub-vertex algebra $W \subset V$ is the sub-vertex algebra $W' \subset V$ consisting of all elements of V that have trivial operator product expansion with all elements of W. In our case, however, $V^G \subset V$ is a conformal embedding (i.e., the image of V^G contains the Virasoro vector of V) and so, since V is assumed simple, $(V^G)'$ is the trivial VOA.

In order to have a meaningful commutant of $V^G \subset V$, we must leave the world of VOAs, which encode the vertex operators in a chiral CFT, and also consider topological line operators. One expects on physical grounds that topological line operators for a fixed CFT form a category, because there can be topological junctions between line operators, and that they can be fused, providing the category of line operators with a monoidal structure. (A typical CFT will have infinitely many nonisomorphic simple topological line operators and so the category of line operators is not fusion; c.f. [Hen17], which uses the term "soliton" for what we call "topological line operator".)

For each $g \in G$, consider the category $\mathcal{C}_g$ of topological line operators X such that, for each vertex operator $v \in V$,

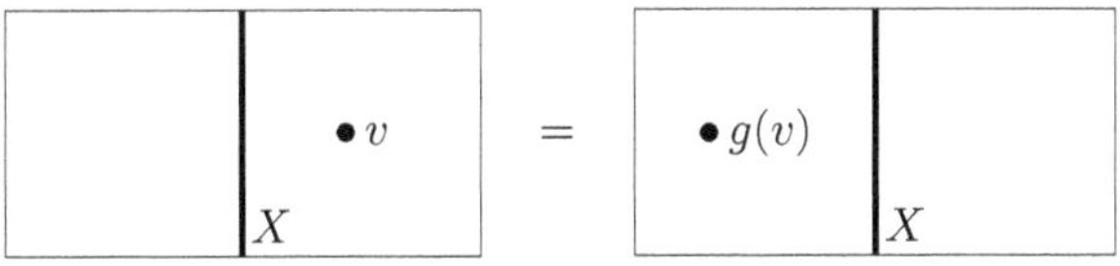

$X \quad \bullet v \quad = \quad \bullet g(v) \quad X$

in the sense that if you take the operator-valued holomorphic function "insert the vertex operator v at a point to the right of the line operator X" and analytically continue it over the X-line, you get the operator-valued holomorphic function "insert the vertex operator $g(v)$ at a point to the left of the line operator X".

There is a natural functor from this category $\mathcal{C}_g$ to what in the VOA literature is called the category of "g-twisted V-modules". Indeed, a typical topological line operator X in a CFT does not admit any topological "endings", i.e., there are usually no topological junctions between X and the trivial line operator. But there is a vector space of chiral conformal endings of X. (Compare: The topological endings of the trivial line operator are the vertex operators in V of conformal dimension 0, which is to say just the multiples of the vacuum. The chiral conformal endings of the trivial line are all of V.) When X is a line operator satisfying $Xv = g(v)X$, the space of chiral conformal endings for X is a g-twisted module.

The functor $\mathcal{C}_g \to$ {g-twisted V-modules}, $X \mapsto$ {endings of X}, is expected to be an equivalence, and so we will conflate the two categories. The only way it could fail to be an equivalence is if there are line operators in $\mathcal{C}_g$ that admit no endings at all. Such line operators can exist in a TQFT (assuming the phrase "TQFT" implicitly restricts just to topological endings of topological line operators) but should not exist for a CFT described by a VOA.

When V is holomorphic, the category $\mathcal{C}_g$ is (noncanonically) equivalent to VECT [DLM00, Theorem 2]. Consider the direct sum of categories $\mathcal{C} = \bigoplus \mathcal{C}_g$. Under the physical expectation that line operators can be fused, $\mathcal{C}$ will be a G-pointed fusion category. The nontrivial data of a G-pointed fusion category is its associator, up to equivalence of G-pointed fusion categories, that data is a cohomology class $\alpha \in \mathrm{H}^3(G; \mathbb{C}^\times)$ [EGNO15, Example 2.3.8 and Proposition 2.6.1]. By construction, V^G is the commutant VOA of $\mathcal{C} \subset V$, since it consists of those vertex operators that commute with the line operators in $\mathcal{C}$, and one expects a "double commutant theorem" realizing $\mathcal{C} = \bigoplus \mathcal{C}_g$ as the commutant of V^G.

Actually, working purely in VOAs, it is not known that one can fuse arbitrary twisted modules. We will therefore impose the following widely believed assumption:

Regularity Assumption *Let V be a holomorphic VOA and $G \subset \mathrm{Aut}(V)$ a finite group. Then the fixed sub-VOA V^G is regular.*

The main result of [CM16] implies that this regularity assumption holds if G is solvable. With the regularity assumption in place, the results of [Kir02] fully justify the story told in this section: in particular, the category $\mathcal{C}$ is fusion and is the commutant of V^G, and so the gauge anomaly $\alpha \in \mathrm{H}^3(G; \mathbb{C}^\times)$ is defined.

2.4 Gauging

To justify the name "gauge anomaly", we must construct, for each trivialization β of α, a *gauging* $V /\!\!/^\beta G$ of V. In the physicist's style, we will first give a recipe for $V /\!\!/^\beta G$. In Sect. 2.5, we will explain the question that this recipe answers. The recipe will imitate the quantum Hamiltonian reduction formula (2) from Sect. 1.4: for Azumaya algebras, the gauging was $A /\!\!/^\beta G := A^G \otimes_{\mathcal{Z}} \mathcal{O}(\mathrm{pt}/G)$, where $\mathcal{Z} \cong \mathcal{Z}(A^G) \cong \mathcal{Z}(B)$ and where β identified the twisted group algebra $B = \bigoplus B_g \cong \mathbb{K}^\alpha[G]$ with the untwisted group algebra $\mathbb{K}[G]$.

In the VOA case, the gauge anomaly α arises as the associator on the G-pointed fusion category $\mathcal{C} = \bigoplus_g \mathcal{C}_g$. The trivialization is therefore the data of an equivalence $\beta : \mathcal{C} \simeq \text{VECT}[G]$, where $\text{VECT}[G]$ means the G-pointed fusion category of G-graded vector spaces with trivial associator. This is directly analogous with the story about $B \cong \mathbb{K}^\alpha[G] \overset{\beta}{\cong} \mathbb{K}[G]$.

We need an analog for the centre $\mathcal{Z}(A^G)$. The *centre* of any algebra is its commutant as a subalgebra of itself. For W a regular vertex algebra, we could take the coset of $W \subset W$, but we would get a trivial answer. Instead, as in Sect. 2.3, we will take its categorical commutant, i.e., the category of the line operators on W that commute with all vertex operators in W. The category of such line operators is precisely $\mathrm{REP}(W)$: each line operator has a W-module worth of chiral conformal endings (and we take for granted the assumption that every line operator has endings, so that the functor from line operators to modules is an equivalence).

A second justification for setting $\mathcal{Z}(W) := \mathrm{REP}(W)$ comes from comparing Sects. 1.2 and 2.2. The centre $\mathcal{Z}(B)$ of a semisimple algebra B arises as the algebra of operators for the $(1+1)$-dimensional topological order constructed from B, whereas $\mathrm{REP}(W)$ are the operators for the $(2+1)$-dimensional topological order constructed from the regular VOA W.

The isomorphism $\mathcal{Z}(A^G) \cong \mathcal{Z}(B)$ used in Sect. 1.4 has an analog: conditional on the regularity assumption, it is shown in [Kir02] that there is a canonical equivalence $\mathcal{Z}(V^G) := \mathrm{REP}(V^G) \simeq \mathcal{Z}(\mathcal{C})$, where the right-hand side denotes the Drinfeld centre of the fusion category $\mathcal{C}$.

Moreover, analogous to the map $\mathcal{Z}(\mathbb{K}[G]) = \mathcal{O}(G/G) \to \mathbb{K} = \mathcal{O}(\mathrm{pt}/G)$, there is a canonical map $\mathcal{Z}(\mathrm{VECT}[G]) \to \mathrm{REP}(G)$, coming from realizing $\mathcal{Z}(\mathrm{VECT}[G])$ and $\mathrm{REP}(G)$ as categories of sheaves on G/G and pt/G, respectively. The map $\mathcal{Z}(\mathrm{VECT}[G]) \to \mathrm{REP}(G)$ extends to an equivalence $\mathcal{Z}(\mathrm{VECT}[G]) \simeq \mathcal{Z}(\mathrm{REP}(G))$ of MTCs [Nik13, Example 3.15]. One can recover the trivialization β from the composition $\mathcal{Z} := \mathcal{Z}(\mathcal{C}) \overset{\beta}{\simeq} \mathcal{Z}(\mathrm{VECT}[G]) \to \mathrm{REP}(G)$.

All together, we can analogize equation (2) to

$$V /\!\!/^{\beta} G := V^G \otimes_{\mathcal{Z}} \mathrm{REP}(G). \tag{1}$$

The last step is to explain how to tensor a VOA with a fusion category. The short answer is that $V /\!\!/^{\beta} G$ is a conformal extension of V^G determined by $\mathcal{Z} \to \mathrm{REP}(G)$: the map $V^G \hookrightarrow V /\!\!/^{\beta} G$ is analogous to the map $A^G \twoheadrightarrow A /\!\!/^{\beta} G := A^G \otimes_{\mathcal{Z}} \mathbb{K}$; in categories, there is no real distinction between quotients and extensions (e.g., extensions of algebras often lead to quotients of module categories).

The long answer is that the data of the map $\mathcal{Z} \to \mathrm{REP}(G)$ is equivalent to the data of an "étale algebra object" $E \in \mathcal{Z}$, i.e., a separable braided-commutative algebra that contains only one copy of the unit object of $\mathcal{Z}$. The equivalence identifies $\mathrm{REP}(G)$ as the category of E-module objects in $\mathcal{Z}$ and identifies E as the $\mathcal{Z}$-object worth of "internal endomorphisms" of the unit in $\mathrm{REP}(G)$. But étale algebra objects in $\mathrm{REP}(V^G)$ are precisely conformal extensions of V^G. All together, equation (1) says that $V /\!\!/^{\beta} G$ is the conformal extension of V^G corresponding to the étale algebra $E \in \mathrm{REP}(V^G)$ corresponding to the map $\mathrm{REP}(V^G) \overset{\beta}{\simeq} \mathcal{Z}(\mathrm{VECT}[G]) \to \mathrm{REP}(G)$.

The étale algebra object $E \in \mathcal{Z}$ is "lagrangian" in the sense that it realizes $\mathcal{Z}$ as the Drinfeld centre of $\mathrm{REP}(G)$. It follows that $V /\!\!/^{\beta} G$ is holomorphic.

2.5 Linear Algebraic Description

The previous sections provided a recipe that took in a holomorphic VOA V and a finite group $G \subset \mathrm{Aut}(V)$ and a trivialization β of the anomaly and produced (assuming the regularity assumption) a VOA $V /\!\!/^{\beta} G$ that we called the "gauging" of V by G. We did not explain abstractly what "gauging" means, nor did we prove that one needs precisely a trivialization of β. We will do so now. Specifically, we will arrive at the answer by explaining how $V /\!\!/^{\beta} G$ decomposes as a module over its subalgebra V^G. We call this a "linear algebraic description" of $V /\!\!/^{\beta} G$ since we will not try to describe the operator product.

The first step is to describe the category $\mathrm{REP}(V^G)$ in which $V /\!\!/^{\beta} G$ lives. Suppose that G acts with anomaly $\alpha \in \mathrm{H}^3(G; \mathbb{C}^\times)$, which we do not yet assume to be trivializable; then $\mathrm{REP}(V^G) \simeq \mathcal{Z}(\mathrm{VECT}^{\alpha}[G])$. Slightly abusively, let us write G/G for the set of conjugacy classes in G, and for each $g \in G/G$, choose a representative also called g. Let $C(g) \subset G$ denote its centralizer. There is a *slant product* map $\iota_g : \mathrm{H}^3(G; \mathbb{C}^\times) \to \mathrm{H}^2(C(g); \mathbb{C}^\times)$, given on cocycle representatives by

$$(\iota_g \alpha)(x, y) = \alpha(g, x, y) - \alpha(x, g, y) + \alpha(x, y, g).$$

(Here and throughout, we write the group law on $\mathrm{H}^\bullet(G; \mathbb{C}^\times)$ additively.)

Write $\mathrm{REP}^{\iota_g \alpha}(C(g))$ for the $\iota_g \alpha$-block in the category of projective $C(g)$-modules. Then, as a linear category,

$$\mathcal{Z}(\mathrm{VECT}^{\alpha}[G]) \simeq \bigoplus_{g \in G/G} \mathrm{REP}^{\iota_g \alpha}(C(g)).$$

For each $g \in G/G$, let $V_g \in \mathcal{C}_g$ denote (a choice for) the simple g-twisted V-module. Then, V_g is naturally a $\iota_g \alpha$-projective $C(g)$-module, i.e., $V_g \in \mathrm{REP}^{\iota_g \alpha}(C(g))$. In fact, under the equivalence $\mathrm{REP}(V^G) \simeq \mathcal{Z}(\mathrm{VECT}^{\alpha}[G])$, the simple objects of $\mathrm{REP}^{\iota_g \alpha}(C(g))$ are identified with the V^G-submodules of V_g.

Suppose now that we choose a trivialization β of α. Then, $\iota_g \beta$ (represented by the cocycle $(\iota_g \beta)(x) = \beta(g, x) - \beta(x, g)$) is a trivialization of $\iota_g \alpha$, i.e., it determines a one-dimensional $\iota_g \alpha$-projective $C(g)$-module that we will call $\mathbb{C}_{\iota_g \beta} \in \mathrm{REP}^{\iota_g \alpha}(C(g))$. Given $M \in \mathrm{REP}^{\iota_g \alpha}(C(g))$, let $M^{\iota_g \beta} := \hom(\mathbb{C}_{\iota_g \beta}, M)$. It is the space of "fixed points" of the $C(g)$-action if we use $\iota_g \beta$ to identify $\mathrm{REP}^{\iota_g \alpha}(C(g))$ with $\mathrm{REP}(C(g))$.

Then, as a representation of V^G, we have

$$V /\!\!/^{\beta} G \cong \bigoplus_{g \in G/G} V_g^{\iota_g \beta}. \tag{2}$$

One can show this by inspecting the Lagrangian étale algebra object in $\mathrm{REP}(\mathrm{VECT}[G])$ corresponding to the fusion category $\mathrm{REP}(G)$.

An immediate consequence of equation (2) is that the only common V^G-submodule of V and $V /\!\!/^{\beta} G$ is $V^G = V_e^{\iota_e \alpha}$ itself (where of course $e \in G$ denotes the unit). Based on this, we take:

Definition Let V be a holomorphic VOA and $G \subset \mathrm{Aut}(V)$ a finite group of automorphisms such that the regularity assumption holds. A *gauging of V by G* is any holomorphic VOA $V /\!\!/ G$ extending the fixed subalgebra V^G such that the only common V^G-submodule shared by V and $V /\!\!/ G$ is V^G itself. The action of G on V is *nonanomalous* if a gauging exists.

We have seen that if the anomaly admits a trivialization β, then $V /\!\!/^{\beta} G$ is a gauging in this sense. Conversely, by studying Lagrangian algebras in $\mathcal{Z}(\mathrm{VECT}^{\alpha}[G])$, one can show that the action of G on V is nonanomalous if and only if the anomaly is trivializable, and that gaugings are in bijection with trivializations.

3 Main Result

Most of Sect. 2 requires working over the complex numbers $\mathbb{C}$ for a simple reason: there is no purely algebraic way to talk about "line operators", because there is no purely algebraic way to talk about real submanifolds of $\mathbb{C}$. Said another way, the MTC structure on $\mathrm{REP}(W)$, for W a regular VOA, requires a priori that we work over $\mathbb{C}$: algebraically, one can contemplate intertwining operators parameterized by configurations of points on $\mathbb{P}^1$; only after taking $\mathbb{C}$-points can one identify configuration spaces on $\mathbb{P}^1$ with classifying spaces for braid groups. Indeed, the construction of the MTC structure on $\mathrm{REP}(W)$ given in [Hua08] repeatedly chooses parameters living on real intervals in $\mathbb{C}$.

Thus, the definition of the anomaly α of an action of G on a holomorphic VOA V requires that V be defined over $\mathbb{C}$. But suppose that γ is a field automorphism of $\mathbb{C}$. Then there is a Galois-conjugate VOA ${}^{\gamma}V$ equal to V as a VOA over $\mathbb{Q}$ but with the $\mathbb{C}$-structure twisted by γ. It is also holomorphic, since its linear category of representations is defined algebraically, and also has a G-action, hence its own anomaly. Our goal in this section is to prove:

Main Theorem *Let V be a holomorphic VOA over $\mathbb{C}$ and $G \subset \mathrm{Aut}(V)$ a finite group of automorphisms, and let γ be a field automorphism of $\mathbb{C}$ and ${}^{\gamma}V$ the Galois-conjugate VOA. Conditional on the regularity assumption, let $\alpha \in \mathrm{H}^3(G; \mathbb{C}^{\times})$ denote the anomaly of the action of G on V. Then the anomaly of the action of G on ${}^{\gamma}V$ is $\gamma^2(\alpha)$.*

3.1 Recollection on the Cyclotomic Galois Group

Let $\boldsymbol{\mu} \subset \mathbb{C}^\times$ denote the group of roots of unity. We will write $\boldsymbol{\mu}$ additively; it is isomorphic to $\mathbb{Q}/\mathbb{Z}$ via the exponential map $x \mapsto \exp(2\pi i x)$. If G is a finite group, then the map $\mathrm{H}^\bullet(G; \boldsymbol{\mu}) \to \mathrm{H}^\bullet(G; \mathbb{C}^\times)$ is an isomorphism for $\bullet \geq 1$.

If $\gamma \in \mathrm{Aut}(\mathbb{C})$ and $x \in \boldsymbol{\mu}$ has order N, then $\gamma(x) \in \boldsymbol{\mu}$ also has order N. It follows that γ acts on $\boldsymbol{\mu}$ by multiplication by some $n \in \widehat{\mathbb{Z}}^\times$, where $\widehat{\mathbb{Z}}$ denotes the profinite completion of $\mathbb{Z}$ and $\widehat{\mathbb{Z}}^\times$ is its group of units. The map $\gamma \mapsto n : \mathrm{Aut}(\mathbb{C}) \to \widehat{\mathbb{Z}}^\times$ is a surjection of $\mathrm{Aut}(\mathbb{C})$ onto the cyclotomic Galois group $\mathrm{Gal}(\mathbb{Q}^{\mathrm{cyc}}/\mathbb{Q}) \cong \mathrm{Aut}(\boldsymbol{\mu}) \cong \widehat{\mathbb{Z}}^\times$.

3.2 Nonanomalous Actions

There are two purely algebraic parts of Sect. 2. First, the definitions in Sect. 2.1 of the words "holomorphic" and "regular" refer only to the structure of the linear category of representations, and so if V is holomorphic, then ${}^\gamma V$ is as well. Second, the definition of a "gauging" of V by G given in Sect. 2.5 is purely algebraic. To see this, first note that $({}^\gamma V)^G = {}^\gamma(V^G)$, and so the expression ${}^\gamma V^G$ is unambiguous. Second, suppose that $V /\!\!/ G$ is a gauging of V by G, i.e., a holomorphic extension of V^G that, as a V^G-module, shares only the submodule V^G with V. Then ${}^\gamma(V /\!\!/ G)$ is a gauging of ${}^\gamma V$ by G, since the only common ${}^\gamma V^G$-submodule shared by ${}^\gamma V$ and ${}^\gamma(V /\!\!/ G)$ is ${}^\gamma V^G$. In summary:

Lemma *If the action of G on V is nonanomalous, then so is the action of G on ${}^\gamma V$.* □

It is not hard to prove that if G acts on V_1 with anomaly α_1 and on V_2 with anomaly α_2, then it acts on $V_1 \otimes V_2$ with anomaly $\alpha_1 + \alpha_2$, where we have written the group law on $\mathrm{H}^3(G; \mathbb{C}^\times)$ additively.

Corollary *Suppose that G acts on V_1 with anomaly α and on V_2 with anomaly $-\alpha$, and suppose that the main theorem holds for $V = V_1$. Then, it also holds for $V = V_2$.* □

3.3 A Construction of Evans and Gannon

Conditional on the regularity assumption, the paper [EG18] constructs, for each finite group G and each class $\alpha \in \mathrm{H}^3(G; \boldsymbol{\mu})$, a VOA V with a faithful G-action with anomaly α. Given the corollary in Sect. 3.2, it suffices to prove the main theorem for the VOAs constructed by Evans and Gannon. This section reviews their construction.

Let G be a finite group. Then, there is a finite abelian extension $1 \to A \to E \to G \to 1$ such that the restriction map $\mathrm{H}^3(G; \boldsymbol{\mu}) \to \mathrm{H}^3(E; \boldsymbol{\mu})$ is the zero map. This fact has been rediscovered a number of times, most recently in [WWW18]. The Lyndon–Hochschild–Serre spectral sequence enhances the usual 5-term inflation-restriction exact sequence (for $\boldsymbol{\mu}$-cohomology) to one of the form:

$$1 \to \mathrm{H}^1(G; \boldsymbol{\mu}) \to \mathrm{H}^1(E; \boldsymbol{\mu}) \to H^1(A; \boldsymbol{\mu}) \to \mathrm{H}^2(G; \boldsymbol{\mu}) \to \mathrm{H}^2(E; \boldsymbol{\mu})$$
$$\to \Lambda(G; A) \stackrel{\delta}{\to} \mathrm{H}^3(G; \boldsymbol{\mu}) \stackrel{0}{\to} \mathrm{H}^3(E; \boldsymbol{\mu}) \to \dots,$$

where $\Lambda(G; A)$ maps naturally to $\mathrm{H}^2(A; \boldsymbol{\mu})$, with kernel $\mathrm{H}^1(G; \mathrm{H}^1(A; \boldsymbol{\mu})$. It follows that for each $\alpha \in \mathrm{H}^3(G; \boldsymbol{\mu})$, we can choose $\beta \in \Lambda(G; A)$ with $\delta\beta = \alpha$.

The strategy of [EG18], then, is to start with a holomorphic VOA W with a faithful action by E with trivialized anomaly. (They prove that the permutation action on $W = (V^\natural)^{\otimes |E|}$, where $V^\natural$ denotes the moonshine VOA, works. One can always arrange a nonanomalous E-action by starting with some holomorphic VOA W on which E acts with anomaly of order N and then taking $W' = W^{\otimes N}$. It could happen that W' is nonanomalous but not canonically so: the anomaly may not have a distinguished trivialization. But all choices of trivialization of the anomaly for W' will lead to the same trivialization of the anomaly for $W'' = (W')^{\otimes |\mathrm{H}^2(E;\boldsymbol{\mu})|}$.) In particular, the induced A-action on W is nonanomalous, and so gaugings of W by A are parameterized by classes in $\mathrm{H}^2(A; \boldsymbol{\mu})$. The central result underlying the construction is:

Proposition ([EG18]) *Conditional on the regularity assumption, $W /\!\!/^{\beta} A$ carries a faithful G-action with anomaly $\alpha = \delta\beta$.* □

In the proposition, the VOA $W /\!\!/^{\beta} A$ depends only on the image of $\beta \in \Lambda(G; A)$ under the map $\Lambda(G; A) \to \mathrm{H}^2(A; \boldsymbol{\mu})$, but the information in β in the kernel of this map is used to determine the action of G. (The group $\mathrm{H}^1(A; \boldsymbol{\mu})$ acts on any orbifold $W /\!\!/ A$, and so an action of G on W can be adjusted by an element of $\mathrm{H}^1(G; \mathrm{H}^1(A; \boldsymbol{\mu})) = \hom(G, \mathrm{H}^1(A; \boldsymbol{\mu}))$; compare the "electromagnetic duality" from [FT18] or the "T-duality" from [JF19].)

3.4 Completion of the Proof

Thus to prove the main theorem, it suffices to study the case $V = W /\!\!/^{\beta} A$ from Sect. 3.3. Specifically, we know that G acts on $W /\!\!/^{\beta} A$ with anomaly $\alpha = \delta\beta$. Suppose that $\gamma \in \mathrm{Aut}(\mathbb{C})$ acts on $\boldsymbol{\mu}$ by multiplication by $n \in \widehat{\mathbb{Z}}^{\times}$. (As in the previous sections, we will write $\boldsymbol{\mu}$ additively.) Then we wish to show that G acts on

${}^{\gamma}(W /\!\!/^{\beta} A)$ with anomaly $n^2\alpha$. Since $\delta : \Lambda(G; A) \to \mathrm{H}^3(G; \boldsymbol{\mu})$ is linear, it suffices to prove:

Claim ${}^{\gamma}(W /\!\!/^{\beta} A) \cong {}^{\gamma}W /\!\!/^{n^2\beta} A$.

To prove the claim, we begin with the linear algebraic description of $W /\!\!/^{\beta} A$ from Sect. 2.5. Using that A is an abelian group, equation (2) says that as a W^A-module,

$$W /\!\!/^{\beta} A \cong \bigoplus_{a \in A} W_a^{\iota_a \beta},$$

and so

$${}^{\gamma}(W /\!\!/^{\beta} A) \cong \bigoplus_{a \in A} {}^{\gamma}(W_a^{\iota_a \beta}).$$

We must now understand the Galois-conjugate W^A-module ${}^{\gamma}(W_a^{\iota_a\beta})$. By assumption, γ acts by multiplication by $n \in \widehat{\mathbb{Z}}^{\times}$ on $\boldsymbol{\mu}$, and so

$${}^{\gamma}(W_a^{\iota_a\beta}) \cong ({}^{\gamma}(W_a))^{\iota_a n\beta}.$$

What is ${}^{\gamma}(W_a)$? It is not $({}^{\gamma}W)_a$. Indeed, the definition of "a-twisted W-module" refers explicitly to the eigenvalues of a. These eigenvalues (or rather their logarithms, so that we can work additively) get multiplied by $n \in \widehat{\mathbb{Z}}^{\times}$ under Galois conjugation by γ. To compensate for this, one must divide a by n. The result is

$${}^{\gamma}(W_a) \cong ({}^{\gamma}W)_{n^{-1}a}.$$

Taking the direct sum over $a \in A$, we find

$${}^{\gamma}(W /\!\!/^{\beta} A) \cong \bigoplus_{a \in A} {}^{\gamma}(W_a^{\iota_a\beta}) \cong \bigoplus_{a \in A} ({}^{\gamma}W)_{n^{-1}a}^{\iota_a n\beta} \cong \bigoplus_{a' \in A} ({}^{\gamma}W)_{a'}^{\iota_{na'} n\beta} \cong \bigoplus_{a' \in A} ({}^{\gamma}W)_{a'}^{n^2\iota_{a'}\beta}. \tag{1}$$

Here, we have reindexed $a' = n^{-1}a$. We also used that $\iota_{nx}\beta = n\iota_x\beta$. To see this, note that, for fixed y, $\iota_x\beta(y) = \beta(x, y) - \beta(y, x) = -\iota_y\beta(x)$ is a 1-cocycle in the x variable.

Equation (1) establishes the claim at the linear algebraic level, i.e., as modules over ${}^{\gamma}W^A$. But, since A is an abelian group, étale algebras in $\mathcal{Z}(\mathrm{VECT}[A]) \simeq \mathrm{REP}({}^{\gamma}W^A)$ are determined by their underlying modules. Thus, the linear algebraic isomorphism implies a VOA isomorphism. Finally, the VOA isomorphism is compatible with the G-action. This completes the proof of the claim and hence of the main theorem.

4 Questions and Conjectures

4.1 Moonshine for Every Group

Since $\mathrm{Aut}(\boldsymbol{\mu}) \cong \widehat{\mathbb{Z}}^{\times}$ is abelian, its action on $\boldsymbol{\mu}$ can be "twisted" by any power. Given $i \in \mathbb{Z}$, we will write $\boldsymbol{\mu}^{\otimes i}$ for the abelian group $\boldsymbol{\mu}$ made into a $\widehat{\mathbb{Z}}^{\times}$-, and hence $\mathrm{Aut}(\mathbb{C})$-, module via the action $n \triangleright x := n^i x$. The main theorem can then be summarized as saying that as an $\mathrm{Aut}(\mathbb{C})$-module, gauge anomalies of VOAs live in $\mathrm{H}^3(-; \boldsymbol{\mu}^{\otimes 2})$.

Corollary of the Main Theorem *Suppose that the action of G on V preserves a $\mathbb{K}$-form for some field $\mathbb{K} \subset \mathbb{C}$. Then the anomaly of the action lives in* $\mathrm{H}^0\big(\mathrm{Gal}(\bar{\mathbb{K}}/\mathbb{K}); \mathrm{H}^3\big(G; \boldsymbol{\mu}^{\otimes 2}\big)\big)$.

Consider the case $\mathbb{K} = \mathbb{Q}$. The "defining property of 24" from [CN79]—that $n^2 \equiv 1 \bmod 24$ for $n \in \widehat{\mathbb{Z}}^{\times}$—implies that for any group G, $\mathrm{H}^0(\mathrm{Gal}(\bar{\mathbb{Q}}/\mathbb{Q}); \mathrm{H}^\bullet(G; \boldsymbol{\mu}^{\otimes 2}))$ is entirely 24-torsion. Let $V^\natural$ denote the moonshine VOA. The monster group $\mathbb{M}$ acts on $V^\natural$ preserving a $\mathbb{Q}$-form [DG12]. Taking for granted the regularity assumption, let $\omega^\natural$ denote the anomaly of this action. In unpublished work [Mas07], Mason computed the image of $\omega^\natural$ under the total slant product map

$$\prod_{g \in \mathbb{M}} \iota_g : \mathrm{H}^3(\mathbb{M}; \boldsymbol{\mu}) \to \prod_{g \in \mathbb{M}} \mathrm{H}^2(C(g); \boldsymbol{\mu})$$

and showed that the image has exact order 24. Together with the above corollary, we find:

Corollary *The anomaly $\omega^\natural$ of the action of $\mathbb{M}$ on $V^\natural$ has exact order* 24.

The exact order of $\omega^\natural$ was first computed in [JF19] using a different argument; that paper modelled CFTs in terms of conformal nets (rather than VOAs), for which the regularity assumption is known [Xu00] to hold.

The construction from [EG18], reviewed in Sect. 3.3, establishes that there is a weak form of "moonshine" for every finite group G and every anomaly $\alpha \in \mathrm{H}^3(G; \boldsymbol{\mu})$. The resulting VOA is by no means canonical and has astronomically large central charge (of order $|G| \times |\mathrm{H}^3(G; \boldsymbol{\mu})|$). Various authors have speculated versions of the following (I heard it from I. Frenkel), while knowing that, as stated, it is surely too naive:

Conjecture *For each quasisimple group G, there is a distinguished "moonshine" VOA representation V_G of G. For the Monster group, $V_{\mathbb{M}} = V^\natural$, and for compact Lie groups, V_G is the level-one WZW model. The VOAs V_G can be used in the classification of finite simple groups analogously to the way Lie algebras are used in the classification of simple Lie groups.*

The words “level-one” in the phrase “level-one WZW model” of a simple Lie group G refer to the fact that the anomaly for these models is a generator of the infinite cyclic group $\mathrm{H}^4(BG; \mathbb{Z})$. Theorem 8.6 of [JFT19] comes close to showing that $\omega^\natural$ generates $\mathrm{H}^3(\mathbb{M}; \boldsymbol{\mu}) \cong \mathrm{H}^4(B\mathbb{M}; \mathbb{Z})$. For all but finitely many finite simple groups G of Lie type, $\mathrm{H}^3(G; \boldsymbol{\mu})$ is cyclic, and the calculations in [JFT19] suggest that cyclicity also holds for “most” of the sporadic simple groups. Thus, the above conjecture may be refined by speculating that one way V_G should be distinguished is that its anomaly should generate $\mathrm{H}^3(G; \boldsymbol{\mu})$.

The calculations in [JFT19] also show that for “most” sporadic simple groups G, $\mathrm{H}^3(G; \boldsymbol{\mu})$ is entirely 24-torsion. (The known exceptions are the Janko groups J_1, J_2, and J_3.) Given our main theorem, this increases the odds that the distinguished representation V_G posited by the conjecture might be defined over $\mathbb{Q}$. For comparison, according to unpublished work of J. Grodal [Gro], if G is a quasisimple group of Lie type defined over $\mathbb{F}_q$, then $\mathrm{H}^3(G; \boldsymbol{\mu}) \approx \mathbb{Z}/(q^2 - 1)$. (The “$\approx$” sign indicates that there are some systematic adjustments needed to accommodate twisted forms and Schur multipliers of G, and that there are finitely many groups G for which further corrections are necessary.) Note that letting $\boldsymbol{\mu} \subset \bar{\mathbb{F}}_q^\times$ now denote the group of roots of unity in positive characteristic,

$$\mathbb{Z}/(q^2 - 1) \cong \mathrm{H}^0(\mathrm{Gal}(\bar{\mathbb{F}}_q/\mathbb{F}_q); \boldsymbol{\mu}^{\otimes 2}).$$

This is consistent with the construction of [GL15] of a VOA over $\mathbb{F}_q$ for each group of Lie type defined over $\mathbb{F}_q$. Missing is a good theory of “gauge anomalies” in positive characteristic.

4.2 A Vertex Brauer Group

Recall the story of the cohomological Brauer group outlined in Sect. 1.5. When a group G acts on an Azumaya algebra A over $\bar{\mathbb{K}}$, the anomaly lives not just in $\mathrm{H}^2(G; \bar{\mathbb{K}}^\times)$ thought of as an abstract group: under the $\mathrm{Gal}(\bar{\mathbb{K}}/\mathbb{K})$-action on {Azumaya algebras over $\bar{\mathbb{K}}$}, the anomaly transforms according to the action on (the coefficients of) $\mathrm{H}^2(G; \bar{\mathbb{K}}^\times)$. This control over the Galois action implies that there is an anomaly-type map for Galois-twisted actions, which in the case of descent data for $\mathrm{Gal}(\bar{\mathbb{K}}/\mathbb{K})$ provides a map

$$\{\text{Azumaya algebras over } \mathbb{K}\} \to \mathrm{H}^2\big(\mathrm{Gal}(\bar{\mathbb{K}}/\mathbb{K}); \bar{\mathbb{K}}^\times\big).$$

Moreover, this map provides an isomorphism of the right-hand side with the Brauer group $\mathrm{Br}(\mathbb{K})$.

Our main theorem suggests that there may be a similar story for VOAs. For G a finite group, our main theorem identifies anomalies of actions of G on holomorphic VOAs as living in $\mathrm{H}^3(G; \boldsymbol{\mu}^{\otimes 2})$. This suggests:

Conjecture *For each field* $\mathbb{K}$*, there is a* vertex Brauer group *of Morita-equivalence classes of holomorphic VOAs over* $\mathbb{K}$ *modulo tensoring with* $E_{8,1}$*. When* $\mathbb{K}$ *is of characteristic* 0*, then the vertex Brauer group is isomorphic to the Galois cohomology group* $\mathrm{H}^3(\mathrm{Gal}(\bar{\mathbb{K}}/\mathbb{K}); \boldsymbol{\mu}^{\otimes 2})$.

We restrict to characteristic zero for two reasons. First, there is so far no theory of 't Hooft anomalies of actions on VOAs in positive characteristic. Second, in the Azumaya case, the map

$$\mathrm{H}^2\big(\mathrm{Gal}(\bar{\mathbb{K}}/\mathbb{K}); \boldsymbol{\mu}\big) \to \mathrm{H}^2\big(\mathrm{Gal}(\bar{\mathbb{K}}/\mathbb{K}); \bar{\mathbb{K}}^\times\big) \cong \mathrm{Br}(\mathbb{K})$$

is an isomorphism only in characteristic 0; in characteristic $p > 0$, it is an isomorphism onto the prime-to-p part of the right-hand side (see, e.g., [NSW08, Chapter VI, Corollary 6.3.6]).

We now outline, in parallel with Sect. 1.5, how one could hope to prove the conjecture. Suppose that V and W are holomorphic VOAs over $\mathbb{K}$ such that $V \otimes_{\mathbb{K}} \bar{\mathbb{K}} \cong W \otimes_{\mathbb{K}} \bar{\mathbb{K}}$. If V is fixed, then the data of the $\mathbb{K}$-form W is equivalent to the data of a 1-cocycle with twisted coefficients

$$\kappa \in \mathrm{H}^1\big(\mathrm{Gal}(\bar{\mathbb{K}}/\mathbb{K}); \mathrm{Aut}_{\bar{\mathbb{K}}}(V \otimes_{\mathbb{K}} \bar{\mathbb{K}})\big).$$

Suppose that the 't Hooft anomalies of finite group actions on $V \otimes_{\mathbb{K}} \bar{\mathbb{K}}$ are the restriction of a universal anomaly living on $\mathrm{Aut}_{\bar{\mathbb{K}}}(V \otimes_{\mathbb{K}} \bar{\mathbb{K}})$; according to our main theorem, this universal anomaly should live in

$$\alpha \in \mathrm{H}^3\big(\mathrm{Aut}_{\bar{\mathbb{K}}}(V \otimes_{\mathbb{K}} \bar{\mathbb{K}}); \boldsymbol{\mu}^{\otimes 2}\big).$$

Then, the composition $\alpha \circ \kappa$ would be a twisted cohomology class

$$\alpha \circ \kappa \in \mathrm{H}^3\big(\mathrm{Gal}(\bar{\mathbb{K}}/\mathbb{K}); \boldsymbol{\mu}^{\otimes 2}\big)$$

that depends only on the $\mathbb{K}$-form W of V.

Unfortunately, we lack direct access to this universal class α: with current understanding of VOAs, we can only define anomalies for actions of finite groups, but $\mathrm{Aut}_{\bar{\mathbb{K}}}(V \otimes_{\mathbb{K}} \bar{\mathbb{K}})$ is typically an infinite (in fact, affine algebraic) group. One could salvage this as follows. Suppose that there is a finite-degree Galois extension $\mathbb{K} \subset \mathbb{L}$ such that $V \otimes_{\mathbb{K}} \mathbb{L} \cong W \otimes_{\mathbb{K}} \mathbb{L}$. Then, still fixing V, the data of the $\mathbb{K}$-form W is given by a twisted cohomology class $\kappa_{\mathbb{L}} \in \mathrm{H}^1(\mathrm{Gal}(\mathbb{L}/\mathbb{K}); \mathrm{Aut}_{\mathbb{L}}(V \otimes_{\mathbb{K}} \mathbb{L}))$. Fix a cocycle for $\kappa_{\mathbb{L}}$, and let $G \subset \mathrm{Aut}_{\mathbb{L}}(V \otimes_{\mathbb{K}} \mathbb{L})$ denote the union of the image of $\kappa_{\mathbb{L}}$ together with its Galois-conjugates. Then, G is a finite group and so does have a well-defined anomaly $\alpha_{\mathbb{L}} \in \mathrm{H}^0(\mathrm{Gal}(\bar{\mathbb{K}}/\mathbb{L}); \mathrm{H}^3(G; \boldsymbol{\mu}^{\otimes 2}))$. Our proof of the main theorem does not provide sufficient control over the Galois action on cocycles—at too many steps we worked instead simply with cohomology classes—but it is reasonable to expect that there is a valid composition

$$\alpha_{\mathbb{L}} \circ \kappa_{\mathbb{L}} \in \mathrm{H}^3\big(\mathrm{Gal}(\mathbb{L}/\mathbb{K}); \mathrm{H}^0\big(\mathrm{Gal}(\bar{\mathbb{L}}/\mathbb{L}); \boldsymbol{\mu}^{\otimes 2}\big)\big) \subset \mathrm{H}^3\big(\mathrm{Gal}(\bar{\mathbb{K}}/\mathbb{K}); \boldsymbol{\mu}^{\otimes 2}\big),$$

providing the class $\alpha \circ \kappa$ above.

Thus, we expect that $\mathbb{K}$-forms of a fixed holomorphic VOA lead to classes in the putative "cohomological vertex Brauer group" $\mathrm{H}^3(\mathrm{Gal}(\bar{\mathbb{K}}/\mathbb{K}); \boldsymbol{\mu}^{\otimes 2})$. In the associative case, the next step is to classify Azumaya algebras over $\bar{\mathbb{K}}$: they are all matrix algebras. In the vertex case, it is hopeless to try to classify holomorphic VOAs up to isomorphism, since in particular the set of even unimodular lattices injects into the set of holomorphic VOAs. (In small central charge, there is an expected classification [Sch93], but it is not elementary.)

Instead, we should recognize the Azumaya classification as being about Morita-equivalence classes: over $\bar{\mathbb{K}}$, all Azumaya algebras are Morita-equivalent. We can hope that a similar statement holds for VOAs. There is no consensus definition of "Morita-equivalence" of VOAs. We will take the following definition:

Definition Suppose that V_1 and V_2 are regular VOAs that are both conformal extensions of the same regular VOA $W \subset V_1, V_2$ (in particular, all three of V_1, V_2, and W have the same central charge). Then, V_1 and V_2 are *Morita-equivalent* if the corresponding étale algebras in $\mathrm{REP}(W)$ are E_2-Morita-equivalent.

Implicit in the definition is the fact that, given a braided monoidal category $\mathcal{C}$, there is a "Morita 3-category" $\mathrm{ALG}_2(\mathcal{C})$ whose objects are braided-commutative algebra objects in $\mathcal{C}$; compare [JFS17, §8]. In the case of a braided fusion category $\mathcal{C}$, an E_2-Morita-equivalence of étale algebras A and B is the same as a $\mathcal{C}$-balanced Morita-equivalence of fusion categories $\mathrm{MOD}(A) \simeq \mathrm{MOD}(B)$. It follows in particular that $\mathrm{REP}(V_1) \simeq \mathrm{REP}(V_2)$ as MTCs.

Suppose that V_1 and V_2 are holomorphic VOAs. Then, using that two fusion categories are Morita-equivalent if and only if they have equivalent centres [ENO11, Theorem 3.1], one finds that V_1 and V_2 are Morita-equivalent as soon as they share a common regular subalgebra.

The analog of the fact that all Azumaya algebras over $\mathbb{C}$ are Morita-trivial would then be the conjecture that the central charge is the only Morita invariant of a holomorphic VOA. Equivalently:

Conjecture *Suppose V_1 and V_2 are two holomorphic VOAs over $\mathbb{C}$ of the same central charge. Then there is a regular VOA W of the same central charge embedding into both V_1 and V_2.*

It is not at all clear whether to believe this conjecture because we do not have enough understanding of the set of all VOAs of large central charge: the conjecture holds for the few holomorphic VOAs that are known simply because the only available methods to construct holomorphic VOAs are all of the form "pass to a regular subalgebra, then extend to a holomorphic algebra in a different way" and so always produce Morita-equivalent VOAs. (For instance, the conjecture holds for all holomorphic VOAs of central charge $c \leq 24$ by [vEMS17, MS19]. Holomorphic VOAs of larger central charge have not been well studied; the conjecture applies, by construction, to those from [GK18].) For lattice VOAs, the conjecture follows from a $\mathbb{Z}$-version of the Graham–Schmidt process, which produces, for any two unimodular lattices of the same rank, a common finite-index sublattice, and which provides the foundation of Kneser's neighbourhood method. Physically, the conjecture would follow from the expectation that every full conformal field theory with $c_L = c_R$ admits a gapped boundary condition.

Finally, if this conjecture holds, then we could take the VOAs $E_{8,1}^{\otimes n}$ as our standard Morita-class representatives, where "$E_{8,1}$" means the level-one WZW theory for E_8 equipped with its "Chevalley" $\mathbb{Q}$-form constructed in [DG12]. This is why "modulo tensoring with $E_{8,1}$" appears in the statement of the conjecture at the beginning of this subsection.

4.3 K-Theoretic Interpretation of Anomalies

The existence of a vertex Brauer group isomorphic to $\mathrm{H}^3(\mathrm{Gal}(\bar{\mathbb{K}}/\mathbb{K}); \boldsymbol{\mu}^{\otimes 2})$ would imply the existence of a "refined anomaly" whenever the action of G on a holomorphic vertex algebra V preserves a $\mathbb{K}$-form of V. Specifically, it would provide an anomaly map

$$\{\text{holomorphic VOAs over } \mathbb{K} \text{ with } G\text{-action}\} \to \mathrm{H}^3\big(G \times \mathrm{Gal}(\bar{\mathbb{K}}/\mathbb{K}); \boldsymbol{\mu}^{\otimes 2}\big). \tag{1}$$

The ordinary gauge anomaly and the vertex Brauer class would be the restrictions to $\mathrm{H}^3(G; \boldsymbol{\mu}^{\otimes 2})$ and $\mathrm{H}^3(\mathrm{Gal}(\bar{\mathbb{K}}/\mathbb{K}); \boldsymbol{\mu}^{\otimes 2})$, respectively.

The group $\mathrm{H}^3(G \times \mathrm{Gal}(\bar{\mathbb{K}}/\mathbb{K}); \boldsymbol{\mu}^{\otimes 2})$ is built, via a spectral sequence, from components of the form $\mathrm{H}^{3-i}(G; \mathrm{H}^i(\mathrm{Gal}(\bar{\mathbb{K}}/\mathbb{K}; \boldsymbol{\mu}^{\otimes 2})))$ for $0 \leq i \leq 3$. Suppose that $\mathbb{K}$ is a number field (i.e., a finite-degree extension of $\mathbb{Q}$). Then, the proposed vertex Brauer group is not itself particularly interesting: it follows from a theorem of Tate and Poitou [Tat63, Poi67] (see [Ser97, §II.6.3, Theorem B]) that $\mathrm{H}^3(\mathrm{Gal}(\bar{\mathbb{K}}/\mathbb{K}); \boldsymbol{\mu}^{\otimes 2})$ is isomorphic to $(\mathbb{Z}/2)^N$, where N is the number of field embeddings $\mathbb{K} \hookrightarrow \mathbb{R}$. But the Galois cohomology groups $\mathrm{H}^i(\mathrm{Gal}(\bar{\mathbb{K}}/\mathbb{K}); \boldsymbol{\mu}^{\otimes 2})$ for $i \leq 2$ are quite interesting. A theorem of Tate in the number-field case [Tat71, Tat76], and of Merkurjev and Suslin in general [MeS82], provides a close relationship between $\mathrm{H}^2(\mathrm{Gal}(\bar{\mathbb{K}}/\mathbb{K}); \boldsymbol{\mu}^{\otimes 2})$ and the second K-group $\mathrm{K}_2(\mathbb{K})$:

$$\mathrm{H}^2(\mathrm{Gal}(\bar{\mathbb{K}}/\mathbb{K}); \boldsymbol{\mu}_n^{\otimes 2}) \cong \mathrm{K}_2(\mathbb{K}) \otimes \mathbb{Z}/n\mathbb{Z}, \tag{2}$$

where $\boldsymbol{\mu}_n^{\otimes 2}$ denotes the n-torsion subgroup of $\boldsymbol{\mu}^{\otimes 2}$. (See [GS06, Chapter 8] and [NSW08, §VI.4].)

Suppose that G is a finite group acting on a holomorphic VOA V over $\mathbb{K}$. The refined anomaly (1) would map to a class in $\mathrm{H}^1(G; \mathrm{H}^2(\mathrm{Gal}(\bar{\mathbb{K}}/\mathbb{K}); \boldsymbol{\mu}^{\otimes 2}))$. The isomorphism (2) then suggests:

Conjecture *Each action of a finite group G on a holomorphic VOA V over $\mathbb{K}$ determines a homomorphism $G \to \mathrm{K}_2(\mathbb{K})$.*

Let n denote the exponent of G, i.e., $g^n = 1$ for all $g \in G$. Then, a map $G \to \mathrm{K}_2(\mathbb{K})$ will land in the n-torsion subgroup of $\mathrm{K}_2(\mathbb{K})$. Suppose that $\mathbb{K}$ contains a primitive nth root of unity ζ. The choice of ζ determines an isomorphism between the n-torsion subgroup of $\mathrm{K}_2(\mathbb{K})$ and the n-torsion subgroup of $\mathrm{Br}(\mathbb{K})$ because it identifies both with the untwisted cohomology group $\mathrm{H}^2(\mathrm{Gal}(\bar{\mathbb{K}}/\mathbb{K}); \mathbb{Z}/n)$. The choice of ζ is also what is needed to define algebraically the category $\mathcal{C}_g$ of g-twisted V-modules. Over an algebraically closed field, $\mathcal{C}_g$ is (noncanonically) equivalent to VECT, but such an equivalence need not survive Galois descent: rather, $\mathcal{C}_g$ will be equivalent to the category of modules of some Azumaya algebra A_g. The map $g \mapsto [A_g] \in \mathrm{Br}(\mathbb{K})$ presumably agrees with the map $G \to \mathrm{K}_2(\mathbb{K})$ predicted in the conjecture.

4.4 Galois and Grothendieck–Teichmuller Groups and Modular Data

Let W be a regular VOA over $\mathbb{C}$. We emphasized at the start of Sect. 3 that the MTC structure on $\mathrm{REP}(W)$ depends, a priori, on the $\mathbb{C}$-structure of W. To control this a priori $\mathbb{C}$-dependence, one should study the following question. Choose $\gamma \in \mathrm{Aut}(\mathbb{C})$. As a $\mathbb{C}$-linear category, $\mathrm{REP}({}^{\gamma}W) \simeq {}^{\gamma}\mathrm{REP}(W)$, since the notion of "representation" is defined algebraic. How do the MTC structures compare?

Conjecture *Fix a semisimple $\mathbb{C}$-linear category $\mathcal{C}$ with finitely many simples objects. The set of MTC structures on $\mathcal{C}$ is permuted by a canonical action of* $\mathrm{Aut}(\mathbb{C})$.

Up to issues of completion, this action of $\mathrm{Aut}(\mathbb{C})$ should come from the injection of $\mathrm{Gal}(\bar{\mathbb{Q}}/\mathbb{Q})$ into the Grothendieck–Teichmuller group, which is again up to issues of completion, the automorphism group of the E_2 operad. MTCs are a particular type of (categorical) algebra for the E_2 operad.

Each MTC $\mathcal{C}$ determines a projective representation of $\mathrm{SL}(2, \mathbb{Z})$ called the *modular data* of $\mathcal{C}$. We will refer to this representation as $\int_{T^2} \mathcal{C}$; the notation follows the "factorization algebra" notation used for example in [Wal06, Sch14]. The representation $\int_{T^2} \mathcal{C}$ has many nice properties. Among them are: (i) $\int_{T^2} \mathcal{C}$ has a basis indexed by the simple objects in $\mathcal{C}$ and (ii) the representation $\int_{T^2} \mathcal{C}$ is defined over the cyclotomic field $\mathbb{Q}^{\mathrm{cyc}}$. Our main theorem strongly suggests that in the case when $\mathcal{C} = \mathrm{REP}(W)$ for a regular VOA W,

Conjecture *At the level of modular data, we have* $\int_{T^2} \mathrm{REP}({}^{\gamma}W) \cong \gamma^2\left(\int_{T^2} \mathrm{REP}(W)\right)$.

This conjecture is enticingly close to the following result of [CG94, DLN15]. Let $\mathcal{C}$ be an arbitrary MTC, and choose $\gamma \in \mathrm{Gal}(\mathbb{Q}^{\mathrm{cyc}}/\mathbb{Q}) \cong \widehat{\mathbb{Z}}^{\times}$. Then, there is a permutation σ of the simple objects of $\mathcal{C}$, and hence of the basis of $\int_{T^2}\mathcal{C}$, which intertwines $\int_{T^2}\mathcal{C}$ with $\gamma^2(\int_{T^2}\mathcal{C})$. Thus, the conjecture predicts that $\int_{T^2}\mathrm{REP}(W)$ and $\int_{T^2}\mathrm{REP}({}^{\gamma}W)$ are isomorphic as projective $\mathrm{SL}(2,\mathbb{Z})$-modules.

4.5 Cyclotomicity of MTCs

Suppose one has some arbitrary module X of $\mathrm{Gal}(\bar{\mathbb{Q}}/\mathbb{Q})$ such that the formula $\gamma \triangleright x := \gamma^2(x)$ also defines an action of $\mathrm{Gal}(\bar{\mathbb{Q}}/\mathbb{Q})$ on X. Then, the action on X factors through $\mathrm{Gal}(\mathbb{Q}^{\mathrm{cyc}}/\mathbb{Q})$. Indeed, if $(\gamma_1\gamma_2)^2 = \gamma_1^2\gamma_2^2$, then $\gamma_1\gamma_2 = \gamma_2\gamma_1$, and the Abelianization of $\mathrm{Gal}(\bar{\mathbb{Q}}/\mathbb{Q})$ is precisely $\mathrm{Gal}(\mathbb{Q}^{\mathrm{cyc}}/\mathbb{Q})$.

The repeated occurrences of a squared Galois action suggest that perhaps the entire theory of MTCs is cyclotomic. For comparison, finite group theory is cyclotomic—all characters are cyclotomic, and all representations split over $\mathbb{Q}^{\mathrm{cyc}}$—and MTCs are a version of "finite quantum groups". The following conjectures have been contemplated by various authors, e.g., [MS12, EG18]:

Conjectures

(1) *If $\mathcal{C}$ is an MTC, then $\mathcal{C}$ has a (canonical?) form over* $\mathbb{Q}^{\mathrm{cyc}}$.
(2) *If $\mathcal{C}$ is an MTC, then $\mathcal{C}$ arises as* $\mathrm{REP}(W)$ *for some (canonical?) regular VOA W.*
(3) *If W is a regular VOA, then* $\mathrm{REP}(W)$ *has a (canonical?) form over* $\mathbb{Q}^{\mathrm{cyc}}$.

References

[BDSPV15] Bruce Bartlett, Christopher L. Douglas, Christopher J. Schommer-Pries, and Jamie Vicary. Modular categories as representations of the 3-dimensional bordism 2-category. 2015. arXiv:1509.06811.
[Bor86] Richard E. Borcherds. Vertex algebras, Kac-Moody algebras, and the Monster. *Proc. Nat. Acad. Sci. U.S.A.*, 83(10):3068–3071, 1986.
[CG94] A. Coste and T. Gannon. Remarks on Galois symmetry in rational conformal field theories. *Phys. Lett. B*, 323(3–4):316–321, 1994. DOI:10.1016/0370-2693(94)91226-2. MR1266785.
[CM16] Scott Carnahan and Masahiko Miyamoto. Regularity of fixed-point vertex operator subalgebras. 2016. arXiv:1603.05645.
[CN79] J. H. Conway and S. P. Norton. Monstrous moonshine. *Bull. London Math. Soc.*, 11(3):308–339, 1979. DOI:10.1112/blms/11.3.308. MR554399.
[DG12] Chongying Dong and Robert L. Griess, Jr. Integral forms in vertex operator algebras which are invariant under finite groups. *J. Algebra*, 365:184–198, 2012. DOI:10.1016/j.jalgebra.2012.05.006. MR2928458. arXiv:1201.3411.

[DLM00] Chongying Dong, Haisheng Li, and Geoffrey Mason. Modular-invariance of trace functions in orbifold theory and generalized Moonshine. *Comm. Math. Phys.*, 214(1):1–56, 2000. DOI:10.1007/s002200000242. MR1794264. arXiv:q-alg/9703016.

[DLN15] Chongying Dong, Xingjun Lin, and Siu-Hung Ng. Congruence property in conformal field theory. *Algebra Number Theory*, 9(9):2121–2166, 2015. DOI:10.2140/ant.2015.9.2121. MR3435813.

[DPR90] R. Dijkgraaf, V. Pasquier, and P. Roche. Quasi Hopf algebras, group cohomology and orbifold models. *Nuclear Phys. B Proc. Suppl.*, 18B:60–72 (1991), 1990. Recent advances in field theory (Annecy-le-Vieux, 1990).

[DVVV89] Robbert Dijkgraaf, Cumrun Vafa, Erik Verlinde, and Herman Verlinde. The operator algebra of orbifold models. *Comm. Math. Phys.*, 123(3):485–526, 1989. MR1003430.

[EG18] David E. Evans and Terry Gannon. Reconstruction and local extensions for twisted group doubles, and permutation orbifolds. 2018. arXiv:1804.11145.

[EGNO15] Pavel Etingof, Shlomo Gelaki, Dmitri Nikshych, and Victor Ostrik. *Tensor categories*, volume 205 of *Mathematical Surveys and Monographs*. American Mathematical Society, Providence, RI, 2015. http://www-math.mit.edu/~etingof/egnobookfinal.pdf. MRnumber3242743. DOI:10.1090/surv/205.

[ENO11] Pavel Etingof, Dmitri Nikshych, and Victor Ostrik. Weakly group-theoretical and solvable fusion categories. *Adv. Math.*, 226(1):176–205, 2011. DOI:10.1016/j.aim.2010.06.009. MR2735754.

[FT14] Daniel S. Freed and Constantin Teleman. Relative quantum field theory. *Comm. Math. Phys.*, 326(2):459–476, 2014. arXiv:1212.1692. DOI:10.1007/s00220-013-1880-1. MR3165462.

[FT18] Daniel S. Freed and Constantin Teleman. Topological dualities in the Ising model. 2018. arXiv:1806.00008.

[FZ92] Igor B. Frenkel and Yongchang Zhu. Vertex operator algebras associated to representations of affine and Virasoro algebras. *Duke Math. J.*, 66(1):123–168, 1992. DOI:10.1215/S0012-7094-92-06604-X. MR1159433.

[GG09] Matthias R. Gaberdiel and Terry Gannon. Zhu's algebra, the C_2 algebra, and twisted modules. In *Vertex operator algebras and related areas*, volume 497 of *Contemp. Math.*, pages 65–78. Amer. Math. Soc., Providence, RI, 2009. DOI:10.1090/conm/497/09769. arXiv:0811.3892. MR2568399.

[GJF19] Davide Gaiotto and Theo Johnson-Freyd. Symmetry protected topological phases and generalized cohomology. *J. High Energy Phys.*, 7, 5 2019. DOI:10.1007/JHEP05(2019)007. arXiv:1712.07950.

[GK18] Thomas Gemünden and Christoph A. Keller. Orbifolds of lattice vertex operator algebras at $d = 48$ and $d = 72$. 2018. arXiv:1802.10581.

[GKO85] P. Goddard, A. Kent, and D. Olive. Virasoro algebras and coset space models. *Phys. Lett. B*, 152(1–2):88–92, 1985. DOI:10.1016/0370-2693(85)91145-1. MR778819.

[GL15] Robert L. Griess, Jr. and Ching Hung Lam. Groups of Lie type, vertex algebras, and modular moonshine. *Int. Math. Res. Not. IMRN*, (21):10716–10755, 2015. DOI:10.1093/imrn/rnv003. MR3456026.

[Gro] Jesper Grodal. Low dimensional homology of finite groups of Lie type, away from the characteristic. Unpublished.

[GS06] Philippe Gille and Tamás Szamuely. *Central simple algebras and Galois cohomology*, volume 101 of *Cambridge Studies in Advanced Mathematics*. Cambridge University Press, Cambridge, 2006. DOI:10.1017/CBO9780511607219. MR2266528.

[Hen17] André Henriques. Bicommutant categories from conformal nets. 2017. arXiv:1701.02052.

[Hua08] Yi-Zhi Huang. Rigidity and modularity of vertex tensor categories. *Commun. Contemp. Math.*, 10(suppl. 1):871–911, 2008. DOI:10.1142/S0219199708003083. MR2468370. arXiv:math.QA/0502533.

[JF16] Theo Johnson-Freyd. The quaternions and Bott periodicity are quantum Hamiltonian reductions. *SIGMA Symmetry Integrability Geom. Methods Appl.*, 12:Paper No. 116, 6, 2016. DOI:10.3842/SIGMA.2016.116. MR3581593. arXiv:1603.06603.

[JF19] Theo Johnson-Freyd. The moonshine anomaly. *Comm. Math. Phys.*, 365(3):943–970, 2019. arXiv:1707.08388. DOI:10.1007/s00220-019-03300-2. MR3916985.

[JF20] Theo Johnson-Freyd. Heisenberg-picture quantum field theory. In *Representation Theory, Mathematical Physics and Integrable Systems*, Progr. Math. Birkhäuser Boston, 2020. Volume in honor of Kolya Reshetikhin. arXiv:1508.05908.

[JFS17] Theo Johnson-Freyd and Claudia Scheimbauer. (Op)lax natural transformations, twisted field theories, and "even higher" Morita categories. *Adv. Math.*, 307:147–223, 2 2017. DOI:10.1016/j.aim.2016.11.014. arXiv:1502.06526.

[JFT19] Theo Johnson-Freyd and David Treumann. Third homology of some sporadic finite groups. *SIGMA Symmetry Integrability Geom. Methods Appl.*, 15:059, 38 pages, 2019. arXiv:1810.00463. DOI:10.3842/SIGMA.2019.059. MR3990846.

[Kir02] Alexander Kirillov, Jr. Modular categories and orbifold models. *Comm. Math. Phys.*, 229(2):309–335, 2002. DOI:10.1007/s002200200650. MR1923177. arXiv:math/0104242.

[LL04] James Lepowsky and Haisheng Li. *Introduction to vertex operator algebras and their representations*, volume 227 of *Progress in Mathematics*. Birkhäuser Boston, Inc., Boston, MA, 2004. DOI:10.1007/978-0-8176-8186-9. MR2023933.

[Mas07] Geoffrey Mason. Some cohomology classes associated with the monster simple group and the monster orbifold. 2007.

[MeS82] A. S. Merkur′ ev and A. A. Suslin. K-cohomology of Severi-Brauer varieties and the norm residue homomorphism. *Izv. Akad. Nauk SSSR Ser. Mat.*, 46(5):1011–1046, 1135–1136, 1982. MR675529.

[MS12] Scott Morrison and Noah Snyder. Non-cyclotomic fusion categories. *Trans. Amer. Math. Soc.*, 364(9):4713–4733, 2012. DOI:10.1090/S0002-9947-2012-05498-5. MR2922607. arXiv:1002.0168.

[MS19] Sven Möller and Nils R. Scheithauer. Dimension formulae and generalised deep holes of the Leech lattice vertex operator algebra. 2019. arXiv:1910.04947.

[Nik13] Dmitri Nikshych. Morita equivalence methods in classification of fusion categories. In *Hopf algebras and tensor categories*, volume 585 of *Contemp. Math.*, pages 289–325. Amer. Math. Soc., Providence, RI, 2013. DOI:10.1090/conm/585/11607. MR3077244. arXiv:1208.0840.

[NSW08] Jürgen Neukirch, Alexander Schmidt, and Kay Wingberg. *Cohomology of number fields*, volume 323 of *Grundlehren der Mathematischen Wissenschaften [Fundamental Principles of Mathematical Sciences]*. Springer-Verlag, Berlin, second edition, 2008. MR2392026. DOI:10.1007/978-3-540-37889-1.

[Poi67] Georges Poitou. *Cohomologie galoisienne des modules finis: Séminaire de l'Institut de Mathématiques de Lille*, volume 13 of *Travaux et recherches mathématiques*. Paris: Dunod, 1967. Textes des séminaires hebdomadaires (1962–1963).

[RT91] N. Reshetikhin and V. G. Turaev. Invariants of 3-manifolds via link polynomials and quantum groups. *Invent. Math.*, 103(3):547–597, 1991. MR1091619. DOI:http://dx.doi.org/10.1007/BF01239527.

[Sch93] A. N. Schellekens. Meromorphic $c = 24$ conformal field theories. *Comm. Math. Phys.*, 153(1):159–185, 1993. MR1213740.

[Sch14] Claudia Scheimbauer. *Factorization Homology as a Fully Extended Topological Field Theory*. PhD thesis, ETH Zürich, 2014. https://folk.ntnu.no/claudiis/ScheimbauerThesis.pdf.

[Ser97] Jean-Pierre Serre. *Galois cohomology*. Springer-Verlag, Berlin, 1997. Translated from the French by Patrick Ion and revised by the author. DOI:10.1007/978-3-642-59141-9. MR1466966.

[SP09] Christopher J. Schommer-Pries. *The Classification of Two-Dimensional Extended Topological Field Theories*. PhD thesis, University of California, Berkeley, 2009. arXiv:1112.1000. MRMR2534210.

[Tat63] John Tate. Duality theorems in Galois cohomology over number fields. In *Proc. Internat. Congr. Mathematicians (Stockholm, 1962)*, pages 288–295. Inst. Mittag-Leffler, Djursholm, 1963. MR0175892.

[Tat71] John Tate. Symbols in arithmetic. pages 201–211, 1971. MR0422212.

[Tat76] John Tate. Relations between K_2 and Galois cohomology. *Invent. Math.*, 36:257–274, 1976. DOI:10.1007/BF01390012. MR0429837.

[vEMS17] Jethro van Ekeren, Sven Möller, and Nils R. Scheithauer. Construction and classification of holomorphic vertex operator algebras. *J. Reine Angew. Math.*, 2017. arXiv:1507.08142.

[Wal06] Kevin Walker. TQFTs [early incomplete draft]. http://canyon23.net/math/tc.pdf, 2006.

[WWW18] Juven Wang, Xiao-Gang Wen, and Edward Witten. Symmetric gapped interfaces of SPT and SET states: Systematic constructions. *Phys. Rev. X*, 8:031048, Aug 2018. DOI:10.1103/PhysRevX.8.031048. arXiv:1705.06728.

[Xu00] Feng Xu. Algebraic orbifold conformal field theories. *Proc. Natl. Acad. Sci. USA*, 97(26):14069–14073, 2000. arXiv:math/0004150. DOI:10.1073/pnas.260375597. MR1806798.

[Zhu96] Yongchang Zhu. Modular invariance of characters of vertex operator algebras. *J. Amer. Math. Soc.*, 9(1):237–302, 1996. DOI:10.1090/S0894-0347-96-00182-8. MR1317233.

Heisenberg-Picture Quantum Field Theory

Theo Johnson-Freyd

For Kolya Reshetikhin on the occasion of his 60th birthday.

Abstract What we should mean by "Heisenberg-picture quantum field theory"? Atiyah–Segal-type axioms do a good job of capturing the "Schrödinger picture": these axioms define a "d-dimensional quantum field theory" to be a symmetric monoidal functor from an (∞, d)-category of "spacetimes" to an (∞, d)-category which at the second-from-top level consists of vector spaces, so at the top level consists of numbers. This paper argues that the appropriate parallel notion "Heisenberg picture" should also be defined in terms of symmetric monoidal functors from the category of spacetimes, but the target should be an (∞, d)-category that in top dimension consists of pointed vector spaces instead of numbers; the second-from-top level can be taken to consist of associative algebras or of pointed categories. The paper ends by outlining two sources of such Heisenberg-picture field theories: factorization algebras and skein theory.

I would like to thank Alexandru Chirvasitu, Owen Gwilliam, and Claudia Scheimbauer for many ongoing discussions about these and related ideas, to the referee for their helpful comments, and to Stephan Stolz and Peter Teichner for sharing the construction developed in Sect. 8. Most importantly, my thanks go to Kolya Reshetikhin, whose work on quantum field theory and ribbon categories (e.g. [RT90, RT91, Res10a, Res10b]) inspired this paper. This work is supported by the grant DMS-1304054.

T. Johnson-Freyd (✉)
Perimeter Institute for Theoretical Physics, Waterloo, ON, Canada
e-mail: theojf@pitp.ca

A. Alekseev et al. (eds.), *Representation Theory, Mathematical Physics, and Integrable Systems*, Progress in Mathematics 340,
https://doi.org/10.1007/978-3-030-78148-4_13

1 Introduction and Motivation

Open the nearest book called *Introduction to Quantum Mechanics*—you probably have one lying around somewhere. Almost certainly somewhere in it is a discussion of the two so-called pictures of quantum mechanics, named after Erwin Schrödinger and Werner Heisenberg. The "Schrödinger picture" has a natural generalization to quantum field theory via Atiyah–Segal-type axioms. My goal in this paper is to motivate, propose, and expound upon axioms for a "Heisenberg picture" of quantum field theory. Let us, then, begin by learning what we can from that *Quantum Mechanics* textbook. The discussion of the Schrödinger picture translates well into a mathematical definition:

Definition 1.1 A *Schrödinger-picture quantum mechanical system* consists of the following data:

(1) A vector space V (over some ground field $\mathbb{K}$) called *the space of states*.
(2) For each "time" $t \in \mathbb{R}_{>0}$, a linear map $u_t : V \to V$ called *time evolution*. These should satisfy a *group law* $u_{t_1+t_2} = u_{t_1} \circ u_{t_2}$.
(3) Some distinguished *states* $v_i \in V$ depending on some labeling set I which our laboratory friends know how to prepare, and some distinguished *costates* $w_j : V \to \mathbb{K}$ that we know how to "post-pare" (or is it "post-pair"?). There may also be some distinguished *observables* $a_k \in \mathrm{End}(V)$ that we know how to test by making some manipulation while the experiment runs. ◊

Remark 1.2 The axioms for a group (a set with a binary operation satisfying ...) do not communicate what questions one might want to know about a group. Moreover, the axioms for a group are often too liberal: groups "in nature" are usually Lie or algebraic or finite or hyperbolic or otherwise more structured than the axioms would suggest. Similarly, Schrödinger-picture quantum mechanical systems are usually defined over $\mathbb{C}$ specifically, the spaces of states are usually Hilbert spaces, time evolution operators are usually unitary, the sets of distinguished states and costates usually agree, and observables are usually self-adjoint.

But none of these extra conditions are what that textbook of yours claims is the essence of quantum mechanics, which is the linearity illustrated by the two-slit experiment. So I will not put it in the definition. Indeed, Hilbert structures and unitarity are closely related to time-reversal symmetry (recent discussions are available in [FH16, JF17]); our axiomatics should accommodate non-symmetric examples as well. General questions one might want to ask about a Schrödinger-picture quantum mechanical system include the spectrum of time evolution and the (perhaps asymptotic) values of various compositions of the data. ◊

Remark 1.3 The discussion of the Schrödinger picture in that textbook of yours probably also includes an important comment emphasizing that the space V itself is not "physical": only its projectivization $\mathbb{P}V = (V \smallsetminus \{0\})/\mathbb{K}^\times$ is. It is often tempting to ignore this comment. ◊

The discussion of the Heisenberg picture begins by emphasizing that in physics, what is most physical are the *observables* in a given system, and that in quantum mechanics these form an associative but generally noncommutative algebra. (Just like how physically interesting Schrödinger-picture quantum mechanics deals with Hilbert spaces rather than general vector spaces, physically interesting Heisenberg-picture quantum mechanics deals with algebras equipped with extra structure—C-star or von Neumann, for example—but we should not build such structure into the basic axiomatics.) Let us recall the main examples and then try to extract a definition:

Example 1.4 Let $(V, u_t, \dots)$ be a Schrödinger-picture quantum mechanical system such that the time evolution operators u_t are all isomorphisms. The corresponding *Heisenberg-picture quantum mechanical system* consists of the following data:

(1) The associative algebra $A = \mathrm{End}(V)$ of all endomorphism of V is the *algebra of observables* of the system.
(2) For each $t \in \mathbb{R}_{>0}$, *evolution for time t* is encoded by the "conjugate by u_t" algebra automorphism $a \mapsto u_t a u_t^{-1}$.
(3) The distinguished observables $a_k \in A$ already do not reference V. To remove V from the data of the distinguished states and costates, we can encode them as ideals: given $v_i \in V$, remember the left ideal $\mathrm{Ann}(v_i) = \{a \in A \text{ s.t. } av_i = 0\}$; for $w_j : V \to \mathbb{K}$, use the right ideal $\mathrm{Ann}(w_j) = \{a \in A \text{ s.t. } w_j \circ a = 0\}$.

This construction is discussed in [Sch09], which does not answer the question of what to do when the operators u_t are not isomorphisms. I will propose an answer in Proposition 3.4. ◊

Remark 1.5 Some contravariance in the constructions later in this paper is unavoidable. Just to fix a convention, let us declare (as is most standard) that $\mathrm{End}(V)$ acts on V from the left. For a general category $\mathcal{C}$ and object $C \in \mathcal{C}$ therein, the algebra $\mathrm{End}_{\mathcal{C}}(C)$ of endomorphisms of C has multiplication $fg = f \circ g = (C \overset{g}{\to} C \overset{f}{\to} C)$. My hope is that the reader will largely ignore left/right questions. ◊

This example illustrates already one feature of the Heisenberg picture: it solves the problem from Remark 1.3 that only $\mathbb{P}V$ is physical, not V. Indeed, multiplying u_t by a non-zero constant does not change the conjugation map $a \mapsto u_t a u_t^{-1}$, and similarly the ideals $\mathrm{Ann}(v_i)$ and $\mathrm{Ann}(w_j)$ depend only on the lines spanned by v_i and w_j. Moreover, in many situations a vector or Hilbert space V can be recovered up to non-unique isomorphism from the algebra $\mathrm{End}(V)$; the ambiguity in recovering V is exactly the group of invertible scalars, so that $\mathrm{End}(V)$ and $\mathbb{P}V$ encode exactly the same information.

The following example illustrates the other main feature of the Heisenberg picture: it is flexible enough to accommodate classical as well as quantum mechanical systems (and systems intermediate between these extremes, corresponding to noncommutative algebras that are not as fully noncommutative as $\mathrm{End}(V)$):

Example 1.6 A *classical mechanical system* consists of a symplectic manifold X and a family of symplectomorphisms $\{\varphi_t : X \to X\}_{t\in\mathbb{R}_{>0}}$ satisfying a group law, and perhaps some other distinguished data. Such a system can be encoded in the Heisenberg picture as follows:

(1) The *algebra of observables* is the commutative algebra $\mathcal{O}(X)$ of smooth functions on $\mathcal{M}$.
(2) *Time evolution* is encoded by the algebra isomorphisms $f_t = \varphi_t^* : \mathcal{O}(X) \to \mathcal{O}(X)$.
(3) One type of distinguished data might be submanifolds of "boundary conditions." These can be encoded by ideals in $\mathcal{O}(X)$. ◊

Examples 1.4 and 1.6 suggest the following definition of the Heisenberg picture:

Definition 1.7 (Tentative) A *Heisenberg-picture quantum mechanical system* consists of:

(1) An associative *algebra of observables* A.
(2) For each time $t \in \mathbb{R}_{>0}$, a *time evolution* homomorphism $f_t : A \to A$. These should satisfy a *group law* $f_{t_1+t_2} = f_{t_1} \circ f_{t_2}$.
(3) Some distinguished right ideals encoding initial states/boundary conditions, some distinguished left ideals encoding terminal costates/boundary conditions, and some distinguished elements of A encoding available experimental manipulations. ◊

Definition 1.7 is only tentative, and will be revised in the next section. For instance, it is not clear how (or whether?) to accommodate Schrödinger-picture systems with non-invertible time evolution.

2 Modulation

The Schrödinger picture of Definition 1.1 generalizes naturally to an Atiyah–Segal style axiomatics for quantum field theory. Indeed, Definition 1.1, and in particular the group law, already defines quantum mechanical systems as a type of functor:

Definition 2.1 Let I, J, K be label sets. The category QMSPACETIMES of *quantum mechanical spacetimes* with "point defect" label sets I, J, K has:

Objects: Two objects, called $\{\mathrm{pt}\}$ and $\emptyset$.
Generating morphisms: For each $t \in \mathbb{R}_{>0}$, there is a morphism $u_t : \{\mathrm{pt}\} \to \{\mathrm{pt}\}$, which can be thought of as a metrized interval of length t. These are required to satisfy the group law relation $u_{t_1} \circ u_{t_2} = u_{t_1+t_2}$.

There are also morphisms $v_i : \emptyset \to \{\text{pt}\}$ for each $i \in I$, $w_j : \{\text{pt}\} \to \emptyset$ for each $j \in J$, and $a_k : \{\text{pt}\} \to \{\text{pt}\}$, called *point defects*. We impose no relations on these.

The full category then consists of compositions of the generating morphisms (modulo the group law relation). ♢

Let VECT denote the category of $\mathbb{K}$-vector spaces and linear maps. Definition 1.1 can then be rephrased to state that a Schrödinger-picture quantum mechanical system is a functor V : QMSPACETIMES $\to$ VECT along with an isomorphism $V(\emptyset) \cong \mathbb{K}$.

Remark 2.2 VECT and QMSPACETIMES are examples of *pointed categories*: categories equipped with distinguished objects (namely, $\mathbb{K} \in$ VECT and $\emptyset \in$ QMSPACETIMES). Then V is precisely a *strong pointed functor* from VECT to QMSPACETIMES (compare Remark 4.3).

One can alternately freely generate a symmetric monoidal category from QMSPACETIMES under the condition that $\emptyset$ becomes the monoidal unit. Calling the monoidal structure $\sqcup$, the objects of this freely generated category are finite sets—formal disjoint unions of copies of {pt}—and its morphisms are disjoint unions of directed metrized intervals with "point defects" labeled by elements of I, J, and K. ♢

This functorial point of view can be applied to higher-dimensional spacetimes:

Definition 2.3 ([Ati88, Seg04]) Given a geometry $\mathcal{G}$, write $\text{SPACETIMES}^{\mathcal{G}}_{d-1,d}$ for the symmetric monoidal category whose objects are $(d-1)$-dimensional manifolds equipped with geometry of type $\mathcal{G}$ and whose morphisms are isomorphism types of d-dimensional cobordisms equipped with geometry of type G. A *Schrödinger-picture (d-dimensional, non-extended, for geometry $\mathcal{G}$) quantum field theory* is a symmetric monoidal functor $\text{SPACETIMES}^{\mathcal{G}}_{d-1,d} \to \text{VECT}$. ♢

Remark 2.4 The details of, and potential difficulty in constructing, the category $\text{SPACETIMES}^{\mathcal{G}}_{d-1,d}$ depends on the choice of geometry $\mathcal{G}$. The notion of "geometry" should be interpreted quite liberally: it could include metrics, background fields, labeled defects, etc. For some discussion and examples of geometric cobordism categories, see [Aya08, Res10b, ST11]. When $\mathcal{G}$ is sufficiently topological, $\text{SPACETIMES}^{\mathcal{G}}_{d-1,d}$ can be built from the category $\text{BORD}^{\text{smooth}}_{d-1,d}$ of smooth but otherwise unstructured spacetimes as in [Lur09, Section 3.2], elaborated upon in [JF17]. ♢

Ignoring for the moment the boundary-condition ideals and distinguished manipulations (item (3) in Definition 1.7), a Heisenberg-picture quantum mechanical system is also a functor: its source is the category QMSPACETIMES of metric segments from Definition 2.1, and its target is the category ALG^{homo} whose objects are associative algebras and whose morphisms are algebra homomorphisms (we will soon introduce a different category called ALG, hence the name for this one). Let VECT^{inv} denote the maximal subgroupoid of VECT: it has the same objects, but morphisms must be isomorphisms. There is a functor End : $\text{VECT}^{\text{inv}} \to \text{ALG}^{\text{homo}}$ which sends a vector space V to its endomorphism algebra $\text{End}(V)$ and an isomorphism $f : V \to W$ to the algebra homomorphism $a \mapsto faf^{-1} : \text{End}(V) \to \text{End}(W)$. Example 1.4 then consists of taking in a Schrödinger-picture quantum mechanical system $V : \text{QMSPACETIMES} \to \text{VECT}^{\text{inv}}$ and producing the composition $\mathcal{A} = \text{End} \circ V : \text{QMSPACETIMES} \to \text{ALG}^{\text{homo}}$. This functor is used in [Sch09] to turn Schrödinger-picture quantum field theories valued in VECT^{inv} into nets of algebras in the sense of "axiomatic" or "algebraic" quantum field theory.

The question then arises: can we similarly compose arbitrary Schrödinger-picture quantum field theories with this End functor to produce Heisenberg-picture quantum field theories? My goal, of course, is to explain that the answer is "yes." But the main obstruction, hinted at already in Example 1.4 and after Definition 1.7, must be confronted: in most Schrödinger-picture quantum field theories, the linear maps associated to spacetimes are not invertible. To resolve this obstruction, let us seek guidance from another example:

Example 2.5 Typical classical field theories correspond to partial differential equations. Let D be some partial differential equation for fields on d-dimensional $\mathcal{G}$-geometric manifolds. Then the classical field theory $\mathcal{F}$ assigns to a $(d-1)$-dimensional manifold N the space $\mathcal{F}(N) = \{$germs of solutions to the PDE D on a small d-dimensional neighborhood of $N\}$ and to a d-dimensional cobordism M the space $\mathcal{F}(M) = \{$solutions to D on $M\}$. These data package into a *span of spaces*: to any cobordism $(M : N_1 \to N_2) \in \text{SPACETIMES}^{\mathcal{G}}_{d-1,d}$, we get the span $\mathcal{F}(N_1) \leftarrow \mathcal{F}(M) \to \mathcal{F}(N_2)$.

The corresponding Heisenberg-picture TQFT should assign to N the algebra $\mathcal{O}(\mathcal{F}(N))$ of functions on $\mathcal{F}(N)$. If the span $\mathcal{F}(N_1) \leftarrow \mathcal{F}(M) \to \mathcal{F}(N_2)$ were the graph of a function $f_M : \mathcal{F}(N_1) \leftarrow \mathcal{F}(N_2)$, then we could associative in the Heisenberg picture the homomorphism $f_M^* : \mathcal{O}(\mathcal{F}(N_1)) \to \mathcal{O}(\mathcal{F}(N_2))$. But typically it is not the graph of a function. Applying $\mathcal{O}$ to all pieces produces a cospan of algebras $\mathcal{O}(\mathcal{F}(N_1)) \to \mathcal{O}(\mathcal{F}(M)) \leftarrow \mathcal{O}(\mathcal{F}(N_2))$. This is about as far as abstract nonsense can take us. $\diamond$

Perhaps we should follow Example 2.5 and expect that Heisenberg-picture TQFTs should assign to cobordisms cospans of associative algebras? This is a tempting answer, but has a few problems. It does not fully accommodate the "boundary ideals" from item (3) of Definition 1.7—if these were two-sided ideals, we could instead use the quotient rings, but generically they are only one-sided. Moreover, the canonical way to compose cospans is via a push-out square. These exist in the category of associative algebras, but are too large (about as large as a free associative algebra). Considering further the classical field theory of Example 2.5, one notes that the composition of spans of spaces, given by a pullback of spaces, corresponds to the push-out of commutative algebras, which is also the tensor product. Given cospans of associative algebras $A_0 \to B_1 \leftarrow A_1 \to B_2 \leftarrow A_2$, one can give B_1 a right A_1-module structure and B_2 a left A_1-module structure, and consider the composition $B_1 \otimes_{A_1} B_2$. But this generically is not an associative algebra.

On the other hand, it is a module, as are ideals. And cospans of commutative algebras are examples of bimodules. Perhaps (bi)modules are the appropriate target of Heisenberg-picture field theory? This will not quite be our final definition, but it is sufficiently important as to merit its own name:

Definition 2.6 The *Morita* bicategory $\mathrm{MOR} = \mathrm{MOR}_1(\mathrm{VECT}_{\mathbb{K}})$ over the field $\mathbb{K}$ has:

Objects: Associative algebras over $\mathbb{K}$.

1-morphisms: A 1-morphism between associative algebras A and B is an A-B-bimodule ${}_AM_B$. These compose by tensor product.

2-morphisms: A 2-morphism is a homomorphism of bimodules.

This bicategory is symmetric monoidal with the usual tensor product over $\mathbb{K}$ [Shu08, Shu10].

A *(non-extended) Morita-picture quantum field theory* for the spacetime category SPACETIMES is a symmetric monoidal functor SPACETIMES $\to$ MOR. ◊

How can we translate a field theory in the sense of Definition 1.7 into a Morita-picture field theory? It is not yet clear how to accommodate the distinguished elements of $\mathcal{A}$ from item (3), but the distinguished ideals are already modules, hence easy to handle. As for time evolution, one can always turn homomorphisms into modules:

Definition 2.7 Let $f : A \to B$ be a homomorphism of associative algebras. The *modulation* of f is the A-B-bimodule $\mathcal{M}(f) = {}_fB$, which as a left B-module is just B, and has a right A-action via f:

$$a \triangleright b' \triangleleft b = f(a)\, b'\, b.$$ ◊

One may easily check that modulation defines a functor $\mathcal{M} : \mathrm{ALG}^{\mathrm{homo}} \to \mathrm{MOR}$. The name is from [TWZ07].

3 The Point of Pointings

Unfortunately, the modulation functor of Definition 2.7 loses too much information:

Lemma 3.1 *Let $f, g : A \to B$ be homomorphisms of associative algebras. Then $\mathcal{M}(f) \cong \mathcal{M}(g) \in \mathrm{MOR}$ if and only if there exists an invertible element $b \in B$ such that for each $a \in A$, $f(a) = b^{-1}\, g(a)\, b$.*

Proof Given such a $b \in B$, the map $\mathcal{M}(f) \to \mathcal{M}(g)$ given by multiplication by b on the left is an isomorphism of A-B-bimodules. The converse is an easy exercise for the reader. □

In particular, one can fully recover from its modulation the Heisenberg-picture encoding of a classical mechanical system in the sense of Example 1.6. However, if one starts with a Schrödinger-picture quantum mechanical system, applies Example 1.4, and then modulates the output, all information is lost: the functor QMSPACETIMES $\to$ MOR produced in this way does not depend (up to isomorphism) on the time evolution operators U_t.

To fix this requires breaking the multiplication by b in the proof of Lemma 3.1. A minimal way to do this is to remember one extra bit of data: which element of $\mathcal{M}(f)$ corresponds to $1_B \in B$. With this extra information, homomorphisms $f : A \to B$ can be recovered. Indeed, given the A-B-bimodule $\mathcal{M}(f) = {}_fB$ and the vector $1_B \in {}_fB$, the element $f(a) \in B$ for a given $a \in A$ is the unique solution to the equation $a \triangleright 1_B = 1_B \triangleleft f(a)$. This suggests that we revise Definition 2.7: the output of modulation is not just a bimodule, but a *pointed* bimodule.

Definition 3.2 The bicategory $\text{ALG} = \text{ALG}_1(\text{VECT}_{\mathbb{K}})$ of *algebras and pointed bimodules* has:

Objects: An object of ALG is an associative algebra over $\mathbb{K}$.

1-morphisms: A 1-morphism from A to B is an A-B-bimodule ${}_AM_B$ along with a *pointing* $1_M \in M$. The composition of $({}_AM_B, 1_M \in M)$ with $({}_BN_C, 1_N \in N)$ is the A-C-bimodule $M \otimes_B N$ pointed by the class of $1_M \otimes 1_N$.

2-morphisms: A 2-morphism $({}_AM_B, 1_M \in M) \to ({}_AN_B, 1_N \in N)$ is a bimodule homomorphism $f : M \to N$ such that $f(1_M) = 1_N$.

Like MOR, this bicategory is symmetric monoidal with the usual tensor product over $\mathbb{K}$.

There is an obvious forgetful functor ALG $\to$ MOR which forgets all pointings. The modulation functor factors through it: abusing notation, we let $\mathcal{M} : \text{ALG}^{\text{homo}} \to \text{ALG}$ denote the modulation functor that sends a homomorphism $f : A \to B$ to the pointed bimodule $({}_fB, 1_B)$. $\lozenge$

Consider modulating the Heisenberg-picture quantum mechanical system from Example 1.4 corresponding to a Schrödinger-picture system in which time evolution u_t is invertible. By Lemma 3.1, the modulation of $a \mapsto u_t a u_t^{-1}$ is isomorphic in MOR to the modulation of the identity $\mathcal{M}(\text{id}_A) = {}_AA_A$, where $A = \text{End}(V)$. They are not isomorphic in ALG when $u_t \neq 1$. The isomorphism in MOR consists of multiplication by u_t. It follows that:

Lemma 3.3 *Given $V \in$ VECT and $u : V \xrightarrow{\sim} V$ an isomorphism, let $A = \text{End}(V)$. The modulation $\mathcal{M}(a \mapsto uau^{-1})$ of conjugation by u is isomorphic in* ALG *to the trivial bimodule ${}_AA_A$ pointed not by 1_A but by the element $u \in A$.* $\square$

This suggests that even when time evolution is not an isomorphism, we can nevertheless encode it as a pointed bimodule: the identity bimodule, pointed by time evolution. Indeed:

Proposition 3.4 *There is a contravariant functor* End : VECT $\to$ ALG *taking a vector space V to its endomorphism algebra* End(V) *and taking a linear map $f \in$*

$\hom(V, W)$ *to the pointed bimodule* $(\hom(V, W), f)$, *where* $\mathrm{End}(V)$ *and* $\mathrm{End}(W)$ *act on* $\hom(V, W)$ *by pre- and post-composition.*

This functor is symmetric monoidal when restricted to finite-dimensional vector spaces or when VECT *(and, correspondingly,* ALG*) is replaced by an appropriate category of topological vector spaces. For example,* End *is symmetric monoidal if* VECT *is replaced by the category of Hilbert spaces and bounded operators,* $\mathrm{End}(V)$ *is topologized as a von Neumann algebra in the usual way, and* ALG *consists of von Neumann algebras and pointed Hilbert-space bimodules with the von Neumann tensor product.* □

Remark 3.5 The functor End is contravariant because, following Remark 1.5, $\hom(V, W)$ carries a *left* action by $\mathrm{End}(W)$ and a *right* action by $\mathrm{End}(V)$. ◊

We leave checking details to the reader. Lemma 3.3 assures that the functor End extends to all of VECT the composition "conjugate, then modulate" from $\mathrm{VECT}^{\mathrm{inv}}$ via $\mathrm{ALG}^{\mathrm{homo}}$ to ALG. With Proposition 3.4 in place, we define:

Definition 3.6 A *(non-extended, affine) Heisenberg-picture quantum field theory* is a symmetric monoidal functor SPACETIMES $\to$ ALG. ◊

Example 3.7 Any Schrödinger-picture quantum field theory $\mathcal{Z}$: SPACETIMES $\to$ VECT determines a Heisenberg-picture quantum field theory $\mathrm{End} \circ \mathcal{Z}$: SPACETIMES $\to$ ALG via Proposition 3.4, provided the values of the functor $\mathcal{Z}$ are in some subcategory (e.g. finite-dimensional vector spaces or Hilbert spaces) for which the functor End : VECT $\to$ ALG is symmetric monoidal. ◊

Example 3.8 Let SPANS denote the category of spans of spaces. Then a *classical field theory* as in Example 2.5 is a symmetric monoidal functor $\mathcal{F}$: SPACETIMES $\to$ SPANS. Any such classical field theory determines a Heisenberg-picture quantum field theory $\mathcal{O} \circ \mathcal{F}$: SPACETIMES $\to$ ALG, where $\mathcal{O}(X)$ is the algebra of functions on the space X. When N is an object of SPACETIMES (i.e. a $(d-1)$-dimensional manifold with appropriate geometry) we regard $\mathcal{O}(\mathcal{F}(N))$ as an associative algebra. When M is a morphism in SPACETIMES (i.e. a d-dimensional manifold with appropriate geometry) we regard $\mathcal{O}(\mathcal{F}(M))$ just as a pointed vector space (pointed by its unit element $1_{\mathcal{O}(\mathcal{F}(M))}$). ◊

Finally we can update Definition 1.7:

Definition 3.9 A *Heisenberg-picture quantum mechanical system* is a Heisenberg-picture quantum field theory for the spacetime category QMSPACETIMES of one-dimensional spacetimes with point defects from Definition 2.1 (or, rather, the symmetric monoidal envelope thereof).

Given a system as in Definition 1.7, we define the contravariant symmetric monoidal functor $\mathcal{H}$: QMSPACETIMES $\to$ ALG by declaring:

(1) $\mathcal{H}(\{\mathrm{pt}\}) = A$. $\mathcal{H}(\emptyset) = \mathbb{K}$ by symmetric monoidality.
(2) $\mathcal{H}(u_t) = \mathcal{M}(f_t)$ is the modulation of time evolution.

(3) Given a distinguished experimental manipulation $a \in A$, the corresponding point defect is sent to $({}_A A_A, a)$, the identity bimodule pointed by a. Given a distinguished left ideal ${}_A I \subseteq {}_A A$, the corresponding point defect is sent to the left module A/I, pointed by the class of 1_A. Given a distinguished right ideal $J_A \subseteq A_A$, the corresponding point defect is sent to the right module $J\backslash A$, pointed by the class of 1_A.

The main thing to note is the treatment of the distinguished ideals and elements in item (3): all determine pointed (bi)modules. ◊

Remark 3.10 I have discussed pointed modules as a way of assigning algebras of quantum observables to codimension-one spaces without losing too much information about the algebra. But pointed modules themselves have a direct interpretation in quantum field theory as the "purely algebraic part" of path integrals.

Indeed, suppose we are given an n-dimensional affine variety X of "field configurations over M" along with a polynomial "action" functional $s \in \mathcal{O}(X)$ and a volume form on X. The Feynman path integral invites us to consider the values of "oscillating" integrals $\langle f \rangle = \int_X f\, e^s\, \mathrm{dVol}$ for polynomial "observables" f. This integral is insufficiently defined: to define it requires choosing a contour in X along which integrals against the measure $e^s\, \mathrm{dVol}$ converge; up to homotopies of contours that leave all integrals unchanged, the space of contours is parameterized by the relative cohomology group $\mathrm{H}_n(X; \{\Re(s) \ll 0\})$.

The purely algebraic part of integration is the calculation of the class of f in the quotient $\mathcal{O}(X)/(\text{total derivatives})$, as the integrals of total derivative vanish on any contour. This vector space is naturally pointed by the class of 1, and is isomorphic (via multiplication by dVol) to the nth cohomology group of the twisted de Rham complex for s. In good situations, the class of f in $\mathcal{O}(X)/(\text{total derivatives})$ (or of $f\,\mathrm{dVol}$ in the twisted de Rham complex) can be calculated using homotopy algebra [JF15]. In a general Heisenberg-picture quantum field theory $\mathcal{Z}$, the pointed vector space $\mathcal{Z}(M)$ assigned to a closed top-dimensional manifold M can be interpreted as "$\mathcal{O}(X)/(\text{total derivatives})$" for an ill-defined path integral. ◊

4 Non-affine Field Theory

Do you still have that *Introduction to Quantum Mechanics* textbook? In one of its more philosophical sections, it is likely to discuss the following basic premise of experimental science: the universe consists only of things that are in principle measurable; if no experiment can distinguish two states, then those states are equal. The tautologous version of this philosophy is the mathematicians' Yoneda lemma. But there is a non-tautologous version, which asserts that by "measurement" and "experiment" we should mean "element of the algebra of observables." For classical phase spaces, for example, the non-tautologous version asserts that points can be separated by real-valued functions—that all spaces are *affine* in the sense of algebraic geometry. This assertion is true for many types of spaces (smooth manifolds; locally compact Hausdorff spaces) but by no means all spaces: there are many roles in physics for non-affine schemes and stacks.

A piece of philosophy that probably is not in your textbook is that "most of 0-algebraic geometry is 1-affine." Said another way, although schemes and stacks usually are not determined by their algebras (0-categories) of global functions, they are often determined by their symmetric monoidal (1-)categories of quasicoherent sheaves of modules—such a category should be understood as "the algebra of VECT-valued global functions." (In general, one could call a space *k-affine* if it is determined by its symmetric monoidal k-category of global maps to the k-categorical analogue of VECT.) Most algebrogeometric objects one comes across are known to be 1-affine [Lur09, BZFN10, Bra11, CJF13, BC14, HR19, BCJF15, HR19] also records a few objects which are known to be non-1-affine.

It is not my intention to develop here the theory of 1-affine algebraic geometry, but it is worth making a few remarks. The Gabriel–Rosenberg theorem [Gab62, Ros98b] reconstructs a scheme up to isomorphism from its category of quasicoherent modules with no extra structure: only the category itself is used. Here is a baby case of this result: let A be a commutative ring and MOD_A the category of right A-modules. Then A can be reconstructed as the algebra of natural endomorphisms of the identity functor $\mathrm{id} : \mathrm{MOD}_A \to \mathrm{MOD}_A$.

This has suggested to many workers in "noncommutative algebraic geometry" that "abelian category" is a good definition of "noncommutative scheme" (e.g. [Ros98a]). I would argue, however, that this misunderstands the variability of categories. Consider, for example, the stacks $\{\mathrm{pt}\} \sqcup \{\mathrm{pt}\}$ and $\{\mathrm{pt}\}/(\mathbb{Z}/2)$, the latter being the classifying stack of the group $\mathbb{Z}/2$. Provided we work over a ring in which 2 is invertible, these have equivalent categories of modules. But they are honestly different as stacks. To fully recover $\{\mathrm{pt}\} \sqcup \{\mathrm{pt}\}$ and $\{\mathrm{pt}\}/(\mathbb{Z}/2)$ from their categories of modules it suffices to remember additionally the symmetric monoidal structures on those categories. Moreover, the natural homomorphisms of abelian categories are the exact functors, but these do not have direct geometric meaning as morphisms of schemes. Thus the papers [Gab62, Ros98b] do not reconstruct a scheme *functorially* from its category of modules. For comparison, the papers [Lur09, Bra11, CJF13, BC14, HR19, BCJF15] do provide functorial reconstruction of various algebrogeometric objects by remembering their module categories' symmetric monoidal structures and demanding that functors be symmetric monoidal.

By a similar token, an associative algebra A is not determined by the equivalence type of the category MOD_A, which encodes only the class of A in the Morita bicategory MOR. But A can be recovered if MOD_A is equipped with a *pointing*—a distinguished object—in MOD_A, namely the rank-one free module A_A. I therefore propose:

Definition 4.1 A *noncommutative 1-affine stack X over* $\mathbb{K}$ is a $\mathbb{K}$-linear cocomplete category $\mathrm{QCOH}(X)$ equipped with a distinguished object $\mathbb{1}_X = \mathcal{O}_X \in \mathrm{QCOH}(X)$.

Let $\mathrm{COCOMP}_{\mathbb{K}}$ denote the bicategory whose objects are $\mathbb{K}$-linear cocomplete categories, whose 1-morphisms are cocontinuous $\mathbb{K}$-linear functors, and whose 2-

morphisms are natural transformations. It is symmetric monoidal for a version of Deligne's tensor product $\boxtimes$; the monoidal unit is VECT [Kel05, Section 6.5]. Noncommutative 1-affine stacks are the objects of a bicategory which we will call $\mathrm{ALG}_0^{\mathrm{lax}}(\mathrm{COCOMP}_{\mathbb{K}})$. It has:

Objects: An object of $\mathrm{ALG}_0^{\mathrm{lax}}(\mathrm{COCOMP}_{\mathbb{K}})$ is a noncommutative stack, i.e. a pair $(\mathcal{C}, \mathbb{1}_{\mathcal{C}} \in \mathcal{C})$ where $\mathcal{C} \in \mathrm{COCOMP}_{\mathbb{K}}$ is a cocomplete $\mathbb{K}$-linear category and $\mathbb{1}_{\mathcal{C}}$ is a pointing thereof.

1-morphisms: A 1-morphism $(\mathcal{A}, \mathbb{1}_{\mathcal{A}}) \to (\mathcal{B}, \mathbb{1}_{\mathcal{B}})$ is a pair $(F, 1_F)$ where $F : \mathcal{A} \to \mathcal{B}$ is a $\mathbb{K}$-linear cocontinuous functor and $1_F : \mathbb{1}_{\mathcal{B}} \to F(\mathbb{1}_{\mathcal{A}})$ is a homomorphism in $\mathcal{B}$.

2-morphisms: A 2-morphism $(F, 1_F) \to (G, 1_G)$ is a natural transformation $\eta : F \to G$ such that $\eta_{\mathbb{1}_{\mathcal{A}}} \circ 1_F = 1_G : \mathbb{1}_{\mathcal{B}} \to G(\mathbb{1}_{\mathcal{A}})$.

The bicategory $\mathrm{ALG}_0^{\mathrm{lax}}(\mathrm{COCOMP}_{\mathbb{K}})$ is symmetric monoidal for $\boxtimes$. ◊

Remark 4.2 Recall that a category is *cocomplete* if it is closed under colimits, and a functor is *cocontinuous* if it preserves colimits. For set theoretic reasons it is often preferable to work just with locally presentable categories rather than all cocomplete categories. Actually, [Kel05, Section 6.5] works with small categories closed under some small set of colimit shapes, and so to extract the tensor product $\boxtimes$ on $\mathrm{COCOMP}_{\mathbb{K}}$ requires the type of judicious Grothendieck-universe jumping standard in category theory. ◊

Remark 4.3 The 1-morphisms $(F, 1_F) : (\mathcal{A}, \mathbb{1}_{\mathcal{A}}) \to (\mathcal{B}, \mathbb{1}_{\mathcal{B}})$ in $\mathrm{ALG}_0^{\mathrm{lax}}(\mathrm{COCOMP}_{\mathbb{K}})$ are *lax homomorphisms of pointed categories*, a.k.a. *lax pointed functors*, as opposed to the *strong* pointed functors of Remark 2.2. The use of "lax" here is consistent with the general notion of "lax homomorphism" in [JFS17]. ◊

Example 4.4 The Eilenberg–Watts theorem [Eil60, Wat60] asserts that the functor $\mathrm{MOR} \to \mathrm{COCOMP}_{\mathbb{K}}$ sending an algebra A to the category MOD_A of right A-modules and a bimodule ${}_AM_B$ to the functor $(-) \otimes_A M : \mathrm{MOD}_A \to \mathrm{MOD}_B$ is a fully faithful inclusion of bicategories in the sense that it induces an equivalence of categories

$$\hom_{\mathrm{COCOMP}_{\mathbb{K}}}(\mathrm{MOD}_A, \mathrm{MOD}_B) \cong \hom_{\mathrm{MOR}}(A, B).$$

Given an algebra A, consider the pointed category $(\mathrm{MOD}_A, A_A) \in \mathrm{ALG}_0^{\mathrm{lax}}(\mathrm{COCOMP}_{\mathbb{K}})$. What is the category of homomorphisms $(F, f) : (\mathrm{MOD}_A, A_A) \to (\mathrm{MOD}_B, B_B)$? By the Eilenberg–Watts theorem, the data of a cocontinuous linear functor $F : \mathrm{MOD}_A \to \mathrm{MOD}_B$ consists (up to canonical isomorphism) of an A-B-bimodule ${}_AF_B$. What about the homomorphism $f : B_B \to F(A) \cong A_A \otimes_A {}_AF_B$? It is nothing but an element of the underlying vector space of F.

Thus the Eilenberg–Watts inclusion $\mathrm{MOR} \hookrightarrow \mathrm{COCOMP}_{\mathbb{K}}$ lifts to an inclusion

$$\mathrm{EW} : \mathrm{ALG}_1(\mathrm{VECT}_{\mathbb{K}}) \hookrightarrow \mathrm{ALG}_0^{\mathrm{lax}}(\mathrm{COCOMP}_{\mathbb{K}})$$

sending $A \mapsto (\text{MOD}_A, A_A)$. It is reasonable to declare therefore that a noncommutative stack is *(0-)affine* if it is in the essential image of EW.

It is worth noting that EW is symmetric monoidal. ◊

Remark 4.5 The composition $\text{EW} \circ \mathcal{M} : \text{ALG}^{\text{homo}} \to \text{ALG}_0^{\text{lax}}(\text{COCOMP}_{\mathbb{K}})$ of the Eilenberg–Watts and modulation functors picks out precisely those 1-morphisms $(F, 1_F) : (\mathcal{A}, \mathbb{1}_{\mathcal{A}}) \to (\mathcal{B}, \mathbb{1}_{\mathcal{B}})$ which are *strong* pointed functors in the sense that $1_F : \mathbb{1}_{\mathcal{B}} \to F(\mathbb{1}_{\mathcal{A}})$ is an isomorphism. ◊

With Definition 4.1 in place, we may study quantum field theories valued in "noncommutative stacks":

Definition 4.6 A *(non-extended) 1-affine Heisenberg-picture quantum field theory* over $\mathbb{K}$ is a contravariant symmetric monoidal functor $\text{SPACETIMES} \to \text{ALG}_0^{\text{lax}}(\text{COCOMP}_{\mathbb{K}})$. ◊

Any 0-affine quantum field theory $\text{SPACETIMES} \to \text{ALG}_1(\text{VECT}_{\mathbb{K}})$ provides an example of a 1-affine quantum field theory, by composing with the Eilenberg–Watts inclusion from Example 4.4.

Remark 4.7 One-affine Heisenberg-picture quantum field theory is one possible formalization of *twisted* [ST11] or *relative* [FT14] quantum field theory: Schrödinger-picture quantum field theory valued in a categorified Schrödinger-picture quantum field theory. Indeed, given a Heisenberg-picture quantum field theory $Z : \text{SPACETIMES} \to \text{ALG}_0^{\text{lax}}(\text{COCOMP}_{\mathbb{K}})$, one can forget to a "categorified" quantum field theory $Z_0 : \text{SPACETIMES} \to \text{COCOMP}_{\mathbb{K}}$. The extra data of the field theory Z is a *symmetric monoidal oplax natural transformation* from the trivial field theory to Z_0 in the sense of [JFS17]. ◊

Example 4.8 Given a vector space $V \in \text{VECT}_{\mathbb{K}}$, there are two natural ways to produce an object of $\text{ALG}_0^{\text{lax}}(\text{COCOMP}_{\mathbb{K}})$:

(1) Following Proposition 3.4 and Example 4.4, apply $\text{End} : \text{VECT} \to \text{ALG}_1(\text{VECT}_{\mathbb{K}})$ to produce the algebra $A = \text{End}(V)$ and then apply $\text{EW} : \text{ALG}_1(\text{VECT}_{\mathbb{K}}) \hookrightarrow \text{ALG}_0^{\text{lax}}(\text{COCOMP}_{\mathbb{K}})$ to produce the pointed category $\big(\text{MOD}_{\text{End}(V)}, \text{End}(V)_{\text{End}(V)}\big)$.
(2) Recognize that $(\text{VECT}_{\mathbb{K}}, V)$ is already a pointed category, pointed by V rather than $\mathbb{K}$.

Both constructions are contravariantly functorial. Functoriality of the first we have already addressed. For the second, given a linear map $f : V \to W$, the pair $(\text{id}_{\text{VECT}}, f)$ is a pointed functor $(\text{VECT}_{\mathbb{K}}, W) \to (\text{VECT}_{\mathbb{K}}, V)$. (The Eilenberg–Watts theorem identifies cocontinuous functors $\text{VECT} \to \text{VECT}$ with vector spaces. So a general pointed functor $(\text{VECT}_{\mathbb{K}}, V) \to (\text{VECT}_{\mathbb{K}}, W)$ consists of a pair (X, f) where $X \in \text{VECT}$ is a vector space and $f \in \hom(W, V \otimes X)$ is a linear map.)

But these constructions are not honestly different for the most important V. Suppose that $V \neq 0$ is finite-dimensional. Then $\text{End}(V)$ is Morita-equivalent to $\mathbb{K}$: there is an equivalence of categories $\text{MOD}_{\text{End}(V)} \simeq \text{VECT}_{\mathbb{K}}$. The choice of V provides a canonical such equivalence $\otimes_{\text{End}(V)} V$. Under this equivalence, the object $V^*_{\mathrm{End}(V)} \in \mathrm{MOD}_{\mathrm{End}(V)}$ is identified with $\mathbb{K} \in \mathrm{VECT}_{\mathbb{K}}$ and the rank-one free module $\mathrm{End}(V)_{\mathrm{End}(V)} \in \mathrm{MOD}_{\mathrm{End}(V)}$ is identified with $V \in \mathrm{VECT}_{\mathbb{K}}$. So, at least for non-zero finite-dimensional V, the pointed categories $(\mathrm{MOD}_{\mathrm{End}(V)}, \mathrm{End}(V)_{\mathrm{End}(V)})$ and $(\mathrm{VECT}_{\mathbb{K}}, V)$ are equivalent in $\mathrm{ALG}_0^{\mathrm{lax}}(\mathrm{COCOMP}_{\mathbb{K}})$. Similar remarks apply when V is a separable Hilbert space and one works in an analytic context in which the algebra of bounded operators on V is Morita-equivalent to $\mathbb{K}$.

The construction $V \mapsto (\mathrm{VECT}_{\mathbb{K}}, V)$ is fully symmetric monoidal, whereas, as mentioned in Proposition 3.4, $V \mapsto \mathrm{End}(V)$ is only symmetric monoidal when $\dim V < \infty$. So in some sense construction (2) above is the more natural one: it turns *any* Schrödinger-picture quantum field theory, independent of dimension, into a Heisenberg-picture one. But it also explains construction (1): most Schrödinger-picture quantum field theories (including all topological ones) are valued in vector spaces V for which $(\mathrm{VECT}_{\mathbb{K}}, V)$ is (0-)affine in the sense of Example 4.4. ◊

5 Extended Affine Field Theory

Atiyah [Ati88] and Segal [Seg04] introduced their functorial axioms for quantum field theory in an attempt to capture the *locality* of physics. In the decades since, it has become clear that locality is stronger than a functor that takes values on $(d-1)$-dimensional "spaces" and d-dimensional "spacetimes": a quantum field theory should also assign algebraic data to k-dimensional manifolds for lower k, and to spacetimes with such corners, since manifolds can be cut and glued along such manifolds [Law93, Fre94, BD95].

The modern consensus (see e.g. [Lur09, Sch14]) is that to fully capture locality, the source $\mathrm{SPACETIMES}_d$ of a d-dimensional quantum field theory should not be just a (symmetric monoidal) category, but a (symmetric monoidal) (∞, d)-category: k-dimensional morphisms for $k \leq d$ should be k-dimensional manifolds with corners, equipped with the appropriate geometric structure; "higher" morphisms for $k > d$ should be isomorphisms of spacetimes and isotopies (of isotopies of ...) thereof. The target of a quantum field theory must also be an (∞, d)-category.

Definition 5.1 A *delooping* of a symmetric monoidal (∞, n)-category $\mathcal{C}$ is a choice of $(\infty, n+1)$-category $\mathcal{D}$ along with an equivalence $\mathcal{C} \cong \hom(\mathbb{1}_{\mathcal{D}}, \mathbb{1}_{\mathcal{D}})$, where $\mathbb{1}_{\mathcal{D}}$ is the unit object in $\mathcal{D}$.

A *d-dimensional fully extended Schrödinger-picture quantum field theory over* $\mathbb{K}$ is a symmetric monoidal functor $\mathrm{SPACETIMES}_d^{\mathcal{G}} \to d\mathrm{VECT}$, where:

- $\mathrm{SPACETIMES}_d^{\mathcal{G}}$ is some symmetric monoidal (∞, d)-category whose d-morphisms are d-dimensional cobordisms with corners equipped with geometric structure of type $\mathcal{G}$;
- $d\mathrm{VECT}_{\mathbb{K}}$ is some $(d-1)$-fold delooping of $\mathrm{VECT}_{\mathbb{K}}$.

For example, the bicategories $\mathrm{MOR}_{\mathbb{K}}$ and $\mathrm{COCOMP}_{\mathbb{K}}$ are reasonable choices for $2\mathrm{VECT}_{\mathbb{K}}$. ◊

Remark 5.2 Definition 5.1 is incomplete because, depending on the geometry $\mathcal{G}$, constructing an (∞, d)-category deserving the name $\mathrm{SPACETIMES}_d^{\mathcal{G}}$ may be quite difficult. When $\mathcal{G}$ is sufficiently topological, $\mathrm{SPACETIMES}_d^{\mathcal{G}}$ can be built from the "topological" bordism category $\mathrm{BORD}_d^{\mathrm{smooth}}$ following [Lur09, Section 3.2] or [SP09, Section 3.3] (elaborated upon in [JF17]). A detailed outline of the construction of $\mathrm{BORD}_d^{\mathrm{smooth}}$ itself is given in [Lur09] and clarified in [CS19].

Although the modern consensus is that $\mathrm{SPACETIMES}_d^{\mathcal{G}}$ should be a symmetric monoidal higher category, there is some evidence that the usual notions of "higher category" (e.g. those in [BSP11]) may not provide the correct framework in which to organize cobordisms, and that other related versions are needed [MW11]. $\Diamond$

For a similar generalization of Heisenberg-picture quantum field theories, we should look for interesting deloopings of $\mathrm{ALG}_1(\mathrm{VECT}_{\mathbb{K}})$ or $\mathrm{ALG}_0^{\mathrm{lax}}(\mathrm{COCOMP}_{\mathbb{K}})$. Various options are available, but we will use one which has a natural interpretation in terms of the pictures drawn in quantum field theory of insertion of local observables. To wit, consider choosing a vector space V and placing on $\mathbb{R}$ (considered just as an oriented manifold) some "beads" labeled by vectors in V. These "beads" can slide back and forth but cannot pass through each other. Further assume that the beads can "fuse" in a nuclear reaction. Imposing linearity in the labels, this "fusion" is an operation $V \otimes V \to V$. At microscopic scales, fusion is not really a discrete process: instead, when beads become very close to each other, they can become bonded and behave like a single particle. But suppose that all physics of beads and fusion is "topological" in the sense that it is independent of distances on $\mathbb{R}$. Then the fusion operation $V \otimes V \to V$ is associative. The "invisible bead"— no bead at all—is a unit for this fusion. So such a system is the same as an associative algebra structure on V. The beads are nothing but "local observables" drawn from the "algebra of observables" V. The theory of local observables for a general quantum field theory has been formulated in terms of *factorization algebras* in [CG16], and restricts to "beads on $\mathbb{R}$" in the case of one-dimensional topological theories.

More generally, consider dividing $\mathbb{R}$ into intervals separated by "point defects." One can imagine a situation in which the different regions follow different physical laws: one might be filled with water, for example, and another air. Each defect might also have its own physics: perhaps there are waves that live only where water and air meet. Each region or defect has its own vector space of observables. As observables move around, they can fuse, and we will require that the universe be topological. An observable in a region can move onto a defect, but not otherwise. There should be "invisible observables" that can be inserted at any point. These rules together comprise (1) an associative algebra assigned to each region, and (2) a pointed bimodule assigned to each defect. These are nothing but the objects and morphisms of $\mathrm{ALG}_1(\mathrm{VECT}_{\mathbb{K}})$.

The natural generalization is to consider systems of regions and defects on higher-dimensional spaces—vector spaces of observables assigned to each stratum, with operations describing the ways points can collide. Let us impose a topological condition. The Eckmann–Hilton argument [EH62] then says that the algebra assigned to any two-dimensional region is commutative:

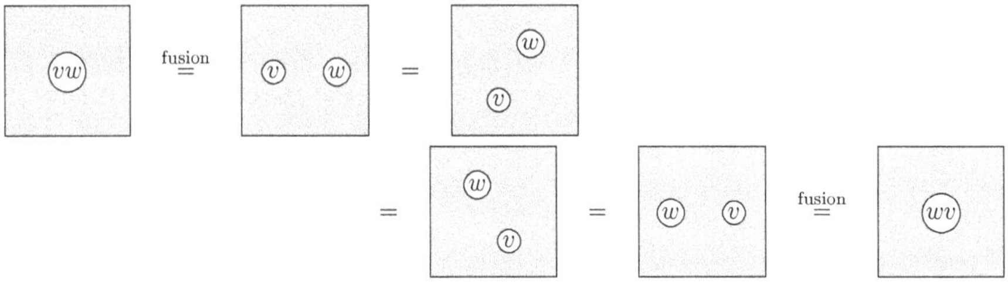

What about a one-dimensional defect separating regions whose commutative algebras of functions are A and B? Its local observables form an associative $(A \otimes B)$-algebra.

Commutative algebras are mildly disappointing because they are not particularly "quantum." Fortunately, there is more to the Eckmann–Hilton argument than the answer "the algebra is commutative." Indeed, the Eckmann–Hilton argument actually gives *two* proofs of the commutativity of multiplication, the above one and:

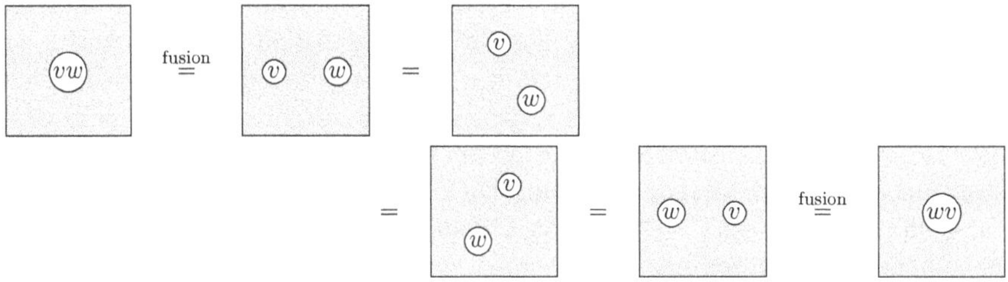

Linear maps are either equal or unequal, but in a homotopical or higher categorical world, *how* two operations are "the same" is a type of data. One can therefore define a nontrivial notion of *n-algebra* in a symmetric monoidal category $\mathcal{S}$ to be an object of $\mathcal{S}$ with operations parameterized by labeled configurations of points in $\mathbb{R}^n$ such that homotopies between configurations correspond to homotopies between operations. (A precise definition, along with much discussion, is in [Lur14, Chapter 5], where n-algebras are called $\mathbb{E}_n$*-algebras*.) For example, a 1-algebra is a homotopy-coherent version of an associative algebra. A 2-algebra in the $(\infty, 1)$-category of (usual) categories is a braided monoidal category. A 0-algebra is a *pointed object*, i.e. an object $X \in \mathcal{S}$ along with a map $\mathbb{1} \to X$, where $\mathbb{1} \in \mathcal{S}$ is the monoidal unit.

The bicategory $\mathrm{ALG}_1(\mathrm{VECT}_{\mathbb{K}})$ of associative algebras and pointed bimodules can then be generalized to higher algebra by working, as discussed above, with systems of algebras of observables that are topological but vary at prescribed lower-dimensional strata in $\mathbb{R}^n$. The following summarizes the main results of [Sch14]:

Theorem 5.3 ([Sch14]) *Let $\mathcal{S}$ be a symmetric monoidal $(\infty, 1)$-category admitting filtered colimits and such that the symmetric monoidal structure distributes over filtered colimits.*

(1) *For each $n \in \mathbb{N}$, there is a symmetric monoidal (∞, n)-category* $\text{ALG}_n(\mathcal{S})$ *with:*

Objects: *n-algebras in $\mathcal{S}$.*
1-morphisms: *$(n-1)$-algebra bimodules between n-algebras.*
2-morphisms: *$(n-2)$-algebra bimodules between $(n-1)$-algebras, for which the various actions of the ambient n-algebras are compatible.*
...: ...
***n*-morphisms**: *pointed bimodules between 1-algebras, for which the various actions of lower-dimensional morphisms are compatible.*
$(n+1)$**-morphisms**: *equivalences of n-morphisms.*
$(n+2)$**-morphisms**: *equivalences of equivalences.*
... ...

(1) *Two n-algebras in $\mathcal{S}$ are equivalent as objects in* $\text{ALG}_n(\mathcal{S})$ *if and only if they are equivalent as n-algebras (i.e. via homomorphisms, rather than via bimodules).*
(2) *Let* $\text{BORD}_d^{\text{fr}}$ *denote the "spacetime" (∞, d)-category of framed smooth cobordisms. For each n-algebra X in $\mathcal{S}$, there is a unique (up to contractible choices) symmetric monoidal functor* $\text{BORD}_d^{\text{fr}} \to \text{ALG}_n(\mathcal{S})$ *assigning X to the standard-framed point* $\{\text{pt}\} \in \text{BORD}_d^{\text{fr}}$. *This functor is called* factorization homology *or* topological chiral homology *with coefficients in X, written $M \mapsto \int_M X$.* □

Remark 5.4 Part (3) of Theorem 5.3 is proved by explicitly constructing the functor $\int_\square X$, rather than by appealing to Lurie's celebrated classification theorem from [Lur09]. Lurie called [Lur09] an "outline," and while it seems the consensus among experts is that it is in every important way correct and complete, it also seems best to prove results without appealing to an "outline." ◊

As I said earlier, it is mildly disappointing that 2-algebras in VECT are automatically commutative. But there is a close cousin to VECT that admits honestly noncommutative n-algebras for all n. The category DGVECT of chain complexes of vector spaces ("derived" vector spaces) has a model category structure making it into an $(\infty, 1)$-category in an interesting way: objects are chain complexes and 1-morphisms are chain maps as in the usual category, but 2-morphisms are chain homotopies between 1-morphisms, 3-morphisms are homotopies between homotopies, and so on. It satisfies the conditions of Theorem 5.3. We can therefore define:

Definition 5.5 A *d-dimensional fully extended derived affine Heisenberg-picture quantum field theory* based on the spacetime category $\text{SPACETIMES}_d^{\mathcal{G}}$ is a contravariant symmetric monoidal functor of (∞, d)-categories $\text{SPACETIMES}_d^{\mathcal{G}} \to \text{ALG}_d(\text{DGVECT})$. ◊

Then part (3) of Theorem 5.3 asserts that each d-algebra (among chain complexes) defines a d-dimensional fully extended framed topological derived affine Heisenberg-picture quantum field theory.

6 Extended Non-Affine Field Theory

Of course, category theory accommodates affine *non*-derived quantum field theories—functors to $\mathrm{ALG}_d(\mathrm{VECT})$—but these will not exhibit truly "quantum" behavior except in codimensions 0 and 1. If the derived world is to be avoided, another option for capturing honestly "quantum" examples is to give up on affineness. (Indeed, as we will see in Sect. 9, important examples are not affine.) Let $\mathcal{S}$ be a symmetric monoidal (∞, k)-category, say the bicategory $\mathrm{COCOMP}_{\mathbb{K}}$. The notion of "$d$-algebra in $\mathcal{S}$" never uses non-invertible 2- or higher morphisms, and so depends only on the maximal $(\infty, 1)$-category inside of $\mathcal{S}$. Thus one can throw away that data and define $\mathrm{ALG}_d(\mathcal{S})$ as in Theorem 5.3. But the higher morphisms in $\mathcal{S}$ can be used to enriched $\mathrm{ALG}_d(\mathcal{S})$:

Theorem 6.1 ([JFS17]) *Let $\mathcal{S}$ be a symmetric monoidal $(\infty, k+1)$-category satisfying conditions analogous to those of Theorem 5.3 (for the details, see [JFS17]). Then the (∞, n)-category $\mathrm{ALG}_n(\mathcal{S})$ from Theorem 5.3 is the n-dimensional truncation of an $(\infty, n+k+1)$-category $\mathrm{ALG}_n^{\mathrm{lax}}(\mathcal{S})$. This extended version has the same 0- through n-morphisms as its non-extended cousin. In particular, the n-morphisms are pointed objects of $\mathcal{S}$ which are acted upon by their sources and targets. The $(n+1)$-morphisms are lax homomorphisms of pointed bimodules; the $(n+2)$-morphisms are lax homomorphisms thereof; etc.* □

Example 6.2 Let $\mathcal{S}$ be a symmetric monoidal $(\infty, k+1)$-category with unit object $\mathbb{1} \in \mathcal{S}$. Generalizing Definition 4.1, the $(\infty, k+1)$-category $\mathrm{ALG}_0^{\mathrm{lax}}(\mathcal{S})$ has:

Objects: An object of $\mathrm{ALG}_0^{\mathrm{lax}}(\mathcal{S})$ consists of a pair $(X, 1_X)$ where object $X \in \mathcal{S}$ and $1_X : \mathbb{1} \to X$ is a *pointing* of X.

$$\mathbb{1} \xrightarrow{\quad 1_X \quad} X$$

1-morphism: A 1-morphism $(X, 1_X) \to (Y, 1_Y)$ is a *lax homomorphism* of pointed objects. It consists of a pair $(F, 1_F)$ where $F : X \to Y$ is a 1-morphism in $\mathcal{S}$ and $1_F : 1_Y \to F \circ 1_X$ is a 2-morphism.

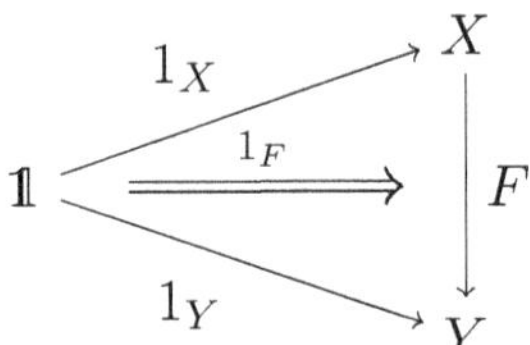

2-morphism: A 2-morphism $(F, 1_F) \to (G, 1_G)$ is pair $(T, 1_T)$ where $T : F \to G$ is a 2-morphism in $\mathcal{S}$ and $1_T : 1_G \to T \circ 1_F$ is a 3-morphism.

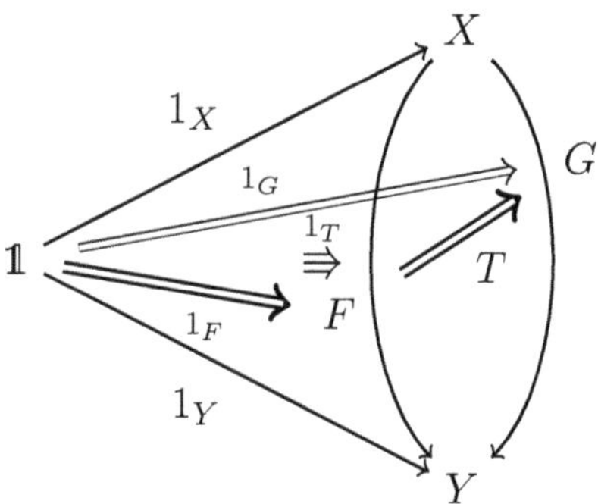

...: ... ◊

Example 6.3 Up to a contractible space of choices, one can identify objects of $\text{ALG}_1^{\text{lax}}(\text{COCOMP}_{\mathbb{K}})$ with *monoidal* $\mathbb{K}$-linear cocomplete categories $\mathcal{C}$, by which I mean that the monoidal functor is $\mathbb{K}$-linear and cocontinuous in each variable, so that it extends to a functor $\otimes : \mathcal{C} \boxtimes_{\mathbb{K}} \mathcal{C} \to \mathcal{C}$. Recall that to be monoidal, in addition to the monoidal functor, we should have distinguished "associator" and "unitor" natural isomorphisms satisfying standard "pentagon" and "triangle" axioms. I will generally suppress these auxiliary data.

Let $\mathcal{C}$ be a monoidal $\mathbb{K}$-linear cocomplete category and $\mathcal{X}$ a $\mathbb{K}$-linear cocomplete category. A *left action* of $\mathcal{C}$ on $\mathcal{X}$ consists of a 1-morphism (i.e. $\mathbb{K}$-linear cocontinuous functor) $\triangleright : \mathcal{C} \boxtimes_{\mathbb{K}} \mathcal{X} \to \mathcal{X}$ and an associator $(C_1 \otimes C_2) \triangleright X \xrightarrow{\sim} C_1 \triangleright (C_2 \triangleright X)$ and a unitor $\mathbb{1}_{\mathcal{C}} \triangleright X \xrightarrow{\sim} X$ (natural in $X \in \mathcal{X}$ and $C_1, C_2 \in \mathcal{C}$, of course) satisfying the appropriate pentagon and triangle equations. A *right action* is similar: there should be a functor $\triangleleft : \mathcal{X} \boxtimes_{\mathbb{K}} \mathcal{C} \to \mathcal{X}$ and an associator and a unitor. Suppose that $(\mathcal{B}, \otimes_{\mathcal{B}})$ and $(\mathcal{C}, \otimes_{\mathcal{C}})$ are monoidal $\mathbb{K}$-linear cocomplete category and $\mathcal{X}$ is a $\mathbb{K}$-linear cocomplete category equipped with a left action $\triangleright : \mathcal{B} \boxtimes_{\mathbb{K}} \mathcal{X} \to \mathcal{X}$ and a right action $\mathcal{X} \boxtimes_{\mathbb{K}} \mathcal{C} \to \mathcal{X}$. A *compatibility* between these actions consists of an associator $\alpha_{B,X,C} : (B \triangleright X) \triangleleft C \xrightarrow{\sim} B \triangleright (X \triangleleft C)$ that satisfies the appropriate pentagon and triangle equations. Such an $\mathcal{X}$ is called a *$\mathcal{B}$-$\mathcal{C}$-bimodule*.

A 1-morphism in $\text{ALG}_1^{\text{lax}}(\text{COCOMP}_{\mathbb{K}})$ between monoidal $\mathbb{K}$-linear cocomplete categories $\mathcal{B}$ and $\mathcal{C}$ is a *pointed* $\mathcal{B}$-$\mathcal{C}$-bimodule, i.e. a $\mathcal{B}$-$\mathcal{C}$-bimodule $\mathcal{X}$ with a distinguished object $\mathbb{1}_{\mathcal{X}} \in \mathcal{X}$. Let $(\mathcal{X}, \mathbb{1}_{\mathcal{X}})$ be a pointed $\mathcal{B}$-$\mathcal{C}$-bimodule and $(\mathcal{Y}, \mathbb{1}_{\mathcal{Y}})$ a pointed $\mathcal{C}$-$\mathcal{D}$-bimodule. The *balanced tensor product* $\mathcal{X} \boxtimes_{\mathcal{C}} \mathcal{Y}$ is the $\mathbb{K}$-linear cocomplete category which is universal for the following:

- There is a 1-morphism $P : \mathcal{X} \boxtimes_{\mathbb{K}} \mathcal{Y} \to \mathcal{X} \boxtimes_{\mathcal{C}} \mathcal{Y}$.
- There is an "associator"

$$\alpha_{X,C,Y} : P((X \triangleleft C) \boxtimes Y) \xrightarrow{\sim} P(X \boxtimes (C \triangleright Y))$$

depending naturally on $X \in \mathcal{X}$, $C \in \mathcal{C}$, and $Y \in \mathcal{Y}$.

- The associator α satisfies triangle and pentagon equations. Suppressing the associators and unitors from $\mathcal{X}, \mathcal{C}, \mathcal{Y}$, these say that $\alpha_{X,\mathbb{1},Y} = \mathrm{id}$ and $\alpha_{X,A\otimes B,Y} = \alpha_{X,A,B\triangleright Y} \circ \alpha_{X\triangleleft A,B,Y}$.

That the balanced tensor product exists follows from [Kel05, Theorem 6.23], which implies (modulo Remark 4.2) that $\mathrm{COCOMP}_{\mathbb{K}}$ contains all colimits. Indeed, $\mathcal{X} \boxtimes_{\mathcal{C}} \mathcal{Y}$ is the colimit of the following diagram:

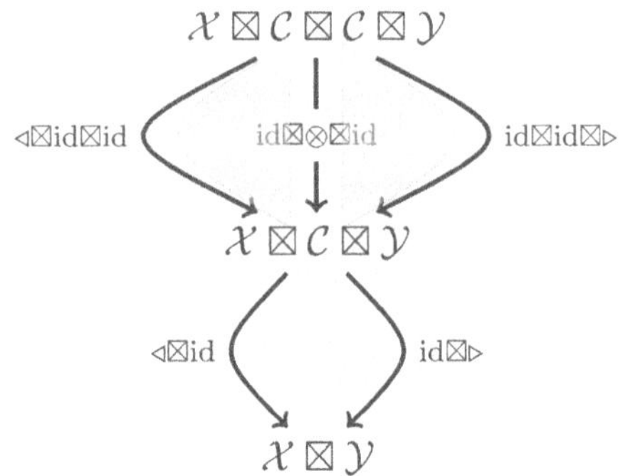

The three 2-cells are:

= associator for $\mathcal{X}$, = associator for $\mathcal{Y}$,

= the commutativity $(\mathrm{id} \boxtimes \triangleright) \circ (\triangleleft \boxtimes \mathrm{id} \boxtimes \mathrm{id}) = (\triangleleft \boxtimes \mathrm{id}) \circ (\mathrm{id} \boxtimes \mathrm{id} \boxtimes \triangleright)$.

The category $\mathcal{X} \boxtimes_{\mathcal{C}} \mathcal{Y}$ is naturally a $\mathcal{B}$-$\mathcal{D}$-bimodule, and the balanced tensor product is coherently associative and unital. If $\mathcal{X}$ is pointed by $\mathbb{1}_{\mathcal{X}}$ and $\mathcal{Y}$ by $\mathbb{1}_{\mathcal{Y}}$, then $\mathcal{X} \boxtimes_{\mathcal{C}} \mathcal{Y}$ is pointed by $P(\mathbb{1}_{\mathcal{X}} \boxtimes \mathbb{1}_{\mathcal{Y}})$. This categorifies the 0- and 1-morphisms of the bicategory $\mathrm{ALG}_1(\mathrm{VECT}_{\mathbb{K}})$ from Definition 3.2.

Let $(\mathcal{X}, \mathbb{1}_{\mathcal{X}})$ and $(\mathcal{Y}, \mathbb{1}_{\mathcal{Y}})$ now be two $\mathcal{B}$-$\mathcal{C}$-bimodules. A 2-morphism $(\mathcal{X}, \mathbb{1}_{\mathcal{X}}) \to (\mathcal{Y}, \mathbb{1}_{\mathcal{Y}})$ in $\mathrm{ALG}_1^{\mathrm{lax}}(\mathrm{COCOMP}_{\mathbb{K}})$ is a *lax pointed bimodule homomorphism*, which is to say a $\mathbb{K}$-linear cocontinuous functor $F : \mathcal{X} \to \mathcal{Y}$, natural transformations $B \triangleright F(X) \to F(B \triangleright X)$ and $F(X) \triangleleft C \to F(X \triangleleft C)$ intertwining the various associators and unitors, and a homomorphism $1_F : \mathbb{1}_{\mathcal{Y}} \to F(\mathbb{1}_{\mathcal{X}})$ in $\mathcal{Y}$.

A 3-morphism $(F, 1_F) \to (G, 1_G)$ is a natural transformation $\eta : F \to G$ intertwining the various natural transformations and homomorphisms. All together this makes $\text{ALG}_1^{\text{lax}}(\text{COCOMP}_{\mathbb{K}})$ into a weak 3-category. ◊

Remark 6.4 I know of no examples of monoidal categories appearing in quantum field theory that are not generated under colimits by their (1-)dualizable objects (see Definition 7.1). Suppose that $\mathcal{B}$ and $\mathcal{C}$ are monoidal $\mathbb{K}$-linear cocomplete categories generated under colimits by their (1-)dualizable objects, and that $\mathcal{X}$ and $\mathcal{Y}$ are $\mathcal{B}$-$\mathcal{C}$-bimodules. Then in fact all lax bimodule homomorphisms $F : \mathcal{X} \to \mathcal{Y}$ are *strong*, which is when the natural transformations $B \triangleright F(X) \to F(B \triangleright X)$ and $F(X) \triangleleft C \to F(X \triangleleft C)$ are isomorphisms; the proof is the same as that of [DSPS19, Lemma 2.10]. The pointing $1_F : \mathbb{1}_{\mathcal{Y}} \to F(\mathbb{1}_{\mathcal{X}})$ is not necessarily an isomorphism. ◊

Suppose that $\mathcal{S}$ is a k-fold delooping of VECT—"the $(\infty, k+1)$-category of cocomplete $\mathbb{K}$-linear (∞, k)-categories," for example. Then $\text{ALG}_n^{\text{lax}}(\mathcal{S})$ is an $(n+k)$-fold delooping of $\text{ALG}_0(\text{VECT})$; i.e. in dimension $(n+k)$, $\text{ALG}_n^{\text{lax}}(\mathcal{S})$ consists of pointed vector spaces. This is what a fully extended $(n+k)$-dimensional Heisenberg-picture quantum field theory would assign to top-dimensional spacetimes. Therefore a *d-dimensional fully extended k-affine Heisenberg-picture quantum field theory*, for $d \geq k$, should be a symmetric monoidal functor $\text{SPACETIMES}_d \to \text{ALG}_{d-k}^{\text{lax}}(\mathcal{S})$, where $\mathcal{S}$ is a k-fold delooping of VECT. For the "derived" version, replace VECT by DGVECT. My favorite delooping of $\text{VECT}_{\mathbb{K}}$ is $\text{COCOMP}_{\mathbb{K}}$, and so I generally choose:

Definition 6.5 A *$\mathcal{G}$-geometric d-dimensional fully extended* 1*-affine Heisenberg-picture quantum field theory* is a symmetric monoidal functor $\text{SPACETIMES}_d^{\mathcal{G}} \to \text{ALG}_{d-1}^{\text{lax}}(\text{COCOMP}_{\mathbb{K}})$, where $\text{SPACETIMES}_d^{\mathcal{G}}$ is an (∞, d)-category of 0- through d-dimensional manifolds equipped with geometric structures of type $\mathcal{G}$. ◊

7 A (Non-)dualizability Result

Let us focus on the case of d-dimensional fully extended framed topological quantum field theories, which are based on the spacetime category $\text{SPACETIMES} = \text{BORD}_d^{\text{fr}}$. The famous *cobordism hypothesis*, whose proof is outlined in detail in [Lur09], asserts that for any symmetric monoidal (∞, N)-category $\mathcal{C}$, symmetric monoidal functors $\text{BORD}_d^{\text{fr}} \to \mathcal{C}$ are the same as d-dualizable objects in $\mathcal{C}$. (It is not too hard to convince oneself that every such functor picks out a d-dualizable object. The deep part of [Lur09] is that each d-dualizable object determines uniquely such a functor, which is in turn a statement about the topology of manifolds.)

Definition 7.1 ([Lur09]) A 1-morphism f in an (∞, N)-category $\mathcal{C}$ is *1-dualizable* if it has left and right adjoints in the homotopy bicategory $\text{h}_2\mathcal{C}$, each of which have left and right adjoints, ad infinitum. This homotopy bicategory has the same 0- and 1-morphisms as has $\mathcal{C}$, but its 2-morphisms are the equivalence classes of 2-morphisms in $\mathcal{C}$. (Any data from non-invertible 3- and higher morphisms in $\mathcal{C}$ is discarded.) Thus 1-dualizability asserts the existence of various "evaluation" and "coevaluation" 2-morphisms in $\mathcal{C}$, which must satisfy certain relations up to non-unique equivalence.

For $k > 1$, let f be a k-morphism in $\mathcal{C}$ with source and target $(k-1)$-morphisms x and y. Then x and y are *parallel*: they have the same source and target $(k-2)$-morphisms. The morphism f is *1-dualizable* if it is 1-dualizable in the $(\infty, N-k)$-category whose objects are all $(k-1)$-morphisms parallel to x and y. Thus 1-dualizability of a k-morphism asserts the existence of various "evaluation" and "coevaluation" $(k+1)$-morphisms in $\mathcal{C}$, which must satisfy certain relations up to non-unique equivalence.

For $d > 1$, a k-morphism f is *d-dualizable* if it is 1-dualizable and the evaluation and coevaluation $(k+1)$-morphisms witnessing such 1-dualizability are themselves $(d-1)$-dualizable.

If $\mathcal{C}$ is a symmetric monoidal (∞, N)-category, an object $X \in \mathcal{C}$ is *d-dualizable* if the functor $\otimes X : \mathcal{C} \to \mathcal{C}$ is d-dualizable as a 1-morphism in the $(\infty, N+1)$-category with a single object $\star$ and $\hom(\star, \star) = \mathcal{C}$ and composition given by $\otimes$. $\lozenge$

In general, given a symmetric monoidal (∞, N)-category $\mathcal{C}$, it is interesting to ask what are the d-dualizable objects for various values of d. First, a trivial observation: 1-dualizable 1-morphisms in an $(\infty, 1)$-category are equivalences, and so $(N+1)$-dualizable objects in a symmetric monoidal (∞, N)-category are invertible. Thus to have non-invertible examples, we should let $d \leq N$. Well-known examples include:

Example 7.2 The 1-dualizable objects in VECT are the finite-dimensional vector spaces. The 2-dualizable objects in MOR are the finite-dimensional semisimple algebras. $\lozenge$

These examples are typical of deloopings of VECT, in which such "full" dualizability is a strong "finiteness" condition which nevertheless admits interesting examples. In terms of field theories, this suggests that d-dimensional fully extended Schrödinger field theories are fairly rigid but can be quite nontrivial. Similar results about 3-dimensional Schrödinger-picture topological field theories are in [DSPS20].

What about Heisenberg-picture field theories? Theorem 5.3 implies that, for any $(\infty, k+1)$-category $\mathcal{S}$ and any n, *every* object of the $(\infty, n+k+1)$-category $\mathrm{ALG}_n^{\mathrm{lax}}(\mathcal{S})$ is n-dualizable. Thus 0-affine Heisenberg-picture topological field theories (when $n = d$ and $k = 0$) are quite flexible. But they are not "fully dualizable": even when $\mathcal{S}$ is an $(\infty, 1)$-category, $\mathrm{ALG}_n^{\mathrm{lax}}(\mathcal{S})$ is an $(\infty, n+1)$-category. In fact, asking for any more dualizablity collapses the whole enterprise:

Theorem 7.3 ([GS18]) *Let $\mathcal{S}$ be a symmetric monoidal $(\infty, k+1)$-category. Then the groupoid of $(n+k+1)$-dualizable objects in $\mathrm{ALG}_n^{\mathrm{lax}}(\mathcal{S})$ is contractible—every $(n+k+1)$-dualizable object is canonically equivalent to the unit object $\mathbb{1}$.*

Theorem 7.3 arose in conversations joint with Scheimbauer. I will give the idea of the proof. Providing all details of the second part of the argument would require working in too much detail for this paper with some particular model of (∞, n)-categories. An alternate complete proof is available in [GS18], which appeared after a preprint of this paper circulated.

Proof We begin by proving the claim when $k = 0$. Let $(X, 1_X)$ be a 1-dualizable object in $\mathrm{ALG}_0^{\mathrm{lax}}(\mathcal{S})$ with dual $(X, 1_X)^* = (Y, 1_Y)$. Denote the evaluation and coevaluation maps by

$$(F, 1_F) : (\mathbb{1}, \mathrm{id}_{\mathbb{1}}) \to (X, 1_X) \otimes (Y, 1_Y) \cong (X \otimes Y, 1_X \otimes 1_Y),$$
$$(G, 1_G) : (Y \otimes X, 1_Y \otimes 1_X) \cong (Y, 1_Y) \otimes (X, 1_X) \to (\mathbb{1}, \mathrm{id}_{\mathbb{1}}).$$

The compatibility conditions between the evaluation and coevaluation necessary for 1-dualizability assert that two equations, the first of which reads

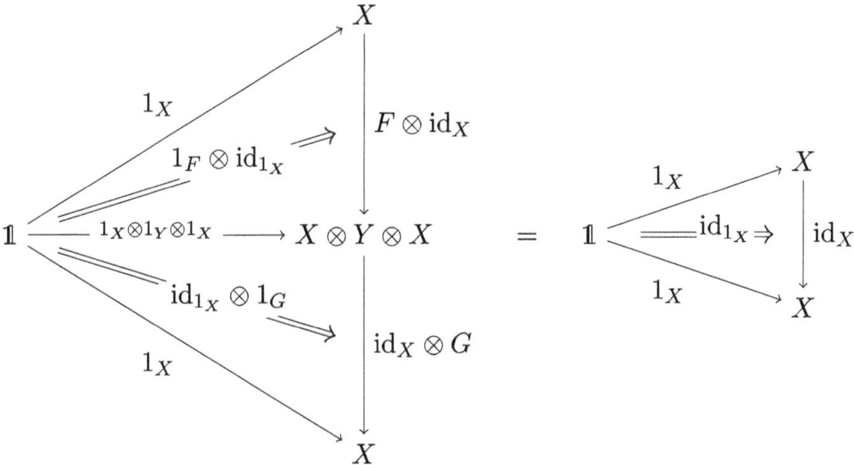

Considering just X, Y, F, G, one finds that X and Y are duals in $\mathcal{S}$. Unpacking the extra conditions coming from 1_F and 1_G gives two commuting triangles, the first of which reads

$$\begin{array}{ccccc} & & 1_X \otimes \big(G \circ (1_Y \otimes 1_X)\big) & & (\mathrm{id}_X \otimes G) \circ (F \otimes \mathrm{id}_X) \circ 1_X \\ & & \wr\!\! \| & & \wr\!\! \| \\ 1_X & \xRightarrow{\mathrm{id}_{1_X} \otimes 1_G} & 1_X \circ \big(G \circ (1_Y \otimes \mathrm{id}_X)\big) \circ 1_X & \xRightarrow{1_F \otimes \mathrm{id}_{1_X}} & 1_X \end{array}$$
$$1_X \xRightarrow{\mathrm{id}_{1_X}} 1_X$$

These triangles precisely assert that $1_X : \mathbb{1} \to X$ and $G \circ (1_Y \otimes \mathrm{id}_X) : X \to \mathbb{1}$ are dual 1-morphisms. In particular, 1_X is 1-dualizable.

An induction implies more generally that a d-dualizable object $(X, 1_X) \in \mathrm{ALG}_0^{\mathrm{lax}}(\mathcal{S})$ consists of a d-dualizable object $X \in \mathcal{S}$ along with a d-dualizable 1-morphism $1_X : \mathbb{1} \to X$. But, as remarked above, an $(n+1)$-dualizable 1-morphism in an $(\infty, n+1)$-category is necessarily invertible. Thus the pointing 1_X furnishes a canonical equivalence between $(X, 1_X)$ and $(\mathbb{1}, \mathrm{id}_{\mathbb{1}})$.

Now let k be arbitrary. I will describe a canonical symmetric monoidal functor $\mathrm{ALG}_n^{\mathrm{lax}}(\mathcal{S}) \to \mathrm{ALG}_0^{\mathrm{lax}}(\mathrm{ALG}_n^{\mathrm{lax}}(\mathcal{S}))$ which splits the forgetful map $\mathrm{ALG}_0^{\mathrm{lax}}(\mathrm{ALG}_n^{\mathrm{lax}}(\mathcal{S})) \to \mathrm{ALG}_n^{\mathrm{lax}}(\mathcal{S})$. (This splitting is a version of the Eilenberg–Watts functor considered in Example 4.4.) With such a functor in hand, any $(n+k+1)$-dualizable object $X \in \mathrm{ALG}_n^{\mathrm{lax}}(\mathcal{S})$ determines an $(n+k+1)$-dualizable object in $\mathrm{ALG}_0^{\mathrm{lax}}(\mathrm{ALG}_n^{\mathrm{lax}}(\mathcal{S}))$, which by above is trivial, and so X is trivial.

First, let M be an m-morphism in $\mathrm{ALG}_n^{\mathrm{lax}}(\mathcal{S})$ for $m < n$; i.e. an $(n-m)$-algebra in $\mathcal{S}$, with $n - m \geq 1$, with some compatible actions by some $(n-m+1)$-algebras. We map it to the pointed m-morphism (M, M_M), where M_M is the "right regular M-module," which is M treated as an $(n-m-1)$-algebra with the canonical right M-action. In terms of the factorization algebra pictures used in [Sch14] to define $\mathrm{ALG}_n^{\mathrm{lax}}(\mathcal{S})$, the m-morphism M is described by a factorization algebra on $\mathbb{R}^n$ which is locally constant with respect to the stratification given by the subspaces $\{x_1 = 0\}$, $\{x_1 = x_2 = 0\}, \ldots, \{x_1 = \cdots = x_m = 0\}$. The pointed m-morphism (M, M_M) is given by pushing M forward along the constructible map

$$\vec{x} \mapsto \begin{cases} \vec{x} & \text{if } x_1 + \cdots + x_n \geq 0; \\ (x_1, \ldots, x_{n-1}, -x_1 - \cdots - x_{n-1}) & \text{if } x_1 + \cdots + x_n \leq 0. \end{cases}$$

Let M now by an n-morphism with source S and target T. Then S and T are 1-algebras and $M = ({}_S M_T, 1_M)$ is a pointed S-T-bimodule (subject to compatibility conditions coming from the common source and target of S and T). The composition $M(S_S) = S_S \otimes_S {}_S M_T$ is the module M_T in given by forgetting the S-action. The pointing 1_M then determines a (strong, and in particular lax) pointed T-module map $T_T \to M_T$ given by $1_T \mapsto 1_M$. Thus $(M, 1_M)$ "is" an n-morphism in $\mathrm{ALG}_0^{\mathrm{lax}}(\mathrm{ALG}_n^{\mathrm{lax}}(\mathcal{S}))$. The construction $M \mapsto (M, 1_M)$ extends naturally to lax pointed bimodule morphisms, and completes the description of the splitting $\mathrm{ALG}_n^{\mathrm{lax}}(\mathcal{S}) \to \mathrm{ALG}_0^{\mathrm{lax}}(\mathrm{ALG}_n^{\mathrm{lax}}(\mathcal{S}))$. □

In summary, the interesting dualizability questions related to k-affine Heisenberg-picture topological quantum field theory are about d-dualizability in $\mathrm{ALG}_n(\mathcal{S})$ for $\mathcal{S}$ a k-fold delooping of VECT and $n + 1 \leq d \leq n + k$.

8 From Factorization Algebra to Heisenberg-Picture Field Theory

I would like now to explain an example which is based on unpublished work of Dwyer, Stolz, and Teichner. I will outline my interpretation of, and elaboration upon, some parts of their construction, but details will need to wait for future work.

Definition 8.1 Let $\mathcal{G}$ be a not-necessarily topological local geometry for d-dimensional manifolds, and $\text{EMB}_d^{\mathcal{G}}$ the symmetric monoidal category of possibly open d-dimensional $\mathcal{G}$-geometric manifolds, with symmetric monoidal structure given by disjoint union. Given a target category $\mathcal{S}$, a *$\mathcal{G}$-geometric prefactorization algebra valued in $\mathcal{S}$* is a symmetric monoidal functor $F : \text{EMB}_d^{\mathcal{G}} \to \mathcal{S}$. In particular, for every $M \in \text{EMB}_d^{\mathcal{G}}$, $F(M)$ is pointed by $F(\emptyset \hookrightarrow M) : \mathbb{1}_{\mathcal{S}} \to F(M)$, and each embedding $M_1 \sqcup M_2 \hookrightarrow M$ determines a "multiplication" map $F(M_1) \otimes F(M_2) \to F(M)$. A prefactorization algebra is a *factorization algebra* if it is local for the Weiss topology (see [Gin13, CG16, Sch14] for details, but note that this usage of the phrase "(pre)factorization algebra" is not quite the same as the one in [CG16]). $\Diamond$

We will need two variations of the notion of "factorization algebra." Let X be a topological space. A *factorization algebra on X* is a precosheaf F on X, local for the Weiss topology, such that if U and V are disjoint opens, then $F(U \sqcup V) \cong F(U) \otimes F(V)$. Let $(X, * \in X)$ be a pointed topological space. An *unpointed point-module on* $(X, * \in X)$ is similar, but whereas in a factorization algebra there are maps $F(U) \to F(V)$ for every inclusion $U \subseteq V$ of open sets in X, in an unpointed point-module we do not ask for such maps in the special case where $U \ni *$ but $V \not\ni *$. These are called "unpointed" because in the special case when $X = \{*\}$, an unpointed point-module is just an object of $\mathcal{S}$, whereas a factorization algebra is a pointed object. The restriction of a $\mathcal{G}$-geometric factorization algebra to a space with $\mathcal{G}$-geometry is a factorization algebra on that space; given a factorization algebra on a pointed space, one can forget to an unpointed point-module.

When $\mathcal{G}$ is a not-necessarily topological "rigid supergeometry," a construction of the unextended spacetime category $\text{SPACETIMES}_{d-1,d}^{\mathcal{G}}$ is described in [ST11]. In that construction, the objects are $(d-1)$-dimensional closed manifolds N equipped with germs of one-sided collars diffeomorphic to $N \times [0, \epsilon)$ with $\mathcal{G}$-structures on $N \times (0, \epsilon)$. A morphism from N_2 to N_1 is a d-dimensional manifold M with boundary $\partial M \cong N_1 \sqcup N_2$ and a germ of an extension of M past N_2:

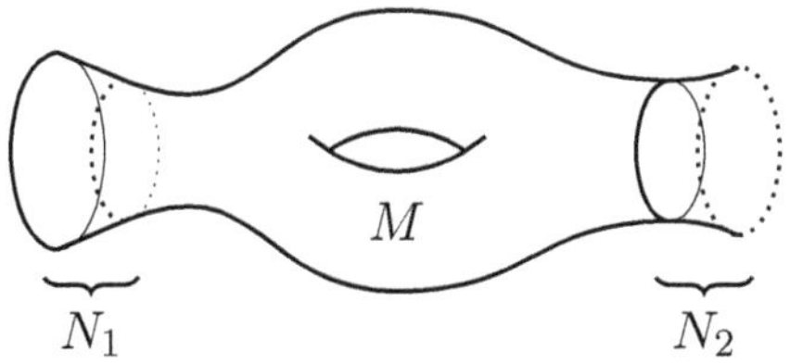

Given a $\mathcal{G}$-geometric factorization algebra F, we will build a Heisenberg-picture field theory $\mathcal{Z}_F : \text{SPACETIMES}^{\mathcal{G}}_{d-1,d} \to \text{ALG}_0(\text{COCOMP}_{\mathbb{K}})$ for this version of $\text{SPACETIMES}^{\mathcal{G}}_{d-1,d}$.

Definition 8.2 Fix a $\mathcal{G}$-geometric factorization algebra $F : \text{EMB}^{\mathcal{G}}_d \to \text{VECT}_{\mathbb{K}}$. Let M be a (possibly open) $\mathcal{G}$-manifold with boundary. Let $M/\partial M$ denote the quotient (in topological spaces) in which ∂M is contracted to a point; it is naturally pointed by that point. The category $\text{BC}_F(M)$ of *boundary conditions* for F on M is

$$\text{BC}_F(M) = \left\{\text{unpointed point-modules } \tilde{F} \text{on} M/\partial M \text{s.t. } \tilde{F}|_{M \smallsetminus \partial M} = F|_{M \smallsetminus \partial M}\right\}.$$

The *free boundary condition* is the boundary condition given by pushing forward of $F|_{M \smallsetminus \partial M}$ along $M \smallsetminus \partial M \to M/\partial M$.

Lemma 8.3 *The category of boundary conditions for F on M depends only on the germ of a $\mathcal{G}$-manifold around ∂M.*

Proof This follows from descent/locality of factorization algebras [Sch14]. Indeed, suppose that $M' \hookrightarrow M$ is a submanifold with boundary such that the inclusion maps $\partial M' \overset{\sim}{\to} \partial M$. One can restrict unpointed point-modules on $M/\partial M$ to unpointed point-modules on $M'/\partial M'$. Conversely, suppose we are given an unpointed point-module $\tilde{F}$ on $M'/\partial M'$ whose restriction to $M' \smallsetminus \partial M'$ is $F|_{M' \smallsetminus \partial M'}$. Then it can be canonically extended to an unpointed point-module on $M/\partial M$ by gluing with $F|_{M \smallsetminus \partial M}$. □

Proposition 8.4 *Let M be a manifold with boundary. Its category* $\text{BC}_F(M)$ *of boundary conditions is cocomplete and $\mathbb{K}$-linear.*

The proof will show indeed that $\text{BC}_F(M)$ is locally presentable.

Proof I will give an alternate characterization of $\text{BC}_F(M)$, closer to the construction of Dwyer, Stolz, and Teichner. Consider the poset $\mathcal{P}(M)$ whose objects are open neighborhoods of ∂M, ordered by inclusion. We build a linear category $\mathcal{C}_F(M)$ whose objects are the elements of $\mathcal{P}(M)$ and whose morphisms are:

$$\hom(A, B) = \begin{cases} 0, & \text{if } A \not\subseteq B \\ F(B \smallsetminus A), & \text{if } A \subseteq B. \end{cases}$$

Note in particular that $\hom(A, A) = F(\emptyset) = \mathbb{K}$. The composition is given by the factorization algebra structure maps $F(A \smallsetminus B) \otimes F(B \smallsetminus C) \cong F((A \smallsetminus B) \cup (B \smallsetminus C)) \to F(A \smallsetminus C)$.

Let $\text{FUN}_{\mathbb{K}}(\mathcal{A}, \mathcal{B})$ denote the category of $\mathbb{K}$-linear functors between $\mathbb{K}$-linear categories $\mathcal{A}$ and $\mathcal{B}$. There is a fully faithful embedding $\text{BC}_F(M) \hookrightarrow \text{FUN}_{\mathbb{K}}(\mathcal{C}_F(M), \text{VECT}_{\mathbb{K}})$ that sends each unpointed point-module $\tilde{F}$ to the functor $A \mapsto \tilde{F}(A \smallsetminus \partial M)$. Functoriality is given by the factorization algebra structure. Since $\text{FUN}_{\mathbb{K}}(\mathcal{C}_F(M), \text{VECT}_{\mathbb{K}})$ is $\mathbb{K}$-linear, so is $\text{BC}_F(M)$.

The essential image of this embedding consists of those $\mathbb{K}$-linear functors $\mathcal{C}(N) \to \text{VECT}_{\mathbb{K}}$ satisfying an appropriate "locality" condition coming from the locality condition for factorization algebras. This locality condition consists of a series of assertions of the following form: $\tilde{F}(A)$ arises as a weighted colimit of a diagram each term of which is $\tilde{F}(B)$ for some $B \subsetneq A$, with the weightings being tensor products of $F(A \smallsetminus B)$s.

Thus $\text{BC}_F(M)$ is the category of models of a colimit sketch structure on $\mathcal{C}_F(M)$, and is therefore locally presentable and in particular cocomplete. □

With Lemma 8.3 and Proposition 8.4 in hand, we may define:

Definition 8.5 Given an object $N \in \text{SPACETIMES}^{\mathcal{G}}_{d-1,d}$ and a factorization algebra $F : \text{EMB}^{\mathcal{G}}_d \to \text{VECT}_{\mathbb{K}}$, we set $\mathcal{Z}_F(N) = \text{BC}_F(N \times [0,\epsilon)) \in \text{ALG}_0^{\text{lax}}(\text{COCOMP}_{\mathbb{K}})$, where the pointing is via the free boundary condition. ◊

Lemma 8.6 *The assignment $N \mapsto \mathcal{Z}_F(N)$ is symmetric monoidal in the sense that there are canonical equivalences $\mathcal{Z}_F(\emptyset) = \text{VECT}_{\mathbb{K}}$ and $\mathcal{Z}_F(N_1 \sqcup N_2) = \mathcal{Z}_F(N_1) \boxtimes \mathcal{Z}_F(N_2)$.*

Proof The category of boundary conditions for the empty set is the category of unpointed point-modules on {pt}, which is easily seen to be $\text{VECT}_{\mathbb{K}}$. Assume N_1 and N_2 are nonempty. Given $\mathbb{K}$-linear categories $\mathcal{A}$ and $\mathcal{B}$, define their *naive tensor product* $\mathcal{A} \otimes_{\mathbb{K}} \mathcal{B}$ to be the category with object set $\text{ob}(\mathcal{A}) \times \text{ob}(\mathcal{B})$ and morphisms $\hom((A_1, B_1), (A_2, B_2)) = \hom_{\mathcal{A}}(A_1, A_2) \otimes \hom_{\mathcal{B}}(B_1, B_2)$. Then there is a natural equivalence $\text{FUN}_{\mathbb{K}}(\mathcal{A}, \text{VECT}_{\mathbb{K}}) \boxtimes \text{FUN}_{\mathbb{K}}(\mathcal{B}, \text{VECT}_{\mathbb{K}}) = \text{FUN}_{\mathbb{K}}(\mathcal{A} \otimes_{\mathbb{K}} \mathcal{B}, \text{VECT}_{\mathbb{K}})$. From this and the description of boundary conditions given in the proof of Proposition 8.4 one may derive the natural equivalence $\mathcal{Z}_F(N_1 \sqcup N_2) = \mathcal{Z}_F(N_1) \boxtimes \mathcal{Z}_F(N_2)$. □

Definition 8.7 We now extend $\mathcal{Z}_F$ to morphisms in $\text{SPACETIMES}^{\mathcal{G}}_{d-1,d}$. Let $M : N_2 \to N_1$ be a 1-morphism, where we have chosen representative collars $N_1 \times [0, \epsilon_1)$ around N_1 and $N_2 \times [0, \epsilon_2)$ around N_2. Since Heisenberg-picture field theories are assumed to be contravariantly functorial, we want to build a functor $\mathcal{Z}_F(M) : \mathcal{Z}_F(N_1) \to \mathcal{Z}_F(N_2)$.

By Lemma 8.3, $\mathcal{Z}_F(N_1) = \text{BC}_F(M \sqcup_{N_2} N_2 \times [0, \epsilon_2))$, as the inclusion $N_1 \times [0, \epsilon_1) \to M \sqcup_{N_2} N_2 \times [0, \epsilon_2)$ is an isomorphism near the boundary. Given a boundary condition $\tilde{F} \in \text{BC}_F(M \sqcup_{N_2} N_2 \times [0, \epsilon_2))$, we may push it forward along the map $(M \sqcup_{N_2} N_2 \times [0, \epsilon_2))/N_1 \to (N_2 \times [0, \epsilon_2))/N_2$ that is the identity on $N_2 \times (0, \epsilon)$ and collapses M to the point. The resulting pushed-forward factorization algebra then restricts over $N_2 \times (0, \epsilon)$ to $F|_{N_2 \times (0,\epsilon)}$ and so defines a boundary condition on $N_2 \times [0, \epsilon_2)$. We set $\mathcal{Z}_F(M) : \mathcal{Z}_F(N_1) \to \mathcal{Z}_F(N_2)$ to be this push-forward operation.

Cocontinuity of $\mathcal{Z}_F(M)$ follows from Remark 8.8. Its structure as a lax pointed functor comes from the canonical maps $F(A_2 \smallsetminus N_2) \to F((M \smallsetminus N_1) \sqcup_{N_2} A_2)$, where $A_2 \subseteq N_2 \times [0, \epsilon)$ is an open neighborhood of the boundary N_2. This defines the functor $\mathcal{Z}_F : \text{SPACETIMES}^{\mathcal{G}}_{d-1,d} \to \text{ALG}_0^{\text{lax}}(\text{COCOMP}_{\mathbb{K}})$; functoriality follows from the functoriality of push-forward of factorization algebras, and symmetric monoidality can be proven by extending Lemma 8.6. ◊

Remark 8.8 One can give $\mathcal{Z}_F(N)$ a much more explicit description, completing the comparison to the construction described by Dwyer, Stolz, and Teichner. I will continue with the notation from the proof of Proposition 8.4.

Set $M = N \times [0, \epsilon)$. For each object $A \in \mathcal{C}_F(M)$, consider the full subcategory $\mathcal{C}_F(M)_{\subsetneq A}$ of $\mathcal{C}_F(M)$ on the objects B with $B \subsetneq A$. There are restriction functors $\mathrm{FUN}_{\mathbb{K}}(\mathcal{C}_F(M)_{\subsetneq A}, \mathrm{VECT}_{\mathbb{K}}) \to \mathrm{FUN}_{\mathbb{K}}(\mathcal{C}_F(M)_{\subsetneq B}, \mathrm{VECT}_{\mathbb{K}})$ for $B \subsetneq A$ whose left adjoints define *extension* functors

$$\mathrm{FUN}_{\mathbb{K}}(\mathcal{C}_F(M)_{\subsetneq B}, \mathrm{VECT}_{\mathbb{K}}) \to \mathrm{FUN}_{\mathbb{K}}(\mathcal{C}_F(M)_{\subsetneq A}, \mathrm{VECT}_{\mathbb{K}}).$$

Unpacking the locality condition for factorization algebras reveals

$$\mathcal{Z}_F(N) = \mathrm{BC}_F(M) = \lim_{A \to \partial M} \mathrm{FUN}_{\mathbb{K}}(\mathcal{C}_F(M)_{\subsetneq A}, \mathrm{VECT}_{\mathbb{K}}),$$

where the limit is taken over $\mathcal{P}(M)$ along the extension functors. It is worth noting that the colimit along restriction functors vanishes because $\hom(A, B) = 0$ if $A \not\subseteq B$.

Let M now be a $\mathcal{G}$-geometric cobordism between N_1 and N_2. There is a canonical inclusion $\mathcal{C}_F(N_2 \times [0, \epsilon_2)) \hookrightarrow \mathcal{C}_F(M)$ given by $B \mapsto M \sqcup_{N_2} B$, and hence a restriction functor

$$\mathrm{FUN}_{\mathbb{K}}(\mathcal{C}_F(M \sqcup_{N_2} N_2 \times [0, \epsilon_2))_{\subsetneq M \sqcup_{N_2} B}, \mathrm{VECT}_{\mathbb{K}}) \to$$
$$\mathrm{FUN}_{\mathbb{K}}(\mathcal{C}_F(N_2 \times [0, \epsilon_2))_{\subsetneq B}, \mathrm{VECT}_{\mathbb{K}}).$$

Composing this with the projection $\mathcal{Z}_F(N_1) = \mathrm{BC}_F(M \sqcup_{N_2} N_2 \times [0, \epsilon_2)) \to \mathrm{FUN}_{\mathbb{K}}(\mathcal{C}_F(M \sqcup_{N_2} N_2 \times [0, \epsilon_2))_{\subsetneq M \sqcup_{N_2} B}, \mathrm{VECT}_{\mathbb{K}})$ gives a sequence of cocontinuous functors

$$\mathcal{Z}_F(N_1) \to \mathrm{FUN}_{\mathbb{K}}(\mathcal{C}_F(N_2 \times [0, \epsilon_2))_{\subsetneq B}, \mathrm{VECT}_{\mathbb{K}}).$$

One may directly check that these commute with the extension functors for $B \hookrightarrow B'$, and so define a cocontinuous functor $\mathcal{Z}_F(N_2) \to \mathcal{Z}_F(N_2)$, which is nothing but the functor $\mathcal{Z}_F(M)$ from Definition 8.7.

As a final remark, note that limits can be computed by passing to cofinal subcategories, and so $\mathcal{Z}_F(N)$ can be presented as a limit over categories just for those opens that contract onto N. Moreover, consider pushing forward $F|_{N \times (0, \epsilon)}$ along the projection $N \times (0, \epsilon) \to (0, \epsilon)$ to produce a factorization algebra on $(0, \epsilon)$. One can then consider the category of boundary conditions on $[0, \epsilon)$ for this factorization algebra. By comparing descriptions in terms of limits of functor categories along extension functors, one can show that this category is nothing but $\mathcal{Z}_F(N)$. With this description and some careful study of limits of locally presentable categories, one can show in particular that when F is locally constant, $\mathcal{Z}_F(N) = \mathrm{MOD}_{\int_N F}$, and so the construction above matches the factorization homology functor from [Sch14]. ◊

The construction of $\mathcal{Z}_F(N)$ did not require that N was compact. Suppose that N is an open $(n-1)$-dimensional manifold along with a germ of a $\mathcal{G}$-structure on $N \times [0,\epsilon)$; then there is still a category $\mathrm{BC}_F(N \times [0,\epsilon))$ of boundary conditions for $F|_{N\times(0,\epsilon)}$, and it is still described as in the proof of Proposition 8.4.

Theorem 8.9 *The assignment* $N \mapsto \mathcal{Z}_F(N)$ *defines a symmetric monoidal precosheaf (i.e. a prefactorization algebra) valued in* $\mathrm{COCOMP}_{\mathbb{K}}$ *on the category* $\mathrm{EMB}^{\mathcal{G}}_{d-1}$ *of* $(d-1)$*-dimensional manifolds equipped with germs of* $\mathcal{G}$*-structures.*

In fact, $\mathcal{Z}_F$ is a factorization algebra, but the proof will await future work.

Proof Suppose that $N_1 \hookrightarrow N_2$ is an inclusion compatible with germs of $\mathcal{G}$-structures. The corresponding map $N_1 \times [0,\epsilon) \hookrightarrow N_2 \times [0,\epsilon)$ extends to an inclusion

$$\bigl(N_1 \times [0,\epsilon)\bigr) \bigsqcup_{N_1\times(0,\epsilon)} \bigl(N_2 \times (0,\epsilon)\bigr) \hookrightarrow \bigl(N_2 \times [0,\epsilon)\bigr),$$

which in turn descends to a continuous function

$$\bigl(N_1 \times [0,\epsilon)/N_1\bigr) \bigsqcup_{N_1\times(0,\epsilon)} \bigl(N_2 \times (0,\epsilon)\bigr) \hookrightarrow \bigl(N_2 \times [0,\epsilon)/N_2\bigr)$$

which is a bijection on points, although not in general a homeomorphism.

Any boundary condition on $N_1 \times [0,\epsilon)$ extends canonically to an unpointed point-module on $\bigl(N_1 \times [0,\epsilon)/N_1\bigr) \sqcup_{N_1\times(0,\epsilon)} \bigl(N_2 \times (0,\epsilon)\bigr)$ by gluing with $F|_{N_2\times(0,\epsilon)}$. Pushing forward along the bijection above defines the functor $\mathcal{Z}_F(N_1) \to \mathcal{Z}_F(N_2)$.

Symmetric monoidality of $\mathcal{Z}_F$ follows from Lemma 8.6. □

Nowhere in the construction of $\mathcal{Z}_F$ did we need that the factorization algebra F took values in $\mathrm{VECT}_{\mathbb{K}}$—any locally presentable target category would have sufficed. Let $\mathrm{PRES}_{\mathbb{K}}$ denote the bicategory of $\mathbb{K}$-linear locally presentable categories; it is a full subbicategory of $\mathrm{COCOMP}_{\mathbb{K}}$. $\mathrm{PRES}_{\mathbb{K}}$ seems in many ways like it is a "locally presentable bicategory": it is cocomplete [Bir84]; every object has a presentation [AR94, Theorem 1.46].

The theory of locally presentable higher categories is under active development; cf. [GH15]. Eventually, one should expect to be able to iterate the above construction $F \rightsquigarrow \mathcal{Z}_F$ to produce an extended field theory $\mathcal{Z}_{\mathcal{Z}_F}$ valued in "locally presentable $\mathrm{PRES}_{\mathbb{K}}$-linear bicategories." Iterating further would produce a fully extended Heisenberg-picture field theory from a factorization algebra.

9 From Skein Theory to Heisenberg-Picture Field Theory

In this final example I would like to describe a family of topological Heisenberg-picture field theories which one can prove are not affine. The construction is my interpretation of the Reshetikhin–Turaev invariants of knots and links [RT90]. The extension to a local field theory is inspired by Walker's work [Wal06, MW11]. It is the three-dimensional extension of the two-dimensional quantum field theories studied in [BZBJ18a, BZBJ18b] and is expected to match the answer given by [AFR18]. Recall some standard definitions:

Definition 9.1 Let $\mathcal{C} = (\mathcal{C}, \otimes, \dots)$ be a small $\mathbb{K}$-linear monoidal category. (The "$\dots$" denote the auxiliary data of an associator, unit, and unitors that I will generally suppress.) A *strong monoidal functor* (F, f) consists of a functor F and a natural isomorphism $f : F(-) \otimes F(-) \overset{\sim}{\to} F(- \otimes -)$ expressing compatibility with the monoidal structure, which must itself be compatible with associators (and also an isomorphism expressing compatibility with the units, but that isomorphism can always be canonically strictified to an identity, and so I will suppress it from the notation). Let $\otimes^{\mathrm{op}}$ denote the opposite monoidal structure on $\mathcal{C}$.

The monoidal category $\mathcal{C}$ is *braided* if it is equipped with a strong monoidal functor $(\mathrm{id}_{\mathcal{C}}, \beta) : (\mathcal{C}, \otimes, \dots) \to (\mathcal{C}, \otimes^{\mathrm{op}}, \dots)$ whose underlying functor is the identity functor $\mathrm{id}_{\mathcal{C}}$. (This is equivalent to the usual hexagon relations for β.) A *full twist* for $\mathcal{C}$ is an isomorphism $\theta : (\mathrm{id}_{\mathcal{C}}, \beta^{-1}) \overset{\sim}{\to} (\mathrm{id}_{\mathcal{C}}, \beta)$ of monoidal functors between $(\mathcal{C}, \otimes)$ and $(\mathcal{C}, \otimes^{\mathrm{op}})$, or equivalently an isomorphism of monoidal endofunctors $\theta : (\mathrm{id}_{\mathcal{C}}, \mathrm{id}) \overset{\sim}{\to} (\mathrm{id}_{\mathcal{C}}, \beta^2)$ of $(\mathcal{C}, \otimes)$. Spelled out, a full twist consists of a natural automorphism $\theta_X : X \overset{\sim}{\to} X$ such that

$$\beta_{X,Y} \circ (\theta_X \otimes \theta_Y) = \beta_{Y,X}^{-1} \circ \theta_{X \otimes Y}$$

for all $X, Y \in \mathcal{C}$.

The name "full twist" comes from interpreting θ_X as a $360°$ twist of a ribbon labeled by X:

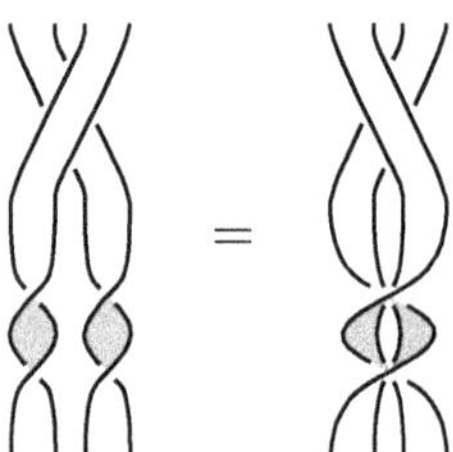

Categories equipped with a full twist are called *balanced* in [JS91], but I will not use this word so as not to conflict with the notion of "balanced tensor product" from Example 6.3 and Definition 9.4.

A small monoidal category $\mathcal{C}$ *has duals* if for every object $X \in \mathcal{C}$ the functor $X\otimes : \mathcal{C} \to \mathcal{C}$ has a right adjoint of the form ${}^*X\otimes$ and a left adjoint of the form $X^*\otimes$ for some objects $X^*, {}^*X \in \mathcal{C}$. (These objects are unique up to unique isomorphism if they exist, in which case $\otimes X^*$ and $\otimes {}^*X$ are the left and right adjoints, respectively, to $\otimes X$. Compare Definition 7.1.) If $\mathcal{C}$ has duals, then there is a canonically defined *double dual* functor $X \mapsto X^{**}$ which is a monoidal equivalence $(\mathcal{C}, \otimes, \dots) \to (\mathcal{C}, \otimes, \dots)$. Suppose that $\mathcal{C} = (\mathcal{C}, \otimes, \beta, \dots)$ is a small $\mathbb{K}$-linear braided monoidal category with duals. The braiding determines isomorphisms

$$\tau_\beta : (-)^{**} \overset{\sim}{\to} (\mathrm{id}, \beta^2) \quad \text{and} \quad \tau_{\beta^{-1}} : (-)^{**} \overset{\sim}{\to} (\mathrm{id}, \beta^{-2})$$

of monoidal functors $\mathcal{C} \to \mathcal{C}$, via

$$X^{**} \xrightarrow{\text{unit of adjunction}} X^{**} \otimes X \otimes X^* \underset{\beta^{-1}_{X,X^{**}} \otimes \mathrm{id}_{X^*}}{\overset{\beta_{X^{**},X} \otimes \mathrm{id}_{X^*}}{\rightrightarrows}} X \otimes X^{**} \otimes X^* \xrightarrow{\text{counit of adjunction}} X$$

(the upper composite being $\tau_\beta(X)$ and the lower composite $\tau_{\beta^{-1}}(X)$) with inverses

$$X \xrightarrow{\text{unit of adjunction}} X^{*} \otimes X^{**} \otimes X \underset{\mathrm{id}_{X^*} \otimes \beta_{X^{**},X}}{\overset{\mathrm{id}_{X^*} \otimes \beta^{-1}_{X,X^{**}}}{\rightrightarrows}} X^* \otimes X \otimes X^{**} \xrightarrow{\text{counit of adjunction}} X^{**}$$

(the upper composite being $\tau_\beta^{-1}(X)$ and the lower composite $\tau_{\beta^{-1}}^{-1}(X)$).

A braided monoidal category $\mathcal{C}$ with duals is *ribbon* if it is equipped with a full twist θ satisfying the quadratic equation $\theta^2 = \tau_{\beta^{-1}}^{-1} \circ \tau_\beta$:

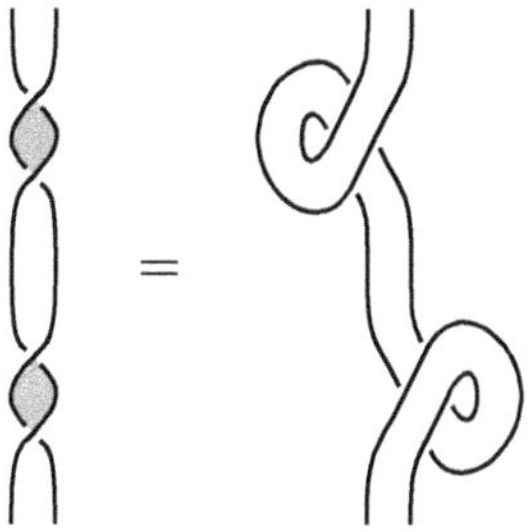

Let $(\mathcal{C}, \otimes, \beta, \theta, \dots)$ be a ribbon category. A *$\mathcal{C}$-labeled ribbon tangle in* $[0, 1]^3$ is a piecewise-smooth embedding (except for as described in item (2)) into $[0, 1]^3$ of a finite collection of rectangles subject to the following conditions:

(1) The rectangles come in two kinds, "long thin" *ribbons* and "short squat" *coupons*. The *bottom end* of a ribbon is the subset $[0, 1] \times \{0\}$ of its boundary,

and the *top end* if $[0, 1] \times \{1\}$. The *bottom side* of a coupon is the subset $[0, 1] \times \{0\}$ of its boundary, and the *top side* is $[0, 1] \times \{1\}$.

(2) Each bottom end of each ribbon is a subset either of the *bottom* of the cube $[0, 1]^2 \times \{0\}$ or of a top side of a coupon; in the latter case the map $[0, 1] \times \{0\} \hookrightarrow [0, 1] \times \{1\}$ mapping the end of the ribbon into the top or bottom of the coupon is orientation-preserving. The top end of each ribbon is a subset either of the *top* of the cube $[0, 1]^2 \times \{1\}$ or of a bottom side of a coupon; again in the latter case orientations should be preserved. These are the only intersections of ribbons and coupons with each other or with the boundary of $[0, 1]^3$.

(3) Each ribbon is labeled with an object of $\mathcal{C}$. Each coupon is labeled with a morphism in $\mathcal{C}$ compatibly with the ribbons that end on it: if the ribbons ending on its bottom are labeled in order $X_1, \dots, X_m$ and the ribbons ending on its top are labeled in order $Y_1, \dots, Y_n$, then the coupon itself is labeled with an element of $\hom(X_1 \otimes \cdots \otimes X_m, Y_1 \otimes \cdots \otimes Y_n)$. ◊

The main coherence theorem about ribbon categories is the following:

Theorem 9.2 ([RT90]) *There is a well-defined* interpretation *map from $\mathcal{C}$-labeled ribbon tangles in* $[0, 1]^3$ *to morphisms in $\mathcal{C}$. The source and target of the morphism depend only on the intersections of the tangle with the bottom and top, respectively, of* $[0, 1]^3$. *The interpretation of a tangle depends only on the isotopy-rel-boundary class of the tangle. Suppose that* $[0, 1]^3 \hookrightarrow [0, 1]^3$ *is an orientation-preserving nesting of a smaller cube into the interior of a larger cube. Suppose further that we are given two $\mathcal{C}$-labeled ribbon tangles in the larger cube such that (i) the two tangles agree outside the smaller cube, (ii) the tangles intersect the boundary of smaller cube only along its top and bottom, and (iii) the intersections of the two tangles with the smaller cube have the same interpretation. Then the two tangles have the same interpretation in the larger cube.* □

Theorem 9.2 allows us to use $\mathcal{C}$ to study tangles in more complicated manifolds. The following construction is particularly important:

Definition 9.3 Let $\mathcal{C}$ be a ribbon category and let N be a possibly open oriented surface admitting a finite cover by disks. The *skein category* $\int_N \mathcal{C}$ is the $\mathbb{K}$-linear category described as follows:

objects: configurations of disjoint "short" oriented intervals in N, each labeled by an object of $\mathcal{C}$.

There is a natural extension of the notion of "ribbon tangle labeled by $\mathcal{C}$" to tangles in $N \times [0, 1]$. Note that for any ribbon tangle, its intersections with $N \times \{0\}$ and with $N \times \{1\}$ are naturally objects of $\int_N \mathcal{C}$.

morphisms: the space of morphisms from X_0 to X_1 is spanned by the set of $\mathcal{C}$-labeled ribbon tangles that intersect $N \times \{i\}$ at X_i, modulo the following relation: for every orientation-preserving embedding $[0, 1]^3 \hookrightarrow N \times [0, 1]$ of a "small" cube into the interior of $N \times [0, 1]$, we set to zero any linear combination of tangles $\sum T_a$ if (i) all the T_as are equal outside the small cube, (ii) they intersect the small cube only along its top and bottom sides, and the intersection is transverse, and (iii) $\sum \text{interpret}(T_a \cap (\text{small cube})) = 0$.

It follows from Theorem 9.2 that isotopic-rel-boundary tangles represent the same morphism in $\int_N \mathcal{C}$. Composition is by stacking of tangles in $N \times [0,1] \cup_N N \times [0,1] \simeq N \times [0,1]$. The skein category $\int_N \mathcal{C}$ is naturally pointed by the empty object. If N has boundary, we set $\int_N \mathcal{C} = \int_{\mathring{N}} \mathcal{C}$. ◊

Definition 9.4 Let $\mathcal{A}$ be a small $\mathbb{K}$-linear monoidal category with a right module $\mathcal{X}$ and a left module $\mathcal{Y}$. Recall from the proof of Lemma 8.6 that the *naive tensor product* of $\mathcal{X}$ with $\mathcal{Y}$ is the category $\mathcal{X} \otimes_{\mathbb{K}} \mathcal{Y}$ with object set $\text{ob}(\mathcal{X}) \times \text{ob}(\mathcal{Y})$ and morphisms $\hom((X_1, Y_1), (X_2, Y_2)) = \hom_{\mathcal{X}}(X_1, X_2) \otimes \hom_{\mathcal{Y}}(Y_1, Y_2)$. The *naive balanced tensor product* $\mathcal{X} \otimes_{\mathcal{A}} \mathcal{Y}$ is the $\mathbb{K}$-linear category formed from $\mathcal{X} \otimes_{\mathbb{K}} \mathcal{Y}$ by adding, for each object $(X, A, Y) \in \mathcal{X} \otimes_{\mathbb{K}} \mathcal{A} \otimes_{\mathbb{K}} \mathcal{Y}$, a natural-in-$(X, A, Y)$ isomorphism $\alpha_{X,A,Y} : (X \triangleleft A) \otimes Y \xrightarrow{\sim} X \otimes (A \triangleright Y)$, subject to a pentagon relation (and also adding some unitor relations that we will not write down). Compare Example 6.3. ◊

The first part of the following result is straightforward. The excision property is tedious, and is worked out in full in [Coo19]. (That paper was based, in turn, on an earlier preprint of this paper.)

Theorem 9.5 ([Coo19]) *The skein category $\int_{N_1 \sqcup N_2} \mathcal{C}$ of a disjoint union is the naive tensor product of skein categories $(\int_{N_1} \mathcal{C}) \otimes_{\mathbb{K}} (\int_{N_2} \mathcal{C})$.*

The construction $\int_\square \mathcal{C} : \{surfaces\} \to \{categories\}$ is functorial on embeddings. It follows that for any oriented one-manifold P, $\int_P \mathcal{C} = \int_{P \times [0,1]} \mathcal{C}$ is naturally a monoidal category and for any oriented zero-manifold P, $\int_P \mathcal{C} = \int_{P \times [0,1]^2} \mathcal{C}$ is naturally a braided monoidal category. It follows also that if N is a surface with boundary ∂N, then $\int_N \mathcal{C}$ is naturally a module category for $\int_{\partial N} \mathcal{C}$. The interpretation map from Theorem 9.2 provides a braided monoidal equivalence $\int_{\{pt\}} \mathcal{C} \xrightarrow{\sim} \mathcal{C}$.

Moreover, the functor $\int_\square \mathcal{C} : \{surfaces\} \to \{categories\}$ satisfies the following version of excision. *Suppose that N is a surface with a decomposition as $N = N_1 \sqcup_P N_2$ along a one-manifold P. Then $\int_{N_1} \mathcal{C}$ is naturally a right $\int_P \mathcal{C}$-module and $\int_{N_2} \mathcal{C}$ is naturally a left module, and $\int_N \mathcal{C}$ is equivalent to the naive balanced tensor product*

$$\int_N \mathcal{C} \simeq \left(\int_{N_1} \mathcal{C}\right) \underset{\int_P \mathcal{C}}{\otimes} \left(\int_{N_2} \mathcal{C}\right).$$

□

Combining results from [AF15, Sch14] gives:

Corollary 9.6 *The assignment $N \mapsto \int_N \mathcal{C}$ defines a symmetric monoidal functor* $\mathrm{BORD}_2^{\mathrm{or}} \to \mathrm{ALG}_2^{\mathrm{lax}}(\mathrm{CAT}_{\mathbb{K}})$, *where* $\mathrm{BORD}_2^{\mathrm{or}}$ *is the fully extended oriented 2-bordism category and* $\mathrm{CAT}_{\mathbb{K}}$ *is the bicategory of small linear categories and linear functors.* □

This is not the preferred target of a Heisenberg-picture field theory—as in Definition 6.5, I would much rather land in $\mathrm{ALG}_2^{\mathrm{lax}}(\mathrm{COCOMP}_{\mathbb{K}})$. But this is not a problem. The following is proved in [Kel05]:

Lemma 9.7 *Let $\mathcal{X}$ be a small $\mathbb{K}$-linear category. The* free cocompletion *$\widehat{\mathcal{X}}$ of $\mathcal{X}$ is $\mathrm{FUN}_{\mathbb{K}}(\mathcal{X}^{\mathrm{op}}, \mathrm{VECT}_{\mathbb{K}})$. It satisfies the following universal property: for any $\mathcal{E} \in \mathrm{COCOMP}_{\mathbb{K}}$, there is a natural equivalence $\mathrm{FUN}_{\mathbb{K}}(\mathcal{X}, \mathcal{E}) \simeq \hom_{\mathrm{COCOMP}_{\mathbb{K}}}(\widehat{\mathcal{X}}, \mathcal{E})$. Free cocompletion is symmetric monoidal: $\widehat{\mathcal{X} \otimes_{\mathbb{K}} \mathcal{Y}} \simeq \widehat{\mathcal{X}} \boxtimes_{\mathbb{K}} \widehat{\mathcal{Y}}$. The essential image of $\widehat{(-)}$ consists of those cocomplete categories generated under colimits by some set of compact projective objects. There is a natural equivalence*

$$\hom_{\mathrm{COCOMP}_{\mathbb{K}}}(\widehat{\mathcal{X}}, \widehat{\mathcal{Y}}) \simeq \mathrm{FUN}_{\mathbb{K}}(\mathcal{X} \otimes_{\mathbb{K}} \mathcal{Y}^{\mathrm{op}}, \mathrm{VECT}_{\mathbb{K}}).$$

□

Linear functors $\mathcal{X} \otimes_{\mathbb{K}} \mathcal{Y}^{\mathrm{op}} \to \mathrm{VECT}_{\mathbb{K}}$ should be thought of as *bimodules* between the categories $\mathcal{X}$ and $\mathcal{Y}$, where $\mathcal{X}$ and $\mathcal{Y}$ themselves are "many object associative algebras."

Corollary 9.8 *Free cocompletion takes naive balanced tensor products to balanced tensor products:*

$$\widehat{\mathcal{X} \otimes_{\mathcal{A}} \mathcal{Y}} \simeq \widehat{\mathcal{X}} \boxtimes_{\widehat{\mathcal{A}}} \widehat{\mathcal{Y}}$$

if $\mathcal{A}$ is a small $\mathbb{K}$-linear monoidal category with a left module $\mathcal{X}$ and a right module $\mathcal{Y}$.

Proof Both sides satisfy the same universal property. □

It follows that $N \mapsto \widehat{\int_N \mathcal{C}}$ defines a symmetric monoidal functor $\mathrm{BORD}_2^{\mathrm{or}} \to \mathrm{ALG}_2^{\mathrm{lax}}(\mathrm{COCOMP}_{\mathbb{K}})$. This is not yet a Heisenberg-picture field theory: it assigns pointed linear categories, not pointed vector spaces, to top-dimensional manifolds. To build a Heisenberg-picture field theory requires extending this functor to three-manifolds.

Definition 9.9 Let M be a compact three-dimensional manifold with boundary decomposed as $\partial M = N_{\mathrm{bot}} \sqcup_{P \times \{0\}} P \times [0, 1] \sqcup_{P \times \{1\}} N_{\mathrm{top}}$, where P is a one-manifold and N_{bot} and N_{top} are the "bottom" and "top" parts of the boundary. Such manifolds are the three-morphisms in $\mathrm{BORD}_3^{\mathrm{or}}$. Given a ribbon category $\mathcal{C}$, the *skein module* $\int_M \mathcal{C}$ of M is (the span of) ribbon tangles in M modulo local relations for little cubes $[0, 1]^3 \hookrightarrow M$ just as in the morphisms of Definition 9.3. By stacking on cylinders over N_{bot} and N_{top}, $\int_M \mathcal{C}$ is naturally an $\int_{N_{\mathrm{bot}}} \mathcal{C}$–$\int_{N_{\mathrm{top}}} \mathcal{C}$–bimodule, i.e. a functor

$$\left(\int_{N_{\text{bot}}} \mathcal{C}\right) \otimes_{\mathbb{K}} \left(\int_{N_{\text{top}}} \mathcal{C}\right)^{\text{op}} \to \text{VECT}_{\mathbb{K}},$$

and so defines a cocontinuous functor

$$\widehat{\int_M \mathcal{C}} : \widehat{\int_{N_{\text{bot}}} \mathcal{C}} \to \widehat{\int_{N_{\text{top}}} \mathcal{C}}.$$

The empty tangle gives $\widehat{\int_M \mathcal{C}}$ the structure of a lax pointed functor. ◊

Conjecture 9.10 When P is nonempty, for each object $X \in \int_P \mathcal{C}$, there is a natural operation that takes a tangle in M and outputs its union with the identity tangle over X. This operation makes $\widehat{\int_M \mathcal{C}}$ into a lax module functor between the $\widehat{\int_P \mathcal{C}}$-modules $\widehat{\int_{N_{\text{bot}}} \mathcal{C}}$ and $\widehat{\int_{N_{\text{top}}} \mathcal{C}}$. The assignment $\widehat{\int_\square \mathcal{C}}$ defines an oriented topological fully extended 1-affine Heisenberg-picture field theory $\text{BORD}_3^{\text{or}} \to \text{ALG}_2^{\text{lax}}(\text{COCOMP}_{\mathbb{K}})$.

Remark 9.11 Conjecture 9.10 surely follows from the previous results, and is equivalent in spirit to results from [Wal06]. Nevertheless, I have labeled it a Conjecture because there are some subtle details that must be checked (and I invite the reader to convert such a check into a paper). That $P \mapsto \int_P \mathcal{C}$ defines a functor $\text{BORD}_2^{\text{or}} \to \text{ALG}_2^{\text{lax}}(\text{COCOMP}_{\mathbb{K}})$ follows from cocompleting Corollary 9.6. This provides the values of the desired functor $\text{BORD}_3^{\text{or}} \to \text{ALG}_2^{\text{lax}}(\text{COCOMP}_{\mathbb{K}})$ on 0, 1, and 2-morphisms. The values on 3-morphisms are given by the skein modules of Definition 9.9, and functoriality for 3-morphism composition is manifest. The subtle detail that needs checking is the compatibility between the composition of 3-morphisms and the composition of 2-morphisms. ◊

A braided monoidal cocomplete category $\mathcal{A} \in \text{ALG}_2^{\text{lax}}(\text{COCOMP}_{\mathbb{K}})$ is of the form $\widehat{\mathcal{C}}$ for $\mathcal{C}$ a small ribbon category if and only if it satisfies all of the following:

- The unit object $\mathbb{1}_{\mathcal{A}} \in \mathcal{A}$ is compact projective.
- Every object of $\mathcal{A}$ is a colimit of dualizable objects.
- $\mathcal{A}$ is equipped with a full twist whose restriction makes the full subcategory $\mathcal{A}^{\text{fd}}$ on the dualizable objects into a ribbon category.
- $\mathcal{A}^{\text{fd}}$ is equivalent to a small category.

For such $\mathcal{A}$, $\mathcal{A} = \widehat{\mathcal{A}^{\text{fd}}}$. In general, $(\widehat{\mathcal{C}})^{\text{fd}}$ is the *Karoubi envelop* or *Cauchy completion* of $\mathcal{C}$: it is what you get when you take $\mathcal{C}$, add all finite direct sums, and then split all idempotents. These conditions are satisfied, for example, for $\mathcal{A} = \text{REP}_{\mathbb{C}(q)}^{\text{integrable}}(U_q\mathfrak{g})$ the category of ind-finite-dimensional representations of a quantum group at generic quantum parameter q, or for $\mathcal{A} = \text{REP}_{\mathbb{C}}^{\text{algebraic}}(G)$ the category of algebraic representations of a reductive group when $\mathbb{C}$; the dualizable objects are then the finite-dimensional modules. The conditions fail for (non-semisimplified) quantum groups at roots of unity and for finite groups over fields of characteristic dividing the order of the group.

Example 9.12 Let us say that a *trivial* ribbon category is the category $\mathrm{MOD}_A^{\mathrm{fd}}$ of finitely generated projective modules over a commutative $\mathbb{K}$-algebra A with $\otimes = \otimes_A$, or some ribbon subcategory thereof. The monoidal category $(\mathrm{MOD}_A^{\mathrm{fd}}, \otimes_A)$ admits a unique braiding and unique full twist, since it is generated under direct sums and passing to direct summands by the monoidal unit. All ribbon subcategories of $\mathrm{MOD}_A^{\mathrm{fd}}$ determine the same Heisenberg-picture field theory valued in $\mathrm{COCOMP}_{\mathbb{K}}$, since they have canonically identified free cocompletions. The 1-affine Heisenberg-picture field theory $\widehat{\int_\square \mathcal{C}}$ is 0-affine if and only if $\mathcal{C}$ is trivial. ◊

Example 9.13 If G is a reductive group over $\mathbb{K} = \mathbb{C}$ and N is a surface, then $\widehat{\int_N \mathrm{REP}_{\mathbb{C}}^{\mathrm{fd}}(G)} \simeq \mathrm{QCOH}(\mathrm{LOC}_G(N))$, where QCOH is the category of quasicoherent sheaves of $\mathcal{O}$-modules and $\mathrm{LOC}_G(N)$ is the stack of G-local systems, presented by the groupoid $\hom(\pi_1(N), G)/G$ [BZBJ18a, BZBJ18b]. This can be proven by showing that $N \mapsto \mathrm{QCOH}(\hom(\pi_1(N), G)/G)$ satisfies excision.

An object $X = \sqcup\{X_i\} \in \int_N \mathrm{REP}^{\mathrm{fd}}(G)$ represents the following sheaf on $\mathrm{LOC}_G(N)$: given a local system on N, restrict it at each of the small intervals $X_i : [0, 1] \to N$ in X; tensor each restriction with the corresponding G-module to get a flat vector bundle on the ith interval; tensor together the spaces of flat sections of each of these bundles.

Suppose that M is a three-manifold with boundary ∂M, read as a morphism $M : \emptyset \to \partial M$. Then $\widehat{\int_M \mathcal{C}} : \mathrm{VECT}_{\mathbb{K}} \to \widehat{\int_{\partial M} \mathcal{C}} \simeq \mathrm{QCOH}(\mathrm{LOC}_G(\partial))$ is the push-forward of the structure sheaf $\mathcal{O}(\mathrm{LOC}_G(M))$ along the restriction map $\mathrm{LOC}_G(M) \to \mathrm{LOC}_G(\partial M)$. ◊

Example 9.14 Let $\mathcal{C} = \mathrm{TL}$ be the well-known *Temperley–Lieb* category, which is a ribbon category defined over $\mathbb{K} = \mathbb{Z}[q, q^{-1}]$ whose $\mathbb{C}$-linearization is the monoidal subcategory of $\mathrm{REP}_{\mathbb{C}}(U_q\mathfrak{sl}_2)$ generated by the defining module. The Heisenberg-picture field theory $\int_\square \mathrm{TL}$ packages together all of Kauffman-bracket skein theory. ◊

Remark 9.15 Using the same skein-theoretic technology, one can show:

- Every $\mathbb{K}$-linear braided monoidal locally presentable category in which the unit object is compact projective and every object is a colimit of dualizable objects defines a framed Heisenberg-picture field theory $\mathrm{BORD}_3^{\mathrm{fr}} \to \mathrm{ALG}_2^{\mathrm{lax}}(\mathrm{COCOMP}_{\mathbb{K}})$.
- Every $\mathbb{K}$-linear monoidal locally presentable category in which the unit object is compact projective and every object is a colimit of dualizable objects defines a framed Heisenberg-picture field theory $\mathrm{BORD}_2^{\mathrm{fr}} \to \mathrm{ALG}_2^{\mathrm{lax}}(\mathrm{COCOMP}_{\mathbb{K}})$.
- Every $\mathbb{K}$-linear monoidal locally presentable category in which the unit object is compact projective and every object is a colimit of dualizable objects, which is additionally equipped with a "pivotal" structure, defines an oriented Heisenberg-picture field theory $\mathrm{BORD}_2^{\mathrm{or}} \to \mathrm{ALG}_2^{\mathrm{lax}}(\mathrm{COCOMP}_{\mathbb{K}})$.

After a preprint of this paper originally circulated, [BJS18] proved versions of the above results using the Cobordism Hypothesis rather than skein theory. ◊

References

[AF15] David Ayala and John Francis. Factorization homology of topological manifolds. *J. Topol.*, 8(4):1045–1084, 2015. arXiv:1206.5522. https://doi.org/10.1112/jtopol/jtv028. MR3431668.

[AFR18] David Ayala, John Francis, and Nick Rozenblyum. Factorization homology I: Higher categories. *Adv. Math.*, 333:1042–1177, 2018. arXiv:1504.04007. https://doi.org/10.1016/j.aim.2018.05.031. MR3818096.

[AR94] Jiří Adámek and Jiří Rosický. *Locally presentable and accessible categories*, volume 189 of *London Mathematical Society Lecture Note Series*. Cambridge University Press, Cambridge, 1994. MR1294136. https://doi.org/10.1017/CBO9780511600579.

[Ati88] Michael Atiyah. Topological quantum field theories. *Inst. Hautes Études Sci. Publ. Math.*, (68):175–186 (1989), 1988. MR1001453.

[Aya08] David Ayala. *Geometric Cobordism Categories*. PhD thesis, Stanford University, 2008. arXiv:0811.2280.

[BC14] Martin Brandenburg and Alexandru Chirvasitu. Tensor functors between categories of quasi-coherent sheaves. *J. Algebra*, 399:675–692, 2014. MR3144607. arXiv:1202.5147.

[BCJF15] Martin Brandenburg, Alexandru Chirvasitu, and Theo Johnson-Freyd. Reflexivity and dualizability in categorified linear algebra. *Theory Appl. Categ.*, 30:808–835, 2015. MR3361309. arXiv:1409.5934.

[BD95] J.C. Baez and J. Dolan. Higher dimensional algebra and topological quantum field theory. *J. Math. Phys.*, 36:6073–6105, 1995. MR1355899. arXiv:q-alg/9503002. https://doi.org/10.1063/1.531236.

[Bir84] Gregory J. Bird. *Limits in 2-categories of locally presentable categories*. PhD thesis, University of Sydney, 1984. Circulated by the Sydney Category Seminar.

[BJS18] Adrien Brochier, David Jordan, and Noah Snyder. On dualizability of braided tensor categories. 2018. 1804.07538.

[Bra11] Martin Brandenburg. Tensorial schemes. 2011. arXiv:1110.6523.

[BSP11] Clark Barwick and Christopher Schommer-Pries. On the unicity of the homotopy theory of higher categories. 12 2011. arXiv:1112.0040.

[BZBJ18a] David Ben-Zvi, Adrien Brochier, and David Jordan. Integrating quantum groups over surfaces. *J. Topol.*, 11(4):874–917, 2018. arXiv:1501.04652. https://doi.org/10.1112/topo.12072. MR3847209.

[BZBJ18b] David Ben-Zvi, Adrien Brochier, and David Jordan. Quantum character varieties and braided module categories. *Selecta Math. (N.S.)*, 24(5):4711–4748, 2018. arXiv:1606.04769. https://doi.org/10.1007/s00029-018-0426-y. MR3874702.

[BZFN10] David Ben-Zvi, John Francis, and David Nadler. Integral transforms and Drinfeld centers in derived algebraic geometry. *J. Amer. Math. Soc.*, 23(4):909–966, 2010. MR2669705. arXiv:0805.0157.

[CG16] Kevin Costello and Owen Gwilliam. *Factorization algebras in quantum field theory*. Cambridge University Press, 2016. Drafts available at http://people.mpim-bonn.mpg.de/gwilliam/vol1may8.pdf and http://people.mpim-bonn.mpg.de/gwilliam/vol2may8.pdf.

[CJF13] Alexandru Chirvasitu and Theo Johnson-Freyd. The fundamental pro-groupoid of an affine 2-scheme. *Appl. Categ. Structures*, 21(5):469–522, 2013. MR3097055. arXiv:1105.3104. https://doi.org/10.1007/s10485-011-9275-y.
[Coo19] Juliet Cooke. Excision of skein categories and factorisation homology. 2019. arXiv:1910.02630.
[CS19] Damien Calaque and Claudia Scheimbauer. A note on the (∞, n)-category of cobordisms. *Algebr. Geom. Topol.*, 19(2):533–655, 2019. arXiv:1509.08906. https://doi.org/10.2140/agt.2019.19.533. MR3924174.
[DSPS19] Christopher L. Douglas, Christopher Schommer-Pries, and Noah Snyder. The balanced tensor product of module categories. *Kyoto J. Math.*, 59(1):167–179, 2019. arXiv:1406.4204. https://doi.org/10.1215/21562261-2018-0006.
[DSPS20] Christopher L. Douglas, Christopher Schommer-Pries, and Noah Snyder. Dualizable tensor categories. *Memoirs of the AMS*, 2020. arXiv:1312.7188.
[EH62] B. Eckmann and P. J. Hilton. Group-like structures in general categories. I. Multiplications and comultiplications. *Math. Ann.*, 145:227–255, 1961/1962. MR0136642.
[Eil60] Samuel Eilenberg. Abstract description of some basic functors. *J. Indian Math. Soc. (N.S.)*, 24:231–234 (1961), 1960. MR0125148.
[FH16] Daniel S. Freed and Michael J. Hopkins. Reflection positivity and invertible topological phases. 2016. arXiv:1604.06527.
[Fre94] Daniel S. Freed. Higher algebraic structures and quantization. *Comm. Math. Phys.*, 159(2):343–398, 1994. MR1256993. arXiv:hep-th/9212115.
[FT14] Daniel S. Freed and Constantin Teleman. Relative quantum field theory. *Comm. Math. Phys.*, 326(2):459–476, 2014. arXiv:1212.1692. https://doi.org/10.1007/s00220-013-1880-1. MR3165462.
[Gab62] Pierre Gabriel. Des catégories abéliennes. *Bull. Soc. Math. France*, 90:323–448, 1962. MR0232821.
[GH15] David Gepner and Rune Haugseng. Enriched ∞-categories via non-symmetric ∞-operads. *Adv. Math.*, 279:575–716, 2015. https://doi.org/10.1016/j.aim.2015.02.007. arXiv:1312.3178. MR3345192.
[Gin13] Grégory Ginot. Notes on factorization algebras, factorization homology and applications. In *Mathematical aspects of quantum field theories*. Springer, 2013. To appear. arXiv:1307.5213.
[GS18] Owen Gwilliam and Claudia Scheimbauer. Duals and adjoints in higher Morita categories. 2018. arXiv:1804.10924.
[HR19] Jack Hall and David Rydh. Coherent Tannaka duality and algebraicity of Hom-stacks. *Algebra Number Theory*, 13(7):1633–1675, 2019. arXiv:1405.7680. https://doi.org/10.2140/ant.2019.13.1633. MR4009673.
[JF15] Theo Johnson-Freyd. Homological perturbation theory for nonperturbative integrals. *Lett. Math. Phys.*, 2015. https://doi.org/10.1007/s11005-015-0791-9. arXiv:1206.5319.
[JF17] Theo Johnson-Freyd. Spin, statistics, orientations, unitarity. *Algebr. Geom. Topol.*, 17(2):917–956, 2017. https://doi.org/10.2140/agt.2017.17.917. MR3623677. arXiv:1507.06297.
[JFS17] Theo Johnson-Freyd and Claudia Scheimbauer. (Op)lax natural transformations, twisted field theories, and "even higher" Morita categories. *Adv. Math.*, 307:147–223, 2 2017. https://doi.org/10.1016/j.aim.2016.11.014. arXiv:1502.06526.
[JS91] André Joyal and Ross Street. The geometry of tensor calculus. *Adv. Math.*, 88(1):55–112, 1991. https://doi.org/10.1016/0001-8708(91)90003-P. MR1113284. Part II available at http://www.researchgate.net/publication/228929964.
[Kel05] G. M. Kelly. Basic concepts of enriched category theory. *Repr. Theory Appl. Categ.*, (10):vi+137, 2005. Reprint of the 1982 original. MR2177301.
[Law93] R. J. Lawrence. Triangulations, categories and extended topological field theories. In *Quantum topology*, volume 3 of *Ser. Knots Everything*, pages 191–208. World Sci. Publ., River Edge, NJ, 1993. https://doi.org/10.1142/9789812796387_0011. MR1273575.

[Lur09] Jacob Lurie. On the classification of topological field theories. In *Current developments in mathematics, 2008*, pages 129–280. Int. Press, Somerville, MA, 2009. MR2555928. arXiv:0905.0465.

[Lur14] Jacob Lurie. *Higher Algebra*. 2014. http://www.math.harvard.edu/~lurie/papers/HigherAlgebra.pdf.

[MW11] Scott Morrison and Kevin Walker. Higher categories, colimits, and the blob complex. *Proc. Natl. Acad. Sci. USA*, 108(20):8139–8145, 2011. https://doi.org/10.1073/pnas.1018168108. MR2806651. arXiv:http://arxiv.org/abs/1108.5386.

[Res10a] N. Reshetikhin. Lectures on quantization of gauge systems. 2010. arXiv:1008.1411.

[Res10b] Nicolai Reshetikhin. Topological quantum field theory: 20 years later. In *European Congress of Mathematics*, pages 333–377. Eur. Math. Soc., Zürich, 2010. https://doi.org/10.4171/077-1/15. MR2648332.

[Ros98a] Alexander L. Rosenberg. Noncommutative schemes. *Compositio Math.*, 112(1):93–125, 1998. MR1622759.

[Ros98b] Alexander L. Rosenberg. The spectrum of abelian categories and reconstruction of schemes. In *Rings, Hopf algebras, and Brauer groups (Antwerp/Brussels, 1996)*, volume 197 of *Lecture Notes in Pure and Appl. Math.*, pages 257–274. Dekker, New York, 1998. MR1615928.

[RT90] N. Yu. Reshetikhin and V. G. Turaev. Ribbon graphs and their invariants derived from quantum groups. *Comm. Math. Phys.*, 127(1):1–26, 1990. MR1036112.

[RT91] N. Reshetikhin and V. G. Turaev. Invariants of 3-manifolds via link polynomials and quantum groups. *Invent. Math.*, 103(3):547–597, 1991. MR1091619. https://doi.org/10.1007/BF01239527.

[Sch09] Urs Schreiber. AQFT from n-functorial QFT. *Comm. Math. Phys.*, 291(2):357–401, 2009. https://doi.org/10.1007/s00220-009-0840-2. MR2530165.

[Sch14] Claudia Scheimbauer. *Factorization Homology as a Fully Extended Topological Field Theory*. PhD thesis, ETH Zürich, 2014. https://folk.ntnu.no/claudiis/ScheimbauerThesis.pdf.

[Seg04] Graeme Segal. The definition of conformal field theory. In *Topology, geometry and quantum field theory*, volume 308 of *London Math. Soc. Lecture Note Ser.*, pages 421–577. Cambridge Univ. Press, Cambridge, 2004. MR2079383.

[Shu08] Michael Shulman. Framed bicategories and monoidal fibrations. *Theory Appl. Categ.*, 20:No. 18, 650–738, 2008. arXiv:0706.1286. MR2534210.

[Shu10] Michael A. Shulman. Constructing symmetric monoidal bicategories. 2010. arXiv:1004.0993.

[SP09] Christopher J. Schommer-Pries. *The Classification of Two-Dimensional Extended Topological Field Theories*. PhD thesis, University of California, Berkeley, 2009. arXiv:1112.1000. MRMR2534210.

[ST11] Stephan Stolz and Peter Teichner. Supersymmetric field theories and generalized cohomology. In *Mathematical foundations of quantum field theory and perturbative string theory*, volume 83 of *Proc. Sympos. Pure Math.*, pages 279–340. Amer. Math. Soc., Providence, RI, 2011. MR2742432. arXiv:1108.0189.

[TWZ07] Xiang Tang, Alan Weinstein, and Chenchang Zhu. Hopfish algebras. *Pacific J. Math.*, 231(1):193–216, 2007. MR2304628.

[Wal06] Kevin Walker. TQFTs [early incomplete draft]. http://canyon23.net/math/tc.pdf, 2006.

[Wat60] Charles E. Watts. Intrinsic characterizations of some additive functors. *Proc. Amer. Math. Soc.*, 11:5–8, 1960. MR0118757.

Irreducibility of the Wysiwyg Representations of Thompson's Groups

Vaughan F. R. Jones

Abstract We prove irreducibility and mutual inequivalence for certain unitary representations of R. Thompson's groups F and T.

1 Introduction

Let F and T be the Thompson groups as usual. In [10], an action of F was shown to arise from a *functor* from the category $\mathcal{F}$ whose objects are natural numbers and whose morphisms are planar binary forests, to another category $\mathcal{C}$. Forests decorated with cyclic permutations of their leaves give a category $\mathcal{T}$ for which functors from $\mathcal{T}$ give actions of T.

The representations studied in [10] came from functors Φ to a trivalent tensor category (planar algebra) $\mathcal{C}$ in the sense of [12], based on a specific "vacuum vector" Ω in the 1-box space of the tensor category. A Thompson group element g is represented by a pair of binary planar trees (see [4]), drawn in the plane with one tree upside down on top of the other as below for an element of F that we will call D:

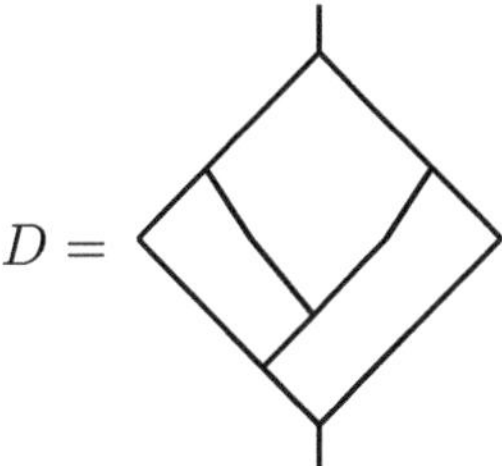

Vaughan F. R. Jones is deceased.

Vaughan F. R. Jones (✉)
Vanderbilt University, Department of Mathematics, Nashville, TN, USA

A. Alekseev et al. (eds.), *Representation Theory, Mathematical Physics, and Integrable Systems*, Progress in Mathematics 340,
https://doi.org/10.1007/978-3-030-78148-4_14

The standard dyadic intervals defined by the leaves of the bottom tree are sent by g (in the only affine way possible) to the corresponding intervals for the top tree.

If π is the unitary representation defined by the (suitably normalised) trivalent vertex in $\mathcal{C}$, the coefficient

$$\langle \pi(g)\Omega, \Omega \rangle$$

is simply equal to the pair of trees of g interpreted as a planar diagram (tangle) for $\mathcal{C}$!! (Or more correctly, the pair of trees as drawn is a multiple of a single vertical straight line and that multiple is $\langle \pi(g)\Omega, \Omega \rangle$.) For this reason, we will call these representations, on the closure of the linear span of $\pi(F)(\Omega)$ or $\pi(T)(\Omega)$, the Wysiwyg (what you see is what you get) representations of the Thompson groups. It is plausible that all the Wysiwyg representations are irreducible; indeed, the entire family of unitary representations defined in [10] and [9] could all be irreducible. Until this paper, the only examples where irreducibility could be shown were when the representation was *induced* from a subgroup. Then, the problem becomes one of calculating the commensurator of that subgroup. This was done in [6] and [13].

In this chapter, we will show that if we change the vacuum vector slightly, then all the Wysiwyg representations are irreducible. The proof will not be difficult but comes from a remarkable piece of luck involving this new vacuum Ψ. To wit, Ψ is fixed by the usual generator A of F and any other fixed vector by A is a multiple of Ψ. This immediately implies that the representation is irreducible on the closure of the linear span of $\pi(F)(\Omega)$. (More information on the spectral measure of elements of F can be found in [1].)

Since these fixed vectors are canonical, any numerical data calculated from them are invariants of the representation. In this way, we are able to show that these Wysiwyg representations are mutually inequivalent for different values of the parameter d of the category $\mathcal{C}$. This was also unknown before, even for the induced representations.

A more "elementary" way to get unitary representations is developed in [3] where the tensor category is Hilbert spaces with direct sum as tensor product. Here too the coefficients of the vacuum vector are given simply by the pair of trees as above, but interpreted as morphisms in the category. Little is known about irreducibility in this situation.

The author would like to thank the referee for carefully reading the manuscript and making suggestions that forced him to give a correct account of some thorny details.

2 Definitions

A binary planar forest is the isotopy classes of a disjoint union of binary trees embedded in $\mathbb{R}^2$ all of whose roots lie on $(\mathbb{R}, 0)$ and all of whose leaves lie on $(\mathbb{R}, 1)$. The isotopies are supported in the strip $(\mathbb{R}, [0, 1])$. Binary planar forests form a category in the obvious way with objects being $\mathbb{N}$ whose elements are identified with isotopy classes of sets of points on a line and whose morphisms are the forests which can be composed by stacking a forest in $(\mathbb{R}, [0, 1])$ on top of another, lining up the leaves of the one on the bottom with the roots of the other by isotopy and then rescaling the y-axis to return to a forest in $(\mathbb{R}, [0, 1])$. The structure is of course actually combinatorial, but it is very useful to think of it in the way we have described.

We will call this category $\mathcal{F}$.

Definition 2.1 Fix $n \in \mathbb{N}$. For each $i = 1, 2, \cdots, n$, let f_i be the planar binary forest with n roots and $n + 1$ leaves consisting of straight lines joining $(k, 0)$ to $(k, 1)$ for $1 \leq k \leq i - 1$ and $(k, 0)$ to $(k + 1, 1)$ for $i + 1 \leq k \leq n$ and a single binary tree with root $(i, 0)$ and leaves $(i, 1)$ and $(i + 1, 1)$; thus,

Note that any element of $\mathcal{F}$ is in an essentially unique way a composition of morphisms f_i, the only relation being $\Phi(f_j)\Phi(f_i) = \Phi(f_i)\Phi(f_{j-1})$ for $i < j-1$. The set of morphisms from 1 to n in $\mathcal{F}$ is the set of binary planar rooted trees $\mathfrak{T}$ and is a *directed set* with $s \leq t$ iff there is and $f \in \mathcal{F}$ with $t = fs$.

Given a functor $\Phi : \mathcal{F} \to \mathcal{C}$ to a category $\mathcal{C}$ whose objects are sets, we define the direct system S_Φ, which associates to each $t \in \mathfrak{T}, t : 1 \to n$, the set $\Phi(target(t)) = \Phi(n)$. For each $s \leq t$, we need to give ι_s^t. For this, observe that there is an $f \in \mathcal{F}$ for which $t = fs$, so we define

$$\iota_s^t(\kappa) = \Phi(f)$$

which is an element of $Mor_{\mathcal{C}}(\Phi(target(s)), \Phi(target(t)))$ as required. The ι_s^t trivially satisfy the axioms of a direct system.

As a slight variation on this theme, given a functor $\Phi : \mathcal{F} \to \mathcal{C}$ to any category $\mathcal{C}$ and an object $\omega \in \mathcal{C}$, form the category $\mathcal{C}^\omega$ whose objects are the sets $Mor_{\mathcal{C}}(\omega, obj)$ for every object obj in $\mathcal{C}$ and whose morphisms are composition with those of $\mathcal{C}$. The definition of the functor $\Phi^\omega : \mathcal{F} \to \mathcal{C}^\omega$ is obvious. Thus, the direct system S_{Φ^ω} associates with each $t \in \mathfrak{T}, t : 1 \to n$, the set $Mor_{\mathcal{C}}(\omega, \Phi(n))$. Given $s \leq t$, let $f \in \mathcal{F}$ be such that $t = fs$. Then, for $\kappa \in Mor_{\mathcal{C}}(\omega, \Phi(target(s)))$,

$$\iota_s^t(\kappa) = \Phi(f) \circ \kappa$$

which is an element of $Mor_{\mathcal{C}}(\omega, \Phi(target(t)))$.

As in [10], we consider the direct limit:

$$\lim_{\to} S_\Phi = \{(t, x) \text{ with } t \in \mathfrak{T}, x \in \Phi(target(t))\}/\sim$$

where $(t, x) \sim (s, y)$ iff there are $r \in \mathfrak{T}, z \in \Phi(target(z))$ with $t = fr, s = gr$ and $\Phi(f)(x) = z = \Phi(g)(y)$.

We use $\frac{t}{x}$ *to denote the equivalence class of* (t, x) *mod* $\sim$.

The limit $\lim\limits_{\rightarrow} S_\Phi$ will inherit structure from the category $\mathcal{C}$. For instance, if the objects of $\mathcal{C}$ are Hilbert spaces and the morphisms are isometries, then $\lim\limits_{\rightarrow} S_\Phi$ will be a pre-Hilbert space that may be completed to a Hilbert space that we will also call the direct limit unless special care is required.

As was observed in [10], if we let Φ be the identity functor and choose ω to be the tree with one leaf, then the inductive limit consists of equivalence classes of pairs $\frac{t}{x}$ where $t \in \mathcal{T}$ and $x \in \Phi(target(t)) = Mor(1, target(t))$. But $Mor(1, target(t))$ is nothing but $s \in \mathcal{T}$ with $target(s) = target(t)$, i.e., trees with the same number of leaves as t. Thus, the inductive limit is nothing but the Thompson group F with group law

$$\frac{r}{s}\frac{s}{t} = \frac{r}{t}$$

Moreover, for any other functor Φ, $\lim\limits_{\rightarrow} S_\Phi$ carries a natural action of F defined as follows:

$$\frac{s}{t}(\frac{t}{x}) = \frac{s}{x}$$

where $s, t \in \mathfrak{T}$ with $target(s) = target(t) = n$ and $x \in \Phi(n)$. A Thompson group element given as a pair of trees with m leaves, and an element of $\lim\limits_{\rightarrow} S_\Phi$ given as a pair (tree with n leaves, element of $\Phi(n)$), may not be immediately composable by the above formula, but they can always be "stabilised" to be so within their equivalence classes.

The Thompson group action preserves the structure of $\lim\limits_{\rightarrow} S_\Phi$, so, for instance, in the Hilbert space case the representations are unitary.

3 The Wysiwyg Representations

We are especially interested in applying the construction of the previous section when $\mathcal{C}$ is the category of bifinite bimodules (correspondences in the sense of Connes) over a von Neumann algebra M - [5]. Here, we begin with a Hilbert space $\mathcal{H}$, which is a binormal bimodule for M, and form the category whose objects are the relative tensor powers $\otimes_M^n \mathcal{H}$, which are also $M - M$ bimodules. The morphisms of the category are $M - M$ bimodule maps. We will for simplicity restrict to certain such bimodule categories that have been described simply diagrammatically in [12]. To be precise, see the following definition:

Definition 3.1 For $d \in \{4cos^2\pi/n - 1|n = 4, 5, 6, \cdots\} \cup [3, \infty)$, let $\mathcal{C}$ be the trivalent (tensor) category described in [12] with objects 1 and $\otimes^n X$ for $n \geq 0$, with $\otimes^0 X = 1$, for a privileged object X.

By definition, $\mathcal{C}_n$ is $Mor(1, \otimes^n X)$, which is the finite-dimensional Hilbert space of linear combinations of tangles with n boundary points and an arbitrary number of trivalent vertices, modulo the skein relations given in [12], and the kernel of the natural positive semidefinite Hermitian form. We have $dim(\mathcal{C}_0) = 1, dim(\mathcal{C}_1) = 0, dim(\mathcal{C}_2 = 1)$ and $dim(\mathcal{C}_3) = 1$.

Here, the object 1 may be realised as $L^2(M)$ for a von Neumann algebra M and X would be some given $M - M$ correspondence.

The category $\mathcal{C}$ is by definition generated by an element in $\mathcal{C}_3$. Although [12] does not address unitarity issues, by realising $\mathcal{C}$ as a cabled Temperley–Lieb category as described in [12], $\mathcal{C}$ has a $*$-structure with a positive definite inner product. To be more explicit, all the pictures that we draw in this chapter can be converted to linear combinations of ordinary Temperley–Lieb pictures (for which positivity is well established) by doubling all the strings and inserting, on the doubled strings, the "second Jones–Wenzl idempotent" jw. Thus, a piece of string | becomes $jw =$ || $- \frac{1}{\delta}$ ∪∩ in Temperley-Lieb, which causes the loop parameter δ to become $d = \delta^2 - 1$. The idempotent property assures us that jw can be inserted as many times as we like. The generator Y of [12] is then

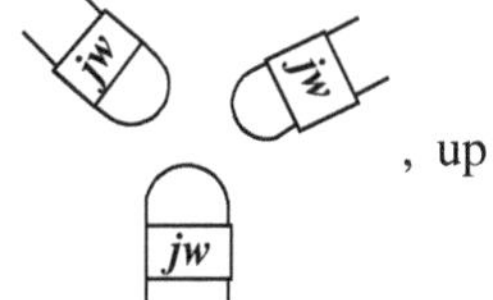

, up to normalisation, which is manifestly invariant under the rotation. $\mathcal{C}$ can also be realised as a category of correspondences over hyperfinite factors [8]. (Indeed, all of our considerations will apply to categories of $M - M$ bimodules generated by $\mathcal{H}$ with suitable irreducibility and self-duality conditions and an appropriate $M - M$ bilinear $Y : \mathcal{H} \to \mathcal{H} \otimes_M \mathcal{H}$ with $Y^*Y = 1$.)

Definition 3.2 $Mor_{\mathcal{C}}(1, X \otimes X)$ is spanned by a single tangle consisting of a string joining the 2 boundary points. We will normalise this tangle by dividing it by $\sqrt{d}$ and call the resulting unit vector Ψ.

$Mor_{\mathcal{C}}(X, X \otimes X)$ is spanned by a single tangle consisting of the trivalent vertex. By our normalisation, it is a unit vector. We will call it Ω.

The above choice of Y gives a functor from the category of binary planar forests to $\mathcal{C}$ as described in [10]. The relation 2.1 of [12] implies that the connecting maps ι_s^t are isometries. So, we obtain two categories $\mathcal{C}^1$ and $\mathcal{C}^X$ and direct systems S_{Φ^1} and

S_{Φ^X}. Thompson's groups F and T act unitarily on the Hilbert spaces $\mathfrak{H}_Y^1$ and $\mathfrak{H}_Y^X$ obtained as the (completions of the) direct limits $\varinjlim S_{\Phi^1}$ and $\varinjlim S_{\Phi^X}$, respectively.

Thus, Φ^X is the functor that sends the object n to the finite dimensional Hilbert space $\mathcal{C}_{n+1} = Mor_{\mathcal{C}}(X, \otimes^n X)$ and a morphism (forest) $f : m \to n \in \mathcal{F}$ to the isometry $\Phi^X(f) : \mathcal{C}_{m+1} \to \mathcal{C}_{n+1}$ obtained by interpreting the forest f as an element of $\mathcal{C}_{m+n}$ by replacing the vertices in f with the element $Y \in \mathcal{C}$. The element $\Phi^X(f)$ then has m ingoing boundary points and n outgoing ones so defines an isometry according to the usual language of planar algebras [11]. We illustrate below, for $x \in \Phi^X(3)$ and f, the forest with 3 roots and 8 leaves which is visible in the diagram (thus, $m = 3andn = 8$):

$$\Phi^X(f)(x) \in Mor_{\mathcal{C}}(X, \otimes^n X) =$$

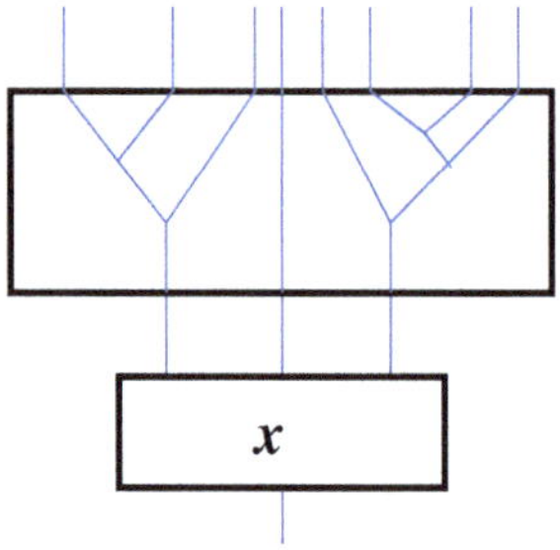

The functor Φ^1 is in some sense even simpler. We have

$$\Phi^1(n) = \begin{cases} 0 & \text{if } n = 1 \\ \mathcal{C}_n & \text{if } n \geq 2 \end{cases}$$

And as an illustration, for $x \in \Phi^1(3)$ and f as above,

$$\Phi^1(f)(x) \in Mor_{\mathcal{C}}(1, \otimes^n X) =$$

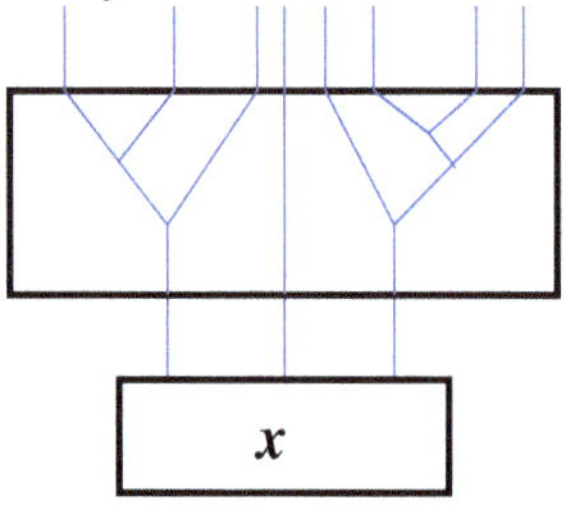

Note that, in the inductive limit, for any tree t with at least two leaves, the equivalence class $[\Omega] = \frac{t}{\Phi^X(f_t)(\Omega)}$, where f_t is the unique forest with $t = f_t \circ y$, y being the tree with two leaves. Similarly, $[\Psi] = \frac{1}{\sqrt{d}} \frac{t}{\Phi^1(f_t)(\Psi)}$. The distinction between t and f_t may seem pedantic, but it is required to make the formalism work.

When representing an element of Thompson's group F as a pair of trees as in the introduction, one can choose to either include or not include the vertical edges at the top and bottom marking the roots. If $g \in F$, let T_g be the version with these vertices and $\mathbb{T}_g$ be the version without, e.g., for the element $D \in F$ of the introduction, the given picture is T_X and $\mathbb{T}_D$ is

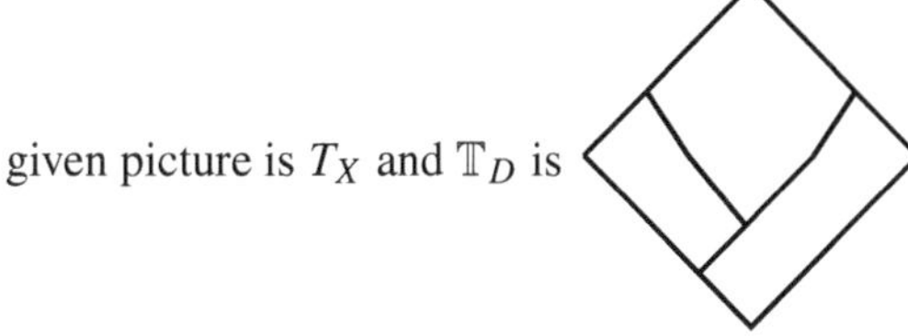

To alleviate notation, we will just use Ψ and Ω for $[\Psi]$ and $[\Omega]$ in the statement of the following proposition and after the proposition.

Proposition 3.1 *Let $g \in F$ be given. Then, the "vacuum expectation value" or "vacuum coefficient" $\langle \pi(g)\Omega, \Omega \rangle$ is $\frac{1}{d}$ times the scalar obtained by tying the top of T_g to the bottom and evaluating in $\mathcal{C}$ by replacing all vertices of the trees by Y.*

And $\langle \pi(g)\Psi, \Psi \rangle$ is $\frac{1}{d}$ times the scalar obtained by replacing all the vertices of $\mathbb{T}_g$ by Y and evaluating in $\mathcal{C}$.

Proof This is an exercise that will be clear if we illustrate using the element $D \in F$ and the representation coming from Φ^1.

If s is the tree and t is the tree then $D = \frac{s}{t}$.

As a vector in the direct limit, $\sqrt{d}[\Psi] = \dfrac{t}{\Phi^1(f_t)(\Psi)}$ so that $\sqrt{d}\pi(D)([\Psi]) = \dfrac{s}{\Phi^1(f_t)(\Psi)}$. But also $\sqrt{d}[\Psi] = \dfrac{s}{\Phi^1(f_s)(\Psi)}$, so by the definition of the inner product in the direct limit Hilbert space, $d\langle \pi(g)[\Psi], [\Psi] \rangle$ is the inner product in $\Phi^1(4)$ of $\Phi^1(f_t)(\Psi)$ with $\Phi^1(f_s)(\Psi)$, which by the definition of Φ is just viewed as an element of $\mathcal{C}_0 = \mathbb{C}$.

The case of Φ^X is proved in the same way.

□

Definition 3.3 We call the representations π_Y of F coming from Φ^1 and Φ^X as above, on the closure of the subspace spanned by $\Phi^1(F)(\Psi)$ and $\Phi^X(F)(\Omega)$, respectively, the *Wysiwyg* representations of F.

Definition 3.4 More generally, one can let p be any minimal projection in one of the algebras $\mathcal{C}_{2n}$ and consider the direct system

$$t \mapsto \begin{cases} \mathcal{C}_m & \text{if } t \text{ is a tree with } m \text{ leaves and } m \geq 2n \\ 0 & \text{if } t \text{ has less than } 2m \text{ leaves} \end{cases}$$

Then, F acts on the direct limit as before, and we let $\pi_{Y,p}$ be the unitary representation of F on the Hilbert space completion, so that $\pi_{Y,\phi}$ contains π_Y, where ϕ is the identity of $\mathcal{C}_0$.

4 The Main Theorem

Let $\mathcal{C}$ and Y be as in the last section. Let π_Y be the Wysiwyg representation of F defined above with vacuum Ψ.

Theorem 4.1 *The unitary representation π_Y is irreducible.*

Our proof will rely heavily on an algorithm for calculating the product of two elements of F, which is to be found in [2] and [7]. Let us describe it.

Given two pairs P and Q of (binary planar rooted) trees represented diagrammatically as in the introduction, place P on top of Q in the plane and join them as below:

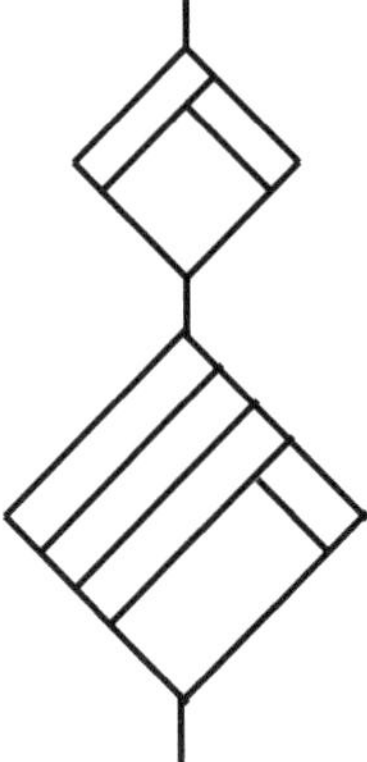

Then, apply the diagrammatic rules → | | and → | (and some isotopy) until they cannot be applied any more:

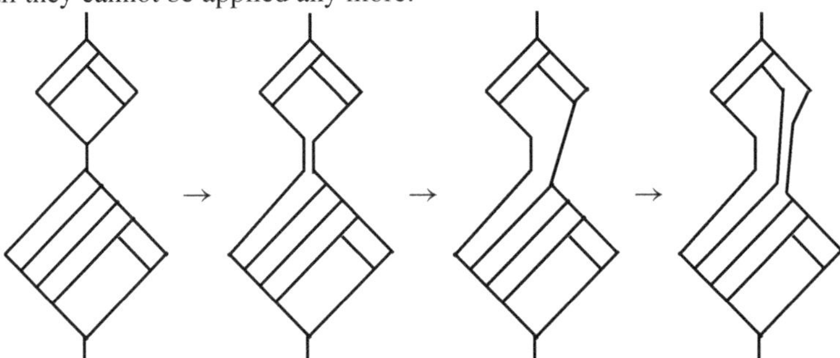

At this point, one may draw a curve through the diagram so that all the vertices below it are Y's and all above it are upside down Y's. Straightening that curve to a horizontal line thus gives a pair of trees that give the product of the two original F elements P and Q:

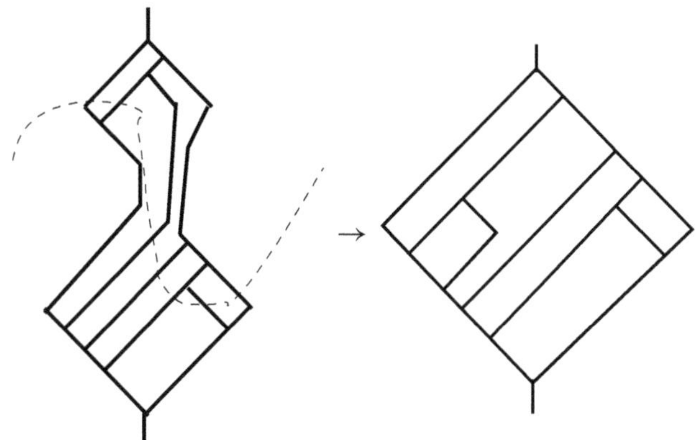

NB It is important to note that, although the diagrams involved in calculating gh make sense in the category $\mathcal{C}$, one may NOT use them to calculate vacuum expectation values until all the cancellations are done. This is because the relation $\times \to |\;|$ does not hold in $\mathcal{C}$.

The following lemma is just a simple calculation, but that calculation leads on to all the lucky accidents that make the proof work. Recall that A is the generator of F in [4].

Lemma 4.1 $\pi_Y(A)\Psi = \Psi$.

Proof The two-tree picture of A is 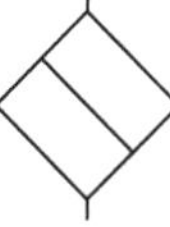 so the diagram in $\mathcal{C}$ giving $\langle A\Psi, \Psi\rangle$ is  which by the rules in $\mathcal{C}$ is the same as $d\langle\Psi, \Psi\rangle$. By the unitarity of π_Y, we are done. □

We leave it to the reader to deduce this result directly from the definition of the action of group elements on vectors in the direct limit so that the result actually holds even in the non-unitary case. (Note that we have suppressed the π_Y in the inner product formula. We will continue in this way.)

Lemma 4.2 *Let g and h be in F. Then,*

$$\langle A^n g\Psi, h\Psi\rangle = \langle g\Psi, \Psi\rangle\langle\Psi, h\Psi\rangle$$

for sufficiently large n.

Proof Observe that $\langle A^n g\Psi, h\Psi\rangle = \langle h^{-1}A^n g\Psi, \Psi\rangle$, so we must calculate the vacuum expectation value of $h^{-1}A^n g$.

The diagram for $h^{-1}A^n g$ before cancellation is (with n sloping lines in the A^n part)

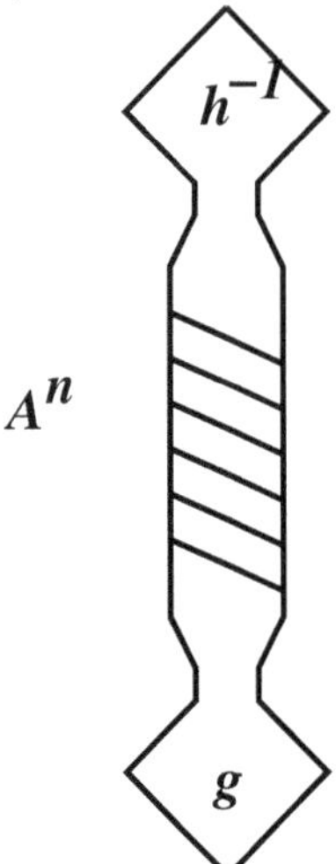

Now take the "pair of trees" picture of g and isolate the vertices on the right branch of the top tree as below:

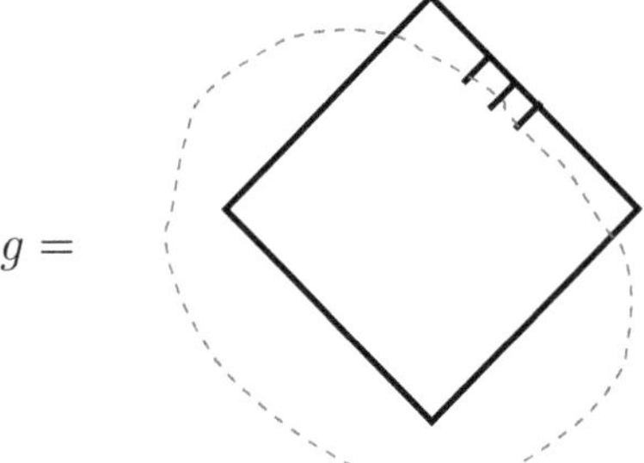

Look near the top of g and the bottom of A^n:

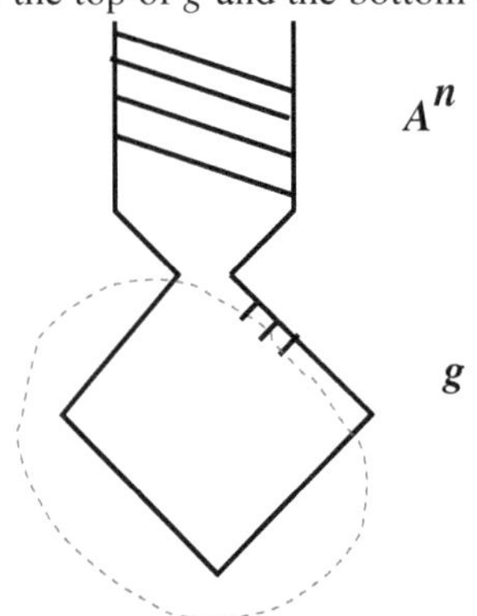

For sufficiently large n, one may apply $\times \to |\ |$ enough times to cancel all the right branch vertices in the top half of g to obtain

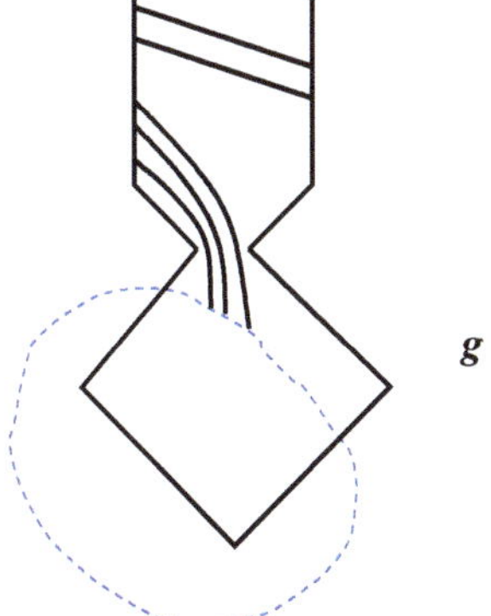

At this stage, all the strings connecting what is left of A^n to g are connected at the bottom as they were in g, which is reduced. Therefore, provided n is large enough, so there is no interference between g and h^{-1}, the only cancellation that can occur in the picture is the leftmost of these strings, marked "a" below, which can only cancel with a vertex of A^n if it is connected directly to the leftmost branch of the top tree of g.

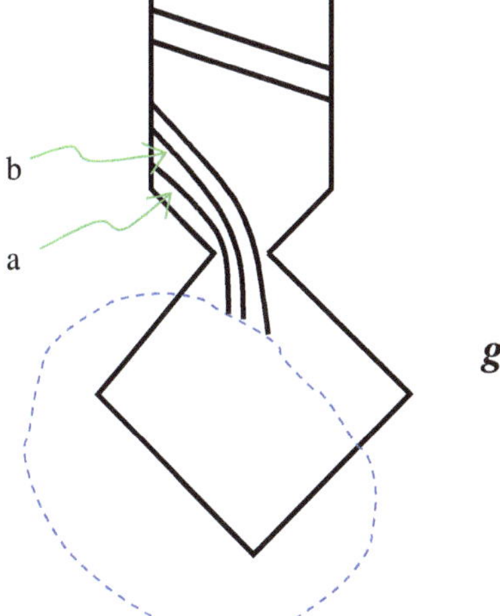

After the string a has been cancelled, the only other cancellation is b and so on. Once all these cancellations are done, no more can happen. The same considerations apply to h^{-1}, so we may calculate $\langle h^{-1}A^n g\Psi, \Psi\rangle$ by looking at the resulting figure in the trivalent category. But before doing that, observe that the last round of cancellations are all of the form ◇ → | which is also true in $\mathcal{C}$! So, we may undo them and obtain the picture below:

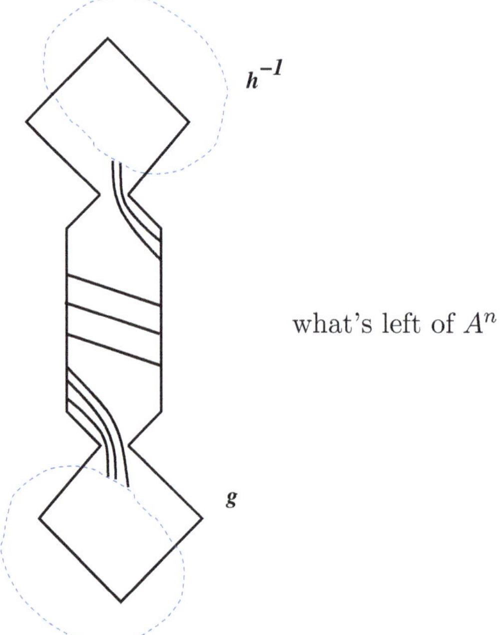

So that $\langle h^{-1}A^n g\Psi, \Psi\rangle$ is $\frac{1}{d}$ times the evaluation of the above picture in $\mathcal{C}$. But now isolate the parts of the picture involving g and h^{-1} as below:

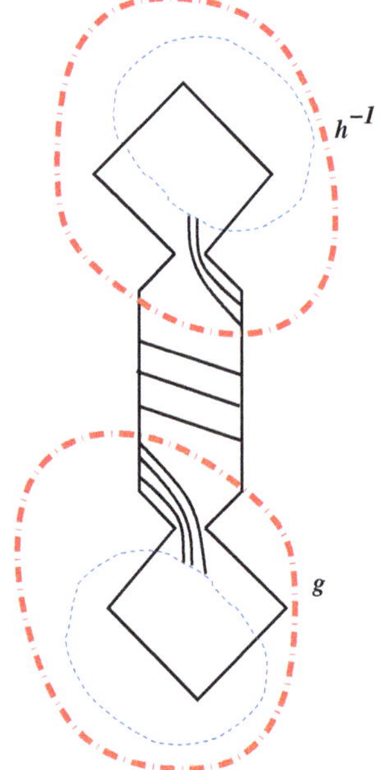

The parts enclosed in the red double dashed regions are actually elements of $\mathcal{C}_2$, which is one dimensional. Hence they are scalar multiples of a single curve joining the boundary points. To find the scalar multiple, cut out the parts inside the red dotted lines and join the boundary points to obtain (for g, h is similar)

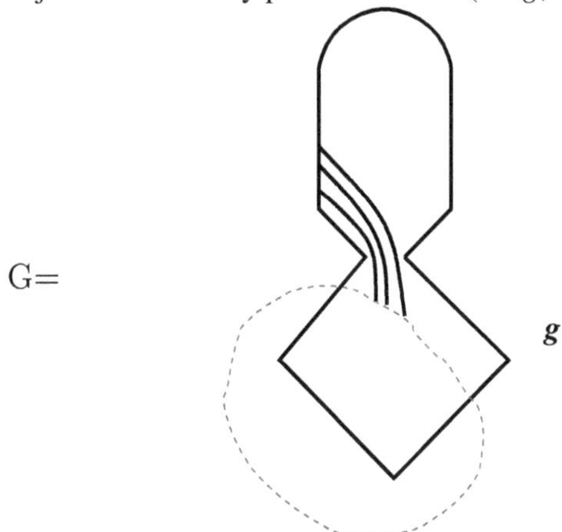

The curves, which started out connected to the top right branch of g, can be moved back there, the top of the picture straightened, and we recognise the picture of g that we started with, and similarly a factor H at the top. By Wysiwyg, we have

$$G = d\langle g\Psi, \Psi\rangle \text{ and } H = d\langle h^{-1}\Psi, \Psi\rangle$$

By easy planar algebra arguments, splitting a "connected sum" of two closed tangles as the product of individual closed tangles incurs a multiplicative factor of $\frac{1}{d}$.

The three pictures resulting from the red double dashed decomposition are G, H and A^k for some k with $0 \leq k \leq n$, to which we may apply $\Diamond \rightarrow \Big|$ k times to obtain just d.

Altogether, $d\langle A^n g\Psi, h\Psi\rangle = \frac{d}{d^2}GH$, so $\langle A^n g\Psi, h\Psi\rangle = \dfrac{G}{d}\dfrac{H}{d} = \langle g\Psi, \Psi\rangle\langle\Psi, h\Psi\rangle$.

It is interesting that the same calculation works for Φ^X right up to this point. But because of the extra string joining the top to the bottom, a multiplicative factor of $(\dfrac{d-2}{d-1})^k$ appears, which means A^n actually tends weakly to zero. Indeed, the projection onto $\mathbb{C}\Omega$ does not even commute with A in this case since if it did, $|\langle A\Omega, \Omega\rangle|$ would be 1. □

Corollary 4.1 *The weak limit of $\pi_Y(A^n)$ as $n \to \infty$ on the closure of the linear span of $\pi_Y(F)(\Psi)$ is equal to P_Ψ, the orthogonal projection onto the linear span of Ψ.*

Proof The formula for P_Ψ is $P_\Psi(\eta) = \langle \Psi, \eta \rangle \Psi$. Linear extension of the result of the previous lemma gives $\langle A^n \xi, \eta \rangle = \langle P_\Psi \xi, \eta \rangle$ for ξ and η in a dense subspace, and the corollary follows from the unitarity of π_Y. □

Proof The proof of the theorem is now easy. Let $\mathfrak{H}$ be the closed linear span of $\pi_Y(F)(\Psi)$ on which π_Y acts. Suppose t is a bounded operator on $\mathfrak{H}$ commuting with $\pi_Y(F)$. Then, by the previous corollary, $t\Psi = \lambda\Psi$ for some scalar λ. Thus, $tg\Psi = \lambda g\Psi$ for all $g \in F$, so taking linear combinations, we see that $t = \lambda id$ on all of $\mathfrak{H}$. □

Corollary 4.2 *The Wysiwyg representations from Φ^1 act values of d.*

Proof The vector Ψ when normalised is unique up to a scalar of modulus one, as a vector spanning the 1-eigenspace of A. But we can calculate, using the relations of [12], $\langle D\Psi, \Psi \rangle = \frac{d-2}{d-1}$. But $\langle \Psi, \Psi \rangle = 1$, so two Wysiwyg representations can only be equivalent if the values of d are equal. □

Remark 4.1 We know that the inductive limit Hilbert space also carries a unitary representation of Thompson's group T, extending that of F and also called π_Y. Exactly the same argument as above shows that T acts irreducibly on the closed T-linear span of Ψ, but we do not know if that is the same Hilbert space as $\mathfrak{H}$.

5 Improvements

We present two improvements on the previous results by extending from the F-linear span of the vacuum vector to the whole direct limit Hilbert space.

(a) Telling the representations apart.

It is curious that all the representations of F the kind we are considering contain the coefficients of the vacuum by weak limits of certain elements, which means that the representations are inequivalent as soon as the corresponding vacuum ones are.

To be more precise, let $\sigma : F \to F$ be the endomorphism which sends an element of F (viewed as a PL homeomorphism of $[0, 1]$) to itself rescaled and acting in $[\frac{1}{2}, 1]$ and by the identity on $[0, \frac{1}{2}]$.

Diagrammatically, we have, for $g = \frac{a}{b}$,

$\sigma(g) =$ 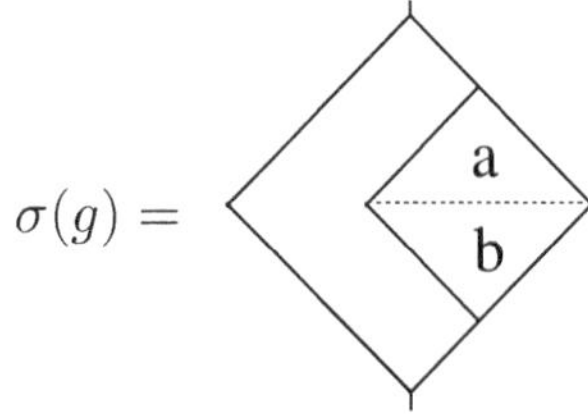

Theorem 5.1 *Let $g \in G$. Then,*

$$\pi_{Y,p}(\sigma^n(g)) \text{ tends weakly to } \langle \pi(g)\Omega, \Omega\rangle id \text{ for any } p \text{ as in } 3.4$$

Proof By unitarity as usual, it suffices to show that

$$\langle \pi_{Y,p}(\sigma^n(g))(\xi), \eta\rangle \to \langle \pi(g)\Omega, \Omega\rangle\langle \xi, \eta\rangle$$

for ξ and η in any finite-dimensional approximant of the direct limit. Let t_m be the full bifurcating planar tree with 2^m leaves (such trees form a cofinite sequence), and let x and y be elements of $\mathcal{C}_{2^m}$. Then, let $\xi = \frac{t_m}{x}$ and $\eta = \frac{t_m}{y}$ be the corresponding elements of the direct limit Hilbert space. We need to calculate $\langle \pi_{Y,p}(\sigma^n(g))(\xi), \eta\rangle$ for large n. For any $k > 0$ and any tree s, let r_s^k be the tree with n branches to the left attached to a single branch to the right and s attached to the end of the right branch; thus,

$r_s^3 =$ 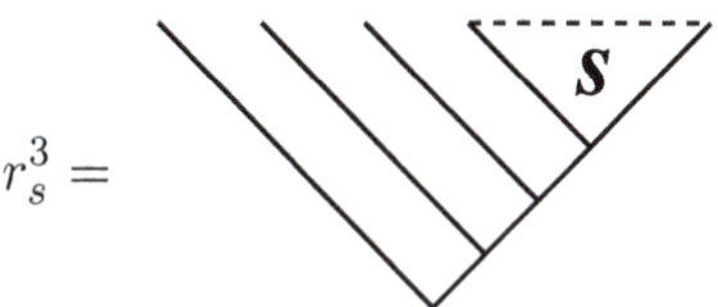

Suppose $n > m$ and that a is a tree. Let t_a be the tree obtained by attaching r_a^{n-m} to the rightmost leaf of t_m. Let g be given by the pair of trees $\frac{a}{b}$. Then, by drawing a diagram and cancelling lots of carets, we see that $\frac{t_a}{t_b} = \sigma^n(\frac{a}{b}) = \sigma^n(g)$. Moreover,

$$\frac{t_m}{x} = \frac{t_b}{x_b}$$

where x_b is the element of $\mathcal{C}$ defined by attaching the diagram r_b^{n-m} to the rightmost string emanating from a disc containing x, and thus (illustrated for $m = 3$, so there are 8 vertical strings emanating from the disc containing x)

$x_b =$ 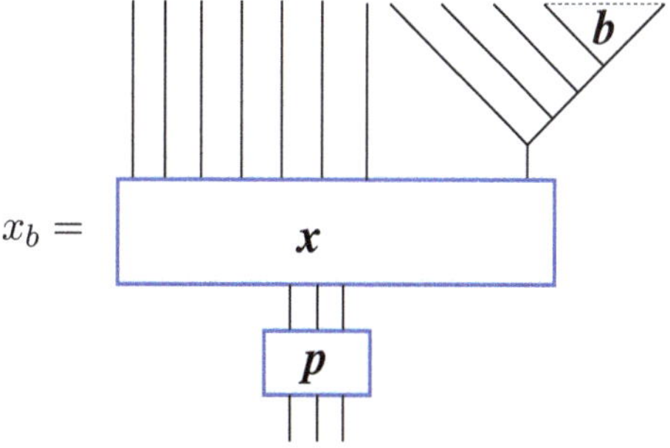

This picture is interpreted as an element of the planar algebra with the trivalent vertices being discs containing Y, as usual.

We have

$$\sigma^n(g)(\xi) = \frac{t_a}{t_b}(\frac{t_b}{x_b}) = \frac{t_a}{x_b}$$

and of course

$$\eta = \frac{t_m}{y} = \frac{t_a}{y_a}$$

so working the $\mathcal{C}$ we see that $\langle \pi_{Y,p}(\sigma^n(g))(\xi), \eta \rangle$ is the value in $\mathcal{C}$ of the following diagram (illustrated for $n = 6$):

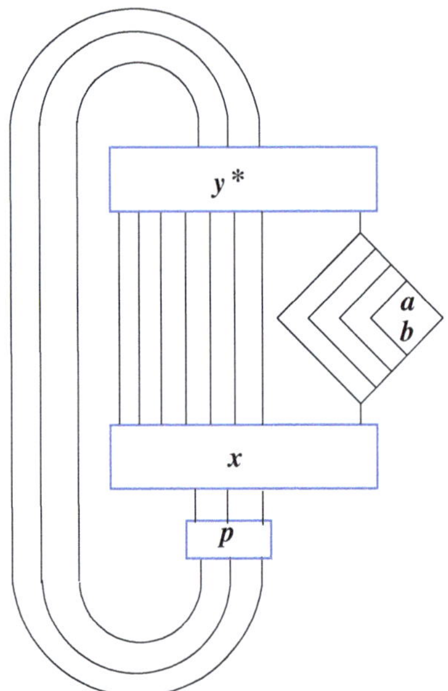

Since the dimension of $\mathcal{C}_1$ is one, we may surround the diamond shape part of the picture by a disc and use the unitarity of Y to conclude that this picture is a multiple of $\langle \pi(g)\Omega, \Omega \rangle \langle \xi, \eta \rangle$. □

Corollary 5.1 *$\pi_{Y,p}$ is inequivalent to $\pi_{Y',q}$ whenever $\pi_{Y,1}$ is inequivalent to $\pi_{Y',1}$.*

(b) In the last section, we showed that A^n tends weakly to the projection onto $\mathbb{C}\Psi$ on the F-linear span of Ψ. Here we extend that to the whole space $\mathfrak{H}^1_Y$. We record also the corresponding result for $\pi_{Y,1}$.

Theorem 5.2

(i) $\pi_{Y,\phi}(A^n)$ *tends weakly to the projection onto* $\mathbb{C}\Psi$ *as* $n \to \infty$.
(ii) $\pi_{Y,1}(A^n)$ *tends weakly to zero as* $n \to \infty$.

Proof (i) The proof will involve a calculation similar to that of the previous theorem, but fortunately the detailed structure of the stabilising forests need not concern us. This is because we know already that there is a fixed vector for $\pi_{Y,\phi}(A)$, so if we can show the weak limit of $\pi_{Y,\phi}(A^n)$ exists and has rank 1, it can only be orthogonal projection onto the subspace spanned by the fixed point.

So, let a_n and b_n be the trees with n leaves illustrated below for $n = 7$:

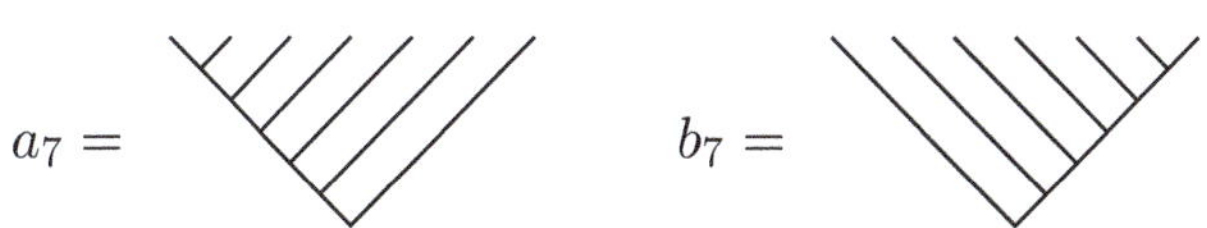

Then, $A^n = \dfrac{a_{n-2}}{b_{n-2}}$.

Suppose we are given $\xi = \dfrac{t_m}{x}$ and $\eta = \dfrac{t_m}{y}$ as in the last theorem in the direct limit Hilbert space on which $\pi_{Y,\phi}(A^n)$ acts. As before, our first job is to stabilise b_n and t_m, so they are equal. Provided n is much larger than 2^m, this will be achieved for b_n by a forest of the form:

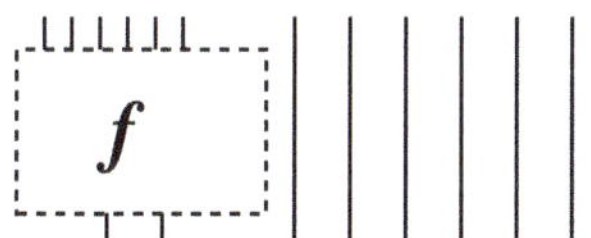

where f is the forest with the $m - 2$ trees $t_{m-1}, t_{m-2}, \cdots$ from left to right. To stabilise the numerator t_m of ξ, simply attach a copy of b_k for some large k (approximately equal to $n - 2^m$) to the rightmost leaf of t_m. Thus, we have

$$\pi_{Y,\phi}(A^n)(\xi) = \frac{\tilde{a}_n}{\tilde{x}}$$

with $\tilde{x}$ given by the following tangle (illustrated with $m = 3$ and $k = 7$):

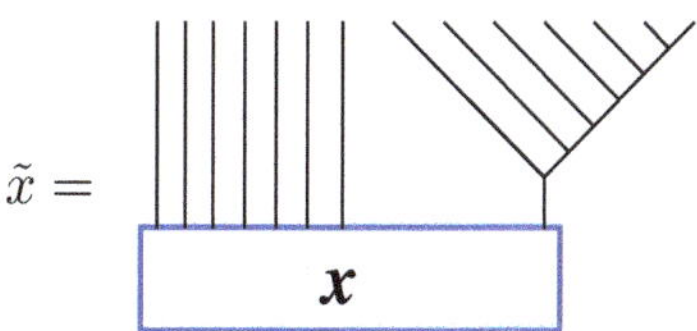

and $\tilde{a}_n$ is the tree a_n stabilised by the forest above; thus,

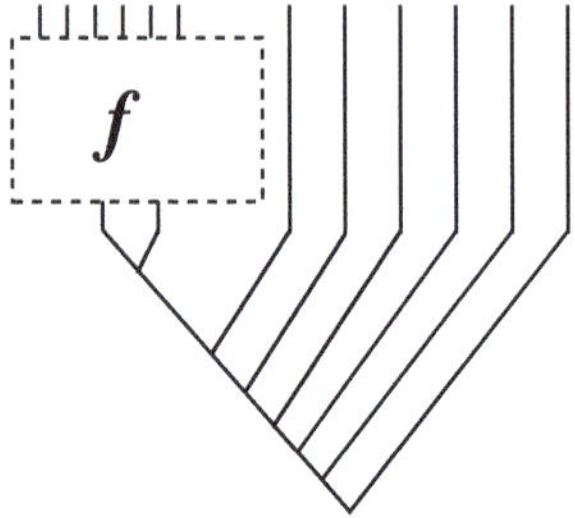

In order to calculate the inner product $\langle \pi_{Y,\phi}(A^n)(\xi), \eta \rangle$, we need to stabilise $\tilde{a}_n$ and t_m so that they are equal. This involves attaching a forest g, which is a reflected version of f on the right hand side of $\tilde{a}$, and attaching a copy of $\tilde{a}_k$ for some k to the left hand side of t_m. Applying these stabilisations to the denominators, we see that we want the inner product in $\mathcal{C}$ of

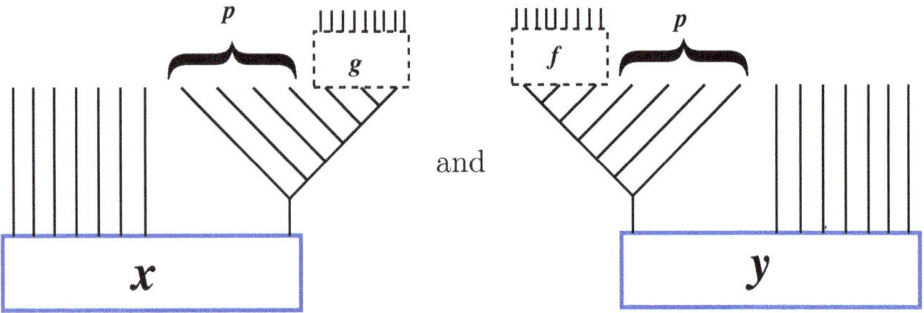

Observe that p (the number of strings as indicated) can be made arbitrarily large by increasing n, whereas the sizes of f and g depend only on m. We see that $\langle \pi_{Y,\phi}(A^n)(\xi), \eta \rangle$ is given by an element of $\mathcal{C}_0$ that looks like

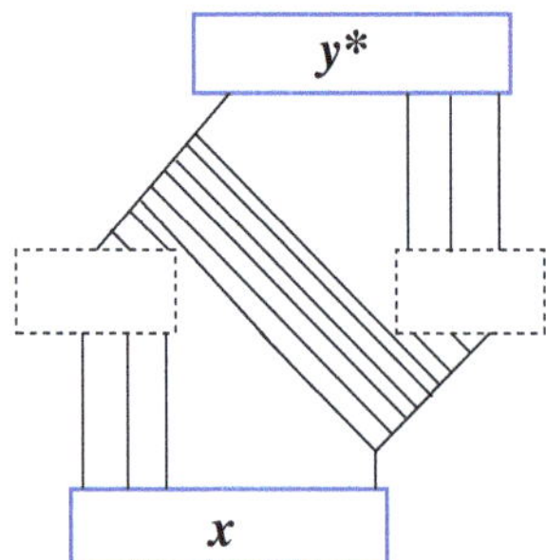

which can be redrawn in $\mathcal{C}$ as

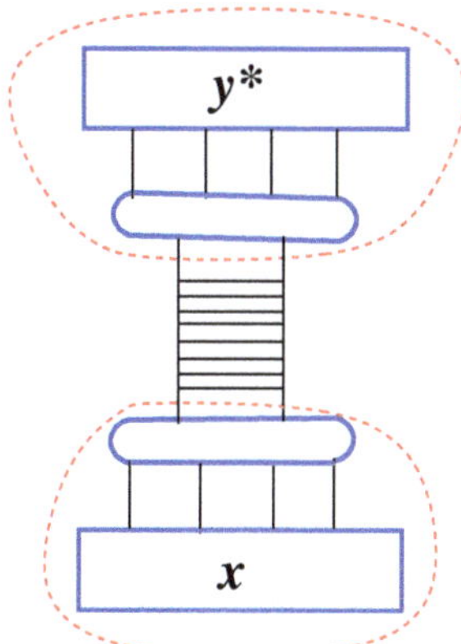

where we observe that, since the dimension of $\mathcal{C}_2$ is one, the contributions of the top and bottom dotted circles are just linear functionals of x and y^*, which do not change as soon as n is large enough. By the unitarity property of Y, we see that the weak limit of $\pi_{Y,\phi}(A^n)$ exists and has rank 1.

Very little changes in the argument for $\pi_{Y,1}(A^n)$. The only difference in the final diagram is a string joining the top of the picture to the bottom so that the dotted circles meet three strings instead of two. Thus, what is inside them contributes a multiple of Y and the ladder in the middle introduces a term t^p, which tends to zero as n, and hence p tends to infinity. □

References

1. Aiello, V. and Jones, V. (2019) On spectral measures for certain unitary representations of R. Thompson's group. arXiv:1905.05806
2. J. Belk, Thompson's group F. Ph.D. Thesis (Cornell University). arXiv preprint:0708.3609 (2007).
3. Brothier, A. and Jones, V. Pythagorean representations of Thompson's groups. arXiv:1807.06215
4. Cannon, J.W., Floyd, W.J. and Parry, W.R.(1996) Introductory notes on Richard Thompson's groups. *L'Enseignement Mathématique* **42** 215–256
5. Connes, A. (1994). Noncommutative geometry. *Academic Press.*
6. G. Golan and M. Sapir, On Jones' subgroup of Thompson group F, *Journal of Algebra* **470** (2017), 122–159.
7. V Guba, M Sapir, Diagram groups, *Mem. Amer. Math. Soc.***130**(1997)
8. Hayashi, T. and Yamagami, S. (2000). Amenable tensor categories and their realizations as AFD bimodules. *Journal of Functional Analysis*, **172**, 19–75.
9. Jones, V.F.R.(2017) Some unitary representations of Thompson's groups F and T. *J. Comb. Algebra* ,**1** , 1–44.
10. Jones, V.F.R. (2018) A no-go theorem for the continuum limit of a periodic quantum spin chain. *Communications in Mathematical Physics*, **357**, 295–317

11. V.F.R. Jones, Planar Algebras I, preprint. math/9909027
12. S. Morrison, E.Peters, N. Snyder, (2017) Categories generated by a trivalent vertex. *Selecta Mathematica,* **23** 817–868 arXiv:1501.06869
13. J. Nikkel, Y. Ren, (2018) On Jones Subgroup of R. Thompson's Group T *International Journal of Algebra and Computation* **28** 877–903 arXiv:1710.06972

Invariants of Long Knots

Rinat Kashaev

Dedicated to Nikolai Reshetikhin on the occasion of his 60th birthday

Abstract By using the notion of a rigid R-matrix in a monoidal category and the Reshetikhin–Turaev functor on the category of tangles, we review the definition of the associated invariant of long knots. In the framework of the monoidal categories of relations and spans over sets, by introducing racks associated with pointed groups, we illustrate the construction and the importance of consideration of long knots. Else, by using the restricted dual of algebras and Drinfeld's quantum double construction, we show that to any Hopf algebra H with invertible antipode, one can associate a universal long knot invariant $Z_H(K)$ taking its values in the convolution algebra $((D(H))^o)^*$ of the restricted dual Hopf algebra $(D(H))^o$ of the quantum double $D(H)$ of H. This extends the known constructions of universal invariants previously considered mostly either in the case of finite-dimensional Hopf algebras or by using some topological completions.

1 Introduction

This chapter is to a large extent a review of certain aspects of quantum invariants where we restrict ourselves exclusively to the context of long knots. More generally, one can consider also the string links, but it is known that the topological classes of string links are in bijection with the classes of ordinary links only if the number of components is one, i.e., if a string link is a long knot.

We describe in detail the construction of invariants of long knots by using rigid R-matrices (solutions of the quantum Yang–Baxter relation) in monoidal categories.

R. Kashaev (✉)
Section de Mathématiques, Université de Genève, Geneva, Switzerland
e-mail: rinat.kashaev@unige.ch

A. Alekseev et al. (eds.), *Representation Theory, Mathematical Physics, and Integrable Systems*, Progress in Mathematics 340,
https://doi.org/10.1007/978-3-030-78148-4_15

The importance of long knots (as opposed to usual closed knots) is illustrated by considering a general class of group-theoretical R-matrices put into the context of monoidal categories of relations and spans over sets. These R-matrices are indexed by pointed groups, that is, groups with a distinguished element. The underlying racks seem not to be considered previously in the existing literature.

Drinfeld's quantum double construction gives rise to a large class of rigid R-matrices, and the associated invariants get factorized through universal invariants associated with underlying Hopf algebras. Such universal invariants were introduced and studied in a number of works [2, 6, 9–11, 14, 15, 18] mostly either in the context of finite-dimensional Hopf algebras or certain topological completions, for example, by considering formal power series. Here, we define the universal invariants purely algebraically and with minimal assumptions on the underlying Hopf algebras. In particular, we emphasize the case of infinite dimensional Hopf algebras. The distinguishing feature of our approach is the use of the restricted or finite dual of an algebra in conjunction with the quantum double construction.

The outline of this chapter is as follows.

In Sect. 2, we recall the definitions of long knots and their diagrams and introduce the notions of a normal diagram and a normalization of an arbitrary diagram.

In Sect. 3, we recall the definition of a rigid R-matrix in a monoidal category and give a detailed description of a long knot invariant associated with a given rigid R-matrix.

In Sect. 4, we consider a special class of rigid R-matrices in the categories of relations and spans over sets. Each such R-matrix is associated with a pointed group with a canonical structure of a rack, and Theorem 1 identifies the associated invariant with the set of representations of the knot group into the group that underlies the rack. The example of an extended Heisenberg group gives rise to an invariant ideal in the polynomial algebra $\mathbb{Q}[t, t^{-1}, s]$ closely related but not equivalent to the Alexander polynomial, at least if the latter admits higher multiplicity roots.

In Sect. 5, based on the restricted dual of an algebra and Drinfeld's quantum double construction, we describe a universal invariant associated with any Hopf algebra with invertible antipode.

2 Long Knots

Definition 1 An embedding $f: \mathbb{R} \to \mathbb{R}^3$ is called *long knot* if there exist $a, b \in \mathbb{R}$ such that $f(t) = (0, 0, t)$ for any $t < a$ or $t > b$.

Two long knots $f, g: \mathbb{R} \to \mathbb{R}^3$ are called *equivalent* if they are ambient isotopic, that is, if there exists an ambient isotopy

$$H: \mathbb{R}^3 \times [0, 1] \to \mathbb{R}^3 \times [0, 1], \quad H(x, t) = (h_t(x), t), \quad h_0 = \mathrm{id}_{\mathbb{R}^3}, \tag{1}$$

such that, for any $t \in [0, 1]$, $h_t \circ f$ is a long knot and $g = h_1 \circ f$.

A long knot is called *tame or regular* if it is equivalent to a polygonal long knot.

An (oriented) *long knot diagram D* is a (1,1)-tangle diagram in $\mathbb{R}^2$ representing a tame long knot

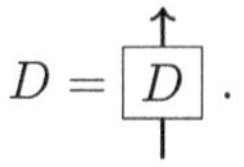

$$D = \boxed{D}\,. \tag{2}$$

Two long knot diagrams are called *(Reidemeister) equivalent* if they can be related to each other by a finite sequence of oriented Reidemeister moves of all types.

Remark 1 The set of long knot diagrams is a monoid with respect to the composition

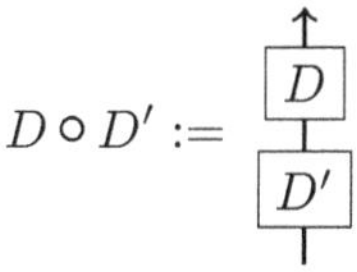

$$D \circ D' := \begin{array}{c}\boxed{D}\\ \boxed{D'}\end{array} \tag{3}$$

Remark 2 In the case of long knots, the Reidemeister theorem states that two long knot diagrams are equivalent if and only if the corresponding long knots are equivalent. Furthermore, a folklore theorem states that the natural map of long knot diagrams to closed oriented knot diagrams

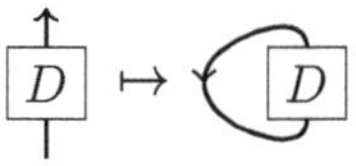

$$\boxed{D} \mapsto \boxed{D} \tag{4}$$

induces a bijection between the respective Reidemeister equivalence classes. In particular, any invariant of long knots is also an invariant of closed knots.

In what follows, we will always assume that a long knot diagram is put into a generic position with respect to the vertical axis so that all crossing have non-vertical strands as in the letter X. We denote by $w(D)$ the writhe of D defined as the number of positive crossings minus the number of negative crossings where a crossing is called positive if the ordered pair of tangent vectors $(v_{\text{overpass}}, v_{\text{underpass}})$ induces the standard orientation of $\mathbb{R}^2$.

Definition 2 A (long knot) diagram is called *normal* if it has no local extrema (with respect to vertical direction) oriented from left to right like ↷ and ↶ .

To any diagram D, we associate its *normalization* $\dot{D}$, the diagram obtained from D by the replacements

$$\curvearrowright \mapsto \ ,\quad \mapsto \ . \tag{5}$$

It will be of special interest for us the normal long knot diagrams

$\xi^+ :=$ , $\xi^- :=$ ,

$$\xi^n := \begin{cases} \uparrow & \text{if } n = 0; \\ \underbrace{\xi^{\operatorname{sgn}(n)} \circ \cdots \circ \xi^{\operatorname{sgn}(n)}}_{|n| \text{ times}} & \text{if } n \neq 0 \end{cases} \tag{6}$$

where $n \in \mathbb{Z}$, $\operatorname{sgn}(n) := n/|n|$ and we identify the signs $\pm$ with the numbers ± 1.

Remark 3 Any normal long knot diagram has an even number of crossings. In particular, we have $w(\xi^n) = 2n$.

3 Invariants of Long Knots from Rigid *R*-Matrices

We say an object G of a monoidal category $\mathcal{C}$ (with tensor product $\otimes$ and unit object $\mathbb{I}$) admits a left *adjoint* if there exists an object F and morphisms

$$\varepsilon\colon F \otimes G \to \mathbb{I}, \quad \eta\colon \mathbb{I} \to G \otimes F \tag{1}$$

such that

$$(\varepsilon \otimes \mathrm{id}_F) \circ (\mathrm{id}_F \otimes \eta) = \mathrm{id}_F, \quad (\mathrm{id}_G \otimes \varepsilon) \circ (\eta \otimes \mathrm{id}_G) = \mathrm{id}_G\,. \tag{2}$$

In that case, the quadruple $(F, G, \varepsilon, \eta)$ is called *adjunction* in $\mathcal{C}$.

Definition 3 Let $\mathcal{C}$ be a monoidal category. An *R-matrix* over an object $G \in \operatorname{Ob}\mathcal{C}$ is an element $r \in \operatorname{Aut}(G \otimes G)$ that satisfies the Yang–Baxter relation

$$(r \otimes \mathrm{id}_G) \circ (\mathrm{id}_G \otimes r) \circ (r \otimes \mathrm{id}_G) = (\mathrm{id}_G \otimes r) \circ (r \otimes \mathrm{id}_G) \circ (\mathrm{id}_G \otimes r). \tag{3}$$

Definition 4 Let $(F, G, \varepsilon, \eta)$ be an adjunction in a monoidal category $\mathcal{C}$. An *R*-matrix r over G is called *rigid* if the morphisms

$$\widetilde{r^{\pm 1}} := (\varepsilon \otimes \mathrm{id}_{G \otimes F}) \circ (\mathrm{id}_F \otimes r^{\pm 1} \otimes \mathrm{id}_F) \circ (\mathrm{id}_{F \otimes G} \otimes \eta) \tag{4}$$

are invertible.

We also denote

$$\widetilde{\widetilde{r^{\pm 1}}} := (\varepsilon \otimes \mathrm{id}_{F \otimes F}) \circ (\mathrm{id}_F \otimes \widetilde{r^{\pm 1}} \otimes \mathrm{id}_F) \circ (\mathrm{id}_{F \otimes F} \otimes \eta). \tag{5}$$

One easily checks the identity

$$\widetilde{\widetilde{r^{-1}}} = \left(\widetilde{\widetilde{r}}\right)^{-1}. \tag{6}$$

Associated with a rigid R-matrix r over G with an adjunction $(F, G, \varepsilon, \eta)$, the *Reshetikhin–Turaev functor* RT_r associates to any normal long knot diagram D the endomorphism $RT_r(D)\colon G \to G$ obtained as follows.

Assuming that the non-trivial part of D is contained in $\mathbb{R} \times [0, 1]$, there exists a finite sequence of real numbers $0 = t_0 < t_1 < \cdots < t_{n-1} < t_n = 1$ such that, for any $i \in \{0, \ldots, n-1\}$, the intersection $D_i := D \cap (\mathbb{R} \times [t_i, t_{i+1}])$ is an ordered (from left to right) finite sequence of connected components each of which is isotopic relative to boundary either to one of the four types of segments

$$\uparrow,\ \downarrow,\ \curvearrowleft,\ \circlearrowleft \tag{7}$$

or to one of the eight types of crossings

$$\tag{8}$$

To such an intersection, we associate a morphism f_i in $\mathcal{C}$ by taking the tensor product (from left to right) of the morphisms associated with the connected fragments of D_i according to the following rules:

$$\uparrow\ \mapsto \mathrm{id}_G,\quad \downarrow\ \mapsto \mathrm{id}_F,\quad \curvearrowleft\ \mapsto \varepsilon,\quad \circlearrowleft\ \mapsto \eta, \tag{9}$$

$$\mapsto r,\quad \mapsto r^{-1}, \tag{10}$$

$$\mapsto \widetilde{r},\quad \mapsto \widetilde{r^{-1}}, \tag{11}$$

$$\mapsto \widetilde{\widetilde{r}},\quad \mapsto \widetilde{\widetilde{r^{-1}}}, \tag{12}$$

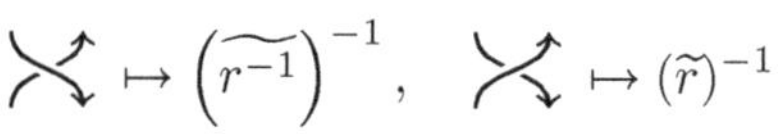

$$\mapsto \left(\widetilde{r^{-1}}\right)^{-1},\quad \mapsto (\widetilde{r})^{-1} \tag{13}$$

The morphism $RT_r(D)\colon G \to G$ associated with D is obtained as the composition

$$RT_r(D) := f_{n-1} \circ \cdots \circ f_1 \circ f_0. \tag{14}$$

Example 1 Let $D = \xi^-$ defined in (6). Then, $RT_r(D) = f_3 \circ f_2 \circ f_1 \circ f_0$, where

$$f_0 = \mathrm{id}_G \otimes \eta, \quad f_1 = \mathrm{id}_G \otimes (\widetilde{r})^{-1}, \quad f_2 = (\widetilde{r})^{-1} \otimes \mathrm{id}_G, \quad f_3 = \varepsilon \otimes \mathrm{id}_G . \tag{15}$$

Theorem 1 ([8, 15, 16]) *Let r be a rigid R-matrix over an object G of a monoidal category $\mathcal{C}$ with an adjunction $(F, G, \varepsilon, \eta)$. Then, for any long knot diagram D, the element*

$$J_r(D) := RT_r(\dot{D} \circ \xi^{-w(\dot{D})/2}) \in \mathrm{End}(G) \tag{16}$$

depends on only the Reidemeister equivalence class of D.

Proof Let $\dot{\sim}$ be the equivalence relation on the set of normal long knot diagrams generated by the oriented versions of the Reidemeister moves RII and RIII and the moves R0$^\pm$ defined by the pictures

$$\xleftrightarrow{\mathrm{R0}^+} \quad , \quad \xleftrightarrow{\mathrm{R0}^-} \tag{17}$$

with two possible orientations for the straight segment and two possibilities for the crossing. The strategy of the proof is first to show the implication

$$\dot{D} \dot{\sim} \dot{D}' \Rightarrow RT_r(\dot{D}) = RT_r(\dot{D}') \tag{18}$$

which, by taking into account the implication

$$\dot{D} \dot{\sim} \dot{D}' \Rightarrow w(\dot{D}) = w(\dot{D}'), \tag{19}$$

ensures the invariance of $J_r(D)$ under all Reidemeister moves RII and RIII and then to verify the invariance under the Reidemeister moves RI. It is in this last part of the proof where the correction of $\dot{D}$ by $\xi^{-w(\dot{D})/2}$ is crucial.

Invariance of the Reshetikhin–Turaev functor with respect to moves R0$^\pm$ follows from the equivalences

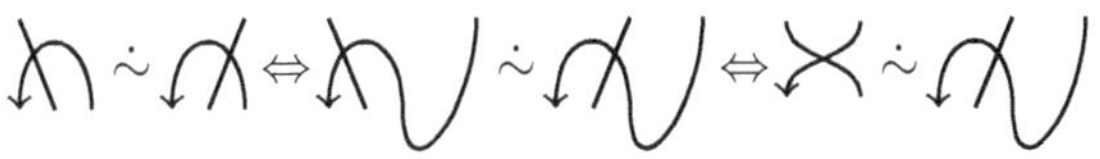

$$\dot{\sim} \quad \Leftrightarrow \quad \dot{\sim} \quad \Leftrightarrow \quad \dot{\sim} \tag{20}$$

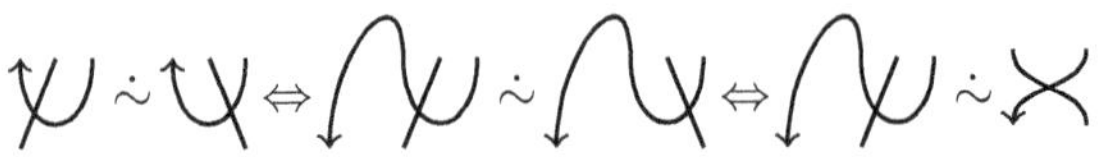

$$\dot{\sim} \quad \Leftrightarrow \quad \dot{\sim} \quad \Leftrightarrow \quad \dot{\sim} \tag{21}$$

and the definitions (4) and (5) of $\widetilde{r^{\pm 1}}$ and $\widetilde{\widetilde{r^{\pm 1}}}$.

Invariance of RT_r with respect to oriented RII moves is easily checked first for eight basic moves

$$\overset{RT_r}{\longmapsto} r^{-1} \circ r = \mathrm{id}_{G\otimes G} = r \circ r^{-1} \tag{22}$$

$$\overset{RT_r}{\longmapsto} \widetilde{r} \circ (\widetilde{r})^{-1} = \mathrm{id}_{G\otimes F} = \widetilde{r^{-1}} \circ \left(\widetilde{r^{-1}}\right)^{-1} \tag{23}$$

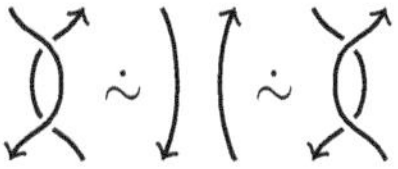

$$\overset{RT_r}{\longmapsto} \left(\widetilde{r^{-1}}\right)^{-1} \circ \widetilde{r^{-1}} = \mathrm{id}_{F\otimes G} = (\widetilde{r})^{-1} \circ \widetilde{r}, \tag{24}$$

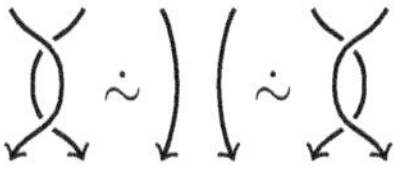

$$\overset{RT_r}{\longmapsto} \widetilde{\widetilde{r^{-1}}} \circ \widetilde{\widetilde{r}} = \mathrm{id}_{F\otimes F} = \widetilde{\widetilde{r}} \circ \widetilde{\widetilde{r^{-1}}}, \tag{25}$$

and then for two composite moves

(26)

with two possible choices for the crossings.

In order to check invariance of RT_r with respect to RIII moves, we remark that altogether there are 48 such moves which can be indexed by the set $\mathrm{Sym}(3) \times \{\pm 1\}^3$ as follows.

Given an RIII move, we enumerate the strands that intervene the move by following their bottom open ends from left to right:

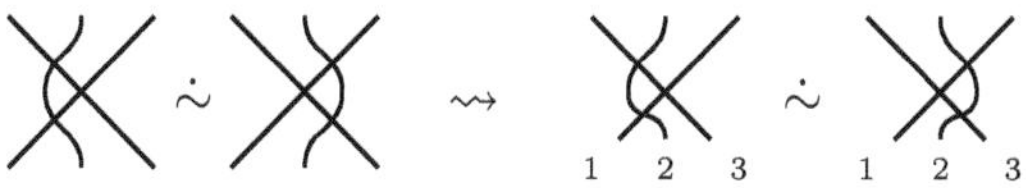

1 2 3 1 2 3 (27)

and we define the associated element $(\sigma, \varepsilon) \in \mathrm{Sym}(3) \times \{\pm 1\}^3$ by the conditions that for any $i \in \{1, 2, 3\}$, $\sigma(i)$ is the number of arcs on the i-th strand and $\varepsilon_i = 1$ if i-th strand is oriented upward. For example, the RIII move

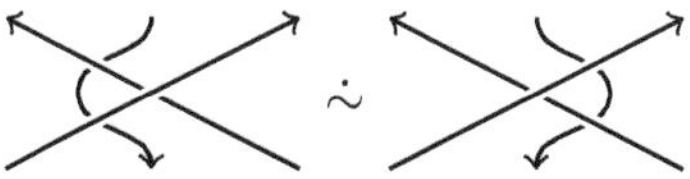

(28)

corresponds to permutation $\sigma = (2, 3) = (1)(2, 3)$ and $\varepsilon = (1, -1, 1)$ while the pair $(\sigma = \mathrm{id}, \varepsilon = (1, 1, 1))$ corresponds to the reference move associated with the Yang–Baxter relation

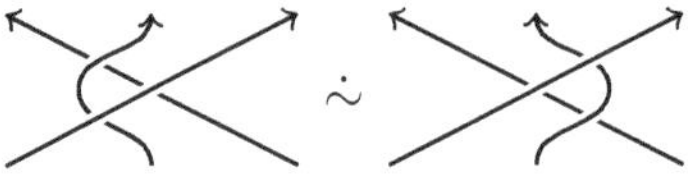

$$\stackrel{RT_r}{\longmapsto} r_1 \circ r_2 \circ r_1 = r_2 \circ r_1 \circ r_2 \tag{29}$$

with the notations $r_1 := r \otimes \mathrm{id}_G$ and $r_2 := \mathrm{id}_G \otimes r$.

One can now show that by using the moves RII and R0$^{\pm}$, any RIII move is equivalent to the reference move (29).

Indeed, in the case $\varepsilon_1 = -1$, we have the equivalences

(30)

which imply the equivalence

$$(\sigma, (-1, \varepsilon_2, \varepsilon_3)) \Leftrightarrow (\sigma \circ (1, 2, 3), (\varepsilon_2, \varepsilon_3, 1)) \tag{31}$$

in the set $\mathrm{Sym}(3) \times \{\pm 1\}^3$, thus allowing to reduce the number of negative components of ε. Additionally, a right action of the permutation group Sym(3) on the set $\mathrm{Sym}(3) \times \{\pm 1\}^3$ is induced by the equivalences

(32)

and

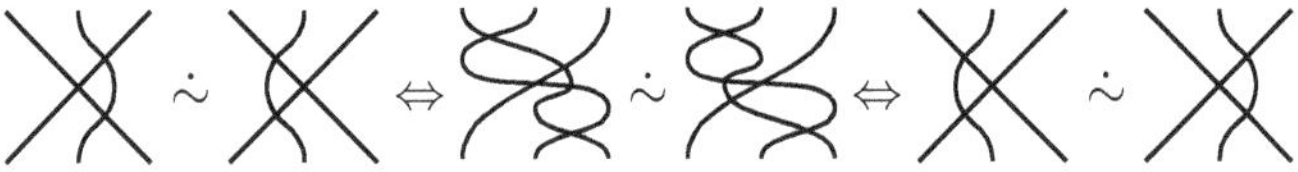

(33)

which correspond to the respective equivalences

$$(\sigma, \varepsilon) \Leftrightarrow (\sigma, \varepsilon) \circ (1, 2) \ \text{ and } \ (\sigma, \varepsilon) \Leftrightarrow (\sigma, \varepsilon) \circ (2, 3) \tag{34}$$

in the set $\mathrm{Sym}(3) \times \{\pm 1\}^3$ where we interpret $(\sigma, \varepsilon) \in \mathrm{Sym}(3) \times \{\pm 1\}^3$ as the map

$$(\sigma, \varepsilon) \colon \{1, 2, 3\} \to \{1, 2, 3\} \times \{\pm 1\}, \quad i \mapsto (\sigma(i), \varepsilon_i). \tag{35}$$

Thus, in conjunction with the equivalence (31), the right action of the group Sym(3) on the set $\mathrm{Sym}(3) \times \{\pm 1\}^3$ establishes the equivalence of any RIII move to the reference move (29) and thereby the invariance of RT_r with respect to all RIII moves.

Finally, in order to prove invariance of J_r with respect to all RI moves, we need to check only the invariance with respect to the four basic moves of the form

(36)

as all others are consequences of the basic ones and the intermediate equivalence relation $\dot{\sim}$ generated by the moves R0$^{\pm}$, RII, and RIII:

(37)

and

(38)

Les us analyze the four cases of (36) separately.

Case 1. If diagrams D and D' differ by the fragments

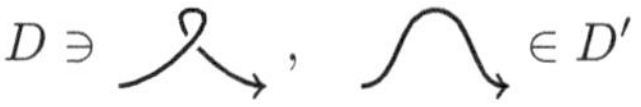

$$D \ni \quad , \quad \in D' \tag{39}$$

then, by the definition of the normalization of a long knot diagram, we have the equality $\dot{D} = \dot{D}'$. Thus, $J_r(D) = J_r(D')$.

Case 2. Diagrams D and D' differ by the fragments

$$D \ni \quad , \quad \in D' \tag{40}$$

so that the normalized diagrams $\dot{D}$ and $\dot{D}'$ differ by the fragments

$$\dot{D} \ni \quad , \quad \in \dot{D}' \tag{41}$$

which imply that

$$w(\dot{D}) = 2 + w(\dot{D}') \Rightarrow \xi^{+} \circ \xi^{-w(\dot{D})/2} = \xi^{-w(\dot{D}')/2}. \tag{42}$$

On the other hand, we have the equivalence

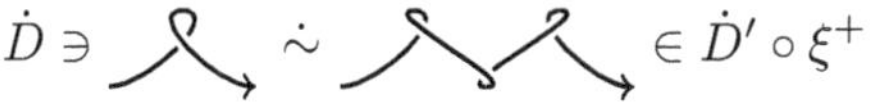

$$\dot{D} \ni \quad \dot{\sim} \quad \in \dot{D}' \circ \xi^{+} \tag{43}$$

which, together with (42), implies that

$$\dot{D} \circ \xi^{-w(\dot{D})/2} \dot{\sim} \dot{D}' \circ \xi^{+} \circ \xi^{-w(\dot{D})/2} = \dot{D}' \circ \xi^{-w(\dot{D}')/2} \Rightarrow J_r(D) = J_r(D'). \tag{44}$$

Case 3. Diagrams D and D' differ by the fragments

$$D \ni \quad , \quad \in D' \tag{45}$$

so that

$$\dot{D} \ni \quad \dot{\sim} \quad \in \dot{D}' \Rightarrow J_r(D) = J_r(D') \tag{46}$$

Case 4. Diagrams D and D' differ by the fragments

$$D \ni \quad , \quad \in D' \tag{47}$$

so that we have for the corresponding normalized diagrams

$$\dot{D} \ni \quad \dot{\sim} \quad \in \dot{D}' \circ \xi^{-} \tag{48}$$

and

$$w(\dot{D}) = w(\dot{D}') - 2 \Rightarrow \xi^{-} \circ \xi^{-w(\dot{D})/2} = \xi^{-w(\dot{D}')/2}. \tag{49}$$

Thus,

$$\dot{D} \circ \xi^{-w(\dot{D})/2} \dot{\sim} \dot{D}' \circ \xi^{-} \circ \xi^{-w(\dot{D})/2} = \dot{D}' \circ \xi^{-w(\dot{D}')/2} \Rightarrow J_r(D) = J_r(D'). \tag{50}$$

□

4 Rigid R-Matrices from Racks

A *binary relation* from a set X to a set Y as a subset of the Cartesian product $X \times Y$. The composition of two binary relations $R \subset X \times Y$ and $S \subset Y \times Z$ is the binary relation $S \circ R \subset X \times Z$ defined by

$$S \circ R := \{(x, z) \in X \times Z \mid \exists y \in Y \colon\ (x, y) \in R,\ (y, z) \in S\}. \tag{1}$$

A *span* from a set X to a set Y is a triple $U = (U, \mathrm{s}_U, \mathrm{t}_U)$, where U is a set and $\mathrm{s}_U : U \to X$ and $\mathrm{t}_U : U \to Y$ are set-theoretical maps. The composition of two spans U from X to Y and V from Y to Z is the span from X to Z defined as the pullback space (fibered product) $V \circ U := U \times_Y V$ together with the natural projections to X and Z.

Two spans U and V from X to Y are called *equivalent* if there exists a bijection $f : U \to V$ such that $\mathrm{s}_V \circ f = \mathrm{s}_U$ and $\mathrm{t}_V \circ f = \mathrm{t}_U$. The composition of spans induces an associative binary operation for the equivalence classes of spans.

Any binary relation $R \subset X \times Y$ is a special case of a span with $\mathrm{s}_R : R \to X$ and $\mathrm{t}_R : R \to Y$ being the canonical projections.

Let **Set** be the monoidal category of sets with the Cartesian product as the monoidal product and **Rel** (respectively, **Span**) the extension of **Set** with morphisms given by binary relations (respectively, equivalence classes of spans). For a morphism $Z : X \to Y$ in **Span**, and any $(x, y) \in X \times Y$, we denote

$$Z(x, y) := \mathrm{s}_Z^{-1}(x) \cap \mathrm{t}_Z^{-1}(y). \tag{2}$$

We have a canonical monoidal functor

$$\varpi : \mathbf{Span} \to \mathbf{Rel}, \tag{3}$$

which is identity on the level of objects and for any morphism $Z : X \to Y$ in **Span**, the corresponding morphism in **Rel** is given by

$$\varpi(Z) = \{(x, y) \in X \times Y \mid Z(x, y) \neq \emptyset\}. \tag{4}$$

Notice that if Z is a relation (as a particular case of spans), then $\varpi(Z) = Z$.

Given a set-theoretical map $f : X \to Y$, its graph

$$\Gamma_f := \{(x, f(x)) \mid x \in X\} \subset X \times Y \tag{5}$$

is naturally interpreted as a morphism $\rho(f)$ in **Rel** and a morphism $\sigma(f)$ in **Span**.

The advantage of the categories **Rel** and **Span** over **Set** is their rigidity, namely, for any set X, the diagonal $\Delta_X := \Gamma_{\mathrm{id}_X}$, interpreted as morphisms $\varepsilon_X : X \times X \to \{0\}$ and $\eta_X : \{0\} \to X \times X$ in **Rel** and **Span**, gives rise to a canonical adjunction $(X, X, \varepsilon_X, \eta_X)$ both in **Rel** and in **Span**.

Let X be a (left) *rack* [4, 7, 13, 17], that is, a set with a map

$$X^2 \to X^2, \quad (x, y) \mapsto (x \cdot y, x * y),$$

such that the binary operation $x \cdot y$ is left self-distributive

$$x \cdot (y \cdot z) = (x \cdot y) \cdot (x \cdot z), \quad \forall (x, y, z) \in X^3,$$

and

$$x * (x \cdot y) = y, \quad \forall (x, y) \in X^2.$$

It is easily verified that for any rack X, the set-theoretical map

$$r\colon X^2 \to X^2, \quad (x, y) \mapsto (x \cdot y, x),$$

is a rigid R-matrix in the categories **Rel** and **Span**. Moreover, all the relevant morphisms are realized by set-theoretical maps. Indeed, introducing the maps

$$r', s, s'\colon X^2 \to X^2, \quad r'(x, y) = (y, y \cdot x), \; s(x, y) = (x * y, x), \; s'(x, y) = (y, y * x),$$

we obtain

$$r^{-1} = \widetilde{r} = s', \quad (\widetilde{r})^{-1} = r, \quad \widetilde{r^{-1}} = \widetilde{\widetilde{r}} = s, \quad \left(\widetilde{r^{-1}}\right)^{-1} = \widetilde{\widetilde{r^{-1}}} = r'.$$

Thus, we obtain two long knot invariants $J_{\rho(r)}(D)$ in **Rel** and $J_{\sigma(r)}(D)$ in **Span**, which are related to each other by the equality

$$J_{\rho(r)}(D) = \varpi(J_{\sigma(r)}(D)). \tag{6}$$

4.1 Racks Associated with Pointed Groups

Let (G, μ) be a *pointed group*, that is, a group G together with a fixed element $\mu \in G$. Then, it is easily verified that the set G with the map

$$G \times G \to G \times G, \quad (g, h) \mapsto (g\mu g^{-1}h, g\mu^{-1}g^{-1}h) \tag{7}$$

is a rack, and, thus, it gives rise to a rigid R-matrix

$$r_{G,\mu}\colon G \times G \to G \times G, \quad (g, h) \mapsto (g\mu g^{-1}h, g), \tag{8}$$

in both the categories **Rel** and **Span**. In this way, we obtain two long knot invariants $J_{\rho(r_{G,\mu})}(D)$ and $J_{\sigma(r_{G,\mu})}(D)$ related to each other by the equality

$$J_{\rho(r_{G,\mu})}(D) = \varpi(J_{\sigma(r_{G,\mu})}(D)). \tag{9}$$

Theorem 1 *There exists a canonical choice of a meridian–longitude pair (m, ℓ) of long knots such that the set $(J_{\sigma(r_{G,\mu})}(D))(1, \lambda)$ is in bijection with the set of group homomorphisms*

$$\{h\colon \pi_1(\mathbb{R}^3 \setminus f(\mathbb{R}), x_0) \to G \mid h(m) = \mu,\ h(\ell) = \lambda\}, \tag{10}$$

where $f : \mathbb{R} \to \mathbb{R}^3$ is a long knot represented by D.

Proof Let $f : \mathbb{R} \to \mathbb{R}^3$ be a long knot whose image under the projection

$$p\colon \mathbb{R}^3 \to \mathbb{R}^2, \quad (x, y, z) \mapsto (y, z). \tag{11}$$

is the diagram $\tilde{D} := \dot{D} \circ \xi^{-w(\dot{D})/2}$ with linearly ordered (from bottom to top) set of arcs $a_0, a_1, \dots, a_n$. As a result, the set of crossings acquires a linear order as well $\{c_i \mid 1 \le i \le n\}$, where c_i is the crossing separating the arcs a_{i-1} and a_i and with the over passing arc a_{κ_i} for a uniquely defined map

$$\kappa\colon \{1, \dots, n\} \to \{0, \dots, n\}. \tag{12}$$

Let $t_0, t_1, \dots, t_n \in \mathbb{R}$ be a strictly increasing sequence such that $f(t) = (0, 0, t)$ for all $t \notin [t_0, t_n]$, and for each $i \in \{1, \dots, n-1\}$, $p(f(t_i))$ belongs to arc a_i and is distinct from any crossing. Choose a base point $x_0 = (s, 0, 0)$ with sufficiently large $s \in \mathbb{R}_{>0}$ and a sufficiently small $\epsilon \in \mathbb{R}_{>0}$, and define the following paths:

$$\begin{aligned} &\alpha_0, \beta_i, \gamma_i\colon [0, 1] \to \mathbb{R}^3,\ i \in \{0, \dots, n\}, \quad \alpha_0(t) = (\epsilon \cos(2\pi t), -\epsilon \sin(2\pi t), t_0), \\ &\beta_i(t) = (1 - t)x_0 + (f(t_i) + (\epsilon, 0, 0))t, \quad \gamma_i(t) = f((1 - t)t_i + tt_0) + (\epsilon, 0, 0). \end{aligned} \tag{13}$$

To each arc a_i of $\tilde{D}$, we associate the homotopy class

$$e_i := [\beta_i \cdot \gamma_i \cdot \bar{\beta}_0] \in \pi_1(\mathbb{R}^3 \setminus f(\mathbb{R}), x_0), \tag{14}$$

so that $e_0 = 1$, and the Wirtinger generator

$$w_i := [\beta_i \cdot \gamma_i \cdot \alpha_0 \cdot \bar{\gamma}_i \cdot \bar{\beta}_i] \in \pi_1(\mathbb{R}^3 \setminus f(\mathbb{R}), x_0). \tag{15}$$

We have the equalities

$$w_i = e_i w_0 e_i^{-1}, \quad \forall i \in \{0, 1, \dots, n\}, \tag{16}$$

$$e_i = w_{\kappa_i}^{\varepsilon_i} e_{i-1}, \quad \forall i \in \{1, \dots, n\}, \tag{17}$$

where $\varepsilon_i \in \{\pm 1\}$ is the sign of the crossing c_i, and

$$e_i = w_{\kappa_i}^{\varepsilon_i} w_{\kappa_{i-1}}^{\varepsilon_{i-1}} \cdots w_{\kappa_1}^{\varepsilon_1}, \quad \forall i \in \{1, \dots, n\}. \tag{18}$$

We define the canonical meridian–longitude pair (m, ℓ) as follows:

$$m := w_0, \quad \ell := e_n. \tag{19}$$

Taking into account the condition $\sum_{i=1}^{n} \varepsilon_i = 0$, we see that ℓ has the trivial image in $H_1(\mathbb{R}^3 \setminus f(\mathbb{R}), \mathbb{Z})$.

Let us show that the following finitely presented groups are isomorphic to the knot group $\pi_1(\mathbb{R}^3 \setminus f(\mathbb{R}), x_0)$:

$$E := \langle m, e_0, \dots, e_n \mid e_0 = 1,\ e_i = e_{\kappa_i} m^{\varepsilon_i} e_{\kappa_i}^{-1} e_{i-1},\ 1 \le i \le n \rangle \tag{20}$$

and

$$W := \langle w_0, \dots, w_n \mid w_{\kappa_i}^{\varepsilon_i} w_{i-1} = w_i w_{\kappa_i}^{\varepsilon_i},\ 1 \le i \le n \rangle. \tag{21}$$

As W is nothing else but the Wirtinger presentation of $\pi_1(\mathbb{R}^3 \setminus f(\mathbb{R}), x_0)$, it suffices to see the isomorphism $E \simeq W$. To see the latter, we remark that there are two group homomorphisms

$$u \colon W \to E, \quad w_i \mapsto e_i m e_i^{-1}, \quad i \in \{0, 1, \dots, n\}, \tag{22}$$

and

$$v \colon E \to W, \quad m \mapsto w_0, \quad e_0 \mapsto 1, \quad e_i \mapsto w_{\kappa_i}^{\varepsilon_i} w_{\kappa_{i-1}}^{\varepsilon_{i-1}} \cdots w_{\kappa_1}^{\varepsilon_1}, \quad \forall i \in \{1, \dots, n\}. \tag{23}$$

Indeed, we have

$$\begin{aligned} u(w_{\kappa_i})^{\varepsilon_i} u(w_{i-1}) = u(w_i) u(w_{\kappa_i})^{\varepsilon_i} &\Leftrightarrow u(w_{\kappa_i})^{\varepsilon_i} e_{i-1} m e_{i-1}^{-1} = e_i m e_i^{-1} u(w_{\kappa_i})^{\varepsilon_i} \\ \Leftrightarrow e_i^{-1} u(w_{\kappa_i})^{\varepsilon_i} e_{i-1} m = m e_i^{-1} u(w_{\kappa_i})^{\varepsilon_i} e_{i-1} &\Leftarrow e_i^{-1} u(w_{\kappa_i})^{\varepsilon_i} e_{i-1} = 1 \\ \Leftrightarrow e_i = u(w_{\kappa_i})^{\varepsilon_i} e_{i-1} &\Leftrightarrow e_i = e_{\kappa_i} m^{\varepsilon_i} e_{\kappa_i}^{-1} e_{i-1} \end{aligned} \tag{24}$$

implying that u is a group homomorphism. We also have

$$\begin{aligned} v(e_i) = v(e_{\kappa_i}) v(m)^{\varepsilon_i} v(e_{\kappa_i})^{-1} v(e_{i-1}) &\Leftrightarrow v(e_i) v(e_{i-1})^{-1} = v(e_{\kappa_i}) w_0^{\varepsilon_i} v(e_{\kappa_i})^{-1} \\ \Leftrightarrow w_{\kappa_i} = v(e_{\kappa_i}) w_0 v(e_{\kappa_i})^{-1} &\Leftarrow \{w_i = v(e_i) w_0 v(e_i)^{-1}\}_{0 \le i \le n}, \end{aligned} \tag{25}$$

and, for all $i \in \{1, \dots, n\}$,

$$\begin{aligned} v(e_i) w_0 &= w_{\kappa_i}^{\varepsilon_i} \cdots w_{\kappa_1}^{\varepsilon_1} w_0 = w_{\kappa_i}^{\varepsilon_i} \cdots w_{\kappa_2}^{\varepsilon_2} w_1 w_{\kappa_1}^{\varepsilon_1} = \dots \\ &= w_{\kappa_i}^{\varepsilon_i} \cdots w_{\kappa_k}^{\varepsilon_k} w_{k-1} w_{\kappa_{k-1}}^{\varepsilon_{k-1}} \cdots w_{\kappa_1}^{\varepsilon_1} = \dots = w_i w_{\kappa_i}^{\varepsilon_i} \cdots w_{\kappa_1}^{\varepsilon_1} = w_i v(e_i) \end{aligned} \tag{26}$$

implying that v is a group homomorphism as well.

It remains to show that $v \circ u = \mathrm{id}_W$ and $u \circ v = \mathrm{id}_E$. Indeed, for all $i \in \{0, \ldots, n\}$, we have

$$v(u(w_i)) = v(e_i m e_i^{-1}) = v(e_i)v(m)v(e_i)^{-1} = v(e_i)w_0 v(e_i)^{-1} = w_i \tag{27}$$

implying that $v \circ u = \mathrm{id}_W$. We prove that $u(v(e_i)) = e_i$ for all $i \in \{0, \ldots, n\}$ by recursion on i. For $i = 0$, we have

$$u(v(e_0)) = u(1) = 1 = e_0. \tag{28}$$

Assuming that $u(v(e_{k-1})) = e_{k-1}$ for some $k \in \{1, \ldots, n-1\}$, we calculate

$$u(v(e_k)) = u(w_{\kappa_k}^{\varepsilon_k} v(e_{k-1})) = u(w_{\kappa_k})^{\varepsilon_k} u(v(e_{k-1})) = e_{\kappa_k} m^{\varepsilon_k} e_{\kappa_k}^{-1} e_{k-1} = e_k. \tag{29}$$

Now, any element g of the set $(J_{\sigma(r_{G,\mu})}(D))(1, \lambda)$ is a map

$$g : \{0, 1, \ldots, n\} \to G \tag{30}$$

such that

$$g_0 = 1, \; g_n = \lambda, \; g_i = g_{\kappa_i} \mu^{\varepsilon_i} g_{\kappa_i}^{-1} g_{i-1}, \quad \forall i \in \{1, \ldots, n\}. \tag{31}$$

This means that g determines a unique group homomorphism

$$h_g : E \to G \tag{32}$$

such that $h_g(m) = \mu$ and $h_g(e_i) = g_i$ for all $i \in \{0, 1, \ldots, n\}$. On the other hand, any group homomorphism $h : E \to G$ such that $h(m) = \mu$ and $h(\ell) = \lambda$ is of the form $h = h_g$ where $g_i = h(e_i)$. Thus, the map $g \mapsto h_g$ is a set-theoretical bijection between $(J_{\sigma(r_{G,\mu})}(D))(1, \lambda)$ and the set of group homomorphisms (10). □

Remark 4 Theorem 1 illustrates the importance of considering long knots as opposed to closed knots. Namely, by closing a long knot, one identifies two open strands and all the associated data. In particular, one has to impose the equality $\lambda = 1$ that corresponds to considering only those representations where the longitude is realized trivially. This means that in the case of closed diagrams one would obtain less powerful invariants.

4.2 An Extended Heisenberg Group

Let R be a commutative unital ring and G the subgroup of $GL(3, R)$ given by upper triangular matrices of the form

$$\begin{pmatrix} a & b & d \\ 0 & 1 & c \\ 0 & 0 & a \end{pmatrix}, \quad a \in U(R), \ (b, c, d) \in R^3, \tag{33}$$

where $U(R)$ is the group of units of R. Then, the elements

$$\mu = \begin{pmatrix} t & 0 & 0 \\ 0 & 1 & 0 \\ 0 & 0 & t \end{pmatrix} \quad \text{and} \quad \lambda = \begin{pmatrix} 1 & 0 & s \\ 0 & 1 & 0 \\ 0 & 0 & 1 \end{pmatrix} \tag{34}$$

commute with each other for any $t \in U(R)$ and $s \in R$. Given the fact that G is an algebraic group, the invariant $(J_{\rho(r_{G,\mu})}(D))(1, \lambda)$ factorizes through an ideal I_D in the polynomial algebra $\mathbb{Q}[t, t^{-1}, s]$. Examples of calculations show that this ideal is often principal and is generated by the polynomial $\Delta_D s$, where $\Delta_D := \Delta_D(t)$ is the Alexander polynomial of D. Nonetheless, there are examples for which this is not the case, at least if the Alexander polynomial has multiple roots. In Table 1, we have collected few selected examples of calculations for knots where the Alexander polynomials of the first three examples of knots 3_1, 4_1, and 6_2 have simple roots, while for all other examples, the respective Alexander polynomials have multiple roots and they are factorized into products of the Alexander polynomials of the first three examples. In particular, one can see that the knots 8_{10} and 8_{20} have different Alexander polynomials but one and the same ideal, while the knots 8_{18} and 9_{24} have one and the same Alexander polynomial but different ideals.

Table 1 Examples of calculation with the extended Heisenberg group

Knot	Δ_D	I_D
3_1	$1 - t + t^2$	$(\Delta_{3_1} s)$
4_1	$1 - 3t + t^2$	$(\Delta_{4_1} s)$
6_2	$1 - 3t + 3t^2 - 3t^3 + t^4$	$(\Delta_{6_2} s)$
8_{10}	$\Delta_{3_1}^3$	$(\Delta_{3_1} s, s^2)$
8_{18}	$\Delta_{4_1} \Delta_{3_1}^2$	$(\Delta_{4_1} \Delta_{3_1} s)$
8_{20}	$\Delta_{3_1}^2$	$(\Delta_{3_1} s, s^2)$
9_{24}	$\Delta_{4_1} \Delta_{3_1}^2$	$(\Delta_{4_1} \Delta_{3_1} s, \Delta_{4_1} s^2)$
10_{99}	$\Delta_{3_1}^4$	$(\Delta_{3_1} s, s^2)$
10_{123}	$\Delta_{6_2}^2$	$(\Delta_{6_2} s)$
10_{137}	$\Delta_{4_1}^2$	$(\Delta_{4_1} s, s^2)$
$11a_5$	$\Delta_{4_1}^3$	$(\Delta_{4_1} s, s^2)$

5 Invariants from Hopf Algebras

Let $\mathbf{Hopf}_{\mathbb{K}}$ be the category of Hopf algebras over a field $\mathbb{K}$ with invertible antipode. We have a contravariant endofunctor $(\cdot)^o\colon \mathbf{Hopf}_{\mathbb{K}} \to \mathbf{Hopf}_{\mathbb{K}}$ that associates to a Hopf algebra H with multiplication ∇ its *restricted dual*

$$H^o := (\nabla^*)^{-1}(H^* \otimes H^*), \tag{1}$$

which can also be identified with the vector subspace of the algebraic dual H^* generated by all matrix coefficients of all finite-dimensional representations of H [3].

Following [12], Drinfeld's *quantum double* of $H \in \operatorname{Ob} \mathbf{Hopf}_{\mathbb{K}}$ is a Hopf algebra $D(H) \in \operatorname{Ob} \mathbf{Hopf}_{\mathbb{K}}$ uniquely determined by the property that there are two Hopf algebra inclusions

$$\imath\colon H \to D(H), \quad \jmath\colon H^{o,\mathrm{op}} \to D(H) \tag{2}$$

such that $D(H)$ is generated by their images subject to the commutation relations

$$\jmath(f)\imath(x) = \langle f_{(1)}, x_{(1)}\rangle\langle f_{(3)}, S(x_{(3)})\rangle\imath(x_{(2)})\jmath(f_{(2)}), \quad \forall(x, f) \in H \times H^o, \tag{3}$$

where we use Sweedler's notation for the comultiplication

$$\Delta(x) = x_{(1)} \otimes x_{(2)}, \quad (\Delta \otimes \mathrm{id})(\Delta(x)) = x_{(1)} \otimes x_{(2)} \otimes x_{(3)}, \ \ldots \tag{4}$$

The restricted dual of the quantum double $(D(H))^o$ is a *dual quasitriangular* Hopf algebra with the *dual universal R-matrix*

$$\varrho\colon (D(H))^o \otimes (D(H))^o \to \mathbb{K}, \quad x \otimes y \mapsto \langle x, \jmath(\imath^o(y))\rangle, \tag{5}$$

which, among other things, satisfies the Yang–Baxter relation

$$\varrho_{1,2} * \varrho_{1,3} * \varrho_{2,3} = \varrho_{2,3} * \varrho_{1,3} * \varrho_{1,2} \tag{6}$$

in the convolution algebra $(((D(H))^o)^{\otimes 3})^*$, and any finite-dimensional right comodule

$$V \to V \otimes (D(H))^o, \quad v \mapsto v_{(0)} \otimes v_{(1)}, \tag{7}$$

gives rise to a rigid R-matrix

$$r_V : V \otimes V \to V \otimes V, \quad u \otimes v \mapsto v_{(0)} \otimes u_{(0)}\langle \varrho, u_{(1)} \otimes v_{(1)}\rangle. \tag{8}$$

This implies that there exists a *universal invariant* of long knots $Z_H(K)$ taking its values in the convolution algebra $((D(H))^o)^*$ such that

$$J_{r_V}(K)v = v_{(0)}\langle Z_H(K), v_{(1)}\rangle, \quad \forall v \in V. \tag{9}$$

Remark 5 By using the coend of the monoidal braided category of finite-dimensional comodules over $(D(H))^o$, the universal invariant $Z_H(K)$ can be interpreted via the Lyubashenko theory [11]. In the case of finite-dimensional quasitriangular Hopf algebras, this coend approach is further developed by Virelizier in [18].

Remark 6 If H is a finite-dimensional Hopf algebra with a linear basis $\{e_i\}_{i\in I}$ and $\{e^i\}_{i\in I}$ is the dual linear basis of H^*, then, the dual universal R-matrix is conjugate to the universal R-matrix

$$R := \sum_{i\in I} \jmath(e^i) \otimes \imath(e_i) \in D(H) \otimes D(H) \tag{10}$$

in the sense that, for any $x, y \in (D(H))^o = (D(H))^*$, we have

$$\begin{aligned}\langle x \otimes y, R\rangle &= \sum_{i\in I}\langle x, \jmath(e^i)\rangle\langle y, \imath(e_i)\rangle = \sum_{i\in I}\langle x, \jmath(e^i)\rangle\langle \imath^o(y), e_i\rangle \\ &= \left\langle x, \jmath\left(\sum_{i\in I}\langle \imath^o(y), e_i\rangle e^i\right)\right\rangle = \left\langle x, \jmath\left(\imath^o(y)\right)\right\rangle = \langle \varrho, x \otimes y\rangle. \end{aligned} \tag{11}$$

In the infinite-dimensional case, formula (10) is formal, but it is a convenient and useful tool for actual calculations.

Remark 7 The algebra inclusion $D(H) \subset ((D(H))^o)^*$ allows to think of the convolution algebra $((D(H))^o)^*$ as a certain algebra completion of the quantum double $D(H)$.

5.1 A Hopf Algebra Associated with a Two-Dimensional Lie Group

Let B be the commutative Hopf algebra over $\mathbb{C}$ generated by an invertible group-like element a and element b with the coproduct

$$\Delta(b) = a \otimes b + b \otimes 1 \tag{12}$$

so that the set of monomials $\{b^m a^n \mid (m, n) \in \mathbb{Z}_{\geq 0} \times \mathbb{Z}\}$ is a linear basis of B.

The restricted dual Hopf algebra B^o is generated by two primitive elements ψ and ϕ and group-like elements

$$u^{\psi},\ e^{v\phi},\quad (u,v)\in\mathbb{C}_{\neq 0}\times\mathbb{C}, \tag{13}$$

which satisfy the commutation relations

$$[\psi,\phi]=\phi,\quad [\psi,e^{v\phi}]=v\phi e^{v\phi},\quad u^{\psi}\phi u^{-\psi}=u\phi,\quad u^{\psi}e^{v\phi}u^{-\psi}=e^{uv\phi},$$
$$[\psi,u^{\psi}]=[\phi,e^{v\phi}]=0,\quad u^{\psi}z^{\psi}=(uz)^{\psi},\quad e^{v\phi}e^{w\phi}=e^{(v+w)\phi}. \tag{14}$$

As linear forms on B, they are defined by the relations

$$\langle\psi,b^m a^n\rangle=\delta_{m,0}n,\quad \langle\phi,b^m a^n\rangle=\delta_{m,1},$$
$$\langle u^{\psi},b^m a^n\rangle=\delta_{m,0}u^n,\quad \langle e^{v\phi},b^m a^n\rangle=v^m,\quad \forall(m,n)\in\mathbb{Z}_{\geq 0}\times\mathbb{Z}. \tag{15}$$

Using the notation $\dot{x}:=\imath(x)$ for $x\in\{a,b\}$, $\dot{y}:=\jmath(y)$ for $y\in\{\psi,\phi\}$ and

$$u^{\dot{\psi}}:=\jmath(u^{\psi}),\quad e^{v\dot{\phi}}:=\jmath(e^{v\psi}), \tag{16}$$

the commutation relations (3) in the case of the quantum double $D(B)$ take the form

$$[\dot{\psi},\dot{b}]=\dot{b},\quad [\dot{\phi},\dot{b}]=1-\dot{a},\quad u^{\dot{\psi}}\dot{b}u^{-\dot{\psi}}=u\dot{b},\quad e^{v\dot{\phi}}\dot{b}e^{-v\dot{\phi}}=\dot{b}+(1-\dot{a})v, \tag{17}$$

and $\dot{a}$ is central. The formal universal R-matrix that reads as

$$R:=(1\otimes\dot{a})^{\dot{\psi}\otimes 1}e^{\dot{\phi}\otimes\dot{b}}=\sum_{m,n\geq 0}\frac{1}{n!}\binom{\dot{\psi}}{m}\dot{\phi}^n\otimes(\dot{a}-1)^m\dot{b}^n \tag{18}$$

should be interpreted as follows.

Any finite-dimensional right comodule V over $(D(B))^o$ is a left module over $D(B)$ defined by

$$xv=v_{(0)}\langle v_{(1)},x\rangle,\quad \forall(x,v)\in D(B)\times V. \tag{19}$$

Thus, it suffices to make sense of formula (18) in the case of an arbitrary finite-dimensional representation of $D(B)$ where the elements $1-\dot{a}$, $\dot{b}$, and $\dot{\phi}$ are necessarily nilpotent, so that the formal infinite double sum truncates to a well-defined finite sum.

Conjecture 16 The universal invariant associated with the Hopf algebra B is of the form $Z_B(K)=(\Delta_K(\dot{a}))^{-1}$, where $\Delta_K(t)$ is the Alexander polynomial of K normalized so that $\Delta_K(1)=1$ and $\Delta_K(t)=\Delta_K(1/t)$.

By direct computation, we were able to prove this conjecture in the case of the trefoil knot. Justification in the general case comes from the following reasoning.

The Hopf algebra B can be q-deformed to a non-commutative Hopf algebra B_q with the same coalgebra structure (12) but with q-commutative relation $ab = qba$. We have $B = B_1$. For q not a root of unity, the quantum double $D(B_q)$ is closely related to the quantum group $U_q(sl_2)$. In particular, for each $n \in \mathbb{Z}_{>0}$, it admits an n-dimensional irreducible representation corresponding to the n-th colored Jones polynomial. In the limit $n \to \infty$ with $q = t^{1/n}$ and fixed t, one recovers an infinite-dimensional representation of the Hopf algebra B where the central element $\dot{a}$ takes the value t. On the other hand, according to the Melvin–Morton–Rozansky conjecture proven by Bar-Nathan and Garoufalidis in [1] and by Garoufalidis and Lê in [5], the n-th colored Jones polynomial in that limit tends to $(\Delta_K(t))^{-1}$.

Acknowledgments I would like to thank Bruce Bartlett, Léo Bénard, Joan Porti, Louis-Hadrien Robert, Arkady Vaintrob, Roland van der Veen, and Alexis Virelizier for valuable discussions. This work is supported in part by the Swiss National Science Foundation, the subsidy no 200020_192081.

References

1. Dror Bar-Natan and Stavros Garoufalidis. On the Melvin-Morton-Rozansky conjecture. *Invent. Math.*, 125(1):103–133, 1996.
2. Alain Bruguières and Alexis Virelizier. Hopf diagrams and quantum invariants. *Algebr. Geom. Topol.*, 5:1677–1710 (electronic), 2005.
3. Sorin Dăscălescu, Constantin Năstăsescu, and Şerban Raianu. *Hopf algebras*, volume 235 of *Monographs and Textbooks in Pure and Applied Mathematics*. Marcel Dekker, Inc., New York, 2001. An introduction.
4. Roger Fenn and Colin Rourke. Racks and links in codimension two. *J. Knot Theory Ramifications*, 1(4):343–406, 1992.
5. Stavros Garoufalidis and Thang T. Q. Lê. Asymptotics of the colored Jones function of a knot. *Geom. Topol.*, 15(4):2135–2180, 2011.
6. Kazuo Habiro. Bottom tangles and universal invariants. *Algebr. Geom. Topol.*, 6:1113–1214, 2006.
7. David Joyce. A classifying invariant of knots, the knot quandle. *J. Pure Appl. Algebra*, 23(1):37–65, 1982.
8. R. M. Kashaev. R-matrix knot invariants and triangulations. In *Interactions between hyperbolic geometry, quantum topology and number theory*, volume 541 of *Contemp. Math.*, pages 69–81. Amer. Math. Soc., Providence, RI, 2011.
9. R. J. Lawrence. A universal link invariant using quantum groups. In *Differential geometric methods in theoretical physics (Chester, 1988)*, pages 55–63. World Sci. Publ., Teaneck, NJ, 1989.
10. H. C. Lee. Tangles, links and twisted quantum groups. In *Physics, geometry, and topology (Banff, AB, 1989)*, volume 238 of *NATO Adv. Sci. Inst. Ser. B Phys.*, pages 623–655. Plenum, New York, 1990.
11. Volodimir Lyubashenko. Tangles and Hopf algebras in braided categories. *J. Pure Appl. Algebra*, 98(3):245–278, 1995.
12. Shahn Majid. *Foundations of quantum group theory*. Cambridge University Press, Cambridge, 1995.

13. S. V. Matveev. Distributive groupoids in knot theory. *Mat. Sb. (N.S.)*, 119(161)(1):78–88, 160, 1982.
14. Tomotada Ohtsuki. Colored ribbon Hopf algebras and universal invariants of framed links. *J. Knot Theory Ramifications*, 2(2):211–232, 1993.
15. N. Yu. Reshetikhin. Quasitriangular Hopf algebras and invariants of links. *Algebra i Analiz*, 1(2):169–188, 1989.
16. N. Yu. Reshetikhin and V. G. Turaev. Ribbon graphs and their invariants derived from quantum groups. *Comm. Math. Phys.*, 127(1):1–26, 1990.
17. Mituhisa Takasaki. Abstraction of symmetric transformations. *Tôhoku Math. J.*, 49:145–207, 1943.
18. Alexis Virelizier. Kirby elements and quantum invariants. *Proc. London Math. Soc. (3)*, 93(2):474–514, 2006.

Rigged Configurations and Unimodality

Anatol N. Kirillov

To Professor Nikolai Reshetikhin on the occasion of the 60th anniversary with grateful and admiration from his friend and co-author

Abstract For a given partition λ and a dominant sequence of rectangular shape partitions $\{R_a\}$, we give sufficient conditions that imply that the corresponding parabolic Kostka polynomial $K_{\lambda,\{R\}}(q)$ is symmetric and **unimodal**. Examples found to satisfy these conditions include:

- principal specialization of the internal product of Schur functions $s_\alpha * s_\beta(\frac{q-q^N}{1-q})$, in particular, q-Gaussian polynomials $\left[{N \atop \lambda}\right]_q$ (the case $\alpha = \lambda$, $\beta = (|\lambda|)$;
- generalized q-Narayana numbers/polynomials $N_\ell((\lambda; \{R\}); q)$ associated with a partition λ and a sequence of rectangular shape partitions $\{R\} = (R_1, \ldots, R_n)$;
- symmetry and unimodality of the statistics charge generating function $\mathcal{K}^{[d]}_{\lambda,1^{|\lambda|}}(q)$ defined on the set of standard Young tableaux of a given shape λ and a fixed number of descents d;
- Schröder–Narayana $SchN_k(n.d; q)$, Kirkman–Caley $KC_d(n; q)$, and Motzkin sum $MS_d(n; q)$ polynomials;
- a (q, t)-deformation of the Euler polynomials $A_n(t)$; and
- reduced decomposition polynomials $RD^{[\ell]}(n; q)$.

As a corollary of our general result, we prove symmetry and unimodality of

- classical and rectangular q-Narayana numbers and

A. N. Kirillov (✉)
Research Institute for Mathematical Sciences (RIMS), Kyoto, Japan

The Kavli Institute for the Physics and Mathematics of the Universe (IPMU), Kashiwa, Japan

Department of Mathematics, National Research University Higher School of Economics, Moscow, Russia
e-mail: kirillov@kurims.kyoto-u.ac.jp
www.kurims.kyoto-u.ac.jp/~kirillov

A. Alekseev et al. (eds.), *Representation Theory, Mathematical Physics, and Integrable Systems*, Progress in Mathematics 340,
https://doi.org/10.1007/978-3-030-78148-4_16

- a certain q-deformation of the odd Eulerian numbers $A(2n+1,k)$.
  We also prove the *strict log-concavity* of q-binomial coefficients, and we introduce and initiate the study of double *Liskova polynomials* $L^{\gamma}_{\alpha,\beta}(q,t)$, which are a natural generalization of the Kostka–Macdonald polynomials $K_{\alpha,\beta}(q,t)$.

2000 Mathematics Subject Classifications: 05E05, 05E10, 05A19

1 Introduction

This paper presents a brief exposition of some results obtained by the author, which are devoted to various applications of the Algebraic Bethe Ansatz associated with generalized Heisenberg chains, to combinatorics of Young tableaux. The study of combinatorial aspects of Algebraic Bethe Ansatz was initiated by joint research with Nikolai Reshetikhin, [9]. For more details and proofs of statements that are formulated in our exposition, see [10, 12–14, 16–20].

1.1 A Bit of History

The story of my collaboration with Nikolai Reshetikhin has been started in the early 80s of the last century from traditional at that time in the Leningrad branch of Steklov Mathematical Institute (LOMI), the so-called *Christmas Exercises*, which were suggested to me by Professors L.D. Faddeev and L.A. Takhtajan, namely, to prove the following identity:

$$(1+x)^n = \sum_{2\ell \le n} \sum_{\nu} \prod_{j \ge 0} \binom{P_j^{(\ell)} + \nu_j}{\nu_j} \frac{x^{\ell} - x^{n-\ell+1}}{1-x}, \tag{1.1}$$

where the sum runs over a set of non-negative integers $\nu = (\nu_1, \nu_2, \ldots, \nu_\ell)$ such that $\sum_k k\nu_k = \ell$,

$$P_j^{(\ell)}(\nu) = n - 2\sum_k \min(j,k), \tag{1.2}$$

and for non-negative integers $\binom{P+m}{m} := \frac{(P+m)!}{P!\, m!}$.
This binomial-type identity looks rather mysterious and very different from a wide variety of binomial identities that are spreading in the mathematical literature. On the other hand, it is clearly seen that the identity (1.1) on the level of characters of the Lie algebra $sl(2)$ describes the decomposition of the nth tensor power of the spin $1/2$ irreducible representation of Lie algebra $sl(2)$ into irreducible $sl(2)$-modules.

In other words, the number

$$Z_n(1/2|\ell) = \sum_{\nu} \prod_{j \geq 0} \binom{P_j^{(\ell)} + \nu_j}{\nu_j} \tag{1.3}$$

is equal to the tensor product multiplicity

$$Z_n(1/2|\ell) = \mathrm{Mult}[V_{(n-2\ell+1)/2} : (V_{1/2})^{\otimes n}],$$

where V_j denotes the irreducible representation of $sl(2)$, $\dim V_j = 2j+1$.

For small n, say $n \leq 5$, the above formula for numbers $Z_n(1/2|\ell)$ can be proved by using the famous Kostant formula for the triple tensor product multiplicities. But I still do not know how to deduce formula (1.3) by reference to Kostant's formula in general.

To prove the identity (1.1), I applied the so-called inverse induction method. More precisely, I proved certain versions of the original identity (1.1) for the case of infinite number of variables, and then, after a certain change of variables (related with solutions to the so-called Q-systems), one can come back to the identity (1.1). This method allows to prove very general identity, namely, the following proposition.

Proposition 1.1 ([10]) *Let $b_1, b_2, \ldots$ be a set of parameters. Then,*

$$(1-x) \prod_{n \geq 1} (1-x^n)^{b_n} = 1 + \sum_{\ell \geq 1} Z(\{b\}|\ell)\, x^{\ell}, \tag{1.4}$$

where

$$Z(\{b\}|\ell)(\nu) = \sum_{\nu_k} \prod_{k} \binom{P_k^{(\ell)} + \nu_k}{\nu_k}', \ \textit{and}\ P_k^{(\ell)}(\nu) = -\sum_{j=1}^{k} (k-j+1) b_j - 2 \sum_{k} \min(j,k).$$

Here, $\binom{P+m}{m}' := P(P+1)\cdots(P+m-1)/m!$.

It is still an open **problem** to describe integer sequences $(b_1, b_2, \ldots)$ and the set $C(\{b\}|\ell)$ consisting of compositions $\nu = (\nu_1, \nu_2, \ldots)$, $\sum_{k \geq 1} k\nu_k = \ell$, such that

(1) the coefficients $Z(\{b\}|\ell)(\nu) \in \mathbb{Z}_{\geq 0}$ for all $k, \ell, \nu \in C(\{b\})|\ell)$, and
(2) $\sum_{\nu \notin C(\{b\})|\ell)} Z(\{b\}|\ell)(\nu) = 0$.

I have proved that conditions (1) and (2) are satisfied for sequences $\{b_k\}$ of the following form: let $\{s_i\}_{1 \leq i \leq n}$ be a sequence of positive integers, define

$$b_1 = -n, b_k = \delta_{k,s_i}, k \geq 2.$$

I proved that such sequences satisfy the conditions (1) and (2) above and give rise to the following identity:

$$(1-x)\prod_{j=1}^{n}\frac{1-x^{s_i}}{1-x}=\sum_{\ell} Z(\{s_i\}|\ell)\,(x^{\ell}-x^{n-2\ell+1}), \tag{1.5}$$

where

$$Z(\{s_i\}|\ell)=\sum_{\nu_k}\prod_{j\geq 1}\binom{\sum_a \min(j,s_a)-2\sum_a \min(j,a)\,\nu_a+\nu_j}{\nu_j},\ \sum k\nu_k=\ell. \tag{1.6}$$

A bit later, I met Leon Takhtajan and showed to him a proof of the identity (1.1) and asked about its origin and known generalizations. He told me that identity (1.1) had been proved by H. Bethe in his famous paper [1] in which he invented a method to diagonalize the Hamiltonian of the spin 1/2 Heisenberg chain of length n. Nowadays, this method is commonly called by *Algebraic Bethe Ansatz* and has many and varied applications in Physics. Leon also recommended me to ask Kolya Reshetikhin about the Bethe Ansatz, since Kolya is one of the leading specialists in this area.

I was really enjoyed very much by discussions with Kolya. He was very clever and accessibly explained me what are the (generalized) Heisenberg models, the Bethe Ansatz Equations, String Conjecture, Yang–Baxter equations, and many other things from Mathematical Physics. And he was always friendly and open for discussions and sharing ideas and results. I am sure that everyone who had, has (and will have) a chance to collaborate or associate with Kolya was (will be) always enjoyed by Kolya's clever lectures, conversations, and collaboration. There are plethora of evidence and examples that confirm this statement. He made (and will do) many important contributions to a wild variety areas of Mathematical Physics and Mathematics.

In this chapter, I want to overview some results concerning combinatorial aspects of the Algebraic Bethe Ansatz (ABA), which I had learned from Kolya. The study combinatorial properties of ABA were initiated in our joint paper [9] and had led to the discovery of Rigged Configuration Bijection, fermionic formula for parabolic Kostka polynomials, Kirillov–Reshetikhin modules, the first combinatorial proof of the unimodality of q-Gaussian polynomials, and several other findings in combinatorics. Other fields of our common interest and research included the study of generalized XXX and XXZ models, Dilogarithm Identities, q-orthogonal polynomials, q-$6j$ symbols and related knot invariants, quantum Weyl groups, etc. Professor Nikolai Reshetikhin made (and will do) great contribution to Mathematical Physics and Mathematics.

1.2 Introduction

The main objective of this chapter is to prove unimodality of certain polynomials related to the parabolic Kostka polynomials. More precisely, we state and prove sufficient conditions for ℓ-partial parabolic Kostka polynomials $\mathcal{K}_{\lambda,R}^{[\ell]}(q)$ to be symmetric and unimodal. As a corollary, we prove unimodality of generalized q-Narayana numbers and generalized q-Gaussian polynomials, and give new proof of the strict unimodality of "general" q-binomial coefficients. All our proofs are based on the use of a fermionic formula for parabolic Kostka polynomials discovered by the author of this chapter in the middle of 80s of the last century. In essence, our fermionic formula gives rise to the decomposition of a given parabolic Kostka polynomial $K_{\lambda,\{R\}}(q)$ into the sum of symmetric and unimodal polynomials. Our main strategy is to find data $(\lambda, \{R\})$ such that a given symmetric polynomial can be identified[1] with some symmetric and unimodal part of the decomposition of certain parabolic Kostka polynomials of type $(\lambda\{R\})$. We illustrate this method by several examples, including, among others, the q-Gaussian polynomials, the Rectangular and Staircase Narayana, the Schröder–Narayana, a q-deformed Riordan, and a (q,t)-deformed Euler polynomials.

Theorem 1.2 *Let λ be partitions, then ℓ-part $\mathcal{K}_{\lambda,1^{|\lambda|}}^{[\ell]}(q)$ of the Kostka polynomial $K_{\lambda,1^{|\lambda|}}(q)$ is a symmetric and unimodal polynomial for $\ell(\lambda)-1 \le \ell \le |\lambda| - \lambda_1'$ (where λ' denotes the transpose partition of λ).*

Theorem 1.3 *Let λ be partition and $R = \{R_a = (\mu_a^{\eta_a})\}$ be dominant sequence of rectangular shape partitions. If $\mu_a \ge (\nu^{\eta_a})_1'$ $\forall a$ and for all (admissible) configurations of type (λ, R), then the parabolic Kostka polynomial $K_{\lambda,R}(q)$ is symmetric and unimodal.*

For example, one can take $\mu_a \ge |\lambda| - \lambda_1$, for all $a \ge 1$. Under these assumptions, the first row λ_1 of λ must be "very long" in comparison with other parts of λ, namely, $\lambda_1 \ge \lambda[1](-1+\sum_a \eta_a)$, where we set $\lambda[1] = \sum_{j\ge 2}\lambda_j$. Examples satisfying conditions of Theorem 1.3 include the principal specialization of Schur functions, i.e., q-Gaussian polynomials, the internal product of Schur polynomials.

Another objective of this chapter is to draw attention of the readers to a natural generalization of the Kostka–Foulkes polynomials, which we named by *Liskova* and double *Liskova* polynomials, see [14, 18, 19]. Namely, let α, β, and γ be three partitions of the same size n.

Definition 1.4 Define Liskova polynomial $L_{\alpha,\beta}^{\gamma}(q)$ from the decomposition

$$s_\alpha * s_\beta(x) = \sum_\gamma L_{\alpha,\beta}^{\gamma}(q) P_\gamma(x;q),$$

[1] Up to multiplication by some power of q.

where $P_\gamma(x;q)$ denotes the Hall–Littlewood polynomial [23], and $s_\alpha * s_\beta(x)$ denotes the internal product of Schur functions, see Sect. 4 for details.

It is easy to see that the constant term of the Liskova polynomials $L^\gamma_{\alpha,\beta}(q)$ is equal to the multiplicity of the symmetric group $\mathbb{S}_n$ character χ^γ in the Kronecker product of characters χ^α and χ^β. In other words, if

$$\chi^\alpha \chi^\beta = \sum_\gamma g_{\alpha\beta\gamma}\chi^\gamma \Longrightarrow L^\gamma_{\alpha,\beta}(0) = g_{\alpha\beta\gamma},$$

and it is easy to see that one has also

$$L^\gamma_{\alpha,\beta}(q) = \sum_\eta g_{\alpha\beta\gamma} K_{\eta,\gamma}(q).$$

Note that if partition β consists of only one part, i.e., $\beta = (|\alpha|)$, then

$$L^\gamma_{\alpha,\beta}(q) = K_{\alpha,\gamma}(q).$$

In other words, the Liskova polynomials are a natural generalization of the Kostka–Foulkes polynomials.
The number $L^\gamma_{\alpha,\beta}(1)$ admits a combinatorial interpretation as the number of the Littlewood–Richardson sequences of tableaux $\mathcal{T} = (T_1, \ldots, T_{\ell(\gamma)})$ of type $(\alpha, \beta; \gamma)$, see [18, Definition 6.4]; cf. [23, pp. 184–185]. We denote by $\boldsymbol{\nu}^\gamma(\alpha,\beta)$ the set of the Littlewood–Richardson sequences of tableaux that are defined in [18, Definition 6.4].

Problem 1.5 *Let $\mathcal{T} \in \boldsymbol{\nu}^\gamma(\alpha,\beta)$ be an LR sequence of type $(\alpha, \beta; \gamma)$. Define statistics $c(\mathcal{T})$ for an LR sequence $\mathcal{T}$ such that*

$$L^\gamma_{\alpha,\beta}(q) = \sum_{\mathcal{T}\in\boldsymbol{\nu}^\gamma(\alpha,\beta)} q^{c(\mathcal{T})}.$$

Therefore,

$$g_{\alpha,\beta,\gamma} = \left|\{\mathcal{T} \in \boldsymbol{\nu}^\gamma(\alpha,\beta)\ c(\mathcal{T}) = 0\right|. \tag{1.7}$$

We consider conjectural formula (1.7) as a "charge version," see [18], or Remark 1.6, of the classical LR formula for the decomposition of the product of Schur functions.

Remark 1.6 (Charge and the Littlewood–Richardson Rule) Let λ, μ, and ν be three partitions such that $|\lambda|+|\mu| = |\nu|$, then the Littlewood–Richardson number $c^\nu_{\lambda,\mu} := \mathrm{Mult}\big[V_\nu\colon V_\lambda \otimes V_\mu\big]$ is equal to the number of semistandard Young tableaux T of skew shape $\nu \setminus \mu$ and weight λ such that the right-to-left and top-to-bottom reading word $w(T)$ corresponding to tableaux T in question, is the *Yamanouchi* word. It is clearly seen that the word $w(T)$ obtained is a the Yamanouchi word iff the charge of $w(T)$ is equal to zero. In other words,

$$c^{\nu}_{\lambda,\mu} = \{T \in SYT(\nu \setminus \mu, \lambda) \mid c(T) = 0\}.$$

We expect a similar rule for computation of the numbers $g_{\alpha,\beta,\gamma}$, as it was suggested in Problem 1.5, formula (1.7).
Note also that if we set $\alpha_N := (N - |\lambda|, \lambda)$, $\beta_N := (N - |\mu|, \mu)$ and $\gamma_N := (N - |\nu|, \nu)$, and $N \geq |\nu| + \nu_1$, then [22]

$$g_{\alpha_N,\beta_N,\gamma_N} = c^{\nu}_{\lambda,\mu}.$$

- (**Big Littlewood–Richardson numbers [19]**)

Let α, β, and γ be three partitions of the same size n. For integer N, consider partitions $\alpha^{(N)} := (N + \alpha_1, \alpha[1])$, $\beta^{(N)} := (N + \beta_1, \beta[1])$, and $\gamma^{(N)} := (N + \gamma_1, \gamma[1])$, where for any partition $\lambda = (\lambda_1, \ldots, \lambda_r)$ and integer $1 \leq k < r$, we set $\lambda[k] := (\lambda_{k+1}, \ldots, \lambda_r)$.

Proposition 1.7 *The sequence of polynomials $\{L_{\alpha_{N,\beta_N}}^{\gamma_N}(q)\}_N$ is stabilized to the polynomial $\mathcal{L}^{\nu}_{\lambda,\mu}(q)$, i.e., for sufficiently large N, the polynomial $L^{\gamma_N}_{\alpha_N,\beta_N}(q)$ does not depend on N and equal to $\mathcal{L}^{\nu}_{\lambda,\mu}(q)$.*

We call numbers $C^{\nu}_{\lambda,\mu} := \mathcal{L}^{\nu}_{\lambda,\mu}(0)$ by the big LR coefficients and polynomials $\mathcal{L}^{\nu}_{\lambda,\mu}(q)$ by stable Liskova polynomials.

For example, $\mathcal{L}^{21}_{21,21}(0) = 9$. Note that if $|\lambda| + |\mu| > |\nu|$, then $\mathcal{L}^{\nu}_{\lambda,\mu}(q) = 0$.

- (**Domino, nspin, and t-LR numbers**)

Let λ, μ, and η be three partitions such that $|\lambda| + |\mu| = |\eta|$. Consider rectangular shape partition $(N^{\lambda'_1})$ such that $N \geq \mu_1$, and define partition $\Lambda_N := ((N^{\lambda'_1}) + \lambda, \mu)$ and dominant sequence of rectangular shape partitions $R_N := \{(N^{\lambda'_1}), \eta^+\}$.

Theorem 1.8

(1) $K_{\Lambda_N,R}(q) \stackrel{\bullet}{=} c^{\eta}_{\lambda,\mu} + \ldots + q^{n(\eta)-n(\lambda)-n(\mu)}$;
(2)

$$K_{\Lambda_N,R}(1) = \Big|\{T \in STY^{(2)}((2\,\lambda) \,\vee\, (2\,\mu), \eta)\}\Big|.$$

Here, $STY^{(2)}((2\,\lambda) \vee (2\,\mu), \eta)$ denotes the set of domino tableaux of shape $(2\,\lambda) \vee (2\,\mu)$ and weight η.

Finally, define polynomials

$$K^{\eta}_{\lambda,\mu}(q,t) := \sum_{\{\nu\}} t^{ns(\nu)} K_{\Lambda_N,R_N}(q),$$

where sum runs over the set of (admissible) configurations of type (Λ_N, R_N), and $ns(\nu)$ stands for the nips of configuration $\{\nu\}$.

Note that if partition β consists of only one part, then the set $\boldsymbol{\nu}^{\gamma}(\alpha, \beta) = STY(\alpha, \gamma)$, that is, an LR sequence $\mathcal{T}$ of type $(\alpha, (|\beta|); \gamma)$, defines a semistandard tableau T of shape α and weight γ, and moreover $charge(\mathcal{T}) = charge(T)$.

- (**Double Liskova polynomials**)

Definition 1.9 Define double Liskova polynomials as follows:

$$L^{\gamma}_{\alpha,\beta}(q,t) = \sum_{\eta} g_{\alpha\beta\eta} K_{\eta,\gamma}(q,t),$$

where $K_{\eta,\gamma}(q,t)$ denotes the Kostka–Macdonald polynomial [23].

It is easy to see [23] that

$$L^{\gamma}_{\alpha,\beta}(0,t) = L^{\gamma}_{\alpha,\beta}(t),\ L^{\gamma}_{\alpha,\beta}(1,1) = f^{\alpha} f^{\beta},\ L^{\gamma}_{\alpha,\beta}(0,0) = g_{\alpha\beta\gamma}.$$

Note also that if partition β consists only one part, then

$$L^{\gamma}_{\alpha,\beta}(q,t) = K_{\alpha,\gamma}(q,t).$$

Problem 1.10 *For a pair of standard Young tableaux $(T_1, T_2) \in STY(\alpha) \times STY(\beta)$, define a pair of statistics $u_{\mu}(T_1, T_2)$ and $v_{\mu}(T_1, T_2)$ such that*

$$L^{\mu}_{\alpha,\beta}(q,t) = \sum_{(T_1,T_2)\in STY(\alpha)\times STY(\beta)} q^{u_{\mu}(T_1,T_2)}\, t^{v_{\mu}(T_1,T_2)}.$$

We also include a fermionic formula for generalized exponents polynomial $G_N(V_{\alpha} \otimes V^{*}_{\beta})(q)$ associated with mixed tensor representation $V_{\alpha} \otimes V^{*}_{\beta}$ of the Lie algebra $sl(N)$.

Let us say few words about the content of this chapter.

In Sect. 2, we collect some basic definitions and notation we will use in this chapter. Section 3 contains our main results. In Sect. 4, we introduce and study some properties of the internal product of Schur functions and define Liskova polynomials. We review our old result concerning a fermionic formula for principal specialization of the internal product of Schur functions. In Sect. 5, we state a fermionic formula for the mixed tensor product generalized exponents polynomial. In Sect. 6, for the reader's convenience, we give brief review of basic facts about rigged configurations.

2 Basic Notation and Definitions

To start with, let us recall definitions of symmetric, unimodal, and strictly unimodal polynomials.

Definition 2.1

(1) A sequence of non-negative real numbers $\mathbf{a} = \{a_k\}_{0 \leq k \leq n}$ is called *unimodal* if for some integer d,

$$a_k \leq a_{k+1},\ 1 \leq k \leq d,\ and\ a_k \geq a_{k+1},\ d \leq k \leq n;$$

(2) A such sequence is called *strictly unimodal* if the inequalities in the above definition are all strict.

(3) A sequence $\{a_k\}_{0 \leq k \leq n}$ is called *symmetric* if

$$a_i = a_{n-i},\ i \leq n/2.$$

For a symmetric sequence $\{a_k\}_{0 \leq k \leq n}$, the number $n/2$ is called *centerline* or *unimodality index* and denoted by $un(\mathbf{a})$.

Definition 2.2 A polynomial $p(x) = x^k(a_0 + a_1x + \cdots + a_nx^n)$ is called *symmetric* (and/or *unimodal* or *strictly unimodal*) iff the sequence of its coefficients $\{a_0, \ldots, a_n\}$ is *symmetric* (and/or *unimodal* or *strictly unimodal*).

Definition 2.3 (Strong Log-Concavity of q-Binomial Coefficients) The sequence of polynomials $\{r_n(q) \in \mathbb{N}[q]\}_{n \geq 0}$, $r_0(q) = 1, r_n(q) = 0,\ if\ n \in \mathbb{Z}_{<0}$, is called strongly log-concave, if the difference

$$r_n(q)r_m(q) - r_{n-k}(q)r_{m+k}(q)$$

is a unimodal polynomial with non-negative coefficients for all $n \geq m$ and $k \leq m$.

Proposition 2.4

(1) *([28]) The product of symmetric and unimodal (respectively, strictly unimodal) polynomials is again symmetric and unimodal (respectively, strictly unimodal) polynomial.*

(2) *([4, 6]) Let $k \geq 3$ and $n_1, \ldots, n_k$ be a sequence of pairwise distinct positive integers. Then, the polynomial*

$$\prod_{a=1}^{k} [n_a + 1]_q$$

is symmetric and strictly unimodal.

2.1 Polynomials Associated with Configurations of Type (λ, R)

Now, we are going to review some basic notation and definitions concerning the rigged configurations. For more details and examples, see, e.g., the Appendix or [18, 20].

For a given data (λ, R) as before,[2] let us define the set of admissible configurations of type (λ, R) and then introduce a few polynomials, which are the main object to study in this chapter.

Thus, let λ be a partition and $R = (R_1, \ldots, R_p)$ be a sequence of rectangular shape partitions, say, $R = \{R_a := (\mu_a^{\eta_a}), a = 1, \ldots, p\}$. We assume that the size of λ is equal to $\sum_a \mu_a \eta_a$. With such data (λ, R), we associate a set of admissible configurations of type (λ, R), which is denoted by $C(\lambda, R)$.

Definition 2.5 An admissible configuration $\{\nu\}$ of type (λ, R) is a sequence of partitions $\{\nu\} = (\nu^{(1)}, \nu^{(2)}, \ldots)$, which satisfies the following conditions:

- (Size of partition $\nu^{(k)}$)

$$|\nu^{(k)}| = \sum_{j>k} \lambda_j - \sum_{a \geq 1} \mu_a \max(\eta_a - k, 0) = -\sum_{j \leq k} \lambda_j + \sum_{a \geq 1} \mu_a \min(k, \eta_a).$$

  We set $\nu^{(0)} = \emptyset$.
- (Non-negativity constraints or positivity of vacancy numbers $P_j^{(k)}(\{\nu\})$)

$$P_j^{(k)}(\{\nu\}) := Q_j(\nu^{(k-1)}) - 2Q_j(\nu^{(k)}) + Q_j(\nu^{(k+1)}) + \sum_{a \geq 1} \min(\mu_a, j)\delta_{\eta_a,k} \geq 0, \ \forall k, j \geq 1,$$

  where $\delta_{a,b}$ stands for the Kronecker delta function; for any Young diagram ν, we set

$$Q_j(\nu) = \sum_{a \geq 1} \min(\nu_j, a) = \sum_{a \leq j} \nu'_a.$$

Note that if $k \geq l(\lambda)$ and $k \geq \eta_a$ for all a, then partition $\nu^{(k)} = \emptyset$. Therefore, there are only a finite number of admissible configurations of type (λ, R). It might be well to point out that one needs to check a non-negativity of a vacancy number $P_j^{(k)}(\{\nu\})$ even if the corresponding number of rows $m_j(\nu^{(k)})$ is equal to zero.

[2] That is, λ is a partition and R is a sequence of rectangular shape partitions.

The main quantities, which we associate with a configuration $\{\nu\} \in C(\lambda; R)$ we are planning to investigate, are

- $n(R) = \sum_{i,k \geq 1} \binom{\sum_a \theta(\eta_a - k)\theta(\mu_a - i)}{2} = \sum_{a<b} \min(\mu_a, \mu_b) \min(\eta_a, \eta_b)$, where for a real number x, we set $\theta(x) = 1$, if $x \geq 0$, and $\theta(x) = 0$, if $x < 0$;
- (Matrix $M(\{\nu\})$ associated with a configuration $\{\nu\}$)
  $M(\{\nu\}) := (m_{ij}(\{\nu\})$, where

$$m_{ij}(\{\nu\}) = ((\nu^{(i-1)})_j' - (\nu^{(i)})_j') + \sum_{a \geq 1} \theta(\eta_a - i)\theta(\mu_a - j);$$

- (The number of parts of size j in diagram/partition $\nu^{(k)}$)

$$m_j^{(k)}(\{\nu\}) := (\nu^{(k)})_j' - (\nu^{(k)})_{j+1}' = \sum_{a \geq 1} \min(\eta_a, k)\delta_{\mu_a, j} - \sum_{i \leq k} (m_{ij} - m_{i,j+1});$$

- (Vacancy numbers)

$$P_i^{(k)}(\{\nu\}) = \sum_{i \geq j} (m_{ki} - m_{k+1,i}),$$

- (Charge $c(\{\nu\})$ and cocharge $\overline{c}(\{\nu\})$ of configuration)

$$c(\{\nu\}) = \sum_{i,j \geq 1} \binom{m_{ij}(\{\nu\})}{2}, \ \overline{c}(\{\nu\}) = \sum_{j,k \geq 1} \binom{(\nu^{(k)})_j' - (\nu^{(k+1)})_j'}{2};$$

- (Centerline or unimodality index)

$$un(\nu) = n(R) - 1/2\Big(\sum_{a \geq 1} Q_{\mu_a}(\nu^{(\eta_a)})\Big);$$

  in the case $\eta_a = 1, 1 \leq a \leq p$, our formula for $un(\{\nu\})$ can be rewritten as follows:

$$2\, un(\{\nu\}) = 2\, n(\mu) - \sum_{a \geq 1} \mu_a' (\nu_1')_a,$$

  where we set $\mu = (\mu_1 \geq \mu_2 \geq \cdots)$.
- (Nips of configuration)

$$2\, ns(\{\nu\}) = \#\Big((ij) \, \Big| m_{ij}(\{\nu\}) \equiv 1 \bmod 2\Big),$$

- (Descent index) $des(\{\nu\}) = \nu_1'$.

Definition 2.6 A sequence of rectangular shape partitions $\{R_a = (\mu_a^{\eta_a})\}_{1 \le a \le p}$ is called *dominant*, if $\mu_1 \ge \mu_2 \ge \ldots \ge \mu_p$.

Now, we are going to define ℓ-partial parabolic Kostka polynomials.
Let ℓ be a positive integer, using the quantities associated with a configuration $\{\nu\}$ of type (λ, R); we define the following polynomials:

- (ℓ-partial $\mathcal{K}$-polynomials $\mathcal{K}_{\lambda,R}^{[\ell]}(q)$). Define

$$\mathcal{K}_{\lambda,R}^{[\ell]}(q) = \sum_{\substack{\{\nu\} \in C(\lambda,R) \\ un(\{\nu\})=\ell}} q^{c(\{\nu\})} \prod_{k,i \ge 1} \begin{bmatrix} P_i^{(k)}(\{\nu\}) + \mu_i^{(k)}(\{\nu\}) \\ \mu_i^{(k)}(\{\nu\}) \end{bmatrix}_q,$$

  where the sum runs over the set of type (λ, R) configurations with the unimodality index equals to ℓ,
- ($\mathcal{K}$-polynomials $\mathcal{K}_{\lambda,R}(q,t)$)

$$\mathcal{K}_{\lambda,R}(q,t) = \sum_{\{\nu\} \in C(\lambda,R)} q^{c(\{\nu\})}\, t^{un(\{\nu\})} \prod_{k,i \ge 1} \begin{bmatrix} P_i^{(k)}(\{\nu\}) + \mu_i^{(k)}(\{\nu\}) \\ \mu_i^{(k)}(\{\nu\}) \end{bmatrix}_q.$$

  Therefore, $\mathcal{K}_{\lambda,R}(q,t) = \sum_{\ell} t^{un(\{\nu\})} \mathcal{K}_{\lambda,R}^{(\ell)}(q)$, and the polynomial $\mathcal{K}_{\lambda,R}(q,1)$ coincides with the parabolic Kostka polynomial corresponding to the data (λ, R).

Definition 2.7 Let $d \in \mathbb{N}$ and (λ, R) be a partition and a dominant sequence of rectangular shape partitions. Define generalized Narayana polynomials of type (λ, R) and degree d, denoted by $N_d(\lambda, R; q)$, as follows:

$$N_d(\lambda, R; q) := \mathcal{K}_{\lambda,R}^{[d]}(q).$$

2.2 Formulas for Unimodality Index

To state our main result, we need to compute the "centerline," that is, the sum

$$un(\{\nu\}) := c(\{\nu\}) + 1/2 \sum_{i,k \ge 1} P_i^{(k)}(\{\nu\})\, m_i^{(k)}(\{\nu\})$$

for any admissible configuration $\{\nu\}$ of type $(\nu; R)$. First of all, let us define an analog of the number $n(\lambda) := \sum_{i \ge 1} (i-1)\, \lambda_i$, where λ is a partition, now for a sequence of rectangles $R = \{(\mu_a^{\eta_a})\}$; namely, let us set

$$n(R) := \sum_{i,k \ge 1} \binom{\sum_a \theta(\eta_a - k)\theta(\mu_a - i)}{2}.$$

It is not difficult to see that

$$n(R) = \sum_{a<b} \min(\mu_a, \mu_b)\ \min(\eta_a, \eta_b).$$

Proposition 2.8 *One has the following relations:*

•

$$c(\{\nu\}) = n(R) + \sum_{i,k\geq 1} \Big(((\nu^{(k)})_i^{'})^2 - (\nu^{(k)})_i^{'} \times (\nu^{(k+1)})_i^{'} \Big) -$$

$$\sum_{k,i\geq 1} \Big(\sum_{a\geq 1} \theta(\mu_a - i)\ \delta_{\eta_a,k} \Big) \nu_i^{(k)}.$$

•

$$\sum_{i,k\geq 1} P_i^{(k)}(\{\nu\})\ m_i^{(k)}(\{\nu\}) = 2\Big((\nu^{(k)})_i^{'} \times (\nu^{(k+1)})_i^{'} - ((\nu^{(k)})_i^{'})^2 \Big) +$$

$$\sum_{k,i\geq 1} \Big(\sum_{a} \theta(\mu_a - i)\ \delta_{\eta_a,k} \Big) \nu_i^{(k)}.$$

A proof is a bit lengthy work with definitions of the charge and vacancy numbers of a configuration $\{\nu\}$.

As the result of these computations, one obtains the following theorem.

Theorem 2.9

$$un(\{\nu;\, R\}) = n(R) - 1/2 \Big(\sum_{i,k\geq 1} \big(\sum_{a} \theta(\mu_a - i)\ \delta_{\eta_a,k} \big)\ \nu_i^{(k)} \Big) = n(R) - 1/2 \sum_{a} Q_{\mu_a}(\nu^{(\eta_a)}).$$

Therefore, one has

$$un(\{\nu\}) = n(R) - 1/2 \sum_{a} Q_{\mu_a}(\nu^{(\eta_a)}).$$

In the special case $\eta_a = 1,\ \forall a \geq 1$, one has

$$2\, un(\{\nu\}) = 2\, n(\mu) - \sum_{a\geq 1} \mu_a^{'} (\nu_1^{'})_a.$$

Corollary 2.10 *Assume that $\eta_a = 1, \forall a$, and denote by μ a unique partition corresponding to the sequence $\{\mu_1, \mu_2, \ldots\}$. Then, for any admissible configuration*

$\{\nu\}$ *of type* $(\lambda;\mu)$, *one has*

$$un(\{\nu\}) = n(\mu) - 1/2 \sum_{a \geq 1} Q_{\mu_a}(\nu^{(1)}) = n(R) - 1/2 \sum_{i \geq 1} \mu_i^{'} \times (\nu^{(1)})_i^{'};$$

$$c(\{\nu\}) + \sum_{i,k \geq 1} P_i^{(k)}(\{\nu\})\, m_i^{(k)}(\{\nu\}) = n(R) - \sum_{i,k \geq 1} (\nu_i^{(k)})' \Big((\nu_i^{(k)})' - (\nu_i^{(k+1)})' \Big).$$

3 Main Results

- **(Unimodality of q-Gaussian polynomials)**

Theorem 3.1

(1) *Let* λ *be a strict partition,* $\ell(\lambda) \geq 3$. *If* $N > \ell(\lambda)$, *then the* q-*Gaussian polynomial* $\left[\begin{smallmatrix} N \\ \lambda \end{smallmatrix}\right]_q$ *is a symmetric and strictly unimodal.*

(2) *Let* λ *be a partition such that* $\lambda_i^{'} - \lambda_{i+1}^{'} \geq 8$. *If* $N > \ell(\lambda) + 8$, *then the* q-*Gaussian polynomial* $\left[\begin{smallmatrix} N \\ \lambda \end{smallmatrix}\right]_q$ *is a symmetric and strictly unimodal.*

Indeed, first of all, let us recall that [15, 18]

$$\begin{bmatrix} N \\ \lambda \end{bmatrix}_q \stackrel{\bullet}{=} K_{\Lambda,\mu}(q),$$

where $\Lambda = (N|\lambda|, \lambda)$ and $\mu = (|\lambda|^{N+1})$.
Now, let us consider the maximal configuration $\Delta := \Delta_{\Lambda,\mu}$ of type (Λ, μ), see Sect. 6. By definition, $\Delta = (\lambda, \lambda[1], \ldots, \lambda[\ell(\lambda) - 1])$, where $\lambda[k] := (\lambda_{k+1}, \ldots, \lambda_{\ell(\lambda)})$. By construction, one can see that if $k \geq 2$, then all the vacancy numbers $P_j^{(k)}(\Delta)$ are equal to zero, whereas in the case (1),

$$P_j^{(1)}(\Delta) = jN - Q_j(\lambda) = \sum_{1 \leq a \leq j} (N - \lambda_a) > 0.$$

Observe that if $a < b$, then

$$P_b^{(1)}(\Delta) - P_a^{(1)}(\Delta) = (b-a)N - \sum_{j=a+1}^{b} \lambda_j^{'} \geq (b-a)(N - \lambda_1^{'}) > 0.$$

Therefore, it follows from Sect. 6, Corollary 6.6. (fermionic formula), assumptions (1) in Theorem 3.1, and Proposition 2.4, (2) that

$$K_{\Lambda,\mu}(\Delta) := \prod_{j \geq 1} [P_j^{(1)}(\Delta) + 1]_q$$

is strictly unimodal and symmetric. Since the polynomials $K_{\Lambda,\mu}(\{\nu\})$ have the same center of symmetry and unimodal for all configurations $\{\nu\} \in C(\Lambda, \mu)$, and

$$\sum_{\{\nu\}} K_{\Lambda,\mu}(\{\nu\}) = K_{\Lambda,\mu}(q),$$

one concludes that the q-Gaussian polynomial $\left[\begin{smallmatrix} N \\ \lambda \end{smallmatrix}\right]_q \doteq K_{\Lambda,\mu}(q)$ is also symmetric and strictly unimodal.
Similarly, in the second case, one has $m_j^{(1)}(\Delta) := \lambda_j^{'} - \lambda_{J=1}^{'} \geq 8$, and $P_j^{(1)}(\Delta) \geq 8j$. Thus, it follows from Proposition 2.4,(1) that all the q-binomial coefficients $\left[\begin{smallmatrix} P_j^{(1)}(\Delta)+m_j^{(1)}(\Delta) \\ m_j^{(1)}(\Delta) \end{smallmatrix}\right]$ are symmetric and unimodal. Therefore, their product, i.e., $K_{\Lambda,\mu}(\Delta)$, is also symmetric and strictly unimodal. Hence, under the assumptions of Proposition 2.4,(2) and assumptions of Theorem 3.1, (2), the polynomial $\left[\begin{smallmatrix} N \\ \lambda \end{smallmatrix}\right]_q$ is also symmetric and strictly unimodal. ■

It is still an open question to describe all the exceptions to Theorem 3.1.
In the case when partition λ consists of one row (or one column), Theorem 3.1 has been proved first in [25]. The first combinatorial proof of unimodality of q-binomial coefficients $\left[\begin{smallmatrix} n+m \\ m \end{smallmatrix}\right]_q$ has been done in [24]. In [11], one can find rigged configuration interpretation of statistics introduced in K. O'Hara's combinatorial proof presented in [24]; see also [33].

Theorem 3.2 (Strong Log-Concavity of q-Binomial Coefficients) *For $0 \leq k \leq m \leq n < N - k$, the polynomial*

$$BS_{n,m}(N, k; q) := \begin{bmatrix} N \\ m \end{bmatrix}_q \begin{bmatrix} N \\ n \end{bmatrix}_q - q^{k(n-m+k)} \begin{bmatrix} N \\ m-k \end{bmatrix}_q \begin{bmatrix} N \\ n+k \end{bmatrix}_q$$

is a symmetric and unimodal polynomial in q with non-negative coefficients.

Indeed, one can show (see, e.g.,[3], Corollary 3.2, or [23], p.44) that

$$\sum_{j=0}^{k} s_{2^{m-j} 1^{n-m+2j}}(1, q, \ldots, a^{N-1}) = q^{\binom{m}{2}+\binom{n}{2}} \left(\begin{bmatrix} m \\ m \end{bmatrix}_q \begin{bmatrix} n \\ n \end{bmatrix}_q - q^{k(n-m+k)} \begin{bmatrix} N \\ m-k \end{bmatrix}_q \begin{bmatrix} N \\ n+k \end{bmatrix}_q \right).$$

Now, let us observe that the unimodality index of polynomial

$$s_{2^{n-j}1^{n-m+2j}}(1, q, \ldots, q^{N-1}),\ 0 \le j \le k,$$

is the same for all j and equal to $\frac{1}{2}(n+m)(N-1)$. According to our result that the principal specialization $s_\alpha(q, q^2, \ldots, q^N)$ of a Schur polynomial $s_\alpha(X)$ is symmetric and unimodal, see, e.g., [15], and as it is clearly seen that the polynomial from our Theorem 1.2 is the sum of symmetric and unimodal polynomials with the same symmetry and unimodality indexes, we conclude that the polynomial from Theorem 3.2 is symmetric and unimodal, and therefore the q-binomial coefficients $\begin{bmatrix} N \\ m \end{bmatrix}_q$ obey the strong log-concavity property.

- (**Unimodality of** *Schröder–Narayana polynomials*)
  Let d and N be positive integers, $N \ge 2d$. For each k, $1 \le k \le d$, define the kth Schröder–Narayana polynomial as follows: $ScN_k(n, d; q) =$

$$\frac{1}{[d-k_q![d-k+1]_q!} \begin{bmatrix} N-2d+k-1 \\ k-1 \end{bmatrix}_q \prod_{j=0}^{d-k} [n-d-j]_q^{\min(2, j+1, d-k-j+1+\delta_{k,d})}.$$

Theorem 3.3 *The Schröder–Narayana polynomials are symmetric and unimodal for all $N, d, N \ge 2d \ge 2$.*

Our proof is based on the observation that

$$ScN_k(n, d; q) = \mathcal{K}^{[d]}_{(d^2, 1^{N-2d}), (1^N)}(q)$$

and our result that, for any partition λ, the truncated Kostka–Foulkes polynomial $\mathcal{K}^{[d]}_{\lambda, (1^{|\lambda|})}(q)$ is symmetric and unimodal.

- (**Unimodality of almost rectangular Narayana polynomials**)
  Let d, k, and N be positive integers, $N \ge kd$. Defined almost rectangular Narayana polynomials $N_p(k, d, N; q)$ to be

$$N_p(k, d, N; q) = \mathcal{K}^{[p]}_{(d^k, 1^{N-kd})}(q).$$

Theorem 3.4

(1) *Polynomials $N_p(k, d, N; q)$ are symmetric and unimodal.*
(2) *(Combinatorial formula)*

$$N_p(k, d, N; q) = \sum_{\substack{\{\nu\} \\ (\nu^{(1)})'_1 = p}} q^{c(\{\nu\})} \prod_{j,\ell} \begin{bmatrix} P_j^{(\ell)}(\{\nu\}) + m_j^{(\ell)} \\ m_j^{(\ell)} \end{bmatrix}_q,$$

where the sum runs over a sequence of partitions ($\nu_0 = \emptyset$)

$$\{\nu^{(1)}, \nu^{(2)}, \dots, \nu^{(k)} = 1^{N-kd}, \nu^{(k+1)} = 1^{N-kd-1}\}, \ |\nu^{(\ell)}| = (k-\ell)d, \ 1 \le \ell \le k-1;$$

$$P_j^{(\ell)}(\{\nu\}) = min(N, j)\delta_{\ell,1} + Q_j(\nu^{(\ell-1)} - 2Q_j(\nu^{(\ell)}) + Q_j(\nu^{(\ell+1)}), \ j \ge 1, \ 1 \le \ell \le k;$$

$m_j(\nu^{(\ell)}) = ((\nu^{(\ell)})')_j - ((\nu^{(\ell)})')_{j+1}$; $c(\{\nu\})$ stands for the charge of configuration $\{\nu\}$.

In the case $k = 2$, we have some explicit product formula for polynomials $N_p(k = 2, d, N; q)$, see Theorem 3.4 stated above. As for the Kostka polynomials $K_{(d^k, 1^{N-kd})}(q)$ thereof, they are not symmetric and unimodal in general. In fact, the number $K_{(d^2, 1^{N-2d})}(1)$ is equal to the so-called *Kirkman–Caley* number

$$KC(N, d) := \frac{1}{N-d+1} \binom{N}{d} \binom{N-d-1}{d-1}$$

and is equal to the number the so-called *short bushes* with N ends and d branching nodes, see [27, $A108263$] for more details and references. Moreover, one has

$$KC(n + d - 1, d + 1) = T(n, d),$$

where $T(n, d)$ denotes the number of dissections of a convex n-gon into $d + 1$ regions, see, e.g., [27, $A033282$]. Note also that the polynomial

$$MS_d(n; q) := \sum_{\substack{d=1 \\ 2d \le n}} K_{(d^2, 1^{n-2d})}(q)$$

is a q deformation of the Motzkin sum (or Riordan) number $MS(n)$, see, e.g., [27, $A005043$] for definition, examples, and references concerning the Motzkin sum numbers; for example,

$$MS_2(8; q) = (1, 0, 1, 1, 2, 2, 4, 3, 6, 5, 7, 6, 9, 6, 8, 6, 7, 4, 5, 2, 3, 1, 1, 0, 1)_q.$$

Finally, we define rectangular Motzkin polynomials $M_d(n; q) := MS_d(n; q) + qMS_d(n + 1; q)$.

Problem 3.5 *Define statistics on the set of Motzkin paths of length n with no horizontal steps at level* 0 *(i.e., the set of Riordan paths) such that the generating function of that statistics is equal to the polynomial $MS_d(n; q)$.*

For example, $M_2(8; q) = (1, 1, 1, 2, 3, 4, 6, 7, 10, 11, 14, 15, 18, 19, 20, 21, 22, 20, 20, 19, 17, 15, 13, 11, 9, 8, 5, 4, 3, 2, 1, 1)$.
We **expect** that the Motzkin polynomials $M_2(n; q)$ are unimodal for $n \ge 2$.

The case $N = dk$ corresponds to rectangular Narayana polynomials studied in [4, 18, 30].

- *(q-***Deformed Eulerian polynomials***)*

Definition 3.6 Define a q-deformation of the Euler polynomials[3] $A_\ell(n,m;q)$, denoted by $A_\ell(n,m;q)$ as follows:

$$A_\ell(n,m;q) = \sum_{\{\nu\}} q^{c(\{\nu\})} \prod_{j,k\geq 1} \begin{bmatrix} P_j^{(k)}(\{\nu\}) + m_j(\nu^{(k)}) \\ m_j(\nu^{(k)}) \end{bmatrix}_q,$$

where summation runs over the set of configurations $\{\nu\}$ of type $(((n^n), 1^m), ((n-1)^{n-1}, 1^{n+m})), (\nu^{(k)})' = n+m+k-1, k = 1, \ldots, n+m-1$. Other notations are the same as in Theorem 3.7 below.

Theorem 3.7 *If n is odd, then the polynomial $E(n;q) := K_{(n^n),((n-1)^{n-1},1^{n-1})}(q)$ is* **symmetric**, *and the polynomial $E(n;q) - q^n - q^{n^2}$ is* **unimodal**.
Moreover, set $E^{[d]}(n;q) := \mathcal{K}^{[d]}_{(n^n),((n-1)^{n-1},1^{n-1})}(q)$. Then,

$$E^{[d]}(n;q) \doteq E^{[3n-3-d]}(n;q^{-1}),\ n-1 \leq d \leq 2(n-1),$$

where for any two polynomials $P_1(q)$ and $P_2(q)$, the expression $P_1(q) \doteq P_2(q)$ means that the ratio $P_1(q)/P_2(q)$ is an integer power of q.

Theorem 3.8 *Polynomials $\mathcal{K}^{[d]}_{(n^n),((n-1)^{n-1},1^{n-1})}(q)$ are unimodal for $n-1 \leq d \leq 2(n-1)$ and any n.*

Note that $K_{(n^n),((n-1)^{n-1},1^{n-1})}(q=1) = n!$ and $\mathcal{K}^{d}_{(n^n),((n-1)^{n-1},1^{n-1})}(q=1) = A(n,d)$.

■

Example 3.9 Take $n = 5$, so we deal with partition $\lambda = (5,5,5,5,5)$ and weight $\mu = (4,4,4,4,4,1,1,1,1,1)$. Clearly, $\mu' = (10,5,5,5)$. The corresponding Euler Polynomial is $A_5(q) = (1,26,66,26,1)_q$. On the other hand, there are 13 (admissible) configurations of type $((5^5),(4^5,1^5))$, namely,

(1) $\nu'_{[1]} = \{(4^5),(3^5),(2^5),(1^5)\}$ with charge $c(\nu_{[1]}) = 25$; all vacancy numbers are equal to zero;

(2)
- $\nu'_{[2]} = \{(5,4^3,3),(3^5),(2^5),(1^5)\}$, with charge $c(\nu_{[2]}) = 17$; the only non-zero vacancy numbers are $P_1^{(1)} = 3 = P_3^{(1)}$;
- $\nu_{[3]}\{(5^2,4^2,2),(3^5),(2^5),(1^5)\}$, with charge $c(\nu_{[3]}) = 16$; the only non-zero vacancy numbers are $P_2^{(1)} = 1 = P_4^{(1)}$;

---

[3]See, e.g., [29], [27, *A*008292] and the literature quoted therein for definition, basic properties, and applications of the Euler numbers $A9n,k)$ and Euler polynomials $E_n(t) = \sum_{k=0}^{n} A(n,k)\,t^k$.

- $\nu_{[4]}\{(5^2, 4, 3^2), (3^5), (2^5), (1^5)\}$, with charge $c(\nu_{[4]}) = 19$; the only non-zero vacancy numbers are $P_2^{(1)} = 1 = P_3^{(1)}$; the contribution to the Kostka polynomial from these configurations is equal to

$$\mathcal{K}_{\lambda,\mu}^{[5]}(q) = q^{16}(1, 3, 4, 5, 6, 4, 2, 1)_q, \ \mathcal{K}_{\lambda,\mu}^{[5]}(1) = 26.$$

(3)

- $\nu_{[5]}\{(6, 4^3, 3^2), (3^5), (2^5), (1^5)\}$, with charge $c(\nu_{[5]}) = 16$; the only non-zero vacancy numbers are $P_1^{(1)} = 1 = P_3^{(1)}$;
- $\nu_{[6]}\{(6, 4^2, 3^2), (3^5), (2^5), (1^5)\}$, with charge $c(\nu_{[6]}) = 13$; the only non-zero vacancy numbers are $P_1^{(1)} = 2 = P_4^{(1)}$;
- $\nu_{[7]}\{(6, 4^3, 2), (4, 3^3, 2), (2^5), (1^5)\}$, with charge $c(\nu_{[7]}) = 11$; the only non-zero vacancy numbers are $P_1^{(1)} = 2 = P_4^{(1)}$;
- $\nu_{[8]}\{(6, 5, 4^2, 1), (3^5), (2^5), (1^5)\}$, with charge $c(\nu_{[8]}) = 11$; the only non-zero vacancy numbers are $P_1^{(1)} = 2, \ P_4^{(2)} = 1$;
- $\nu_{[9]}\{(6, 5, 4, 3, 2), (4, 3^3, 2), (2^5), (1^5)\}$, with charge $c(\nu_{[9]}) = 13$; the only non-zero vacancy numbers are $P_1^{(1)} = 2 = P_4^{(1)}$;
  the contribution to the Kostka polynomial from these configurations is equal to

$$\mathcal{K}_{\lambda,\mu}^{[6]}(q) = q^{11}(2, 4, 9, 11, 14, 11, 9, 4, 2)_q, \ \mathcal{K}_{\lambda,\mu}^{[6]}(1) = 66.$$

(4)

- $\nu_{[10]}\{(7, 4^4, 1), (5, 3^3, 1), (3, 2^3, 1), (1^5)\}$, with charge $c(\nu_{[10]}) = 7$; the only non-zero vacancy numbers are $P_1^{(1)} = 1 = P_4^{(1)}$;
- $\nu_{[11]}\{(7, 4^2, 3, 2), (4, 3^3, 2), (3, 2^3, 1), (1^5)\}$, with charge $c(\nu_{[11]}) = 11$; the only non-zero vacancy numbers are $P_4^{(1)} = 2, \ P_1^{(2)} = 1$;
- $\nu_{[12]}\{(7, 4^3, 3, 1), (4, 3^3, 2), (3, 2^3, 1), (1^5)\}$, with charge $c(\nu_{[12]}) = 9$; the only non-zero vacancy numbers are $P_1^{(2)} = 1 = P_4^{(2)}$;
  the contribution to the Kostka polynomial from these configurations is equal to

$$\mathcal{K}_{\lambda,\mu}^{[7]}(q) = q^{7}(1, 2, 4, 6, 5, 4, 3, 1)_q, \ \mathcal{K}_{\lambda,\mu}^{[7]}(1) = 26.$$

(5)

- $\nu_{[13]}\{(8, 4^3), (6, 3^3), (4, 2^3), (2, 1^3)\}$, with charge $c(\nu_{[12]}) = 5$; all the vacancy numbers are equal to zero;
  Finally, one can easily check that

$$\sum_{d=4}^{8} \mathcal{K}_{\lambda,\mu}^{[d]}(q) = q^5(1, 0, 2, 4, 6, 7, 8, 12, 12, 14, 12, 12, 8, 7, 6, 4, 2, 0, 1)_q$$

$$= K_{(5^5),(4^5,1^5)}(q),$$

as well as to see that the polynomial $K_{(5^5),(4^5,1^5)}(q) - q^5 - q^{25}$ is symmetric and unimodal.

Exercise 3.10 *Consider a set of configurations* $C_n^{[d]} = \{\nu\} \in C((n^n),((n-1)^{n-1},1^{n-1})) \mid \nu_1^{'} = d\}$. *Construct bijection between the sets* $C_n^{[d]}$ *and that* $C_n^{[3n-3-d]}$.

It is an interesting **task** to find connections between our polynomials and a q-analog of Eulerian polynomials studied in [26].

Definition 3.11 (Symmetric and unimodal polynomials associated with reduced decompositions of the longest element $w_0^{(n)} \in \mathbb{S}_n$) Define polynomials

$$RD^{[\ell]}(n;q) := \mathcal{K}^{[\ell]}_{\delta_n,1^{\binom{n}{2}}}(q), \ n-1 \le \ell \le 2(n-1),$$

where $\delta_n := (n-1, n-2, \ldots, 2, 1)$ denotes the staircase partition of size $\binom{n}{2}$. Clearly, $RD(n;q) := \sum_\ell RD^{[\ell]}(q) = K_{\delta_n,1^{\binom{n}{2}}}(q)$. It is well known, see, e.g., [29], that the number $K_{\delta_n,1^{\binom{n}{2}}}(q=1)$ is equal to the number of standard Young tableaux of the staircase shape δ_n and is also equal to the number of reduced decompositions of the longest permutation $w_0^{(n)} := [n, n-1, \ldots, 2, 1] \in \mathbb{S}_n$.

Proposition 3.12

- *Polynomials* $RD^{[\ell]}(n;q) := \mathcal{K}^{[\ell]}_{\delta_n,1^{\binom{n}{2}}}(q), n-1 \le \ell \le 2(n-1)$, *are all symmetric and unimodal;*
- $RD^{[n-1]}(n;q) = \prod_{j=1}^{n-1}(1+q^j)^{n-j}$, *and the ratio* $RD^{[\ell]}(n;q)/RD^{[n-1]}(n;q)$ *is a polynomial with non-negative coefficients.*

We **expect** that the ratio is a unimodal polynomial.

Example 3.13 Take $n = 4$. It is not difficult to check that there are 8 configurations, and

$$RD(4,q) = q^{10}\,(1+q)^3(1+q^2)^2(1+q^3)(q^{15}, q^8\,[5]_q, q^3\,[5]_q, 1).$$

For $n = 5$, one can check that $\{RD^{[\ell]}(5, q = 1), \ell = 4, \ldots, 8\} = 2^{10}(1, 16, 252, 16, 1)$.

3.1 Generalized Catalan Polynomials

Definition 3.14 Define generalized Catalan polynomials $Cat(n,m;q,t)$ and Narayana polynomials $N_d(\lambda, R)$ to be

$$Cat(n,m;q,t) := \mathcal{K}_{(n^m,1^{nm})}(q,t), \ N_d(\lambda,R)(q) := \mathcal{K}^{[d]}_{\lambda,R}(q).$$

Note that for $n = 2$, the polynomial $Cat(n, 2; q, 1)$ coincides with the so-called Carlitz–Riordan q-analog of the Catalan number C_n, as well as the rectangular Narayana polynomial $N_d(n, 2; q)$ corresponding to partitions $\lambda = (n^2)$ and $R = (1^{2n})$ gives rise to a q-analog of the classical Narayana number $N(d, n)$. Note that, in the special case when $\lambda = (n^m)$ is a rectangular shape partition and $R = (1^{nm})$, the value of polynomial $N_d((n^m), 1^{nm}; q)$ at $q = 1$ coincides with the so-called *rectangular* Narayana number studied in [20, 30, 31].

Proposition 3.15 *Let λ be a partition and R be a dominant sequence of rectangular shape partitions. The generalized Narayana polynomials $N_d(\lambda, R; q)$ are symmetric and unimodal.*

Indeed, using the well-known fact that the product of two polynomials, which are both symmetric and unimodal, is again symmetric and unimodal one, see, e.g., [28], we conclude that any d-partial $\mathcal{K}$-polynomial associated with (λ, R) is symmetric and unimodal. Therefore, we have proved Proposition 3.15 and the following corollary.

Corollary 3.16 *Let λ be a partition of size n and $\mu = (1^n)$. Then, generalize q-Narayana numbers of type $(\lambda, 1^n)$ are symmetric and unimodal polynomials of q.*

Indeed, for any admissible configuration of type $(\lambda, 1^m)$, we have

$$un(\{\nu\}) = n(\mu) - \frac{1}{2} m \, (\nu^{(1)})_1',$$

that is, the unimodality indices are the same for all configurations $\{\nu\}$ with the fixed length of the first column $(\nu^{(1)})_1'$ of the first partition $\nu^{(1)}$. It is well known that for a rigged configuration $(\{\nu\}, J)$ of type (λ, μ), the number $(\nu^{(1)})_1'$ is equal to the number of descents of the semistandard Young tableau which corresponds to that rigged configuration under the Rigged Configuration Bijection, see, e.g., [16, 17].

In the particular case when $\lambda = (n^m)$ and $\mu = 1^{nm}$, the q-Narayana numbers of type $(n^m, 1^{nm})$ coincide with the rectangular Narayana polynomials that have been introduced and studied[4] in [11, 18, 20, 30, 31]. More combinatorial proof of Corollary 3.16 has been done in [4].

Moreover, for a dominant sequence of rectangular shape partitions R such that the unimodality index $un(\{\nu\})$ is the same for *all* configurations $\{\nu\} \in C(\lambda, R)$, one can deduce that the parabolic Kostka polynomial $K_{\lambda,R}(q)$ is symmetric and unimodal. For example,

Theorem 3.17 *Let λ be a partition and R be a dominant sequence of rectangular shape partitions. Assume that*

$$\mu_a \geq \left|\nu^{(\eta_a)}\right| = \left|\lambda[\eta_a]\right| - \sum_b \mu_b \, \max(\eta_b - \eta_a, 0), \forall a,$$

[4]Erroneously, it was stated as Conjecture 2.3 in [20], even though more general statement has been proved in [18, 20].

where for any partition $\lambda = (\lambda_1, \lambda_2, \ldots)$, *we used notation* $\lambda[k]$ *for partition* $(\lambda_{k+1}, \lambda_{k+2}, \ldots)$. *Then, the parabolic Kostka polynomial* $K_{\lambda,R}(q)$ *is symmetric and unimodal.*

Indeed under the above assumptions, the unimodality index

$$un(\{\nu\}) = n(R) - 1/2 \sum_a |\nu^{(\eta_a)}|$$

is the same for all configurations $\{\nu\} \in C(\lambda; R)$. Therefore, under the assumptions stated in Theorem 3.17, the parabolic Kostka polynomial $K_{\lambda,R}(q)$ is symmetric and unimodal.

3.2 Some Remarks on Strict Unimodality

The first algebra-combinatorial proof of strict unimodality of q-binomial coefficients has been appeared in [25]. A combinatorial proof of more general statement concerning the strict unimodality of q-binomial coefficients has been done in [6]. Some results concerning strict unimodality of the product of at most two q-binomial coefficient have been done in [4]. Both of the proofs given in [4, 6] use basically the following result.

Proposition 3.18 *Let* $a_1 \geq a_2 \cdots \geq a_n$. *Then, the product* $\prod_{j=1}^{n} [a_j + 1]_q$ *is strictly unimodal iff*

$$a_1 \leq a_2 + \cdots + a_n + 1.$$

To investigate strict unimodality of q-binomial coefficients, we apply the result of Proposition 3.18 to the fermionic formula for parabolic Kostka polynomials discovered by the author in the middle of 80s of the twentieth century. It was observed by the author that some special cases of that fermionic formula give rise to combinatorial/fermionic formulas for q-Gaussian polynomials $\begin{bmatrix} N \\ \lambda \end{bmatrix}_q$, the principal specialization of internal product of Schur functions, generalized exponents polynomial of mixed tensor modules denoted by $G_N(V_\alpha \otimes V_\beta)(q)$, among several others, [20]. In the case of q-binomial coefficients, we have proved, in particular, that the q-binomial $\begin{bmatrix} m+n \\ n \end{bmatrix}_q$ is equal to the Kostka–Foulkes polynomial

$$\begin{bmatrix} m+n \\ n \end{bmatrix}_q = q^{-n\binom{n+m}{2}} K_{((m+n)n,n),(n^{n+m+1})}(q).$$

Therefore, a fermionic formula for q-binomial coefficients $\begin{bmatrix} m+n \\ n \end{bmatrix}_q$ has the following form:

$$\begin{bmatrix} n+m \\ n \end{bmatrix}_q = \sum_{\nu \vdash n} q^{\hat{c}(\nu)} \prod_{j \geq 1} \begin{bmatrix} P_j(\nu) + (\nu)'_j - (\nu)'_{j+1} \\ (\nu)'_j - (\nu)'_{j+1} \end{bmatrix}_q,$$

where summation runs over all partitions on n with at most $[\frac{n+m+1}{2}]$ parts; $\hat{c}(\nu) = c(\nu) - n\binom{m+n}{2}$.

It follows from Theorem 3.1 and Proposition 3.18 that if among the partitions have been involved in the fermionic formula displayed above for q-binomial $\begin{bmatrix} m+n \\ n \end{bmatrix}_q$, there exists at least one strict partition ν, $\ell(\nu) \leq [\frac{n+1}{2}]$ and with at least three non-zero vacancy numbers, then such a partition contributes a *strict unimodality* summand in the fermionic formula for q-binomials, and therefore the total sum, i.e., the q-binomial in question, appears to be also strict unimodal. As a corollary, we get the following.

Corollary 3.19 *Write* $\begin{bmatrix} m+n \\ n \end{bmatrix}_q := \sum_{d=0}^{mn} c_d(n,m)\, q^d$. *Then,*

$$c_{[\frac{mn}{2}}(n,m) - c_{[\frac{mn-2}{2}}(n,m) \geq \widehat{p}([\frac{n+1}{2}],[\frac{n+1}{2}]),$$

where $\widehat{p}(m,d)$ *stands for the number of strict partitions of* m *with at most* d *parts and with positive (i.e.,* ≥ 1*) vacancy numbers.*

The first non-trivial case appears when $n = m = 8$. In this case, $\widehat{p}(8,4) = 1$ and the corresponding configuration is $(4, 3, 1)$. Note that conditions $n, m \geq 8$ guarantee that all the vacancy numbers for configuration $\nu = ([\frac{n+1}{2}],[\frac{n-1}{2}],1)$ are positive. Recall that by the vacancy numbers associated with partition/configuration ν, we mean the following collection of numbers:

$$\{(n+m+1)((\nu')_j - (\nu')_{j+1}) - 2\sum_{a \geq 1} \min\left((\nu')_j - (\nu')_{j+1}, \nu_a\right),\ j = 1,\ldots,\nu_1.$$

Note that if $n, m \geq 8d$, then $\widehat{p}([\frac{n+1}{2}],[\frac{n+1}{2}]) \geq \widehat{p}(4d,4d) \geq 2d - 1$. This observation is a generalization of a similar result has been proved first in [6]. Note that in [32], the author has used another configurations/partitions to find another lower bonds for n and m, which guarantee the strict unimodality of q-binomial polynomials.

Comments 3.20 For a review of different proofs of the unimodality of q-binomial coefficients, and more generally that for q-Gaussian polynomials, see, e.g.,[28]. The first combinatorial proof of unimodality of the q-Gaussian polynomials $\begin{bmatrix} N \\ \lambda \end{bmatrix}_q$ that has been done in [15]. The first proof of the unimodality of q-binomials goes back to J. Sylvester paper of the year 1878 . In the modern notation, it can be interpreted as a consequence of the following identity:

$$\begin{bmatrix} m \\ \lambda' \end{bmatrix} = \sum_{\ell} g_{m-1,\ell}^{\lambda} \, q^{\frac{(m-1)|\lambda|}{2} - n(\lambda)} \, \frac{1-q^{\ell+1}}{1-q},$$

where $g_{m-1,\ell}^{\lambda}$ stands for the so-called $sl(2)$-*plethysm* coefficient; see, e.g., [11, 18, 19, 22, 23]. Writing as before

$$\begin{bmatrix} N \\ \lambda' \end{bmatrix}_q = \sum_{k=0}^{m|\lambda|} c_k(m,\lambda) \, q^k,$$

we find that

$$c_j(m,\lambda) - c_{j-1}(m,\lambda) = g_{m-1,j}^{\lambda} \geq 0, \; 2 \leq j \leq \frac{m|\lambda|}{2}.$$

In other words, strict unimodality of q-binomial coefficient is equivalent to non-vanishing of certain $sl(2)$-plethysm coefficients. Recall that $sl(2)$-plethysm coefficients $c_j(m.\lambda)$ are defined from the decomposition of the Nth tensor power of $sl(2)$-irreducible representation V_m into the direct sum of irreducible $\mathbb{S}_N \times sl(2)$-models, namely

$$V_m^{\otimes N} = \bigoplus_{\lambda,\ell} g_{m,\ell}^{\lambda} \, \mathbb{S}_N^{\lambda} \times V_{\ell}^{sl(2)},$$

where $\mathbb{S}_N^{\lambda}$ denotes the irreducible representation of the symmetric group $\mathbb{S}_N$ corresponding to partition λ.

Similarly, if α and β are partitions of the same size and the both have length $\leq N$, then one can consider the decomposition

$$V_{\alpha}^{sl(N)} \otimes V_{\beta}^{sl(N)}|_{sl(2)} = \bigoplus_{\ell} g_{\alpha\beta\ell} \, V_{\ell}^{sl(2)}.$$

In terms of characters of representations involved, we came to identity

$$s_{\alpha} * s_{\beta}(q, q^2, \ldots, q^{N-1}) = \sum_{\ell} g_{\alpha\beta\ell} \frac{1-q^{\ell+1}}{1-q}.$$

Now let us write

$$s_{\alpha} * s_{\beta}(q, q^2, \ldots, q^{N-1} = \sum_{k=0}^{N|\alpha|} r_{\alpha\beta k}^{(n)} \, q^k.$$

Therefore,

$$g_{\alpha\beta k} = r^{(N)}_{\alpha\beta k} - r^{(N)}_{\alpha\beta(k-1)} \geq 0, \ 1 \leq k \leq \frac{N|\alpha|}{2} \geq 0.$$

Recall that for three partitions α, β, and γ of the same size $|\alpha| = N$, the coefficient $g_{\alpha\beta\gamma}$ is equal to the multiplicity of the symmetric group $\mathbb{S}_N$ representation $\mathbb{S}_N^{\gamma}$ in the Kronecker product of representations $\mathbb{S}_N^{\alpha}$ and $\mathbb{S}_N^{\beta}$. In other words,

$$s_\alpha * s_\beta = \sum_{\gamma} g_{\alpha\beta\gamma}\ s_\gamma.$$

See Sect. 4 for definition and basic properties of the internal product of Schur functions.

Note that as it has been proved in [15], the principal specialization of the Schur function, namely, $s_\alpha(q, q^2, \ldots, q^{N-1})$ is symmetric and unimodal (with unimodality index $\frac{N|\alpha|}{2}$) for any partition α and $N \geq 2$. Therefore the polynomial $s_\alpha * s_\beta(q, q^2, \ldots, q^{N-1})$ is also symmetric and unimodal. As a corollary, we see that the polynomial $s_\alpha * s_\beta(q, q^2, \ldots, q^{N-1})$ is strictly unimodal iff $g_{\alpha\beta k} > 0$ for $2 \leq k \leq \frac{N|\alpha|}{2}$.

3.3 Generalized q-Narayana and Carlitz–Riordan Polynomials

Let λ be a partition and $R = (\{(\mu_a)^{\eta_a}\}_{1 \leq a \leq n}$ be a sequence of rectangular shape partitions such that $|\lambda| = \sum_a \mu_a\, \eta_a$.

Definition 3.21 Define generalized q-Narayana numbers $N_\ell((\lambda; R); q)$ to be

$$N_\ell((\lambda; R); q) = q^{c(\{\nu\})} \sum_{\substack{\{\nu\} \\ un(\{\nu\}))=\ell}} \prod_{i,k \geq 1} \begin{bmatrix} P_i^{(k)}(\{\nu\}) + m_i^{(k)}(\{\nu\}) \\ m_i^{(k)}(\{\nu\}) \end{bmatrix}_q.$$

Theorem 3.22 *The generalized q-Narayana numbers are symmetric and unimodal polynomials in q.*

Examples 3.23

(1) (Rectangular q-Narayana numbers)
Let λ be a partition. Denote by $SYT(\lambda)$ the set of standard Young tableaux of shape λ and by $STY_\ell(\lambda)$ a subset of $STY(\lambda)$ consisting of Young tableaux which have the descent set of cordiality ℓ.

Proposition 3.24 *The polynomial*

$$D_\ell(\lambda;q) := \sum_{T \in STY_\ell(\lambda)} q^{charge(T)}$$

is symmetric and unimodal.

Indeed, if $|\lambda| = n$, then

$$un((\lambda;1^{|\lambda|})) = \binom{n}{2} - n\,(\nu_1^{(1)})',$$

and it is well known that $(\nu_1^{(1)})'$ is equal to the cordiality of the descent set of the corresponding Young tableau under the Rigged Configuration Bijection.

For example, take $\lambda = (n,n)$, and then one can check that

$$N_\ell((n,n),1^{2n});q) \stackrel{\bullet}{=} \frac{1}{[\ell]_q} \begin{bmatrix} n \\ \ell \end{bmatrix}_q \begin{bmatrix} n \\ \ell-1 \end{bmatrix}_q = N(n,\ell;q),$$

where $N(n,\ell;q) = \frac{1}{[\ell]_q} \left[\begin{smallmatrix} n \\ \ell \end{smallmatrix}\right]_q \left[\begin{smallmatrix} n \\ \ell-1 \end{smallmatrix}\right]$ stands for a q-deformed Narayana number; we have used a notation $A(q) \stackrel{\bullet}{=} B(q)$ to stress that the ratio $A(q)/B(q)$ is equal to q^r for some $r \in \mathbb{Z}$.

More generally, for positive integers n and k, define rectangular q-Narayana number $N_\ell(n,k;q)$ to be the generalized q-Narayana number $N_\ell(((n^k),1^{nk});q)$ of type $((n^k),1^{nk})$. It follows from Theorem 3.22 that rectangular q-Narayana numbers are symmetric and unimodal polynomials of q.

(2) (Generalized Gaussian polynomials)
Let λ be a partition of size nr. Assume that $r \geq \lambda_2$, and set $\mu = \underbrace{(n\ldots,n)}_{r}$. Then,
the Kostka polynomial $K_{\lambda,\mu}(q)$ is symmetric and unimodal.
Indeed, for any admissible configuration $\{\nu\}$ of type (λ,μ), one has (taking into account that $\sum_a \nu_a^{(1)} = |\nu^{(1)}| = nr - \lambda_1$)

$$un(\{\nu\}) = n\binom{r}{2} - r\sum_a \nu_a^{(1)} = r\,|\nu^{(1)}| = n\binom{r}{2} - r(nr-\lambda_1).$$

That means that the unimodality centerlines are the same for all admissible configurations in question. Therefore, the Kostka polynomial $K_{\lambda,\mu}(q)$ is symmetric and unimodal. Since one can show that for any partition λ, the generalized q-Gaussian polynomial $\left[\begin{smallmatrix} \lambda' \\ N \end{smallmatrix}\right]$ (up to a power of q) is equal to $K_{\Lambda_N,\mu_N}(q)$, where

$\Lambda_N = (N\ |\lambda|, \lambda)\ \mu_N = \underbrace{(|\lambda|, \ldots, |\lambda)}_{N+1}$, we conclude that the generalized q-Gaussian coefficients are symmetric and unimodal polynomials of q.

Now let λ and μ be partitions such that $\lambda \geq \mu$ with respect to the dominant order on the set of partitions. Let $\{\nu\} = (\nu_i^{(k)})$ be an admissible configuration of type (λ, μ). We define *spin*, denoted by $spin(\{\nu\})$, of an admissible configuration $\{\nu\}$ to be

$$spin(\{\nu\}) = \# \left| (i, k) : (\nu_i^{(k)})' - (\nu_i^{(k+1)})' \equiv 1 \ (\mathrm{mod}\ 2) \right|.$$

Finally, we define (q, p)-deformation of Kostka polynomial $K_{\lambda,\mu}(q)$ to be

$$\mathcal{K}_{\lambda,\mu}(q, t, p) := \sum_{\{\nu\}\in C(\lambda,\mu)} q^{c(\{\nu\})}\, t^{nu(\{\nu\})}\, p^{spin(\{\nu\})} \prod_{i,k\geq 1} \begin{bmatrix} P_i^{(k)}(\{\nu\}) + m_i^{(k)}(\{\nu\}) \\ m_i^{(k)}(\{\nu\}) \end{bmatrix}_q .$$

Definition 3.25 Let $\{\nu\}$ be an admissible configuration of type $C(\lambda, \mu)$. We say that $\{\nu\}$ is a *Yamanuchi configuration* configuration of type (λ, μ) if

$$0 \leq (\nu_i^{(k-1)})' - (\nu_i^{(k)})' \leq 2, \ \forall i, k \geq 1.$$

Recall that we set by definition $\nu_i^{(0)} := \mu$. We denote the set of Yamanuchi configurations of type (λ, μ) by $Yam(\lambda, \mu)$. We define polynomial $Yam_{\lambda,\mu}(q, t, p)$ as follows:

$$Yam_{\lambda,\mu}(q, t, p) := \sum_{\{\nu\}\in Yam(\lambda,\mu)} q^{c(\{\nu\})}\, t^{un(\{\nu\})}\, p^{spin(\{\nu\})} \prod_{i,k\geq 1} \begin{bmatrix} P_i^{(k)}(\{\nu\}) + m_i^{(k)}(\{\nu\}) \\ m_i^{(k)}(\{\nu\}) \end{bmatrix}_q .$$

It is clearly seen that $\mathcal{K}_{\lambda,\mu}(q, t = 1, p = 1) = K_{\lambda,\mu}(q)$, and if $core(\lambda) = \emptyset$, and $\mu = \beta \vee \beta$ for a composition β, then the polynomial $\mathcal{K}_{\lambda,\mu}(q = -1, t = 1, p)$ is equal to the *spin* generating function for the set of semistandard domino tableaux of shape λ and weight β, see, e.g., [5].

For a Yamanuchi configuration $\{\nu\}$ of type (λ, μ), we set $\epsilon(\nu\}) = \epsilon^{c(\{\nu\})}$, and for all others, we set $\epsilon(\{\nu\}) = 1$.

Define 5-parameter deformation of the Kostka polynomial $K_{\lambda,\mu}(q)$ as follows:

$$\mathcal{K}_{\lambda,\mu}(q, t, p, v, \epsilon) := \sum_{\{\nu\}\in C(\lambda,\mu)} q^{c(\{\nu\})}\, t^{(\nu^{(k)})'_1}\, v^{un(\{\nu\})}\, p^{spin(\{\nu\})}\, \epsilon(\{\nu\})$$

$$\prod_{i,k\geq 1} \begin{bmatrix} P_i^{(k)}(\{\nu\}) + m_i^{(k)}(\{\nu\}) \\ m_i^{(k)}(\{\nu\}) \end{bmatrix}_q .$$

Consider deformed rectangular Narayana polynomials $\mathcal{K}_{(n^m),1^{nm}}(t,q,p)$ more closely. For example, $q^{-30}\ \mathcal{K}_{(6^2),(1^{12})}(t,q,p) =$

$$tq^{30}p^6+t^2\left(q^{20}\begin{bmatrix}9\\1\end{bmatrix}_q p^4+q^{22}\begin{bmatrix}5\\1\end{bmatrix}_q p^2+q^{24}\right)+t^3\left(q^{12}\begin{bmatrix}8\\2\end{bmatrix}_q p^4+q^{14}\begin{bmatrix}7\\1\end{bmatrix}_q\begin{bmatrix}3\\1\end{bmatrix}_q p^2+q^{18}p^2\right)+$$

$$t^4\left(q^6\begin{bmatrix}7\\3\end{bmatrix}_q p^2+q^8\begin{bmatrix}6\\2\end{bmatrix}_q\right)+t^5\ q^2\begin{bmatrix}6\\2\end{bmatrix}_q p^2+t^6.$$

Clearly,

$$\mathcal{K}_{(6^2),(1^{12})}(t=1,q,p=1) = K_{(6^2),(1^{12})} = q^{30}\ \frac{\begin{bmatrix}12\\6\end{bmatrix}_q}{[7]_q} = q^{30}\ Cat_6^{CR}(q),$$

where $Cat_n^{CR}(q) := \frac{\begin{bmatrix}2n\\n\end{bmatrix}_q}{[n+1]_q}$ stands for the Carlitz–Riordan q-deformation of the Catalan number C_n.

$$\mathcal{K}_{(6^2),(1^{12})}(t,q=1,p=1) = t(1,10,50,50,10,1)_t\ ,$$

and two-variable polynomial $\mathcal{K}_{(n^2),(1^{2n})}(t,q,p = 1)$ coincides with generating function for (normalized) q-Narayana numbers.

$$\mathcal{K}_{(6^2),(1^{12})}(t=1,q=-1,p) = (5,9,5,1)_p$$

coincides with the *spin*-generating function for the number of standard *domino* tableaux of shape (6^2).

Theorem 3.26

(*a*) $\mathcal{K}_{((2n)^2),(1^{4n})}(t=1,q=-1,p=1) = \binom{2n}{n}$, *and*

$$\mathcal{K}_{((2n)^2),(1^{4n})}(t=1,q=-1,p) = \sum_{k=0}^{n} \#\left|STY(n+k,n-k)\right|\ p^k,$$

see also [27], A039599 *for additional information concerning these polynomials.*

(*b*)

$$\mathcal{K}_{((3)^n),(1^{3n})}(t=1,q=\zeta_3,p=1) = \frac{n!}{[n/3]!\ [(n+1)/3]!\ [(n+2)/3]!},$$

4 Internal Product of Schur Functions

The irreducible characters χ^{λ} of the symmetric group S_n are indexed in a natural way by partitions λ of n. If $w \in S_n$, then define $\rho(w)$ to be the partition of n whose parts are the cycle lengths of w. For any partition λ of m of length l, define the power–sum symmetric function $p_{\lambda} = p_{\lambda_1} \dots p_{\lambda_l}$, where $p_n(x) = \sum x_i^n$. For brevity write $p_w := p_{\rho(w)}$. The Schur functions s_{λ} and power–sums p_{μ} are related by a famous result of Frobenius

$$s_{\lambda} = \frac{1}{n!} \sum_{w \in S_n} \chi^{\lambda}(w) p_w. \tag{4.1}$$

For a pair of partitions α and β, $|\alpha| = |\beta| = n$, let us define the internal product $s_{\alpha} * s_{\beta}$ of Schur functions s_{α} and s_{β}:

$$s_{\alpha} * s_{\beta} = \frac{1}{n!} \sum_{w \in S_n} \chi^{\alpha}(w) \chi^{\beta}(w) p_w. \tag{4.2}$$

It is well known that $s_{\alpha} * s_{(n)} = s_{\alpha}$, $s_{\alpha} * s_{(1^n)} = s_{\alpha'}$, where α' denotes the conjugate partition to α.

Let α, β, and γ be partitions of a natural number $n \geq 1$, and consider the following numbers:

$$g_{\alpha\beta\gamma} = \frac{1}{n!} \sum_{w \in S_n} \chi^{\alpha}(w) \chi^{\beta}(w) \chi^{\gamma}(w). \tag{4.3}$$

The numbers $g_{\alpha\beta\gamma}$ coincide with the structural constants for multiplication of the characters χ^{α} of the symmetric group S_n:

$$\chi^{\alpha} \chi^{\beta} = \sum_{\gamma} g_{\alpha\beta\gamma} \chi^{\gamma}. \tag{4.4}$$

Hence, $g_{\alpha\beta\gamma}$ are non-negative integers. It is clear that

$$s_{\alpha} * s_{\beta} = \sum_{\gamma} g_{\alpha\beta\gamma} s_{\gamma}. \tag{4.5}$$

Theorem 4.1 (A.N. Kirillov) *Let α and β be partitions of the same size, and $\ell(\alpha) = r$, then*

$$s_{\alpha} * s_{\beta}(q, q^2, \dots, q^{N-1}) = \sum_{\{\nu\}} q^{\nu} \prod_{k,j \geq 1} \begin{bmatrix} P_j^{(k)}(\nu) + m_j(\nu^{(k)}) + N(k-1)\delta_{j,\beta_1}\theta(r-k) \\ P_j^{(k)}(\nu) \end{bmatrix}_q ,$$

where ν runs over the set of all admissible configurations $\{\nu\}$ of the type $([\alpha,\beta]_N,\beta_1^N)$; $c(\{\nu\})$ stands for the charge of configuration $\{\nu\}$; $\ell(\alpha)$ denotes the length of partition α; for a real number x we set $\theta(x) = 1$, if $x \geq 0$ and $\theta(x) = 0$, if $x < 0$; $P_j^{(k)}(\{\nu\}) = N\min(j,\beta_1)\delta_{k,1} + Q_j(\nu^{(k-1)}) - 2\,Q_j(\nu^{(k)}) + Q_j(\nu^{(k+1)})$; $m_j(\nu^{(k)}) = \nu_j^{(k)} - \nu_{j+1}^{(k)}$; and for the same size partitions α and β, we have used notation $s_\alpha * s_\beta(X)$ to denote the internal product of the corresponding Schur functions.

In particular, if $\beta = (|\alpha|)$ has only one part, when $s_\alpha * s_\beta = s_\alpha$, and Theorem 4.1 gives the so-called *fermionic formula* for the principal specialization of Schur function s_α, [9, 11, 16–19]. For definition of admissible configurations of type $(\lambda;\{R\})$, see Sect. 2.

It follows from Theorem 4.1 that polynomials $s_\alpha * s_\beta(q,\ldots,q^{N-1})$ are symmetric and unimodal, as it was stated in the Introduction section.

4.1 Liskova Polynomials

Let us introduce polynomials $L_{\alpha\beta}^{\mu}(q)$ via the decomposition of the internal product of Schur functions $s_\alpha * s_\beta(x)$ in terms of Hall–Littlewood functions $P_\mu(x;q)$:

$$s_\alpha * s_\beta(x) = \sum_\mu L_{\alpha\beta}^{\mu}(q) P_\mu(x;q). \tag{4.6}$$

It follows from (4.5) and definition of Kostka polynomials that

$$L_{\alpha\beta}^{\mu}(q) = \sum_\gamma g_{\alpha\beta\gamma} K_{\gamma\mu}(q). \tag{4.7}$$

Thus, the polynomials $L_{\alpha\beta}^{\mu}(q)$ have non-negative integer coefficients, and $L_{\alpha\beta}^{\mu}(0) = g_{\alpha\beta\mu}$. The polynomials $L_{\alpha\beta}^{\mu}(q)$ can be considered as a generalization of the Kostka polynomials. Indeed, if partition β consists of only one part, $\beta = (|\alpha|)$, and then

$$L_{\alpha\beta}^{\mu}(q) = K_{\alpha,\mu}(q).$$

Example 4.2 Take partitions $\alpha = (4,2)$ and $\beta = (3,2,1)$, then

$$s_\alpha * s_\beta = s_{51} + 2s_{42} + 2s_{41^2} + s_{3^2} + 3s_{321} + 2s_{31^3} + s_{2^3} + 2s_{2^21^2} + s_{21^4}.$$

Below we list all (non-zero) polynomials $L_{(4,2),(3,2,1)}^{\gamma}(q)$:

$$L_{\alpha\beta}^{(51)}(q) = 1,\; L_{\alpha\beta}^{(42)}(q) = (12),\; L_{\alpha\beta}^{(411)}(q) = (132),\; L_{\alpha\beta}^{(33)}(q) = (121),$$

$$L_{\alpha\beta}^{(321)}(q) = (1353),\; L_{\alpha\beta}^{(31^3)}(q) = (135752),$$

$$L_{\alpha\beta}^{(2^3)}(q) = (135531), \ L_{\alpha\beta}^{(2^21^2)} = (1369962),$$

$$L_{\alpha\beta}^{(21^4)}(q) = (1, 3, 6, 10, 13, 14, 12, 8, 4, 1)$$
$$= (1+q)^2(1+q^2)(1+q+q^2)(1+q^2+q^3),$$
$$L_{\alpha\beta}^{(1^6)}(q) = q(1, 3, 6, 10, 14, 18, 20, 20, 18, 14, 10, 6, 3, 1)$$
$$= q(1+q)^3(1+q^2)(1+q^2+q^4)^2.$$

Exercise 4.3 *Let $N \geq 1$, $M \geq 1$ be integer numbers. For any pair of partitions α and β of the same size n, consider the following polynomial:*

$$S_{\alpha\beta;N,M}(q) = \tag{4.8}$$
$$\frac{1}{n!} \sum_{w \in S_n} \chi^{\alpha}(w) \chi^{\beta}(w) \prod_{k \geq 1} \left(\frac{q^k - (1+(-1)^k) q^{kN} + (-1)^k q^{k(N+M-1)}}{1-q^k} \right)^{\rho_k(w)}.$$

It is clear that

$$S_{\alpha\beta;N,1}(q) = s_{\alpha} * s_{\beta}(q, \ldots, q^{N-1}),$$
$$S_{\alpha\beta;2,2}(q) = (1+q) \sum_{\mu=(a|b)} g_{\alpha\beta\mu} q^b,$$

summed over all hook partitions $\mu = (a+1, 1^b)$, $a+b = n-1$.

- *Show that $S_{\alpha\beta;N,M}(q)$ is a polynomial with non-negative integer coefficients.*

Example 4.4 Take partitions $\alpha = (31)$ and $\beta = (22)$. Using the character table for the symmetric group S_4, one can easily find the following expression for the internal product of Schur functions in question:

$$s_{\alpha} * s_{\beta} = \frac{1}{4!}(6p_1^4 - 6p_2^2),$$

and therefore,

$$S_{\alpha\beta;n,m}(q) = q^5 \begin{bmatrix} n+m-2 \\ 1 \end{bmatrix}_q \begin{bmatrix} 2n+2m-5 \\ 1 \end{bmatrix}_q + q^7 \begin{bmatrix} 3 \\ 1 \end{bmatrix}_q \begin{bmatrix} n+m-1 \\ 4 \end{bmatrix}_q$$
$$+ q^9 \begin{bmatrix} 3 \\ 1 \end{bmatrix}_q \begin{bmatrix} n+m-2 \\ 4 \end{bmatrix}_q + q^{2n+2} \begin{bmatrix} n-1 \\ 1 \end{bmatrix}_{q^2} \begin{bmatrix} m-1 \\ 1 \end{bmatrix}_{q^2}.$$

In particular,

$$S_{\alpha\beta;2,2}(q) = q^5(1+q+q^2) + q^6 = q^5(1+q)^2.$$

Problem 4.5 *Find a fermionic formula for polynomials $S_{\alpha\beta;N,M}(q)$, which generalizes that from Theorem 4.1.*

4.2 Two-Variable Liskova Polynomials $L^{\mu}_{\alpha\beta}(q,t)$

Let α, β, and μ be partitions, $|\alpha| = |\beta| = |\mu| = n$, and define

$$L^{\mu}_{\alpha\beta}(q,t) = \sum_{\gamma} g_{\alpha\beta\gamma} K_{\gamma\mu}(q,t).$$

Polynomials $L^{\mu}_{\alpha\beta}(q,t)$ may be considered as a generalization of the double Kostka–Macdonald polynomials $K_{\alpha\mu}(q,t)$. Indeed, if $\beta = (|\alpha|)$, then $L^{\mu}_{\alpha\beta}(q,t) = K_{\alpha\mu}(q,t)$, and $L^{\mu}_{\alpha(1^n)}(q,t) = K_{\alpha'\mu}(q,t)$. Polynomials $L^{\mu}_{\alpha\beta}(q,t)$ have properties similar to those of $K_{\alpha\mu}(q,t)$.

Exercises 4.6 *Show that*

i) $L^{\mu}_{\alpha\beta}(0,t) = L^{\mu}_{\alpha\beta}(t)$;
ii) $L^{\mu}_{\alpha\beta}(0,0) = g_{\alpha\beta\mu}$, $L^{\mu}_{\alpha\beta}(1,1) = f^{\alpha} f^{\beta}$, *where f^{α} denotes the number of standard (i.e., weight $(1^{|\alpha|})$) Young tableaux of shape α;*
iii) $L^{\mu}_{\alpha\beta}(q,t) = L^{\mu'}_{\alpha'\beta'}(t,q)$;
iv) $L^{\mu}_{\alpha\beta}(q,t) = q^{n(\mu')} t^{n(\mu)} L^{\mu}_{\alpha'\beta}(q^{-1},t^{-1})$;
v) $L^{1^n}_{\alpha\beta}(q,t) = K_{\alpha'\beta}(t,t)\widetilde{K}_{\beta,(1^n)}(t) = K_{\beta'\alpha}(t,t)\widetilde{K}_{\alpha,(1^n)}(t)$; *and*
vi) let λ and μ be partitions and $\lambda \geq \mu$ with respect to the dominance ordering on the set of partitions and set $\widetilde{L}^{\mu}_{\alpha,\beta}(q) = q^{d(\alpha\beta\mu)} L^{\mu}_{\alpha,\beta}(q^{-1})$, where $d(\alpha\beta\mu) = deg_q(L^{\mu}_{\alpha,\beta}(q))$.

- *Show that $L^{\mu}_{\alpha\beta}(q) \geq L^{\lambda}_{\alpha\beta}(q)$, i.e., the difference $L^{\mu}_{\alpha\beta}(q) - L^{\lambda}_{\alpha\beta}(q)$ is a polynomial with non-negative coefficients.*
  More generally, show that for any interval $I = [\mu_1, \mu_2]$ in the Young graph $(\mathbb{Y}, \leq)$, one has

$$\sum_{\tau \in [\mu_1,\mu_2} \mu(\tau; I) \widetilde{L}^{\tau}_{\alpha,\beta}(q) \geq 0,$$

where $\mu(\tau; I)$ denotes the Möbius function of the interval I.

- *Construct a natural embedding of the sets*

$$\boldsymbol{\nu}^{\mu}(\alpha, \beta) \hookrightarrow \boldsymbol{\nu}^{\lambda}(\alpha, \beta).$$

Hence, for any partition μ, there exists a natural embedding (standardization map)

$$i_{\mu} : \boldsymbol{\nu}^{\mu}(\alpha, \beta) \hookrightarrow STY(\alpha) \times STY(\beta). \tag{4.9}$$

5 Generalized Exponents and Mixed Tensor Representations

Let $\mathfrak{g} = sl(N, \mathbb{C})$ denote the Lie algebra of all $N \times N$ complex matrices of trace 0 and $G = SL(N, \mathbb{C})$ denote the Lie group of all invertible $N \times N$ complex matrices. Let ad : $G \to \mathrm{Aut}(\mathfrak{g})$ denote the adjoint representation of G, defined by $(adX)(A) = XAX^{-1}$, where $X \in G$, and $A \in \mathfrak{g}$.

The adjoint action of $SL(N, \mathbb{C})$ extends to an action on the symmetric algebra $S^{\bullet}(\mathfrak{g}) = \oplus_{k \geq 0} S^k(\mathfrak{g})$, where $S^k(\mathfrak{g})$ denotes the kth symmetric power. It is well known [21] that the ring $I = S^{\bullet}(\mathfrak{g})^G = \{f \in S^{\bullet}(\mathfrak{g}) | X \cdot f = f, \forall x \in G\}$ of invariants of this action is a polynomial ring in $N-1$ variables $f_2, \ldots, f_{N-1}$, where $f_i \in S^i(\mathfrak{g})^G$. By a theorem of Kostant [21], $S^{\bullet}(\mathfrak{g}) = I \otimes H$ is a free module over G-invariants I generated by harmonics H. Moreover, $H = \oplus_{p \geq 0} H^p$ is a graded (so $H^p = H \cap S^p(\mathfrak{g})$) locally finite $\mathfrak{g}$-representation. The graded character ch_q of the symmetric algebra of adjoint representation is given by the following formal power series:

$$\mathrm{ch}_q(S^{\bullet}(\mathfrak{g})) = \sum_{k \geq 0} q^k \mathrm{ch}(S^k(\mathfrak{g})) = \prod_{1 \leq i, j \leq n} (1 - q x_i / x_j)^{-1}.$$

For any finite dimensional $\mathfrak{g}$-representation V, let us define

$$G_N(V) := \sum_{p \geq 0} \langle G, H^p \rangle q^p, \tag{5.1}$$

where $\langle V_1, V_2 \rangle = \dim_{\mathfrak{g}} \mathrm{Hom}(V_1, V_2)$ is the standard pairing on the representation ring of the Lie algebra $\mathfrak{g}$. By a theorem of Kostant [21], $G_N(V)|_{q=1} = \dim V(0)$, where $V(0)$ denotes the zero-weight subspace of representation V. Hence, $G_N(V)$ is a polynomial in q with non-negative integer coefficients. Following Kostant [21], the integers $e_1, \ldots, e_s$ with $G_N(V) = \sum_{i=1}^{s} q^{e_i}$ are called *generalized exponents* of the representation V. Kostant's problem [*ibid*] is to determine/compute these numbers for a given representation V.

Let $V_{\lambda} := V_{\lambda}^{[N]}$ denote the irreducible highest weight λ representation of the Lie algebra $\mathfrak{g} := sl(N, \mathbb{C})$. Theorem 5.1 below together with the fermionic formula (6.2) for the Kostka–Foulkes polynomials, see Sect. 6, gives an effective method for computing the generalized exponents of *irreducible* representation of the Lie algebra $\mathfrak{g} = sl(N)$.

Theorem 5.1 ([8]) *Let λ be a partition, and then*

$$G_N(V_\lambda) = \begin{cases} K_{\lambda, \left(\left(\frac{|\lambda|}{N}\right)^N\right)}(q), & \text{if } |\lambda| \equiv 0 (\text{mod } N), \\ 0, & \text{otherwise.} \end{cases}$$

For an "elementary" proof of Theorem 5.1, which is based only on the theory of symmetric functions, see [7].

Using Theorem 5.1, one can compute—in principal—the generalized exponents for any finite dimensional $\mathfrak{gl}(N)$-module V. What seems to be very interesting is that for certain representations—see below—there exist alternative expressions for the generalized exponents polynomials which have independent interest and more convenient for computations. Before turning to our main results of this section, it is useful to recall a few definitions and results from [2] and [8].

Let α and β be partitions and $V_\alpha^{(N)}$, $V_\beta^{(N)}$ be the highest weight α and β (respectively) irreducible representations of the Lie algebra $\mathfrak{gl}(N)$. For any finite dimensional $\mathfrak{gl}(N)$-module V, let V^* denote its dual. If $l(\alpha) + l(\beta) \leq N$, denote by $V_{\alpha,\beta}^{(N)}$ the Cartan piece in the tensor product $V_\alpha^{(N)} \otimes V_\beta^{(N)*}$, i.e., the irreducible submodule generated by the tensor product of the highest weight vectors of each component. Following [8], we call a representation obtained in this way as *mixed tensor representation*. Clearly, $V_{\alpha,\beta}^{(N)}$ is the dual of $V_{\beta,\alpha}^{(N)}$.

Since it is irreducible, $V_{\alpha,\beta}^{(N)}$ is equal to $V_\lambda^{(N)}$ for a unique partition λ of less than N rows. Let us write $[\alpha, \beta]_N$ for this λ. It is well known [8], and goes back to D. Littlewood [22], that

$$[\alpha, \beta]_N = (\alpha_1 + \beta_1, \ldots, \alpha_s + \beta_1, \underbrace{\beta_1, \ldots, \beta_1}_{N-s-r}, \beta_1 - \beta_r, \ldots, \beta_2 - \beta_1, 0),$$

where $s = l(\alpha)$, $r = l(\beta)$, and we assume that $s + r \leq N$. For example, $V_{0,0}^{(N)} = V_0^{(N)} \simeq \mathbb{C}$, $V_{(1),(1)}^{(N)} = V_{(21^{N-2})}^{(N)} = \mathfrak{g}$ is the adjoint representation.

Theorem 5.2 *Let α and β be partitions, $|\alpha| = |\beta|$, $l(\alpha) \leq r$. Then,*

$$G_N(V_\alpha \otimes V_\beta^*) \doteq K_{[\alpha,\beta]_{N+r}, R_N}(q), \tag{5.2}$$

where $R_N = \{\underbrace{(\beta_1), \ldots, (\beta_1)}_{N}, (\beta_1^r)\}$.

Corollary 5.3 (Fermionic Formula for the Generalized Exponents Polynomial $G_N(V_\alpha \otimes V_\beta^*)$**)** *Let α and β be partitions, $|\alpha| = |\beta|$, $l(\alpha) \le r$. Then,*

$$q^{|\alpha|} G_N(V_\alpha \otimes V_\beta^*) = \sum_\nu q^{c(\nu)} \prod_{k,j \ge 1} \begin{bmatrix} P_j^{(k)}(\nu) + m_j(\nu^{(k)}) + k\delta_{j,\beta_1}\theta(r-k) \\ P_j^{(k)}(\nu) \end{bmatrix}_q, \tag{5.3}$$

summed over all admissible configurations ν of type $([\alpha,\beta]_N; (\beta_1^N))$.

Remark 5.4 Let α and β be partitions such that $l(\alpha) + l(\beta) \le N$, $l(\alpha) \le r$. Then, $G_N(V_\alpha \otimes V_\beta^*) \ne 0$ if and only if $|\alpha| \equiv |\beta| \bmod N$ and $\widetilde{\beta}_1 = \beta_1 + \dfrac{|\alpha| - |\beta|}{N} \ge 0$; if so, then

$$G_N(V_\alpha \otimes V_\beta^*) = K_{[\alpha,\beta]_{N+r}, \widetilde{R}_N}(q), \tag{5.4}$$

where $\widetilde{R}_N = \{\underbrace{(\widetilde{\beta}_1), \ldots, (\widetilde{\beta}_1)}_{N}, (\beta_1^r)\}$.

6 Rigged Configurations: A Brief Review

Let λ be a partition and $R = ((\mu_a^{\eta_a}))_{a=1}^p$ be a sequence of rectangular shape partitions such that $|\lambda| = \sum_a |R_a| = \sum_a \mu_a \eta_a$.

Definition 6.1 The configuration of type (λ, R) is a sequence of **partitions** $\{\nu\} = (\nu^{(1)}, \nu^{(2)}, \ldots)$ such that

$$|\nu^{(k)}| = \sum_{j>k} \lambda_j - \sum_{a \ge 1} \mu_a \max(\eta_a - k, 0) = -\sum_{j \le k} \lambda_j + \sum_{a \ge 1} \mu_a \min(k, \eta_a)$$

for each $k \ge 1$.

Note that if $k \ge l(\lambda)$ and $k \ge \eta_a$ for all a, then a partition $\nu^{(k)}$ is the empty one. As in the previous section, in the sequel, we make the convention that $\nu^{(0)}$ is the empty partition.[5]

For a partition μ and an integer $j \ge 1$, define the number

$$Q_j(\mu) = \mu_1' + \cdots + \mu_j',$$

which is equal to the number of cells in the first j columns of μ.

[5]However, in the case when $\eta_{i_a} = 1, \forall a \ge 1$, it is more convenient to set $\nu^{(0)} = \mu := (\mu_1, \ldots, \mu_p)$.

Definition 6.2 The *vacancy* numbers $P_j^{(k)}(\{\nu\}) := P_j^{(k)}(\nu; R)$ of a configuration $\{\nu\}$ of type (λ, R) are defined by

$$P_j^{(k)}(\{\nu\}) = Q_j(\nu^{(k-1)}) - 2Q_j(\nu^{(k)}) + Q_j(\nu^{(k+1)}) + \sum_{a \geq 1} \min(\mu_a, j)\delta_{\eta_a, k}$$

for $k, j \geq 1$, where $\delta_{a,b}$ is the Kronecker delta.

Definition 6.3 The configuration $\{\nu\}$ of type (λ, R) is called **admissible**, if

$$P_j^{(k)}(\nu; R) \geq 0 \text{ for all } k, j \geq 1.$$

We denote by $C(\lambda; R)$ the set of all admissible configurations of type (λ, R) and call the vacancy number $P_j^{(k)}(\nu, R)$ to be *essential*, if $m_j(\nu^{(k)}) > 0, \forall j \geq 1$.

Definition 6.4 For a configuration $\{\nu\}$ of type (λ, R), let us define its **charge**

$$c(\{\nu\}) = \sum_{k,j \geq 1} \binom{\alpha_j^{(k-1)} - \alpha_j^{(k)} + \sum_a \theta(\eta_a - k)\theta(\mu_a - j)}{2},$$

and **cocharge**

$$\bar{c}(\nu) = \sum_{k,j \geq 1} \binom{\alpha_j^{(k-1)} - \alpha_j^{(k)}}{2},$$

where $\alpha_j^{(k)} = (\nu^{(k)})'_j$ denotes the size of the jth column of the kth partition $\nu^{(k)}$ of the configuration $\{\nu\}$; for any real number $x \in \mathbb{R}$, we put $\theta(x) = 1$, if $x \geq 0$, and $\theta(x) = 0$, if $x < 0$.

Theorem 6.5 (Fermionic Formula for Parabolic Kostka Polynomials [11]) *Let λ be a partition and R be a dominant*[6] *sequence of rectangular shape partitions. Then,*

$$K_{\lambda R}(q) = \sum_{\nu} q^{c(\nu)} \prod_{k,j \geq 1} \begin{bmatrix} P_j^{(k)}(\nu; R) + m_j(\nu^{(k)}) \\ m_j(\nu^{(k)}) \end{bmatrix}_q, \tag{6.1}$$

summed over all admissible configurations ν of type $(\lambda; R)$; $m_j(\lambda)$ denotes the number of parts of a partition λ of size j.

[6]That is, $m_1 \geq \mu_2 \geq \ldots \geq \mu_p$.

Corollary 6.6 (Fermionic Formula for Kostka–Foulkes Polynomials [9]) *Let λ and μ be partitions of the same size. Then,*

$$K_{\lambda\mu}(q) = \sum_{\nu} q^{c(\nu)} \prod_{k,j\geq 1} \begin{bmatrix} P_j^{(k)}(\nu,\mu) + m_j(\nu^{(k)}) \\ m_j(\nu^{(k)}) \end{bmatrix}_q, \tag{6.2}$$

summed over all sequences of partitions $\nu = \{\nu^{(1)}, \nu^{(2)}, \ldots\}$ such that

- $|\nu^{(k)}| = \sum_{j>k} \lambda_j$, $k = 1, 2, \ldots$;
- $P_j^{(k)}(\nu,\mu) := Q_j(\nu^{(k-1)}) - 2Q_j(\nu^{(k)}) + Q_j(\nu^{(k+1)}) \geq 0$ *for all* $k, j \geq 1$, *where by definition, we* put $\nu^{(0)} = \mu$;

$$\bullet\ c(\nu) = \sum_{k,j\geq 1} \binom{(\nu^{(k-1)})'_j - (\nu^{(k)})'_j}{2}. \tag{6.3}$$

It is frequently convenient to represent an admissible configuration $\{\nu\}$ by a matrix $m(\{\nu\}) = (m_{ij})$, $m_{ij} \in \mathbb{Z}, \forall i, j \geq 1$, which must meet certain conditions. Namely, starting from a collection of partitions $\{\nu\} = (\nu^{(1)}, \nu^{(2)}, \ldots, \ldots)$ corresponding to a configuration $\{\nu\}$ of type (λ, R), define matrix

$$m(\{\nu\}) := (m_{ij}),\ m_{ij} = (\nu^{(i-1)})'_j - (\nu^{(i)})'_j + \sum_{a\geq 1} \theta(\eta_a - i)\theta(\mu_a - j),\ \nu^{(0)} := \emptyset,$$

where, as before, we set by definition $\theta(x) = 1,\ if\ x \in \mathbb{R}_{\geq 0}$ and $\theta(x) = 0,\ x \in \mathbb{R}_{<0}$.
One can check that a configuration $\{\nu\}$ of type (λ, R) is **admissible** if and only if the matrix $m(\{\nu\})$ meets the following conditions:

(0) $m_{ij} \in \mathbb{Z}$,
(1) $\sum_{i\geq 1} m_{ij} = \sum_{a\geq 1} \eta_a \theta(\mu_a - j)$,
(2) $\sum_{j\geq 1} m_{ij} = \lambda_i$,
(3) $\sum_{j\leq k} (m_{ij} - m_{i+1,j}) \geq 0$, for all i, j, k, and
(4) $\sum_{a\geq 1} \min(\eta_a, k)\delta_{\mu_a, j} \geq \sum_{i\leq k} (m_{ij} - m_{i,j+1})$, for all i, j, k.

One can check that if the matrix (m_{ij}) satisfies the conditions (0)–(4), then the set of partitions $\{\nu\} = (\nu^{(1)}, \nu^{(2)}, \ldots, \ldots)$, where

$$(\nu^{(k)})'_j := \sum_{i>k} m_{ij} - \sum_{a} \max(\eta_a - k, 0)\theta(\mu_a - j)$$

defines an admissible configuration of type $(\lambda, R = \{\mu_a^{\eta_a}\})$.

Example 6.7 Take $\lambda = (44332)$, $R = \{(2^3), (2^2), (2^2), (1), (1)\}$, so that

$$\{\mu_a\} = (2, 2, 2, 1, 1)\ and\ \{\eta_a\} = (3, 2, 2, 1, 1),\ a = 1, \ldots, 5.$$

Therefore $|\nu^{(1)}| = 12-2\times 2-2\times 1-2\times 1 = 4$, $|\nu^{(2)}| = 8-2\times 1 = 6$, $|\nu^{(3)}| = 5$, and $|\nu^{(4)}| = 2$. It is not hard to check that there exist 6 admissible configurations. They are

(1) $\{\nu^{(1)} = (3,1), \nu^{(2)} = (3,3), \nu^{(3)} = (3,2), \nu^{(4)} = (2)\}$,
(2) $\{\nu^{(1)} = (3,1), \nu^{(2)} = (3,2,1), \nu^{(3)} = (3,2), \nu^{(4)} = (2)\}$,
(3) $\{\nu^{(1)} = (2,2), \nu^{(2)} = (2,2,2), \nu^{(3)} = (3,2), \nu^{(4)} = (2)\}$,
(4) $\{\nu^{(1)} = (4), \nu^{(2)} = (3,3), \nu^{(3)} = (3,2), \nu^{(4)} = (2)\}$,
(5) $\{\nu^{(1)} = (3,1), \nu^{(2)} = (2,2,1,1), \nu^{(3)} = (2,2,1), \nu^{(4)} = (2)\}$, and
(6) $\{\nu^{(1)} = (3,1), \nu^{(2)} = (2,2,1,1), \nu^{(3)} = (3,1,1), \nu^{(4)} = (2)\}$.

Let us compute the matrix $m(\{\nu\}) := (m_{ij})$ corresponding to the configuration (2). Let us write,
$(m_{ij}) = ((\nu^{(i-1)})'_j - (\nu^{(i)})'_j) + (\sum_{a\geq 1}\theta(\eta_a - i)\theta(\mu_a - j)) := U + W.$
One can check that

$$U = \begin{pmatrix} -2 & -1 & -1 & 0 & 0 \\ -1 & -1 & 0 & 0 & 0 \\ 1 & 0 & 0 & 0 & 0 \\ 1 & 1 & 1 & 0 & 0 \\ 1 & 1 & 0 & 0 & 0 \end{pmatrix}, \quad W = \begin{pmatrix} 5 & 3 & 0 \\ 3 & 3 & 0 \\ 1 & 1 & 0 \end{pmatrix}.$$

Therefore,

$$m(\{\nu\}) = \begin{pmatrix} 3 & 2 & -1 & 0 & 0 \\ 2 & 2 & 0 & 0 & 0 \\ 2 & 1 & 0 & 0 & 0 \\ 1 & 1 & 1 & 0 & 0 \\ 1 & 1 & 0 & 0 & 0 \end{pmatrix}.$$

One can read off directly from the matrix $m(\{\nu\}) = (m_{ij})$ all the additional quantities, which is needed to compute the parabolic Kostka polynomial corresponding to λ and a (dominant) sequence of rectangular shape partitions R; namely,

$$P_j^{(k)}(\{\nu\}) = \sum_{i\geq j}(m_{ki} - m_{k+1,i}),\ m_j(\nu^{(k)}) = \sum_{a\geq 1}\min(\eta_a, k)\delta_{\mu_a,j} - \sum_{i\leq k}(m_{ij} - m_{i,j+1}),$$

$$c(\{\nu\}) = \sum_{i,j\geq 1}\binom{m_{ij}}{2}.$$

For example, in our example, we have

$$c(\nu) = 8,\ \mathit{and}\ P_1^{(1)} = 1,\ P_2^{(2)} = 1,\ P_3^{(2)} = 1,\ P_2^{(3)} = 1$$

are all non-zero vacancy numbers. Therefore the contribution of the configuration in question to the parabolic Kostka polynomial is equal to

$$q^8 \begin{bmatrix} 2 \\ 1 \end{bmatrix}_q^4.$$

Treating in a similar fashion all the other configurations, we come to a fermionic formula

$$K_{44332,\{(2^3),(2^2),(2^2),(1),(1)\}}(q) = q^{10}\begin{bmatrix} 3 \\ 1 \end{bmatrix} + q^8 \begin{bmatrix} 2 \\ 1 \end{bmatrix}^4 + q^8 \begin{bmatrix} 3 \\ 2 \end{bmatrix} + q^{12} + q^6 \begin{bmatrix} 2 \\ 1 \end{bmatrix}\begin{bmatrix} 3 \\ 2 \end{bmatrix} + q^8.$$

■

Note that in the case when $\eta_a = 1, \ \forall a$, one can show that $\sum_{a \geq 1} \eta_a \theta(\mu_a - j) = \mu'_j$, and we *set* $\nu^{(0)} = \mu$. In this case, one can rewrite the conditions (1) – (4) as follows:

$(1')$ $\sum_{i \geq 1} m_{ij} = \mu'_j$,
$(2')$ $\sum_{j \geq 1} m_{ij} = \lambda_i$,
$(3')$ $\sum_{j \leq k} (m_{ij} - m_{i+1,j}) \geq 0$, for all i, j, k, and
$(4')$ $\sum_{i > k} (m_{ij} - m_{i,j+1}) \geq 0$, for all i, j, k.

Let us remark that if $m_{ij} \in \mathbb{Z}_{\geq 0}, \forall i, j$, then the matrix $m(\{\nu\}) = (m_{ij})$ defines a *low-lattice plane partition* of shape λ. For example, take $\lambda = (6, 4, 2, 2, 1, 1)$, $\mu = (2^8)$ and admissible configuration $\{\nu\} = \{(5, 5), (4, 2), (3, 1), (2), (1)\}$. The corresponding matrix and low-lattice[7] plane partition of shape λ are

$$(m_{ij}) = \begin{pmatrix} 3\,3\,0\,0\,0\,0 \\ 1\,3\,0\,0\,0\,0 \\ 1\,1\,0\,0\,0\,0 \\ 1\,1\,0\,0\,0\,0 \\ 1\,0\,0\,0\,0\,0 \\ 1\,0\,0\,0\,0\,0 \end{pmatrix}, \text{ and } \textit{plane partition} \begin{array}{l} 2\,2\,2\,1\,1\,1 \\ 2\,2\,2\,1 \\ 2\,1 \\ 2\,1 \\ 1 \\ 1 \end{array}.$$

The corresponding low-lattice word is 1.1.12.12.1222.111222.

In the case when $\eta_a = 1, \ \forall a$, there exists a unique admissible configuration of type (λ, μ), denoted by $\Delta(\lambda, \mu)$, such that $\max(c((\Delta(\lambda, \mu), J))) = n(\mu) - n(\lambda)$, where the maximum is taken over all *rigged configurations* associated with configuration $\Delta(\lambda, \mu)$. Recall that for any partition λ,

[7]Let π be a plane partition. We associate with π a word $w(\pi)$ as follows: let us begin reading of the entices of π right to left starting from the most right and bottom cell of π till the first cell of the last row of π. Then, do the same reading of the next row above the previous one, and so on till the same reading of the first row of plane partition π. As a result, we obtain a word $w(\pi)$. A plane partition is called *low-lattice*, if the corresponding word $w(\pi)$ is a lattice word.

$$n(\lambda) = \sum_{j \geq 1} \binom{\lambda'_a}{2}.$$

If $\lambda \geq \mu$ with respect to the *dominance order* on the set of partitions, then the degree of the Kostka polynomial $K_{\lambda,\mu}(q)$ is equal to $n(\mu) - n(\lambda)$, see, e.g. [23], Chapter 1, for details. Note that the maximal configuration $\Delta(\lambda, \mu)$ corresponds to the following matrix:

$$m_{1j} = \mu'_j - \max(\lambda'_j - 1, 0),\ j \geq 1,\ m_{ij} = 1,\ if\ (i, j) \in \lambda,\ i \geq 2,\ m_{ij} = 0,\ if\ (i, j) \notin \lambda.$$

In other words, the configuration $\Delta(\lambda, \mu)$ consists of the following partitions $(\lambda[1], \lambda[2], \ldots)$, where we set $\lambda[k] = (\lambda_{k+1}, \lambda_{k+2}, \ldots)$. It is not difficult to see that the contribution to the Kostka polynomial $K_{\lambda,\mu}(q)$ coming from the maximal configuration is equal to

$$K_{max}(\Delta(\lambda, \mu)) := q^{c(\Delta(\lambda,\mu))} \prod_{j=1}^{\lambda_2} \begin{bmatrix} Q_j(\mu) - Q_j(\lambda) + \lambda'_j - \lambda'_{j+1} \\ \lambda'_j - \lambda'_{j+1} \end{bmatrix}_q,$$

where $c(\Delta(\lambda, \mu)) = n(\lambda) + n(\mu) - \sum_{j \geq 1} \mu'_j(\lambda'_j - 1)$. Therefore,

$$K_{\lambda,\mu}(q) \geq K_{max}(\Delta(\lambda, \mu)). \tag{6.4}$$

It is clearly seen that if $\lambda \geq \mu$, then $Q_j(\mu) \geq Q_j(\lambda)$, $\forall\ j \geq 1$, and thus, $K_{q=1}(\Delta(\lambda, \mu)) \geq 1$, and the inequality (2.4) can be considered as a "quantitative" generalization of the Gale–Ryser theorem, see, e.g. [23], Chapter I, Section 7, or [13] for details.

Now let us **stress** that for a fixed k, all the partitions $\nu^{(k)}$ that contribute to the set of admissible configurations of type (λ, μ), have the same size equal to $\sum_{j \geq k+1} \lambda_j$, and thus the size of each $\nu^{(k)}$ does not depend on μ. However, the Rigged Configuration Bijection

$$RC_{\lambda,\mu} : STY(\lambda, \mu) \longrightarrow RC(\lambda, \mu)$$

happens to be essentially depend on μ. One can check that the map $RC_{\lambda,\mu}$ is compatible with the familiar *Bender–Knuth* transformations on the set of semistandard Young tableaux of a fixed shape. More precisely, let $\mu = (\mu_1, \ldots, \mu_i, \mu_{i+1}, \ldots)$ be a composition. We set $\mu^{(i)} = (\mu_1, \ldots, \mu_{i+1}, \mu_i, \ldots)$, and denote by κ_i the Bender–Knuth transformation, namely a bijection $\kappa_i : STY(\lambda, \mu) \longrightarrow STY(\lambda, \mu^{(i)})$. Then, $\kappa_i(T) = RC_{\lambda,\mu^{(i)}}(T)$, $T \in STY(\lambda, \mu)$. Let us stress again that $RC(\lambda, \mu) = RC(\lambda, \mu^{(i)})$.

As it was mentioned above, for a fixed k, all the (admissible) configurations have the same size. Therefore, the set of admissible configurations admits a partial ordering denoted by "$\succcurlyeq$." Namely, if $\{\nu\}$ and $\{\xi\}$ are two admissible configurations of the same type (λ, μ), we will write $\{\nu\} \succcurlyeq \{\xi\}$, if either $\{\nu\} = \{\xi\}$ or there exists an integer ℓ such that $\nu^{(a)} = \xi^{(a)}$ if $1 \le a \le \ell$, and $\nu^{(\ell+1)} > \xi^{(\ell+1)}$ with respect to the dominance order on the set of the same size partitions. It seems an interesting **Problem** to study poset structures on the set of admissible configurations of type (λ, μ), especially to investigate the posets of admissible configurations associated with the multidimensional Catalan numbers (work in progress).

■

Theorem 6.8 (Duality Theorem for Parabolic Kostka Polynomials [11]) *Let λ be partition and $R = \{(\mu_a^{\eta_a})\}$ be a dominant sequence of rectangular shape partitions. Denote by λ' the conjugate of λ and by R' a dominant rearrangement of a sequence of rectangular shape partitions $\{(\eta_a^{\mu_a})\}$. Then,*

$$K_{\lambda,R}(q) = q^{n(R)} K_{\lambda',R'}(q^{-1}),$$

where

$$n(R) = \sum_{a<b} \min(\mu_a, \mu_b) \min(\eta_a, \eta_b).$$

A technical proof is based on checking of the statement that the map

$$\iota : m_{ij} \longrightarrow \hat{m}_{ij} = -m_{ji} + \theta(\lambda_j - i) + \sum_{a \ge 1} \theta(\mu_a - j)\theta(\eta_a - i)$$

establishes bijection between the set of admissible configurations of type (λ, R) and that (λ', R') and the equality $\iota(c(m_{ij})) = c((\hat{m}_{ij}))$.

6.1 Example

Let $n = 6$, and consider, for example, a standard Young tableau

$$T = \begin{matrix} 1\,2\,3 & 6 & 8 & 9 \\ 4\,5\,7 & 10 & 11 & 12 \end{matrix}, \quad c(T) = 48.$$

The corresponding rigged configuration (ν, J) is

$$\nu = (321),\ J = (J_3 = 0,\ J_2 = 2, J_1 = 6),\ (m_{ij})(\nu) = \begin{pmatrix} 9 & -2 & -1 \\ 3 & 2 & 1 \end{pmatrix},\ c(\nu) = 44.$$

Recall that $c(T)$ and $c(\nu)$ denote the charge of tableau T and configuration ν correspondingly.

- One can see that $c(T) = c(\nu) + J_3 + J_2 + J_1$, as it should be in general.
- Next, the descent set and the descent number of tableau T are $Des(T) = \{3, 6, 9\}$, $des(T) = 3$. One can see that $des(T) = 3 = \nu_1^{'}$, as it should be in general.[8]
- One can check that our tableau T is invariant under the action of the Schützenberger involution[9] on the set of standard Young tableaux of a shape λ. It is clearly seen from the set of riggings J[10] that the rigged configuration (ν, J) corresponding to tableau T is invariant under the Flip involution.[11]

Acknowledgments I would like to thank the Research Institute for Mathematical Sciences(RIMS), Kyoto University, Japan, the Kavli Institute for the Physics and Mathematics of the Universe (IPMU), Tokyo University, Japan, and Department of Mathematics, the National Research University Higher School of Economics (HES), Moscow, Russia, for hospitality and financial support. This work was also supported by JSPS KAKEHI 16K05057.

References

1. H. Bethe, *On the theory of metals. 1. Eigenvalues and eigenfunctions for the linear atomic chain* (In German), Z. Phys. 71 (1931) 205–226 https://doi.org/10.1007/BF01341708.
2. R.K. Brylinski, *Matrix concomitants with mixed tensor model*, Adv. in Math., 100 (1993), 28–52.

[8] In fact the shape of the first configuration $\nu^{(1)}$ of type (λ, μ) can be read off from the set of "secondary" descent sets $\{Des^{(1)}(T) = Des(T), Des^{(2)}(T), \ldots, \ldots)$, cf. [12].

[9] http://en.wikipedia.org/wiki/Jeu_de_taquin.

[10] In our example, $J = (0, 1, 3)$.

[11] Recall that a rigging of an admissible configuration ν is a collection of integers

$$J = (\{J_{s,r}^{(k)}\}, \ 1 \le s \le m_r(\nu^{(k)})$$

such that for a given k, r one has

$$0 \le J_{1,r}^{(k)} \le J_{2,r}^{(k)} \le \ldots \le J_{m_r(\nu^{(k)}),r}^{(k)} \le P_r^{(k)}(\nu).$$

The Flip involution κ is defined as follows:

$$\kappa(\nu, \{J_{s,r}^{(k)}\}) = (\nu, \{J_{m_r(\nu^{(k)})-s+1,r}^{(k)}\}).$$

On the set of rigged configurations of type $(\lambda, 1^{|\lambda|})$, as it should be in general, see [17] for a complete proof of the statement that the action of the Schützenberger transformation on a Littlewood–Richardson tableau $T \in LR(\lambda, R)$, under the Rigged Configuration Bijection transforms tableau T to a Littlewood–Richardson tableau corresponding to the rigged configuration $\nu\kappa(J))$, where (νJ) is the rigged configuration corresponding to tableau T we are started with.

3. L. M. Butler, The q-Log-Concavity of q-Binomial Coefficients, Journal of Combinatorics, Serie A 54, (1990), 54–63.
4. H. Z.Q. Chen, A.L.B. Yang, P.B. Zhang, Kirillov's unimodality conjecture for the rectangular Narayana polynomials. Preprint `arXiv:1601.05863` [math.CO].
5. C. Carré and B. Leclerc, *Splitting the square of a Schur function into its symmetric and antisymmetric parts*, J. Alg. Combin. 4 (1995), 201–231
6. V. Dhand, A Combinatorial proof of strict unimodality for q-binomial coefficients; preprint `arXiv:1402.1199` [mathCO].
7. J. Desarmenien, B. Leclerc and J.-Y. Thibon, *Hall–Littlewood functions and Kostka–Foulkes polynomials in representation theory,* Seminaire Lotharingien de Combinatoire, v.32, 1994.
J. Fürlinger and J. Hofbauer, *q–Catalan numbers*, J. of Comb. Theory A **40** (1985), 248–264.
8. R.K. Gupta, *Generalized exponents via Hall-Littlewood symmetric functions,* Bull. Amer. Math. Soc, 16 (1987), 287–291.
9. A.N. Kirillov and N. Reshetikhin, *The Bethe ansatz and the combinatorics of Young tableaux*, (Russian), Zap. Nauch. Sem. LOMI **155** (1986), 65–115, translation in Journal of Soviet Math. **41** (1988), 925–955.
10. A.N. Kirillov, *Combinatorial identities and completeness of states for the Heisenberg magnet*, (Russian), Zap. Nauch. Sem. LOMI **131** (1983), 88–105, translation in Journal of Soviet Math. **30** (1985), 2298–3310.
11. A.N. Kirillov, *Combinatorics of Young tableaux and rigged configurations* (Russian) Proceedings of the St.Petersburg Math. Soc., 7 (1999), 23–115; translation in Proceedings of the St.Petersburg Math. Soc. volume VII, Amer. Math. Soc. Transl. Ser.2, 203, 17–98, AMS, Providence, RI, 2001.
12. A.N. Kirillov, *On some properties of the Robinson–Schensted correspondence*, Proceedings of Hayashibara conference on special functions, 1990, 122–126.
13. A.N. Kirillov, *Generalization of the Gale-Ryser theorem*, European Journal of Combinatorics, **21** (2000), 1047–1055; $hep-th/9304099$.
14. A.N. Kirillov, *Decomposition of symmetric and exterior powers of the adjoint representation of* $\mathfrak{gl}(N)$. *1. Unimodality of principal specialization of the internal product of the Schur functions*, Jour. Mod. Phys. A **7** (1992), 545–579.
15. A.N. Kirillov, *Unimodality of generalized Gaussian coefficients*, C.R. Acad. Sci. Paris, Ser, I, **315** (1992), no, 5, 497–501.
16. A.N. Kirillov, *Bijective correspondences for rigged configurations*, Algebra i Analiz **12** (2000), 204–240, translation in St.Petersburg Math. J. **12** (2001), no.1, 161–190.
17. A.N. Kirillov, A. Schilling and M. Schimozono, A bijection between Littlewood-Richardson tableaux and rigged configurations, Selecta Mathematica **8** (2002), 67–135.
18. A.N. Kirillov, *Ubiquity of Kostka polynomials,* Physics and Combinatorics 1999 (Proceedings of the Nagoya 1999 International Workshop on Physics and Combinatorics, Nagoya University, August 23–27, 1999, ed. A. Kirillov, A. Tsuchiya and H. Umemura), 85–200, World Scientific, Singapore, 2001.
19. A.N. Kirillov, *An invitation to the generalized saturation conjecture*, Publications of RIMS Kyoto University **40** (2004), 1147–1239.
20. A.N. Kirillov, *Kostka, and higher dimensional Catalan and Narayana numbers*, unpublished preprint (2001). Updated version: *Rigged Configurations and Catalan, Stretched Parabolic Kostka Numbers and Polynomials: Polynomiality, Unimodality and Log-concavity*, Preprint RIMS-1824, 2015, and arXiv:1505.01542; Advanced Studies in Pure Mathematics **76** (2018) Representation Theory, Special Functions and Painlevé Equations. RIMS 2015 pp. 303–3346.
21. B. Kostant, *Lie group representations on polynomial rings,* Amer. J. Math., 85 (1963), 327–404.
22. D.E. Littlewood, *On the Kronecker product of symmetric group representations,* J. London Math. Soc., 31 (1956), 89–93.
23. I.G. Macdonald, *Symmetric functions and Hall polynomials,* 2nd ed., Oxford, 1995.
24. K. O'Hara, *Unimodality of Gaussian coefficients: a constructive proof,* J. Comb. Theory A, **53** (1990), 29–52.

25. I. Pak, G. Panova, Strict Unimodality of q-Binomial Coefficients, C. R. Acad. Sci. Paris, Ser. I, **351** (2013), 415–418.
26. J. Shareshian, M. Wachs, q-Eulerian Polynomials: excedence number and major index, `arXiv:math/0608274` [math/CO].
27. N.J.A. Sloan, *The on-line encyclopedia of integer sequences,* (2004), http://www.researche.att.com/njas/sequences/ .
28. R. Stanley, *Log-concave and unimodal sequences in algebra, combinatorics, and geometry,* Annals of the New York Academy of Sciences **576** (1989), 500–535.
29. R. Stanley, *Enumerative Combinatorics*, vols. 1 and 2, Cambridge University Press, 1999.
30. R. Sulanke, *Generalizing Narayana and Schröder numbers to higher dimensions,* Electr. Journ. Comb. **11** (2004), #R54.
31. R. Sulanke, *Three dimensional Narayana and Schröder numbers,* Theor. Comp. Sci. **346(2–3)** (2005), 455–468.
32. F. Zanello, Zeilberger's KOH Theorem and the strict unimodality of q-binomial coefficients, Proc. of AMS, **143**, v.7, 2795–2799.
33. D. Zeilberger, *Kathy O'Hara's constructive proof of the unimodality of the Gaussian polynomials,* Amer. Math. Monthly, 96 (1989), 590–602.

Turning Point Processes in Plane Partitions with Periodic Weights of Arbitrary Period

Sevak Mkrtchyan

To Nicolai Reshetikhin on the occasion of his 60th birthday.

Abstract I study random plane partitions with respect to volume measures with periodic weights of arbitrarily high period. I show that near the vertical boundary the system develops up to as many turning points as the period of the weights and that these turning points are separated by vertical facets which can have arbitrary rational slope. In the lozenge tiling formulation of the model, the facets consist of only two types of lozenges arranged in arbitrary periodic deterministic patterns. We compute the correlation functions near turning points and show that the point processes at the turning points can be described as several GUE-corners processes which are non-trivially correlated.

The weights we study introduce a first-order phase transition in the system. We compute the limiting correlation functions near this phase transition and obtain a process which is translation invariant in the vertical direction but not the horizontal.

1 Introduction

Recall that a plane partition confined to a $c \times d$ box is a two-dimensional array $\pi = (\pi_{i,j})_{1\le i\le c, 1\le j\le d}$ of non-negative integers such that $\pi_{i,j} \ge \pi_{i+1,j}$ and $\pi_{i,j} \ge \pi_{i,j+1}$ for all i, j. If c or d is infinite, then we also require that $\pi_{i,j} = 0$ if $i + j$ is large enough. We will denote the set of all plane partitions confined to a $c \times d$ box by $\Pi^{c,d}$.

A natural measure on plane partitions is the so-called volume measure, where the probability $\mathbb{P}_q(\pi)$ of a plane partition π is proportional to $q^{|\pi|}$, where q is

S. Mkrtchyan (✉)
University of Rochester, Rochester, NY, USA
e-mail: sevak.mkrtchyan@rochester.edu

A. Alekseev et al. (eds.), *Representation Theory, Mathematical Physics, and Integrable Systems*, Progress in Mathematics 340,
https://doi.org/10.1007/978-3-030-78148-4_17

a parameter in $(0, 1)$ and $|\pi| = \sum_{i,j} \pi_{i,j}$ is the volume of π. Scaling limits of plane partitions and their generalizations under volume measures and with various boundary conditions have been studied extensively (see e.g. [NHB84, KO07, OR03, OR07, BMRT12, Mkr11], and references therein). In particular, in [OR03], a broad family of measures called the Schur process was introduced, and it was shown that the "volume" measure is a special case of the Schur process. Using the vertex operator formalism, it was shown in [OR03] that the Schur process is determinantal and an explicit contour integral representation of the kernel was obtained, paving the way for studying the asymptotics of the local correlation functions for the volume measure.

In the formulation of the model as a Schur process, a plane partition is thought of as a sequence of (interlacing) Young diagrams. More precisely, given a plane partition $\pi = (\pi_{i,j})$ and an integer $-c < t < d$, let $\pi(t)$ be the tth diagonal slice of π defined by

$$\pi(t) = \begin{cases} (\pi_{1,k+1}, \pi_{2,k+2}, \pi_{3,k+3}, \dots), & k \geq 0, \\ (\pi_{-k+1,1}, \pi_{-k+2,2}, \pi_{-k+3,3}, \dots), & k \leq 0. \end{cases}$$

This decomposes π into a sequence of interlacing Young diagrams

$$\pi \Leftrightarrow \pi(-c+1) \prec \cdots \prec \pi(-1) \prec \pi(0) \succ \pi(1) \succ \cdots \succ \pi(d-1),$$

where the notation $\mu \succ \nu$ for Young diagrams $\mu = (\mu_1, \mu_2, \dots)$ and $\nu = (\nu_1, \nu_2, \dots)$ stands for the interlacing condition

$$\mu_1 \geq \nu_1 \geq \mu_2 \geq \nu_2 \dots$$

The volume measure can be expressed as

$$\mathbb{P}_q(\pi) \propto \prod_{-c<t<d} q^{|\pi(t)|},$$

where $|\pi(t)|$ stands for the sum of the entries in $\pi(t)$. In this formulation, it is natural to consider a generalization of the volume measure where the weight q depends on the "time" parameter t (see [OR07]). Namely, given a set of positive real numbers $\bar{q} = (q_t)_{-c<i<d}$, define the measure $\mathbb{P}_{\bar{q}}$ on the space of plane partitions $\Pi^{c,d}$ by

$$\mathbb{P}_{\bar{q}}(\pi) \propto \prod_{-c<t<d} q_t^{|\pi(t)|}, \tag{1}$$

i.e. give weight q_t to the boxes on slice t of the plane partition. The case $q_t = q, \forall t$, corresponds to the volume measure introduced earlier, and we will call it the homogeneous case. The measure (1) with inhomogeneous weights is still a Schur process. In [OR07], Okounkov and Reshetikhin showed that it is a determinantal point process, gave a contour integral representation for the correlation kernel, and studied the asymptotic behavior in the case of homogeneous weights. In this chapter, we will analyze the scaling limit of this measure with inhomogeneous periodic weights.

To study the scaling limit, it is convenient to use a different formulation of the model. Plane partitions can be visualized through their height functions, which can be obtained by stacking $\pi_{i,j}$ identical cubes at position (i, j) for all (i, j) (see Fig. 1). It is evident from the figure that if we regard it as a two-dimensional object, plane partitions can be thought of as a tiling of a certain region of the plane by rhombi of three different orientations, called lozenges.

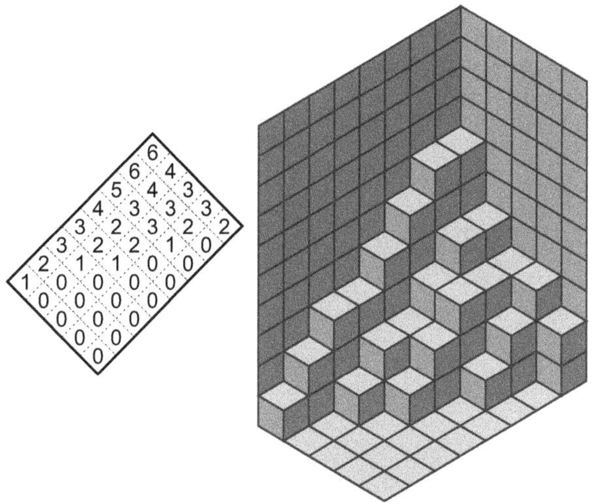

Fig. 1 A plane partition π and the corresponding stack of cubes/lozenge tiling

In the appropriate scaling limit, the system develops frozen and liquid regions separated by a curve called the arctic curve or the frozen boundary. In a frozen region, randomness disappears and the lozenges form a deterministic pattern, whereas in a liquid region any possible local configuration of lozenges appears with positive probability. Each frozen region that forms in the scaling limit of random plane partitions distributed with respect to the homogeneous volume measure consists of only one type of lozenges, and in the 3-dimensional formulation of the model via its height function, it corresponds to a facet with a normal vector pointing in one of the three lattice directions.

Along the frozen boundary, there are special points called turning points. These are the points where two different frozen regions and the liquid region meet. In [OR06], Okounkov and Reshetikhin conjectured that the point processes at turning points should converge to the Gaussian Unitary Ensemble (GUE)-corners process and proved that this is indeed the case for the homogeneous volume measure with certain boundary conditions. For results toward proving the conjecture in the case of boxed plane partitions with respect to the uniform measure, see [JN06, GP15,

Nov15], for Gelfand–Tsetlin patterns with the homogeneous volume measure, see [MP17], and for more general Gaussian orbital beta processes, see [Cue18].

In this chapter, we will mainly concentrate on the behavior of the turning points that arise in random plane partitions with respect to the non-homogeneous measure (1) when the weights are k-periodic and the new types of frozen regions that arise. We show that the point processes at turning points are no longer the GUE-corners process, but rather several copies of the GUE-corners process non-trivially interlaced. Moreover, we show that the turning points are separated by new types of frozen regions. In the 3-dimensional formulation, these are facets that are vertical and whose projections on the horizontal plane give a line of arbitrary rational slope. In the lozenge tiling formulation, each facet consists of two types of lozenges, arranged in an arbitrary periodic deterministic pattern (see Fig. 2 for examples). The exact value of the slope, or the exact pattern, is determined by the specific weights taken.

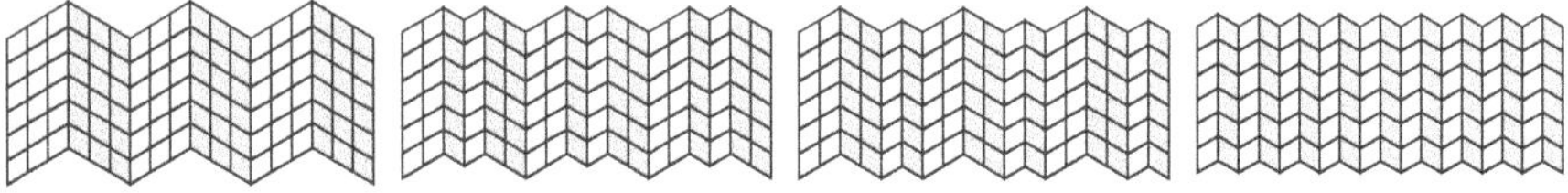

Fig. 2 All possible frozen regions of slope 0 that can arise when weights are 6-periodic

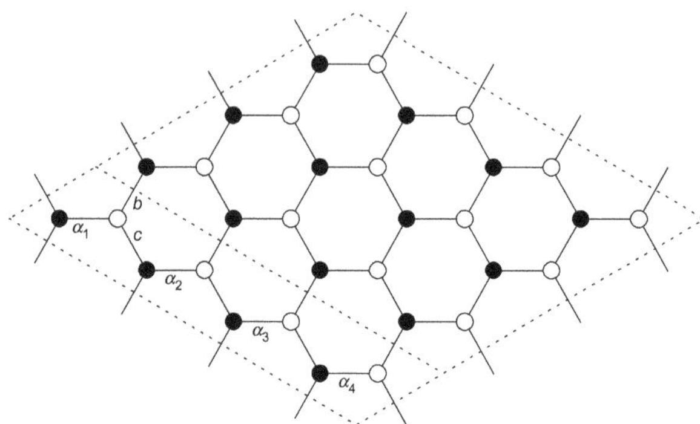

Fig. 3 The smaller region is the fundamental domain when weights are 4-periodic

In the formulation of the model as a dimer model, the weights we take essentially correspond to taking a larger fundamental domain in the language of [Ken09] (see Fig. 3) as opposed to the smaller fundamental domain from Fig. 4 in the case of homogeneous weights. The larger fundamental domain results in a larger Newton polygon (see [KO07, KOS06]), which has integer points on the boundary. As was discussed in [KO07, KOS06], these integer points are responsible for the new types of frozen facets. Because the weights we take are non-homogeneous only in the horizontal direction, the Newton polygon does not have any integer points in its interior, which means we will not see gaseous regions in our setup.

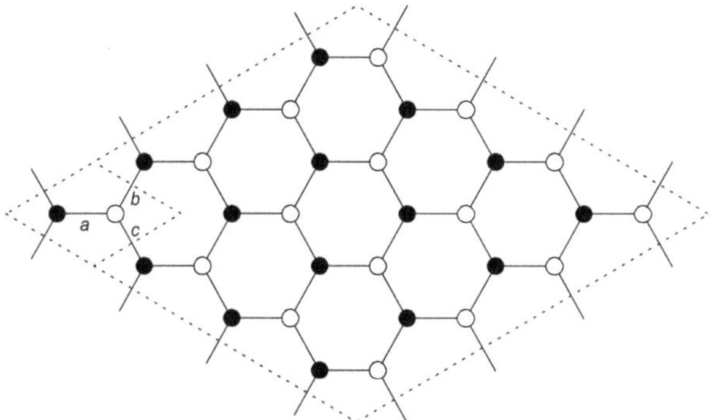

Fig. 4 The smaller region is the fundamental domain when weights are homogeneous

Certain doubly periodic models have been studied in the literature; see, for example, [CJ16] or [BD19]. One significant difference they have is that the systems under doubly periodic weights can develop gaseous regions as well.

When dealing with the measure (1) with appropriate non-homogeneous weights, care should be taken to ensure that the measure is indeed well defined, as the partition function becomes infinite. The case of 2-periodic weights was studied in [Mkr14], and in that case, this issue was resolved by modifying the boundary and studying certain skew plane partitions instead. This approach, however, does not work for $k > 2$. Instead, what we do here is to modify the weight in one position, namely, the 0th slice at the corner, so weights are periodic everywhere except this slice. This should have no effect on the processes at turning points, as they are macroscopically far away from that slice. However, the modification we introduce does affect the behavior of the system at the 0th slice. Namely, it introduces a first-order phase transition, and the local point process loses translation invariance in the horizontal direction. For a precise description of this process, see Theorem 5.1.

2 Notation and Description of Main Results

The Boundary We will work with the formulation of the model in terms of lozenge tilings. Knowing the positions of lozenges of only one type completely determines the tiling. In particular, the positions of the horizontal tiles (those at the top of a stack of cubes) determine the complete tiling. Introduce planar coordinates (t, h), so that the centers of horizontal tiles are on the lattice $\mathbb{Z} \times \frac{1}{2}\mathbb{Z}$ and position the point $(0, 0)$ as in Fig. 5.

Fig. 5 Coordinate axes

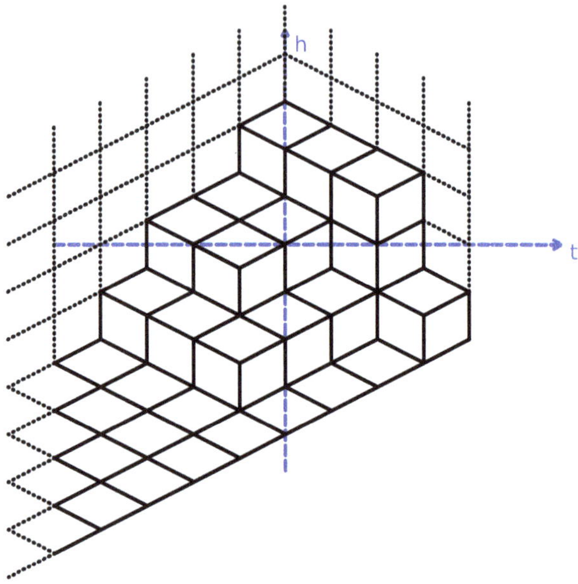

To simplify the computations, we will consider the case $c = \infty$, so the plane partition will consist of the slices $\pi(t)$ for $t \in (-\infty, d-1] \cap \mathbb{Z}$.

Weights We will consider a weight sequence $\bar{q}$, which is periodic of period $k \in \mathbb{N}$, i.e. $q_i = q_{i+k}$ for all $i \in \mathbb{Z}$. The thermodynamic limit in the homogeneous case corresponds to the limit $q_t = q \to 1^-$. In the periodic case, one could consider the limit when $q_i \to 1^-, \forall i$; however, in that regime, the behavior of the system is the same as that of the homogeneous measure with the weight given by the geometric average of the periodic weights. The more interesting limit to study is when $q_i \to \alpha_i$, where $\alpha_0, \ldots, \alpha_{k-1}$ are positive numbers. When $\prod_{i=0}^{k-1} \alpha_i < 1$, the expected size of the system is finite, whereas when $\prod_{i=0}^{k-1} \alpha_i > 1$, the measure does not make sense as the partition function is infinite. Thus, the appropriate setup is when $\prod_{i=0}^{k-1} \alpha_i = 1$. However, having any $\alpha_i > 1$ still makes the partition function infinite, so the measure $P_{\bar{q}}$ is not well defined. To see this, consider cases based on the value of α_{k-1}.

It is convenient to set $q_i = \alpha_i q$ with $0 < q < 1$ and consider the limit $q \to 1^-$.

First, suppose $\alpha_{k-1} < 1$. Then, $\alpha_0 \cdot \ldots \cdot \alpha_{k-2} > 1$. Let π^m be the plane partition given by $\pi^m_{1,j} = m$ for $1 \le j \le k-1$ and $\pi^m_{i,j} = 0$ otherwise (see Fig. 6). Then,

$$P_{\bar{q}}(\pi^m) = (q_0 \ldots q_{k-2})^m \approx (\alpha_0 \ldots \alpha_{k-2})^m,$$

so we have $\sum_{m=1}^{\infty} P_{\bar{q}}(\pi^m) = \infty$, which means the measure $P_{\bar{q}}$ is not well defined.

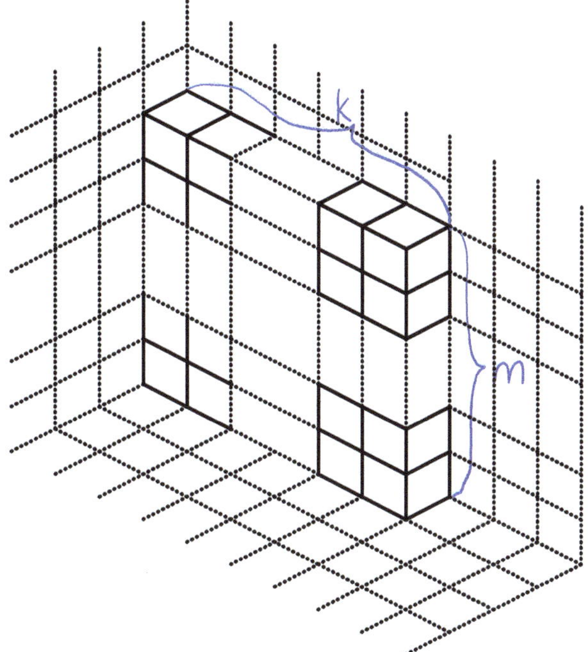

Fig. 6 The configuration π^m when $\alpha_{k-1} < 1$

Now, suppose $\alpha_{k-1} > 1$. Then, $q_{-1} = q_{k-1} \approx \alpha_{k-1} > 1$. Let π^m be the plane partition given by $\pi^m_{1,j} = m$ for $1 \le j \le k$, $\pi^m_{2,1} = m$ and $\pi^m_{i,j} = 0$ otherwise (see Fig. 7). Then,

$$P_{\bar{q}}(\pi^m) = (q_{-1}q_0 \dots q_{k-1})^m \approx (\alpha_{k-1}\alpha_0 \dots \alpha_{k-1})^m = \alpha^m_{k-1},$$

so we again have $\sum_{m=1}^{\infty} P_{\bar{q}}(\pi^m) = \infty$.

Lastly, if $\alpha_{k-1} = 1$, then we can consider α_{k-2}, and so on. This shows that unless $\alpha_0 = \cdots = \alpha_{k-1} = 1$, the measure $P_{\bar{q}}$ is not well defined if q is close enough to 1.

One way to fix the issue is to modify the boundary. This was the approach taken in [Mkr14], where skew plane partitions with 2-periodic weights with a staircase cut out at the corner were considered. Such an approach does not work when the period $k > 2$. Instead, in this chapter, we consider ordinary plane partitions but modify the weights, so they are periodic everywhere except the slice containing the corner. More precisely, let $\alpha_0, \alpha_1, \dots, \alpha_{k-1} > 0$ with

$$\alpha_0 \cdot \dots \cdot \alpha_{k-1} = 1,$$

and let

$$\gamma = \prod_{j:\alpha_j < 1} \alpha_j.$$

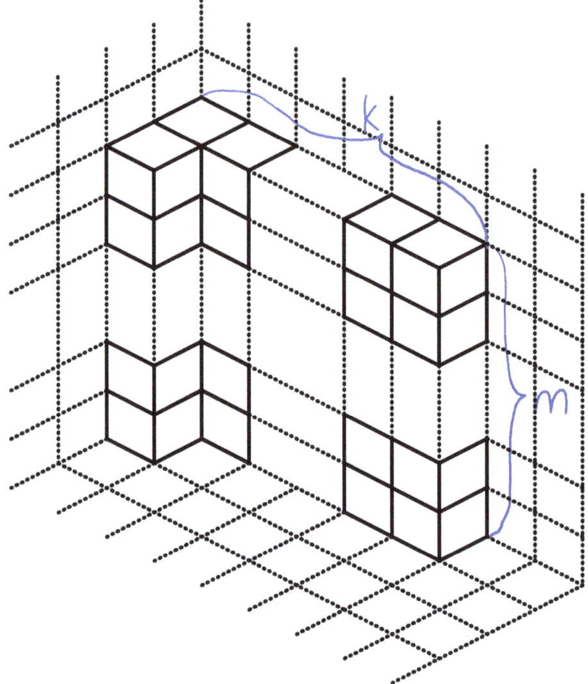

Fig. 7 The configuration π^m when $\alpha_{k-1} > 1$

Define

$$q_i := q_{i,r} := \begin{cases} q\alpha_{i \bmod k}, & i \neq 0 \\ q\alpha_0\gamma, & i = 0, \end{cases} \tag{1}$$

where $q = e^{-r}$. We will study random plane partitions distributed according to the measure (1) with weights given by (1), in the limit when $r \to 0^+$ and d grows at the rate $d = V/r$ for some constant $V > 0$. To simplify the formulas, we will assume that d mod k is independent of r. For an illustration of the weight assignments, see Fig. 8.

Note that the modification of the weight q_0 by γ will not have any effect on the processes observed at turning points since the turning points will be macroscopically away from the 0th slice. Indeed, when $k = 2$, the process observed here matches with that obtained in [Mkr14], where the weights are completely 2-periodic. The addition of γ to the slice at zero does and, however, has an effect on the local processes near that slice. Whereas the processes appearing in the bulk are translation invariant in two directions, the addition of γ causes a first-order phase transition with a local point process only translation invariant in the vertical direction near the special slice.

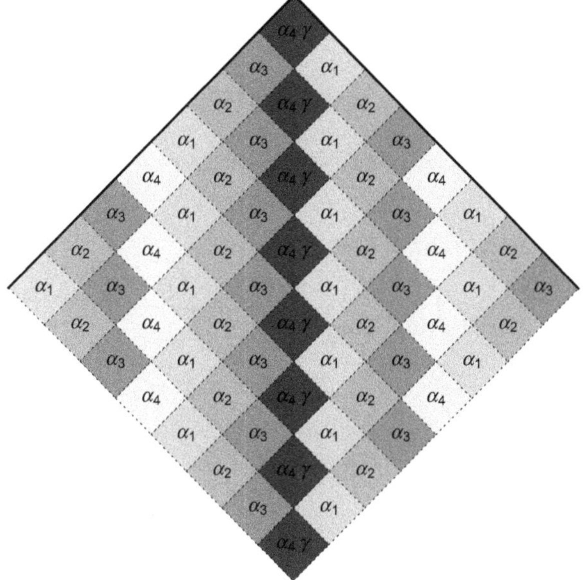

Fig. 8 Weight assignments of period 4. The number in a cell indicates the weight of the boxes above the cell

All fully $\mathbb{Z} \times \mathbb{Z}$ translation invariant ergodic Gibbs measures were classified in [She05]. The process we obtain away from the phase transition is translation invariant under translations by elements of $k\mathbb{Z} \times \mathbb{Z}$ only. As a result, it is not given by the incomplete beta kernel, but rather a deformation of it, which is a special case of processes studied previously in [Bor07] and [BS10].

Main Results We will study the scaling limit of our model via the local correlation functions. Given a subset $U = \{(t_1, h_1), \dots, (t_n, h_n)\} \subset \mathbb{Z} \times \frac{1}{2}\mathbb{Z}$, define the corresponding local correlation function $\rho_{\bar{q}}(U)$ as the probability for a random tiling taken from the above probability space to have horizontal tiles centered at all positions $(t_i, h_i)_{i=1}^n$.

The main theorems of the chapter describe the point processes that appear in the thermodynamic limit of the system in terms of their correlation functions. Theorem 4.2 gives the limiting point process in the bulk near the edge, Theorem 5.1 gives the non-translation invariant point process in the bulk along the slice at the corner where we observe a first-order phase transition, and Theorem 4.6 gives the point processes near turning points. Section 4.6 explains the connection with the GUE-corners process, and Sect. 4.7 describes the new frozen regions that arise.

3 The Correlation Kernel and Its Leading Asymptotics

Okounkov and Reshetikhin [OR07] showed that for arbitrary $\bar{q}$, the correlation functions of the Schur process are determinantal.

Theorem 3.1 (Theorem 2, Part 3 [OR07]) *The correlation functions $\rho_{\bar{q}}$ are determinants*

$$\rho_{\bar{q}}(U) = \det(K_{\bar{q}}((t_i, h_i), (t_j, h_j)))_{1 \le i, j \le n},$$

where the correlation kernel $K_{\bar{q}}$ is given by the double integral

$$K_{\bar{q}}((t_1, h_1), (t_2, h_2)) = \frac{1}{(2\pi \mathrm{i})^2} \int_{z \in C_z} \int_{w \in C_w} \frac{\Phi_{\bar{q}}(z, t_1)}{\Phi_{\bar{q}}(w, t_2)} \frac{1}{z - w} z^{-h_1 - \frac{1}{2}|t_1| + \frac{1}{2}} w^{h_2 + \frac{1}{2}|t_2| + \frac{1}{2}} \frac{dz\, dw}{zw}, \tag{1}$$

where

$$\Phi_{\bar{q}}(z, t) = \frac{\Phi_{\bar{q}}^{-}(z, t)}{\Phi_{\bar{q}}^{+}(z, t)},$$
$$\Phi_{\bar{q}}^{+}(z, t) = \prod_{\max(0,t) < m < d, m \in \mathbb{Z} + \frac{1}{2}} (1 - z x_m^{+}), \tag{2}$$
$$\Phi_{\bar{q}}^{-}(z, t) = \prod_{m < \min(0,t), m \in \mathbb{Z} + \frac{1}{2}} (1 - z^{-1} x_m^{-}),$$

the parameters $x_m^{\pm}$ satisfy the conditions

$$\frac{x_{m+1}^{+}}{x_m^{+}} = q_{m+\frac{1}{2}}, \; 0 < m < d - 1,$$
$$x_{-\frac{1}{2}}^{-} x_{+\frac{1}{2}}^{+} = q_0, \tag{3}$$
$$\frac{x_m^{-}}{x_{m+1}^{-}} = q_{m+\frac{1}{2}}, \; m < -1,$$

and C_z (respectively, C_w) is a simple positively oriented contour around 0 such that its interior contains none of the poles of $\Phi_{\bar{q}}(\cdot, t_1)$ (respectively, all of $\Phi_{\bar{q}}(\cdot, t_2)^{-1}$). Moreover, if $t_1 < t_2$, then C_z is contained in the interior of C_w, and otherwise, C_w is contained in the interior of C_z.

The conditions in (3) can be rewritten in the form

$$x_m^{-} = a^{-1} q_{m+\frac{1}{2}} q_{m+\frac{3}{2}} \cdot \ldots \cdot q_{d-1}, \tag{4}$$
$$x_m^{+} = a q_{m+\frac{1}{2}}^{-1} q_{m+\frac{3}{2}}^{-1} \cdot \ldots \cdot q_{d-1}^{-1},$$

where $a > 0$ is an arbitrary parameter. We will choose $a = e^{-V} = e^{-rd}$.

3.1 *The Asymptotically Leading Term in the Integral for the Correlation Kernel*

It can be shown that as $r \to 0$, the volume of a typical plane partition grows at the rate of $1/r^3$, so we scale the system by r in all directions. Let (τ, χ) be the scaled coordinates. First, given a point (τ, χ), we will study the asymptotics of the correlation kernel (1) when $r \to 0$, $rt_i \to \tau$, $rh_i \to \chi$, and $\Delta t = t_1 - t_2$, $\Delta h = h_1 - h_2$ are fixed.

Definition 3.2 For $i \geq 1$, let $\beta_i = \alpha_{d-1}\alpha_{d-2} \cdot \ldots \cdot \alpha_{d-i}$. Let l be the number of distinct β's, and let $\tilde{\beta}_1 < \cdots < \tilde{\beta}_l$ be the ordered list of distinct β's. For $i = 1, \ldots, l$, let m_i be the number of times $\tilde{\beta}_i$ appears among $\beta_1, \ldots, \beta_k$. We have $m_1 + \cdots + m_l = k$.

Note that since $d \bmod k$ is constant, the values $\beta_1, \beta_2, \ldots$ do not depend on r. Moreover, since $a_0 \cdot \ldots \cdot a_{k-1} = 1$, the sequence $\beta_1, \beta_2, \ldots$ is also periodic of period k. We might, however, have some of the numbers $\beta_1, \ldots, \beta_k$ be equal, so l can be less than k and $m_i > 1$.

We have

$$q_{m+\frac{1}{2}} \cdot \ldots \cdot q_{d-1} = q^{d-m-\frac{1}{2}} \beta_{(d-m-\frac{1}{2}) \bmod k} \gamma^{c_m},$$

where $c_m = 0$ if $m > 0$ and $c_m = 1$ if $m < 0$. Thus, we have

$$\lim_{r \to 0} r \ln \Phi_{\bar{q}}(z, t) = \lim_{r \to 0} \left(-\sum_{i=1}^{k} \sum_{\substack{\max(0,t) < m < d \\ m \in \mathbb{Z}+\frac{1}{2} \\ \left(d-m-\frac{1}{2}\right) \bmod k = i}} r \ln(1 - zae^{r(d-m-\frac{1}{2})}\beta_i^{-1}) \right.$$

$$\left. + \sum_{i=1}^{k} \sum_{\substack{m < \min(0,t) \\ m \in \mathbb{Z}+\frac{1}{2} \\ \left(d-m-\frac{1}{2}\right) \bmod k = i}} r \ln(1 - z^{-1}a^{-1}e^{-r(d-m-\frac{1}{2})}\beta_i \gamma) \right)$$

$$= \frac{1}{k} \sum_{i=1}^{k} \left(\int_{-\infty}^{\min(\tau,0)} \ln(1 - z^{-1}\beta_i \gamma e^{M}) dM - \int_{\max(\tau,0)}^{V} \ln(1 - z\beta_i^{-1}e^{-M}) dM \right),$$

where in the last step we replaced a sum of $2k$ Riemann sums by the sum of the corresponding integrals.

It follows that the integrand in formula (1) for the correlation kernel has the leading asymptotics given by

$$K_{\bar{q}}((t_1,h_1),(t_2,h_2)) = \frac{1}{(2\pi \mathrm{i})^2}\int_{z\in C_z}\int_{w\in C_w} e^{\frac{1}{r}(S_{\tau,\chi}(z)-S_{\tau,\chi}(w))+O(1)}\frac{dzdw}{z-w}, \tag{5}$$

where

$$S_{\tau,\chi}(z) = \frac{1}{k}\sum_{i=1}^{k}\left(\int_{-\infty}^{\min(\tau,0)}\ln(1-z^{-1}\beta_i\gamma e^{M})dM - \int_{\max(\tau,0)}^{V}\ln(1-z\beta_i^{-1}e^{-M})dM\right) - \left(\chi+\frac{1}{2}|\tau|\right)\ln z. \tag{6}$$

4 The Point Processes in the Bulk Near the Edge and at the Turning Points

To compute the asymptotics of the correlation kernel, we use the steepest descent method, so we need to study the critical points of $S_{\tau,\chi}$. We have

$$z\frac{dS_{\tau,\chi}}{dz} = -\frac{1}{k}\sum_{i=1}^{k}\ln\left(\frac{z-\beta_i\gamma e^{\min(\tau,0)}}{z}\right)+\frac{1}{k}\sum_{i=1}^{k}\ln\left(\frac{z-\beta_i e^{V}}{z-\beta_i e^{\max(\tau,0)}}\right)-V-\chi+\frac{\tau}{2}$$

and

$$\frac{d}{dz}\left(z\frac{dS_{\tau,\chi}(z)}{dz}\right) = -\frac{1}{k}\sum_{i=1}^{k}\left(\frac{1}{z-e^{\min(\tau,0)}\beta_i\gamma}-\frac{1}{z}\right)+\frac{1}{k}\sum_{i=1}^{k}\left(\frac{1}{z-\beta_i e^{V}}-\frac{1}{z-\beta_i e^{\max(\tau,0)}}\right).$$

We are interested in the behavior of the system near the turning points that appear when $\tau = V$, so we set $\tau = V-\varepsilon$ with $\varepsilon > 0$ and small. We need ε to be smaller than a constant that only depends on d and the weights $\alpha_0,\ldots,\alpha_{k-1}$. In particular, we have $\tau \geq 0$, so

$$z\frac{dS_{\tau,\chi}}{dz}(z) = -\frac{1}{k}\sum_{i=1}^{l}m_i\ln\left(\frac{z-\tilde{\beta}_i\gamma}{z}\right)+\frac{1}{k}\sum_{i=1}^{l}m_i\ln\left(\frac{z-\tilde{\beta}_i e^{V}}{z-\tilde{\beta}_i e^{\tau}}\right)-V-\chi+\frac{\tau}{2} \tag{1}$$

and

$$\frac{d}{dz}\left(z\frac{dS_{\tau,\chi}(z)}{dz}\right) = -\frac{1}{k}\sum_{i=1}^{l}m_i\left(\frac{1}{z-\tilde{\beta}_i\gamma}-\frac{1}{z}\right)+\frac{1}{k}\sum_{i=1}^{l}m_i\left(\frac{1}{z-\tilde{\beta}_i e^{V}}-\frac{1}{z-\tilde{\beta}_i e^{\tau}}\right), \tag{2}$$

where we have grouped terms with equal β's together. To simplify the notation, we will denote $Sp(z) := z\frac{dS_{\tau,\chi}}{dz}(z)$.

4.1 *The Possible Locations of Double Real Critical Points*

First, we look for double real critical points of $S_{\tau,\chi}$. If z is such a point, then from (6) or (1), it is easy to see that $z \notin [0, \tilde{\beta}_i\gamma]$ and $z \notin [\tilde{\beta}_i e^{\tau}, \tilde{\beta}_i e^{V}]$ for any i.

Since the $\tilde{\beta}$'s are distinct, by picking the bound on ε to be small enough, we will have that the intervals $[\tilde{\beta}_i e^{\tau}, \tilde{\beta}_i e^{V}]$ are disjoint. Also, we have

$$\tilde{\beta}_i\gamma < \tilde{\beta}_l\gamma < \tilde{\beta}_1 e^{\tau} < \tilde{\beta}_i e^{\tau}. \tag{3}$$

Here, the only non-trivial inequality is the middle one. It holds, since $e^{\tau} \geq 1$, and $\tilde{\beta}_1/\tilde{\beta}_l$ is either the product of consecutive α's or the inverse of the product of consecutive α's. Since the product of all the α's is one, we have that $\tilde{\beta}_1/\tilde{\beta}_l$ is always the product of consecutive α's. However, γ is the product of all α's that are less than 1, so we have $\gamma < \tilde{\beta}_1/\tilde{\beta}_l$. It follows that the intervals $[0, \tilde{\beta}_i\gamma]$ and $[\tilde{\beta}_i e^{\tau}, \tilde{\beta}_i e^{V}]$ are situated on the real line as in Fig. 9.

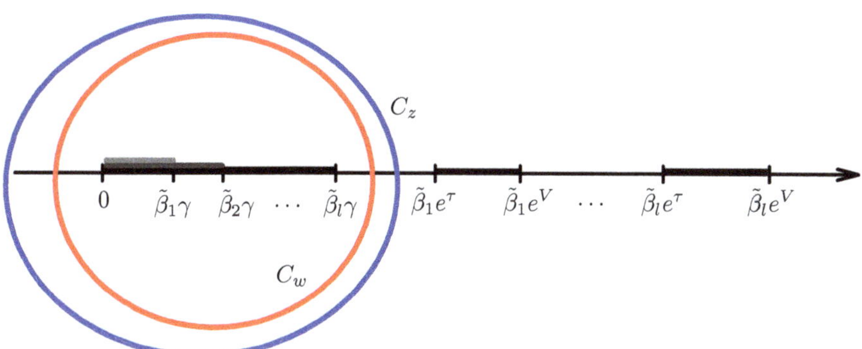

Fig. 9 The integration contours and the intervals where there cannot be real critical points

Note also that the C_z contour from (1) contains no zeros of $\Phi^{+}_{\bar{q}}(z, t_1)$, i.e. none of $e^{r(m+\frac{1}{2})}\tilde{\beta}_i$ for $t_1 < m < d$, and the C_w contour from (1) contains all zeros of $\Phi^{-}_{\bar{q}}(z, t_1)$, i.e. all of $e^{r(m+\frac{1}{2})}\tilde{\beta}_i\gamma$ for $m < 0$. This means that the contours C_z and C_w cross the positive part of the real line between $\tilde{\beta}_l\gamma$ and $\tilde{\beta}_1 e^{\tau}$ as in Fig. 9.

4.2 Computing the Double Real Critical Points

It follows from the computations in the previous section that if z is a double real critical point of $S_{\tau,\chi}$, then the summands in the first sum in (2) are all of the same sign, whereas the summands in the second sum in (2) are all small (they converge to 0 as ε goes to 0), unless z is close to $\tilde{\beta}_i e^{\tau}$ or $\tilde{\beta}_i e^{V}$, so if z is a double real critical point, z must be close to $\tilde{\beta}_i e^{\tau}$ or $\tilde{\beta}_i e^{V}$ for some i. Let $z = \tilde{\beta}_j e^V + \delta$, where $\delta \to 0$ as $\varepsilon \to 0$. All the summands in the second sum in (2) will be of size $o(1)$ except the summand with $i = j$. The latter one should be of order 1. Thus, we have

$$\frac{\tilde{\beta}_j(e^V - e^{\tau})}{(z - \tilde{\beta}_j e^V)(z - \tilde{\beta}_j e^{\tau})} = \theta(1),$$

where $\theta(1)$ means of constant order as $\varepsilon \to 0$. Plugging in the values of τ and z, we have

$$\frac{\tilde{\beta}_j e^V(\varepsilon + O(\varepsilon))}{\delta(\delta + \tilde{\beta}_j e^V \varepsilon + O(\varepsilon^2))} = \theta(1),$$

which implies that $\delta = \theta(\sqrt{\varepsilon})$. From (2), we must have

$$0 = -\frac{1}{k}\sum_{i=1}^{l} \frac{m_i \tilde{\beta}_i \gamma}{\tilde{\beta}_j e^V(\tilde{\beta}_j e^V - \tilde{\beta}_i \gamma)} + O(\delta) + O(\varepsilon) + \frac{1}{k}\frac{m_j \tilde{\beta}_j e^V \varepsilon}{\delta^2 + O(\varepsilon^{3/2})}.$$

Solving for δ, we get

$$\delta = \pm \frac{\sqrt{m_j}\tilde{\beta}_j e^V}{\sqrt{\sum_{i=1}^{l} \frac{m_i \tilde{\beta}_i \gamma}{\tilde{\beta}_j e^V - \tilde{\beta}_i \gamma}}}\sqrt{\varepsilon} + O(\varepsilon). \tag{4}$$

Note that (3) implies that the summands in the denominator are all positive, so δ is real. Setting

$$f_j := \frac{\sqrt{m_j}\tilde{\beta}_j e^V}{\sqrt{\sum_{i=1}^{l} \frac{m_i \tilde{\beta}_i \gamma}{\tilde{\beta}_j e^V - \tilde{\beta}_i \gamma}}},$$

let $z_{j,\pm} = \tilde{\beta}_j e^V \pm f_j\sqrt{\varepsilon} + g_\varepsilon \varepsilon + o(\varepsilon)$. By setting $Sp(z_{j,\pm}) = 0$ and solving for χ from (1) we obtain

$$\chi_{j,\pm} = -\frac{V}{2} - \frac{1}{k}\sum_{i=1}^{l} m_i \ln\left(\frac{e^V\tilde{\beta}_j - \tilde{\beta}_i\gamma}{e^V\tilde{\beta}_j}\right) \mp \sum_{i=1}^{l} f_j \frac{\tilde{\beta}_i\gamma}{e^V\tilde{\beta}_j(e^V\tilde{\beta}_j - \tilde{\beta}_i\gamma)}\sqrt{\varepsilon} \mp \frac{1}{k} m_j \frac{e^V\tilde{\beta}_j}{f_j}\sqrt{\varepsilon} + O(\varepsilon). \tag{5}$$

Note that to evaluate $\chi_{j,\pm}$ up to order $\sqrt{\varepsilon}$, we do need an expansion of $z_{j,\pm}$ up to order ε, even though the coefficient g_ε of ε in $z_{j,\pm}$ cancels out in the computation of $\chi_{j,\pm}$.

Also, note that we have

$$z_{l,+} > z_{l,-} > z_{l-1,+} > z_{l-1,-} > \cdots > z_{2,+} > z_{2,-1} > z_{1,+} > z_{1,-}$$

and

$$\chi_{l,+} < \chi_{l,-} < \chi_{l-1,+} < \chi_{l-1,-} < \cdots < \chi_{2,+} < \chi_{2,-1} < \chi_{1,+} < \chi_{1,-}.$$

Now, fix χ and consider the function $Sp(z)$. We have $Sp(z) < 0$ if $z < 0$.

It is easy to see that

- when z ranges from $-\infty$ to 0, the function $Sp(z)$ decreases from $-V - \chi + \frac{\tau}{2}$ to $-\infty$,
- when z ranges from $\tilde{\beta}_l \gamma$ to $\tilde{\beta}_1 e^\tau$, the function $Sp(z)$ decreases from $-\infty$ to $\chi_{1,-} - \chi$ at $z_{1,-}$ and increases to ∞,
- when z ranges from $\tilde{\beta}_i e^V$ to $\tilde{\beta}_{i+1} e^\tau$, the function $Sp(z)$ increases from $-\infty$ to $\chi_{i,+} - \chi$ at $z_{i,+}$, decreases to $\chi_{i+1,-} - \chi$ at $z_{i+1,-}$, and increases to ∞, and
- when z ranges from $\tilde{\beta}_l e^V$ to ∞, the function $Sp(z)$ increases from $-\infty$ to $\chi_{l,+} - \chi$ at $z_{l,+}$ and decreases to $-V - \chi + \frac{\tau}{2}$.

These properties of $Sp(z)$ are visualized in Fig. 10.

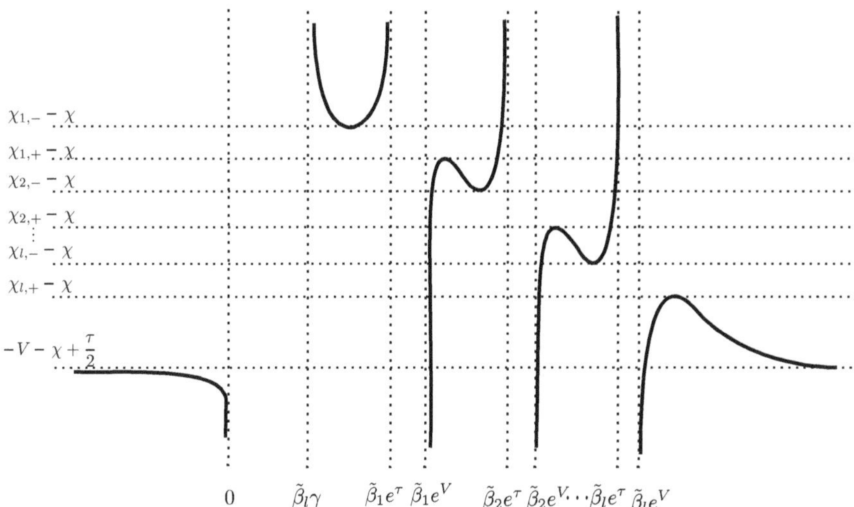

Fig. 10 A plot of the function $Sp(z)$

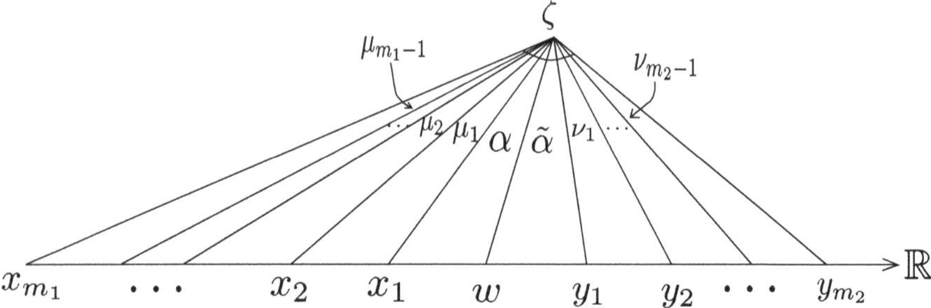

Fig. 11 Setup of Lemma 4.1

4.3 The Number of Real and Complex Critical Points

First, we count the critical points of $S_{\tau,\chi}$ when χ is large. We use the following lemma proven in [Mkr11].

Lemma 4.1 (Lemma 2.3 in [Mkr11]) *Let* $\{x_i\}_{i=1}^{m_1}$, $\{y_i\}_{i=1}^{m_2}$, w, *and* $\mathfrak{w} \neq 0$ *be real numbers such that*

$$x_{m_1} < \ldots < x_2 < x_1 < w < y_1 < y_2 < \ldots < y_{m_2}.$$

Let ζ *be a complex number with* $\Im\zeta \geq 0$. *Define*

$$\begin{aligned} \mu_i &= \text{angle}(\zeta - x_{i+1}, \zeta - x_i), \ \forall i = 1, 2, \ldots, m_1 - 1, \\ \nu_i &= \text{angle}(\zeta - y_i, \zeta - y_{i+1}), \ \forall i = 1, 2, \ldots, m_2 - 1, \end{aligned}$$

and

$$\alpha = \text{angle}(\zeta - x_1, \zeta - w).$$

(See Fig. 11 for an illustration of the setup on the complex plane.) Given any real numbers $\{\mathfrak{x}_i\}_{i=1}^{m_1-1}$ *and* $\{\mathfrak{y}_i\}_{i=1}^{m_2-1}$, *there exists* $\varepsilon > 0$ *such that if* $|\zeta - w| < \varepsilon$, *then*

$$\sum_{i=1}^{m_1-1} \mathfrak{x}_i \mu_i + \mathfrak{w}\alpha + \sum_{i=1}^{m_2-1} \mathfrak{y}_i \nu_i = 0 \tag{6}$$

if and only if $\zeta \in \mathbb{R}$ *and* $\zeta > w$.

The same holds if α *is replaced by* $\tilde{\alpha} :=$ angle$(\zeta - w, \zeta - y_1)$ *and* $\zeta > w$ *is replaced by* $\zeta < w$.

It follows from (1) that if χ is large, then z is close to either $\tilde{\beta}_l\gamma$ or $\tilde{\beta}_i e^{\tau}$ for some $i = 1, \ldots, l$. Setting the imaginary part of $Sp(z)$ to zero, it follows from Lemma 4.1 that all the critical points of $S_{\tau,\chi}$ are real if χ is large. For example, if z is close to

$\tilde{\beta}_l\gamma$, then setting $\zeta = z$, $m_1 = l$, $x_{m_1} = 0$, $x_i = \tilde{\beta}_{l-i}\gamma$, $\mathfrak{x}_i = m_{l-i} + \cdots + m_l$ for $i = 1, \ldots, l-1$, $w = \tilde{\beta}_l\gamma$, $\mathfrak{w} = m_1$, $m_2 = 2l$, $y_{2j-1} = \tilde{\beta}_j e^{\tau}$, $y_{2j} = \tilde{\beta}_j e^{V}$, $\mathfrak{y}_{2j-1} = m_j$, and $\mathfrak{y}_{2j} = 0$ for $j = 1, \ldots, l$ in Lemma 4.1 gives us that z must be real.

Thus, when χ is large enough, $S_{\tau,\chi}$ has exactly $l+1$ real critical points and no non-real complex critical points. As χ varies, the number of real or complex non-real critical points of $S_{\tau,\chi}$ can only change when it has a double real critical point. It follows from Fig. 10 that if $\chi_{i,+} < \chi < \chi_{i,-}$ for some $i = 1, \ldots, l$, then $S_{\tau,\chi}$ has exactly two non-real complex critical points. Otherwise, it has only real critical points. Let $z_{cr,i}$ be the critical point with positive imaginary part when $\chi_{i,+} < \chi < \chi_{i,-}$. We moreover have that as χ increases from $\chi_{i,+}$ to $\chi_{i,-}$, the real part of z_{cr} decreases from $\tilde{\beta}_i e^V + f_i\sqrt{\varepsilon}$ to $\tilde{\beta}_i e^V - f_i\sqrt{\varepsilon}$.

4.4 Deformation of Contours and the Limiting Correlation Kernel in the Bulk

We carry out the steepest descent analysis by deforming the contours of integration in (1) appropriately. Let $\chi_{i,+} < \chi < \chi_{i,-}$ and $\tau = V - \varepsilon$ with ε small. Then, as we saw, the function $S_{\tau,\chi}$ has exactly one pair of complex conjugate critical points. Let z_{cr} be the one with positive imaginary part. It follows that the curve $\Re S_{\tau,\chi}(z) = \Re S_{\tau,\chi}(z_{cr})$ looks as in Fig. 12. We deform the contours C_z and C_w to contours C_z' and C_w' such that they pass through $z_{cr}, \bar{z}_{cr}$ and $\Re(S_{\tau,\chi}(z)) \leq \Re(S_{\tau,\chi}(z_{cr})) \leq \Re(S_{\tau,\chi}(w))$ for all $z \in C_z'$ and $w \in C_w'$, with equality only at the critical points.

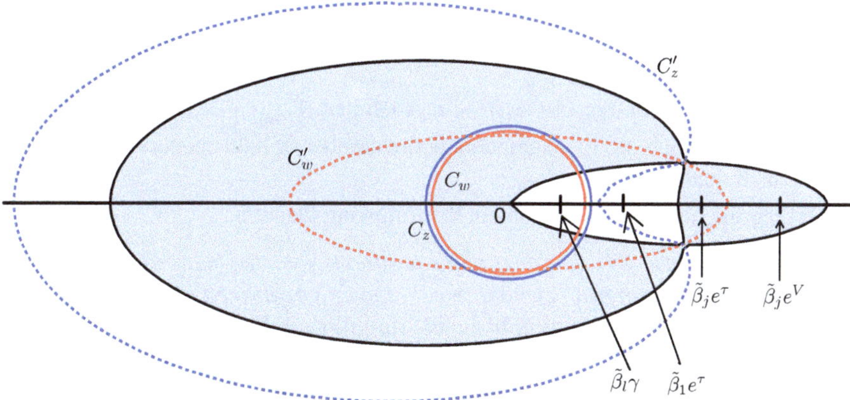

Fig. 12 The shaded region is the region $\Re(S_{\tau,\chi}(z)) > \Re(S_{\tau,\chi}(z_{cr}))$. The solid curves are the original contours, whereas the dotted ones are the new contours

This deformation of contours results in the z and w contours crossing, which, together with the presence of the $\frac{1}{z-w}$ term in (1), means that as a result of the deformation we pick up residues where the curves cross, i.e. along an arc from $\bar{z}_{cr}$ to z_{cr} which crosses the real axes to the right of 0 if $t_1 \geq t_2$ and to the left of 0 if $t_1 < t_2$. We have

$$K_{\bar{q}}((t_1, h_1), (t_2, h_2)) = \int_{z\in C'_z} \int_{w\in C'_w} e^{\frac{S_{\tau,\chi}(z)-S_{\tau,\chi}(w)}{r}+O(1)} \frac{dzdw}{z-w} + \frac{1}{2\pi \mathrm{i}} \int_{\bar{z}_{cr}}^{z_{cr}} \frac{\Phi_{\bar{q}}(z,t_1)}{\Phi_{\bar{q}}(z,t_2)} z^{h_2-h_1-\frac{1}{2}(t_1-t_2)-1} dz.$$

Since $\Re(S_{\tau,\chi}(z)) \leq \Re(S_{\tau,\chi}(z_{cr})) \leq \Re(S_{\tau,\chi}(w))$ along the contours C'_z and C'_w, the first integral converges to 0 as $r \to 0$. Thus,

$$\lim_{r\to 0} K_{\bar{q}}((t_1, h_1), (t_2, h_2)) = \lim_{r\to 0} \frac{1}{2\pi \mathrm{i}} \int_{\bar{z}_{cr}}^{z_{cr}} \frac{\Phi_{\bar{q}}(z,t_1)}{\Phi_{\bar{q}}(z,t_2)} z^{h_2-h_1-\frac{1}{2}(t_1-t_2)-1} dz.$$

Since $t_1, t_2 \approx \frac{\tau}{r} > 0$, from (2), we have

$$\frac{\Phi_{\bar{q}}(z,t_1)}{\Phi_{\bar{q}}(z,t_2)} = \frac{\Phi^-_{\bar{q}}(z,t_1)\,\Phi^+_{\bar{q}}(z,t_2)}{\Phi^-_{\bar{q}}(z,t_2)\,\Phi^+_{\bar{q}}(z,t_1)} = \frac{\Phi^+_{\bar{q}}(z,t_2)}{\Phi^+_{\bar{q}}(z,t_1)} = \prod_{\substack{\min(t_1,t_2)<m<\max(t_1,t_2)\\ m\in\mathbb{Z}+\frac{1}{2}}} (1 - zx^+_m)^{\Delta t},$$

$$= \prod_{\substack{\min(t_1,t_2)<m<\max(t_1,t_2)\\ m\in\mathbb{Z}+\frac{1}{2}}} \left(1 - ze^{r(-m-\frac{1}{2})}\beta^{-1}_{(d-m-\frac{1}{2}) \bmod k}\right)^{\Delta t}. \tag{7}$$

Let $N_{t,i}$ be the number of $t < m < d_r$ such that $\beta_{d-m-\frac{1}{2}} = \tilde{\beta}_i$, and let $\Delta N_i := N_{t_1,i} - N_{t_2,i}$, which, up to a sign, counts the number of half-integers m between t_1 and t_2 such that $\beta_{d-m-\frac{1}{2}} = \tilde{\beta}_i$.

Taking the limit $r \to 0$, we obtain the following theorem.

Theorem 4.2 *Let $r > 0$. Consider plane partitions confined to an $\infty \times d_r$ box, with periodic weights given by* (1) *with $q = e^{-r}$. Let $\tilde{\beta}_i$'s be defined as in Definition 3.2. Assume d_r mod k is independent of r and $\lim_{r\to 0} rd_r = V$. In the limit when r goes to 0 along a vertical line $\tau = V - \varepsilon$ near the edge of the system, the liquid region consists of l intervals, one for each $\tilde{\beta}_i$. The correlation functions of the system near a point (τ, χ) in the bulk where χ is in the interval $(\chi_{j,+}, \chi_{j,-})$ (see* (5)*) are given by determinants of the kernel*

$$\lim_{r\to 0} K_{\bar{q}}((t_1, h_1), (t_2, h_2)) = \frac{1}{2\pi \mathrm{i}} \int_C \prod_{i=1}^{l} \left(1 - ze^{-\tau}\tilde{\beta}_i^{-1}\right)^{\Delta N_i} z^{-\Delta h-\frac{1}{2}\Delta t-1} dz,$$

where $\Delta t = t_1 - t_2$, $\Delta h = h_1 - h_2$, $rt_i \to \tau$, $rh_i \to \chi$, *and the integration contour* C *connects the two non-real critical points of* $S_{\tau,\chi}(z)$, *passing through the real line in the interval* $(\tilde{\beta}_l\gamma, e^{\tau}\tilde{\beta}_1)$ *if* $\Delta(t) \geq 0$ *and through* $(-\infty, 0)$ *otherwise.*

Remark 4.3 Notice that $\sum_{i=1}^{l} \Delta N_i = \Delta t$. In particular, in the homogeneous case $\alpha_1 = \cdots = \alpha_k = 1$, we have $l = 1$ and $\tilde{\beta}_1 = 1$, so we recover the incomplete beta kernel. Otherwise, the process we have is translation invariant under translations by $k\mathbb{Z} \times \mathbb{Z}$ only and is a special case of processes studied previously in [Bor07] and [BS10].

Remark 4.4 When χ is such that $S_{V-\varepsilon,\chi}$ has only real critical points, then during the deformation of the contours, they do not cross each other and we do not pick up any residues from the $\frac{1}{z-w}$ term. It follows that the correlation functions for horizontal lozenges decay exponentially fast as $r \to \infty$, so in the scaling limit, almost surely there will be no horizontal lozenges near such a point $(V - \varepsilon, \chi)$.

Remark 4.5 The argument in Sect. 4.3 showing that $S_{V-\varepsilon,\chi}$ has only real critical points when χ is large and also applies for $S_{0,\chi}$ without modification if the sets $\{\tilde{\beta}_i\}_{i=1}^{l}$ and $\{\tilde{\beta}_i\gamma\}_{i=1}^{l}$ are disjoint. Thus, in that case, we also obtain that when $\tau = 0$ and χ is large, almost surely there will be no horizontal lozenges near the macroscopic point $(0, \chi)$. It follows that the height of the scaled plane partition is actually bounded, unlike the cases of homogeneous weights studied in [OR03, OR07, BMRT12, Mkr11].

4.5 Deformation of Contours and the Limiting Correlation Kernel Near Turning Points

We now turn our attention to the point processes at the turning points. Let $rd_r = V$, $t_i = d_r - \hat{t}_i$, $h_i = \lfloor \frac{\chi_i}{r} \rfloor + \frac{\hat{h}_i}{r^{1/2}}$ with $\hat{t}_i, \hat{h}_i$ fixed, independent of r, and χ_j the height of the jth turning point, which, from (5), is

$$\chi_j = -\frac{V}{2} - \frac{1}{k}\sum_{i=1}^{l} m_i \ln\left(\frac{e^V\tilde{\beta}_j - \tilde{\beta}_i\gamma}{e^V\tilde{\beta}_j}\right). \tag{8}$$

From the earlier computations of critical points, we see that the critical point corresponding to the turning point at height χ_i is $z_j = \tilde{\beta}_j e^V$.

We have

$$S_{\tau,\chi}(z) = \frac{1}{k}\sum_{i=1}^{k}\int_{-\infty}^{0} \ln(1 - z^{-1}\beta_i\gamma e^M)dM - \left(\chi + \frac{V}{2}\right)\ln z,$$

$$\left.\frac{dS_{V,\chi_j}(z)}{dz}\right|_{z=z_j} = 0,$$

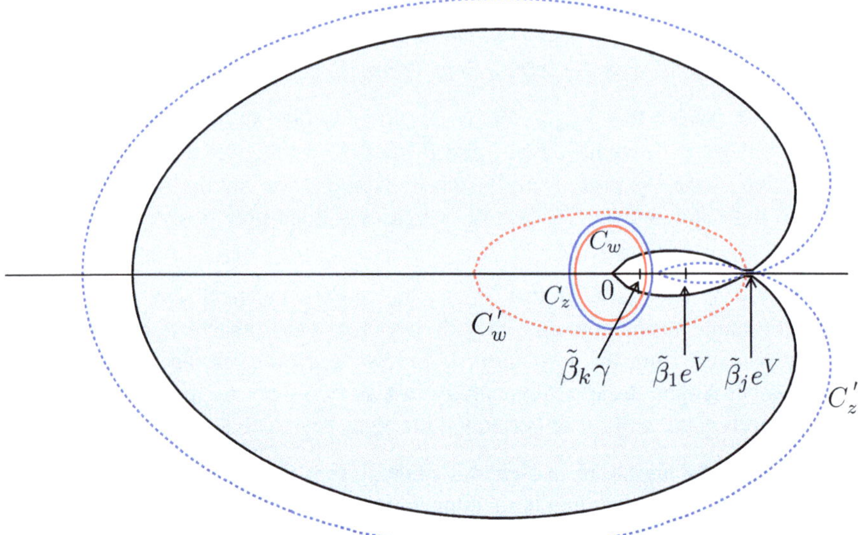

Fig. 13 The shaded region is the region $\Re(S_{V,\chi_j}(z)) > \Re(S_{V,\chi_j}(z_j))$. The solid curves are the original contours, whereas the dotted ones are the new deformed contours

and

$$\frac{d}{dz}\left(z\frac{dS_{V,\chi}(z)}{dz}\right) = -\frac{1}{k}\sum_{i=1}^{l} m_i\left(\frac{1}{z-\tilde{\beta}_i\gamma} - \frac{1}{z}\right) = -\frac{1}{k}\sum_{i=1}^{l}\frac{m_i\tilde{\beta}_i\gamma}{(z-\tilde{\beta}_i\gamma)z}.$$

It follows that

$$\frac{d}{dz}\left(z\frac{dS_{V,\chi}(z)}{dz}\right)\bigg|_{z=z_j} < 0,$$

which implies $S''_{V,\chi}(z_j) < 0$ for all j.

Now, the contours $\Re(S_{V,\chi_j}(z)) = \Re(S_{V,\chi_j}(z_j))$ look as in Fig. 13, and we deform the integration contours of the correlation kernel to pass through z_j as shown in Fig. 13. When $\Delta t \geq 0$, the contours do not pass through each other, so we do not pick up any resides from the $\frac{1}{z-w}$ term. Along the new contours, away from the critical point, we have $\Re(S_{V,\chi_j}(w)) > \Re(S_{V,\chi_j}(z))$, so the contribution to the correlation kernel is exponentially small, and the main contribution comes from the vicinity of the critical point. We make the change of variable $z = z_j e^{r^{1/2}\xi}$, $w = z_j e^{r^{1/2}\omega}$. In these coordinates, the steepest descent contours are as in Fig. 14 (similarly to [OR06]).

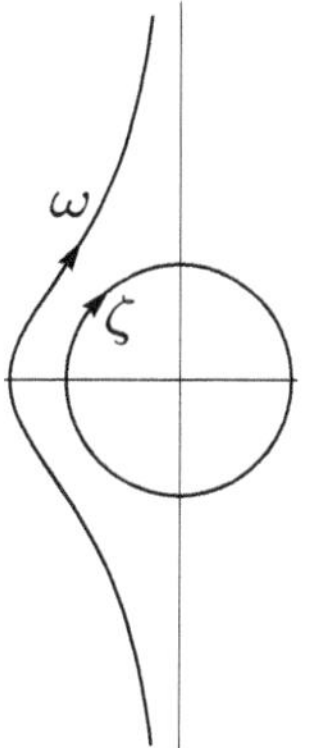

Fig. 14 The ξ and ω contours

We have

$$S_{V,\chi_j}(z) = S_{V,\chi_j}(z_j) + S'_{V,\chi_j}(z_j)(z - z_j) + \frac{S''_{V,\chi_j}(z_j)}{2}(z - z_j)^2 + \cdots .$$

Since

$$z - z_j = -\sqrt{r}\xi z_j(1 + O(\sqrt{r}),$$

we get

$$S_{V,\chi_j}(z) = S_{V,\chi_j}(z_j) + \frac{S''_{V,\chi_j}(z_j)}{2} r\xi^2 z_j^2 + O(r^{\frac{3}{2}}).$$

Using $z_j = \tilde{\beta}_j e^V$, a similar computation to (7) gives

$$\Phi^+_{\bar{q}}(z, t_1) = \prod_{\substack{t_1 < m < d_r \\ m \in \mathbb{Z}+\frac{1}{2}}} \left(1 - \tilde{\beta}_j e^V e^{r^{1/2}\xi} e^{r(-m-\frac{1}{2})} \beta^{-1}_{(d-m-\frac{1}{2}) \bmod k}\right).$$

Since $t_1 = d_r - \hat{t}_1$ with $\hat{t}_1$ fixed, this is a finite product. Moreover,

$$e^V e^{r(-m-\frac{1}{2})} = e^{r(d_r-m-\frac{1}{2})} = 1 + O(r),$$

from which it follows that

$$\begin{aligned}\Phi^+_{\bar{q}}(z, t_1) &= \prod_{\substack{t_1 < m < d_r \\ m \in \mathbb{Z}+\frac{1}{2}}} \left(1 - \tilde{\beta}_j e^{r^{1/2}\xi} \beta^{-1}_{(d-m-\frac{1}{2}) \bmod k}\right)(1 + O(r)) \\ &= (-r^{1/2}\xi)^{N_{t_1,j}} \prod_{\substack{i=1 \\ i \neq j}}^{l} (1 - \tilde{\beta}_j \tilde{\beta}_i^{-1})^{N_{t_1,i}} (1 + O(\sqrt{r})).\end{aligned}$$

Thus, we obtain

$$K_{\bar{q}}((t_1,h_1),(t_2,h_2)) = z_j^{-\Delta h-\frac{\Delta}{t}}(-r^{1/2})^{-\Delta N_j}\prod_{\substack{i=1\\ i\neq j}}^{l}(1-\tilde{\beta}_j\tilde{\beta}_i^{-1})^{-\Delta N_i}$$

$$\times\frac{1}{(2\pi\mathrm{i})^2}\int_{z\in C_z'}\int_{w\in C_w'} e^{\frac{S''_{V,\chi_j}(z_j)}{2}z_j^2(\xi^2-\omega^2)}\frac{\omega^{N_{t_2,j}}}{\xi^{N_{t_1,j}}}\frac{1}{\sqrt{r}(\xi-\omega)}\frac{e^{\hat{h}_2\omega}}{e^{\hat{h}_1\xi}}r(1+O(\sqrt{r}))\,d\xi\,d\omega.$$

When computing $\det K_{\bar{q}}((t_i,h_i),(t_j,h_j))_{i,j=1}^n$, the terms

$$z_j^{-\Delta h-\frac{\Delta}{t}}(-r^{1/2})^{-\Delta N_j}\prod_{\substack{i=1\\ i\neq j}}^{l}(1-\tilde{\beta}_j\tilde{\beta}_i^{-1})^{-\Delta N_i}$$

cancel out. Thus, we obtain the following theorem.

Theorem 4.6 *Let $r>0$. Consider plane partitions confined to an $\infty\times d_r$ box, with periodic weights given by* (1) *with $q=e^{-r}$. Let $\tilde{\beta}_i$'s be defined as in Definition 3.2. Assume d_r* mod *k is independent of r and* $\lim_{r\to 0} rd_r = V$. *In the limit when r goes to* 0 *near the vertical boundary $\tau=V$, the system develops l turning points, one for each $\tilde{\beta}_i$. The jth highest turning point is at position (V,χ_j), where χ_j is given by* (8), *and the correlation functions of the system near it are given by determinants of the kernel*

$$K_{\bar{q}}^{j}((\hat{t}_1,\hat{h}_1),(\hat{t}_2,\hat{h}_2)) = \frac{\sqrt{r}}{(2\pi\mathrm{i})^2}\int_{z\in C_z'}\int_{w\in C_w'} e^{\frac{S''_{V,\chi_j}(z_j)}{2}z_j^2(\xi^2-\omega^2)}\frac{\omega^{N_{\hat{t}_2,j}}}{\xi^{N_{\hat{t}_1,j}}}\frac{e^{\hat{h}_2\omega}}{e^{\hat{h}_1\xi}}\frac{d\xi\,d\omega}{(\xi-\omega)}, \tag{9}$$

where $\hat{t}_i=d_r-t_i$ is the horizontal distance from the boundary, $\frac{\hat{h}_i}{r^{1/2}}$ is the vertical distance from χ_j, $N_{t,i}$ is the number of $m\in\mathbb{Z}+\frac{1}{2}$ such that $d_r-t<m<d_r$ and $\beta_{d-m-\frac{1}{2}}=\tilde{\beta}_i$, and the integration contours are as in Fig. 14.

Remark 4.7 When $\Delta t<0$, the w contour crosses completely over the z contour, so we pick up residues from the $\frac{1}{z-w}$ term along a simple contour going once around 0. However, as explained in particular in [OR07] or [Mkr14, Section 5.0.3], this does not alter the expression of the correlation kernel given in (9).

Remark 4.8 While the limiting process is discrete in the horizontal coordinates t, it is continuous in the vertical coordinates h, so the right-hand side of (9) converges to 0 as $r\to 0$. In fact, if we condition on the vertical slices $\hat{t}_i$, then the point process in the limit is a continuous determinantal point process whose correlation measures have densities given by determinants

$$\det\left(\frac{K_{\bar{q}}^{j}((\hat{t}_i,\hat{h}_i),(\hat{t}_j,\hat{h}_j))}{\sqrt{r}}\right)_{i,j}.$$

4.6 The GUE-Corners Process

Fix j with $1 \leq j \leq l$ and choose integers $\bar{j} = (j_1, j_2, \dots)$ such that $N_{j_i,j} = i$ for all $i \in \mathbb{N}$. Note that we must have $j_1 < j_2 < \cdots$. Given $\tilde{t}_i \in \mathbb{N}$ and $\tilde{h}_i \in \mathbb{R}$, let $\tilde{\rho}_r((\tilde{t}_1,\tilde{h}_1),\dots,(\tilde{t}_n,\tilde{h}_n))$ be the probability that there are horizontal lozenges at positions $(t_i,h_i) = (d_r - j_{\tilde{t}_i}, \lfloor\frac{\chi_i}{r}\rfloor + \frac{\tilde{h}_i}{r^{1/2}})$. It follows that as $r \to 0$, we have

$$\tilde{\rho}_r((\tilde{t}_1,\tilde{h}_1),\dots,(\tilde{t}_n,\tilde{h}_n)) \to \det(K_{\bar{q}}^{\bar{j}}((\tilde{t}_i,\tilde{h}_i),(\tilde{t}_l,\tilde{h}_l)))_{i,l=1}^{n},$$

with

$$K_{\bar{q}}^{\bar{j}}((\tilde{t}_1,\tilde{h}_1),(\tilde{t}_2,\tilde{h}_2)) = \frac{\sqrt{r}}{(2\pi \mathrm{i})^2}\int_{z\in C_z'}\int_{w\in C_w'} e^{\frac{S''_{V,\chi_j}(z_j)}{2} z_j^2(\xi^2-\omega^2)}\frac{\omega^{\tilde{t}_2}}{\xi^{\tilde{t}_1}}\frac{e^{\tilde{h}_2\omega}}{e^{\tilde{h}_1\xi}}\frac{d\xi\, d\omega}{(\xi-\omega)}. \tag{10}$$

This is the correlation kernel for the GUE-corners process (see e.g. [OR06]). In other words, if we restrict the point process at the jth turning point to the slices $j_1, j_2, \dots$, then we get the GUE-corners process. In particular, since the ith slice of the GUE-corners process has i points, it follows that near the jth turning point the ith slice from the boundary has $N_{i,j}$ points.

We can describe the connection with the GUE-corners process differently. The number of horizontal lozenges on vertical slices near a turning point weakly increases as you move further away from the boundary. Pick any turning point and any collection of slices, such that the ith slice has i horizontal lozenges near the turning point. Then, the point process near the turning point restricted to the selected slices will be the GUE-corners process. Note that the GUE-corners process has i particles on slice i, so to obtain the GUE-corners process, it is obviously necessary to pick slices so that the ith has i particles. The key is that doing this is also sufficient, so as long as the number of particles is selected correctly, we obtain the GUE-corners process.

4.7 The Frozen Regions Separating the Turning Points

It follows from Theorems 4.2 and 4.6 that near the boundary at $\tau = V$, the system develops $l + 1$ frozen regions separated by l liquid regions. In the scaling limit the correlation functions for horizontal lozenges converge to zero in these frozen regions, so no horizontal lozenges appear in these regions. Let us call the lozenges of orientation left lozenges and the lozenges of orientation right lozenges. Since the frozen regions consist of only left and right lozenges, along any vertical line inside a frozen region only one type of lozenge can appear. Thus, the frozen regions that appear can be characterized by specifying a horizontal cross section, which can be done via a sequence of the letters R and L standing for right and left lozenges, respectively. It is obvious that the top frozen region has only left lozenges, whereas the bottom frozen region has only right lozenges.

First, let us consider the case when $l = k$, i.e. when $\beta_1, \ldots, \beta_k$ are distinct, and let $\tilde{\beta}_j = \beta_{i_j}$ for $j = 1, \ldots, k$. From the discussion in Sect. 4.6, it follows that at the top turning point, the first $i_1 - 1$ slices will have no horizontal lozenges, the next k slices one horizontal lozenge each, the next k slices two each, etc. It follows that the second frozen region from the top will have profile $(L^{k-1}R)L^{i_1-1}$, where $*^m$ stands for $*$ repeated m times, and $(*)$ stands for $*$ repeated indefinitely.

At the second turning point, the first $i_2 - 1$ slices will have no horizontal lozenges, the next k slices one horizontal lozenge each, the next k slices two each, etc. It follows that the third frozen region from the top will have profile $(L^{k-|i_1-i_2|-1}RL^{|i_1-i_2|-1}R)L^{\min(i_1,i_2)-1}$. It can inductively be shown that after "passing through" a turning point, exactly one of the L's in the repeating pattern of the frozen region changes to an R.

Thus, all frozen regions near the right boundary are vertical facets. If we vertically project them to the floor and introduce polar coordinates so that the projection of the top facet (L^k) has angle 0 and the projection of the bottom facet (R^k) has angle $\pi/2$, then the ith facet from the top will project to angle $\frac{(i-1)\pi}{2k}$ and, ignoring the first few slices, will be formed by a sequence of $i - 1$ R's and $k - i + 1$ L's repeated periodically. The particular pattern of L's and R's will depend on the order of $\beta_1, \ldots, \beta_k$ and can be arbitrary.

For example, for $k = 4$, ignoring the first few slices, the frozen regions from top to bottom will have profiles

$$(L^4), (L^3R), (L^2R^2), (LR^3), (R^4),$$

$$\text{or} \qquad (11)$$

$$(L^4), (L^3R), (LRLR), (LR^3), (R^4),$$

depending on the order of the β's (see Fig. 15).

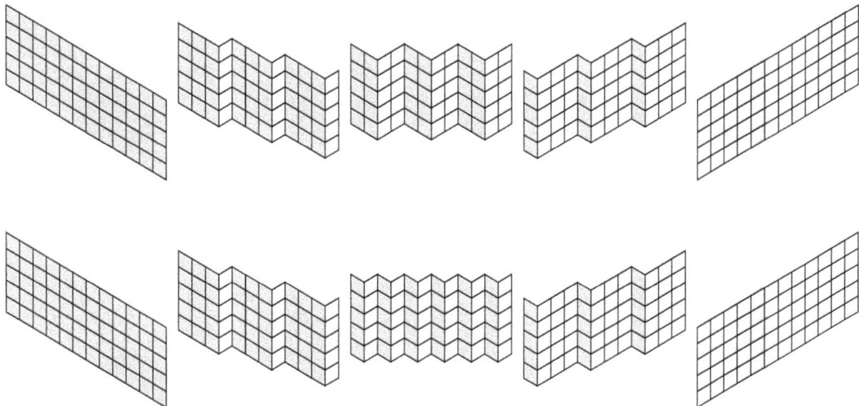

Fig. 15 The frozen regions corresponding to the profiles (11)

The $l < k$ case is similar, with the only difference being that some facets are missing. For example, if $k = 4$ and $\beta_1 = \beta_2 > \beta_3 > \beta_4$, then there will only be 3 turning points and the (L^3R) facet will be missing as the top two turning points will have merged.

An exactly sampled plane partition with 3-periodic weights is shown in Fig. 16.

5 Point Processes at the First-Order Phase Transition

It can be seen from Fig. 16 that the system develops a first-order phase transition at the slice $\tau = 0$: the limit surface is not differentiable along $\tau = 0$. The goal of this section is to study this phase transition and obtain the point processes along it, which are not translation invariant in the horizontal direction.

The nature of the fluctuations at $\tau = 0$ should not depend on the precise boundary. To simplify the computations, we set $V = \infty$ and restrict the derivations to the $k = 2$ case. In this case, the computations can be made explicitly. We have

$$\alpha_0 = \alpha > 1, \alpha_1 = \frac{1}{\alpha}, \gamma = \frac{1}{\alpha}, \tilde{\beta}_1 = 1, \tilde{\beta}_2 = \alpha. \tag{1}$$

Since $d = \infty$, the form of $x_m^{\pm}$ from (4) is not suitable anymore. Instead, we rewrite (3) in the form

$$x_m^+ = q^m y_m, x_{-m}^- = q^m y_{-m} = q^m y_m, \forall m \in \frac{1}{2} + \mathbb{Z}_{\geq 0}, \tag{2}$$

where y_m is as in Fig. 17.

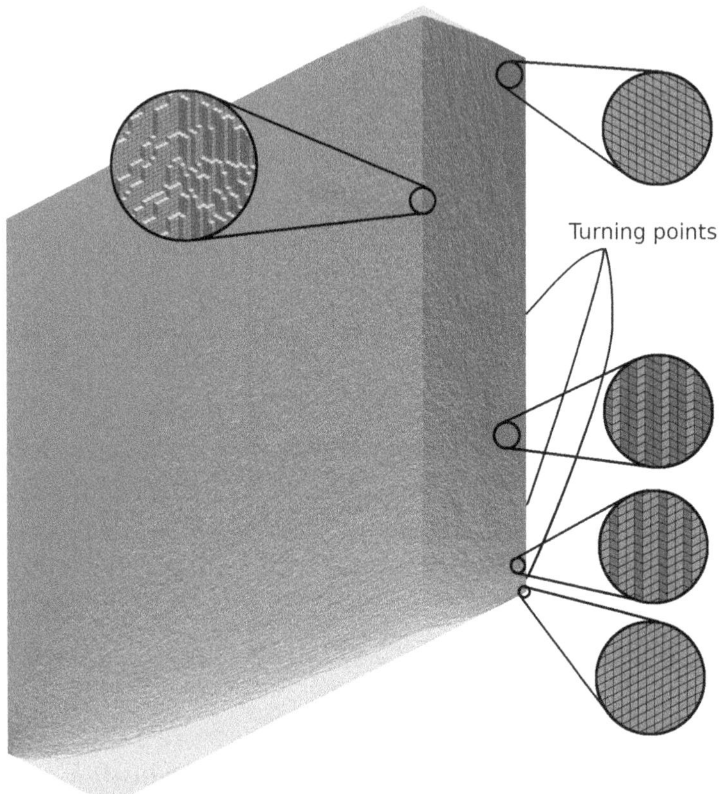

Fig. 16 An exact sample (generated using the algorithm in [Bor11]) of plane partitions with 3-periodic weights, modified at the slice through the corner. Near the right edge of the figure, the 3 turning points and the frozen regions separating them are visible

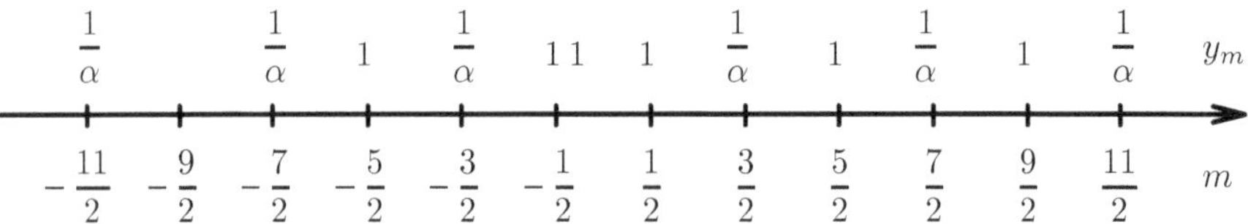

Fig. 17 The sequence y_m for $m \in \mathbb{Z} + \frac{1}{2}$

From (1) and (2) for $\tau \geq 0$, we have

$$2z\frac{dS_{\tau,\chi}}{dz}(z) = -\ln\left(\frac{z-1}{z}\right) - \ln\left(\frac{z-\frac{1}{\alpha}}{z}\right) + \ln\left(\frac{-1}{z-e^{\tau}}\right) + \ln\left(\frac{-\alpha}{z-\alpha e^{\tau}}\right) - 2\chi + \tau \tag{3}$$

and

$$2\frac{d}{dz}\left(z\frac{dS_{\tau,\chi}(z)}{dz}\right) = -\frac{1}{z-1} + \frac{2}{z} - \frac{1}{z-\alpha^{-1}} - \frac{1}{z-e^{\tau}\alpha} - \frac{1}{z-e^{\tau}}. \quad (4)$$

5.1 The Double Real Critical Points

As in Sect. 4, we first determine the points (τ, χ) for which the function $S_{\tau,\chi}$ has double real critical points. Given $z \in \mathbb{R}$, we ask for which (τ, χ) it is a double real critical point of $S_{\tau,\chi}$. If z is a double real critical point of $S_{\tau,\chi}$, then τ can be obtained from

$$\frac{d}{dz}\left(z\frac{dS_{\tau,\chi}(z)}{dz}\right) = 0, \quad (5)$$

and χ then can be easily computed from $z\frac{dS_{\tau,\chi}(z)}{dz} = 0$.

Suppose $\tau \geq 0$. Equation (5) is quadratic in e^{τ}. Solving it, we obtain

$$e^{\tau} = z^2, \text{ or } e^{\tau} = \frac{z + \alpha z - 2\alpha z^2}{2\alpha - \alpha z - \alpha^2 z}. \quad (6)$$

If

$$e^{\tau} = \frac{z + \alpha z - 2\alpha z^2}{2\alpha - \alpha z - \alpha^2 z}, \quad (7)$$

then $z > 0$, since otherwise $e^{\tau} \leq 0$. It follows from (3) that χ is real if and only if

$$1 - z^{-1} > 0,\ 1 - \alpha^{-1}z^{-1} > 0,\ 1 - e^{-\tau}\alpha^{-1}z > 0, \text{ and } 1 - e^{-\tau}z > 0,$$

or, equivalently, since $z > 0$, that

$$1 < z < e^{\tau} = \frac{z + \alpha z - 2\alpha z^2}{2\alpha - \alpha z - \alpha^2 z},$$

which, using (7), becomes

$$1 < z \text{ and } 1 < \frac{1 + \alpha - 2\alpha z}{\alpha(2 - z - \alpha z)}.$$

Since $\alpha > 1$, we have $1 < \alpha z$, which implies the denominator in the formula above is negative. Rewriting the second inequality, we obtain $(\alpha - 1)(1 - \alpha z) > 0$, which is a contradiction. Hence, if (τ, χ) is a double real critical point and $\tau \geq 0$, we must have $e^{\tau} = z^2$. Similarly, it can be established that when $\tau < 0$, we still must have

$e^{\tau} = z^2$. Thus, at the real critical points, we have $z = \pm e^{\frac{\tau}{2}}$. Solving $\frac{dS_{\tau,\chi}}{dz}(\pm e^{\frac{\tau}{2}}) = 0$ for χ we obtain that independently of the sign of τ we have

$$\chi_{\pm}(\tau) = -\ln(1 \mp e^{-\frac{|\tau|}{2}}) - \ln(1 \mp \alpha^{-1} e^{-\frac{|\tau|}{2}}) - \frac{1}{2}|\tau|. \tag{8}$$

Figure 18 gives a plot of the curve $\chi_{\pm}(\tau)$. Note that the curve $\chi_{\pm}(\tau)$ near $\tau = 0$ is not differentiable. We have

$$\frac{d\chi}{d\tau} = \frac{\mp \frac{\text{sign}(\tau)}{2} e^{-\frac{|\tau|}{2}}}{1 \mp e^{-\frac{|\tau|}{2}}} + \frac{\mp \frac{\text{sign}(\tau)}{2} \alpha^{-1} e^{-\frac{|\tau|}{2}}}{1 \mp \alpha^{-1} e^{-\frac{|\tau|}{2}}} - \frac{\text{sign}(\tau)}{2}.$$

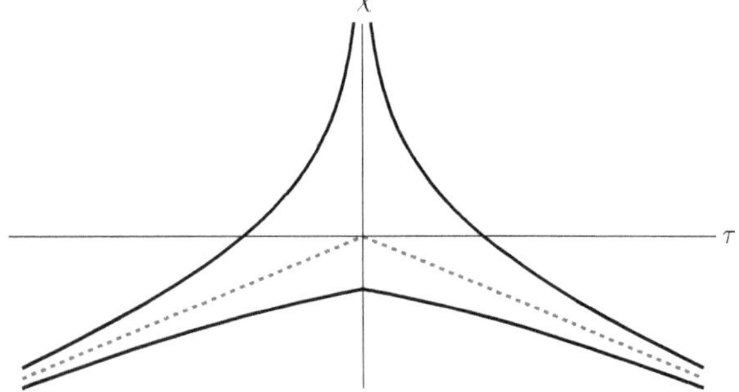

Fig. 18 A plot of the curve $\chi_{\pm}(\tau)$ when $a = 9$

If $z = e^{\frac{\tau}{2}}$, we have

$$\lim_{\tau \to 0\pm} \frac{d\chi}{d\tau} = \mp\infty.$$

If $z = -e^{\frac{\tau}{2}}$, we have

$$\lim_{\tau \to 0\pm} \frac{d\chi}{d\tau} = \pm \frac{1-\alpha}{4(1+\alpha)},$$

which implies that the frozen boundary is not differentiable at $\tau = 0$.

5.2 The Number of Critical Points

To carry out the steepest descent analysis, we need to know the number of complex non-real critical points of $S_{\tau,\chi}$. To obtain the number, we exponentiate the equation $2z\frac{dS_{\tau,\chi}(z)}{dz} = 0$ to obtain

$$\left(z - e^{\min\{\tau,0\}}\right)\left(z - e^{\min\{\tau,0\}}\alpha^{-1}\right)\left(1 - e^{-\max\{\tau,0\}}\alpha^{-1}z\right)\left(1 - e^{-\max\{\tau,0\}}z\right) - e^{-2\left(\chi - \frac{1}{2}V(\tau)\right)}z^2 = 0. \quad (9)$$

The left-hand side is a degree 4 polynomial with positive leading coefficient. Since its value is negative at $e^{\min\{\tau,0\}}$, the equation must have at least two distinct real solutions. Hence, the number of non-real complex solutions is at most 2.

If we divide the (τ, χ) plane into regions according to the number of critical points, $S_{\tau,\chi}$ will have double real critical points at the boundaries of those regions. Thus, the boundary is given by the curve $\chi_{\pm}(\tau)$ from (8). It is easy to see that for any $\tau \neq 0$, if χ is very large, then Eq. (9) has four distinct real solutions, whence $S_{\tau,\chi}$ has no non-real critical points. It follows that in the region bounded by the curve $\chi_{\pm}(\tau)$, the function $S_{\tau,\chi}$ has a pair of complex conjugate critical points, whereas in the region outside the curve all critical points are real.

Carrying out the saddle point analysis is straightforward. When $\tau \neq 0$, we obtain that the limiting correlation functions at any point in the interior of the curve $\chi_{\pm}(\tau)$ are of the same type as in Theorem 4.2 and outside the curve we have frozen regions.

Let $\tau = 0$ and $\chi > \chi_{-}(\tau)$. Then, $S_{\tau,\chi}$ has exactly one pair of complex conjugate critical points. The deformation of contours and the saddle point analysis work exactly as in Theorem 4.2. The only difference is in the residue terms picked up when the z and w contours cross each other. We have t_1, t_2 fixed, $\Delta t = t_1 - t_2$ and $rh_i \to \chi$ with $\Delta h = h_1 - h_2$ fixed as before. We obtain

$$\lim_{r\to 0} K_{\bar{q}}((t_1, h_1), (t_2, h_2)) = \lim_{r\to 0} \frac{1}{2\pi \mathrm{i}} \int_C \frac{\Phi_{\bar{q}}(z, t_1)}{\Phi_{\bar{q}}(z, t_2)} z^{-\Delta h - \frac{1}{2}|t_1| + \frac{1}{2}|t_2| - 1} dz,$$

where the contour C is as before. From (2), it follows that if $|m| < \max(|t_1|, |t_2|)$, then $x_m^{\pm} \to y_m$ as $r \to 0$. Given integers t_1 and t_2, let $A(t_1, t_2)$ be the number of half integers m between t_1 and t_2 such that $y_m = 1$, taken with a positive sign if $t_1 \geq t_2$ and with a negative sign otherwise. Define $g(t_1, t_2)$ by

$$g(t_1, t_2) = \begin{cases} 0, & t_1, t_2 > 0, \\ \Delta t, & t_1, t_2 < 0, \\ t_2, & t_2 < 0 < t_1, \\ t_1, & t_1 < 0 < t_2. \end{cases}$$

We obtain

$$\frac{\Phi_{\bar{q}}(z,t_1)}{\Phi_{\bar{q}}(z,t_2)} z^{-\Delta h-\frac{1}{2}|t_1|+\frac{1}{2}|t_2|-1} \to (-1)^{g(t_1,t_2)}(1-z)^{A(t_1,t_2)}(1-z/\alpha)^{\Delta t-A(t_1,t_2)} z^{-\Delta h-\frac{\Delta t}{2}-1}.$$

When taking determinants, the terms $(-1)^{g(t_1,t_2)}$ cancel out and we obtain the following result.

Theorem 5.1 *Let $r > 0$. Consider plane partitions confined to an $\infty \times \infty$ box, with periodic weights given by* (1),(1) *with $q = e^{-r}$. In the limit when r goes to 0 along the middle slice $\tau = 0$ when $\chi > \chi_-(0)$, the correlation functions of the system (see* (5)*) are given by determinants of the kernel*

$$\lim_{r\to 0} K_{\bar{q}}((t_1,h_1),(t_2,h_2)) = \frac{1}{2\pi \mathrm{i}} \int_C (1-z)^{A(t_1,t_2)}(1-z/\alpha)^{\Delta t-A(t_1,t_2)} z^{-\Delta h-\frac{1}{2}\Delta t-1} dz,$$

where $\Delta t = t_1 - t_2$, $\Delta h = h_1 - h_2$, t_1, t_2 are fixed, $rh_i \to \chi$ with Δh fixed, and the integration contour C connects the two non-real critical points of $S_{0,\chi}(z)$, passing through the real line in the interval $(1,\alpha)$ if $\Delta(t) \geq 0$ and through $(-\infty, 0)$ otherwise.

Remark 5.2 Note that this is a point process that is invariant under translations in the vertical direction but not in the horizontal direction. This is the case because the sequence y_m is not translation invariant (see Fig. 17). When $\alpha = 1$, the process becomes the incomplete beta process from [OR03].

Acknowledgments I would like to thank the referee for a careful reading of the manuscript and in particular for bringing to my attention the work of Borodin and Shlosman [BS10].

The work was partially supported by the Simons Foundation Collaboration Grant No. 422190.

References

[BD19] T. Berggren and M. Duits. Correlation functions for determinantal processes defined by infinite block Toeplitz minors. *arXiv e-prints*, page arXiv:1901.10877, Jan 2019.

[BMRT12] Cedric Boutillier, Sevak Mkrtchyan, Nicolai Reshetikhin, and Peter Tingley. Random skew plane partitions with a piecewise periodic back wall. *Ann. Henri Poincaré*, 13(2):271–296, 2012.

[Bor07] Alexei Borodin. Periodic Schur process and cylindric partitions. *Duke Math. J.*, 140(3):391–468, 2007.

[Bor11] Alexei Borodin. Schur dynamics of the Schur processes. *Adv. Math.*, 228(4):2268–2291, 2011.

[BS10] Alexei Borodin and Senya Shlosman. Gibbs Ensembles of Nonintersecting Paths. *Communications in Mathematical Physics*, 293(1):145–170, Jan 2010.

[CJ16] Sunil Chhita and Kurt Johansson. Domino statistics of the two-periodic Aztec diamond. *Advances in Mathematics*, 294:37–149, 2016.

[Cue18] Cesar Cuenca. Universal behavior of the corners of Orbital Beta Processes. *arXiv e-prints*, page arXiv:1807.06134, Jul 2018.

[GP15] Vadim Gorin and Greta Panova. Asymptotics of symmetric polynomials with applications to statistical mechanics and representation theory. *Ann. Probab.*, 43(6):3052–3132, 11 2015.

[JN06] Kurt Johansson and Eric Nordenstam. Eigenvalues of GUE minors. *Electron. J. Probab.*, 11:no. 50, 1342–1371, 2006.

[Ken09] Richard Kenyon. Lectures on dimers. *Statistical Mechanics*, 16, 10 2009.

[KO07] Richard Kenyon and Andrei Okounkov. Limit shapes and the complex Burgers equation. *Acta Math.*, 199(2):263–302, 2007.

[KOS06] Richard Kenyon, Andrei Okounkov, and Scott Sheffield. Dimers and amoebae. *Ann. of Math. (2)*, 163(3):1019–1056, 2006.

[Mkr11] Sevak Mkrtchyan. Scaling limits of random skew plane partitions with arbitrarily sloped back walls. *Comm. Math. Phys.*, 305(3):711–739, 2011.

[Mkr14] Sevak Mkrtchyan. Plane partitions with two-periodic weights. *Letters in Mathematical Physics*, 104(9):1053–1078, 2014.

[MP17] Sevak Mkrtchyan and Leonid Petrov. Gue corners limit of q-distributed lozenge tilings. *Electron. J. Probab.*, 2017. accepted. arXiv:1703.07503.

[NHB84] Bernard Nienhuis, Hendrik Jan Hilhorst, and Henk W. J. Blöte. Triangular SOS models and cubic-crystal shapes. *J. Phys. A*, 17(18):3559–3581, 1984.

[Nov15] Jonathan Novak. Lozenge tilings and hurwitz numbers. *Journal of Statistical Physics*, 161(2):509–517, Oct 2015.

[OR03] Andrei Okounkov and Nikolai Reshetikhin. Correlation function of Schur process with application to local geometry of a random 3-dimensional Young diagram. *J. Amer. Math. Soc.*, 16(3):581–603 (electronic), 2003.

[OR06] Andrei Okounkov and Nicolai Reshetikhin. The birth of a random matrix. *Mosc. Math. J.*, 6(3):553–566, 588, 2006.

[OR07] Andrei Okounkov and Nicolai Reshetikhin. Random skew plane partitions and the Pearcey process. *Comm. Math. Phys.*, 269(3):571–609, 2007.

[She05] Scott Sheffield. Random surfaces. *Astérisque*, (304):vi+175, 2005.

The Skein Category of the Annulus

K. Al Qasimi and J. V. Stokman

Dedicated to Nicolai Reshetikhin on the occasion of his 60th birthday

Abstract We construct the skein category $\mathcal{S}$ of the annulus and show that it is equivalent to the affine Temperley-Lieb category of Graham and Lehrer. It leads to a skein theoretic description of the extended affine Temperley-Lieb algebras. We construct an endofunctor of $\mathcal{S}$ that corresponds, on the level of tangle diagrams, to the insertion of an arc connecting the inner and outer boundary of the annulus. We use it to define and construct towers of extended affine Temperley-Lieb algebra modules. It allows us to construct a tower of modules acting on spaces of link patterns on the punctured disc which play an important role in the study of loop models. In case of trivial Dehn twist we show that the direct sum of the representation spaces of the link pattern tower defines a graded algebra that may be regarded as a relative version of the Roger-Yang skein algebra of arcs and links on the punctured disc. We also describe the link pattern tower in terms of fused extended affine Temperley-Lieb algebra modules.

1 Introduction

In [16, 17] Kauffman constructed a knot invariant based on an elementary combinatorial rule for eliminating crossings in the associated knot diagram. In doing so he introduced the two skein relations, the Kauffman skein relation (3.1) and the loop

K. Al Qasimi (✉)
Department of Mathematics and Statistics, American University of Sharjah, Sharjah, United Arab Emirates
e-mail: kalqasemi@aus.edu

J. V. Stokman
KdV Institute for Mathematics, University of Amsterdam, Amsterdam, The Netherlands
e-mail: j.v.stokman@uva.nl

A. Alekseev et al. (eds.), *Representation Theory, Mathematical Physics, and Integrable Systems*, Progress in Mathematics 340,
https://doi.org/10.1007/978-3-030-78148-4_18

removal relation (3.2). It has led to skein theory, the study of knots and links in 3-manifolds modulo the Kauffman skein relation and the loop removal relation, see, e.g., [1, 20, 22, 26]. The current paper deals with the 3-manifold $A \times [0, 1]$, with A the annulus in the complex plane.

We introduce and study the skein category $\mathcal{S}$ of the annulus A. It is the linear category with objects the nonnegative integers and morphisms $\mathrm{Hom}_{\mathcal{S}}(m, n)$ the linear skein of the annulus with m marked points on the inner boundary and n marked points on the outer boundary. In other words, $\mathrm{Hom}_{\mathcal{S}}(m, n)$ consists of the ambient isotopy classes of (m, n)-tangle diagrams on the annulus modulo the equivalence relation generated by the Kauffman skein relation and the loop removal relation. The skein category $\mathcal{S}$ is in fact a strict monoidal category with the tensor product obtained from the relative Kauffman bracket skein product introduced in Przytycki and Sikora [24].

Following closely Przytycki [23, §3], we construct a relative version of the Kauffman bracket to prove that the skein category $\mathcal{S}$ is equivalent to Graham and Lehrer's [11] affine Temperley-Lieb category. As a consequence it follows that the endomorphism algebra $\mathrm{End}_{\mathcal{S}}(n) := \mathrm{Hom}_{\mathcal{S}}(n, n)$ is isomorphic to Green's [13] n-affine diagram algebra, also known as the (extended) affine Temperley-Lieb algebra.

We will define an endofunctor $\mathcal{I}$ of $\mathcal{S}$ called the arc insertion functor, which on the level of morphisms inserts a new arc connecting the inner and outer boundary of the annulus in a particular way while undercrossing all arcs it meets along the way. On the level of endomorphisms it provides a tower of algebras

$$\mathrm{End}_{\mathcal{S}}(0) \xrightarrow{\mathcal{I}_0} \mathrm{End}_{\mathcal{S}}(1) \xrightarrow{\mathcal{I}_1} \mathrm{End}_{\mathcal{S}}(2) \xrightarrow{\mathcal{I}_2} \cdots$$

with connecting maps $\mathcal{I}_n$ the algebra maps $\mathcal{I}|_{\mathrm{End}_{\mathcal{S}}(n)} : \mathrm{End}_{\mathcal{S}}(n) \to \mathrm{End}_{\mathcal{S}}(n+1)$. This tower was considered before in the context of knot theory [2] and in the context of fusion of extended affine Temperley-Lieb algebra modules [8], respectively. It differs from the arc-tower from, e.g., [4, 8], which is defined with respect to the two-step algebra embedding $\mathrm{End}_{\mathcal{S}}(n) \to \mathrm{End}_{\mathcal{S}}(n+2)$ that corresponds to the identification of an idempotent subalgebra of $\mathrm{End}_{\mathcal{S}}(n+2)$ with $\mathrm{End}_{\mathcal{S}}(n)$.

We introduce and study towers

$$V_0 \xrightarrow{\phi_0} V_1 \xrightarrow{\phi_1} V_2 \xrightarrow{\phi_2} V_3 \xrightarrow{\phi_3} \cdots$$

of extended affine Temperley-Lieb algebra modules. These are chains of left $\mathrm{End}_{\mathcal{S}}(n)$-modules V_n ($n \in \mathbb{Z}_{\geq 0}$) connected by morphisms $\phi_n : V_n \to \mathrm{Res}^{\mathcal{I}_n}(V_{n+1})$ of $\mathrm{End}_{\mathcal{S}}(n)$-modules, where $\mathrm{Res}^{\mathcal{I}_n}(V_{n+1})$ is the $\mathrm{End}_{\mathcal{S}}(n+1)$-module V_{n+1} viewed as $\mathrm{End}_{\mathcal{S}}(n)$-module via the algebra map $\mathcal{I}_n : \mathrm{End}_{\mathcal{S}}(n) \to \mathrm{End}_{\mathcal{S}}(n+1)$.

Our motivation for studying such towers stems from integrable models in statistical physics with extended affine Temperley-Lieb algebra symmetry. Examples are inhomogeneous dense loop models and inhomogeneous XXZ spin-$\frac{1}{2}$ chains with quasi-periodic boundary conditions, see, e.g., [5, 10, 15] and the references therein. In this context the representation space V_n of the tower represents the state space of the model at system size n and the connecting maps relate the models of different system sizes.

We introduce a special tower of extended affine Temperley-Lieb algebra modules, which we will call the *link pattern tower*. It depends on a free parameter v, called the *twist weight* of the tower. For even n the representation space is spanned by ambient isotopy classes of $(0, n)$-tangle diagrams in A without crossings and without loops, connecting n marked points on the outer boundary of A. For odd n the tangle diagrams include a defect line connecting the outer boundary to the inner boundary, and we add the rule that Dehn twists of the defect line may be removed by v (see (8.3)). The $\mathrm{End}_{\mathcal{S}}(n)$-action is described as follows. The skein class of an (n, n)-tangle diagram on the annulus acts on a diagram $D \in V_n$ by placing D inside the (n, n)-tangle diagram, removing crossings and contractible loops by the skein relations, and removing noncontractible loops by a particular weight factor depending on v (see (8.2)).

For even n the connecting maps $\phi_n : V_n \to V_{n+1}$ of the link pattern tower correspond, from the skein theoretic perspective, to the insertion of a defect line, undercrossing all arcs it meets along the way. These maps were considered before in [5] in the study of the inhomogeneous dense loop model on the half-infinite cylinder. The connecting maps $\phi_n : V_n \to V_{n+1}$ for odd n are more subtle. From a skein theoretic perspective they can be described as follows. The connecting map ϕ_n acts on a diagram by detaching the defect line from the inner boundary and reconnecting it to the outer boundary in two different ways, either encircling the hole of the annulus before reattaching it to the outer boundary, or not. These two contributions are given explicit weights depending on the twist weight v and on the Temperley-Lieb algebra parameter, see Theorem 8.3.

We show that the link pattern tower is nondegenerate for generic parameter values, in the sense that the induced morphisms $\widehat{\phi}_n : \mathrm{Ind}^{\mathcal{I}_n}(V_n) \to V_{n+1}$ of $\mathrm{End}_{\mathcal{S}}(n+1)$-modules are surjective, where $\mathrm{Ind}^{\mathcal{I}_n}(V_n)$ is the $\mathrm{End}_{\mathcal{S}}(n+1)$-module obtained by inducing V_n along the algebra map $\mathcal{I}_n : \mathrm{End}_{\mathcal{S}}(n) \to \mathrm{End}_{\mathcal{S}}(n+1)$. We relate the link pattern tower to the recently introduced fusion [8] of extended affine Temperley-Lieb algebra modules. We construct for each $n \in \mathbb{Z}_{\geq 0}$ a fused $\mathrm{End}_{\mathcal{S}}(n+1)$-module W_{n+1} and a morphism $\psi_n : W_{n+1} \to V_{n+1}$ of $\mathrm{End}_{\mathcal{S}}(n+1)$-modules such that $\widehat{\phi}_n$ factorizes through ψ_n. The $\mathrm{End}_{\mathcal{S}}(n+1)$-module W_{n+1} is obtained by fusing the $\mathrm{End}_{\mathcal{S}}(n)$-module V_n with a one-dimensional $\mathrm{End}_{\mathcal{S}}(1)$-module.

For twist weight $v = 1$ the representation spaces of the link pattern tower may be naturally identified with spaces of link patterns on the punctured disc $\mathbb{D}^* = \mathbb{D} \setminus \{0\}$ by shrinking the hole of the annulus to a point. The resulting modules play an important role in the description of the inhomogeneous dense loop models on the half-infinite cylinder (see, e.g., [5, 15]). We show that in this case the direct sum of the representation spaces of the link pattern tower is a graded algebra, and as such may be viewed as a relative version of Roger's and Yang's [25, Def. 2.3] skein algebra of arcs and links on the punctured disc $\mathbb{D}^*$. In this skein algebra perspective multiple endpoints of arcs in $\mathbb{D}^* \times [0, 1]$ may connect to the pole $\{0\} \times [0, 1]$ but each line segment $\{\xi\} \times [0, 1]$ above the marked points ξ on the outer boundary of $\mathbb{D}^*$ is met by only one endpoint. The number of endpoints on $\partial\mathbb{D} \times [0, 1]$ is the grading of the associated element in the algebra. In this identification our connecting maps ϕ_n for n odd relate to the *puncture-skein relation* in [25], which is the skein theoretic reduction rule when multiple arcs connect to the center pole $\{0\} \times [0, 1]$. In fact, the connecting map ϕ_n, for each $n \in \mathbb{Z}_{\geq 0}$, just becomes right multiplication by the class of the identity of $\mathrm{End}_{\mathcal{S}}(1)$ on the nth graded piece of the algebra.

In our future work [3] we use the link pattern tower to construct a tower of solutions to quantum Knizhnik-Zamolodchikov (qKZ) equations. The tower consists of V_n-valued solutions of qKZ equations ($n \in \mathbb{Z}_{\geq 0}$) which are compatible with respect to the connecting maps ϕ_n. At the stochastic/combinatorial value of the extended affine Temperley-Lieb parameter, the V_n-valued solution in the tower reduces to the ground state of the dense loop model on the half-infinite cylinder with perimeter n. In that case the tower structure gives explicit recursion relations of the ground states with respect to the system size, leading to a refinement of the results in [5].

The structure of the paper is as follows. In Sects. 2 and 3 we define the category of tangle diagrams, $\mathcal{T}$ and the skein category of the annulus, $\mathcal{S}$, respectively. We explain in Sect. 3 that $\mathcal{S}$ is a monoidal category, with the tensor product obtained from the relative Kauffman bracket skein product from [24]. In Sect. 4 we introduce Graham's and Lehrer's [11] affine Temperley-Lieb category $\mathcal{TL}$, whose morphisms are defined in terms of affine diagrams, and we show that the affine Temperley-Lieb category is equivalent to $\mathcal{S}$. In Sect. 5 we define the extended affine Temperley-Lieb algebra, TL_n, algebraically and we recall Green's [13] result that TL_n is equivalent to the endomorphism algebra $\mathrm{End}_{\mathcal{TL}}(n)$ of $\mathcal{TL}$. Combined with the result from Sect. 4 it leads to three different realizations of the extended affine Temperley-Lieb algebra (skein theoretic, combinatorial and algebraic). In Sect. 6 we define the arc insertion functor $\mathcal{I} : \mathcal{S} \to \mathcal{S}$. In Sect. 7 we introduce the notion of towers of extended affine Temperley-Lieb algebra modules. In Sect. 8 we construct the link pattern tower and we explain how it gives rise to a relative version of the Roger-Yang [25] skein algebra on the punctured disc. We show in Sect. 9 how the link pattern tower is related to fusion. Finally in the appendix we discuss how the resulting tower of extended affine Temperley-Lieb algebras lifts to extended affine braid groups and extended affine Hecke algebras, and we discuss a B-type presentation of the extended affine Temperley-Lieb algebra.

2 The Category of Tangle Diagrams

Consider the three-manifold $\Sigma := A \times [0, 1]$ with A the annulus

$$A := \{z \in \mathbb{C} \mid 1 \leq |z| \leq 2\}$$

in the complex plane. We think of Σ as a thickened cylinder in $\mathbb{R}^3$, as depicted below.

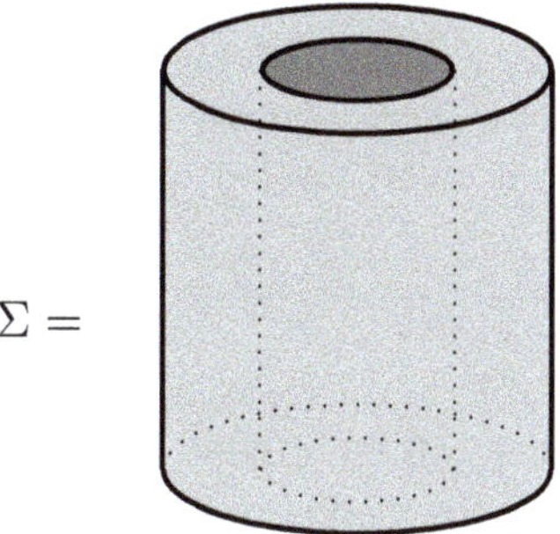

Write $\partial A = C_i \cup C_o$ for the boundary of A with $C_i := S^1 = \{z \in \mathbb{C} \mid |z| = 1\}$ and $C_o = \{z \in \mathbb{C} \mid |z| = 2\}$ (the indices "i" and "o" stand for inner and outer, respectively). Give A the counterclockwise orientation.

Set $\zeta_n := \exp(2\pi i/n)$ for $n \in \mathbb{Z}_{>0}$. Let $m, n \in \mathbb{Z}_{\geq 0}$ with $m+n$ even. A (framed) (m, n)-tangle T in Σ is a disjoint union of smooth framed loops and $\frac{1}{2}(m+n)$ framed arcs in Σ satisfying:

(a) The loops are in the interior of Σ.
(b) The $m+n$ marked points $(2\xi_m^{j-1}, 1)$ $(1 \leq j \leq m)$ and $(2\xi_n^{i-1}, 0)$ $(1 \leq i \leq n)$, framed along $\partial A \times \{1\}$ and $\partial A \times \{0\}$ with the orientation induced from A, are the endpoints of the framed arcs.

Let proj : $\Sigma \to A$ be the map obtained by projecting radially on the outer wall $C_o \times [0, 1]$ of Σ and identifying $C_o \times [0, 1] \simeq A$ by collapsing the wall $C_o \times [0, 1]$ inwards onto the floor $A \times \{0\}$ of Σ. The projection $D := \text{proj}(T)$ of an (m, n)-tangle T in general position with respect to proj, together with the crossing data at the crossing points in the diagram, is called an (m, n)-tangle diagram in A.

If we draw a picture of an (m, n)-tangle diagram, then we label the inner points ξ_m^{i-1} on the diagram by i $(i = 1, \ldots, m)$ and the outer points $2\xi_n^{j-1}$ by j $(j = 1, \ldots, n)$. An example of a tangle diagram is given in Fig. 1.

We say that two (m, n)-tangle diagrams D and D' in A are equivalent if we can transform D to D' by a planar isotopy of the annulus that fixes the boundary. That

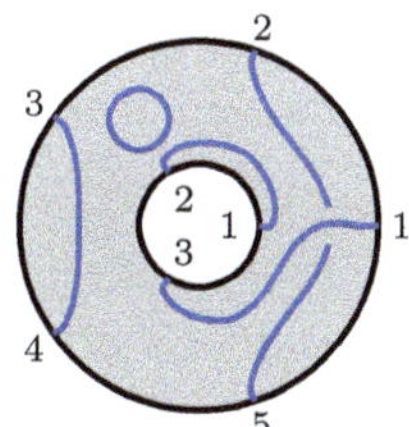

Fig. 1 An example of a (3,5)-tangle diagram in A

is, there exists a smooth ambient isotopy $h : A \times [0, 1] \to A$ fixing ∂A pointwise, satisfying $h(D, 1) = D'$ and respecting the crossing data. If D is an (m, n)-tangle diagram we write $\overline{D} \in \mathrm{Hom}_{\mathcal{T}}(m, n)$ for its equivalence class.

Definition 2.1 The category $\mathcal{T}$ of tangle diagrams in A is the category with objects $\mathbb{Z}_{\geq 0}$ and morphisms $\mathrm{Hom}_{\mathcal{T}}(m, n)$ the equivalence classes of (m, n)-tangle diagrams in A if $m + n$ is even, and the empty set if $m + n$ is odd. The composition map

$$\mathrm{Hom}_{\mathcal{T}}(k, m) \times \mathrm{Hom}_{\mathcal{T}}(m, n) \to \mathrm{Hom}_{\mathcal{T}}(k, n), \qquad (\overline{D}, \overline{D'}) \mapsto \overline{D'} \circ \overline{D}$$

is defined as follows: $\overline{D'} \circ \overline{D} := \overline{D' \circ D}$ with $D' \circ D$ the (k, n)-tangle diagram obtained by rescaling D to $\{z \in \mathbb{C} \mid 1 \leq |z| \leq \frac{3}{2}\}$, D' to $\{z \in \mathbb{C} \mid \frac{3}{2} \leq |z| \leq 2\}$ and placing D inside D'. The identity morphism $\mathrm{Id}_n \in \mathrm{End}_{\mathcal{T}}(n)$ is the equivalence class of the tangle diagram with straight line arcs from ξ_n^{j-1} to $2\xi_n^{j-1}$ for $j = 1, \ldots, n$ and no loops (it is the empty diagram for $n = 0$).

An example of the composition of two tangle diagrams is given in (2.1).

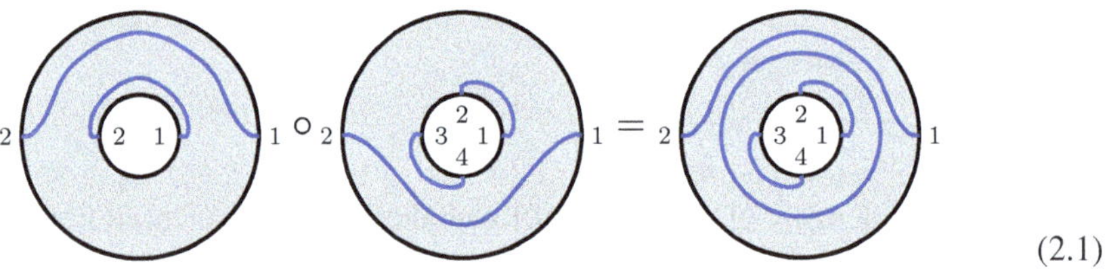

(2.1)

3 The Skein Category of the Annulus

It is well known that skein modules on the strip $\mathbb{R} \times [0, 1]$ form the morphisms of a strict monoidal, linear category called the skein category, see, e.g., [27, Chpt. XII]. In this section we extend this result to skein modules on the annulus.

Write $\mathbb{C}[\mathrm{Hom}_{\mathcal{T}}(m, n)]$ for the complex vector space with linear basis the equivalence classes of (m, n)-tangle diagrams in A. We take it to be $\{0\}$ if $m + n$ is odd. Extend the category $\mathcal{T}$ of tangle diagrams in A to a linear category $\mathrm{Lin}(\mathcal{T})$ with objects $\mathbb{Z}_{\geq 0}$, morphisms $\mathrm{Hom}_{\mathrm{Lin}(\mathcal{T})}(m, n) := \mathbb{C}[\mathrm{Hom}_{\mathcal{T}}(m, n)]$, and composition map the complex bilinear extension of the composition map of $\mathcal{T}$. The skein category on the annulus is now defined as the quotient category obtained from $\mathrm{Lin}(\mathcal{T})$ by modding out the *Kauffman skein relations* [16, 17]:

Definition 3.1 Let $t^{\frac{1}{4}}$ be a nonzero complex number. The skein category $\mathcal{S} = \mathcal{S}(t^{\frac{1}{4}})$ of the annulus A is the quotient of $\mathrm{Lin}(\mathcal{T})$ by the equivalence relation obtained by taking the linear and transitive closure of the following local relations on tangle diagrams:

a. The *Kauffman skein relation* $\overline{D} \sim t^{\frac{1}{4}}\overline{D'} + t^{-\frac{1}{4}}\overline{D''}$ with D, D', D'' three tangle diagrams that are identical except in a small open disc in A where they are as shown

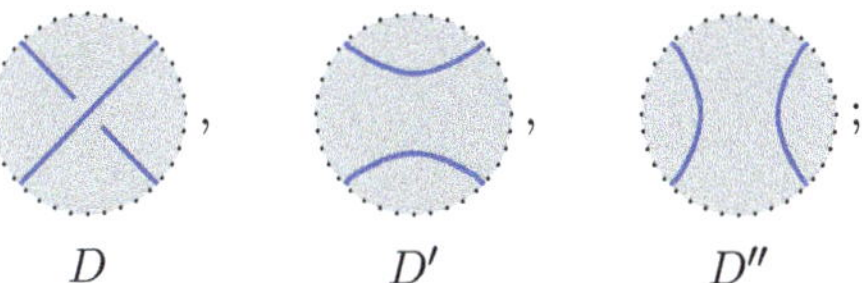

b. The *loop removal relation* $\overline{D} \sim -(t^{\frac{1}{2}} + t^{-\frac{1}{2}})\overline{D'}$ with D, D' two tangle diagrams that are identical except in a small open disc in A where they are as shown

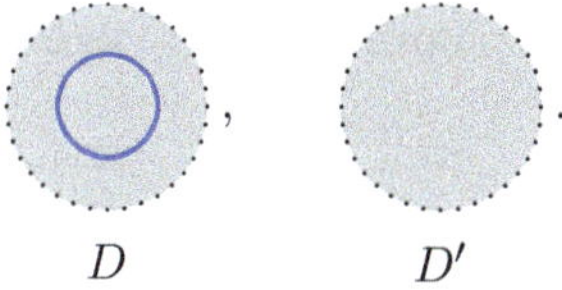

Note that if $m+n$ is odd, then $\mathrm{Hom}_{\mathcal{S}}(m,n) = \{0\}$. If D is a tangle diagram in A, then we will write $[D]$ for the corresponding element in $\mathrm{Hom}_{\mathcal{S}}(m,n)$. We write $\mathbf{1}_n = [\mathrm{Id}_n] \in \mathrm{End}_{\mathcal{S}}(n)$ for the identity morphism ($n \in \mathbb{Z}_{\geq 0}$).

As is customary in skein theory, we write the Kauffman skein relation in $\mathrm{Hom}_{\mathcal{S}}(m,n)$ as

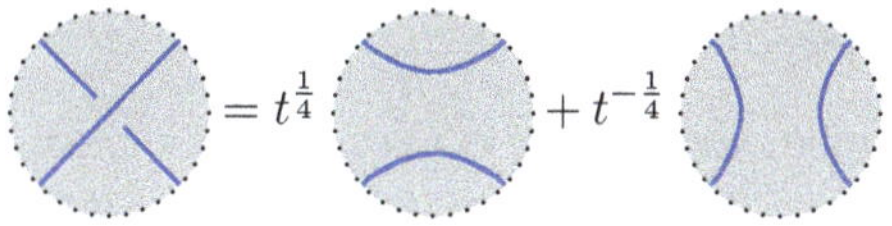

$$= t^{\frac{1}{4}} \qquad + t^{-\frac{1}{4}} \tag{3.1}$$

and the loop removal relation in the skein module $\mathrm{Hom}_{\mathcal{S}}(m,n)$ as

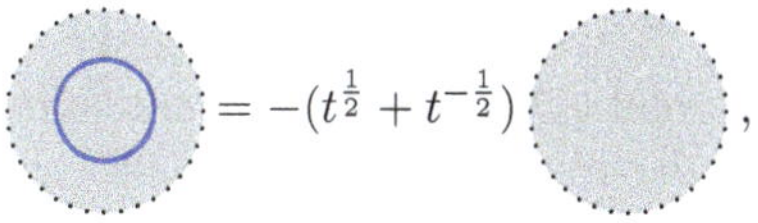

$$= -(t^{\frac{1}{2}} + t^{-\frac{1}{2}}) \qquad , \tag{3.2}$$

with the disc showing the local neighborhood in A where the tangle diagrams differ. We will also write down identities in skein modules by depicting both sides of the equation as linear combinations of the tangle diagrams D representing $[D]$.

Remark 3.2 The important observation, due to Kauffman [16, 17], is that $[D] \in \mathrm{Hom}_{\mathcal{S}}(m,n)$ is invariant under the Reidemeister moves R1', R2 and R3 (see Fig. 2) and their mirror versions, applied to the (m,n)-tangle diagram D in A. Hence $[D]$ represents the ambient isotopy class of the associated framed (m,n)-tangle in Σ.

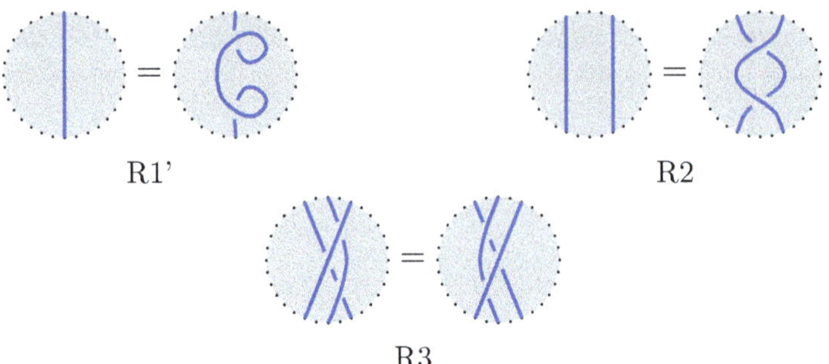

Fig. 2 Reidemeister moves

Note that the Reidemeister move R1 is only satisfied up to a scalar multiple,

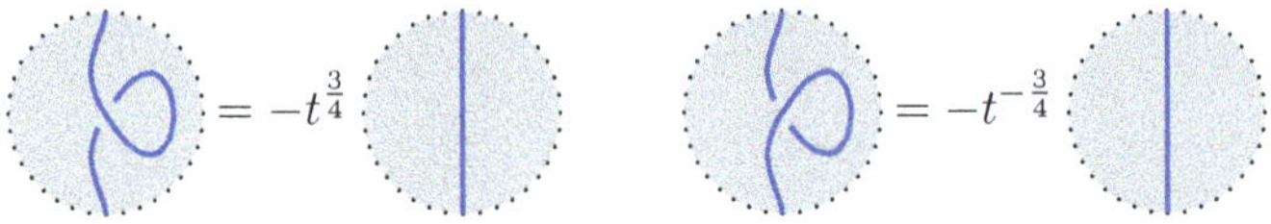

$= -t^{\frac{3}{4}}$ $= -t^{-\frac{3}{4}}$

Remark 3.3 The morphism space $\mathrm{Hom}_{\mathcal{S}}(m, n)$ can be identified with a relative Kauffman bracket skein module on the thickened cylinder Σ with (framed) marked points $(2\xi_m^{i-1}, 1)$ $(1 \leq i \leq m)$ and $(2\xi_n^{j-1}, 0)$ $(1 \leq j \leq n)$, cf. [23]. The identification goes through the projection map proj. In this 3-dimensional description of the hom-spaces the composition rule turns into the vertically stacking of the thickened cylinders.

We now show that the skein category $\mathcal{S}$ is a strict monoidal, linear category.[1] The tensor functor $\times_{\mathcal{S}} : \mathcal{S}\times\mathcal{S} \to \mathcal{S}$ on objects $m, n \in \mathbb{Z}_{\geq 0}$ is given by $m \times_{\mathcal{S}} n := m+n$. On morphisms the tensor product is defined through Przytycki's and Sikora's [24, §3] relative version of the skein algebra multiplication on the associated relative Kauffman bracket skein modules from Remark 3.3. On the level of tangles T, T' on the thickened cylinder Σ, the Kauffman bracket skein product $T \cdot T'$ amounts to placing T' inside the solid cylindrical hole of the thickened cylinder of T and moving the endpoints of the arcs to the marked points on $C_o \times \{1\}$ and $C_o \times \{0\}$ in a specific way. The exact rule regarding the repositioning of the endpoints is determined as follows.

Before putting T' inside T, fix the parametrizations $\gamma_\ell(s) := (2\exp(2\pi i s), \ell)$ $(s \in [0, 1])$ of $C_o \times \{\ell\} \subset A \times \{\ell\}$ $(\ell = 0, 1)$. Place the endpoints of the two tangles T and T' on the line segments $(1, 2]\times\{\ell\} \subset A\times\{\ell\}$ $(\ell = 0, 1)$ using an isotopy of Σ which, for $\ell \in \{0, 1\}$, stabilizes $A \times \{\ell\}$, fixes the endpoint $(2, \ell) \in \partial\Sigma$ and pushes, for $\epsilon > 0$ sufficiently small, the boundary arc $\gamma_\ell([0, 1 - \epsilon])$ into the line segment

[1] We thank an anonymous referee for this observation.

$(1, 2] \times \{\ell\}$. The skein algebra multiplication rule then produces a new tangle with endpoints on the two line segments $(1, 2] \times \{\ell\}$ ($\ell = 0, 1$), which is converted back to a tangle with endpoints on the marked points on $C_o \times \{\ell\}$ ($\ell = 0, 1$) by applying a reverse isotopy of the type as described above (see [24, §3] for further details).

Through the projection map proj the relative skein algebra multiplication rule as described in the previous paragraph gives bilinear operations

$$\mathrm{Hom}_{\mathcal{S}}(k, \ell) \times \mathrm{Hom}_{\mathcal{S}}(m, n) \xrightarrow{\times_{\mathcal{S}}} \mathrm{Hom}_{\mathcal{S}}(k+m, \ell+n), \quad ([D], [D']) \mapsto [D] \times_{\mathcal{S}} [D']$$

for $k, \ell, m, n \in \mathbb{Z}_{\geq 0}$. They are explicitly described as follows. Let D be a (k, ℓ)-tangle diagram on A and D' an (m, n)-tangle diagram on A. Then $[D] \times_{\mathcal{S}} [D'] = [D * D']$ with $D * D'$ the following $(k + m, \ell + n)$-tangle diagram.

Let $D_{\curvearrowright}$ be a diagram on A obtained from D by applying a planar isotopy of A which

1. rotates the endpoints $\xi_k^{i-1} \in C_i$ clockwise to ξ_{k+m}^{i-1} $(1 \leq i \leq k)$,
2. rotates the endpoints $2\xi_\ell^{i-1} \in C_o$ clockwise to $\xi_{\ell+n}^{i-1}$ $(1 \leq i \leq \ell)$,
3. fixes some straight line segment between the inner and outer boundary of A.

Similarly, let $D'_{\curvearrowleft}$ be the diagram on A obtained from D' by applying a planar isotopy of A which

1. rotates the endpoints $\xi_m^{i-1} \in C_i$ counterclockwise to ξ_{k+m}^{k+i-1} $(1 \leq i \leq m)$,
2. rotates the endpoints $2\xi_n^{i-1} \in C_o$ counterclockwise to $2\xi_{\ell+n}^{n+i-1}$ $(1 \leq i \leq n)$,
3. fixes some straight line segment between the inner and outer boundary of A.

Then $D * D'$ is the $(k + m, \ell + n)$-tangle diagram obtained by placing $D_{\curvearrowright}$ on top of $D'_{\curvearrowleft}$.

In the following picture we give an example of the $*$-product of two tangle diagrams on A. We use a different color for the (2,2)-tangle diagram to assist comprehension.

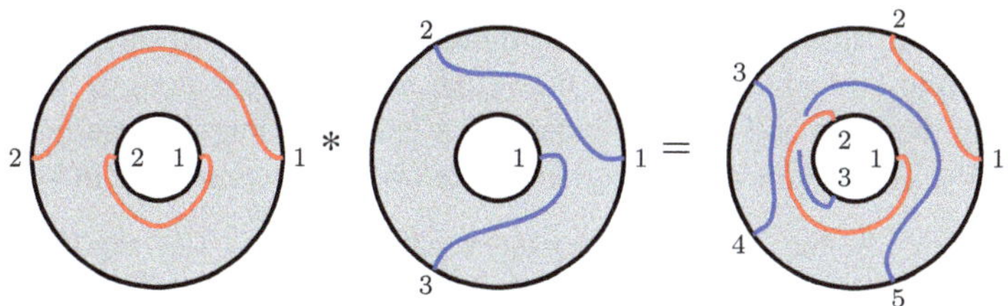

Example 3.4 The tensor product maps $\mathrm{End}_{\mathcal{S}}(0) \times \mathrm{Hom}_{\mathcal{S}}(m, n) \to \mathrm{Hom}_{\mathcal{S}}(m, n)$ and $\mathrm{Hom}_{\mathcal{S}}(m, n) \times \mathrm{End}_{\mathcal{S}}(0) \to \mathrm{Hom}_{\mathcal{S}}(m, n)$ correspond to placing knot diagrams on top or below tangle diagrams within $\mathrm{Hom}_{\mathcal{S}}(m, n)$. The resulting $\mathrm{End}_{\mathcal{S}}(0)$-bimodule structure on $\mathrm{Hom}_{\mathcal{S}}(m, n)$ has been described and studied in the more general context of relative Kauffman skein modules over surfaces, see, e.g., [22, 23]. See also [18, §4.1] for a discussion of $\mathrm{Hom}_{\mathcal{S}}(0, 2)$ as $\mathrm{End}_{\mathcal{S}}(0)$-bimodule.

Proposition 3.5 *The skein category $\mathcal{S}$ of the annulus is a strict monoidal linear category with tensor functor $\times_{\mathcal{S}} : \mathcal{S} \times \mathcal{S} \to \mathcal{S}$ as defined above, and unit object* 0.

Proof By the remarks preceding the proposition, the only nontrivial check is the compatibility of $\times_{\mathcal{S}}$ with composition of morphisms. For the first tensor component this follows from the fact that all the endpoints of D in $D * D'$ are rotated clockwise, while over-rotation by angles $\geq 2\pi$ cannot occur due to the third property of the planar isotopy transforming D into $D_{\curvearrowright}$. A similar remark applies for the second tensor component. □

We write $\otimes$ for the usual tensor product of complex vector spaces.

Corollary 3.6 *For $m, n \in \mathbb{Z}_{\geq 0}$ we have algebra morphisms*

$$\epsilon_{m,n} : \mathrm{End}_{\mathcal{S}}(m) \otimes \mathrm{End}_{\mathcal{S}}(n) \to \mathrm{End}_{\mathcal{S}}(m+n)$$

defined by $\epsilon_{m,n}\big([D] \otimes [D']\big) := [D] \times_{\mathcal{S}} [D'] = [D * D']$.

As we shall see in Remark 5.4, the algebra $\mathrm{End}_{\mathcal{S}}(m)$ is isomorphic to the mth extended affine Temperley-Lieb algebra. Under this identification, the algebra maps $\epsilon_{m,n}$ were considered before in [8, §3.3].

4 Equivalence with the Affine Temperley-Lieb Category

The affine Temperley-Lieb category was introduced by Graham and Lehrer [11]. In this category the morphisms are affine diagrams, which are defined as follows.

Definition 4.1 Let $m, n \in \mathbb{Z}_{\geq 0}$. An affine (m, n)-diagram is an (m, n)-tangle diagram in A with no crossings and without contractible loops in A. We write $\mathcal{D}_{m,n}$ for the subclass of $\mathrm{Hom}_{\mathcal{T}}(m, n)$ consisting of equivalence classes $\overline{D}$ of affine (m, n)-diagrams D.

Remark 4.2 An affine diagram on the annulus can be viewed as a periodic diagram on the infinite horizontal strip by cutting the annulus open along a line segment connecting the inner and outer boundary of A and extending the resulting diagram periodically. This is how affine diagrams were originally considered in [11, 13].

Let $\mathrm{Lin}_c(\mathcal{T})$ be the quotient of the linear category $\mathrm{Lin}(\mathcal{T})$ by the loop removal relation (3.2) (compare with Definition 3.1). The sublabel "c" stands for contractible, signifying that in $\mathrm{Lin}_c(\mathcal{T})$ contractible loops in tangle diagrams may be removed by the multiplicative factor $-(t^{\frac{1}{2}} + t^{-\frac{1}{2}})$. If D is an (m, n)-tangle diagram, then we write $\langle D \rangle$ for its equivalence class in $\mathrm{Hom}_{\mathrm{Lin}_c(\mathcal{T})}(m, n)$.

Note that the skein category $\mathcal{S}$ is the quotient of $\mathrm{Lin}_c(\mathcal{T})$ by the Kauffman skein relation (3.1). Graham's and Lehrer's [11] affine Temperley-Lieb category, which is closely related to Jones' [14] annular Temperley-Lieb category, is the following subcategory of $\mathrm{Lin}_c(\mathcal{T})$.

Definition 4.3 ([11]) The affine Temperley-Lieb category $\mathcal{TL} = \mathcal{TL}(t^{\frac{1}{2}})$ is the linear subcategory of $\mathrm{Lin}_c(\mathcal{T})$ with objects $\mathbb{Z}_{\geq 0}$ and morphisms $\mathrm{Hom}_{\mathcal{TL}}(m,n)$ the subspace of $\mathrm{Hom}_{\mathrm{Lin}_c(\mathcal{T})}(m,n)$ spanned by the equivalence classes $\langle D\rangle$ of affine (m,n)-diagrams D.

If D is an affine (m,n)-diagram and D' is an affine (k,m)-diagram, then

$$\langle D\rangle \circ \langle D'\rangle = \big(-t^{\frac{1}{2}} - t^{-\frac{1}{2}}\big)^{l(D'')}\langle D''_c\rangle$$

in $\mathrm{Hom}_{\mathcal{TL}}(k,n)$, with D'' the (k,n)-tangle diagram in A obtained by inserting D' inside D (in the same way as in Definition 2.1), with $l(D'')$ the number of loops in D'' contractible in A, and with $D''_c \in \mathcal{D}_{k,n}$ the affine (k,n)-diagram obtained from D'' by removing the contractible loops.

Note that $\mathrm{Hom}_{\mathcal{TL}}(m,n) = \{0\}$ if $m+n$ is odd, and

$$\{\langle D\rangle \mid D \text{ affine } (m,n)\text{-diagram}\}$$

is a linear basis of $\mathrm{Hom}_{\mathcal{TL}}(m,n)$.

Next we show that the linear categories $\mathcal{S}$ and $\mathcal{TL}$ are equivalent. The subtle point is to show that the obvious linear functor from $\mathcal{TL}$ to $\mathcal{S}$ is faithful. The proof uses a relative version of the Kauffman bracket for (m,n)-tangle diagrams in A, compare with the proof of [23, Thm. 3.1].

Theorem 4.4 *The linear categories $\mathcal{S}$ and $\mathcal{TL}$ are equivalent.*

Proof Consider the essentially surjective linear functor $\mathcal{F}: \mathcal{TL} \to \mathcal{S}$ which is the identity on objects and maps $\langle D\rangle$ to $[D]$ for an affine (m,n)-diagram D. It is clearly well-defined since the loop removal relation holds in $\mathcal{S}$ as well as in $\mathcal{TL}$.

Let D be an (m,n)-tangle diagram in A. The Kauffman skein relation and the loop removal relation allow us to write $[D]$ as a linear combination of classes $[D'] \in \mathrm{Hom}_{\mathcal{S}}(m,n)$ with the D''s being affine (m,n)-diagrams. It follows that the functor $\mathcal{F}$ is full. It remains to show that $\mathcal{F}$ is faithful.

Suppose that $m+n$ is even and let D be an (m,n)-tangle diagram in A with k crossing points. Let $\mathcal{S}_D$ be the set of cardinality 2^k containing the (m,n)-tangle diagrams S without crossings that are obtained from D by removing each crossing in D by either or . For $S \in \mathcal{S}_D$ let $h_D(S)$ (respectively, $v_D(S)$) be the number of crossing points at which is replaced by (respectively,). Let $c_D(S)$ be the number of loops in S that are contractible in A, and write $\widetilde{S}$ for the affine (m,n)-diagram obtained from S by removing these contractible loops.

It is easy to see that there exists a well-defined linear map

$$\widehat{\Psi}: \mathbb{C}[\mathrm{Hom}_{\mathcal{T}}(m,n)] \to \mathrm{Hom}_{\mathcal{TL}}(m,n)$$

satisfying

$$\widehat{\psi}(\overline{D}) := \sum_{S\in\mathcal{S}_D} \big(-t^{\frac{1}{2}} - t^{-\frac{1}{2}}\big)^{c_D(S)} t^{(h_D(S)-v_D(S))/4} \langle \widetilde{S} \rangle \tag{4.1}$$

for all (m, n)-tangle diagrams D in A. Direct computations show that the map $\widehat{\psi}$ respects the Kauffman skein relation (3.1) and the loop removal relation (3.2), so it gives rise to a linear map

$$\psi : \mathrm{Hom}_{\mathcal{S}}(m, n) \to \mathrm{Hom}_{\mathcal{TL}}(m, n)$$

satisfying $\psi([D]) = \widehat{\psi}(\overline{D})$ for (m, n)-tangle diagrams D in A.

By the Kaufmann skein relation (3.1) and the loop removal relation (3.2), the linear map ψ is the inverse of the linear map $\mathcal{F} : \mathrm{Hom}_{\mathcal{TL}}(m, n) \to \mathrm{Hom}_{\mathcal{S}}(m, n)$. This shows that $\mathcal{F}$ is faithful. □

Remark 4.5 The special case $\mathrm{End}_{\mathcal{S}}(0) \simeq \mathrm{End}_{\mathcal{TL}}(0)$ was established for general surfaces in [23, Lem. 3.3].

Definition 4.6 We call $\psi([D]) = \widehat{\psi}(\overline{D}) \in \mathrm{Hom}_{\mathcal{TL}}(m, n)$ (see (4.1)) the *relative Kauffman bracket* of the (m, n)-tangle diagram D in A.

Remark 4.7 Note that for $(m, n) = (0, 0)$, the relative Kauffman bracket $\psi([D])$ of a link diagram D in A lands in the algebra $\mathrm{End}_{\mathcal{TL}}(0)$, which is isomorphic to the algebra of polynomials in one variable (the variable corresponds to the equivalence class of a noncontractible loop in A). Evaluating the resulting polynomial at $-t^{\frac{1}{2}} - t^{-\frac{1}{2}}$ can be thought of as closing the hole of the annulus and viewing the link diagram as an element in the skein module of the disc (or equivalently, of the plane). As a result one obtains the usual Kauffman [16] bracket of D, viewed as a link diagram in the plane (see [19] and [20, §1.7]).

5 The Extended Affine Temperley-Lieb Algebra

Write $\mathrm{TL}_0 := \mathbb{C}[X]$ for the algebra of complex polynomials in one variable X and $\mathrm{TL}_1 := \mathbb{C}[\rho, \rho^{-1}]$ for the algebra of complex Laurent polynomials in the variable ρ. Let TL_2 be the complex associative unital algebra with generators $e_1, e_2, \rho, \rho^{-1}$ and defining relations

$$e_i^2 = \big(-t^{\frac{1}{2}} - t^{-\frac{1}{2}}\big)e_i,$$

$$\rho e_i = e_{i+1}\rho,$$

$$\rho\rho^{-1} = 1 = \rho^{-1}\rho,$$

$$\rho^2 e_1 = e_1,$$

where the indices are taken modulo two. Finally, for $n \geq 3$ let TL_n be the complex associative unital algebra with generators $e_1, e_2, \ldots, e_n, \rho, \rho^{-1}$ and defining relations

$$\begin{aligned}
e_i^2 &= \big(-t^{\frac{1}{2}} - t^{-\frac{1}{2}}\big)e_i, \\
e_i e_j &= e_j e_i \qquad\qquad \text{if } i - j \neq \pm 1, \\
e_i e_{i\pm 1} e_i &= e_i, \\
\rho e_i &= e_{i+1}\rho, \\
\rho\rho^{-1} &= 1 = \rho^{-1}\rho, \\
\big(\rho e_1\big)^{n-1} &= \rho^n(\rho e_1),
\end{aligned} \tag{5.1}$$

where the indices are taken modulo n. Observe that the last defining relation $(\rho e_1)^{n-1} = \rho^n(\rho e_n)$ in (5.1) can be replaced by

$$\rho^2 e_{n-1} = e_1 e_2 \cdots e_{n-1}.$$

Note that $\mathrm{TL}_n = \mathrm{TL}_n(t^{\frac{1}{2}})$ for $n \geq 2$ depends on the nonzero complex parameter $t^{\frac{1}{2}}$, which we omit from the notations if no confusion can arise.

Remark 5.1 The definition for $n = 2$ and $n \geq 3$ can be placed at the same footing by describing TL_n in terms of the smaller set $e_1, e_2, \ldots, e_{n-1}, \rho, \rho^{-1}$ of algebraic generators. The defining relations then are

$$\begin{aligned}
e_i^2 &= \big(-t^{\frac{1}{2}} - t^{-\frac{1}{2}}\big)e_i,, && 1 \leq i < n, \\
e_i e_j &= e_j e_i && 1 \leq i, j < n \text{ and } i - j \neq \pm 1, \\
e_i e_{i\pm 1} e_i &= e_i, && 1 \leq i, i \pm 1 < n, \\
\rho e_i &= e_{i+1}\rho, && 1 \leq i < n-1, \\
\rho^2 e_{n-1} &= e_1 \rho^2, \\
\rho\rho^{-1} &= 1 = \rho^{-1}\rho, \\
\rho^2 e_{n-1} &= e_1 e_2 \cdots e_{n-1}.
\end{aligned} \tag{5.2}$$

Definition 5.2 ([13]) TL_n is called the *(nth) extended affine Temperley-Lieb algebra*.

Denote $\mathcal{TL}_n := \mathrm{End}_{\mathcal{TL}}(n)$ for the algebra of endomorphisms of n in the affine Temperley-Lieb category $\mathcal{TL}$. The following result is essentially due to Green [13].

Theorem 5.3

a. $\mathrm{TL}_0 \simeq \mathcal{TL}_0$ *with the algebra isomorphism* $\mathrm{TL}_0 \to \mathcal{TL}_0$ *defined by*

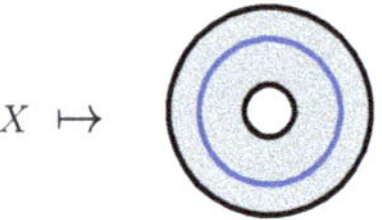

b. $\mathrm{TL}_1 \simeq \mathcal{TL}_1$ *with the algebra isomorphism* $\mathrm{TL}_1 \to \mathcal{TL}_1$ *defined by*

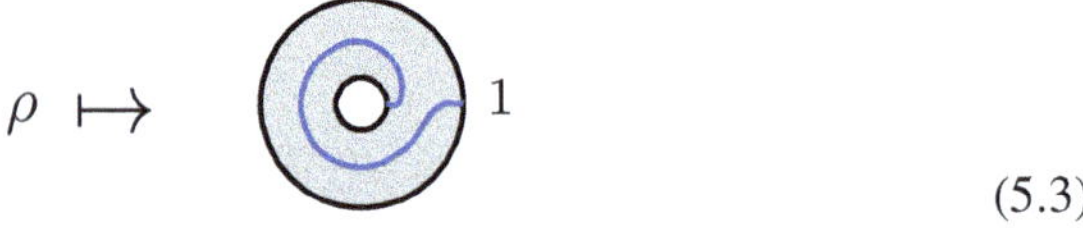

(5.3)

c. $\mathrm{TL}_2 \simeq \mathcal{TL}_2$ *with the algebra isomorphism* $\mathrm{TL}_2 \to \mathcal{TL}_2$ *defined by*

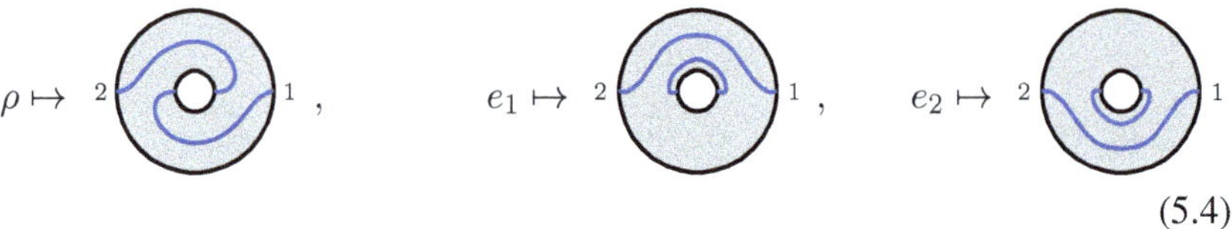

(5.4)

d. *If* $n \geq 3$, *then* $\mathrm{TL}_n \simeq \mathcal{TL}_n$ *with the algebra isomorphism* $\mathrm{TL}_n \to \mathcal{TL}_n$ *defined by*

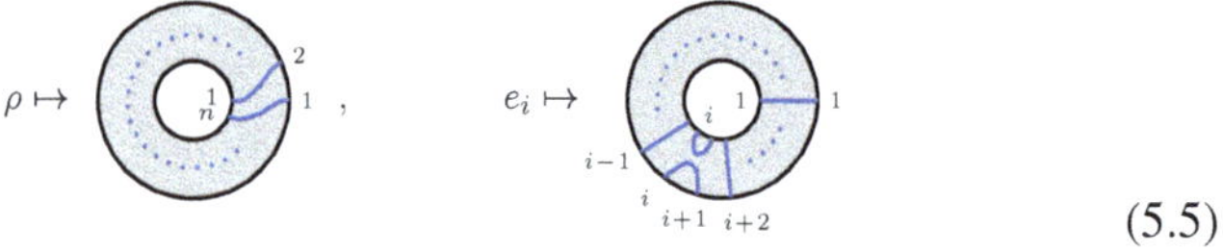

(5.5)

for $i = 1, \ldots, n$ *(with the indices and the labels of the marked points taken modulo* n*).*

Proof **a** and **b** are well known (see, for instance, [20, §1.7] and [21, §4]), while **d** is due to Green [13, Prop. 2.3.7].

Proof of **c** A direct check shows that there exists a unique unital algebra homomorphism $\phi : \mathrm{TL}_2 \to \mathcal{TL}_2$ satisfying (5.4).

Recall that the set $\mathcal{D}_2$ of affine $(2, 2)$-diagrams form a linear basis of $\mathcal{TL}_2$. The affine $(2, 2)$-diagrams can be described explicitly as follows.

The affine $(2, 2)$-diagram $\phi(\rho^m)$ for $m \in \mathbb{Z}$ is obtained from the identity element of $\mathcal{TL}_2$ by winding the outer boundary counterclockwise by an angle of $m\pi$. It follows that the pairwise distinct affine $(2, 2)$-diagrams $\phi(\rho^m)$ $(m \in \mathbb{Z})$ form the subset of $\mathcal{D}_2$ consisting of affine $(2, 2)$-diagrams whose arcs all connect the inner boundary with the outer boundary. The remaining affine $(2, 2)$-diagrams are the diagrams of the form

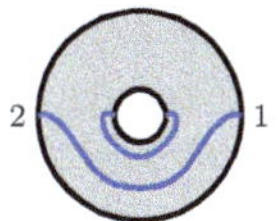

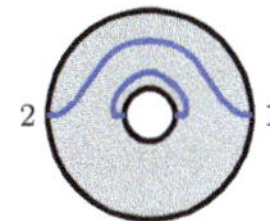

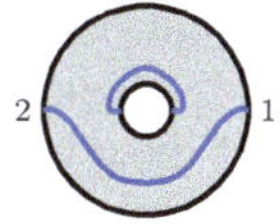

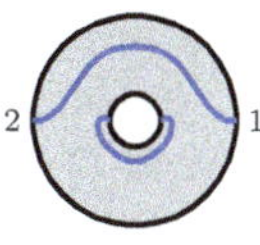

in which r nonintersecting, noncontractible loops are inserted for some $r \in \mathbb{Z}_{\geq 0}$. For $r = 2k$ the resulting four types of affine (2, 2)-diagrams are

$$\phi(e_2(e_1e_2)^k), \quad \phi(e_1(e_2e_1)^k), \quad \phi(\rho e_1(e_2e_1)^k), \quad \phi(\rho e_2(e_1e_2)^k).$$

For $r = 2k + 1$ they are

$$\phi(\rho(e_1e_2)^k), \quad \phi(\rho(e_2e_1)^k), \quad \phi((e_2e_1)^k), \quad \phi((e_1e_2)^k).$$

Hence ϕ maps the subset

$$\begin{aligned} \{\rho^m\}_{m\in\mathbb{Z}} \cup \{(e_2e_1)^k, \rho(e_2e_1)^k, e_1(e_2e_1)^k, \rho e_1(e_2e_1)^k\}_{k\in\mathbb{Z}_{\geq 0}} \\ \cup \{(e_1e_2)^k, \rho(e_1e_2)^k, e_2(e_1e_2)^k, \rho e_2(e_1e_2)^k\}_{k\in\mathbb{Z}_{\geq 0}} \end{aligned} \tag{5.6}$$

of TL_2 bijectively onto the linear basis $\mathcal{D}_2$ of $\mathcal{TL}_2$. By the defining relations in TL_2 we see that (5.6) spans TL_2. We conclude that ϕ is an isomorphism of algebras. □

Remark 5.4 By Theorem 4.4 we now also have a skein theoretic description $\mathrm{End}_{\mathcal{S}}(n)$ of the nth extended affine Temperley-Lieb algebra,

$$\mathrm{TL}_n \simeq \mathcal{TL}_n \simeq \mathrm{End}_{\mathcal{S}}(n). \tag{5.7}$$

The skein theoretic description of the finite Temperley-Lieb algebra is described in [16, 17, 19, 20].

6 The Arc Insertion Functor

Definition 6.1 The arc insertion functor $\mathcal{I} : \mathcal{S} \to \mathcal{S}$ is the endofunctor

$$\mathcal{I} := - \times_{\mathcal{S}} 1,$$

defined concretely by

$$\mathcal{I}(m) := m \times_{\mathcal{S}} 1 = m + 1,$$
$$\mathcal{I}([D]) := [D] \times_{\mathcal{S}} \mathbf{1}_1 = [D * \mathrm{Id}_1]$$

for $m \in \mathbb{Z}_{\geq 0}$ and for tangle diagrams D.

Let D be an (m, n)-tangle diagram and write $D^{ins} = D * \mathrm{Id}_1$, so that $\mathcal{I}([D]) = [D^{ins}]$. The $(m+1, n+1)$-tangle diagram D^{ins} is obtained from D by inserting an arc in D connecting the inner boundary of A with its outer boundary and going underneath all arcs it meets. See Sect. 3 for the specific requirements on the location of the endpoints and on the winding of the inserted arc. We give two examples.

Example 6.2 For the $(0, 2)$-tangle diagram D_1 and the $(1, 3)$-tangle diagram D_2 given by

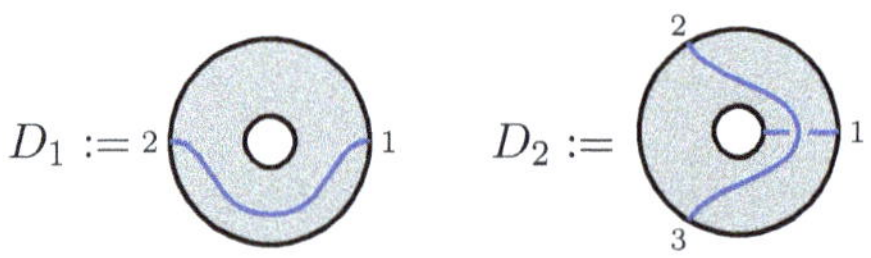

$$D_1 := \qquad\qquad D_2 :=$$

we get

$$D_1^{ins} = \qquad\qquad D_2^{ins} =$$

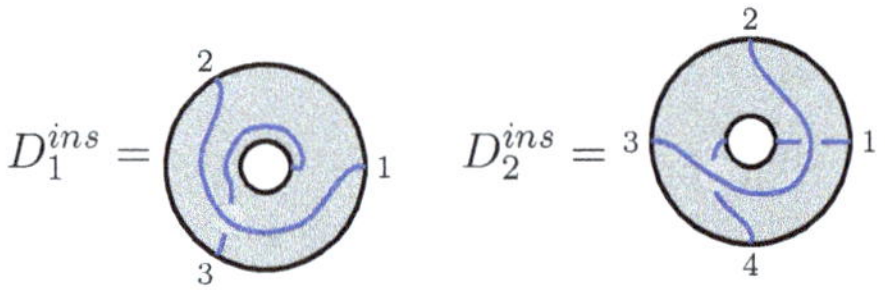

For $n \in \mathbb{Z}_{\geq 0}$ consider the unit preserving algebra map

$$\mathcal{I}_n := \mathcal{I}|_{\mathcal{S}_n} : \mathrm{End}_{\mathcal{S}}(n) \to \mathrm{End}_{\mathcal{S}}(n+1).$$

In terms of the algebra maps $\epsilon_{m,n}$ (see Corollary 3.6) we have

$$\mathcal{I}_n([D]) = \epsilon_{n,1}([D] \otimes \mathbf{1}_1).$$

The map $\mathcal{I}_n$ can be interpreted as an algebra map $\mathcal{I}_n : \mathrm{TL}_n \to \mathrm{TL}_{n+1}$ since $\mathrm{TL}_n \simeq \mathrm{End}_{\mathcal{S}}(n)$ (see Theorem 5.3 and Remark 5.4). In the following proposition we explicitly compute $\mathcal{I}_n$ on the algebraic generators of TL_n.

Proposition 6.3

a. $\mathcal{I}_0(X) = t^{\frac{1}{4}}\rho + t^{-\frac{1}{4}}\rho^{-1}$.
b. $\mathcal{I}_1(\rho) = \rho(t^{-\frac{1}{4}}e_1 + t^{\frac{1}{4}})$ *and* $\mathcal{I}_1(\rho^{-1}) = (t^{\frac{1}{4}}e_1 + t^{-\frac{1}{4}})\rho^{-1}$.
c. *For* $n \geq 2$ *we have*

$$\mathcal{I}_n(e_i) = e_i, \qquad i = 1, \ldots, n-1,$$

$$\mathcal{I}_n(e_n) = (t^{\frac{1}{4}}e_n + t^{-\frac{1}{4}})e_{n+1}(t^{-\frac{1}{4}}e_n + t^{\frac{1}{4}}),$$

$$\mathcal{I}_n(\rho) = \rho(t^{-\frac{1}{4}}e_n + t^{\frac{1}{4}}),$$

$$\mathcal{I}_n(\rho^{-1}) = (t^{\frac{1}{4}}e_n + t^{-\frac{1}{4}})\rho^{-1}.$$

Proof These are direct computations in the skein module.

*Proof of **a*** We have

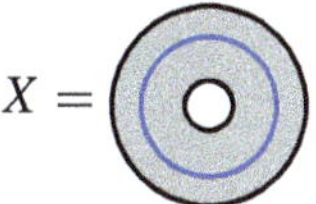

so

$$\mathcal{I}_0(X) = \quad {}^{1} = t^{\frac{1}{4}} \quad {}^{1} + t^{-\frac{1}{4}} \quad {}^{1} = t^{\frac{1}{4}}\rho + t^{-\frac{1}{4}}\rho^{-1}$$

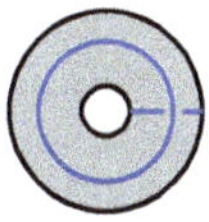

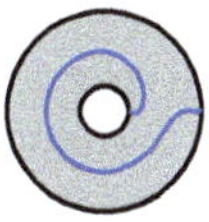

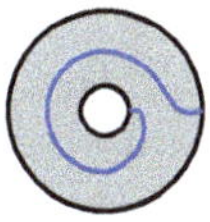

by the Kauffman skein relation (3.1).

*Proof of **b*** We have

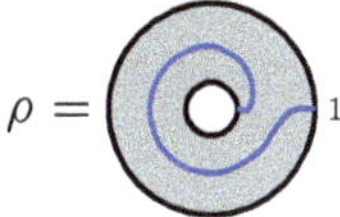

so

$$\mathcal{I}_1(\rho) = {}^{2} \quad {}^{1} = \rho(t^{-\frac{1}{4}}e_1 + t^{\frac{1}{4}})$$

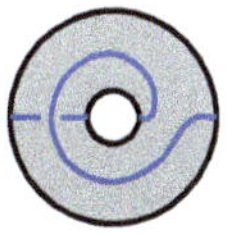

by applying the Kauffman skein relation (3.1) to the crossing and rewriting the resulting expressions in terms of the generators of TL_2 (compare with the proof of Theorem 5.3**c**). In a similar way one proves the explicit formula for $\mathcal{I}_1(\rho^{-1}) \in \mathrm{TL}_2$.

Proof of **c** The formulas for $\mathcal{I}_n(\rho^{\pm 1}) \in \mathrm{TL}_{n+1}$ are obtained by a similar computation as in **b**.

For $1 \leq i < n$, applying the arc insertion functor to $e_i \in \mathrm{TL}_n$ does not introduce crossings. The resulting $(n+1, n+1)$-affine diagram represents the generator e_i in TL_{n+1}, so $\mathcal{I}_n(e_i) = e_i$.

Note that applying the arc insertion functor to $e_n \in \mathrm{TL}_n$ introduces two crossings. Resolving both crossings with the Kauffman skein relation (3.1) and expressing the resulting linear combination of four $(n+1, n+1)$-affine diagrams in terms of the generators of TL_{n+1} yield the formula

$$\mathcal{I}_n(e_n) = t^{\frac{1}{2}}e_n e_{n+1} + t^{-\frac{1}{2}}e_{n+1}e_n + e_{n+1} + e_n = (t^{\frac{1}{4}}e_n + t^{-\frac{1}{4}})e_{n+1}(t^{-\frac{1}{4}}e_n + t^{\frac{1}{4}}).$$

□

Remark 6.4 The calculation for part **a** in the proposition above has also been done in [18, Prop. 2.2] where a similar skein algebra on the annulus is used to prove centrality of certain skeins when $t^{\frac{1}{2}}$ is a root of unity.

7 Towers of Extended Affine Temperley-Lieb Algebra Modules

In [2] the sequence $\{\mathcal{I}_n\}_{n\in\mathbb{Z}_{\geq 0}}$ of algebra maps $\mathcal{I}_n : \mathrm{TL}_n \to \mathrm{TL}_{n+1}$ was used to study affine Markov traces. In [8] it was used to study fusion of affine Temperley-Lieb modules. In the next two sections we use the sequence $\{\mathcal{I}_n\}_{n\in\mathbb{Z}_{\geq 0}}$ of algebra maps to introduce the notion of towers of extended affine Temperley-Lieb modules. We construct examples that are relevant for understanding the dependence of dense loop models and Heisenberg XXZ spin-$\frac{1}{2}$ chains on their system size (cf. [3, 5, 15]).

We first introduce some notations. Let A be a $\mathbb{C}$-algebra. Write $\mathcal{C}_A$ for the category of left A-modules and $\mathrm{Hom}_A(M, N)$ for the space of morphisms $M \to N$ in $\mathcal{C}_A$, which we will call intertwiners. Suppose that $\eta : A \to B$ is a (unit preserving) morphism of $\mathbb{C}$-algebras and $\mathrm{Ind}^\eta : \mathcal{C}_A \to \mathcal{C}_B$ and $\mathrm{Res}^\eta : \mathcal{C}_B \to \mathcal{C}_A$ for the corresponding induction and restriction functor. Concretely, if M is a left A-module, then

$$\mathrm{Ind}^\eta(M) := B \otimes_A M$$

with B viewed as right A-module by $b \cdot a := b\eta(a)$ for $b \in B$ and $a \in A$. If N is a left B-module, then $\mathrm{Res}^\eta(N)$ is the complex vector space N viewed as A-module by $a \cdot n := \eta(a)n$ for $a \in A$ and $n \in N$. The restriction functor Res^η is right adjoint to Ind^η. If M is a left A-module and N a left B-module, then the corresponding linear isomorphism

$$\mathrm{Hom}_A\big(M, \mathrm{Res}^\eta(N)\big) \xrightarrow{\sim} \mathrm{Hom}_B\big(\mathrm{Ind}^\eta(M), N\big)$$

is $\phi \mapsto \widehat{\phi}$ with $\widehat{\phi} \in \mathrm{Hom}_B\big(\mathrm{Ind}^\eta(M), N\big)$ defined by

$$\widehat{\phi}\big(Z \otimes_A m\big) := Z\phi(m)$$

for $Z \in B$ and $m \in M$.

For a left TL_{n+1}-module V_{n+1} we use the shorthand notation $V_{n+1}^{\mathcal{I}}$ for the left TL_n-module $\mathrm{Res}^{\mathcal{I}_n}(V_{n+1})$.

Definition 7.1 We call

$$V_0 \xrightarrow{\phi_0} V_1 \xrightarrow{\phi_1} V_2 \xrightarrow{\phi_2} V_3 \xrightarrow{\phi_3} \cdots$$

with V_n a left TL_n-module and $\phi_n \in \mathrm{Hom}_{\mathrm{TL}_n}\big(V_n, V_{n+1}^{\mathcal{I}}\big)$ a tower of extended affine Temperley-Lieb algebra modules. We will sometimes denote the tower by $\{(V_n, \phi_n)\}_{n\in\mathbb{Z}_{\geq 0}}$.

Example 7.2 The interpretation of the extended affine Temperley-Lieb algebras as the endomorphism spaces of the skein category $\mathcal{S}$ immediately produces examples of towers of extended affine Temperley-Lieb algebra modules. For example, for $m \in \mathbb{Z}_{\geq 0}$ we have the tower $\{(V_n^{(m)}, \phi_n^{(m)})\}_{n\in\mathbb{Z}_{\geq 0}}$ with

$$V_n^{(m)} := \mathrm{Hom}_{\mathcal{S}}(m+n, n)$$

viewed as a left module over $\mathrm{TL}_n \simeq \mathrm{End}_{\mathcal{S}}(n)$ with representation map

$$\pi_n^{(m)}(Y)Z := Y \circ Z$$

for $Y \in \mathrm{End}_{\mathcal{S}}(n)$ and $Z \in V_n^{(m)} = \mathrm{Hom}_{\mathcal{S}}(m+n, n)$, and with intertwiners

$$\phi_n^{(m)} := \mathcal{I}|_{\mathrm{Hom}_{\mathcal{S}}(m+n,n)} : V_n^{(m)} \to V_{n+1}^{(m)}.$$

There are other intertwiners $V_n^{(m)} \to V_{n+1}^{(m)}$ one can take here; for instance, $Z \mapsto \mathcal{I}(Z) \circ R$ for some $R \in \mathrm{End}_{\mathcal{S}}(m+n+1)$. A refinement of this example will play an important role in the construction of the link pattern tower in the next section.

In the definition of towers $\{(V_n, \phi_n)\}_{n\in\mathbb{Z}_{\geq 0}}$ of extended affine Temperley-Lieb algebra modules we do not require conditions on the intertwiners ϕ_n, in particular allowing trivial intertwiners. The interesting towers are the nondegenerate ones, which are defined as follows.

Definition 7.3 We say that the tower $\{(V_n, \phi_n)\}_{n\in\mathbb{Z}_{\geq 0}}$ of extended affine Temperley-Lieb algebra modules is nondegenerate if $\widehat{\phi}_n : \mathrm{Ind}^{\mathcal{I}_n}(V_n) \to V_{n+1}$ is surjective for all $n \in \mathbb{Z}_{\geq 0}$.

In particular, for a nondegenerate tower $\{(V_n, \phi_n)\}_{n\in\mathbb{Z}_{\geq 0}}$ of extended affine Temperley-Lieb algebra modules, the module V_{n+1} is a quotient of $\mathrm{Ind}^{\mathcal{I}_n}(V_n)$,

$$V_{n+1} \simeq \mathrm{coim}(\widehat{\phi}_n).$$

We give an important example of a nondegenerate tower of extended affine Temperley-Lieb algebra modules in the next section.

8 The Link Pattern Tower

Motivated by applications to integrable models in statistical physics [3, 5, 15], in particular to the dense loop model and the Heisenberg XXZ spin$-\frac{1}{2}$ chain, we construct in this section a family of towers of extended affine Temperley-Lieb algebra modules acting on spaces of link patterns on the punctured disc. We use the skein categorical context to build the tower.

The composition in the skein category $\mathcal{S}$ turns the hom-space $\mathrm{Hom}_{\mathcal{S}}(m,n)$ into a $\mathrm{End}_{\mathcal{S}}(n)$-$\mathrm{End}_{\mathcal{S}}(m)$-bimodule. We regard this as a TL_n-TL_m-bimodule structure on $\mathrm{Hom}_{\mathcal{S}}(m,n)$ using the isomorphism $\mathrm{End}_{\mathcal{S}}(n) \simeq \mathrm{TL}_n$ from Remark 5.4. Note that for a left TL_m-module W_m,

$$\mathrm{Hom}_{\mathcal{S}}(m,n) \otimes_{\mathrm{TL}_m} W_m$$

is naturally a left TL_n-module.

For $n = 2k$ with $k \in \mathbb{Z}_{\geq 0}$ and $u \in \mathbb{C}$ we define the left TL_{2k}-module $V_{2k}(u)$ by

$$V_{2k}(u) := \mathrm{Hom}_{\mathcal{S}}(0, 2k) \otimes_{\mathrm{TL}_0} \mathbb{C}_0^{(u)},$$

with $\mathbb{C}_0^{(u)}$ the one-dimensional module over $\mathrm{TL}_0 = \mathbb{C}[X]$ satisfying $X \mapsto u$. For $Y \in \mathrm{Hom}_{\mathcal{S}}(0, 2k)$ we write Y_u for the element $Y \otimes_{\mathrm{TL}_0} 1$ in $V_{2k}(u)$.

For $n = 2k+1$ with $k \in \mathbb{Z}_{\geq 0}$ and $v \in \mathbb{C}^*$ we define the left TL_{2k+1}-module $V_{2k+1}(v)$ by

$$V_{2k+1}(v) := \mathrm{Hom}_{\mathcal{S}}(1, 2k+1) \otimes_{\mathrm{TL}_1} \mathbb{C}_1^{(v)},$$

with $\mathbb{C}_1^{(v)}$ the one-dimensional module over $\mathrm{TL}_1 = \mathbb{C}[\rho^{\pm 1}]$ satisfying $\rho \mapsto v$. For $Z \in \mathrm{Hom}_{\mathcal{S}}(1, 2k+1)$ we write Z_v for the element $Z \otimes_{\mathrm{TL}_1} 1$ in $V_{2k+1}(v)$.

Remark 8.1 The left TL_{2k}-module $V_{2k}(u)$ and the left TL_{2k+1}-module $V_{2k+1}(v)$ are examples of the so-called standard TL_N-modules $\mathcal{W}_{j,z}[N]$ from [8, §4.2] (the extended affine Temperley-Lieb algebra TL_N is denoted by TL_N^a in [8]). Concretely, writing $u = x + x^{-1}$ with $x \in \mathbb{C}^*$, we have

$$V_{2k}(u) = \mathcal{W}_{0,x}[2k], \qquad V_{2k+1}(v) = \mathcal{W}_{\frac{1}{2},v}[2k+1].$$

Next we study towers having the modules $V_{2k}(u)$ and $V_{2k+1}(v)$ as building blocks. For this we need special elements in the skein modules $\mathrm{End}_{\mathcal{S}}(0)$, $\mathrm{End}_{\mathcal{S}}(1)$ and $\mathrm{Hom}_{\mathcal{S}}(0,2)$. Let $\emptyset \in \mathrm{End}_{\mathcal{S}}(0)$ be the skein class of the empty tangle diagram in A and write $\mathbf{1} := \mathbf{1}_1$ for the identity morphism in $\mathrm{End}_{\mathcal{S}}(1)$. Then $V_0(u) = \mathbb{C}\emptyset_u$ and $V_1(v) = \mathbb{C}\mathbf{1}_v$. For $V_2(u)$, note that the skein module $\mathrm{Hom}_{\mathcal{S}}(0,2)$ is a free right $\mathrm{TL}_0 = \mathbb{C}[X]$-module with TL_0-basis $\{[c_+], [c_-]\}$, where

$c_+ =$ 2 1 , $c_- =$ 2 1 .

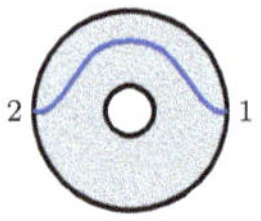

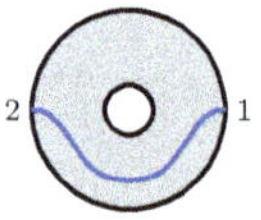

In particular, $V_2(u)$ is two-dimensional with linear basis $\{(c_+)_u, (c_-)_u\}$. Write $U := t^{\frac{1}{4}}[c_+] + v[c_-] \in \mathrm{Hom}_{\mathcal{S}}(0, 2)$. In pictures,

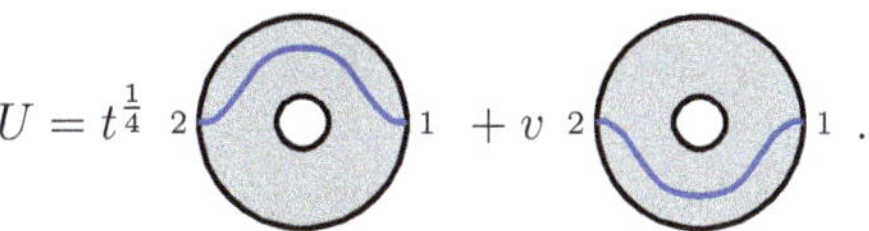

Lemma 8.2 *Let $u \in \mathbb{C}$ and $v \in \mathbb{C}^*$.*

(i) *Define the linear map $\phi_0 : V_0(u) \to V_1(v)$ by $\phi_0(\emptyset_u) := \mathbf{1}_v$. Then*

$$\mathrm{Hom}_{\mathrm{TL}_0}\big(V_0(u), V_1(v)^{\mathcal{I}}\big) = \begin{cases} \mathbb{C}\phi_0 & \text{if } u = t^{\frac{1}{4}}v + t^{-\frac{1}{4}}v^{-1}, \\ \{0\} & \text{otherwise.} \end{cases}$$

(ii) *Let $u = t^{\frac{1}{4}}v + t^{-\frac{1}{4}}v^{-1}$. Define the linear map $\phi_1 : V_1(v) \to V_2(u)$ by $\phi_1(\mathbf{1}_v) := U_u$. Then*

$$\mathrm{Hom}_{\mathrm{TL}_1}\big(V_1(v), V_2(u)^{\mathcal{I}}\big) = \mathbb{C}\phi_1.$$

Proof

(i) Note that $\phi_0 \in \mathrm{Hom}_{\mathrm{TL}_0}\big(V_0(u), V_1(v)^{\mathcal{I}}\big)$ if and only if $\mathcal{I}_0(X)\mathbf{1}_v = u\mathbf{1}_v$. Proposition 6.3**(a)** gives $\mathcal{I}_0(X)\mathbf{1}_v = (t^{\frac{1}{4}}v + t^{-\frac{1}{4}}v^{-1})\mathbf{1}_v$, hence the result.

(ii) Take an arbitrary element $Z_u \in V_2(u)$ with $Z \in \mathrm{Hom}_{\mathcal{S}}(0, 2)$. The linear map $\chi : V_1(v) \to V_2(u)$ defined by $\chi(\mathbf{1}_v) = Z_v$ is in $\mathrm{Hom}_{\mathrm{TL}_1}\big(V_1(v), V_2(u)^{\mathcal{I}}\big)$ if and only if

$$\mathcal{I}_1(\rho)Z_u = vZ_u$$

in $V_2(u)$. By Proposition 6.3**(b)** we have $\mathcal{I}_1(\rho) = \rho(t^{-\frac{1}{4}}e_1 + t^{\frac{1}{4}})$. A direct computation in $\mathrm{Hom}_{\mathcal{S}}(0, 2)$ shows that

$$\begin{aligned} \rho(t^{-\frac{1}{4}}e_1 + t^{\frac{1}{4}}) \circ [c_+] &= -t^{-\frac{3}{4}}[c_-], \\ \rho(t^{-\frac{1}{4}}e_1 + t^{\frac{1}{4}}) \circ [c_-] &= t^{\frac{1}{4}}[c_+] + t^{-\frac{1}{4}}([c_-] \circ X), \end{aligned} \tag{8.1}$$

where we have used the loop removal relation (3.2) in the derivation of the first identity. Writing $m_{\alpha,\beta} := \alpha(c_+)_u + \beta(c_-)_u \in V_2(u)$ with $\alpha, \beta \in \mathbb{C}$ we obtain from (8.1),

$$\mathcal{I}_1(\rho)m_{\alpha,\beta} = \rho(t^{-\frac{1}{4}}e_1 + t^{\frac{1}{4}})m_{\alpha,\beta} = vm_{\alpha',\beta'}$$

with

$$\binom{\alpha'}{\beta'} = M\binom{\alpha}{\beta}, \qquad M := \begin{pmatrix} 0 & t^{\frac{1}{4}}v^{-1} \\ -t^{-\frac{3}{4}}v^{-1} & 1+t^{-\frac{1}{2}}v^{-2} \end{pmatrix}.$$

Since $m_{t^{\frac{1}{4}},v} = U_u$ it remains to show that M has eigenvalue 1 with corresponding eigenspace $\mathbb{C}\binom{t^{\frac{1}{4}}}{v}$. Clearly $\binom{t^{\frac{1}{4}}}{v}$ is an eigenvector of M with eigenvalue 1. The characteristic polynomial of M is

$$p_M(\lambda) = (\lambda - 1)(\lambda - t^{-\frac{1}{2}}v^{-2}),$$

hence the result follows for $t^{-\frac{1}{2}}v^{-2} \neq 1$. If $t^{-\frac{1}{2}}v^{-2} = 1$, then a direct check shows that the geometric multiplication of the eigenvalue 1 of M is still one. □

Note that the intertwiners ϕ_0 and ϕ_1 can alternatively be characterized by the formulas

$$\phi_0(Y_u) = \big(\mathcal{I}(Y)\big)_v, \qquad Y \in \mathrm{End}_{\mathcal{S}}(0),$$
$$\phi_1(Z_v) = \big(\mathcal{I}(Z) \circ U\big)_u, \qquad Z \in \mathrm{End}_{\mathcal{S}}(1)$$

since $\mathcal{I}(\emptyset)_v = 1_v$ and $(\mathcal{I}(\mathbf{1}) \circ U)_u = U_u$.

The following theorem shows that ϕ_0 and ϕ_1 can be extended to a nondegenerate tower

$$V_0(u) \xrightarrow{\phi_0} V_1(v) \xrightarrow{\phi_1} V_2(u) \xrightarrow{\phi_2} \cdots$$

of extended affine Temperley-Lieb modules when $u = t^{\frac{1}{4}}v + t^{-\frac{1}{4}}v^{-1}$.

Theorem 8.3 *Let $v \in \mathbb{C}^*$. Set $u := t^{\frac{1}{4}}v + t^{-\frac{1}{4}}v^{-1}$ and let $k \in \mathbb{Z}_{\geq 0}$.*

(i) *There exist unique intertwiners $\phi_{2k} \in \mathrm{Hom}_{\mathrm{TL}_{2k}}\big(V_{2k}(u), V_{2k+1}(v)^{\mathcal{I}}\big)$ and $\phi_{2k+1} \in \mathrm{Hom}_{\mathrm{TL}_{2k+1}}\big(V_{2k+1}(v), V_{2k+2}(u)^{\mathcal{I}}\big)$ satisfying*

$$\phi_{2k}(Y_u) := \big(\mathcal{I}(Y)\big)_v, \qquad Y \in \mathrm{Hom}_{\mathcal{S}}(0, 2k),$$
$$\phi_{2k+1}(Z_v) := \big(\mathcal{I}(Z) \circ U\big)_u, \qquad Z \in \mathrm{Hom}_{\mathcal{S}}(1, 2k+1).$$

(ii) *The tower*

$$V_0(u) \xrightarrow{\phi_0} V_1(v) \xrightarrow{\phi_1} V_2(u) \xrightarrow{\phi_2} V_3(v) \xrightarrow{\phi_3} \cdots$$

of extended affine Temperley-Lieb algebra modules is nondegenerate if $v^2 \neq t^{\frac{1}{2}}$.

Proof

(i) If the maps ϕ_{2k} and ϕ_{2k+1} are well-defined, then they are obviously intertwiners. To prove that ϕ_{2k} and ϕ_{2k+1} are well-defined we have to show that $\big(\mathcal{I}(Y) \circ \mathcal{I}(X)\big)_v = u\mathcal{I}(Y)_v$ in $V_{2k+1}(v)$ for $Y \in \mathrm{Hom}_{\mathcal{S}}(0, 2k)$ and $\big(\mathcal{I}(Z) \circ (\mathcal{I}(\rho) \circ U)\big)_u = v\big(\mathcal{I}(Z) \circ U\big)_u$ in $V_{2k+2}(u)$ for $Z \in \mathrm{Hom}_{\mathcal{S}}(1, 2k+1)$. This is analogous to the proof of Lemma 8.2.

(ii) Consider the tangle diagrams

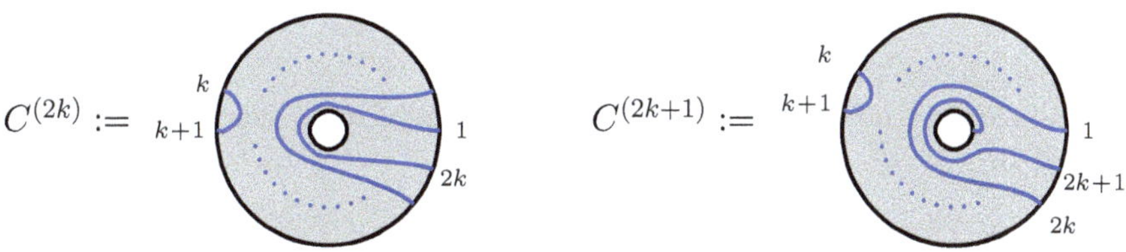

We claim that $V_{2k}(u) = \mathrm{TL}_{2k} \cdot [C^{(2k)}]_u$ and $V_{2k+1}(v) = \mathrm{TL}_{2k+1} \cdot [C^{(2k+1)}]_v$.

To prove this we use the matchmaker representation of the finite Temperley-Lieb algebra TL_{2k}^{fin} (see, e.g., [9, §2.1]). The finite Temperley-Lieb algebra TL_{2k}^{fin} is the subalgebra of TL_{2k} generated by $e_1, \ldots, e_{2k-1}$. The representation space M_{2k} of the matchmaker representation is the vector space with linear basis the non-crossing perfect matchings of $\{1, \ldots, 2k\}$. Such non-crossing perfect matchings are viewed as nonintersecting arcs in a strip with the ordered endpoints $1, \ldots, 2k$ positioned on the bottom line of the strip. We will call such non-crossing perfect matchings link patterns. The e_j acts on link patterns as the matchmaker of j and $j+1$ (see [9, (1)]), with the convention that if j and $j+1$ in the link pattern were already matched, then e_j acts by multiplication by the scalar factor $-(t^{\frac{1}{2}} + t^{-\frac{1}{2}})$.

Let $L^{(2k)} \in M_{2k}$ be the link pattern connecting j to $2k+1-j$ for $j = 1, \ldots, 2k$. By wrapping the link pattern on the annulus in such a way that $\{1, \ldots, 2k\}$ correspond to the marked points $2\xi_{2k}^{j-1}$ $(j = 1, \ldots, 2k)$, we get an injective TL_{2k}^{fin}-module morphism $M_{2k} \hookrightarrow V_{2k}(u)$ mapping $L^{(2k)}$ to $[C^{(2k)}]_u$. If we in addition insert an arc via $- * \mathrm{Id}_1$ before projecting onto the skein, we get an injective TL_{2k}^{fin}-module morphism $M_{2k} \hookrightarrow V_{2k+1}(v)$ mapping $L^{(2k)}$ to $[C^{(2k)} * \mathrm{Id}_1]_v = [C^{(2k+1)}]_v$.

With these observations and the fact that $\rho \in \mathrm{TL}_n$ can be used to turn diagrams in A counterclockwise by an angle of $2\pi/n$, the claim is a consequence of $M_{2k} = \mathrm{TL}_{2k}^{fin} \cdot L^{(2k)}$. This in turn is easy to establish using the alternative description of link patterns in terms of Dyck paths (see, e.g., [10, §2.4]).

Now note that

$$\widehat{\phi}_{2k}\big(\mathbf{1}_{2k+1} \otimes_{\mathrm{TL}_{2k}} [C^{(2k)}]_u\big) = \big(\mathcal{I}([C^{(2k)}])\big)_v = [C^{(2k+1)}]_v$$

hence $\widehat{\phi}_{2k} \in \mathrm{Hom}_{\mathrm{TL}_{2k+1}}\big(\mathrm{Ind}^{\mathcal{I}_{2k}}\big(V_{2k}(u)\big), V_{2k+1}(v)\big)$ is surjective. By a direct computation we have

$$\begin{aligned}\widehat{\phi}_{2k+1}\big(e_{k+1}\cdots e_{2k}e_{2k+1}\otimes_{\mathrm{TL}_{2k+1}}[C^{(2k+1)}]_v\big) &=\\ &= \big(e_{k+1}\cdots e_{2k}e_{2k+1}\mathcal{I}([C^{(2k+1)}])U\big)_u\\ &= \big(t^{\frac{1}{4}}v^2 - t^{\frac{3}{4}}\big)[C^{(2k+2)}]_u,\end{aligned}$$

hence $\widehat{\phi}_{2k+1} \in \mathrm{Hom}_{\mathrm{TL}_{2k+2}}\big(\mathrm{Ind}^{\mathcal{I}_{2k+1}}(V_{2k+1}(v)), V_{2k+2}(u)\big)$ is surjective if $v^2 \neq t^{\frac{1}{2}}$. □

Remark 8.4 The two skein classes $[C^{(2k)}]$ and $[C^{(2k+1)}]$ play an important role in determining the normalization of the ground state of the dense loop model (see [3, 5, 15]).

Fix $v \in \mathbb{C}^*$ and set $u = t^{\frac{1}{4}}v + t^{-\frac{1}{4}}v^{-1}$ for the remainder of this section. Note that for $n = 2k$ the representation space $V_{2k}(u)$ consists of the equivalence classes of the skein module $\mathrm{Hom}_{\mathcal{S}}(0, 2k)$ with respect to the equivalence relation obtained as the linear and transitive closure of the *noncontractible loop removal relation*

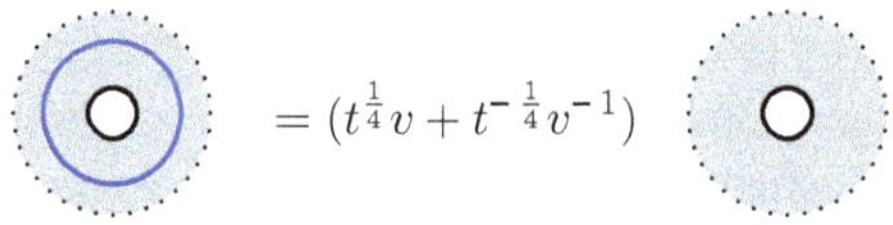

$$= (t^{\frac{1}{4}}v + t^{-\frac{1}{4}}v^{-1}) \tag{8.2}$$

For $n = 2k + 1$ odd, the representation space $V_{2k+1}(v)$ consists of the equivalence classes of the skein module $\mathrm{Hom}_{\mathcal{S}}(1, 2k+1)$ with respect to the equivalence relation obtained as the linear and transitive closure of the following *Dehn twist removal relation*

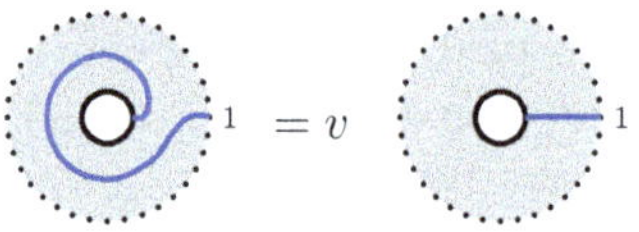

$$1 \quad = v \quad 1 \tag{8.3}$$

Let $\widetilde{\mathcal{C}}_{2k}$ be a set of representatives of the planar isotopy classes of affine $(0, 2k)$-diagrams without noncontractible loops. Let $\widetilde{\mathcal{C}}_{2k+1}$ be a set of representatives of the planar isotopy classes of the affine $(1, 2k + 1)$-diagrams that are planar isotopic to $D * \mathrm{Id}_1$ for some affine $(0, 2k)$-diagram D. We will call the inserted arc connecting the inner boundary of A with the outer boundary of A the *defect line* of the affine $(1, 2k + 1)$-diagram. Observe that $\widetilde{\mathcal{B}}_{2k} := \{[D]_u \mid D \in \widetilde{\mathcal{C}}_{2k}\}$ is a linear basis of $V_{2k}(u)$ and $\widetilde{\mathcal{B}}_{2k+1} := \{[D]_v \mid D \in \widetilde{\mathcal{C}}_{2k+1}\}$ is a linear basis of $V_{2k+1}(v)$.

Definition 8.5 Let $u = t^{\frac{1}{4}}v + t^{-\frac{1}{4}}v^{-1}$. We call the tower

$$V_0(u) \xrightarrow{\phi_0} V_1(v) \xrightarrow{\phi_1} V_2(u) \xrightarrow{\phi_2} V_3(v) \xrightarrow{\phi_3} \cdots$$

of extended affine Temperley-Lieb algebra modules the link pattern tower. We call $v \in \mathbb{C}^*$ the twist weight and $t^{\frac{1}{4}}v + t^{-\frac{1}{4}}v^{-1}$ the noncontractible loop weight of the link pattern tower.

Note that the intertwiners ϕ_{2k} of the link pattern tower are simply given by the insertion of an arc in the underlying $(0, 2k)$-tangle diagrams connecting the outer boundary with the inner boundary. This newly inserted arc is the defect line. The intertwiners ϕ_{2k+1} in the link pattern tower are more subtle. The intertwiner ϕ_{2k+1} acts on the representative of a $(1, 2k+1)$-tangle diagram by detaching the defect line from the inner boundary and reattaching it to the outer boundary in two different ways, corresponding to the two obvious ways that it can pass the hole of the annulus. The two contributions get different weights $t^{\frac{1}{4}}$ and v, respectively. In Theorem 8.3 we have described the operation ϕ_{2k+1} as the composition of arc insertion and composing with the linear combination $U \in \mathrm{Hom}_{\mathcal{S}}(0, 2)$ of the two basic $(0, 2)$-tangle diagrams c_+ and c_-.

Example 8.6

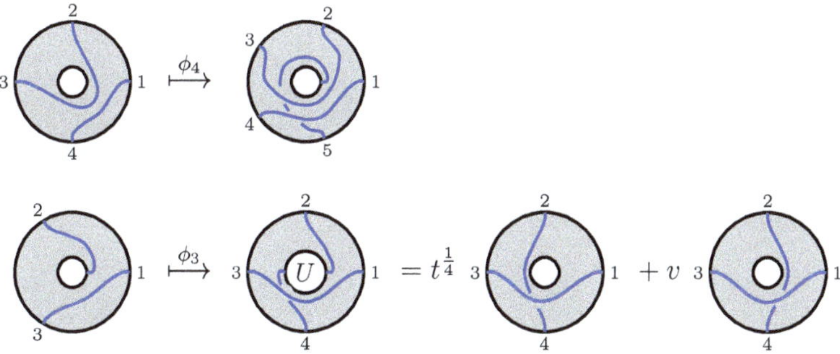

Let $\mathbb{D} := \{z \in \mathbb{C} \mid |z| \leq 2\}$ be the unit disc of radius two and $\mathbb{D}^* := \mathbb{D} \setminus \{0\}$. Let $\mathcal{L}_{2k}$ be the set of link patterns in $\mathbb{D}^*$ connecting the $2k$ marked points $\{2\xi_{2k}^{i-1}\}_{i=1}^{2k}$, i.e., it is the set of perfect non-crossing matchings within $\mathbb{D}^*$ of the marked points $\{2\xi_{2k}^{i-1}\}_{i=1}^{2k}$. For $n = 2k+1$ odd, let $\mathcal{L}_{2k+1}$ be the set of link patterns in $\mathbb{D}$ connecting the $2k+2$ marked points $\{0\} \cup \{2\xi_{2k+1}^{i-1}\}_{i=1}^{2k+1}$. In this context we call the line connecting to 0 the defect line. Since the defect line is now connected to 0 instead of the hole of the annulus we are losing the information about the winding of the defect line. This allows us to realize the link pattern tower for twist weight $v = 1$ on the vector spaces $\mathbb{C}[\mathcal{L}_n]$ with linear basis $\mathcal{L}_n$ as follows.

Consider the map $A \to \mathbb{D} := \{z \in \mathbb{C} \mid |z| \leq 2\}$ given by $re^{i\theta} \mapsto 2e^{\frac{2-r}{r-1}}e^{i\theta}$ for $r \in (1,2]$ and mapping C_i onto 0. Note that the map fixes the outer boundary C_o pointwise. In this way the set $\widetilde{\mathcal{C}_n}$ of affine diagrams in A labelling a basis of the nth representation space in the link pattern tower is identified with $\mathcal{L}_n$ for $n \in \mathbb{Z}_{\geq 0}$. This gives a vector space identification of L_n with $\mathbb{C}[\mathcal{L}_n]$. We now transport the TL_{2k}-module structure on $V_{2k}(u)$ and the TL_{2k+1}-module structure on $V_{2k+1}(v)$ to $\mathbb{C}[\mathcal{L}_{2k}]$ and $\mathbb{C}[\mathcal{L}_{2k+1}]$, respectively, through these linear isomorphisms. It leads to an explicit realization of the link pattern tower with twist weight $v = 1$ as a tower $\{(\phi_n, \mathbb{C}[\mathcal{L}_n])\}_{n\in\mathbb{Z}_{\geq 0}}$ of extended affine Temperley-Lieb algebra modules.

Note that the descriptions of the intertwiners ϕ_{2k} and ϕ_{2k+1} in terms of link patterns are as before: ϕ_{2k} is the insertion of a defect line, and ϕ_{2k+1} is detaching the defect line from the puncture 0 and reattaching it to the outer boundary in two different ways. Note though that the crucial second description of ϕ_{2k+1}, in which a second defect line is added first and then the two defect lines are detached from the puncture 0 and connected to each other in two different ways, requires that one works on the annulus A instead of on the punctured disc $\mathbb{D}^*$. However, there is an analogue to this on the punctured disc using the so-called puncture-skein relations [25], see Remark 8.11 for further details.

Example 8.7 Let $v = 1$.

(1) Example of the action of $e_2 \in \mathrm{TL}_3$ on $\mathbb{C}[\mathcal{L}_3]$:

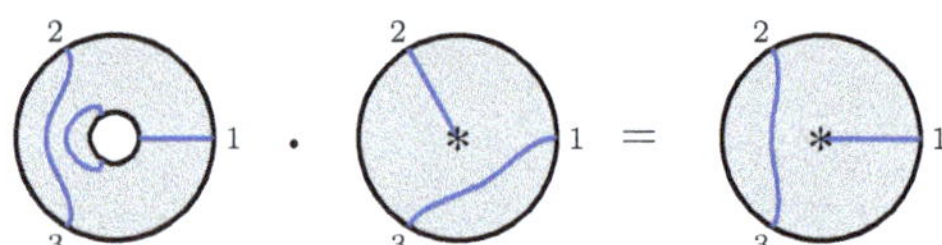

(2) Example of the action of $e_2 \in \mathrm{TL}_4$ on $\mathbb{C}[\mathcal{L}_4]$:

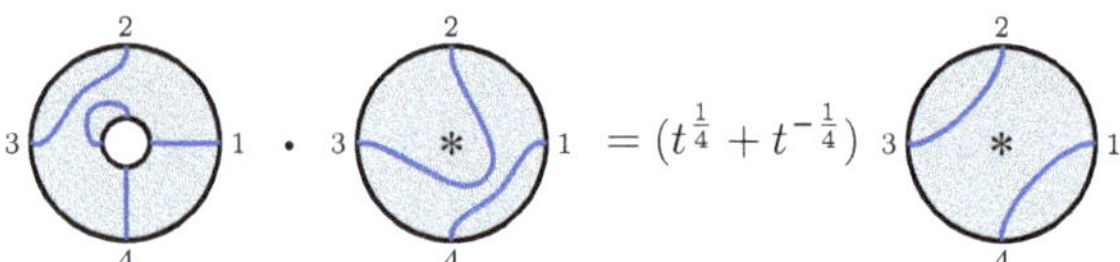

(3) Example of the intertwiner ϕ_2 acting on $\mathbb{C}[\mathcal{L}_2]$:

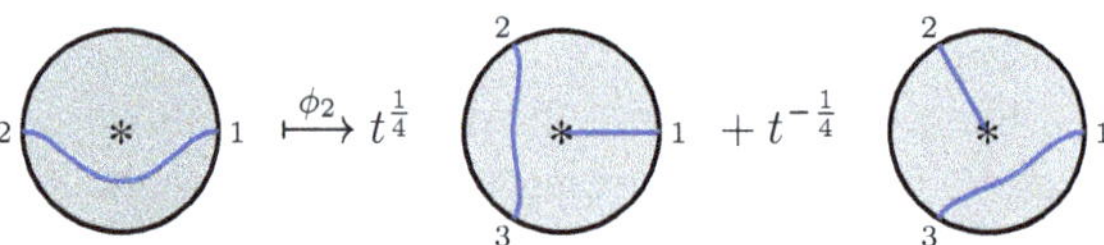

(4) Example of the intertwiner ϕ_3 acting on $\mathbb{C}[\mathcal{L}_3]$ (it corresponds to the second example from Example 8.6 with $v = 1$):

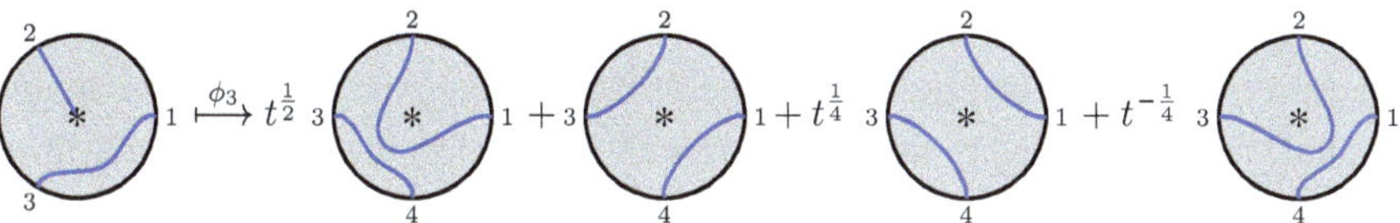

Remark 8.8 The link pattern tower $\{(\mathbb{C}[\mathcal{L}_n], \phi_n)\}_{n \in \mathbb{Z}_{\geq 0}}$ with twist weight $v = 1$ plays an important role in the study of the dense loop model on the semi-infinite cylinder [3, 5, 15]. The representation space $\mathbb{C}[\mathcal{L}_n]$ is the state space of the model of system size n. In [5] the dense loop model of system size $2k + 1$ is related to the dense loop model of system size $2k$ through the map ϕ_{2k}. The results in this paper allows one to also relate the dense loop model of system size $2k + 2$ to the dense loop model of system size $2k+1$ through the (nontrivial) intertwiner ϕ_{2k+1}. We will return to this in [3], in which we also derive recursion relations for associated ground states and for associated solutions of quantum Knizhnik-Zamolodchikov equations.

We end the section[2] by relating the link pattern tower and the connecting maps to a relative version of Roger's and Yang's [25, Def. 2.3] skein algebra on the punctured disc $\mathbb{D}^*$. For $n \in \mathbb{Z}_{\geq 0}$ write $\overline{n} \in \{0, 1\}$ for the residue of n modulo two and set

$$s_0 := t^{\frac{1}{4}} + t^{-\frac{1}{4}}, \qquad s_1 := 1,$$

which are the noncontractible loop weight and the twist weight of the link pattern tower for $v = 1$. Set

$$\mathcal{A} := \bigoplus_{n \in \mathbb{Z}_{\geq 0}} V_n(s_{\overline{n}}) \tag{8.4}$$

for the direct sum of the representation spaces of the link pattern tower. To simplify notations we write $\overline{Y}_n$ for the element $Y_{s_{\overline{n}}} \in V_n(s_{\overline{n}}) = \mathrm{Hom}_{\mathcal{S}}(\overline{n}, n) \otimes_{\mathrm{TL}_{\overline{n}}} \mathbb{C}_{s_{\overline{n}}}$ associated with $Y \in \mathrm{Hom}_{\mathcal{S}}(\overline{n}, n)$.

Proposition 8.9 *$\mathcal{A}$ is a graded associative complex algebra with multiplication defined by*

$$\overline{Y}_m \cdot \overline{Z}_n := \begin{cases} (Y \times_{\mathcal{S}} Z)_{m+n}, & \text{if } (\overline{m}, \overline{n}) \neq (1, 1), \\ ((Y \times_{\mathcal{S}} Z) \circ U)_{m+n}, & \text{if } (\overline{m}, \overline{n}) = (1, 1) \end{cases}$$

for $Y \in \mathrm{Hom}_{\mathcal{S}}(\overline{m}, m)$ and $Z \in \mathrm{Hom}_{\mathcal{S}}(\overline{n}, n)$. The unit element is $\emptyset_0 \in V_0(s_0)$.

[2] We thank an anonymous referee for pointing us to a possible connection with [25].

Proof We first show that the product is well-defined. If $(\overline{m},\overline{n})=(0,0)$, then clearly

$$\overline{(Y\circ X)}_m\cdot\overline{Z}_n=(t^{\frac{1}{4}}+t^{-\frac{1}{4}})\overline{Y}_m\cdot\overline{Z}_n=\overline{Y}_m\cdot\overline{(Z\circ X)}_n$$

since in both the left and right hand side of the equation, the inserted loop around the hole can be removed by the scalar factor $t^{\frac{1}{4}}+t^{-\frac{1}{4}}$ using the noncontractible loop removal relation. If $(\overline{m},\overline{n})=(0,1)$, then

$$\overline{(Y\circ X)}_m\cdot\overline{Z}_n=(t^{\frac{1}{4}}+t^{-\frac{1}{4}})\overline{Y}_m\cdot\overline{Z}_n$$

from (the proof of) Proposition 6.3(**a**), while $\overline{Y}_m\cdot\overline{(Z\circ\rho)}_n=\overline{Y}_m\cdot\overline{Z}_n$ is a direct consequence of the Dehn twist removal relation since $v=1$. The case $(\overline{m},\overline{n})=(0,1)$ is checked similarly. For $(\overline{m},\overline{n})=(1,1)$ we have

$$\begin{aligned}\overline{(Y\circ\rho)}_m\cdot\overline{Z}_n&=\overline{((Y\circ\rho)\times_{\mathcal{S}}Z)\circ U}_{m+n}\\&=\overline{(Y\times_{\mathcal{S}}Z)\circ(\rho\times_{\mathcal{S}}\mathbf{1}_1)\circ U}_{m+n}\\&=\overline{(Y\times_{\mathcal{S}}Z)\circ\mathcal{I}_1(\rho)\circ U}_{m+n}\\&=\overline{(Y\times_{\mathcal{S}}Z)\circ U}_{m+n}=\overline{Y}_m\cdot\overline{Z}_n,\end{aligned}$$

where we used (the proof of) Lemma 8.2(**ii**) for the fourth equality. In fact, the nontrivial equality we are using here is

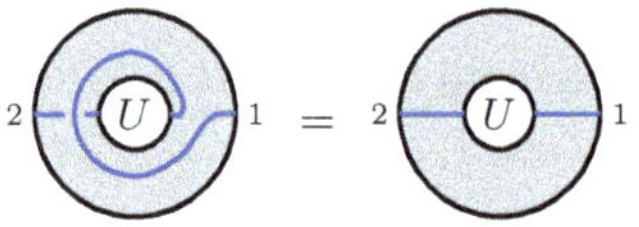

(8.5)

viewed as an identity in $V_2(t^{\frac{1}{4}}+t^{-\frac{1}{4}})$. In a similar manner, it can be shown that $\overline{Y}_m\cdot\overline{Z\circ\rho}_n=\overline{Y}_m\cdot\overline{Z}_n$ if $(\overline{m},\overline{n})=(1,1)$.

Now it remains to show that the product is associative,

$$(\overline{T}_k\cdot\overline{Y}_m)\cdot\overline{Z}_n=\overline{T}_k\cdot(\overline{Y}_m\cdot\overline{Z}_n)$$

for $T\in\mathrm{Hom}_{\mathcal{S}}(\overline{k},k)$, $Y\in\mathrm{Hom}_{\mathcal{S}}(\overline{m},m)$ and $Z\in\mathrm{Hom}_{\mathcal{S}}(\overline{n},n)$. The only nontrivial case is $(\overline{k},\overline{m},\overline{n})=(1,1,1)$. Then we have

$$\begin{aligned}(\overline{T}_k\cdot\overline{Y}_m)\cdot\overline{Z}_n&=\overline{((T\times_{\mathcal{S}}Y)\circ U)\times_{\mathcal{S}}Z}_{k+m+n}\\&=\overline{((T\times_{\mathcal{S}}Y)\times_{\mathcal{S}}Z)\circ(U\times_{\mathcal{S}}\mathbf{1}_1)}_{k+m+n}\\&=\overline{(T\times_{\mathcal{S}}(Y\times_{\mathcal{S}}Z))\circ(\mathbf{1}_1\times_{\mathcal{S}}U)}_{k+m+n}\\&=\overline{T}_k\cdot(\overline{Y}_m\cdot\overline{Z}_n),\end{aligned}$$

where we used in the third equality that $\overline{U \times_{\mathcal{S}} \mathbf{1}_1} = \overline{\mathbf{1}_1 \times_{\mathcal{S}} U}$ in $V_3(1)$. This follows from a direct calculation in the skein, showing that both sides of the equation are equal to

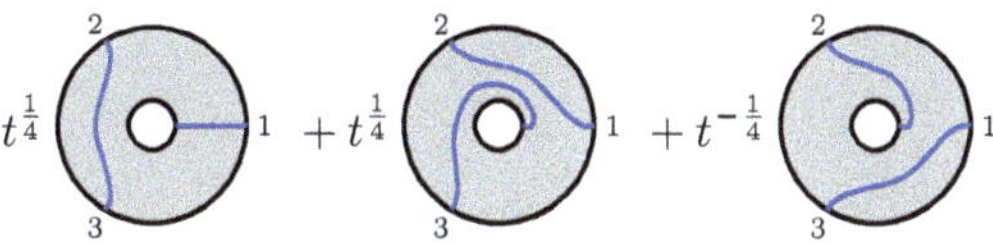

when viewed as identity in $V_3(1)$. □

Corollary 8.10 *Let* $v = 1$. *The connecting map* $\phi_n : V_n(s_{\overline{n}}) \to V_{n+1}(s_{\overline{n+1}})$ *of the link pattern tower is given by*

$$\phi_n(\overline{Y}_n) = \overline{Y}_n \cdot \overline{\mathbf{1}}_1$$

for $\overline{Y}_n \in V_n(s_{\overline{n}})$.

Remark 8.11 We wish to point out the connection between Proposition 8.9 and the work of Roger and Yang [25]. We view the representation space $V_n(s_{\overline{n}})$ of the link pattern tower with $v = 1$ as the following relative version of Roger's and Yang's [25, Def. 2.3] skein algebra of arcs and links on $\Sigma = \mathbb{D}^*$ with puncture $V := \{0\}$. In the relative version we consider, besides the puncture, also the set $\{2\xi_n^{j-1} \mid 1 \leq j \leq n\}$ of n marked points on the outer boundary of $\mathbb{D}^*$. The associated relative skein module $\mathcal{M}_n$ is a quotient of the vector space generated by the isotopy classes of framed arcs and links in $\mathbb{D}^* \times [0, 1]$ such that each pole $\{2\xi_n^{j-1}\} \times [0, 1]$ is met by exactly one endpoint ($1 \leq j \leq n$), and multiple endpoints may connect to the internal pole $\{0\} \times [0, 1]$ at different heights. The quotient is the linear and transitive closure of the Kauffman skein relation (3.1), the nullhomotopic loop removal relation (3.2), the noncontractible loop removal relation (8.2) for $v = 1$ with the hole shrunk to the puncture (it is called the puncture-framing relation in [25]), and finally the Roger-Yang *puncture-skein relation*

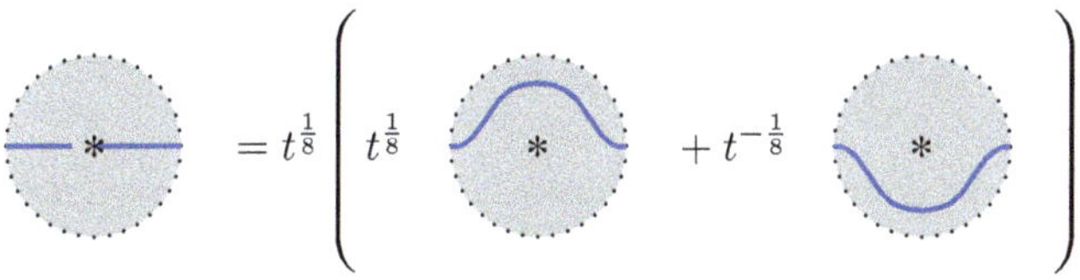

$$= t^{\frac{1}{8}} \left(t^{\frac{1}{8}} \quad + t^{-\frac{1}{8}} \right) \tag{8.6}$$

where at the left, the right curve lies above the left when meeting at the internal pole $\{0\} \times [0, 1]$ (the parameters $v, q^{\frac{1}{2}}$ in [25] is set to the specific values $t^{-\frac{1}{8}}, t^{-\frac{1}{8}}$

to match up with our conventions). Then we have a natural linear isomorphism $V_n(s_{\overline{n}}) \simeq \mathcal{M}_n$ such that

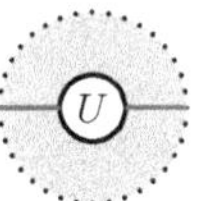

in $V_{2k}(s_{\overline{2k}})$ corresponds to the left hand side of (8.6) in $\mathcal{M}_{2k}$. With this identification our graded algebra structure on

$$\mathcal{A} = \bigoplus_{n=0}^{\infty} V_n(s_{\overline{n}}) \simeq \bigoplus_{n=0}^{\infty} \mathcal{M}_n$$

is the natural relative version of the skein algebra multiplication (cf. the definition of the tensor functor $\times_{\mathcal{S}}$ from Sect. 3). Note that under this identification (8.5) is one of the Reidemeister II' relations from [25], and the proof of Proposition 8.9 is a direct generalization of the proof of [25, Thm. 2.4]. In fact, $\mathcal{M}_0 \simeq \mathbb{C}$ is the Roger-Yang skein algebra of arcs and links on $\mathbb{D}^*$ with puncture $V = \{0\}$.

By Corollary 8.10, we can describe ϕ_n on $\mathcal{M}_n$ as inserting an arc connecting the punctured cylinder to the pole $\{0\} \times [0, 1]$ that passes underneath all other arcs and links and then modding out by the Kauffman skein, loop removal and noncontractible loop removal relations, as well as the puncture-skein relation.

9 The Link Pattern Tower and Fusion

The algebra maps $\epsilon_{n,m}$ (see Corollary 3.6) are used in [8] to define the following fusion product of extended affine Temperley-Lieb modules.

Definition 9.1 ([8]) The fusion product of a left TL_n-module M_1 and a left TL_m-module M_2 is the left TL_{n+m}-module

$$M_1 \widehat{\times}_f M_2 := \mathrm{Ind}^{\epsilon_{n,m}}\big(M_1 \otimes M_2\big).$$

We will show that the consecutive constituents $V_{2k}(u)$ and $V_{2k+1}(v)$ (respectively, $V_{2k+1}(v)$ and $V_{2k+2}(u)$) in the link pattern tower are naturally related by fusion with TL_1-modules. An important role in the analysis is played by the element $d_n \in \mathrm{End}_{\mathcal{S}}(n)$ $(n \geq 1)$, defined by

$$d_n := \epsilon_{n-1,1}(\mathbf{1}_{n-1} \otimes \rho) = \mathbf{1}_{n-1} \times_{\mathcal{S}} \rho$$

with $\rho \in \mathrm{End}_{\mathcal{S}}(1)$ given by (5.3). Note that d_n is the skein class of the (n, n)-tangle diagram

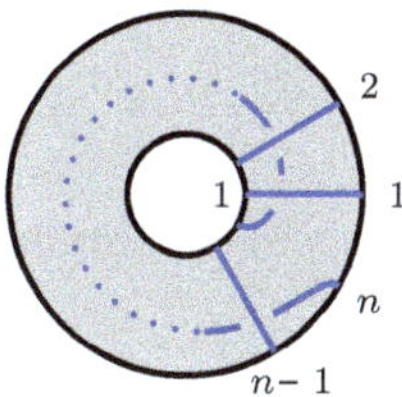

Furthermore, $d_n \in \mathrm{TL}_n$ is invertible and

$$d_{j+1} \circ \mathcal{I}(Z) = Z \times_{\mathcal{S}} \rho = \mathcal{I}(Z) \circ d_{i+1} \qquad \forall\, Z \in \mathrm{Hom}_{\mathcal{S}}(i, j). \tag{9.1}$$

In particular, d_n lies in the centralizer of $\mathcal{I}_{n-1}(\mathrm{TL}_{n-1})$ in TL_n. In terms of the algebraic generators of $\mathrm{TL}_n \simeq \mathrm{End}_{\mathcal{S}}(n)$, the element d_n can be expressed as

$$d_n = (t^{\frac{1}{4}} e_{n-1} + t^{-\frac{1}{4}}) \cdots (t^{\frac{1}{4}} e_1 + t^{-\frac{1}{4}})\rho.$$

Consider now the link pattern tower

$$V_0(u) \xrightarrow{\phi_0} V_1(v) \xrightarrow{\phi_1} V_2(u) \xrightarrow{\phi_2} V_3(v) \xrightarrow{\phi_3} \cdots$$

where $u = t^{\frac{1}{4}} v + t^{-\frac{1}{4}} v^{-1}$. Consider the surjective intertwiners

$$\pi_{2k} : \mathrm{Ind}^{\mathcal{I}_{2k}}\big(V_{2k}(u)\big) \twoheadrightarrow V_{2k}(u) \widehat{\times}_f V_1(v),$$
$$\pi_{2k+1} : \mathrm{Ind}^{\mathcal{I}_{2k+1}}(V_{2k+1}(v)) \twoheadrightarrow V_{2k+1}(v) \widehat{\times}_f V_1(v^{-1}),$$

of TL_{2k+1}-modules and TL_{2k+2}-modules, respectively, defined by

$$\pi_{2k}\big(Y \otimes_{\mathrm{TL}_{2k}} w_{2k}\big) := Y \otimes_{\mathrm{TL}_{2k} \otimes \mathrm{TL}_1} (w_{2k} \otimes 1_v),$$
$$\pi_{2k+1}\big(Z \otimes_{\mathrm{TL}_{2k+1}} w_{2k+1}\big) := Z \otimes_{\mathrm{TL}_{2k+1} \otimes \mathrm{TL}_1} (w_{2k+1} \otimes 1_{v^{-1}})$$

for $Y \in \mathrm{TL}_{2k+1}$, $w_{2k} \in V_{2k}(u)$ and $Z \in \mathrm{TL}_{2k+2}$, $w_{2k+1} \in V_{2k+1}(v)$.

Proposition 9.2 *Let $u = t^{\frac{1}{4}} v + t^{-\frac{1}{4}} v^{-1}$. Let*

$$V_0(u) \xrightarrow{\phi_0} V_1(v) \xrightarrow{\phi_1} V_2(u) \xrightarrow{\phi_2} V_3(v) \xrightarrow{\phi_3} \cdots$$

be the link pattern tower. Then the intertwiner $\widehat{\phi}_n$ *factors through* π_n*. In other words, there exist unique intertwiners*

$$\psi_{2k} : V_{2k}(u)\widehat{\times}_f V_1(v) \longrightarrow V_{2k+1}(v),$$
$$\psi_{2k+1} : V_{2k+1}(v)\widehat{\times}_f V_1(v^{-1}) \longrightarrow V_{2k+2}(u)$$

of TL_{2k+1}*-modules and* TL_{2k+2}*-modules, respectively, such that* $\psi_n \circ \pi_n = \widehat{\phi}_n$ *for all* $n \in \mathbb{Z}_{\geq 0}$.

Proof We need to show that

$$\psi_{2k}\big(Y \otimes_{\mathrm{TL}_{2k}\otimes \mathrm{TL}_1} (w_{2k} \otimes \mathbf{1}_v)\big) := Y\phi_{2k}(w_{2k}),$$
$$\psi_{2k+1}\big(Z \otimes_{\mathrm{TL}_{2k+1}\otimes \mathrm{TL}_1} (w_{2k+1} \otimes \mathbf{1}_{v^{-1}})\big) := Z\phi_{2k+1}(w_{2k+1})$$

for $Y \in \mathrm{TL}_{2k+1}$, $w_{2k} \in V_{2k}(u)$ and $Z \in \mathrm{TL}_{2k+2}$, $w_{2k+1} \in V_{2k+1}(v)$ are well-defined linear maps. The balancing condition of the tensor product for the first tensor component of the algebra $\mathrm{TL}_n \otimes \mathrm{TL}_1$ is respected because of the intertwining properties of ϕ_{2k} and ϕ_{2k+1}. For instance, for $X \in \mathrm{TL}_{2k}$,

$$Y\epsilon_{2k,1}(X \otimes \mathbf{1})\phi_{2k}(w_{2k}) = Y\mathcal{I}_{2k}(X)\phi_{2k}(w_{2k}) = Y\phi_{2k}(Xw_{2k}).$$

For the balancing condition of the tensor product for the second tensor component of $\mathrm{TL}_n \otimes \mathrm{TL}_1$ we need to show that

$$\begin{aligned} d_{2k+1}\phi_{2k}(w_{2k}) &= v\,\phi_{2k}(w_{2k}), \\ d_{2k+2}\phi_{2k+1}(w_{2k+1}) &= v^{-1}\phi_{2k+1}(w_{2k+1}) \end{aligned} \tag{9.2}$$

for all $w_{2k} \in V_{2k}(u)$ and $w_{2k+1} \in V_{2k+1}(v)$. Write $w_{2k} = Y_u$ with $Y \in \mathrm{Hom}_{\mathcal{S}}(0, 2k)$ and $w_{2k+1} = Z_v$ with $Z \in \mathrm{Hom}_{\mathcal{S}}(1, 2k+1)$. Then

$$d_{2k+1}\phi_{2k}(Y_u) = \big(d_{2k+1}\mathcal{I}(Y)\big)_v = \big(\mathcal{I}(Y)\rho\big)_v = v\,\phi_{2k}(Y_u),$$

where we used (9.1) and the fact that $d_1 = \rho \in \mathrm{TL}_1$ for the second equality. To prove the second equality of (9.2), first note that

$$d_{2k+2}\phi_{2k+1}(Z_v) = \big(d_{2k+2}\mathcal{I}(Z)U\big)_u = \big(\mathcal{I}(Z)d_2U\big)_u$$

by (9.1). Now $d_2 = (t^{\frac{1}{4}}e_1 + t^{-\frac{1}{4}})\rho = \mathcal{I}_1(\rho^{-1})\rho^2$ in TL_2 by Proposition 6.3(**b**). Furthermore we have $\rho^2U = U$ in $\mathrm{Hom}_{\mathcal{S}}(0, 2)$, so we conclude that

$$d_{2k+2}\phi_{2k+1}(Z_v) = \big(\mathcal{I}(Z)\mathcal{I}_1(\rho^{-1})U\big)_u = v^{-1}\big(\mathcal{I}(Z)U\big)_u = v^{-1}\phi_{2k+1}(Z_v),$$

where the second step follows from the proof of Lemma 8.2. □

Corollary 9.3 *Let $u = t^{\frac{1}{4}}v + t^{-\frac{1}{4}}v^{-1}$ with $v^2 \neq t^{\frac{1}{2}}$. Let*

$$V_0(u) \xrightarrow{\phi_0} V_1(v) \xrightarrow{\phi_1} V_2(u) \xrightarrow{\phi_2} V_3(v) \xrightarrow{\phi_3} \cdots$$

be the link pattern tower. The left TL_{2k+1}*-module* $V_{2k+1}(v)$ *is a quotient of* $V_{2k}(u)\widehat{\times}_f V_1(v)$ *and the left* TL_{2k+2}*-module* $V_{2k+2}(u)$ *is a quotient of* $V_{2k+1}(v)\widehat{\times}_f V_1(v^{-1})$ *for all* $k \geq 0$.

Proof By Theorem 8.3(**ii**) the link pattern tower is nondegenerate, i.e., the $\widehat{\phi}_n$ are surjective for all $n \in \mathbb{Z}_{\geq 0}$. By the previous proposition we conclude that the intertwiners ψ_n are surjective for all $n \in \mathbb{Z}_{\geq 0}$. □

Example 9.4 Let $u = t^{\frac{1}{4}}v + t^{-\frac{1}{4}}v^{-1}$ with $v^2 \neq t^{\frac{1}{2}}$. Recall the identification of

$$V_{2k}(u) = \mathcal{W}_{0,t^{\frac{1}{4}}v}[2k], \qquad V_{2k+1}(v) = \mathcal{W}_{\frac{1}{2},v}[2k+1]$$

with the standard modules from [8] (see Remark 8.1). Then [8, (4.26)] shows that

$$V_1(v)\widehat{\times}_f V_1(v^{-1}) \simeq V_2(u).$$

Appendix: Relation to Affine Hecke Algebras and Affine Braid Groups, and type *B* Presentations

We first show how the algebra maps $\mathcal{I}_n$ can be lifted to extended affine Hecke algebras and to the group algebras of extended affine braid groups. We give the constructions below for $n \geq 3$. The adjustments needed for $n = 1, 2$ are left to the reader as long as they are obvious.

The affine Temperley-Lieb algebra $\overline{\mathrm{TL}}_n$ of type $\widehat{A}_{n-1}$ is the subalgebra of TL_n generated by $e_1, e_2, \ldots, e_n$, see [6]. The defining relations of $\overline{\mathrm{TL}}_n$ are given by the first three lines in (5.1). Note that $\mathbb{Z}$ acts on $\overline{\mathrm{TL}}_n$ by algebra automorphisms with $m \in \mathbb{Z}$ acting by $e_i \mapsto e_{i+m}$ (with the indices modulo n). Let TL_n^e be the corresponding crossed product algebra $\mathbb{Z} \ltimes \overline{\mathrm{TL}}_n$. Note that TL_n^e is isomorphic to the algebra generated by $e_1, \ldots, e_n, \rho^{\pm 1}$ with defining relations all but the last relation in (5.1). It follows that

$$\mathrm{TL}_n \simeq \mathrm{TL}_n^e / \langle \rho^2 e_{n-1} - e_1 e_2 \cdots e_{n-1} \rangle$$

with $\langle \rho^2 e_{n-1} - e_1 e_2 \cdots e_{n-1} \rangle$ the two-sided ideal generated by the element $\rho^2 e_{n-1} - e_1 e_2 \cdots e_{n-1}$.

In [6] the affine Temperley-Lieb algebra $\overline{\mathrm{TL}}_n$ is realized as a quotient of the affine Hecke algebra of type $\widehat{A}_{n-1}$. We recall this here, and give the extension to TL_n^e.

Definition A.1 The extended affine Hecke algebra H_n of type $\widehat{A}_{n-1}$ is the unital complex associative algebra with generators $T_1, T_2, \ldots, T_n, \rho, \rho^{-1}$ and defining relations

$$\begin{aligned} &(T_i - t^{-\frac{1}{2}})(T_i + t^{\frac{1}{2}}) = 0, \\ &T_iT_j = T_jT_i \qquad \text{if } i-j \neq \pm 1, \\ &T_iT_{i+1}T_i = T_{i+1}T_iT_{i+1}, \\ &\rho T_i = T_{i+1}\rho, \\ &\rho\rho^{-1} = 1 = \rho^{-1}\rho, \end{aligned} \tag{A.1}$$

where the indices are taken modulo n.

Note that $T_i \in H_n$ is invertible with inverse $T_i^{-1} = T_i - t^{-\frac{1}{2}} + t^{\frac{1}{2}}$. The affine Hecke algebra of type $\widehat{A}_{n-1}$ is the subalgebra $\overline{H}_n$ of H_n generated by $T_1, T_2, \ldots, T_n$. The defining relations of $\overline{H}_n$ are given by the first three lines in (A.1). The extended affine Hecke algebra H_n is isomorphic to the crossed product algebra $\mathbb{Z} \ltimes \overline{H}_n$, where $m \in \mathbb{Z}$ acts on $\overline{H}_n$ by the algebra automorphism $T_i \mapsto T_{i+m}$ (with the indices modulo n).

Proposition A.2 *There exists a unique surjective algebra map* $\psi_n : H_n \to \mathrm{TL}_n^e$ *satisfying* $\rho \mapsto \rho$ *and* $T_i \mapsto e_i + t^{-\frac{1}{2}}$. *The kernel of* ψ_n *is the two-sided ideal in* H_n *generated by the elements*

$$T_iT_{i+1}T_i - t^{-\frac{1}{2}}T_iT_{i+1} - t^{-\frac{1}{2}}T_{i+1}T_i + t^{-1}T_i + t^{-1}T_{i+1} - t^{-\frac{3}{2}}, \qquad i \in \mathbb{Z}/n\mathbb{Z}. \tag{A.2}$$

Proof Fan and Green [6] showed that the kernel of the unique surjective algebra map $\overline{\psi}_n : \overline{H}_n \to \overline{\mathrm{TL}}_n$ satisfying $T_i \mapsto e_i + t^{-\frac{1}{2}}$ is generated by the elements (A.2) (see also [12]). The proposition now follows since the $\mathbb{Z}$-actions on $\overline{H}_n$ and $\overline{\mathrm{TL}}_n$ are intertwined by $\overline{\psi}_n$. □

The extended affine braid group $\mathcal{B}_n$ is the group generated by $\sigma_1, \sigma_2, \ldots, \sigma_n, \widetilde{\rho}$ with defining relations

$$\begin{aligned} &\sigma_i\sigma_j = \sigma_j\sigma_i \qquad \text{if } i-j \neq \pm 1, \\ &\sigma_i\sigma_{i+1}\sigma_i = \sigma_{i+1}\sigma_i\sigma_{i+1}, \\ &\widetilde{\rho}\sigma_i = \sigma_{i+1}\widetilde{\rho}, \end{aligned} \tag{A.3}$$

where the indices are taken modulo n, see, e.g., [12]. Recall that $\mathcal{B}_n$ can be realized topologically in terms of n strands in $\mathbb{C}^* \times [0, 1]$ starting at $\{(2\xi_n^{j-1}, 0)\}_{j=1}^n$ and ending at $\{(2\xi_n^{j-1}, 1)\}_{j=1}^n$,

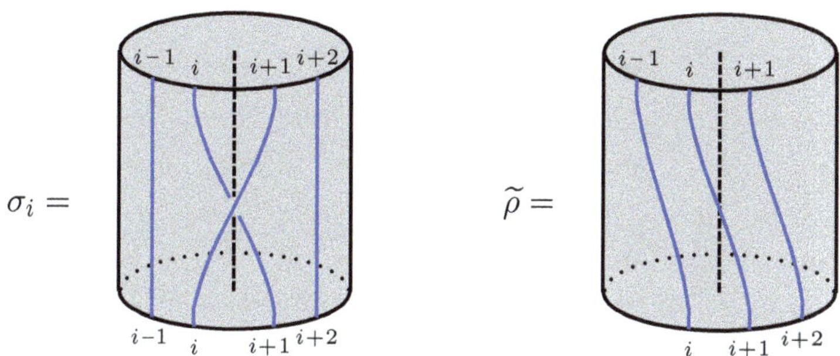

Given a braid in $\mathcal{B}_n$, project it onto the cylinder $C_o \times [0, 1]$ and map $C_o \times [0, 1]$ homeomorphically onto A by collapsing the wall of the cylinder inwards onto the domain $A \times \{0\}$. This results in an (n, n)-tangle diagram in A, which we subsequently interpret as an element in the linear skein $\mathrm{End}_{\mathcal{S}}(n)$. This defines a surjective algebra map $\mu_n : \mathbb{C}[\mathcal{B}_n] \to \mathrm{End}_{\mathcal{S}}(n)$ satisfying

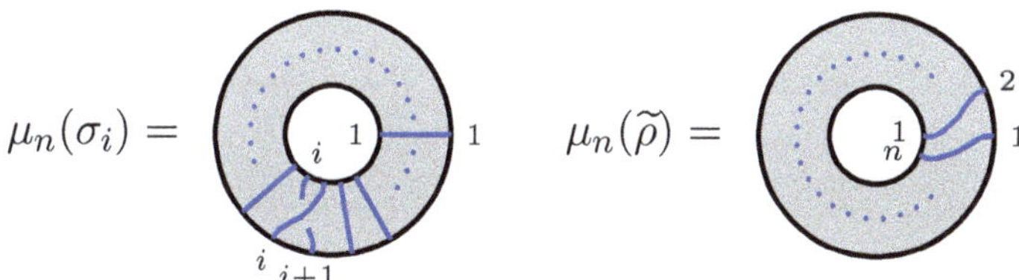

Note that $\mu_n(\widetilde{\rho}) = \rho$ and $\mu_n(\sigma_i) = t^{\frac{1}{4}} e_i + t^{-\frac{1}{4}}$, where the last equality follows from the Kauffman skein relation (3.1). Note also that $\mu_n(\sigma_i^{-1}) = t^{-\frac{1}{4}} e_i + t^{\frac{1}{4}}$.

Remark A.3 Denote by $\nu_n : \mathbb{C}[\mathcal{B}_n] \to H_n$ be the surjective algebra map satisfying $\nu_n(\widetilde{\rho}) = \rho$ and $\nu_n(\sigma_i) = t^{\frac{1}{4}} T_i$, then we have $\psi_n \circ \nu_n = \mu_n$.

Let $\mathcal{I}_n^{br} : \mathcal{B}_n \to \mathcal{B}_{n+1}$ be the group homomorphism that topologically is described by sticking in an additional strand between the nth and the first strand, with the new strand running "behind" all other strands (but not wrapping around the pole). For example,

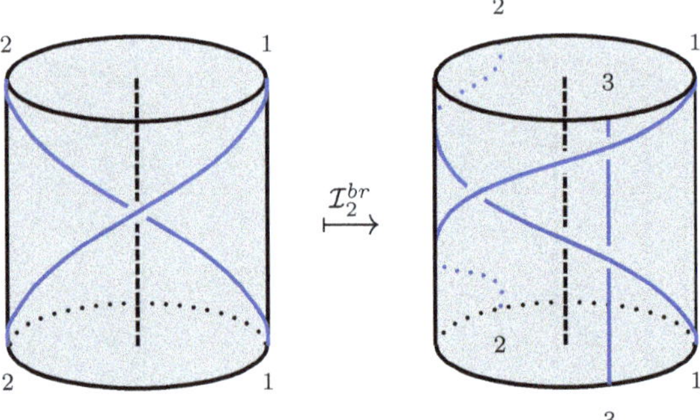

It is the unique group homomorphism satisfying

$$\mathcal{I}_n^{br}(\sigma_i) = \sigma_i, \qquad i = 1, \ldots, n-1,$$
$$\mathcal{I}_n^{br}(\sigma_n) = \sigma_n \sigma_{n+1} \sigma_n^{-1},$$
$$\mathcal{I}_n^{br}(\widetilde{\rho}) = \widetilde{\rho} \sigma_n^{-1}.$$

Extending $\mathcal{I}_n^{br}$ linearly to an algebra map $\mathcal{I}_n^{br} : \mathbb{C}[\mathcal{B}_n] \to \mathbb{C}[\mathcal{B}_{n+1}]$, we have

$$\mu_{n+1} \circ \mathcal{I}_n^{br} = \mathcal{I}|_{\mathcal{S}_n} \circ \mu_n$$

with $\mu_n : \mathbb{C}[\mathcal{B}_n] \to \mathcal{S}_n$ the algebra map as defined in the previous section.

In addition, it is easy to show that there exists a unique unit preserving algebra map $\mathcal{I}_n^{ha} : H_n \to H_{n+1}$ satisfying

$$\mathcal{I}_n^{ha}(T_i) = T_i, \qquad i = 1, \ldots, n-1,$$
$$\mathcal{I}_n^{ha}(T_n) = T_n T_{n+1} T_n^{-1},$$
$$\mathcal{I}_n^{ha}(\rho) = t^{-\frac{1}{4}} \rho T_n^{-1},$$

and

$$\nu_{n+1} \circ \mathcal{I}_n^{br} = \mathcal{I}_n^{ha} \circ \nu_n$$

with $\nu_n : \mathbb{C}[\mathcal{B}_n] \to H_n$ as defined in the previous section.

Remark A.4 The maps $\mathcal{I}_n^{br}$ and $\mathcal{I}_n$ were constructed before in [2, 8].

We end the section by discussing the relation to the braid group and the affine Temperley-Lieb algebra of type B. Let $\mathcal{B}_n^B$ be the braid group of type B_n, i.e., the group with generators $\sigma_0^B, \ldots, \sigma_{n-1}^B$ and defining relations the braid relations associated with the type B Coxeter diagram

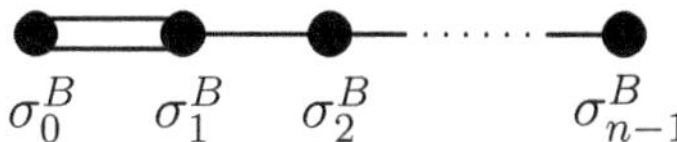

It is known that $\mathcal{B}_n^B$ is isomorphic to the extended affine braid group $\mathcal{B}_n$, with the isomorphism given by

$$\begin{aligned} \sigma_0^B &\mapsto \rho\sigma_{n-1}^{-1}\cdots\sigma_1^{-1} \\ \sigma_i^B &\mapsto \sigma_i \qquad (1 \leq i < n), \end{aligned} \tag{A.4}$$

see [7, Rem. 1.1] and the references therein. We discuss now a similar B-type presentation of the extended affine Temperley-Lieb algebra TL_n.

The B-type affine Temperley-Lieb algebra TL_n^B is defined as follows (see [23, Thm. 3.13]). For $n \geq 2$, TL_n^B is the unital complex associative algebra with generators $\alpha, \tau, e_1, \cdots, e_{n-1}$ and defining relations

$$\begin{aligned} e_i^2 &= \left(-t^{\frac{1}{2}} - t^{-\frac{1}{2}}\right)e_i, \\ e_ie_j &= e_je_i && \text{if } |j-i| \geq 2, \\ e_ie_{i\pm1}e_i &= e_i, \\ \tau e_i &= e_i\tau && \text{if } i > 1, \\ e_1\tau e_1 &= \alpha e_1 = e_1\alpha, \\ \tau^2 &= -t^{\frac{1}{2}}\alpha\tau - t. \end{aligned}$$

For $n = 0$ and $n = 1$ we set $\mathrm{TL}_0^B := \mathbb{C}[\alpha]$ and $\mathrm{TL}_1^B := \mathbb{C}[\tau, \tau^{-1}]$.

Note that τ is invertible with inverse $\tau^{-1} = -t^{-1}\tau - t^{-\frac{1}{2}}\alpha$. It follows that $\alpha = -t^{-\frac{1}{2}}\tau - t^{\frac{1}{2}}\tau^{-1}$, and α is central. For $n = 1$ we define α by this formula.

Note that the assignments

$$\begin{aligned} \sigma_0^B &\mapsto -t^{-\frac{3}{4}}\tau \\ \sigma_i^B &\mapsto t^{\frac{1}{4}}e_i + t^{-\frac{1}{4}} \qquad (1 \leq i < n) \end{aligned}$$

define a surjective algebra map $\mu_n^B : \mathbb{C}[\mathcal{B}_n^B] \to \mathrm{TL}_n^B$. In particular, TL_n^B is isomorphic to a quotient of the group algebra $\mathbb{C}[\mathcal{B}_n^B]$.

Recall the algebra map $\mu_n : \mathbb{C}[\mathcal{B}_n] \to \mathrm{TL}_n$ from Sect. 7. The following result is an algebraic reformulation of [23, Thm. 3.13(a)], see also Remark A.6.

Proposition A.5 *There exists a unique isomorphism* $\mathrm{TL}_n^B \xrightarrow{\sim} \mathrm{TL}_n$ *of algebras such that the diagram*

$$\begin{array}{ccc} \mathbb{C}[\mathcal{B}_n^B] & \xrightarrow{\sim} & \mathbb{C}[\mathcal{B}_n] \\ \downarrow{\scriptstyle \mu_n^B} & & \downarrow{\scriptstyle \mu_n} \\ \mathrm{TL}_n^B & \xrightarrow{\sim} & \mathrm{TL}_n \end{array}$$

of algebra maps is commutative, with the isomorphism $\mathbb{C}[\mathcal{B}_n^B] \xrightarrow{\sim} \mathbb{C}[\mathcal{B}_n]$ *given by* (A.4).

Proof It is to show there exists a well-defined algebra map $f_n : \mathrm{TL}_n^B \to \mathrm{TL}_n$ satisfying $f_n(e_i) = e_i$ $(1 \leq i < n)$ and

$$f_n(\tau) = -t^{\frac{3}{4}} \rho \mu_n(\sigma_{n-1}^{-1}) \cdots \mu_n(\sigma_1^{-1}),$$

and that there exists a well-defined algebra map $g_n : \mathrm{TL}_n \to \mathrm{TL}_n^B$ satisfying $g_n(e_i) = e_i$ $(1 \leq i < n)$ and

$$g_n(\rho) = -t^{-\frac{3}{4}} \tau \mu_n^B(\sigma_1^B) \cdots \mu_n^B(\sigma_{n-1}^B).$$

We omit the proof as it is a straightforward check that all the algebra relations are respected by f_n and g_n. For these checks it is convenient to use the presentation of TL_n in terms of the generators $\rho^{\pm 1}, e_1, \ldots, e_{n-1}$ as given in Remark 5.1. □

Remark A.6 Combining Proposition A.5 with Theorem 5.3 and Remark 5.4 yields an isomorphism $\mathrm{TL}_n^B \xrightarrow{\sim} \mathrm{End}_{\mathcal{S}}(n)$ given by

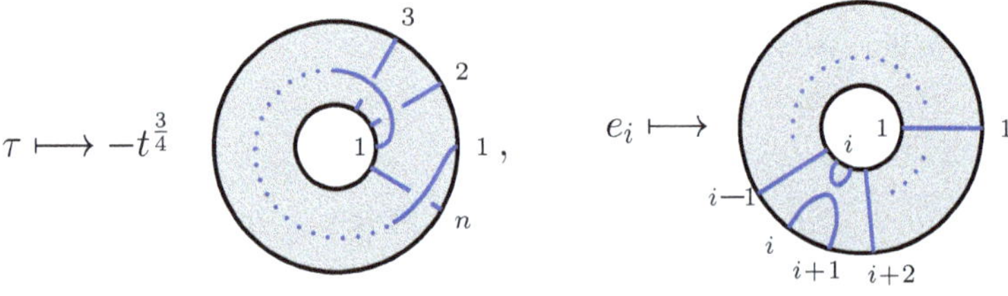

for $1 \leq i < n$. Note that under this isomorphism,

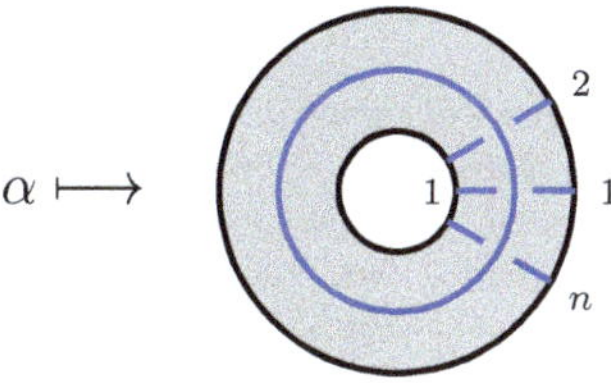

This is the algebra isomorphism $\mathrm{TL}_n^B \xrightarrow{\sim} \mathrm{End}_{\mathcal{S}}(n)$ from [23, Thm. 3.13(a)].

Using the B type presentation of TL_n the algebra maps $\mathcal{I}_n : \mathrm{TL}_n \to \mathrm{TL}_{n+1}$ take on a simple form.

Corollary A.7 *The algebra maps $\mathcal{I}_n : \mathrm{TL}_n \to \mathrm{TL}_{n+1}$, viewed as algebra maps $\mathrm{TL}_n^B \to \mathrm{TL}_{n+1}^B$ via the identification $\mathrm{TL}_n^B \cong \mathrm{TL}_n$, satisfy $\tau \mapsto \tau$ and $e_i \mapsto e_i$ for $1 \leq i < n$.*

Using Corollary A.7 in combination with [23, Thm. 3.13(b)] it follows that the algebra maps $\mathcal{I}_n : \mathrm{TL}_n \to \mathrm{TL}_{n+1}$ are injective.

Acknowledgments The authors would like to thank Bernard Nienhuis as motivation for this paper came from the joint work [3]. We thank two anonymous referees for their comments which led to substantial improvements of the paper. The second author thanks Nicolai Reshetikhin for the wonderful and inspirational collaborations in the past decade. The work by Kayed Al Qasimi is supported by the Ministry of Education of the United Arab Emirates. Diagrams were coded using PSTricks.

References

1. F. Bonahon, H. Wong, *Representations of the Kauffman bracket skein algebra I: invariants and miraculous cancellations*, Invent. Math. **204** (2016), 195–243.
2. S. Al Harbat, *Markov trace on a tower of affine Temperley-Lieb algebras of type A*, J. Knot Theory Ramifications **24** (2015), 1550049, 28pp.
3. K. Al Qasimi, B. Nienhuis, J.V. Stokman, *Towers of solutions of qKZ equations and their applications to loop models*, Ann. Henri Poincaré **20** (2019), 3743–3797.
4. A. Cox, P. Martin, A. Parker, C. Xi, *Representation theory of towers of recollement: theory, notes and examples*, Journal of Algebra **302** (2006), 340–360.
5. P. Di Francesco, P. Zinn-Justin, J.-B. Zuber, *Sum rules for the ground states of the* O(1) *loop model on a cylinder and the XXZ spin chain*, J. Stat. Mech. (2006), P08011, 22pp.
6. C.K. Fan, R.M. Green, *On the affine Temperley-Lieb algebras*, J. London Math. Soc. (2) **60** (1999), 366–380.
7. A. Gadbled, A.-L. Thiel and E. Wagner,*Categorical action of the extended braid group of affine type A*, Commun. Contemp. Math. **19** (2017), 1650024, 39 pp.
8. A.M. Gainutdinov, H. Saleur, *Fusion and braiding in finite and affine Temperley-Lieb categories*, arXiv:1606.04530.
9. J. de Gier, *Loops, matchings and alternating-sign matrices*, Discrete Math. **298** (2005), 365–388,

10. J. de Gier, P. Pyatov, *Factorized solutions of Temperley-Lieb qKZ equations on a segment*, Adv. Theor. Math. Phys. **14** (2010), 795–877.
11. J.J. Graham, G.I. Lehrer, *The representation theory of affine Temperley-Lieb algebras*, Enseign. Math. (2) **44** (1998), 173–218.
12. J.J. Graham, G.I. Lehrer, *Diagram algebras, Hecke algebras and decomposition numbers at roots of unity*, Ann. Sci. École Norm. Sup. (4) **36** (2003), 479–524.
13. R.M. Green, *On representations of affine Temperley-Lieb algebras*, in: “Algebras and modules, II” (Geiranger, 1990), 245–261, CMS Conf. Proc., **24**, Amer. Math. Soc., Providence, RI, 1998.
14. V.F.R. Jones, *Planar algebras. I*, arXiv:math/9909027.
15. M. Kasatani, V. Pasquier, *On polynomials interpolating between the stationary state of a $O(n)$ model and a Q.H.E. ground state*, Comm. Math. Phys. **276** (2007), 397–435.
16. L.H. Kauffman, *State models for knot polynomials*, Topology **26** (1987), 395–407.
17. L.H. Kauffman, *An invariant of regular isotopy*, Trans. Amer. Math. Soc. **318** (1990), 417–471.
18. T. Le, *On Kauffman bracket skein modules at roots of unity*, Algebr. Geom. Topol. **15** (2015), 1093–1117.
19. W.B.R. Lickorish, *The skein method for three-manifold invariants*, J. Knot Theory Ramifications **2** (1993), 171–194.
20. H.R. Morton, *Invariants of links and 3-manifolds from skein theory and from quantum groups*, in: “Topics in knot theory” (Erzurum, 1992), 107–155, NATO Adv. Sci. Inst. Ser. C Math. Phys. Sci., **399**, Kluwer Acad. Publ., Dordrecht, 1993.
21. H.R. Morton, *Skein theory and the Murphy operators*, in: “Knots 2000 Korea, Vol. 2” (Yongpyong), J. Knot Theory Ramifications **11** (2002), 475–492.
22. J.H. Przytycki, *Skein modules of 3-manifolds*, (English summary) Bull. Polish Acad. Sci. Math. **39** (1991), no. 1–2, 91–100.
23. J.H. Przytycki, *Fundamentals of Kauffman bracket skein modules*, Kobe J. Math. **16** (1999), 45–66.
24. J.H. Przytycki, A.S. Sikora, *Skein algebras of surfaces*, Trans. Amer. Math. Soc. **371** (2019), 1309–1332.
25. J. Roger, T. Yang, *The skein algebra of arcs and links and the decorated Teichmüller space*, J. Differential Geom. **96** (2014), no. 1, 95–140.
26. V.G. Turaev, *The Conway and Kauffman modules of a solid torus*, (Russian) Zap. Nauchn. Sem. Leningrad. Otdel. Mat. Inst. Steklov. (LOMI) **167** (1988), Issled. Topol. **6**, 79–89, 190; English translation in J. Soviet Math. **52** (1988), 2799–2805.
27. V.G. Turaev, *Quantum invariants of knots and 3-manifolds*, 3rd edition. De Gruyter Studies in Math., **18**. De Gruyter, Berlin, 2016.

Tensor Product of the Fock Representation with Its Dual and the Deligne Category

Vera Serganova

To Kolya Reshetikhin for his 60th birthday

Abstract We describe $\mathfrak{sl}(\infty)$-module structure of the tensor product of the Fock representation and its shifted dual using action of $\mathfrak{sl}(\infty)$ on the abelian envelope of the Deligne's category $GL(t)$.

1 Introduction

Let $\mathbb{V} := \mathbb{C}^{\mathbb{Z}}$ be a countable dimensional vector space with fixed basis $\{u_i \mid i \in \mathbb{Z}\}$. Consider the Lie algebra $\mathfrak{sl}(\infty)$ of all traceless linear operators in $\mathbb{C}^{\mathbb{Z}}$ annihilating almost all u_i. Clearly, $\mathfrak{sl}(\infty)$ can be identified with the Lie algebra of traceless infinite matrices with finitely many non-zero entries. We consider $\mathfrak{sl}(\infty)$ as a Kac–Moody Lie algebra associated with Dynkin diagram A_∞. The Chevalley–Serre generators $e_a, f_a, a \in \mathbb{Z}$ of $\mathfrak{sl}(\infty)$ act on $\mathbb{V}$ by

$$f_a u_b = \delta_{a,b} u_{b+1}, \quad e_a u_b = \delta_{a+1,b} u_{b-1}.$$

The fermionic Fock space $\mathfrak{F}$ is a simple $\mathfrak{sl}(\infty)$-module with fundamental highest weight ω_{-1}. It has a realization as the "semi-infinite exterior power" $\Lambda^{\infty/2}\mathbb{C}^{\mathbb{Z}}$ that is the span of all formal expressions $u_{i_1} \wedge u_{i_2} \wedge \dots$ satisfying the conditions $i_j > i_{j+1}$ for all $j \geq 1$ and $i_k = -k$ for sufficiently large k. In this way, the highest weight vector is $u_\emptyset := u_{-1} \wedge u_{-2} \wedge \dots$. The famous boson–fermion correspondence identifies $\mathfrak{F}$ with the space of symmetric functions. That in particular implies that $\mathfrak{F}$

V. Serganova (✉)
Department of Mathematics, University of California at Berkeley, Berkeley, CA, USA
e-mail: serganov@math.berkeley.edu

A. Alekseev et al. (eds.), *Representation Theory, Mathematical Physics, and Integrable Systems*, Progress in Mathematics 340,
https://doi.org/10.1007/978-3-030-78148-4_19

has a natural basis $\{u_\lambda\}$ enumerated by partitions λ (this basis corresponds to Schur functions) where

$$u_\lambda := u_{\lambda_1-1} \wedge u_{\lambda_2-2} \wedge u_{\lambda_3-3} \wedge \dots .$$

Let $t \in \mathbb{Z}$. We denote by $\mathfrak{F}_t^\vee$ the simple $\mathfrak{sl}(\infty)$-module with lowest weight $-\omega_{t-1}$. We will use the following realization of $\mathfrak{F}_t^\vee$. Set $\mathbb{V}^\vee = \mathbb{C}^\mathbb{Z}$ with basis $\{w_i \mid i \in \mathbb{Z}\}$ and define the action of e_a, f_a on $\mathbb{V}^\vee$ by

$$e_a w_b = \delta_{a,b} w_{b+1}, \quad f_a w_b = \delta_{a+1,b} w_{b-1}.$$

Then, $\mathfrak{F}_t^\vee$ is the span of all formal expressions $w_{i_1} \wedge w_{i_2} \wedge \dots$ satisfying the conditions $i_j > i_{j+1}$ for all $j \geq 1$ and $i_k = t - k$ for sufficiently large k. We can enumerate the elements of the basis of $\mathfrak{F}_t^\vee$ by partitions

$$w_\mu := w_{\mu_1+t-1} \wedge w_{\mu_2+t-2} \wedge w_{\mu_3+t-3} \wedge \dots .$$

The goal of this chapter is to describe the structure of $\mathfrak{F}_t^\vee \otimes \mathfrak{F}$. Let us consider $(m, n) \in \mathbb{Z}^2$ such that $m - n = t$. As follows from [PS], $\Lambda^m\mathbb{V}^\vee \otimes \Lambda^n\mathbb{V}$ is an indecomposable $\mathfrak{sl}(\infty)$-module with simple socle $S_{m,n}$. To describe this socle, consider the contraction map $c : \mathbb{V} \otimes \mathbb{V}^\vee \to \mathbb{C}$ given by $c(w_i \otimes u_j) = (-1)^j \delta_{i,j}$ and extend it to $c_{m,n} : \Lambda^m\mathbb{V}^\vee \otimes \Lambda^n\mathbb{V} \to \Lambda^{m-1}\mathbb{V}^\vee \otimes \Lambda^{n-1}\mathbb{V}$. Then, $S_{m,n}$ is the kernel of $c_{m,n}$.

Theorem 1.1

(1) *The $\mathfrak{sl}(\infty)$-module $\mathfrak{R} := \mathfrak{F}_t^\vee \otimes \mathfrak{F}$ has an infinite decreasing filtration*

$$\mathfrak{R} := \mathfrak{R}^0 \supset \mathfrak{R}^1 \supset \dots \supset \mathfrak{R}^k \supset \dots$$

such that $\cap_k \mathfrak{R}^k = 0$ *and*

$$\mathfrak{R}^k/\mathfrak{R}^{k+1} \simeq \begin{cases} S_{k+t,k} & \text{if } t \geq 0, \\ S_{k,k-t} & \text{if } t < 0. \end{cases}$$

(2) *Every non-zero submodule of $\mathfrak{R}$ coincides with $\mathfrak{R}^r$ for some $r \geq 0$.*

The proof of this theorem is based on categorification of $\mathfrak{F}_t^\vee \otimes \mathfrak{F}$ by the complexified Grothendieck group $K[\mathcal{V}_t]_\mathbb{C}$ of the abelian envelope $\mathcal{V}_t$ of the Deligne category $\operatorname{Rep} GL_t$ explained in [E] and Brundan categorification of $\Lambda^m\mathbb{V}^\vee \otimes \Lambda^n\mathbb{V}$ via representation theory of the supergroup $GL(m|n)$, [B]. We use the symmetric monoidal functor

$$DS_{m,n} : \mathcal{V}_t \to \operatorname{Rep} GL(m|n)$$

for $m - n = t$. Existence of such functor follows from construction of $\mathcal{V}_t$, see [EHS]. While $DS_{m,n}$ is not exact, it has a certain property, see Lemma 2.3 below, which allows to define the linear map

$$ds_{m,n} : K[\mathcal{V}_t]_{\mathbb{C}} \to K_{red}[\mathrm{Rep}\, GL(m|n)]_{\mathbb{C}}$$

where by K_{red}, we denote the quotient of the Grothendieck group K by the relation $[\mathbb{C}^{0|1}] = -[\mathbb{C}]$ in the category $\mathrm{Rep}\, GL(m|n)$. Furthermore, $ds_{m,n}$ is a homomorphism of rings and also a homomorphism of $\mathfrak{sl}(\infty)$-modules. We prove that the quotients $\mathrm{Ker} ds_{m-1,n-1}/\mathrm{Ker} ds_{m,n}$ form the layers of the radical filtration of $\mathfrak{F}_t^\vee \otimes \mathfrak{F} \simeq K[\mathcal{V}_t]_{\mathbb{C}}$. Let us warn the reader that the image of $ds_{m,n}$ is not $\Lambda^m \mathbb{V}^\vee \otimes \Lambda^n \mathbb{V}$ but another submodule in $K_{red}[\mathrm{Rep}\, GL(m|n)]_{\mathbb{C}}$. While this submodule has the same Jordan–Holder series as $\Lambda^m \mathbb{V}^\vee \otimes \Lambda^n \mathbb{V}$, it is not isomorphic to $\Lambda^m \mathbb{V}^\vee \otimes \Lambda^n \mathbb{V}$ as an $\mathfrak{sl}(\infty)$-module.

The second part of the paper contains calculation of dimensions of certain objects in $\mathcal{V}_t$.

2 The Category Rep $GL(m|n)$ and DS Functors

2.1 *Translation Functors*

Let $\mathrm{Rep}\, GL(m|n)$ denote the category of finite-dimensional $GL(m|n)$-modules. Let $\mu = (a_1, \dots, a_m | b_1, \dots b_n) \in \mathbb{Z}^{m+n}$ satisfy the condition $a_1 \geq a_2 \geq \cdots \geq a_m, b_1 \geq b_2 \geq \cdots \geq b_n$. For every such μ, there are three canonical objects in $\mathrm{Rep}\, GL(m|n)$:

1. The simple module $S(\mu)$ with highest weight μ.
2. The Kac module $K(\mu) := U(\mathfrak{gl}(m|n)) \otimes_{U(\mathfrak{p})} S_0(\mu)$, where $\mathfrak{p}$ is the parabolic subalgebra with Levi subalgebra $\mathfrak{gl}(m|n)_{\bar{0}}$ and $S_0(\mu)$ is the simple $\mathfrak{gl}(m|n)_{\bar{0}}$-module with highest weight μ.
3. The indecomposable projective cover $P(\mu)$ of $S(\mu)$.

The category $\mathrm{Rep}\, GL(m|n)$ is the highest weight category, [Z]. We denote by $K_{red}[\mathrm{Rep}\, GL(m|n)]$ the reduced Grothendieck group of $\mathrm{Rep}\, GL(m|n)$ and set

$$J_{m|n} := K_{red}[\mathrm{Rep}\, GL(m|n)] \otimes_{\mathbb{Z}} \mathbb{C}.$$

It was a remarkable discovery of J. Brundan that $J_{m|n}$ has a natural structure of $\mathfrak{sl}(\infty)$-module, [B]. To define it, let us consider translation functors E_a, F_a : $\mathrm{Rep}\, GL(m|n) \to \mathrm{Rep}\, GL(m|n)$ defined in the following way. There is a canonical $\mathfrak{gl}(m|n)$-invariant map $\omega : \mathbb{C} \to \mathfrak{gl}(m|n) \otimes \mathfrak{gl}(m|n)$ usually called the Casimir element. Let $V_{m|n}$ be the standard $GL(m|n)$-module and M be an arbitrary object of $\mathrm{Rep}\, GL(m|n)$. Let Ω be the composition map

$$\mathbb{C}\otimes M\otimes V_{m|n}\xrightarrow{\omega\otimes\mathrm{id}}\mathfrak{gl}(m|n)\otimes\mathfrak{gl}(m|n)\otimes M\otimes V_{m|n}\xrightarrow{\mathrm{id}\otimes s\otimes\mathrm{id}}$$

$$\mathfrak{gl}(m|n)\otimes M\otimes\mathfrak{gl}(m|n)\otimes V_{m|n}\xrightarrow{a_M\otimes a_{V_{m|n}}} M\otimes V_{m|n},$$

where s is the braiding in Rep $GL(m|n)$ defined by the sign rule and $a_M, a_{V_{m|n}}$ are the action maps. Let $E_a(M)$ be the generalized eigenspace of Ω in $M\otimes V_{m|n}$ with eigenvalue a. Similarly, we define $F_a(M)$ as the generalized eigenspace of Ω' in $M\otimes V^*_{m|n}$ with eigenvalue a, where Ω' is defined as above with substitution of $V^*_{m|n}$ in place of $V_{m|n}$.

The following theorem is a direct consequence of results in [B]:

Theorem 2.1

(1) *E_a, F_a are non-zero only for $a\in\mathbb{Z}$.*
(2) *E_a, F_a are biadjoint exact endofunctors of* Rep $GL(m|n)$.
(3) *Let e_a, $f_a : J_{m|n}\to J_{m|n}$ be the induced $\mathbb{C}$-linear maps. Then, e_a, f_a satisfy the Chevalley–Serre relations for A_∞. Hence, $J_{m|n}$ is an $\mathfrak{sl}(\infty)$-module.*
(4) *The subspace of $\Lambda_{m|n}\subset J_{m|n}$ generated by classes of all Kac modules $[K(\mu)]$ is an $\mathfrak{sl}(\infty)$-submodule isomorphic to $\Lambda^m\mathbb{V}^\vee\otimes\Lambda^n\mathbb{V}$.*

We need the exact description of the socle filtration of $J_{m|n}$ obtained in [HPS], Corollary 29.

Proposition 2.2 *The $\mathfrak{sl}(\infty)$-module $J_{m|n}$ has a finite length. Furthermore, the socle filtration of $J_{m|n}$ is given by the formula*

$$\mathrm{soc}^i(J_{m|n})/\mathrm{soc}^{i-1}(J_{m|n})\simeq S^{\oplus i}_{m-i+1|n-i+1}.$$

In particular, the socle of $J_{m|n}$ is a simple $\mathfrak{sl}(\infty)$-module isomorphic to $S_{m|n}$. It is identified with the subspace generated by classes of all projective modules $[P(\mu)]$.

2.2 *DS-functor*

Fix an odd $x\in\mathfrak{gl}(m|n)_{\bar 1}$ such that $[x,x]=0$ and $\mathrm{rk}\,x=1$. Define a functor DS_x from Rep $GL(m|n)$ to the category of vector superspaces by setting

$$DS_x(M)=\mathrm{Ker}x_M/\mathrm{Im}x_M.$$

It is shown in [DS] that M_x has a natural structure of $GL(m-1|n-1)$-module and DS_x is a symmetric monoidal functor

$$\mathrm{Rep}\,GL(m|n)\to\mathrm{Rep}\,GL(m-1|n-1).$$

Furthermore, although DS_x is not an exact functor, it has the following property pointed out by V. Hinich. For the proof, see [HPS] Lemma 30.

Lemma 2.3 *Every exact sequence* $0 \to N \to M \to K \to 0$ *of* $GL(m|n)$*-modules induces the exact sequence*

$$0 \to E \to DS_xN \to DS_xM \to DS_xK \to E' \to 0$$

for certain $E \in \operatorname{Rep} GL(m-1|n-1)$ *and* $E' \simeq E \otimes \mathbb{C}^{0|1}$.

It follows immediately from Lemma 2.3 that DS_x induces a homomorphism of complexified reduced Grothendieck groups $ds_x : J_{m|n} \to J_{m-1|n-1}$. While DS_x and DS_y are not isomorphic if x and y are not conjugate by the adjoint action of $GL(m) \times GL(n)$, the homomorphism ds_x does not depend on a choice of x. In [HR], the homomorphism ds_x was constructed explicitly in terms of supercharacters and the kernel of ds_x was computed.

Lemma 2.4

(1) DS_x *commutes with translation functors* E_a, F_a, *and hence,* DS_x *induces a homomorphism* $ds_x : J_{m|n} \to J_{m-1|n-1}$ *of* $\mathfrak{sl}(\infty)$*-modules.*
(2) *The kernel of* ds_x *coincides with* $\Lambda_{m|n}$.

Proof For (1), see Lemma 32 in [HPS]. For (2), see [HR]. □

3 The Category $\mathcal{V}_t$, Translation Functors, and Categorification

3.1 The Deligne Category $\mathcal{D}_t$

In [DM], Deligne and Milne constructed a family $\{D_t = \operatorname{Rep} GL_t \mid t \in \mathbb{C}\}$ of symmetric monoidal rigid categories satisfying the following properties:

1. $\mathcal{D}_t$ is a universal additive symmetric monoidal Karoubian category generated by a dualizable object V_t of dimension t.
2. The indecomposable objects of $\mathcal{D}_t$ are in bijection with bipartitions $\lambda = (\lambda^\bullet, \lambda^\circ)$; we denote the corresponding indecomposable objects by $T(\lambda)$.
3. If $t \notin \mathbb{Z}$, then $\dim\operatorname{Hom}(T(\lambda), T(\nu)) = \delta_{\lambda,\mu}$, and hence, the category $\mathcal{D}_t$ is an abelian semisimple category.
4. If $t \in \mathbb{Z}$, and $m-n = t$, then there exists a (unique up to isomorphism) symmetric monoidal functor $F_{m|n} : \mathcal{D}_t \to \operatorname{Rep} GL(m|n)$ that sends V_t to $V_{m|n}$. This functor is full.

The functor $F_{m|n}$ was studied in [CW]. In particular, it was computed on the indecomposable objects of $\mathcal{D}_t$. We call a bipartition $\lambda = (\lambda^\bullet, \lambda^\circ)$ an $(m|n)$-cross if for there exists $0 \leq k \leq m$ such that $\lambda^{\bullet}_{k+1} + (\lambda^{\circ})^T_{m-k+1} \leq n$. Here, μ^T stands for the conjugate of μ. Denote by $C(m|n)$ the set of all $(m|n)$-crosses.

Theorem 3.1

(1) *$F_{m|n} T(\lambda) \neq 0$ if and only if $\lambda \in C(m|n)$.*
(2) *The set $\{F_{m|n} T(\lambda) \mid \lambda \in C(m|n)\}$ is a complete set of pairwise non-isomorphic indecomposable direct summands in tensor powers $V_{m|n}^{\otimes p} \otimes (V_{m|n}^*)^{\otimes q}$ for $p, q \geq 0$.*

Proof The first statement is Theorem 8.7.6 in [CW], and the second is the particular case of Theorem 4.7.1 in [CW]. □

3.2 The Abelian Envelope of $\mathcal{D}_t$

Let $t \in \mathbb{Z}$. Then, $\mathcal{D}_t$ is not abelian. In [EHS], we construct an abelian envelope $\mathcal{V}_t$ of $\mathcal{D}_t$. We need here some particular features of this construction. Let $m - n = t$ and let $\mathrm{Rep}^k\, GL(m|n)$ be the abelian full subcategory of $\mathrm{Rep}\, GL(m|n)$ containing mixed tensor powers $V_{m|n}^{\otimes p} \otimes (V_{m|n}^*)^{\otimes q}$ for $p, q \leq k$. The following statement is crucial for our construction.

Lemma 3.2 *Let $m, n >> k$ and $x \in \mathfrak{gl}(m|n)_{\bar{1}}$ be a self-commuting element of rank 1. Then, the restriction of DS_x to $\mathrm{Rep}^k\, GL(m|n)$ defines an equivalence of the categories $\mathrm{Rep}^k\, GL(m|n) \to \mathrm{Rep}^k\, GL(m-1|n-1)$.*

That allows us to define the abelian category $\mathcal{V}_t^k$ as the inverse limit $\varprojlim \mathrm{Rep}^k\, GL(m|n)$. Then set

$$\mathcal{V}_t := \varinjlim \mathcal{V}_t^k.$$

We have an exact fully faithful functor $I : \mathcal{D}_t \to \mathcal{V}_t$. Slightly abusing notation, we write $T(\lambda) = IT(\lambda)$.

Lemma 3.3 *For every $(m|n)$ such that $m-n = t$, there exists a symmetric monoidal functor $DS_{m|n} : \mathcal{V}_t \to \mathrm{Rep}\, GL(m|n)$. This functor is not exact but satisfies the condition of Lemma 2.3. Moreover, $DS_{m|n} \circ I$ is isomorphic to $F_{m|n}$.*

Proof It suffices to construct $DS_{m|n} : \mathcal{V}_t^k \to \mathrm{Rep}\, GL(m|n)$. We identify V_t^k with $\mathrm{Rep}^k\, GL(m'|n')$ for sufficiently large m', n' and define $DS_{m|n} : \mathrm{Rep}^k\, GL(m'|n') \to \mathrm{Rep}^k\, GL(m|n)$ as a composition of the functors $DS_{x_r} \circ DS_{x_{r-1}} \circ \dots DS_{x_1}$ for some self-commuting rank 1 odd elements $x_i \in \mathfrak{gl}(m+i|n+i)$ with $r = m' - m = n' - n$. Lemma 3.2 ensures that this composition does not depend on the choice of $(m'|n')$ and that passing to the direct limit is well defined. By construction, $DS_{m|n}$ satisfies Lemma 2.3. Finally, $DS_{m|n} \circ I$ is a symmetric monoidal functor from $\mathcal{D}_t$ to $\mathrm{Rep}^k\, GL(m|n)$, which maps V_t to $V_{m|n}$. Hence by (4), it must be isomorphic to $F_{m|n}$. □

Remark 3.4 Construction of $DS_{m|n}$ given in the above proof depends on a choice of $x_s \in \mathfrak{gl}(m+s, n+s)_{\bar{1}}$. A priori there may be several non-isomorphic functors satisfying the condition of Lemma 3.3. We suspect however that all these functors are isomorphic. Anyway as follows from the proof, we can choose the sequence $DS_{m|n}$ so that $DS_{m-1|n-1} = DS_x \circ DS_{m|n}$ for some $x \in \mathfrak{gl}(m|n)_{\bar{1}}$. Note that $DS_{m|n}T(\lambda) \simeq F_{m|n}T(\lambda)$; hence, on tilting objects, the image of $DS_{m|n}$ does not depend on the choice of x_s. Furthermore, $DS_{m|n}$ defines a homomorphism $ds_{m|n} : K[\mathcal{V}_t]_{\mathbb{C}} \to J_{m|n}$, which does not depend on a choice of x_s.

3.3 Objects of $\mathcal{V}_t$

There are three types of objects in $\mathcal{V}_t$ enumerated by bipartitions:

- Simple objects $L(\lambda)$, after identification of $\mathcal{V}_t^k$ with $\mathrm{Rep}^k GL(m|n)$ the highest weight of the corresponding representation is $\sum \lambda_i^\bullet \varepsilon_i - \sum(\lambda_i^\circ)\delta_i$ for the following set of simple roots $\mathfrak{gl}(m|n)$: $\varepsilon_1 - \varepsilon_2, \ldots, \varepsilon_m - \delta_n, \delta_n - \delta_{n-1}, \ldots .\delta_2 - \delta_1$.
- Standard objects $V(\lambda)$, those are maximal quotients of the Kac modules lying in $\mathrm{Rep}^k GL(m|n)$. They can be described as images of the irreducible module in $\mathrm{Rep}\mathfrak{gl}(\infty)$, see [EHS].
- Indecomposable tilting objects $T(\lambda)$.

It is proven in [EHS] that for every $k \geq 0$ the abelian category $\mathcal{V}_t^k$ is a highest weight category. Moreover, simple standard and tilting objects do not depend on k as soon as k is sufficiently large. In particular, $T(\lambda)$ has a filtration by $V(\mu)$ with the property:

$$(T(\lambda) : V(\lambda)) = 1, \quad (T(\lambda) : V(\mu)) \neq 0 \Rightarrow \lambda = \mu \text{ or } \mu \subset \lambda. \tag{3.1}$$

Here, we say $\mu \subset \lambda$ if $\mu^\bullet$ is contained in $\lambda^\bullet$ and μ° is contained in λ°. Furthermore, there is an interesting reciprocity, [E]:

$$(T(\lambda) : V(\mu)) = [V(\lambda) : L(\mu)]. \tag{3.2}$$

It is shown in [EHS] that $[V(\lambda) : L(\mu)] \leq 1$. In [E], all pairs (λ, μ) for which $[V(\lambda) : L(\mu)] = 1$ are described in terms of weight diagrams.

Lemma 3.5 *All three sets* $\{[L(\lambda]\}$, $\{[V(\lambda)]\}$, *and* $\{[T(\lambda)]\}$ *are bases in the Grothendieck group* $K[\mathcal{V}_t]$. *Furthermore, there exists* $K(\lambda, \mu) = 0, 1$ *such that*

$$[V(\lambda)] = \sum_{\mu \subset \lambda} K(\lambda, \mu)[L(\mu)], \quad [T(\lambda)] = \sum_{\mu \subset \lambda} K(\lambda, \mu)[V(\mu)].$$

Proof The second assertion is a consequence of (3.1) and (3.2). The first assertion follows from the fact that $K(\lambda, \mu)$ is upper triangular matrix with respect to the partial order with rank function $|\lambda^\circ| + |\lambda^\bullet|$, and $K(\lambda, \mu)$ has 1-s on the main diagonal. □

3.4 Translation Functors and Categorical Action of $\mathfrak{sl}(\infty)$

One readily sees that $\mathfrak{gl}(V_t) := V_t \otimes V_t^*$ is a Lie algebra object in $\mathcal{V}_t$. Furthermore, there exists a unique canonical morphism $\omega : \mathbf{1} \to \mathfrak{gl}(V_t)$. For every $X \in \mathcal{V}_t$, we do have the action morphism $a_X : \mathfrak{gl}(V_t) \otimes X \to X$. Hence in the same way as for $\mathrm{Rep}GL(m|n)$, we can define the translation functors $E_a X$ and $F_a X$ as generalized eigenspaces with eigenvalue a for

$$\Omega : X \otimes V_t \xrightarrow{\omega\otimes\mathrm{id}} \mathfrak{gl}(V_t) \otimes \mathfrak{gl}(V_t) \otimes X \otimes V_t \xrightarrow{\mathrm{id}\otimes s\otimes\mathrm{id}}$$

$$\mathfrak{gl}(V_t) \otimes X \otimes \mathfrak{gl}(V_t) \otimes V_t \xrightarrow{a_X\otimes a_{V_t}} X \otimes V_t$$

and

$$\Omega' : X \otimes V_t^* \xrightarrow{\omega\otimes\mathrm{id}} \mathfrak{gl}(V_t) \otimes \mathfrak{gl}(V_t) \otimes X \otimes V_t^* \xrightarrow{\mathrm{id}\otimes s\otimes\mathrm{id}}$$

$$\mathfrak{gl}(V_t) \otimes X \otimes \mathfrak{gl}(V_t) \otimes V_t^* \xrightarrow{a_X\otimes a_{V_t^*}} X \otimes V_t^*,$$

respectively.

The following theorem is proven in [E].

Theorem 3.6 *Let $t \in \mathbb{Z}$.*

(1) *E_a, F_a are non-zero only for $a \in \mathbb{Z}$.*
(2) *E_a, F_a are biadjoint exact endofunctors of $\mathcal{V}_t$.*
(3) *Let $e_a, f_a : K[\mathcal{V}_t]_\mathbb{C} \to K[\mathcal{V}_t]_\mathbb{C}$ be the induced $\mathbb{C}$-linear maps. Then, e_a, f_a satisfy the Chevalley–Serre relations for A_∞. Hence, $K[\mathcal{V}_t]_\mathbb{C}$ is an $\mathfrak{sl}(\infty)$-module.*
(4) *There is a unique isomorphism $f : K[\mathcal{V}_t]_\mathbb{C} \to \mathfrak{F}_t^\vee \otimes \mathfrak{F}$ of $\mathfrak{sl}(\infty)$-modules such that $f([V(\lambda)]) = v_\lambda := w_{\lambda^\bullet} \otimes u_{\lambda^\circ}$.*

4 Proof of the Main Theorem

Recall the functor $DS_{m|n}$ defined in Lemma 3.3.

Lemma 4.1 *We have the following commutative diagrams of functors:*

$$\begin{array}{ccc} \mathcal{V}_t & \xrightarrow{E_a(F_a)} & \mathcal{V}_t \\ {\scriptstyle DS_{m|n}}\downarrow & & \downarrow{\scriptstyle DS_{m|n}} \\ \mathrm{Rep}GL(m|n) & \xrightarrow{E_a(F_a)} & \mathrm{Rep}GL(m|n). \end{array}$$

Proof By Lemma 2.4, one has the following commutative diagram:

$$\begin{array}{ccc} \mathrm{Rep}GL(m|n) & \xrightarrow{E_a(F_a)} & \mathrm{Rep}GL(m|n) \\ {\scriptstyle DS_x}\downarrow & & \downarrow{\scriptstyle DS_x} \\ \mathrm{Rep}GL(m-1|n-1) & \xrightarrow{E_a(F_a)} & \mathrm{Rep}GL(m-1|n-1). \end{array}$$

Hence, the statement follows from definition of $\mathcal{V}_t$ and the proof of Lemma 3.3. □

Corollary 4.2 *The induced map $ds_{m|n} : K[\mathcal{V}_t]_{\mathbb{C}} \to K_{red}[\mathrm{Rep}GL(m|n)]_{\mathbb{C}}$ is a homomorphism of $\mathfrak{sl}(\infty)$-modules.*

Lemma 4.3

(1) *$ds_{m|n}([T(\lambda)]) \neq 0$ if and only if $\lambda \in C(m|n)$.*
(2) *If $ds_{m|n}([T(\lambda)]) \neq 0$ and $ds_{m-1|n-1}([T(\lambda)]) = 0$, then $DS_{m|n}T(\lambda)$ is projective in $\mathrm{Rep}GL(m|n)$.*
(3) *The set $\{ds_{m|n}([T(\lambda)]) \mid \lambda \in C(m|n)\}$ is linearly independent in $K_{red}[\mathrm{Rep}GL(m|n)]_{\mathbb{C}}$.*

Proof By Lemma 3.3, we have $DS_{m|n}T(\lambda) = F_{m|n}T(\lambda)$. Therefore, (1) follows from Theorem 3.1 (1).

To prove (2), let $P = DS_{m|n}T(\lambda)$. Then, we have $DS_{m-1|n-1}T(\lambda) = DS_x P$ for any odd self-commuting $x \in \mathfrak{gl}(m|n)$ of rank 1, see Remark 3.4. Since the set $X_P = \{y \mid DS_y P \neq 0\}$ is Zariski closed $GL(m) \times GL(n)$-stable subset, we obtain $X_P = \{0\}$ and therefore P is projective, see [DS].

Now let us prove (3) by induction on m. Consider a linear combination

$$\sum_{\lambda \in C(m|n)} c_\lambda ds_{m|n}([T(\lambda)]) = 0.$$

It can be written as

$$\sum_{\lambda \in C(m-1|n-1)} c_\lambda ds_{m|n}([T(\lambda)]) + \sum_{\lambda \notin C(m-1|n-1)} c_\lambda ds_{m|n}([T(\lambda)]) = 0.$$

Applying ds_x, we get

$$\sum_{\lambda \in C(m-1|n-1)} c_\lambda ds_{m-1|n-1}([T(\lambda)]) = 0.$$

By induction assumption, we obtain $c_\lambda = 0$ for all $\lambda \in C(m-1|n-1)$. On the other hand, $ds_{m|n}([T(\lambda)])$ for all $\lambda \in C(m|n) \setminus C(m-1|n-1)$ is the set of isomorphism classes of all indecomposable projective modules. Hence, this set is linearly independent and all $c_\lambda = 0$. □

Corollary 4.4 *The quotient* $\mathrm{Ker} ds_{m-1|n-1}/\mathrm{Ker} ds_{m|n}$ *is isomorphic to* $S_{m|n}$ *as an* $\mathfrak{sl}(\infty)$*-module.*

Proof Let us write $ds_{m-1|n-1} = ds_x ds_{m|n}$. Then, $\mathrm{Ker} ds_{m-1|n-1}/\mathrm{Ker} ds_{m|n}$ is isomorphic to $\mathrm{Im} ds_{m|n} \cap \mathrm{Ker} ds_x$. Furthermore, Lemma 4.3 implies that $\mathrm{Im} ds_{m|n}$ is spanned by $ds_{m|n}([T(\lambda)])$ for all $\lambda \in C(m|n)$, and $\mathrm{Im} ds_{m|n} \cap \mathrm{Ker} ds_x$ is spanned by classes of all indecomposable projective modules in $\mathrm{Rep} GL(m|n)$. Therefore, the statement follows from Proposition 2.2. □

Lemma 4.5

$$\bigcap_{m-n=t} \mathrm{Ker} ds_{m|n} = 0.$$

Proof Suppose $ds_{m|n}([X]) = 0$ for all m, n such that $m - n = t$. There exists k such that $[X] \in K[\mathcal{V}_t^k]_{\mathbb{C}}$. But $ds_{m|n} : K[\mathcal{V}_t^k]_{\mathbb{C}} \to K[\mathrm{Rep}^k GL(m|n)]_{\mathbb{C}}$ is injective for sufficiently large m, n. Therefore, $[X] = 0$. □

Corollary 4.4 and Lemma 4.5 prove Theorem 1.1(1). Indeed, it suffices to put

$$\mathfrak{R}^k := \begin{cases} \ker ds_{k+t-1,k-1} & \text{if } t \geq 0, \\ \ker ds_{k-1,k-1-t} & \text{if } t < 0. \end{cases}$$

Now let us prove Theorem 1.1(2). We consider the case $t \geq 0$, the case of negative t is similar. Note that $\mathfrak{R}$ satisfies the following property: for any $u \in \mathfrak{R}$, $e_a u = f_a u = 0$ for all but finitely many a. Let $\mathfrak{l}_s^-$ (resp., $\mathfrak{l}_s^+$) be the Lie subalgebra of $\mathfrak{sl}(\infty)$ generated by e_a, f_a for $a < s$ (resp., $a > s$). Let $M_s^+ := M^{\mathfrak{l}_s^-}$. Then M_s^+ is a $\mathfrak{l}_s^+$-module. If M is a submodule of $\mathfrak{R}$, then $M = \bigcup_{s<0} M_s^+$ by the above property. In particular, if M, N are two submodules of $\mathfrak{R}$ such that $M_s^+ = N_s^+$ for all $s < s_0$, then $M = N$. A simple computation shows that for any $s < 0$

$$\mathfrak{R}_s^+ \simeq \Lambda^{-s-1}((\mathbb{V}^\vee)_s^+) \otimes \Lambda^{t-s-1}(\mathbb{V}_s^+).$$

Note that $\mathfrak{l}_s^+$ is isomorphic to $\mathfrak{sl}(\infty)$, and $(\mathbb{V}^\vee)_s^+$ and $\mathbb{V}_s^+$ are isomorphic to the standard and costandard $\mathfrak{l}_s^+$-modules, respectively. A description of the lattice of all submodules of $\mathfrak{R}_s^+$ follows immediately from the socle filtration of $\mathfrak{R}_s^+$, see [PS].

Since every layer of this socle filtration is simple, the only submodules of $\mathfrak{R}_s^+$ are members of the socle filtration $\mathrm{soc}^{r+1}(\mathfrak{R}_s^+)$ for some $0 \leq r \leq -1-s$. Furthermore, $\mathrm{soc}^{r+1}(\Lambda^{-s-1}((\mathbb{V}^\vee)_s^+) \otimes \Lambda^{t-s-1}(\mathbb{V}_s^+)$ is cyclic and is generated by a monomial vector x such that $c^{r+1}(x) = 0$, $c^r(x) \neq 0$ for the contraction map

$$c : \Lambda^k((\mathbb{V}^\vee)_s^+) \otimes \Lambda^{t+k}(\mathbb{V}_s^+) \to \Lambda^{k-1}((\mathbb{V}^\vee)_s^+) \otimes \Lambda^{t+k-1}(\mathbb{V}_s^+).$$

For any $p \geq 0$, set

$$v(p) := (w_{t-1} \wedge w_{t-2} \wedge \ldots) \otimes (u_{t+p} \wedge u_{t+p-1} \wedge \cdots \wedge u_{t+1} \wedge u_{-p-1} \wedge u_{-p-2} \wedge \ldots).$$

By above, $\mathrm{soc}^{r+1}(\mathfrak{R}_s^+)$ is generated by $v(-r-s-1)$. Passing to the direct limit for $s \to -\infty$, we obtain that every submodule of $\mathfrak{R}_s^+$ is generated by $v(p)$ for some $p \geq 0$. Thus, we obtain that every submodule of $\mathfrak{R}$ is generated $v(p)$. On the other hand, it is not difficult to see that $\mathfrak{R}^r$ is generated by $v(r)$. The statement follows.

Remark 4.6 The last argument uses presentation of $\mathfrak{R}$ as a direct limit. Indeed, for the directed system of algebras $\cdots \subset \mathfrak{l}_s^+ \subset \mathfrak{l}_{s-1}^+ \subset \ldots$ (here $s \to -\infty$), we get

$$\mathfrak{R} = \varinjlim \Lambda^{-s+t-1}((\mathbb{V}^\vee)_s^+) \otimes \Lambda^{-s-1}(\mathbb{V}_s^+)$$

for $t \geq 0$, and similarly,

$$\mathfrak{R} = \varinjlim \Lambda^{-s-1}((\mathbb{V}^\vee)_s^+) \otimes \Lambda^{-s-t-1}(\mathbb{V}_s^+)$$

for $t \leq 0$.

5 Blocks in $\mathcal{V}_t$ and Dimensions of Tilting and Standard Objects

The module $\mathfrak{R}$ is a weight $\mathfrak{sl}(\infty)$-module. To simplify bookkeeping, we embed $\mathfrak{sl}(\infty) \hookrightarrow \mathfrak{gl}(\infty)$ and define a $\mathfrak{gl}(\infty)$-action on $\mathfrak{R}$ in the natural way. We fix the Cartan subalgebra $\mathfrak{h}$ of the diagonal matrices in $\mathfrak{gl}(\infty)$, choose the basis $\{E_{i,i} \mid \in \mathbb{Z}\}$, and denote by $\{\theta_i \mid i \in \mathbb{Z}\}$ the dual system in $\mathfrak{h}^*$. It is easy to compute the weight $\mathrm{wt}(v_\lambda)$ of the monomial vector v_λ. Precisely for a bipartition λ, define the sets

$$A(\lambda) := \{\lambda_i^\bullet - i \mid \lambda_j^\circ + t - j \neq \lambda_i^\bullet - i \, \forall j\},$$

$$B(\lambda) := \{\lambda_j^\circ + t - j \mid \lambda_j^\circ + t - j \neq \lambda_i^\bullet - i \, \forall i\}.$$

It follows immediately from definition that $A(\lambda)$ and $B(\lambda)$ are finite subsets of $\mathbb{Z}$ and $|B(\lambda)| - |A(\lambda)| = t$.

Example 5.1 If $\lambda = (\emptyset, \emptyset)$, then $A(\lambda) = \emptyset$, $B(\lambda) = \{0, 1, \ldots, t-1\}$ for $t > 0$ and $A(\lambda) = \{-1, \ldots, t\}$, $B(\lambda) = \emptyset$ for $t < 0$. For $t = 0$ $A(\lambda) = B(\lambda) = \emptyset$.

Then, we have

$$\mathrm{wt}(v_\lambda) = -\sum_{a \in A(\lambda)} \theta_a + \sum_{b \in B(\lambda)} \theta_b. \tag{5.1}$$

Theorem 5.2 *For a weight θ of $\mathfrak{R}$, let $\mathcal{V}_t^\theta$ denote the full subcategory of $\mathcal{V}_t$ consisting of objects with simple constituents isomorphic to $L(\lambda)$ with* $\mathrm{wt}(v_\lambda) = \theta$. *Then, $\mathcal{V}_t$ is the direct sum of $\mathcal{V}_t^\theta$. Moreover, $\mathcal{V}_t^\theta$ is a block in $\mathcal{V}_t$ for every θ.*

Proof Since $\mathcal{V}_t^k$ is a highest weight category for every k, we have

$$\mathrm{Ext}^1(L(\lambda), L(\mu)) \neq 0 \Rightarrow [V(\lambda) : L(\mu)] \neq 0 \text{ or } [V(\mu) : L(\lambda)] \neq 0.$$

On the other hand, since $V(\lambda)$ is indecomposable, all its simple constituents lie in the same block of $\mathcal{V}_t$. Combinatorial description of the multiplicities $[V(\lambda) : L(\mu)] \neq 0$ is given in [E]. It is clear from this description that $[V(\lambda) : L(\mu)] \neq 0$ implies $\mathrm{wt}(v_\lambda) = \mathrm{wt}(v_\mu)$. Let $\sim$ be the equivalence closure of $[V(\lambda) : L(\mu)] \neq 0$. Then, a simple combinatorial argument implies that $\lambda \sim \mu$ if and only if $\mathrm{wt}(v_\lambda) = \mathrm{wt}(v_\mu)$. □

Let us denote by $\dim M$ the categorical dimension of an object M in $\mathcal{V}_t$. Since $DS_{m|n}$ is a symmetric monoidal functor, it preserves categorical dimension. Therefore, for every m, n such that $m - n = t$, we have

$$\dim M = \mathrm{sdim} DS_{m|n} M. \tag{5.2}$$

We call weight θ positive (resp., negative) if $\theta = \sum_{c \in C} \theta_c$ (resp., $\theta = -\sum_{c \in C} \theta_c$). In this definition, $\theta = 0$ is both positive and negative.

Lemma 5.3

(1) *If θ is neither positive nor negative, then* $\dim M = 0$ *for every object M in $\mathcal{V}_t^\theta$.*
(2) *If $t < 0$ and $\theta = \sum_{c \in C} \theta_c$ is positive (resp., $t \geq 0$ and $\theta = -\sum_{c \in C} \theta_c$ is negative), then for every object M in $\mathcal{V}_t^\theta$, we have* $\dim M = \kappa(M) q(\theta)$ *for some integer $\kappa(M)$ and*

$$q(\theta) = \frac{\prod_{a<b, a,b \in C} (b - a)}{\prod_{j=1}^{|t|-1} j!}.$$

Remark 5.4 If $t = 0$, the only positive (and negative) weight θ is zero and $q(\theta) = 1$.

Proof Say $t \geq 0$. All weights of $\Lambda_{t|0}$ are negative. Since $ds_{t|0} : \mathfrak{R} \to \Lambda_{t|0}$ is a homomorphism of $\mathfrak{sl}(\infty)$-modules, $ds_{t|0}[M] = 0$ for every $M \in \mathcal{V}_t^\theta$. Hence, the statement is a consequence of (5.2). Similarly for $t < 0$, we have $ds_{0|-t} : \mathcal{R} \to \Lambda_{0|-t}$ is zero since all weights of $\Lambda_{0|-t}$ are positive. The proof of (1) is complete.

Let us prove (2). Note in $\Lambda_{t|0}$ and $\Lambda_{0|-t}$, all weight spaces are one-dimensional and the corresponding categories of $GL(|t|)$-supermodules are semisimple. Therefore, $DS_{t|0}M$ (resp., $DS_{0|-t}M$) is a direct sum of several copies of a certain irreducible representation $W(\theta)$ of $GL(|t|)$. The highest weight $\nu(\theta)$ of $W(\theta)$ can be easily expressed in terms of $C = \{c_1 > c_2 > \cdots > c_{|t|}\}$. For $t \geq 0$, $\nu(\theta) = (c_1 + 1 - t, c_2 + 2 - t, \ldots, c_t)$, and for $t < 0$ $\nu(\theta) = (c_1 + 1, \ldots, c_{-t} - t)$. Then by the Weyl dimension formula, we have $\operatorname{sdim} W(\theta) = \pm q(\theta)$. This implies (b). □

Remark 5.5 It is proven in [DS] that $DS_x : \operatorname{Rep} GL(m|n) \to \operatorname{Rep} GL(m-k|n-k)$ maps a block to a block corresponding to the same weight of $\mathfrak{gl}(\infty)$. Hence, $DS_{m|n}$ induces a functor from a block $\mathcal{V}_t^\theta$ to the corresponding block $\operatorname{Rep}^\theta GL(m|n)$. In particular, $DS_{t|0}$ (resp., $DS_{0,|t|}$) annihilates any object in $\mathcal{V}_t^\theta$ if θ is not negative (resp., not positive).

Lemma 5.6 *Let $t \geq 0$ (resp., $t < 0$). Then,*

$$\operatorname{Hom}_{\mathfrak{sl}(\infty)}(\mathfrak{R}, \Lambda^t(\mathbb{V}^\vee)) = \mathbb{C}, \text{ respectively, } \operatorname{Hom}_{\mathfrak{sl}(\infty)}(\mathfrak{R}, \Lambda^{-t}(\mathbb{V})) = \mathbb{C}.$$

Proof Immediate consequence of Theorem 1.1. □

Next we are going to construct a homomorphism $\varphi : \mathfrak{R} \to \Lambda^t(\mathbb{V}^\vee)$ (resp., $\varphi : \mathfrak{R} \to \Lambda^{-t}(\mathbb{V})$ by defining it on the monomial basis $v_\lambda = w_{\lambda^\bullet} \otimes u_{\lambda^\circ}$. Let $t > 0$ and

$$u_{\lambda^\circ} = u_{i_1} \wedge u_{i_2} \wedge \ldots, \qquad w_{\lambda^\bullet} = w_{j_1} \wedge w_{j_2} \wedge \ldots.$$

If $\operatorname{wt}(v_\lambda) = -\theta_{a_1} - \cdots - \theta_{a_t}$ is negative, we can write

$$w_{\lambda^\bullet} = (-1)^{s(\lambda)} w_{a_1} \wedge \cdots \wedge w_{a_t} \wedge w_{i_1} \wedge \ldots w_{i_2} \wedge$$

and then set

$$\varphi(v_\lambda) := (-1)^{s(\lambda)} \prod_{i_k \neq -k} (-1)^{i_k} w_{a_1} \wedge \cdots \wedge w_{a_t}.$$

If $\operatorname{wt}(v_\lambda)$ is not negative, we set $\varphi(v_\lambda) := 0$. The easiest way to see that φ commutes with action of $\mathfrak{sl}(\infty)$ is to realize it as the direct limit as in Remark 4.6. Then, φ is the direct limit of contraction maps $\Lambda^{-s+t}(\mathbb{V}^\vee) \otimes \Lambda^{-s}(\mathbb{V}) \to \Lambda^t(\mathbb{V}^\vee)$.

Similarly, for negative t with $\operatorname{wt}(v_\lambda) = \theta_{a_1} + \cdots + \theta_{a_{-t}}$, we write

$$u_{\lambda^\circ} = (-1)^{s(\lambda)} u_{a_1} \wedge \cdots \wedge u_{a_{-t}} \wedge u_{j_1} \wedge \ldots w_{j_2} \wedge,$$

and we set $\varphi(v_\lambda) = (-1)^{s(\lambda)} \prod_{j_k \neq -k} (-1)^{j_k} u_{a_1} \wedge \cdots \wedge u_{a_{-t}}$. In both cases if $\theta = \mathrm{wt}(\lambda)$ is positive or negative, we can write

$$\varphi(v_\lambda) = (-1)^{r(\lambda)}[W(\theta)],$$

for certain $r(\lambda) \in \mathbb{Z}$.

Proposition 5.7 *If $t \geq 0$ and θ is negative, then dimension of $V(\lambda)$ in $\mathcal{V}_t^\theta$ equals $(-1)^{r(\lambda)} q(\theta)$.*

If $t < 0$ and θ is positive, then dimension of $V(\lambda)$ in $\mathcal{V}_t^\theta$ equals $(-1)^{r(\lambda)+\frac{t(t-1)}{2}+\sum_{i=1}^{-t} a_i} q(\theta)$.

Proof First let us see that $ds_{t|0}$ (resp., $ds_{0|-t}$) equals φ. Indeed, if $\mathbf{1}$ denotes the unit object in $\mathcal{V}_t$, then $DS_{t|0}(\mathbf{1})$ (resp., $DS_{0|-t}(\mathbf{1})$) is the trivial module. Hence, $ds_{t|0}$ (resp., $ds_{0|-t}$) coincides with φ on the vacuum vector $v_{\emptyset,\emptyset}$. Then, the statement follows from Lemma 5.6.

Let $t \geq 0$; then $ds_{t|0}(v_\lambda) = (-1)^{r(\lambda)}[W(\theta)]$ and $\mathrm{sdim} W(\theta) = q(\theta)$ since $W(\theta)$ is even. This implies the lemma by (5.2).

Let $t < 0$; then $ds_{0|-t}(v_\lambda) = (-1)^{r(\lambda)}[W(\theta)]$, and the parity of $W(\theta)$ is equal to the parity of the highest weight $\nu(\theta)$. The latter is equal to the parity of $\sum_{i=1}^{-t} a_i + \frac{t(t-1)}{2}$. Hence the lemma. □

Remark 5.8 Let us explain how to compute $r(\lambda)$ in terms of weight diagram f_λ (see Section 4.1 in [E]). Recall that $f_\lambda : \mathbb{Z} \to \{<, >, \times, \circ\}$ is defined as follows:

- $f_\lambda(i) = \circ$ if u_i and w_i do not occur in v_λ.
- $f_\lambda(i) =<$ if u_i occurs in v_λ and w_i does not.
- $f_\lambda(i) =>$ if w_i occurs in v_λ and u_i does not.
- $f_\lambda(i) = \times$ if both u_i and w_i occur in v_λ.

We represent f_λ graphically by putting symbol $f_\lambda(i)$ into position i on the number line. By definition, $f_\lambda(i) = \circ$ for $i >> 0$ and $f_\lambda(i) = \times$ for $i << 0$. If $\theta = \mathrm{wt}(\lambda)$ is positive, then there are no symbols $>$, and if it is negative, there are no symbols $<$. Symbols $<, >$ are called the core symbols. The core diagram is obtained from f_λ by replacing all $\times$-s by $\circ$-s. Furthermore, $L(\lambda)$ and $L(\mu)$ are in the same block if and only if the core diagrams of λ and μ coincide. Then, $s(\lambda)$ equals the sum over all core symbols of the number of $\times$ to the right of that symbol. Now let

$$u(\lambda) = \begin{cases} \sum_{i \geq 0, f_\lambda(i)=\times} i \text{ for } t \geq 0, \\ \sum_{i > -t, f_\lambda(i)=\times} i \text{ for } t < 0 \end{cases}.$$

Then, $r(\lambda) = u(\lambda) + s(\lambda)$.

Proposition 5.9 *Let θ be negative or positive. There is exactly one up to isomorphism tilting object $T(\lambda)$ in the block $\mathcal{V}_t^\theta$ such that $\dim T(\lambda) \neq 0$. This is a unique tilting object in $\mathcal{V}_t^\theta$ such that $T(\lambda) \simeq V(\lambda) \simeq L(\lambda)$.*

Proof We start with proving that $\dim T(\lambda) \neq 0$ implies $T(\lambda) \simeq V(\lambda)$ and deal with the case $t \geq 0$. The other case is similar. Every $T(\lambda)$ is a direct summand in $V_t^{\otimes p} \otimes (V_t^*)^{\otimes q}$; therefore, it is an indecomposable summand in $F_{a_1} \dots F_{a_q} E_{b_1} \dots E_{b_q} \mathbf{1}$. Note that $\mathbf{1} = V(\emptyset, \emptyset)$. An easy computation shows that for every κ $e_a(v_\kappa)$ and $f_a(v_\kappa)$ are zero, v_μ or a sum $v_\mu + v_\nu$. Moreover, the latter case is only possible if $\mathrm{wt}(\kappa)$ is not positive. If $T(\lambda)$ is not isomorphic to $V(\lambda)$, then for some k

$$F_{a_k} \dots F_{a_q} E_{b_1} \dots E_{b_q} \mathbf{1} \in \mathcal{V}_t^\theta$$

for non-positive θ. Then by Remark 5.5 for some $k \geq 1$

$$DS_{t|0} F_{a_k} \dots F_{a_q} E_{b_1} \dots E_{b_q} \mathbf{1} = 0$$

and hence

$$DS_{t|0} F_{a_1} \dots F_{a_q} E_{b_1} \dots E_{b_q} \mathbf{1} = 0.$$

But then $DS_{t|0}(T_\lambda) = 0$, which implies $\dim T(\lambda) = 0$.

From combinatorial description of $K(\lambda, \mu)$ given in [E], we see that if in f_λ, there is $\circ$ to the left of some $\times$, then $K(\lambda, \mu) = 1$ for at least one $\mu \neq \lambda$. If the core diagram is fixed, then there is exactly one diagram such that all $\times$-s lie to the left of all $\circ$-s. That implies uniqueness of λ in every block. We can also characterize λ as the minimal weight in the block. □

Acknowledgments The author was supported by NSF grant DMS-1701532. The author would like to thank Inna Entova-Aizenbud for reading the first version of the paper and pointing out typos and unclear arguments.

References

[B] J. Brundan, *Kazhdan-Lusztig polynomials and character formulae for the Lie superalgebra* $\mathfrak{gl}(m|n)$, J. Amer. Math. Soc. 16 (2003), no. 1, 185–231.

[CW] J. Comes, B. Wilson, *Deligne's category* $\underline{Rep}(GL_\delta)$ *and representations of general linear supergroups*, Represent. Theory **16** (2012), 568–609; arXiv:1108.0652.

[DM] P. Deligne, J.S. Milne, *Tannakian Categories*, Hodge Cycles, Motives, and Shimura Varieties, Lecture Notes in Math. **900** (1982), 101–228.

[DS] M. Duflo, V. Serganova, *On associated variety for Lie superalgebras*, arXiv:Math/0507198.

[E] I. Entova-Aizenbud, *Categorical actions on Deligne's categories*, J. of Algebra **504** (2018), 391–431.

[EHS] I. Entova-Aizenbud, V. Serganova, V. Hinich, *Deligne categories and the limit of categories* $\mathrm{Rep}(GL(m|n))$, to appear in IMRN **15** (2020), 4602–4666. arXiv:1511.07699.

[HR] Crystal Hoyt, Shifra Reif, *Grothendieck rings for Lie superalgebras and the Duflo–Serganova functor*, Algebra Number Theory **12** (2018), no. 9, 2167–2184.

[HPS] Crystal Hoyt, Ivan Penkov, Vera Serganova, *Integrable* $\mathfrak{sl}(\infty)$*-modules and the category* $\mathcal{O}$ *for* $\mathfrak{gl}(m|n)$. Journal LMS. https://doi.org/10.112/jlms.12176.

[PS] I. Penkov, K. Styrkas, *Tensor representations of infinite-dimensional root-reductive Lie algebras*, in Developments and Trends in Infinite-Dimensional Lie Theory, Progr. Math. **288**, Birkhäuser (2011), 127–150.

[Z] Y. M. Zou, *Categories of finite-dimensional weight modules over type I classical Lie superalgebras,* J. of Algebra, 180 (1996),459–482.

Exact Density Matrix for Quantum Group Invariant Sector of XXZ Model

F. Smirnov

Dedicated to Nikolai Reshetikhin on the occasion of his 60th birthday

Abstract Using the fermionic basis, we obtain the expectation values of all $U_q(\mathfrak{sl}_2)$-invariant local operators on 8 sites for the anisotropic six-vertex model on a cylinder with generic Matsubara data. In the case when the $U_q(\mathfrak{sl}_2)$-symmetry is not broken, this computation is equivalent to finding the entire density matrix up to 8 sites. As application, we compute the entanglement entropy without and with temperature and compare the results with CFT predictions.

1 Introduction

This chapter is dedicated to my long-time friend Nikolai Reshetikhin, and its main idea has much in common with our joint paper [1]. In this chapter, it was shown that the restriction of the degrees of freedom for the scattering states of the sine-Gordon model with rational coupling constant is a non-violent procedure: if the quantum group invariant operators are considered, the contributions from the states that do not satisfy the RSOS restriction vanish in the correlation functions. In this chapter, we apply the same logic to rather different situations. To be precise, we are talking here about two different quantum groups that can be combined into the modular double [2].

Membre du CNRS.

F. Smirnov (✉)
Sorbonne Université, UPMC Univ Paris 06, CNRS, UMR 7589, LPTHE, Paris, France
e-mail: smirnov@lpthe.jussieu.fr

A. Alekseev et al. (eds.), *Representation Theory, Mathematical Physics, and Integrable Systems*, Progress in Mathematics 340,
https://doi.org/10.1007/978-3-030-78148-4_20

Consider the XXZ spin chain in critical regime:

$$\mathcal{H}=\tfrac{1}{2}\sum_{k=-\infty}^{\infty}\left(\sigma_k^1\sigma_{k+1}^1+\sigma_k^2\sigma_{k+1}^2+\Delta\sigma_k^3\sigma_{k+1}^3\right),\quad \Delta=\tfrac{1}{2}(q+q^{-1}). \tag{1}$$

We often use the parameter ν:

$$q=e^{\pi i\nu}.$$

It is well known that the XXZ model is closely related to the quantum affine group $U_q(\widehat{\mathfrak{sl}}_2)$. We consider one of its finite-dimensional subgroups $U_q(\mathfrak{sl}_2)$. For finite temperature, consider the modified partition function a correlation function

$$Z=\mathrm{Tr}\left(e^{-\frac{H}{T}}q^{-2S}\right)\quad \langle O\rangle_T=\frac{1}{Z}\mathrm{Tr}\left(e^{-\frac{H}{T}}q^{-2S}O\right),$$

where S is the total spin: $S=\frac{1}{2}\sum\sigma_j^3$. The insertion of q^{-2S} is important: it corresponds to the "quantum group invariant" trace (see [3] for relevant discussion). In some sense, we have a generalisation of the Witten index used in the SUSY models.

In fact, we can consider more general case of generalised Gibbs distribution; in other words, we use rather arbitrary Matsubara data (six-vertex model on a cylinder). Really crucial property for our construction is that the Matsubara maximal vector is quantum group invariant (we call this unbroken quantum group symmetry). We explain this in Sect. 5. The similarity with [1] is in the fact that for $q^r=1$ the states of the lattice model that do not satisfy the RSOS restriction do not contribute.

If the quantum group symmetry is unbroken, only the quantum group invariant operators possess non-vanishing expectation values, and this is similar to the usual Lie group symmetry. We consider a finite interval $[1,n]$ of the lattice and introduce the density matrix ρ. Then for any local operator O localised on this interval, we have

$$\langle O\rangle=\mathrm{Tr}\left(q^{-\sum_{j=1}^{n}\sigma_j^3}\,\rho\, O\right).$$

Again, we use the invariant trace.

For

$$\nu=1-\frac{r}{s}$$

(r and s are positive integers), the scaling theory of the lattice model obtained in this way must coincide with the minimal model $M_{r,s}$. In particular, $\nu=1/s$ corresponds to unitary models, and it can be shown that for them all the contributions to the modified partition function are positive. Otherwise, we have non-unitary models.

In this chapter, we compute the density matrices up to 8 sites. In order to check the agreement with the scaling limit, we compute the Von Neumann entropy. For the non-unitary case, it has strange features: it may be negative getting more negative as temperature increases. This is not very surprising in this case. Nevertheless, we always find very good agreement with the CFT predictions [4–6].

Naïve idea is that for rational ν the restricted case of the six-vertex model on a cylinder coincides completely with the RSOS case. This, however, is not quite simple as we explain in Sect. 5.1. Presumably, for that reason, our results differ from those of papers [7, 8]: they agree with the CFT formulae for usual, not effective, central charge and depend continuously on ν.

Our procedure is the same as in [9]: we compute the expectation values of invariant operators for arbitrary Matsubara data and arbitrary twist $q^{\kappa S}$. However, unbroken quantum group symmetry is possible only for $\kappa = -1$. For broken quantum group symmetry, we do not obtain complete density matrix. Considering all the operators, not only the invariant ones, is possible. We do not do it for two reasons: first, it is technically more complicated, and second, we find it more interesting to have in the scaling limit the central charge $c = 1 - 6\nu^2/(1-\nu)$ then $c = 1$ independently of the coupling constant.

This chapter is organised as follows. In Sect. 2, we briefly recall some definitions concerning the expectation value on a cylinder. Section 3 contains useful information for us from the theory of quantum groups. Section 4 gives an account of our computational procedure. In Sect. 5, we give general explanations regarding the unbroken quantum group symmetry. In Sects. 6 and 7, we expose our numerical results and comparison to the CFT for the cases of zero and non-zero temperature, respectively.

2 Matsubara Expectation Values

Consider the quantum affine group $U_q(\widehat{\mathfrak{sl}}_2)$ and its universal R-matrix $\mathcal{R}$. We use usual Cartan generators. Central charge equals to 0, so $h_1 = -h_0$, denote $H = h_1$. The affine quantum group allows evaluation representations π_ζ^{2s} with s being a spin, and ζ an evaluation parameter. Fix two integers n, L and define two representations:

$$\pi_{\mathbf{S}} = (\pi_1^1)^{\otimes n}, \quad \pi_{\mathbf{M}} = \pi_{\tau_0}^{2s_0} \otimes \cdots \otimes \pi_{\tau_{L-1}}^{2s_{L-1}}$$

with spins s_j and inhomogeneities τ_j. Later we shall use

$$t_j = \tau_j^2 .$$

The index $\mathbf{S}$ stands for “space” and $\mathbf{M}$ stands for Matsubara.

We have the image of the universal R-matrix:

$$T_{\mathbf{S},\mathbf{M}} = (\pi_{\mathbf{S}} \otimes \pi_{\mathbf{M}})(\mathcal{R}) .$$

Consider further the commuting family of Matsubara transfer matrices

$$T_{\mathbf{M}}(\zeta,\kappa)=(\mathrm{Tr}\otimes I)\Big[(\pi^1_\zeta\otimes\pi_{\mathbf{M}})(\mathcal{R}(q^H\otimes I))\Big]\,,$$

and their eigenvector $|\Psi\rangle$.

The main object of our study is a linear functional on operators $O_{\mathbf{S}}$ acting on the representation space of $\pi_{\mathbf{S}}$:

$$Z_\kappa\{O_{\mathbf{S}}\}=\frac{\langle\Psi|T_{\mathbf{S},\mathbf{M}}\,O_{\mathbf{S}}\,q^{2\kappa S_{\mathbf{S}}}|\Psi\rangle}{\langle\Psi|T_{\mathbf{S},\mathbf{M}}\,q^{2\kappa S_{\mathbf{S}}}|\Psi\rangle}\,, \tag{1}$$

where $S_{\mathbf{S}}=1/2\sum_{j=1}^n\sigma^3_j$; in other words, $2S_{\mathbf{S}}$ is the Cartan generator H evaluated on $\pi_{\mathbf{S}}$. If $O_{\mathbf{S}}$ is localised on smaller interval than $[1,n]$, the space shrinks automatically. Clearly, Z_κ can be considered as such automatic reduction for the case $\mathbf{S}=(-\infty,\infty)$ if $|\Psi\rangle$ is the eigenvector with maximal in absolute value eigenvalue of $T_{\mathbf{M}}(1,\kappa)$. This explains importance of Z_κ for physical applications. The main tool of computation is the fermionic basis.

3 Quantum Group Invariant Operators

3.1 Generalities

Let us start with some simple facts concerning the quantum groups. We use Drinfeld's notations [10]. Take a basis e_a, $e_0=1$. The comultiplication is a homomorphism

$$\Delta(e_a)=\mu_a^{bc}e_b\otimes e_c\,.$$

We have the R-matrix

$$\mathcal{R}\mu_a^{bc}e_b\otimes e_c=\mu_a^{bc}e_c\otimes e_b\mathcal{R}\,, \tag{1}$$

and the antipode

$$\mu_a^{bc}s(e_b)e_c=\mu_a^{bc}e_bs(e_c)=\mu_a^{bc}s^{-1}(e_c)e_b=\mu_a^{bc}e_cs^{-1}(e_b)=\delta_{a,0}e_0\,,$$

which is an anti-automorphism. We define two adjoint actions

$$ad_{e_a}(x)=\mu_a^{bc}e_bxs(e_c)\,,\quad ad^*_{e_a}(x)=\mu_a^{bc}s^{-1}(e_c)xe_b\,.$$

The first is a homomorphism, the second is an anti-homomorphism. In any quantum double s^2 is an inner automorphism:

$$s^2(x) = YxY^{-1}\,,$$

for certain Y. The pairing

$$\langle x_1, x_2\rangle = \mathrm{Tr}(Yx_1x_2)$$

is an invariant scalar product of operators for Tr being any cyclic functional, for example usual trace over a finite-dimensional representation if any. We have

$$\langle x_1, ad_{e_a}(x_2)\rangle = \langle ad^*_{e_a}(x_1), x_2\rangle\,.$$

Below we give some additional formulae that will be used only in Sect. 5. From now on, we consider two representations; for economy of space, we do not distinguish between the generators of the quantum group and their representations; everything should be clear from the context.

Consider an invariant vector v. From (1), one derives

$$(I\otimes e_a)\mathcal{R}(I\otimes v) = \left(ad^*_{e_a}\otimes I\right)\mathcal{R}(I\otimes v)\,. \tag{2}$$

Act by $O\otimes I$ on (2) from the left, and then take the invariant trace with respect to the first tensor component obtaining

$$(I\otimes e_a)(\mathrm{Tr}\otimes I)\left[\mathcal{R}(OY\otimes I)\right](I\otimes v) = (\mathrm{Tr}\otimes I)\left[\left(ad^*_{e_a}\otimes I\right)(\mathcal{R})(OY\otimes I)\right](I\otimes v) \tag{3}$$

$$= (\mathrm{Tr}\otimes I)\left[\mathcal{R}(ad_{e_a}(O)Y\otimes I)\right](I\otimes v)\,.$$

In particular, if O is invariant,

$$(I\otimes e_a)(\mathrm{Tr}\otimes I)\left[\mathcal{R}(OY\otimes I)\right](I\otimes v) = \delta_{a,0}(\mathrm{Tr}\otimes I)\left[\mathcal{R}(OY\otimes I)\right](I\otimes v)\,. \tag{4}$$

So, acting on invariant vectors, we obtain invariant ones.

3.2 Invariant Operators

The quantum group $U_q(\widehat{\mathfrak{sl}}_2)$ contains two finite-dimensional subgroups isomorphic to $U_q(\mathfrak{sl}_2)$. Take one of them, for example, the one created by e_0, f_0, h_0, which we denote, respectively, by $E, F, -H$. We use $K = q^H$. The defining relations of $U_q(\mathfrak{sl}_2)$ are well known, so we give them without comments just to fix the notations

$$KE = q^2 EK\,, \quad KF = q^{-2} FK\,, \quad [E, F] = \frac{K - K^{-1}}{q - q^{-1}}\,.$$

The coproduct, counit and antipode are given by

$$\Delta(E) = E \otimes 1 + K \otimes E\,, \quad \Delta(F) = F \otimes K^{-1} + 1 \otimes F\,, \quad \Delta(K) = K \otimes K\,,$$
$$\epsilon(E) = \epsilon(F) = 0\,, \quad \epsilon(K) = 1\,,$$
$$s(E) = -K^{-1}E\,, \quad s(F) = -FK\,, \quad s(K) = K^{-1}\,.$$

Clearly,

$$Y = K^{-1}\,.$$

Consider an invariant under $U_q(\mathfrak{sl}_2)$-operator $O_{\mathbf{S}}$. It is convenient to identify $O_{\mathbf{S}}$ with a vector v in $V_1 \otimes \cdots \otimes V_{2n}$, $V_k \simeq \mathbb{C}^2$ The identification is

$$O_{\mathbf{S}} = \mathcal{I}(v) = v^{t_{n+1},\cdots t_{2n}} c_1 \cdots c_n\,, \tag{5}$$

where c_k acts from V_{2n+1-k} to V_k as

$$c = \begin{pmatrix} 0 & 1 \\ -q^{-1} & 0 \end{pmatrix}.$$

For example for $n = 1$, there is one invariant vector:

$$s_{1,2} = -q e_{-1/2} \otimes e_{1/2} + e_{1/2} \otimes e_{-1/2}\,, \tag{6}$$

which under (5) goes to the unit operator acting in V_1. Generally, we denote the element of the basis for an irreducible representation of spin j by $e_{-j}, e_{-j+1}, \cdots, e_j$.

We are going to give several definitions; for future convenience, they are a bit more general than we need at this point. Bratteli diagram J is a sequence $\{j_0, j_1, \cdots, j_k\}$ such that $j_{p+1} = j_p \pm 1/2$, $j_p \geq 0$. Lexicographically ordered totality of Bratteli diagrams (for fixed k) with $j = j_0$ and $j' = j_k$ will be denoted by $B(k, j, j', \infty)$; the meaning of the latter argument will be clear later when we shall consider restricted case. With every Bratteli diagram $J \in B(k, j, j', \infty)$ and $-j \leq m \leq j$, $-j' \leq m' \leq j'$ (integer of half-integer depending on j), we associate a vector from $\left(\mathbb{C}^2\right)^{\otimes k}$:

$$E_{k,J,m,m'} = \sum_{\substack{\epsilon_1,\cdots\epsilon_k=\pm 1/2 \\ m+\sum_{p=1}^k \epsilon_p = m'}} \prod_{p=1}^{k} \begin{pmatrix} j_{p-1} & 1/2 & j_p \\ m + \sum_{s=1}^{p-1} \epsilon_s & \epsilon_p & m + \sum_{s=1}^{p} \epsilon_s \end{pmatrix} e_{\epsilon_1} \otimes \cdots \otimes e_{\epsilon_k}\,.$$

In order to avoid denominators that are dangerous for q being a root of unity, we use non-normalised $3j$-symbols

$$\begin{pmatrix} j & 1/2 & j+1/2 \\ m & -1/2 & m-1/2 \end{pmatrix} = 1\,, \quad \begin{pmatrix} j & 1/2 & j+1/2 \\ m & 1/2 & m+1/2 \end{pmatrix} = \frac{q^{2(m+1)} - q^{-2j}}{q^2-1}\,,$$
$$\begin{pmatrix} j & 1/2 & j-1/2 \\ m & -1/2 & m-1/2 \end{pmatrix} = 1\,, \quad \begin{pmatrix} j & 1/2 & j-1/2 \\ m & 1/2 & m+1/2 \end{pmatrix} = \frac{q^{2(m+1)} - q^{2(j+1)}}{q^2-1}\,.$$

For $j = 0$, this gives a $U_q(\mathfrak{sl}_2)$ decomposition of the tensor product of k spaces $\mathbb{C}^2$:

$$\left(\mathbb{C}^2\right)^{\otimes k} = \bigoplus_{J \in B(k,0,j,\infty)} M_J \otimes V_j\,. \tag{7}$$

We have for the dimensions of multiplicity spaces

$$\text{card}(B(k,0,j,\infty)) = \binom{k}{\frac{k-2j}{2}} - \binom{k}{\frac{k-2j}{2}-1}\,.$$

We shall use two bases of the space of invariant operators acting on of $\left(\mathbb{C}^2\right)^{\otimes n}$; the dimension of this space is the Catalan number C_n:

1. To have an orthogonal basis, we use the above construction with $3j$-symbols. The basis is given by

$$O(J) = \mathcal{I}(E_{2n,J,0,0})\,.$$

This basis is orthogonal, but not orthonormal with respect to the scalar product

$$\langle O_1, O_2 \rangle = \text{Tr}\left(q^{-\sum \sigma_j^3} O_1 O_2\right)\,.$$

The normalisation is

$$(O(J), O(J')) = \prod_{p=1}^{2n-1} \dim_q(j_p)\,,$$

where the quantum dimension is

$$\dim_q(j) = [2j+1]\,. \tag{8}$$

Here and later

$$[k] = \frac{q^k - q^{-k}}{q - q^{-1}}\,.$$

2. For our computations, it is convenient to use a simpler basis. Take $2n$ points on the real axis: $1, 2, \cdots, 2n$ and connect them pairwise by arcs in the upper half-plain requiring that the arcs do not intersect (Fig. 1),

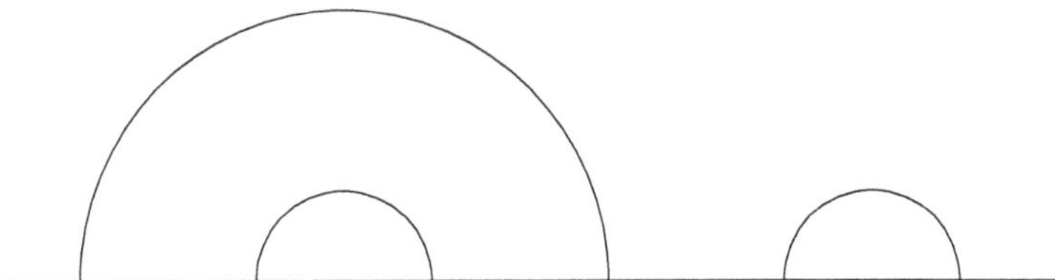

Fig. 1 Simple construction of invariant vectors

With any such design associate, the Bratelli diagram J writes one half of the number of arcs passing over every interval $[k, k+1]$. For instance, Fig. 1 corresponds to $J = \{0, 1/2, 1, 1/2, 0, 1/2, 0\}$. Denote by $i_1, \cdots, i_m$ the beginnings of the arcs, and by $k_1, \cdots, k_n$ their ends. Then, the vector associated with every design is

$$s_{i_1,k_1} s_{i_2,k_2} \cdots s_{i_n,k_n} ,$$

with $s_{i,k}$ given by (6), and define

$$\widetilde{O}(J) = \mathcal{I}(s_{i_1,k_1} s_{i_2,k_2} \cdots s_{i_n,k_n}) .$$

This is another basis.

Recall that $B(2n, 0, 0, \infty)$ is lexicographically ordered. We have two bases of invariant operators $O(J)$ and $\widetilde{O}(J)$. It is easy to see that they are related by a triangular transformation

$$O(J) = U_{J,J'} \widetilde{O}(J') .$$

The matrix U is not hard to compute inductively.

Considering the interval $[1, n]$, we are interested only in the translationally irreducible operators, i.e., the operators that are not localised on subintervals of smaller length. It is easy to figure out that the number of translationally irreducible invariant operators is

$$D_n = C_n - 2C_{n-1} + C_{n-2} .$$

For the second basis, eliminating the translationally reducible operators is simple: it suffices to throw all the vectors containing $s_{1,2n}$ or $s_{n,n+1}$.

4 Procedure of Computation

Fermionic basis for the case of $U_q(\mathfrak{sl}_2)$-invariant operators is parallel to the $\mathfrak{sl}_2$-invariant case for the XXX model that is explained in detail in [9]. So, we shall be brief here. We have two sets of fermionic operators $b_j,\ b^*_j, c_j,\ c^*_j\ (j = 1, 2, 3, \cdots)$, with canonical commutation relations, and use notations $b^*_J,\ c^*_J$ for products, J being a strictly ordered multi-index $\{j_1, \cdots, j_k\}$. For two multi-indices of the same length, we write $I \preccurlyeq J$ if $i_p \le j_p$ for all p. We denote by $|I|$ the sum of elements in I. Our fermionic operators act on the space of local fields; role of vacuum is played by the unit operator I. Consider the space $\mathfrak{H}^{(n)}$ with the basis

$$b^*_{I^+}c^*_{I^-}\cdot \mathrm{I}, \tag{1}$$

with $\#(I^+) = \#(I^-) \le [n/2]$, $\max(I^+ \cup I^-) \le n$, $I^+ \preccurlyeq I^-$ (the difference with the XXX case is that we do not impose $|I^+| + |I^-| \equiv 0 \pmod 2$, since there is no C-symmetry). Define the operators

$$Q_m = \sum_{j=1}^{m-1} c_j b_{m-j}\,, \quad 2, 3, \cdots; \qquad M = \sum_{i=1}^{\infty} c^*_i b_i\,.$$

Introduce the space $\widetilde{\mathfrak{H}}^{(n)}$ defined as above with the condition $I^+ \preccurlyeq I^-$ lifted. The operators Q_m act from $\mathfrak{H}^{(n)}$ to $\widetilde{\mathfrak{H}}^{(n)}$. The operator M acts from the space $\widetilde{\mathfrak{H}}^{(n)}_2$ (space of charge 2), span by the vectors (1) with $\#(I^+)+1 = \#(I^-)-1 \le [n/2]$, $\max(I^+ \cup I^-) \le n$, to $\widetilde{\mathfrak{H}}^{(n)}$. We define the subspace $\mathfrak{V}^{(n)}$ of $\mathfrak{H}^{(n)}$ by

$$\mathfrak{V}^{(n)} = \{v \in \mathfrak{H}^{(n)} \mid Q_m v \in M\widetilde{\mathfrak{H}}^{(n)}_2 \text{ for } m = n+1, n+2, \cdots\}.$$

It is easy to see that $Q_m\mathfrak{V}^{(n)} = 0$ for $m > 2n-1$, so the actual number of requirements is finite.

Denoting basis of $\mathfrak{V}^{(n)}$ by v_α, we have $F = ||F_{\alpha,\{I^+,I^-\}}||$, the first one of several matrices used below:

$$v_\alpha = \sum_{\#(I^+)=\#(I^-)//\max(I^+\cup I^-)\le n,\ I^+\preccurlyeq I^-} F_{\alpha,\{I^+,I^-\}}\, b^*_{I^+}c^*_{I^-}\cdot \mathrm{I}\,.$$

Our goal is to find an analogue of OPE:

$$\widetilde{O}(J) = \sum_{\alpha} c_{J,\alpha}(n) v_\alpha\,, \tag{2}$$

where $\widetilde{O}(J)$ is a $U_q(\mathfrak{sl}_2)$-invariant operator constructed via the Bratteli diagram J and the second of the bases above.

As in the previous paper [9], we fix the OPE coefficients considering finite Matsubara chains. For every Matsubara data, we have an equation

$$\langle\widetilde{O}(J)\rangle=\sum_{\alpha}c_{J,\alpha}(n)\langle v_\alpha\rangle\,. \tag{3}$$

To compute the right-hand side, we use

$$\langle b^*_{I^+}c^*_{I^-}\cdot \mathrm{I}\rangle=\omega_{I^+,I^-}\,,$$

here and later

$$\omega_{I^+,I^-}=\det\left(\omega_{i^+_p,i^-_q}\right)_{p,q=1,\cdots,|I^+|}\,,$$

with $\omega_{i,j}$ are the Taylor series coefficients

$$\omega(x,y)=\sum_{i,j=1}^{\infty}\omega_{i,j}(x-1)^{i-1}(y-1)^{j-1}\,,$$

of a function $\omega(x,y)$ defined below. Here and later, the Latin letters are squares of the evaluation parameters of $U_q(\widehat{\mathfrak{sl}}_2)$ representations.

With every eigenvector of the Matsubara transfer matrix, we associate an eigenvalue of the Q-operator

$$Q(x)=\prod_{j=1}^{m}(1-x/b_j)\,.$$

The information about the Matsubara spin chain encoded in two functions

$$a(x)=x^L+\sum_{j=1}^{L}a_jx^{L-j},\,,\quad d(x)=x^L+\sum_{j=1}^{L}d_jx^{L-j}\,.$$

We have the Bethe equations

$$q^{-m}a(b_j)Q(b_jq^2)+q^md(b_j)Q(b_jq^{-2})=0,\quad j=1,\cdots m\,.$$

In principle, we could introduce a twist multiplying $a(x)$ and $d(x)$ by q^κ and $q^{-\kappa}$, respectively, but our goal will be to obtain equations for the OPE coefficients, and practice shows that twist does not produce independent ones.

Our main trick is to take for the input data

$$\{b_1,\cdots,b_m,a_{m+1},\cdots,a_L,d_1,\cdots,d_L\}\,.$$

Then for the unknowns $a_1,\cdots,a_m$, we have linear equations.

Introduce the measure

$$dm(x)=\frac{1}{1+\mathfrak{a}(x)}\frac{dx}{x},\qquad \mathfrak{a}(x)=q^{-2m}\frac{Q(b_jq^2)}{Q(b_jq^{-2})},$$

the auxiliary function

$$\psi(x)=\frac{x+1}{2(x-1)},$$

and kernels

$$K(x)=\psi(xq^2)-\psi(xq^{-2}),\qquad f_R(x)=\psi(xq^{-2})-\psi(x),\qquad f_L(x)=\psi(xq^2)-\psi(x).$$

We have "integral" equation

$$G(x,z)=f_R(x,z)-\frac{1}{2\pi i}\oint_\Gamma K(x/y)G(y,z)dm(y),\tag{4}$$

with γ going around $y=b_1,\cdots,b_m,z$. For finite m, this is equivalent to a system of linear equations for m functions $G(b_j,z)$. The function $\omega(x,y)$ [15] is

$$\omega(x,y)=-\frac{1}{2\pi i}\oint_{\Gamma'} f_L(x/y)G(y,z)dm(y)-\frac{1}{4}K(x/z),$$

where Γ' encircles, in addition to Γ, the point $y=x$.

The expectation values of invariant operators are computed exactly as in [9]; basically, we rewrite in the basis of Young diagrams formulae of [12–14].

We repeat some definitions. Consider Young diagrams Y_λ where $\lambda=(\lambda_1,\cdots,\lambda_n)$, $\lambda_i\geq\lambda_{i+1}>0$ is a partition. We set $\#(\lambda)=n$. It is called the length of Y_λ. We work in the space H_q whose elements are

$$Y=\sum_{\#(\lambda)\leq q}c_\lambda Y_\lambda.$$

Below we will identify Y_λ with λ. The symbol Ø denotes the empty diagram. Define the operation cut_q that acts from $H_{q'}$ with $q'>q$ to H_q erasing all the terms with $\#(\lambda)>q$. Consider the Grassmann space F_q with the basis $\psi^*_{k_1}\cdots\psi^*_{k_q}$ $(k_1>\cdots>k_q\geq 0)$. We have the usual isomorphism between the spaces H_q and F_q.

$$\begin{aligned}&\psi^*_{k_1}\psi^*_{k_2}\cdots\psi^*_{k_q}\ \mapsto\ (k_1-(q-1),k_2-(q-2),\cdots,k_q)_0,\\&(\lambda_1,\cdots,\lambda_n)\ \mapsto\ \psi^*_{\lambda_1+q-1}\cdots\psi^*_{\lambda_n+q-n}\psi^*_{q-n-1}\cdots\psi^*_0,\ \text{where } n\leq q.\end{aligned}\tag{5}$$

In the above, $()_0$ means removing all entries equal to 0. Schur polynomial $s_\lambda(x_1, \cdots, x_q)$ is the symmetric polynomial

$$s_\lambda(x_1, \cdots, x_q) = \frac{\det ||x_j^{\lambda_i+q-i}||}{\det ||x_j^{q-i}||}.$$

The above formula gives an isomorphism between H_q and P_q.

For a given polynomial of one variable $P(x) = \sum_{j=0}^{d} p_j x^j$, we define the operator $P \wedge F_{q-1} \subset F_q$ multiplying by $\sum_{j=0}^{d} p_j \psi_j^*$; this operator is defined as $P \wedge H_{q-1} \subset H_q$ by the isomorphism (5). This definition can be generalised to polynomials of several variables in obvious way. Certainly, the polynomials anti-symmetrise themselves automatically.

We shall also need the simplest Littlewood–Richardson formula for multiplication of a Schur polynomial by elementary symmetric function σ_j, which translates as action on H_q

$$\sigma_j \circ (\lambda_1, \cdots, \lambda_n) = \sum_{I^-}^{\binom{n+\min(j,q-n)}{j}} \Big((\lambda_1, \cdots, \lambda_n, \underbrace{0, \cdots, 0}_{\min(j,q-n)}) + e_{I^-} \Big)_{\text{order}},$$

where e_{I^-} are all vectors of dimension $n + \min(j, q - n)$ with j elements equal to 1 and other elements being 0, "order" means that we have to drop all the tables in which elements happen to be not ordered, and we also drop all zeros in the final table.

In what follows, we shall also need the operation $\mathrm{cut}_m(Y)$, which erases all the Young diagrams in Y with lengths greater than m.

For a partition λ, define the coefficients $e_{\lambda,\lambda'}$ via

$$s_\lambda(x_1, \cdots, x_k, 1) = \sum_{\lambda'} e_{\lambda,\lambda'}\, s_{\lambda'}(x_1, \cdots, x_k).$$

This gives rise to an operator

$$EY_\lambda = \sum_{\lambda'} e_{\lambda,\lambda'} Y_{\lambda'}.$$

The easiest algorithm for finding $e_{\lambda,\lambda'}$ consists in the following: express $s_\lambda(x_1, \cdots, x_k, 1)$ through Jacobi–Trudi formula using the transposed λ. This formula is given in terms of the elementary symmetric functions $\sigma_k(x_1, \cdots, x_k, 1)$. Using

$$\sigma_k(x_1, \cdots, x_k, 1) = \sigma_k(x_1, \cdots, x_k) + \sigma_{k-1}(x_1, \cdots, x_k),$$

expand the Jacobi–Trudi determinant getting a sum of Schur polynomials.

Slavnov formula for the scalar product of the on-shell and off-shell Bethe vectors $\langle b_1, \cdots, b_m | x_1, \cdots x_m\rangle$ is a symmetric polynomial of the off-shell $x_1, \cdots x_m$ with the following representation in terms of the Young diagrams:

$$\mathcal{N}(b_1, \cdots, b_m) = \frac{\prod_{j=1}^{m} (b_j^{m-2} d(b_j)) q^{m(m-2)} (-1)^{\frac{1}{2}m(m+1)}}{W(b_1, \cdots b_m)} \Big(P_1 \wedge \cdots \wedge P_m \wedge \emptyset\Big),$$

where

$$P_j(x) = \frac{x b_j^2 (q^2 - 1)}{x - b_j} \left(q^{2-2m} a(x) \frac{Q(xq^2)}{xq^2 - b_j} - d(x) \frac{Q(xq^{-2})}{xq^{-2} - b_j} \right).$$

We shall also need the Gaudin formula for normalisation:

$$G(b_1, \cdots, b_m) = (q - q^{-1})^m \prod_{j=1}^{m} a(b_j) d(b_j) \prod_{i \neq j} \frac{b_i q - b_j q^{-1}}{b_i - b_j} \det b_i \partial_{b_i} (\log \mathfrak{a}(b_j)).$$

As usual, we consider the matrix elements $T_{i,j}$ of the Matsubara monodromy matrix. Their action in our framework translates into the action of the operators $\mathcal{T}_{i,j}$ described below.

We begin with the operators $\mathcal{T}_{1,1}$, $\mathcal{T}_{2,2}$ that do not change the charge.

$$\mathcal{T}_{1,1} Y = \sum_{k=0}^{l} \mathrm{cut}_l \Big(\sigma_k \circ E A_k \wedge Y \Big), \qquad A_k(x) = (-1)^k q^{l-2k} x^{l-k} a(x),$$

$$\mathcal{T}_{2,2} Y = \sum_{k=0}^{l} \mathrm{cut}_l \Big(\sigma_k \circ E D_k \wedge Y \Big), \qquad D_k(x) = (-1)^k q^{-l+2k} x^{l-k} d(x).$$

The most complicated operator is $\mathcal{T}_{1,2}$ that raises the number of variables:

$$\mathcal{T}_{1,2} Y = \sum_{j,k=0}^{l} \mathrm{cut}_l \Big(\sigma_j \circ \sigma_k \circ E B_{j,k} \wedge Y \Big) + \sum_{k=0}^{l} \mathrm{cut}_l \Big(\sigma_k \circ B_k \wedge Y \Big), \tag{6}$$

where $B_{j,k}$, B_k are the polynomials of, respectively, two and one variable:

$$B_{j,k}(x, y) = (-1)^{j+k} q d(y) a(x) (q^{-2} y)^{l-k} \frac{(xq^2)^{l-j} - y^{l-j}}{xq^2 - y},$$

$$B_k(x) = (-1)^{k+1} q^{-1} d(1) a(x) \frac{q^{-2(l-k)} - x^{l-k}}{xq^{-2} - x}.$$

Finally, the operator $\mathcal{T}_{2,1}$ that lowers the number of variables is the simplest one:

$$\mathcal{T}_{2,1}Y = EY\,.$$

In order to compute $\langle O \rangle$, we present O as a sum of the operators $E_{i_1,j_1}\cdots E_{i_n,j_n}$ and use

$$\langle E_{i_1,j_1}\cdots E_{i_n,j_n}\rangle = \frac{1}{G(b_1,\cdots,b_m)}\mathrm{Schur}_{b_1,\cdots,b_m}\Bigl(\mathcal{T}_{i_n,j_n}\cdots\mathcal{T}_{i_1,j_1}\mathcal{N}(b_1,\cdots,b_m)\Bigr)\,, \tag{7}$$

where $\mathrm{Schur}_{b_1,\cdots,b_m}$ is a linear functional on the vector space H_0 that maps every Young diagram to the corresponding Schur polynomial of arguments $b_1,\cdots,b_m$. Now we can construct as many equations for the coefficients of OPE as we wish.

The rest of the computations follow closely that of [9]. For example, for $n=8$, we have 324 fermionic vectors v_α, and construct 20 eqs. with $L=1, m=0$, 120 eqs. with $L=2, m=0$; 100 eqs. with $L=3, m=0$; 10 eqs. with $L=4, m=0$; 2 eqs. with $L=2, m=1$; 50 eqs. with $L=3, m=1$; 70 eqs. with $L=4, m=1$; 2 eqs. with $L=5, m=1$. Then, we proceed with Gauss triangularisation. This is easy to do for a numeric value of q, but rather impossible keeping q as a variable. That is why we would like to proceed with interpolation. But the solutions for the coefficients $c_{J,\alpha}(n)$ contain denominators that we have to fix, then interpolating for the numerators is possible, but we also have to estimate the degree of them as functions of q. All these data can be guessed considering several numerical examples.

We have for the lowest common denominators ($\mathrm{den}(n)$) and for the maximal exponent of the numerators ($\mathrm{mn}(n)$):

$$\begin{aligned}
&\mathrm{den}(2) = 4q, && \mathrm{mn}(2) = 2\\
&\mathrm{den}(3) = 4q[2], && \mathrm{mn}(3) = 2\\
&\mathrm{den}(4) = 16q^5[2][3], && \mathrm{mn}(4) = 10\\
&\mathrm{den}(5) = 16q^8[2][3][4], && \mathrm{mn}(5) = 16\\
&\mathrm{den}(6) = 64q^{15}[2][3]^2[4][5], && \mathrm{mn}(6) = 30\\
&\mathrm{den}(7) = 64q^{22}[3][4]^2[5][6], && \mathrm{mn}(7) = 44\\
&\mathrm{den}(8) = 256q^{33}[3]^2[4]^2[5][6][7], && \mathrm{mn}(8) = 74.
\end{aligned} \tag{8}$$

We proceed with interpolation obtaining finally matrices $c_{J,\alpha}(n)$. Now for any Matsubara data, we have

$$\langle O(J)\rangle = U_{J,J'}c_{J',\alpha}F_{\alpha,I^+,I^-}\omega_{I^+,I^-}\,. \tag{9}$$

Here only ω_{I^+,I^-} depends on the Matsubara data.

Let us give an example. For $n=4$, the basis v_α consists of 9 elements

$$v_1=\mathrm{I}\,,\ v_2=b_1^*c_1^*\cdot\mathrm{I}\,,\ v_3=b_1^*c_2^*\cdot\mathrm{I}\,,\ v_4=b_1^*c_3^*\cdot\mathrm{I}\,,\ v_5=(b_1^*c_4^*-b_2^*c_3^*)\cdot\mathrm{I}\,,\ v_6=b_2^*c_2^*\cdot\mathrm{I}\,,$$
$$v_7=(b_2^*c_4^*-b_3^*c_3^*)\cdot\mathrm{I}\,,\ v_8=b_2^*b_1^*c_2^*c_1^*\cdot\mathrm{I}\,,\ v_9=(b_2^*b_1^*c_3^*c_2^*+b_2^*b_1^*c_4^*c_1^*-b_3^*b_1^*c_3^*c_1^*)\cdot\mathrm{I}\,.$$

For $\widetilde{O}(J_1)$ with $J_1=\{1/2,0,1/2,0,1/2,0,1/2,0\}$, which is constructed from $s_{1,2}s_{3,4}s_{5,6}s_{7,8}$, we have

$$\widetilde{O}(J_1)=\frac{(1+q^2)^2}{(16q^2)}v_1+\frac{(1+q^4)(1-q^2)(1+10q^2+q^4)}{4q(1+q^2)(1-q^6)}v_2+\frac{(1-q^2)^4}{2(1+q^2)(1-q^6)}v_3$$
$$+\frac{4(1-q^2)^3q}{(1+q^2)(1-q^6)}\big(-4v_4+3v_6-v_7\big)+\frac{q^2(1-q^2)}{(1-q^6)}\big(2v_8-v_9\big)\,.$$

By triangularity, $O(J_1)=\widetilde{O}(J_1)$. Certainly, for $n>4$, the formulae are getting more complicated, but their structure is more inspiring that in XXX case.

5 The Case of Unbroken Quantum Group Symmetry

Consider the functional Z_κ for particular case $\kappa=-1$:

$$Z_{-1}\{O_{\mathbf{S}}\}=\frac{\langle\Psi|T_{\mathbf{S},\mathbf{M}}\,O_{\mathbf{S}}\,q^{-2S_{\mathbf{S}}}|\Psi\rangle}{\langle\Psi|T_{\mathbf{S},\mathbf{M}}\,q^{-2S_{\mathbf{S}}}|\Psi\rangle}\,. \tag{1}$$

From the identity (4), one concludes that the transfer matrix $T_{\mathbf{M}}(\zeta)=T_{\mathbf{M}}(\zeta,-1)$ preserves the $U_q(\mathfrak{sl}_2)$-invariant subspace of the Matsubara space. Hence, there are invariant eigenvectors. Let $|\Psi\rangle$ be one of them:

$$\pi_{\mathbf{M}}(e_a)|\Psi\rangle=\delta_{a,0}|\Psi\rangle,\quad \langle\Psi|\pi_{\mathbf{M}}(e_a)=\delta_{a,0}\langle\Psi|\,.$$

With such choice, we say that the quantum group symmetry is unbroken. Then, we conclude from (3) that

$$\delta_{a,0}\langle\Psi|\mathrm{Tr}_{\mathbf{S}}\Big[T_{\mathbf{S},\mathbf{M}}\,O_{\mathbf{S}}\,q^{-2S_{\mathbf{S}}}\Big]|\Psi\rangle=\langle\Psi|\mathrm{Tr}_{\mathbf{S}}\Big[\pi_{\mathbf{M}}(e_a)T_{\mathbf{S},\mathbf{M}}\,O_{\mathbf{S}}\,q^{-2S_{\mathbf{S}}}\Big]|\Psi\rangle$$
$$=\langle\Psi|\mathrm{Tr}_{\mathbf{S}}\Big[T_{\mathbf{S},\mathbf{M}}\,\mathrm{ad}_{e_a}\left(O_{\mathbf{S}}\right)\,q^{-2S_{\mathbf{S}}}\Big]|\Psi\rangle\,.$$

So, like in the case of invariance under usual Lie group in the unbroken case, only invariant operators count. Hence, our computations give access to the entire density matrix for the space interval $[1,n]$ with $|\Psi\rangle$ describing the environment. Before going any further, let us see what happens for q being a root of unity. To simplify notations, let us assume that all Matsubara spins equal $1/2$. Certainly, in this case,

L must be even to have an invariant subspace. Construct the basis in the invariant subspace of the Matsubara space using the second basis from Sect. 3 (replacing $2n$ by L). The linear independence is obvious for any $q \neq 0$. Construct the matrix of scalar products of the basis vectors and compute its rank. If $q^r = 1$, the rank is smaller than the maximal. This is well known as well as the fact that the rank equals cardinality of restricted Bratteli diagrams in which all the intermediate spins

$$j_p \leq \frac{r-2}{2}\,. \tag{2}$$

Since action of $\mathrm{Tr}_{\mathbf{S}}\left[T_{\mathbf{S},\mathbf{M}}\, O_{\mathbf{S}}\, q^{-2S_{\mathbf{S}}}\right]$ for invariant $O_{\mathbf{S}}$ preserves the invariant subspace and at the end we take the scalar product with invariant vector $\langle\Psi|$, the states that do not satisfy (2) can be just thrown away since their contributions vanish. We denote the ensemble of restricted Bratteli diagrams by $B(k, 0, j, r)$; certainly, j should also satisfy the restriction.

We have a tautological formula for the density matrix:

$$\rho = \sum_{J\in B(2n,0,0,r)} \frac{Z_{-1}\{O(J)\}}{\big(O(J), O(J)\big)}\, O(J)\,. \tag{3}$$

For any invariant operator $O_{\mathbf{S}}$, we have

$$Z_{-1}\{O_{\mathbf{S}}\} = \mathrm{Tr}_{\mathbf{S}}\left[\rho\, O_{\mathbf{S}}\, q^{-2S_{\mathbf{S}}}\right].$$

Actually, this formula can be used for any, not necessarily invariant, operator: ρ will project on invariant sector. From that point of view, we have a lattice analogue of Dotsenko–Fateev construction; in the scaling limit, this procedure gives the expectation values for the CFT with the central charge

$$c = 1 - \frac{6\nu^2}{1-\nu}\,, \tag{4}$$

and to all the operators of the original $c = 1$ CFT (scaling limit of XXZ), screening procedure is applied.

We are interested in the spectral properties of ρ. For generic q, the space $\left(\mathbb{C}^2\right)^{\otimes n}$ is decomposed into the irreducible representations with multiplicities:

$$\left(\mathbb{C}^2\right)^{\otimes n} = \bigoplus_{j=j_{\min}}^{j_{\max}} M_j \otimes V_j\,,$$

where $j_{\min} = 0$ for $n \equiv 0\ (\mathrm{mod}2)$, $j_{\min} = 1/2$ for $n \equiv 1\ (\mathrm{mod}2)$, $j_{\max} = \frac{n}{2}$, and M_j is the space of multiplicities whose basis is counted by $B(n, 0, j, \infty)$.

For $q^r = 1$, we use the reduced space

$$\left[\left(\mathbb{C}^2\right)^{\otimes n}\right]_{\mathrm{rd}} = \bigoplus_{j=j_{\min}}^{j_{\max}} M_{j,r} \otimes V_j \,,$$

with $j_{\max} = \min\left(\frac{n}{2}, \frac{r-2}{2}\right)$, and the basis of $M_{j,r}$ is counted by $B(n, 0, j, r)$. Being invariant, the density matrix acts as a unit operator on every V_j and reduces effectively to the block-diagonal matrix acting on $\oplus M_j$. Denoting blocks by ρ_j, we obtain

$$\mathrm{Tr}\left[\rho^k q^{-2S_{\mathbf{S}}}\right] = \sum_{j=j_{\min}}^{j_{\max}} \dim_q(j) \mathrm{Tr}\left[\rho_j^k\right] , \tag{5}$$

and interesting property of this formula comparing to the usual $\mathfrak{sl}_2$ is that the dimensions are replaced by quantum dimensions: traces of $q^{-2S_{\mathbf{S}}}$ over irreducible representations.

Let us consider more carefully the restricted case. Recall that the coefficients $c_{J,\alpha}(n)$ have denominators (8) that may vanish for $q^r = 1$. Our first idea was that the singular ones will go when we restrict J, J' in the main formula (9). But the life is not so simple, actually they do not. So, working with q_0 such that $q_0^r = 1$, we have to take $q = q_0 + \epsilon$, compute and send ϵ to 0. Doing that we shall first encounter singularities. Their cancellation is a property of the $\omega_{i,j}$. This is a consequence of our general construction, but direct understanding of such property would be desirable. Then in the finite part of the answer, derivatives $\partial_\nu^k \omega_{i,j}$ appear. We shall return to this point in the next section.

Let us dwell on applications of our construction. Take a long space of length $2N$ and consider the functional

$$\mathbf{Z}(O) = \lim_{N\to\infty} \frac{\mathrm{Tr}_{\mathbf{M}}\mathrm{Tr}_{\mathbf{S}}\left[T_{\mathbf{S},\mathbf{M}} O q^{-2S_{\mathbf{S}}}\right]}{\mathrm{Tr}_{\mathbf{M}}\mathrm{Tr}_{\mathbf{S}}\left[T_{\mathbf{S},\mathbf{M}} q^{-2S_{\mathbf{S}}}\right]} , \tag{6}$$

where in the space direction $\mathbf{S}$ we have the tensor product of $2N$ spaces $\mathbb{C}^2$ counted from $-N+1$ to N. For O being located on the interval $[1, n]$, this functional reduces to the functional Z_{-1} with $|\Psi\rangle$ being the eigenvector of the Matsubara transfer matrix

$$T_{\mathbf{M}} = \mathrm{Tr}_j(T_{j,\mathbf{M}}(1) q^{-\sigma_j^3}) ,$$

with maximal eigenvalue (which we assume to be unique) since

$$\lim_{K\to\infty} \left(\frac{T_{\mathbf{M}}}{T}\right)^K = |\Psi\rangle\langle\Psi| . \tag{7}$$

If $|\Psi\rangle$ is quantum group invariant, we are in the framework of the considerations above. Let us show that the only possibility for $|\Psi\rangle$ to be not invariant is to be orthogonal to the entire invariant subspace. Indeed, using (4), (7), one derives

$$\pi_{\mathbf{M}}(e_a)|\Psi\rangle\langle\Psi|\Phi\rangle = 0\,,$$

for all e_a and all invariant vectors $|\Phi\rangle$. In some physically important cases, this orthogonality can be excluded for continuity reasons near the isotropic point $q = 1$.

We can avoid all these arguments putting from the very beginning the projector on invariant subspaces $\mathcal{P}$ under the traces in (8). This gives rise to another functional

$$\widehat{\mathbf{Z}}(O) = \lim_{N\to\infty} \frac{\mathrm{Tr}_{\mathbf{M}}\mathrm{Tr}_{\mathbf{S}}\left[T_{\mathbf{S},\mathbf{M}}Oq^{-2S_{\mathbf{S}}}\mathcal{P}_{\mathbf{M}}\right]}{\mathrm{Tr}_{\mathbf{M}}\mathrm{Tr}_{\mathbf{S}}\left[T_{\mathbf{S},\mathbf{M}}q^{-2S_{\mathbf{S}}}\mathcal{P}_{\mathbf{M}}\right]}\,. \tag{8}$$

Now the transfer matrices would look for an invariant maximal eigenvector. Personally, the author does not like this violent procedure. The reasoning above shows that in a normal situation $Z = \widehat{Z}$. However, $\widehat{Z}$ is useful for mathematical rigour, in particular, as we shall see in the next subsection.

Finally, by the well-known procedure of taking limit $L \to \infty$ with staggered inhomogeneities [11], one obtains from $\mathbf{Z}$ the temperature average:

$$\langle O\rangle_T = \frac{\mathrm{Tr}\left[e^{-\frac{\mathcal{H}}{T}}q^{-S}O\right]}{\mathrm{Tr}\left[e^{-\frac{\mathcal{H}}{T}}q^{-S}\right]}\,.$$

5.1 *Relation to SOS/RSOS*

For invariant O, consider the functional $\widehat{\mathbf{Z}}(O)$ before the limit $N \to \infty$ is taken:

$$Z_{N,L}(O) = \frac{\mathrm{Tr}_{\mathbf{M}}\mathrm{Tr}_{\mathbf{S}}\left[T_{\mathbf{S},\mathbf{M}}Oq^{-2S_{\mathbf{S}}}\mathcal{P}_{\mathbf{M}}\right]}{\mathrm{Tr}_{\mathbf{M}}\mathrm{Tr}_{\mathbf{S}}\left[T_{\mathbf{S},\mathbf{M}}q^{-2S_{\mathbf{S}}}\mathcal{P}_{\mathbf{M}}\right]}\,.$$

We have seen that for $q^r = 1$ the restriction $j \le (r-2)/2$ for the intermediate spins is automatic. Let us try to rewrite $Z_{N,L}(O)$ in terms of SOS for generic q or RSOS for $q^r = 1$ configurations.

The SOS configurations are counted by spins $j = 0, 1/2, 1, \cdots$ put on the faces of a square lattice; for RSOS case, they are restricted. The spins are subject to the restriction $j' - j = \pm 1/2$ across vertical and horizontal edges. Folklore says that the SOS model is equivalent to the six-vertex model. But what does it mean exactly? On a torus, which is considered here, the number of six-vertex configurations is finite, while the number of SOS configurations is infinite. Let us consider the equivalence carefully rewriting the functional $\widehat{Z}$ in the SOS/RSOS language.

We start by the partition function in the denominator of $Z_{N,L}$. The vertices (i, j) are counted by space coordinate i and Matsubara coordinate j. We count faces by its left-lower vertex. We have seen that $T_{\mathbf{M}}$ acts from the invariant space to itself, so. The projector $\mathcal{P}$ can be freely moved along the space direction. Let us put it between the 0-th and 1-st lines. The natural orthogonal basis in the invariant subspace follows from Sect. 3, it consists of $E_{L,J,0,0}$, $j_0 = j_k = 0$, and we construct our projector out of them. One can rewrite $T_{\mathbf{M}}$ in the basis $E_{L,J,0,0}$; however, combining $(T_{\mathbf{M}})^N$, we do not obtain a local SOS interaction: all the spins on the faces $(\star, 0)$ equal 0, so they do not satisfy the SOS condition along the space direction.

We shall avoid this non-locality problem using that the vectors

$$E_{L,J} = \sum_{l=0}^{2j} q^{2(-j+l)} E_{L,J,-j+l,-j+l}\,, \quad j_1 = j_{2n+1} = j\,, \quad j_j = j_k = j\,,$$

which also happen to be invariant (for $E_{L,J,0,0}$, redundant indices are omitted). For $j_0 > 0$, these vectors are *not linearly independent*; this circumstance will cause some non-locality on the border.

Let us renormalise these vectors

$$|J\rangle = \frac{1}{\prod_{p=1}^{L-1} \sqrt{2j_p + 1}} E_{L,J}\,,$$

in order that for $j_0 = 0$ we have an orthonormal basis. Since for $q^r = 1$ the intermediate spins are restricted, we shall not meet zeros in the denominators.

We observe that the transfer matrix $T_{\mathbf{M}}$ acts locally on E_J:

$$T_{\mathbf{M}}|J\rangle = \sum_{J':\ j'_p - j_p = \pm 1/2} X_{J,J'}|J'\rangle\,, \tag{9}$$

where

$$X_{J,J'} = \prod_{p=0}^{L-1} F(t_p^{-1}, j'_p, j_p, j'_{p+1}, j_{p+1})\,,$$

with the coefficients F obtained using the $6j$-coefficients. Explicitly

$$F(x, j, j+1/2, j+1/2, j+1) = \frac{\sqrt{[2j+1]}}{\sqrt{[2j+2]}}(qx - q^{-1}x^{-1})$$

$$F(x, j, j-1/2, j-1/2, j+1) = \frac{\sqrt{[2j+1]}}{\sqrt{[2j]}}(qx - q^{-1}x^{-1})\,,$$

$$F(x, j, j+1/2, j+1/2, j) = \frac{1}{\sqrt{[2j+1]}\sqrt{[2j+2]}}(x^{-1}q^{2j+1} - xq^{-(2j+1)})\,,$$

$$F(x,j,j-1/2,j-1/2,j)=\frac{1}{\sqrt{[2j+1]}\sqrt{[2j]}}(q^{2j+1}x-q^{-(2j+1)}x^{-1}),$$

$$F(x,j,j+1/2,j-1/2,j)=\frac{\sqrt{[2j+2]}}{\sqrt{[2j+1]}}(x-x^{-1}),$$

$$F(x,j,j-1/2,j+1/2,j)=\frac{\sqrt{[2j]}}{\sqrt{[2j+1]}}(x-x^{-1}),$$

the parameters t_p are the inhomogeneities of the Matsubara chain.

Let us present the identity (9) graphically. We use Kirillov–Reshetikhin notations [16], namely, for the R-matrix, we use cross of solid lines, and for the face R-matrix F, we use cross of dashed lines. The vector $|J\rangle$ is represented by closed dashed line with solid intervals pointed to the left; the initial point is marked by a short fat interval; finally, bullet in the left-hand side stands for the insertion of $q^{-\sigma^3}$ (Fig. 2).

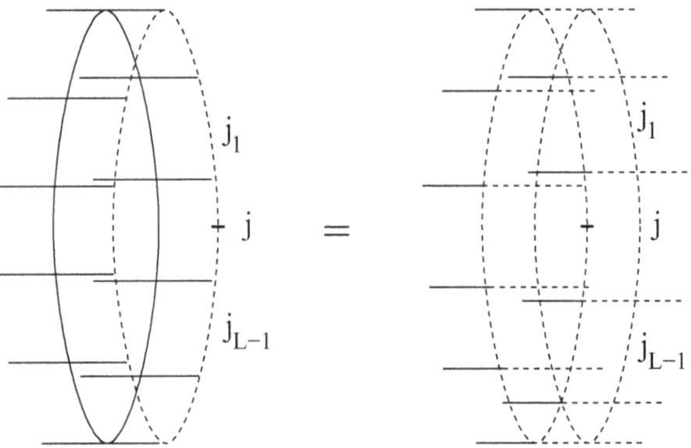

Fig. 2 First step of vertex-face correspondence

Already at this point, we see that the equivalence of the invariant sector of XXZ and SOS/RSOS is far from being clear: in SOS/RSOS formulation different face configurations are supposed to be independent, while from XXZ prospective, they correspond to linearly dependent vectors.

Using this and similar identities we obtain for $T_{\mathbf{S},\mathbf{M}}Oq^{-2S_{\mathbf{S}}}\mathcal{P}_{\mathbf{M}}$ which enters the numerator of $Z_{N,L}$ the equality presented graphically on Fig. 3. For $J=\emptyset$ we get $T_{\mathbf{S},\mathbf{M}}q^{-2S_{\mathbf{S}}}\mathcal{P}_{\mathbf{M}}$ present in the denominator.

Now it is clear that the functional $Z_{N,L}(O)$ can be rewritten in SOS/RSOS formulation, but if we consider different face configurations as independent and take trace over them, then we have to insert under this trace an operator M whose matrix elements are

$$M_{J,J'}=\langle J|J'\rangle\,.$$

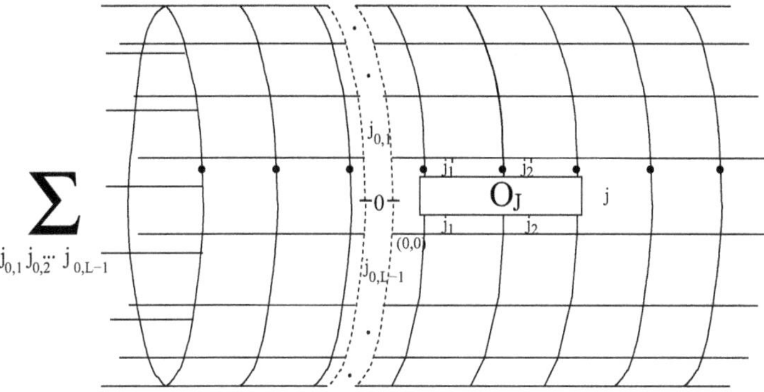

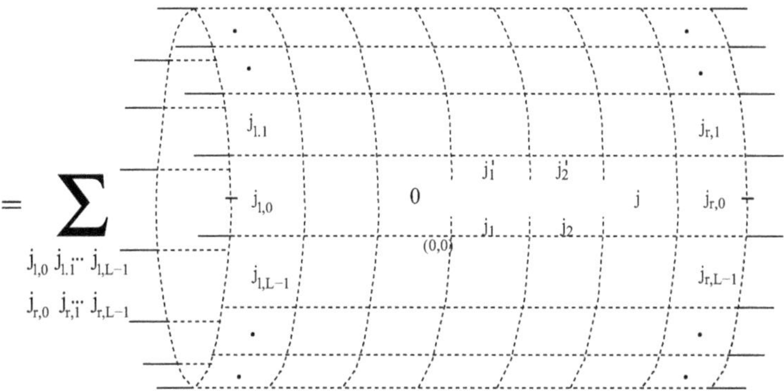

Fig. 3 Global vertex-face correspondence

In our opinion even in the limit $N \to \infty$, this boundary contribution cannot be dropped except, probably for such rough characteristics of the system as the free energy.

6 Numerical Results. Zero Temperature

In the zero temperature limit, the twist q^{-2S} becomes irrelevant, and the Bethe roots for the maximal eigenvector are real, dense, distributed with Lieb density. For the isotropic case, the maximal eigenvector $|\Psi\rangle$ is $\mathfrak{sl}_2$-invariant. Let us take some simple invariant vector $|\Phi\rangle$ to which it is not orthogonal (for example, one from the second basis of Sect. 3.2 basis). The scalar product

$$f(\nu) = \langle \Psi | \Phi \rangle \tag{1}$$

is an analytical function of ν. There is no reason to believe that $f(\nu)$ has branch points on the interval ([0, 1)]. So, as analytical function that is not identically 0, the function $f(\nu)$ has a discrete set of zeros on the interval.

In order to write down the function $\omega(x, y)$, it is convenient to make a change of variables:

$$\lambda = \frac{1}{2\pi\nu}\log x\,.$$

We shall slightly abuse notations using the same letters for functions of different variables:

$$\psi(\lambda) = \pi\nu\tanh(\pi\nu\lambda)\,,\qquad K(\lambda) = \frac{1}{2\pi i}\left(\psi(\lambda+i) - \psi(\lambda-i)\right)\,,$$

but we shall change the notation for ω:

$$\varpi(\lambda,\mu) = \omega\big(e^{2\pi\nu\lambda}, e^{2\pi\nu\mu}\big)\,.$$

The function $\varpi(\lambda,\mu)$ is simple for $T = 0$:

$$\varpi(\lambda,\mu) = \varpi_0(\lambda-\mu)\,, \tag{2}$$

$$\varpi(\lambda) = \int_{-\infty}^{\infty}\cos(2\lambda k)\frac{4\sinh\left(\frac{(1-\nu)}{\nu}k\right)}{\sinh\left(\frac{1}{\nu}k\right)\cosh(k)}dk - \frac{\pi}{2}K(\lambda)\,.$$

It is not difficult to recalculate the Taylor series for $\varpi(\lambda,\mu)$ in λ,μ as the Taylor series for $\omega(x, y)$ in $x-1, y-1$.

We consider the values of ν that correspond in the scaling limit to minimal models $M_{r,s}$ with $s \le 11$. The relation to ν is

$$\frac{r}{s} = 1-\nu\,.$$

The procedure is simple: we take $\omega(x, y)$ given by (2), plug it into (9), compute the density matrix and diagonalise it block-wise. The only problem is that for $s \le 7$ there are singularities in $c_{J,\alpha}$. So we check that they disappear in the final formula, and the final result contains not only the function $\omega(x, y)$, but also its derivatives with respect to ν that are not hard to compute, just a little boring.

What do we do having the spectra of the density matrix? First thing to do is to check that the invariant trace of ρ equals 1. For all ν's under consideration except $\nu = \frac{s-1}{s}$, this is true. Let us give a simple example

$$n = 2,\quad \nu = \frac{2}{11}\,.$$

Anticipating the results of the next section, we consider different temperatures. The density matrix has only two 1×1 blocks with $j = 0, 1$; let ρ_0 and ρ_1 are the corresponding eigenvalues. In the table below, we set the accuracy 10^{-24}, but actually, we compute with 60 digits.

T	ρ_0	ρ_1
0	0.749590171881005195039534	0.136773935626112993010667
1/16	0.749312246998088632583255	0.136925738294284871532951
1/8	0.748472992599050986722480	0.137384139340322728234512
3/16	0.747055192673494195413593	0.138158542155122243436699
1/4	0.745025105413502060299643	0.139267376525958562486133

With these data, one checks that always

$$\rho_0 + \frac{\sin\left(\frac{6\pi}{11}\right)}{\sin\left(\frac{2\pi}{11}\right)}\rho_1 = 1\,.$$

For $\nu = \frac{s-1}{s}$ that corresponds to degenerate minimal models $M_{1,s}$, we get unreasonable result: the invariant trace of ρ is not equal to 1. Presumably, these are the points where the scalar product (1) vanishes for all $|\Phi\rangle$?

All together, we consider

$$\nu \in \left\{\frac{1}{11}, \frac{1}{10}, \frac{1}{8}, \frac{1}{7}, \frac{1}{6}, \frac{2}{11}, \frac{1}{5}, \frac{1}{4}, \frac{3}{11}, \frac{2}{7}, \frac{3}{10}, \frac{4}{11}, \frac{3}{8}, \frac{2}{5}, \frac{3}{7}, \frac{5}{11}, \frac{6}{11}, \frac{4}{7}, \frac{3}{5}, \frac{5}{8}, \right.$$
$$\left. \frac{7}{11}, \frac{11}{16}, \frac{7}{10}, \frac{5}{7}, \frac{8}{11}, \frac{7}{9}\right\}.$$

There is one additional point $\nu = 11/16$, which we included in order to see more clear what happens near the point $\nu = 3/4$.

Now we can compute $\mathrm{Tr}\left[\rho^k q^{-2S}\right]$. For integer k, we observe a combination of monotonous and staggering behaviours, making it hard to compare with the scaling limit with data at hand. But for $k \ll 1$, the behaviour gets smoother and becomes really nice for the Von Neumann entropy. The latter is defined as

$$s(\nu, n) = -\sum_{j=j_{\min}}^{j_{\max}} \dim_q(j)\mathrm{Tr}\left[\rho_j \log(\rho_j)\right], \tag{3}$$

where ρ is computed on n sites.

In the scaling $n \to \infty$ limit $s(\nu, n)$ should behave as $c/3 \log n + B$, with B being some non-universal constant. With our data, we find typically pictures like that shown on Fig. 4.

Analysing these data, we come to the conclusion that the first correction to the CFT behaviour is $O(n^{-2})$. Then, we fit the numerical data for the largest n's at hand,

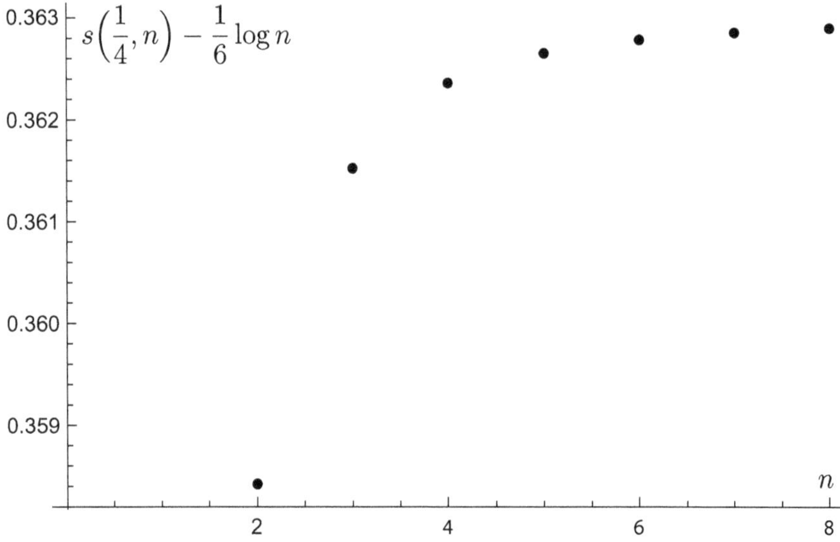

Fig. 4 Entropies for $\nu = 1/4,\ T = 0$

$n = 6, 7, 8$, with the ansatz

$$s(\nu, n) = \frac{c(\nu)}{3}\Big(A(\nu)\log n - G(\nu)n^{-2}\Big) + B(\nu)\,.$$

Our results agree with CFT if $A(\nu)$ is close to 1 and $G(\nu)$ is sufficiently small. The correction $O(n^{-2})$ is experimental. It should come from two places: the contribution of the descendants in the definition of operators in CFT perturbation, and from the perturbation of the CFT Hamiltonian [17].

Here are our results for $A(\nu)$. We see that they are generally good, and sometimes amazingly good having in mind relatively small sizes of our sub-lattices.

ν	1/11	1/10	1/8	1/7	1/6	2/11	1/5	1/4
$A(\nu)$	1.00103	1.00081	1.00032	1.00007	0.99984	0.99974	0.99966	0.99959

3/11	2/7	3/10	4/11	3/8	2/5	3/7	5/11	6/11
0.99959	0.99959	0.99960	0.99962	0.99962	0.99963	0.99965	0.99966	0.99977

4/7	3/5	5/8	7/11	11/16	7/10	5/7	8/11	7/9
0.99982	0.99990	0.99999	1.00004	1.00030	1.00037	1.00044	1.00047	1.00962

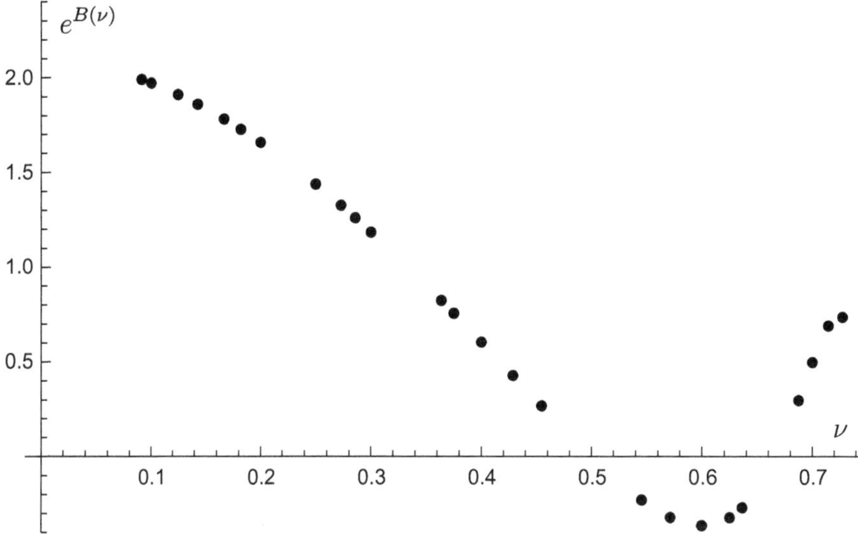

Fig. 5 Behaviour of e^B as a function of ν

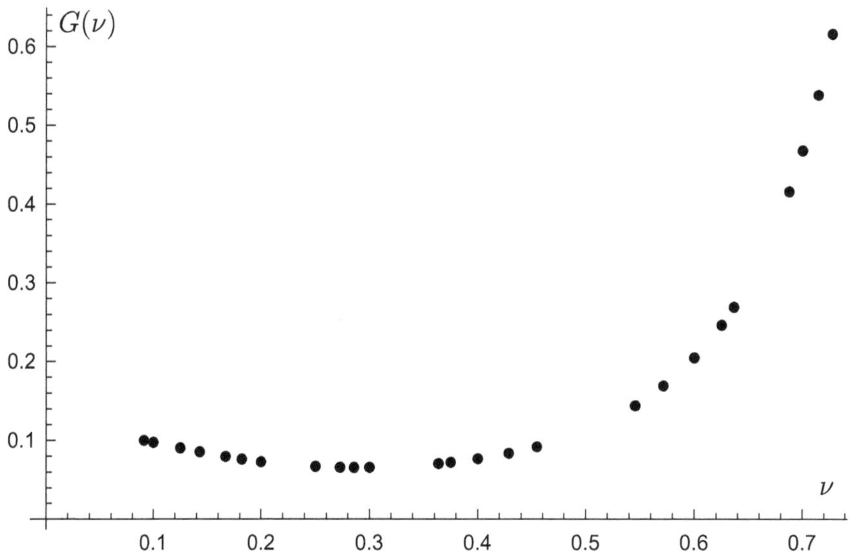

Fig. 6 Behaviour of G as a function of ν

The worst agreement is close to the isotropic case $\nu = 0$.

The constant $B(\nu)$ for $1/2 < \nu < 2/3$ acquires an imaginary part which equals π with a great precision. So, it makes sense to consider $e^{B(\nu)}$. On Fig. 5 we give the graphics with the last point, $\nu = 7/9$ omitted. The latter belongs to the interval $[3/4, 4/5]$, we do not have enough points in this interval to draw conclusions.

Finally, for the constant $G(\nu)$ we obtain Fig. 6 ($\nu = 7/9$ is again omitted).

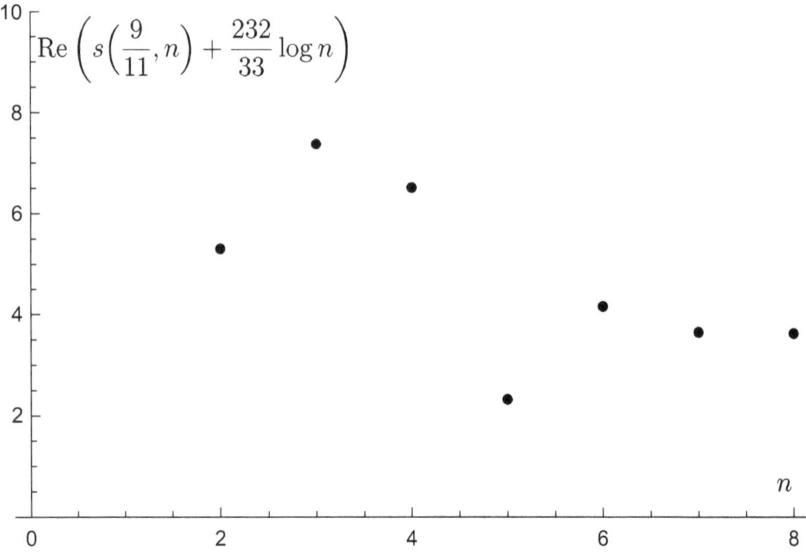

Fig. 7 Entropies for $\nu = 9/11$, $T = 0$

Trying to fit data for $\nu = 9/11$, we get unreasonable results. Let us see what happens. First, $s(9/11, n)$ has an imaginary part. However, the imaginary part is getting relatively small with n growing; here are the arguments of $-s(9/11, n)$ for $n = 6, 7, 8$, respectively: -0.0131820, -0.0006312, -0.000017. For the real part, we have Fig. 7.

This is quite different from the typical behaviour on figure. For small n, the behaviour is complicated, and only the last three points seem to start approaching the CFT behaviour. However, more careful look convinces that the point $n = 6$ is still rather far from the asymptotical trend. It would be desirable to have results for $n = 9$, but this is hard for the moment.

7 Temperature

Computation of the function $\varpi(\lambda, \mu)$ for the case of finite temperature is not very different from the isotropic case [9] that is why we shall be brief. We shall consider ν in the range from 0 to $1/2$. Above the latter, free fermion, point precision of our numerics drops drastically requiring more work.

We assume that in the ground state all Bethe roots are real and situated inside the interval $\lambda \in [-R, R]$. The value of R must be consistent with the requirement

$$|\log(\mathfrak{a}(\pm R))| < \pi, \tag{1}$$

where the counting function $\mathfrak{a}(\lambda)$ will be defined soon. We have

$$|\mathfrak{a}(\lambda)| = 1, \quad \lambda \in \mathbb{R},$$

the Swartz principle asserts

$$\mathfrak{a}(\bar{\lambda}) = \frac{1}{\overline{\mathfrak{a}(\lambda)}}.$$

We have to define several objects. The first one is the resolvent $R(\lambda, \mu)$ that solves the equation

$$R(\lambda, \mu) = K(\lambda - \mu) + \int_{-R}^{R} K(\lambda - \eta) R(\eta, \mu) d\eta. \tag{2}$$

The resolvent is continued analytically by virtue of Eq. (2) itself. We consider the ellipse

$$\lambda(\phi) = -R\cos(\phi) - it\sin\phi, \quad 0 \leq \phi < 2\pi. \tag{3}$$

The parameter t must be between 0 and 1. In numerical solution, we usually take $t = 2/5$. We denote by $C_\pm$ the parts of the ellipse belonging to $\mathbb{C}^\pm$, both oriented from left to right.

We shall need the function

$$F(\lambda, \mu) = f(\lambda - \mu) + \int_{C_+} R(\lambda, \eta) f(\eta - \mu) d\eta, \qquad f(\lambda) = \psi(\lambda - i) - \psi(\lambda).$$

In order to avoid singularities situated close to the integration contour, this equation is rewritten as

$$F(\lambda, \mu) = F_0(\lambda - \mu) + \Delta F(\lambda, \mu),$$

$$\Delta F(\lambda, \mu) = d(\lambda, \mu) + \int_{-R}^{R} R(\lambda, \eta) d(\eta, \mu) d\eta,$$

$$d(\lambda, \mu) = -\Big(\int_{-\infty}^{-R} + \int_{R}^{\infty}\Big) K(\lambda - \eta) F_0(\eta - \mu) d\eta.$$

Then, we have the NLIE:

$$\log \mathfrak{a}(\lambda) = \frac{1}{T}F(\lambda,0) - \int\limits_{C_+} R(\lambda,\mu)\log\left(1+\mathfrak{a}(\mu)\right)d\mu + \int\limits_{C_-} R(\lambda,\mu)\log\left(1+\overline{\mathfrak{a}(\mu)}\right)d\mu\,. \tag{4}$$

The iterations for this equation converge fast because $|\mathfrak{a}(\mu)| \ll 1$ for $\mathrm{Re}(\mu)$ close to 0 (at this point $\mathfrak{a}(\mu)$ develops an essential singularity.

Introduce the measures

$$dm(\lambda) = \frac{d\lambda}{1+\mathfrak{a}(\lambda)}\,, \quad d\overline{m}(\lambda) = \frac{\mathfrak{a}(\lambda)d\lambda}{1+\mathfrak{a}(\lambda)}\,.$$

We need the auxiliary $G(\lambda,\mu)$ function that solves the equation

$$G(\lambda,\mu) = F(\lambda,\mu) - \int\limits_{C_+} R(\lambda,\eta)G(\eta,\mu)d\overline{m}(\eta) - \int\limits_{C_-} R(\lambda,\eta)G(\eta,\mu)dm(\eta)\,.$$

Now we are ready to define the function $\varpi(\lambda,\mu)$:

$$\varpi(\lambda,\mu) = \varpi_0(\lambda-\mu) + \varpi_1(\lambda,\mu) + \varpi_2(\lambda,\mu)\,, \tag{5}$$

$$\varpi_1(\lambda,\mu) = -\frac{1}{2\pi}\Bigl(\int\limits_{-\infty}^{-R} + \int\limits_{R}^{\infty}\Bigr) f(\lambda-\eta)F_0(\eta-\mu)d\eta + \frac{1}{2\pi}\int\limits_{C_-} f(\lambda-\eta)\Delta F(\eta,\mu)d\eta\,,$$

$$\varpi_2(\lambda,\mu) = \frac{1}{\pi}\Bigl(\int\limits_{C_+} F(\lambda,\eta)G(\eta,\mu)d\overline{m}(\eta) + \int\limits_{C_-} F(\lambda,\eta)G(\eta,\mu)dm(z)\Bigr)\,.$$

For small T, the functions $\varpi_1(\lambda,\mu)$, $\varpi_2(\lambda,\mu)$ are small corrections to $\varpi_0(\lambda-\mu)$, and we get rather fast numerical procedure, parallel to that of [9]. We compute up to T from $1/160$ to $1/4$. Smaller is T greater interval we need. To be precise, we use $R = 1, 2, 3$. Certainly, we could do everything with $R = 3$, but it requires more precision and more time consuming. For numerical integration, we use the double exponential method [18] approximating [19]

$$\int_{-R}^{R} f(x)dx \simeq hR\sum_{k=-N}^{N} f(Rg(hk))g'(hk)\,.$$

$$g(t) = -1 + \frac{4}{\pi}\arctan(\exp(c\sinh(t)))\,.$$

Then for different R, we set

$$h = 1/20, \quad N = 200, \quad \text{for} \quad R = 1\,,$$
$$h = 1/25, \quad N = 250, \quad \text{for} \quad R = 2\,,$$
$$h = 1/40, \quad N = 400, \quad \text{for} \quad R = 3$$

for integrals over real axis and

$$h = 1/30, \quad N = 300, \quad \text{for} \quad R = 1$$
$$h = 1/40, \quad N = 400, \quad \text{for} \quad R = 2\,,$$
$$h = 1/60, \quad N = 600, \quad \text{for} \quad R = 3$$

for integration over a half of the ellipse.

For the finite temperature case, we consider only the most advanced case $n = 8$. We want to avoid the complications related to singularities of $c_{J,\alpha}(n)$ because computing derivative with respect to ν of functions defined above presents additional, unnecessary, problem. So, we take ν with denominator 11. Also, we work only below the free fermion point: $\nu < 1/2$, so we take

$$n = 1/11,\ 2/11,\ 3/11,\ 4/11,\ ,\ 5/11\,,$$

which corresponds to the minimal models $M_{r,11}$, $r = 1, 2, 3, 4, 5$. This is quite representative because we have in this list the unitary model $M_{1,11}$ and non-unitary models with positive and negative central charges.

The CFT predicts that the difference between the Von Neumann entropies at temperature T and at zero temperature behaves as

$$s(n, T) - s(n, 0) = \frac{c}{3} \log\left(\frac{\sinh(nT)}{nT}\right) .$$

Fig. 8 presents comparing of our results with the CFT prediction up to $8T = 2$. We see that agreement is good for the unitary case and is getting better when we move further from the isotropic point (Fig. 8). The nice agreement for the non-unitary case requires additional study.

We think that this beautiful picture is a good point to finish this chapter.

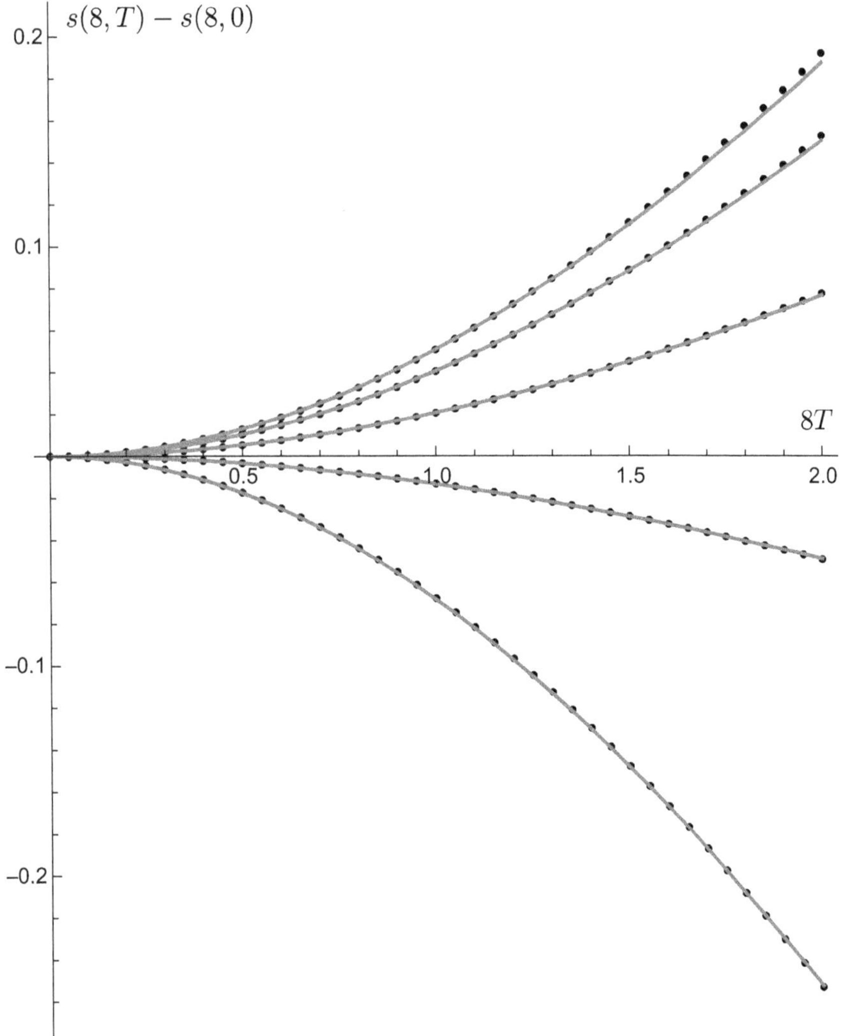

Fig. 8 Comparison with the minimal models $M_{p,11}$ for $p = 1, 2, 3, 4, 5$ (from up to down)

Acknowledgments For the case of finite temperature, I used slightly modified version of a program written together with T. Miwa, I am grateful to him for his effort. I also acknowledge discussions with B. Estienne, Y. Ikhlef and N. Reshetikhin.

References

1. N. Reshetikhin and F. Smirnov. Hidden quantum group symmetry and integrable perturbations of conformal field theories, *Comm. Math. Phys.* **131**(1990), 157–177
2. L. Faddeev. Modular Double of Quantum Group, *Math.Phys.Stud.* **21** (2000)149–156
3. H. Boos, M. Jimbo, T. Miwa, F. Smirnov, Y. Takeyama. Algebraic representation of correlation functions in integrable spin chains, *Annales Henri Poincaré* **7** (2006).1395–1428.
4. C. Holzhey C, F. Larsen, F. Wilczek. Geometric and renormalized entropy in conformal field theory *Nucl. Phys. B* **424** (1994) 443–467
5. V.E. Korepin *Physical Review Letters* **92**, issue 9, electronic identifier 096402, 05 March 2004
6. P. Calabrese, J. Cardy. Entanglement entropy and conformal field theory *J.Phys.A* **42** (2009) 504005–504036
7. D. Bianchini, O. A. Castro-Alvaredo, B. Doyon, E. Levi and F. Ravanini, Entanglement Entropy of Non Unitary Conformal Field Theory, **J. Phys. A 48** (2015) 04FT01
8. T. Dupic, B. Estienne, Y. Ikhlef. Entanglement entropies of minimal models from null-vectors, *SciPostPhys* **4, 031** (2018)
9. T. Miwa, F. Smirnov. New exact results on density matrix for XXX spin chain, *Lett. Math. Phys.* **109** (2019) 675–698
10. V. Drinfeld. Quantum groups, *Journal of Soviet Mathematics* **41** (1988) 898–915
11. A. Klümper. Thermodynamics of the anisotropic spin-1/2 Heisenberg chain and related quantum chains *Zeitschrift für Physik B Condensed Matter* **91**(1993) 507–519
12. L.D. Faddeev, E.K. Sklyanin, L.A. Takhtajan. The quantum inverse problem method *Theoretical and Mathematical Physics* **40** (1980) 688–711
13. V.E. Korepin, N.M. Bogoliubov, A.G. Izergin Quantum inverse scattering method and correlation functions *Cambridge University Press* (1993)
14. N.A. Slavnov. Calculation of scalar products of wave functions and form factors in the framework of the algebraic Bethe ansatz *Theoretical and Mathematical Physics* **79** (1989) 502–508
15. H. Boos, F. Göhmann, A. Klümper and J. Suzuki. Factorization of the finite temperature correlation functions of the XXZ chain in a magnetic field" *J. Phys. A* 40 (2007) 10699- 10727
16. A.N. Kirillov, N.Yu. Reshetikhin. Representations of the algebra $U_q(\mathfrak{sl}_2)$, q-orthogonal polynomials and invariants of links. *Adv. Ser. in Math. Phys.* **7** (1988) 285–339
17. S. Lukyanov. Low energy effective Hamiltonian for the XXZ spin chain. *Nucl.Phys.*, **B522**, (1998) 533–549
18. M. Mori, M. Sugihara. The double-exponential transformation in numerical analysis *Journal of Computational and Applied Mathematics* **127** (2001) 287–296
19. T. Ooura. Double exponential quadratures for various kinds of integral, Talk given at Second International ACCA-JP/UK Workshop, January 19, (2016) Kyoto University

Loops in Surfaces and Star-Fillings

Vladimir Turaev

Abstract We discuss a new approach to computing the standard algebraic operations on homotopy classes of loops in a surface: the homological intersection number, Goldman's Lie bracket, and the author's Lie cobracket. Our approach uses fillings of the surface by certain graphs.

1 Introduction

The homological intersection number of loops in an oriented surface Γ is computed by deforming these loops into a transversal position and counting their intersections with signs determined by the orientation of Γ. This defines a skew-symmetric bilinear form $\cdot_\Gamma : H_1(\Gamma) \times H_1(\Gamma) \to \mathbb{Z}$. A subtler way to count intersections of loops leads to Goldman's Lie bracket in the module generated by free homotopy classes of loops in Γ, see [Go1, Go2]. This bracket was complemented in [Tu1] by a Lie cobracket defined in terms of self-intersections of loops; for recent work on these operations see [AKKN1, AKKN2, Ha1, Ha2, Ka, KK1, KK2, LS, Ma]. For compact Γ, the duality isomorphism $H_1(\Gamma) \approx H^1(\Gamma, \partial\Gamma)$ carries the form $\cdot_\Gamma$ into the composition of the cup-product $\cup : H^1(\Gamma, \partial\Gamma) \times H^1(\Gamma, \partial\Gamma) \to H^2(\Gamma, \partial\Gamma)$ with the linear map $H^2(\Gamma, \partial\Gamma) \to \mathbb{Z}$ evaluating 2-cohomology classes on the fundamental class of Γ. So, the intersection number of two elements of $H_1(\Gamma)$ can be computed by taking the dual cohomology classes, representing them by 1-cocycles on a triangulation of Γ, evaluating the cup-product of these 1-cocycles on all triangles of the triangulation, and summing up the resulting values. This method allows us to compute the algebraic number of intersections of loops without ever considering the intersections themselves. The aim of this paper is to give similar computations of the Lie bracket and cobracket mentioned above. To do it, we switch from triangulations to a more flexible language of graphs in surfaces.

V. Turaev (✉)
Department of Mathematics, Indiana University Bloomington, Bloomington, IN, USA
e-mail: vtouraev@indiana.edu

A. Alekseev et al. (eds.), *Representation Theory, Mathematical Physics, and Integrable Systems*, Progress in Mathematics 340,
https://doi.org/10.1007/978-3-030-78148-4_21

In the rest of the introduction we focus on the case of surfaces with non-void boundary. Let Γ be a compact connected oriented surface with $\partial\Gamma \neq \emptyset$. By a *star* we mean an oriented graph formed by $n \geq 2$ vertices of degree 1 (the leaves), a vertex of degree n (the center), and n edges leading from the center to the leaves. The set of leaves of a star s is denoted by ∂s. A *star* s *in* Γ is a star embedded in Γ so that $s \cap \partial\Gamma = \partial s$. The orientation of Γ at the center of s determines a cyclic order in the set $\mathrm{Edg}(s)$ of edges of s. For $e \in \mathrm{Edg}(s)$ we let $e^+ \in \mathrm{Edg}(s)$ be the next edge with respect to this order. We say that a loop in Γ is s*-generic* if it misses the vertices of s, meets all edges of s transversely, and never traverses a point of s more than once. It is clear that any loop in Γ can be made s-generic by a small deformation. Given an s-generic loop a in Γ and an edge $e \in \mathrm{Edg}(s)$ we let $a \cap e$ be the set of points of e traversed by a. For $p \in a \cap e$, the intersection sign of a and e at p is denoted by $\mu_p(a)$. The integer $a \cdot e = \sum_{p \in a \cap e} \mu_p(a)$ is the algebraic number of intersections of a and e. For s-generic loops a, b in Γ, set

$$a \cdot_s b = \sum_{e \in \mathrm{Edg}(s)} \big((a \cdot e)(b \cdot e^+) - (b \cdot e)(a \cdot e^+)\big).$$

Theorem 1.1 *The map* $(a, b) \mapsto a \cdot_s b$ *defines a skew-symmetric bilinear form* $\cdot_s : H_1(\Gamma) \times H_1(\Gamma) \to \mathbb{Z}$ *depending only on the isotopy class of the star* s *in* Γ.

The idea behind this theorem is to view each pair of points $p \in a \cap e$, $q \in b \cap e^+$ as a pseudo-intersection of a and b, and to consider the (skew-symmetrized) algebraic number of such pairs. To recover the standard homological intersection form $\cdot_\Gamma$ in $H_1(\Gamma)$, we need one more notion. A *star-filling* of Γ is a finite family F of disjoint stars in Γ such that each component of the set $\Gamma \setminus \cup_{s \in F} s$ is a disk meeting $\partial\Gamma$ at one or two open segments. We explain in the body of the paper that Γ has star-fillings. The following theorem computes the intersection form $\cdot_\Gamma$ in terms of star-fillings.

Theorem 1.2 *For any star-filling* F *of* Γ*, we have*

$$2 \cdot_\Gamma = \sum_{s \in F} \cdot_s : H_1(\Gamma) \times H_1(\Gamma) \to \mathbb{Z}. \tag{1.1}$$

So, for any $x, y \in H_1(\Gamma)$, the integer $\sum_{s \in F} x \cdot_s y$ is even and is equal to $2\, x \cdot_\Gamma y$. As a consequence, the form $\cdot_\Gamma$ can be fully recovered from the forms $\{\cdot_s\}_{s \in F}$. Similar remarks apply to our computations of the bracket and cobracket below.

To state an analogue of Theorems 1.1 and 1.2 for Goldman's Lie bracket, we first recall the definition of this bracket. Let $\mathcal{L} = \mathcal{L}(\Gamma)$ be the set of free homotopy classes of loops in Γ and let $M = M(\Gamma)$ be the free abelian group with basis $\mathcal{L}$. For a loop a in Γ, we let $\langle a \rangle \in \mathcal{L} \subset M$ be the free homotopy class of a. For a point $p \in \Gamma$ traversed by a once, we let a_p be the loop starting at p and going along a until the return to p. We say that a pair of loops a, b in Γ is *generic* if these loops are transversal and do not meet at self-intersections of a or b. Then the (finite) set of intersection points of a, b is denoted by $a \cap b$. For $p \in a \cap b$, we let $\varepsilon_p(a, b) = \pm 1$ be the intersection sign of a and b at p. Also, $a_p b_p$ stands for the product of the loops

a_p, b_p based at p. Goldman's bracket is the bilinear form $[\,,\,]_\Gamma : M \times M \to M$ defined on the basis $\mathcal{L} \subset M$ by

$$[\langle a\rangle, \langle b\rangle]_\Gamma = \sum_{p \in a \cap b} \varepsilon_p(a,b) \langle a_p b_p \rangle$$

for any generic pair of loops a, b in Γ. The bracket $[\,,\,]_\Gamma$ is a well-defined homotopy lift of the form $\cdot_\Gamma$ in $H_1(\Gamma)$.

With any star s in Γ we now associate a bilinear map $[\,,\,]_s : M \times M \to M$. For any s-generic loops a, b in Γ and any points $p, q \in s$ traversed, respectively, by a, b, we pick a path c from p to q in s and write $a_p b_q$ for the loop $a_p c b_q c^{-1}$. Clearly, the free homotopy class of this loop does not depend on the choice of c. Set

$$[a,b]_s = \sum_{e \in \mathrm{Edg}(s)} \Big(\sum_{p \in a \cap e,\, q \in b \cap e^+} \mu_p(a)\, \mu_q(b) \langle a_p b_q \rangle - \sum_{p \in a \cap e^+,\, q \in b \cap e} \mu_p(a)\, \mu_q(b) \langle a_p b_q \rangle \Big).$$

Theorem 1.3 *The map $(\langle a\rangle, \langle b\rangle) \mapsto [a,b]_s$ defines a skew-symmetric bilinear form $[\,,\,]_s : M \times M \to M$ depending only on the isotopy class of s in Γ. For any star-filling F of Γ, we have*

$$2[\,,\,]_\Gamma = \sum_{s \in F} [\,,\,]_s. \tag{1.2}$$

One may ask whether the bracket $[\,,\,]_s$ shares the fundamental properties of Goldman's bracket, namely whether it satisfies the Jacobi identity and induces Poisson brackets on the moduli spaces of Γ. In general, the answer to both questions is negative though some weaker results hold true, see [Tu2]. Note also that in analogy with Goldman's bracket, Kawazumi and Kuno [KK1] defined an action of the Lie algebra M on the modules generated by homotopy classes of paths in Γ with endpoints in $\partial\Gamma$. The Kawazumi-Kuno action can be computed similarly to Theorem 1.3 in terms of star-fillings. The same methods work for the double brackets of surfaces defined in [MT], this will be discussed elsewhere.

We next state our results concerning the Lie cobracket on loops in Γ, see [Tu1]. For a loop a in Γ, set $\langle a\rangle_0 = \langle a\rangle \in \mathcal{L} \subset M$ if a is non-contractible and $\langle a\rangle_0 = 0 \in M$ if a is contractible. We say that the loop a is *generic* if all its self-intersections are double transversal intersections. The set of self-intersections of a generic loop a is finite and is denoted $\#a$. The loop a crosses each point $r \in \#a$ twice; we let v_r^1, v_r^2 be the tangent vectors of a at r numerated so that the pair (v_r^1, v_r^2) is positively oriented. For $i = 1, 2$, let a_r^i be the loop starting in r and going along a in the direction of the vector v_r^i until the first return to r. Up to parametrization, $a = a_r^1 a_r^2$ is the product of the loops a_r^1, a_r^2. The cobracket ν_Γ is a linear map $M \to M \otimes M$ defined on the basis $\mathcal{L} \subset M$ by

$$\nu_\Gamma(\langle a\rangle) = \sum_{r \in \#a} \langle a_r^1\rangle_0 \otimes \langle a_r^2\rangle_0 - \langle a_r^2\rangle_0 \otimes \langle a_r^1\rangle_0$$

for any generic loop a in Γ. The cobracket ν_Γ is skew-symmetric in the sense that its composition with the permutation in $M \otimes M$ is equal to $-\nu_\Gamma$.

For any star s in Γ we define a linear map $\nu_s : M \to M \otimes M$. Given an s-generic loop a in Γ and points p_1, p_2 of s traversed by a, we write a_{p_1,p_2} for the loop going from p_1 to p_2 along a and then going back to p_1 along a path in s. The free homotopy class of this loop does not depend on the choice of the latter path. Set

$$\nu_s(\langle a\rangle)$$

$$= \sum_{e\in \mathrm{Edg}(s)} \sum_{p_1\in a\cap e,\, p_2\in a\cap e^+} \mu_{p_1}(a)\, \mu_{p_2}(a) \left(\langle a_{p_1,p_2}\rangle_0 \otimes \langle a_{p_2,p_1}\rangle_0 - \langle a_{p_2,p_1}\rangle_0 \otimes \langle a_{p_1,p_2}\rangle_0\right).$$

Theorem 1.4 *The map $\langle a\rangle \mapsto \nu_s(\langle a\rangle)$ defines a skew-symmetric cobracket $M \to M \otimes M$ depending only on the isotopy class of s in Γ. For any star-filling F of Γ, we have*

$$2\,\nu_\Gamma = \sum_{s\in F} \nu_s. \tag{1.3}$$

The cobracket ν_Γ has a refinement depending on a framing of Γ, see [Tu1], Section 18.1 and [AKKN2]; it would be interesting to extend Theorem 1.4 to this refined cobracket.

Theorems 1.1–1.4 are proved in Sect. 8 where we also discuss the case of closed surfaces. The proofs of Theorems 1.1–1.4 are based on the theory of quasi-surfaces developed in Sects. 2–7.

This work was supported by the NSF grant DMS-1664358.

2 Quasi-Surfaces

2.1 Basics

By a *surface* we mean a smooth 2-dimensional manifold with boundary. A *quasi-surface* is a topological space X obtained by glueing an oriented surface Σ to a topological space Y along a continuous map $\alpha \to Y$ where $\alpha \subset \partial\Sigma$ is the union of a finite number of disjoint segments in $\partial\Sigma$. Clearly, $Y \subset X$ and $X \setminus Y = \Sigma \setminus \alpha$. We call Y the *singular core* of X and call Σ the *surface core* of X. We fix a closed neighborhood of α in Σ and identify it with $\alpha \times [-1, 1]$ so that

$$\alpha = \alpha \times \{-1\} \qquad \text{and} \qquad \partial\Sigma \cap (\alpha \times [-1, 1]) = \alpha \cup (\partial\alpha \times [-1, 1]).$$

The surface

$$\Sigma' = \Sigma \setminus (\alpha \times [-1, 0)) \subset \Sigma \setminus \alpha \subset X$$

is a copy of Σ embedded in X. We provide Σ' with the orientation induced from that of Σ and call Σ' the *reduced surface core* of X.

Set $\pi_0 = \pi_0(\alpha) = \pi_0(\alpha \times \{0\})$. For $k \in \pi_0$, we let α_k be the corresponding segment component of $\alpha \times \{0\} \subset \partial\Sigma' \subset X$. We call α_k the *k-th gate* of X. The gates $\{\alpha_k\}_{k\in\pi_0}$ separate $\Sigma' \subset X$ from the rest of X. A *gate orientation* of X is an orientation of all gates. Gate orientations of X canonically correspond to orientations of the 1-manifold α. For a gate orientation ω of X, we let $\overline{\omega}$ be the gate orientation of X opposite to ω on all gates.

We keep the notation $X, Y, \alpha, \Sigma, \Sigma, \{\alpha_k\}_k, \pi_0$ till the end of Sect. 7.

2.2 Loops in X

By a *loop* in X we mean a continuous map $a : S^1 \to X$. A *generic loop* a in X is a loop in X such that (1) all branches of a in Σ' are smooth immersions meeting $\partial\Sigma'$ transversely at a finite set of points lying in the interior of the gates, and (2) all self-intersections of a in Σ' are double transversal intersections in $\text{Int}(\Sigma') = \Sigma' \setminus \partial\Sigma'$. The set of self-intersections in Σ' of a generic loop a is denoted by $\#a$. This set is finite and lies in $\text{Int}(\Sigma')$. Using cylinder neighborhoods of the gates, it is easy to see that any loop in X may be transformed into a generic loop by a small deformation.

More generally, a finite family of loops in X is *generic* if these loops are generic and all their intersections in Σ' are double transversal intersections in $\text{Int}(\Sigma')$. In particular, these loops cannot meet at the gates. As above, any finite family of loops in X may be transformed into a generic family by a small deformation.

For a loop a in X and any point p of $a(S^1)$ which is not a self-intersection of a (i.e., which is traversed by a only once), we let a_p be the loop which starts at p and goes along a until coming back to p. For any $k \in \pi_0$, we set $a \cap \alpha_k = a(S^1) \cap \alpha_k$. If the loop a is generic then it never traverses a point of $a \cap \alpha_k$ more than once and the set $a \cap \alpha_k$ is finite.

2.3 Local Moves

We define six local moves $L_0 - L_5$ on a generic loop a in X keeping its free homotopy class. The move L_0 is a deformation of a in the class of generic loops. This move preserves the number card$(\#a)$. The moves $L_1 - L_3$ modify a in a small disk in $\text{Int}(\Sigma')$. The move L_1 adds a small curl to a and increases card$(\#a)$ by 1. The move L_2 pushes a branch of a across another branch of a increasing card$(\#a)$ by 2. The move L_3 pushes a branch of a across a double point of a keeping card$(\#a)$. The moves L_4, L_5 modify a in a neighborhood of a gate α_k in X. The move L_4 pushes a branch of a across α_k increasing card$(a \cap \alpha_k)$ by 2 and keeping card$(\#a)$. The move L_5 pushes a double point of a across α_k keeping card$(a \cap \alpha_k)$ and decreasing card$(\#a)$ by 1. We call the moves $L_0 - L_5$ and their inverses *loop moves*. It is clear that generic loops in X are freely homotopic if and only if they can be related by a finite sequence of loop moves.

3 Homological Intersection Forms

We define homological intersection forms in $H_1(X)$. Here and below, by $H_1(X)$ we mean the 1-homology of the underlying topological space of the quasi-surface X.

3.1 *The Intersection Form of* (X, ω)

Given a gate orientation ω of X, we define a bilinear form in $H_1(X)$ called the *homological intersection form of the pair* (X, ω). The idea is to properly position the loops in X near the gates and then to count their algebraic number of intersections in Σ'. We begin with definitions.

For any points p, q of a gate α_k, we say that p *lies on the ω-left of* q and write $p <_\omega q$ if $p \neq q$ and the ω-orientation of α_k leads from p to q. We say that an (ordered) pair of loops a, b in X is *ω-admissible* if it is generic (in the sense of Sect. 2.2) and the crossings of a with any gate lie on the ω-left of the crossings of b with this gate. Taking a generic pair of loops a, b in X and pushing the branches of a crossing the gates to the ω-left of the crossings of b with the gates, we obtain an ω-admissible pair of loops. Thus, any pair of loops in X may be deformed into an ω-admissible pair.

For each generic pair of loops a, b in X, we consider the finite set

$$a \cap b = a(S^1) \cap b(S^1) \cap \Sigma' \subset \mathrm{Int}(\Sigma').$$

For $r \in a \cap b$, set $\varepsilon_r(a, b) = 1$ if the tangent vectors of a and b at r form a positive basis in the tangent space of Σ' at r and set $\varepsilon_r(a, b) = -1$ otherwise.

Lemma 3.1 *For any ω-admissible pair a, b of loops in X, the integer*

$$a \bullet_{X,\omega} b = \sum_{r \in a \cap b} \varepsilon_r(a, b) \tag{3.1}$$

depends only on the homology classes of a, b in $H_1(X)$. The formula $(a, b) \mapsto a \bullet_{X,\omega} b$ defines a bilinear form

$$\bullet_{X,\omega} : H_1(X) \times H_1(X) \to \mathbb{Z}.$$

Proof For each $k \in \pi_0$, one endpoint of the gate α_k lies on the ω-left of the other endpoint. Pick disjoint closed segments $\alpha_k^- \subset \alpha_k$ and $\alpha_k^+ \subset \alpha_k$ containing these two endpoints, respectively. Clearly, $p <_\omega q$ for all $p \in \alpha_k^-$ and $q \in \alpha_k^+$. We say that a loop in X is *ω-left* (respectively, *ω-right*) if it is generic and meets the gates of X only at points of $\cup_k \alpha_k^-$ (respectively, of $\cup_k \alpha_k^+$). Given an ω-admissible pair of loops a, b in X, we can push the branches of a crossing the gates to the left and push the branches of b crossing the gates to the right without creating or destroying intersections between a and b. Consequently, a is homotopic (in fact, isotopic) to an ω-left loop a' and b is homotopic to an ω-right loop b' such that $a \bullet_{X,\omega} b = a' \bullet_{X,\omega} b'$. Since α_k^- is a deformation retract of α_k for all k, any ω-left loops homotopic in X are homotopic in the class of ω-left loops. Similarly, any ω-right loops homotopic in X are homotopic in the class of ω-right loops. Such homotopies of the loops a', b' expand as compositions of loop moves keeping a', b', respectively, ω-left and ω-right. The latter moves obviously preserve $a' \bullet_{X,\omega} b'$. Therefore the integer $a \bullet_{X,\omega} b = a' \bullet_{X,\omega} b'$ depends only on the (free) homotopy classes of a, b in X. Moreover, since $a \bullet_{X,\omega} b$ depends linearly on a and b, it depends only on the homology classes of a, b. This implies the claim of the lemma. □

We stress that the crossings in $X \setminus \Sigma'$ of an ω-admissible pair of loops a, b do not contribute to $a \bullet_{X,\omega} b$. Note also that for such a, b, the pair b, a is $\overline{\omega}$-admissible. Using these pairs to compute $a \bullet_{X,\omega} b$ and $b \bullet_{X,\overline{\omega}} a$, we obtain the same terms with opposite signs. Hence, for any $x, y \in H_1(X)$,

$$x \bullet_{X,\omega} y = -y \bullet_{X,\overline{\omega}} x. \tag{3.2}$$

3.2 The Intersection Form of X

We state the main result of this section.

Theorem 3.2 *The skew-symmetric bilinear form* $\bullet_X : H_1(X) \times H_1(X) \to \mathbb{Z}$ *defined by*

$$x \bullet_X y = x \bullet_{X,\omega} y - y \bullet_{X,\omega} x$$

for all $x, y \in H_1(X)$ *and a gate orientation* ω *of* X *does not depend on* ω.

We will prove this theorem in Sect. 3.3. We call $\bullet_X$ the *homological intersection form of* X. Both $\bullet_{X,\omega}$ and $\bullet_X$ generalize the intersection form in the homology of Σ': the value of $\bullet_{X,\omega}$ (respectively, of $\bullet_X$) on any pair of homology classes of loops in $\Sigma' \subset X$ is equal to the usual intersection number of these loops in Σ' (respectively, twice this number). The image of the inclusion homomorphism $H_1(Y) \to H_1(X)$ annihilates both $\bullet_{X,\omega}$ and $\bullet_X$.

To prove Theorem 3.2, we study the dependence of $\bullet_{X,\omega}$ on ω. We start with notation. For a generic loop a in X, the *sign* $\varepsilon_p(a)$ of a at a point $p \in a \cap \alpha_k$

is $+1$ if a goes near p from $X \setminus \Sigma'$ to $\mathrm{Int}(\Sigma')$ and -1 otherwise. The linear map $v_k : H_1(X) \to \mathbb{Z}$ "dual" to the gate α_k carries the homology class of any generic loop a to $\sum_{p \in a \cap \alpha_k} \varepsilon_p(a)$. For any loops a, b in X, we define the set of triples

$$T(a, b) = \{(k, p, q) \,|\, k \in \pi_0, p \in a \cap \alpha_k, q \in b \cap \alpha_k, p \neq q\}.$$

Given a gate orientation ω of X, we set

$$T_\omega(a, b) = \{(k, p, q) \in T(a, b) \,|\, q <_\omega p\} \subset T(a, b).$$

For $k \in \pi_0$, we set $\varepsilon(\omega, k) = +1$ if the ω-orientation of α_k is compatible with the orientation of Σ', i.e., if the pair (a ω-positive tangent vector of $\alpha_k \subset \partial\Sigma'$, a vector directed inside Σ') is positively oriented in Σ'. Otherwise, $\varepsilon(\omega, k) = -1$.

Lemma 3.3 *For any gate orientation ω and any $x, y \in H_1(X)$ represented by a generic pair of loops a, b in X, we have*

$$x \bullet_{X,\omega} y = \sum_{r \in a \cap b} \varepsilon_r(a, b) + \sum_{(k,p,q) \in T_\omega(a,b)} \varepsilon(\omega, k)\, \varepsilon_p(a)\, \varepsilon_q(b). \tag{3.3}$$

Proof Consider an ω-admissible pair of loops a', b where a' is obtained from a by pushing its branches crossing the gates to the ω-left of the branches of b crossing the gates. This transformation modifies a in a small neighborhood of the gates so that a', b have the same intersections in Σ' as a, b plus one additional intersection $r = r(k, p, q) \in \Sigma'$ for each triple $(k, p, q) \in T_\omega(a, b)$. It is easy to check that $\varepsilon_r(a', b) = \varepsilon(\omega, k)\, \varepsilon_p(a)\, \varepsilon_q(b)$. Consequently,

$$x \bullet_{X,\omega} y = a' \bullet_{X,\omega} b = \sum_{r \in a' \cap b'} \varepsilon_r(a', b)$$

$$= \sum_{r \in a \cap b} \varepsilon_r(a, b) + \sum_{(k,p,q) \in T_\omega(a,b)} \varepsilon(\omega, k)\, \varepsilon_p(a)\, \varepsilon_q(b).$$

□

Formula (3.3) generalizes (3.1) because $T_\omega(a, b) = \emptyset$ for any ω-admissible pair of loops a, b.

3.3 *Proof of Theorem 3.2*

For any $l \in \pi_0$, we let $l\omega$ be the gate orientation obtained from ω by inverting the direction of the gate α_l while keeping the directions of all other gates. We claim that for any $x, y \in H_1(X)$,

$$x \bullet_{X,l\omega} y = x \bullet_{X,\omega} y - \varepsilon(\omega, l)\, v_l(x)\, v_l(y). \tag{3.4}$$

Indeed, pick an ω-admissible pair of loops a, b representing, respectively, x, y. We compute $x \bullet_{X,\omega} y = a \bullet_{X,\omega} b$ from the definition and compute $x \bullet_{X,l\omega} y = a \bullet_{X,l\omega} b$ from Lemma 3.3. The resulting expressions differ in the sum associated with $T_{l\omega}(a, b)$. Since the pair a, b is ω-admissible, the set $T_{l\omega}(a, b)$ consists of all triples (l, p, q) with $p \in a \cap \alpha_l, q \in b \cap \alpha_l$. Therefore

$$x \bullet_{X,l\omega} y = x \bullet_{X,\omega} y + \sum_{p\in a\cap\alpha_l, q\in b\cap\alpha_l} \varepsilon(l\omega, l)\, \varepsilon_p(a)\, \varepsilon_q(b)$$

$$= x \bullet_{X,\omega} y - \varepsilon(\omega, l)\, v_l(x)\, v_l(y).$$

Formula (3.4) implies that

$$x \bullet_{X,l\omega} y - y \bullet_{X,l\omega} x = x \bullet_{X,\omega} y - y \bullet_{X,\omega} x$$

for all $l \in \pi_0$. This implies the claim of the theorem.

3.4 Computation of $\bullet_X$

To compute $x \bullet_X y$ for $x, y \in H_1(X)$ we will use the following method. Pick a generic pair of loops a, b in X representing x, y. Let ω_0 be the orientation of the gates induced by the orientation of $\Sigma' \subset \Sigma$ so that $\varepsilon(\omega_0, k) = 1$ for all $k \in \pi_0$. Lemma 3.3 implies that

$$x \bullet_X y = x \bullet_{X,\omega_0} y - y \bullet_{X,\omega_0} x$$

$$= 2 \sum_{r\in a\cap b} \varepsilon_r(a, b) + \sum_{(k,p,q)\in T(a,b)} \delta(p, q)\, \varepsilon_p(a)\, \varepsilon_q(b)$$

, where $\delta(p, q) = 1$ for $q <_{\omega_0} p$ and $\delta(p, q) = -1$ for $p <_{\omega_0} q$. If $a \cap b = \emptyset$, then

$$x \bullet_X y = \sum_{(k,p,q)\in T(a,b)} \delta(p, q)\, \varepsilon_p(a)\, \varepsilon_q(b). \tag{3.5}$$

4 The Intersection Brackets

We define homotopy intersection brackets refining the homological forms above.

4.1 The Brackets

Let $\mathcal{L} = \mathcal{L}(X)$ be the set of free homotopy classes of loops in the quasi-surface X and let $M = M(X)$ be the free abelian group with basis $\mathcal{L}$. Pick a gate orientation ω of X. By Sect. 3.1, any pair $x, y \in \mathcal{L}$ can be represented by an ω-admissible pair of loops a, b in X. For a point $r \in a \cap b$, consider the loops a_r, b_r which are reparametrizations of a, b based at r. Consider the product loop $a_r b_r$ and set

$$[x, y]_{X,\omega} = \sum_{r \in a \cap b} \varepsilon_r(a, b) \langle a_r b_r \rangle \in M, \tag{4.1}$$

where for a loop c in X, we let $\langle c \rangle \in \mathcal{L} \subset M$ be its free homotopy class. The sum on the right-hand side of (4.1) is an algebraic sum of all possible ways to graft a and b at their intersections in Σ'. It is straightforward to see that this sum is preserved under all loop moves on a, b keeping this pair ω-admissible. Hence, $[x, y]_{X,\omega}$ does not depend on the choice of a, b in the homotopy classes x, y. Extending the map $(x, y) \mapsto [x, y]_{X,\omega}$ by bilinearity, we obtain a bilinear bracket $[-, -]_{X,\omega}$ in M. The proof of Formula (3.2) applies here and shows that for any $x, y \in M$,

$$[x, y]_{X,\omega} = -[y, x]_{X,\overline{\omega}}. \tag{4.2}$$

We now compute the bracket $[x, y]_{X,\omega}$ from an arbitrary generic pair of loops a, b representing x, y. Note that for any points $p \in a \cap \alpha_k$, $q \in b \cap \alpha_k$ on the same gate, we can multiply the loops a_p, b_q based at p, q using a path connecting p, q in α_k. The product loop determines an element of $\mathcal{L}$ denoted $\langle a_p b_q \rangle$.

Lemma 4.1 *Let $x, y \in \mathcal{L}$ be represented by a generic pair of loops a, b. Then*

$$[x, y]_{X,\omega} = \sum_{r \in a \cap b} \varepsilon_r(a, b) \langle a_r b_r \rangle + \sum_{(k,p,q) \in T_\omega(a,b)} \varepsilon(\omega, k)\, \varepsilon_p(a)\, \varepsilon_q(b) \langle a_p b_q \rangle. \tag{4.3}$$

The proof repeats the proof of Lemma 3.3 with obvious modifications. If $a \cap b = \emptyset$, then (4.3) simplifies to

$$[x, y]_{X,\omega} = \sum_{(k,p,q) \in T_\omega(a,b)} \varepsilon(\omega, k)\, \varepsilon_p(a)\, \varepsilon_q(b) \langle a_p b_q \rangle. \tag{4.4}$$

Theorem 4.2 *The skew-symmetric bracket $[-, -]_X$ in M defined by*

$$[x, y]_X = [x, y]_{X,\omega} - [y, x]_{X,\omega}$$

for all $x, y \in M$ and a gate orientation ω of X does not depend on ω.

We prove this theorem in Sect. 4.3 using the content of Sect. 4.2. We call the bracket $[-, -]_X$ the *homotopy intersection bracket of* X. Both brackets $[-, -]_{X,\omega}$

and $[-,-]_X$ generalize Goldman's bracket [Go1, Go2]: the value of $[-,-]_{X,\omega}$ (respectively, $[-,-]_X$) on any pair of free homotopy classes of loops in $\Sigma' \subset X$ is equal to their Goldman's bracket (respectively, twice this bracket). The free homotopy classes of loops lying in $Y \subset X$ annihilate both $[-,-]_{X,\omega}$ and $[-,-]_X$.

4.2 *The Pairing μ_k*

For each $k \in \pi_0$, we define a bilinear form $\mu_k : M \times M \to M$ as follows. It suffices to define $\mu_k(x, y)$ for all $x, y \in \mathcal{L}$. To this end, pick generic loops a, b representing x, y, and set

$$\mu_k(x, y) = \sum_{p \in a \cap \alpha_k, q \in b \cap \alpha_k} \varepsilon_p(a)\, \varepsilon_q(b) \langle a_p b_q \rangle \in M.$$

Lemma 4.3 *The vector $\mu_k(x, y) \in M$ does not depend on the choice of generic loops a, b representing x, y.*

Proof We need only to prove that $\mu_k(x, y)$ is preserved under the loop moves on a, b. For the moves $L_0 - L_3$, this is clear from the definitions. A move L_4 on a creates two additional points $p', p'' \in a \cap \alpha_k$ for some k such that $\varepsilon_{p'}(a) = -\varepsilon_{p''}(a)$. Then the expressions $\varepsilon_{p'}(a)\, \varepsilon_q(b) \langle a_{p'} b_q \rangle$ and $\varepsilon_{p''}(a)\, \varepsilon_q(b) \langle a_{p''} b_q \rangle$ cancel each other for all $q \in b \cap \alpha_k$. So, this move preserves $\mu_k(x, y)$. The move L_5 on a replaces two points $p_1, p_2 \in a \cap \alpha_k$ by two points p'_1, p'_2 such that $\varepsilon_{p'_i}(a) = \varepsilon_{p_i}(a)$ and $\langle a_{p'_i} b_q \rangle = \langle a_{p_i} b_q \rangle$ for $i = 1, 2$ and any $q \in b \cap \alpha_k$. So, this move preserves $\mu_k(x, y)$. Similar computations show that the moves L_4, L_5 on b preserve $\mu_k(x, y)$. □

Lemma 4.3 shows that $\mu_k(x, y)$ depends only on x, y. The identity $\langle a_p b_q \rangle = \langle b_q a_p \rangle$ implies that $\mu_k(x, y) = \mu_k(y, x)$ for all $x, y \in M$.

4.3 *Proof of Theorem 4.2*

It suffices to prove that

$$[x, y]_{X,l\omega} - [y, x]_{X,l\omega} = [x, y]_{X,\omega} - [y, x]_{X,\omega} \tag{4.5}$$

for all $x, y \in \mathcal{L}$ and $l \in \pi_0$. Pick an ω-admissible pair of loops a, b representing, respectively, x, y. We compute $[x, y]_{X,\omega}$ from (4.1) and compute $[x, y]_{X,l\omega}$ applying (4.3) (with ω replaced by $l\omega$) to the pair a, b. The same argument as in the proof of Theorem 3.2 shows that

$$[x, y]_{X,l\omega} - [x, y]_{X,\omega} = \sum_{p \in a \cap \alpha_l, q \in b \cap \alpha_l} \varepsilon(l\omega, l)\, \varepsilon_p(a)\, \varepsilon_q(b) \langle a_p b_q \rangle \tag{4.6}$$

$$= -\varepsilon(\omega, l)\, \mu_l(x, y).$$

Since the form μ_l is symmetric, the expression $[x, y]_{X,l\omega} - [x, y]_{X,\omega}$ is symmetric in x, y. This implies (4.5) and completes the proof of the theorem.

4.4 Computation of $[-, -]_X$

To compute $[x, y]_X$ for $x, y \in \mathcal{L}$ we will use a method parallel to the one in Sect. 3.4. Pick a generic pair of loops a, b in X representing x, y. Let ω_0 be the orientation of the gates induced by the orientation of $\Sigma' \subset \Sigma$. Lemma 4.1 implies that

$$[x, y]_X = [x, y]_{X,\omega_0} - [y, x]_{X,\omega_0}$$

$$= 2 \sum_{r \in a \cap b} \varepsilon_r(a, b) \langle a_r b_r \rangle + \sum_{(k,p,q) \in T(a,b)} \delta(p, q)\, \varepsilon_p(a)\, \varepsilon_q(b) \langle a_p b_q \rangle$$

, where $\delta(p, q) = \pm 1$ is defined in Sect. 3.4. If $a \cap b = \emptyset$, then

$$[x, y]_X = \sum_{(k,p,q) \in T(a,b)} \delta(p, q)\, \varepsilon_p(a)\, \varepsilon_q(b) \langle a_p b_q \rangle. \tag{4.7}$$

4.5 Remarks

1. Applying (4.6) consecutively to all $l \in \pi_0$ and using (4.2), we obtain that for any $x, y \in M$ and any gate orientation ω of X,

$$[x, y]_{X,\omega} + [y, x]_{X,\omega} = \sum_{l \in \pi_0} \varepsilon(\omega, l)\, \mu_l(x, y).$$

Since $[x, y]_X = [x, y]_{X,\omega} - [y, x]_{X,\omega}$, we deduce that

$$2[x, y]_{X,\omega} = [x, y]_X + \sum_{l \in \pi_0} \varepsilon(\omega, l)\, \mu_l(x, y).$$

As a consequence, the form $\bullet_{X,\omega}$ in $H_1(X)$ may be computed from $\bullet_X$ via

$$2\,x \bullet_{X,\omega} y = x \bullet_X y + \sum_{l \in \pi_0} \varepsilon(\omega, l)\, v_l(x)\, v_l(y)$$

for all $x, y \in H_1(X)$.

2. Generally speaking, the brackets $[-,-]_{X,\omega}$ and $[-,-]_X$ do not satisfy the Jacobi identity. Their Jacobiators can be computed in terms of operations associated with the gates of X. Similar results hold for the self-intersection cobrackets defined in the next section; this will be discussed in more detail elsewhere.

5 The Cobrackets

We define self-intersection cobrackets for loops in the quasi-surface X.

5.1 *The Cobracket* $\nu_{X,\omega}$

For a loop a in X, we set $\langle a \rangle_0 = \langle a \rangle \in M = M(X)$ if a is non-contractible and $\langle a \rangle_0 = 0 \in M$ if a is contractible. A generic loop a crosses each point $r \in \#a$ twice; we let v_r^1, v_r^2 be the tangent vectors of a at r numerated so that the pair (v_r^1, v_r^2) is positively oriented. For $i = 1, 2$, let a_r^i be the loop starting in r and going along a in the direction of the vector v_r^i until the first return to r. Up to parametrization, $a = a_r^1 a_r^2$ is the product of the loops a_r^1, a_r^2 based at r. Also, for any $k \in \pi_0$ and any distinct points $p_1, p_2 \in a \cap \alpha_k$ we define a loop a_{p_1,p_2} in X which goes from p_1 to p_2 along a and then goes back to p_1 along the gate α_k. Consider the set of ordered triples

$$T(a) = \{(k \in \pi_0, p_1 \in a \cap \alpha_k, p_2 \in a \cap \alpha_k) \mid p_1 \neq p_2\}.$$

We call a triple $(k, p_1, p_2) \in T(a)$ a *chord of* a *with endpoints* p_1, p_2.

For a gate orientation ω of X, we let $T_\omega(a)$ be the set of chords $(k, p_1, p_2) \in T(a)$ such that $p_1 <_\omega p_2$. Set

$$\nu_{X,\omega}(a) = \sum_{r \in \#a} (\langle a_r^1 \rangle_0 \otimes \langle a_r^2 \rangle_0 - \langle a_r^2 \rangle_0 \otimes \langle a_r^1 \rangle_0)$$

$$+ \sum_{(k,p_1,p_2) \in T_\omega(a)} \varepsilon(\omega, k)\, \varepsilon_{p_1}(a)\, \varepsilon_{p_2}(a)\, \langle a_{p_2,p_1} \rangle_0 \otimes \langle a_{p_1,p_2} \rangle_0 \in M \otimes M.$$

Lemma 5.1 *The cobracket $\nu_{X,\omega}(a)$ is preserved under all loop moves on a.*

Proof The moves $L_0 - L_3$ proceed in Σ' and are treated as in [Tu1]. The move L_4 pushes a branch of a across a gate α_l for some $l \in \pi_0$ creating a loop a' which has two additional crossings $q_1, q_2 \in a' \cap \alpha_l$ such that $q_1 <_\omega q_2$ and $\varepsilon_{q_1}(a) = -\varepsilon_{q_2}(a)$. The contributions to the cobracket of the self-intersection points and of the chords containing neither q_1 nor q_2 are the same before and after the deformation. The contributions to $\nu_{X,\omega}(a')$ of the chords containing exactly one of the points q_1, q_2 cancel each other. The chord (l, q_1, q_2) of a' contributes zero to $\nu_{X,\omega}(a')$ because at least one of the loops a_{q_1,q_2} and a_{q_2,q_1} is contractible. Therefore $\nu_{X,\omega}(a) = \nu_{X,\omega}(a')$. We now prove the invariance of $\nu_{X,\omega}(a)$ under the move L_5 which pushes a self-crossing of a in Σ' across a gate, say α_l, into $X \setminus \Sigma'$. Note that under the inversion of the orientation of Σ, the signs $\varepsilon(\omega, k)$ and the expression $\nu_{X,\omega}(a)$ are multiplied by -1. Therefore, inverting if necessary the given orientation of Σ, we can reduce the proof of the invariance of $\nu_{X,\omega}(a)$ under our move to the case where $\varepsilon(\omega, l) = +1$. Assume that the move changes a in a small disk D by pushing a self-crossing $r_0 \in \#a$ from $D \cap \Sigma'$ to $D \setminus \Sigma'$. For $i = 1, 2$ set $\gamma_i = \langle a^i_{r_0} \rangle_0 \in M$. The branches of a meeting at r_0 intersect the gate α_l in two points lying in D. We label these points q_1, q_2 so that $q_1 <_\omega q_2$. By definition, r_0 contributes $\gamma_1 \otimes \gamma_2 - \gamma_2 \otimes \gamma_1$ to $\nu_{X,\omega}(a)$. The contribution of the chord (l, q_1, q_2) to $\nu_{X,\omega}(a)$ also can be computed from the definitions: it is equal to $\gamma_2 \otimes \gamma_1$ if $\varepsilon_{q_1}(a) = \varepsilon_{q_2}(a)$ and to $-\gamma_1 \otimes \gamma_2$ otherwise. Thus the joint contribution of r_0 and (l, q_1, q_2) to $\nu_{X,\omega}(a)$ is equal to $\gamma_1 \otimes \gamma_2$ if $\varepsilon_{q_1}(a) = \varepsilon_{q_2}(a)$ and to $-\gamma_2 \otimes \gamma_1$ otherwise. The loop, a', produced by the move meets $D \cap \alpha_l$ in two points forming a chord of a'. A similar computation shows that the contribution of this chord to $\nu_{X,\omega}(a')$ also is $\gamma_1 \otimes \gamma_2$ if $\varepsilon_{q_1}(a) = \varepsilon_{q_2}(a)$ and $-\gamma_2 \otimes \gamma_1$ otherwise. All the other self-crossings and chords contribute the same expressions to $\nu_{X,\omega}(a)$ and $\nu_{X,\omega}(a')$. Therefore $\nu_{X,\omega}(a) = \nu_{X,\omega}(a')$. □

Lemma 5.1 implies that $\nu_{X,\omega}(a) \in M \otimes M$ depends only on the free homotopy class $\langle a \rangle$ of a. The map $\langle a \rangle \mapsto \nu_{X,\omega}(a) : \mathcal{L} \to M^{\otimes 2}$ extends uniquely to a linear map $M \to M^{\otimes 2}$ denoted $\nu_{X,\omega}$.

5.2 The Cobracket ν_X

We define a cobracket ν_X independent of ω.

Theorem 5.2 *Let P be the linear automorphism of $M \otimes M$ carrying $x \otimes y$ to $y \otimes x$ for all $x, y \in M$. The skew-symmetric cobracket*

$$\nu_X = \nu_{X,\omega} - P\nu_{X,\omega} : M \to M \otimes M$$

does not depend on the choice of ω.

Proof It suffices to prove that $\nu_{X,\omega} - P\nu_{X,\omega}$ is preserved when ω is replaced with $l\omega$ for $l \in \pi_0$. For any generic loop a in X, we have

$$\nu_{X,\omega}(a) - \nu_{X,l\omega}(a) = \varepsilon(\omega, l) \sum_{(l,p_1,p_2)\in T_\omega(a)} \varepsilon_{p_1}(a)\,\varepsilon_{p_2}(a)\,\langle a_{p_2,p_1}\rangle_0 \otimes \langle a_{p_1,p_2}\rangle_0$$

$$- \varepsilon(l\omega, l) \sum_{(l,p_1,p_2)\in T_{l\omega}(a)} \varepsilon_{p_1}(a)\,\varepsilon_{p_2}(a)\,\langle a_{p_2,p_1}\rangle_0 \otimes \langle a_{p_1,p_2}\rangle_0.$$

Clearly, $\varepsilon(l\omega, l) = -\varepsilon(\omega, l)$. Also, the inclusion $(l, p_1, p_2) \in T_{l\omega}(a)$ holds if and only if $(l, p_2, p_1) \in T_\omega(a)$. Therefore

$$\nu_{X,\omega}(a) - \nu_{X,l\omega}(a) =$$

$$= \varepsilon(\omega, l) \sum_{(l,p_1,p_2)\in T_\omega(a)} \varepsilon_{p_1}(a)\,\varepsilon_{p_2}(a)\,\big(\langle a_{p_2,p_1}\rangle_0 \otimes \langle a_{p_1,p_2}\rangle_0 + \langle a_{p_1,p_2}\rangle_0 \otimes \langle a_{p_2,p_1}\rangle_0\big).$$

The latter expression is, obviously, invariant under the transposition P. So,

$$\nu_{X,\omega}(a) - \nu_{X,l\omega}(a) = P\nu_{X,\omega}(a) - P\nu_{X,l\omega}(a)$$

or, equivalently,

$$\nu_{X,\omega}(a) - P\nu_{X,\omega}(a) = \nu_{X,l\omega}(a) - P\nu_{X,l\omega}(a).$$

□

We call ν_X the *self-intersection cobracket* of X. Both cobrackets $\nu_{X,\omega}$ and ν_X generalize the cobracket ν defined for loops in surfaces in [Tu1]: the value of $\nu_{X,\omega}$ (respectively, ν_X) on any free homotopy class of loops in $\Sigma' \subset X$ is equal to the value of ν on this class (respectively, twice that value). The free homotopy classes of loops lying in $Y \subset X$ are annihilated by both $\nu_{X,\omega}$ and ν_X.

5.3 Computation of ν_X

To compute ν_X we use a method parallel to the one used in Sects. 3.4 and 4.4. Namely, for any generic loop a in X, we have

$$\nu_X(\langle a\rangle) = 2\sum_{r\in\#a} (\langle a_r^1\rangle_0 \otimes \langle a_r^2\rangle_0 - \langle a_r^2\rangle_0 \otimes \langle a_r^1\rangle_0)$$

$$+ \sum_{(k,p_1,p_2)\in T(a)} \delta(p_1, p_2)\,\varepsilon_{p_1}(a)\,\varepsilon_{p_2}(a)\,\langle a_{p_1,p_2}\rangle_0 \otimes \langle a_{p_2,p_1}\rangle_0.$$

If $\#a = \emptyset$, then

$$\nu_X(\langle a \rangle) = \sum_{(k,p_1,p_2)\in T(a)} \delta(p_1, p_2)\, \varepsilon_{p_1}(a)\, \varepsilon_{p_2}(a)\, \langle a_{p_1,p_2} \rangle_0 \otimes \langle a_{p_2,p_1} \rangle_0. \qquad (5.1)$$

5.4 Examples

We give two examples where the operations $\bullet_X$, $[-,-]_X$, and ν_X vanish. Set $I = [0,1]$ and $\Sigma = I^2$ with an arbitrary orientation.

1. Let $\alpha = I \times \{0\} \subset \partial\Sigma$ and $\Sigma' = I \times [1/3, 1] \subset \Sigma$. Then X has only one gate $I \times \{1/3\}$. It is clear that any loop in X can be deformed away from Σ'. Therefore $\bullet_{X,\omega} = 0$, $[-,-]_{X,\omega} = 0$, and $\nu_{X,\omega} = 0$ for both gate orientations ω of X. Consequently, $\bullet_X = 0$, $[-,-]_X = 0$, and $\nu_X = 0$.
2. Let $\alpha = I \times \{0,1\} \subset \partial\Sigma$ and $\Sigma' = I \times [1/3, 2/3] \subset \Sigma$. Then X has two gates $\alpha_1 = I \times \{1/3\}$ and $\alpha_2 = I \times \{2/3\}$. Let ω be the orientation of α_1, α_2 induced by the orientation of I from 0 to 1. Any pair of free homotopy classes of loops in X can be represented by loops a, b such that a meets Σ' at several segments $\{s\} \times [1/3, 2/3]$ with $s \in [0, 1/3]$ and b meets Σ' at several segments $\{t\} \times [1/3, 2/3]$ with $t \in [2/3, 1]$. Then the pair a, b is ω-admissible and $a \cap b = \emptyset$. Hence, $a \bullet_{X,\omega} b = 0$. Consequently, $\bullet_{X,\omega} = 0$ and $\bullet_X = 0$. Similar arguments show that $[-,-]_X = 0$. Next, any free homotopy class of loops in X can be represented by a generic loop a which meets Σ' at the segments $\{s\} \times [1/3, 2/3]$ where s runs over a finite set $S \subset (0,1)$. Clearly, $\#a = \emptyset$. The set $T_\omega(a)$ consists of the triples $(k \in \{1,2\}, s_1, s_2)$ with $s_1, s_2 \in S$ and $s_1 < s_2$. For any such s_1, s_2, the triples $(1, s_1, s_2)$ and $(2, s_1, s_2)$ contribute opposite values to $\nu_{X,\omega}(a)$ because $\varepsilon(\omega, 1) = -\varepsilon(\omega, 2)$ while all other terms of these contributions are the same. Thus, $\nu_{X,\omega}(a) = 0$. Consequently, $\nu_{X,\omega} = 0$ and $\nu_X = 0$. Note that for the gate orientation ω of X which directs one gate from 0 to 1 and the other gate from 1 to 0, the operations $\bullet_{X,\omega}$, $[-,-]_{X,\omega}$, and $\nu_{X,\omega}$ may be non-zero.

6 Transformations of Quasi-Surfaces

We study two transformations of the quasi-surface X: the transformation D (for disjoint unions) and the transformation C (for cuttings). Both D and C preserve the underlying topological space of X but change the structure of a quasi-surface.

6.1 The Transformation D

The transformation D applies when $\Sigma = \sqcup_{j=1}^N \Sigma_j$ is a disjoint union of $N \geq 2$ oriented surfaces. For each $j = 1, \ldots, N$, we let Y_j be the topological space obtained by glueing $N-1$ surfaces $\{\Sigma_i\}_{i\neq j}$ to Y along the maps $\alpha \cap \partial\Sigma_i \to Y$ used in the definition of X. The underlying topological space of X is obtained by glueing Σ_j to Y_j along the map $\alpha \cap \partial\Sigma_j \to Y \subset Y_j$ used in the definition of X. This turns the space in question into a quasi-surface, X_j, with surface core Σ_j and singular core Y_j. For the reduced surface core and the gates of X_j we take $\Sigma' \cap \Sigma_j$ and the gates of X lying in Σ_j.

Lemma 6.1 *We have*

$$\bullet_X = \sum_{j=1}^{N} \bullet_{X_j} : H_1(X) \times H_1(X) \to \mathbb{Z}, \tag{6.1}$$

$$[-,-]_X = \sum_{j=1}^{N} [-,-]_{X_j} : M \times M \to M, \tag{6.2}$$

$$\nu_X = \sum_{j=1}^{N} \nu_{X_j} : M \to M \otimes M. \tag{6.3}$$

Proof Note that each gate orientation ω of X restricts to a gate orientation ω_j of X_j. Formulas (6.1)–(6.3) are direct consequences of the following stronger claim: for any ω, we have

$$\bullet_{X,\omega} = \sum_{j=1}^{N} \bullet_{X_j,\omega_j} : H_1(X) \times H_1(X) \to \mathbb{Z}. \tag{6.4}$$

$$[-,-]_{X,\omega} = \sum_{j=1}^{N} [-,-]_{X_j,\omega_j} : M \times M \to M, \tag{6.5}$$

$$\nu_{X,\omega} = \sum_{j=1}^{N} \nu_{X_j,\omega_j} : M \to M \otimes M. \tag{6.6}$$

To prove (6.4), we pick an ω-admissible pair a, b of loops in X representing $x, y \in H_1(X)$. Then $x \bullet_{X,\omega} y$ is the algebraic number of intersections of a, b in $\Sigma' = \sqcup_j(\Sigma' \cap \Sigma_j)$. Also, for all j, the pair of loops a, b is ω_j-admissible in X_j and $x \bullet_{X_j,\omega_j} y$ is the algebraic number of intersections of a, b in $\Sigma' \cap \Sigma_j$. Hence,

$$x \bullet_{X,\omega} y = \sum_{j=1}^{N} x \bullet_{X_j,\omega_j} y.$$

The proofs of (6.5) and (6.6) are similar. $\square$

6.2 The Transformation C

A submanifold β of a manifold N is said to be *proper* if $\beta \cap \partial N = \partial\beta$. The transformation C applies when we are given a proper compact 1-dimensional submanifold β of Σ' whose components are segments disjoint from the gates of X (which, recall, all lie in $\partial\Sigma'$). Cutting $\Sigma \supset \Sigma' \supset \beta$ along β, we obtain an oriented surface Σ^β. A copy of the 1-manifold $\alpha \subset \partial\Sigma \setminus \Sigma'$ lies in $\partial\Sigma^\beta$ and is denoted α^β. The 1-manifold β gives rise to two copies of itself in $\partial\Sigma^\beta \setminus \alpha^\beta$. The underlying topological space of X can be obtained by glueing the surface Σ^β to the disjoint union $Y^\beta = Y \sqcup \beta$ along the map $\alpha^\beta = \alpha \to Y \subset Y^\beta$ used in the definition of X and along the tautological identity maps of the copies of β in $\partial\Sigma^\beta$ to $\beta \subset Y^\beta$. This turns the underlying topological space of X into a quasi-surface X^β with surface core Σ^β and singular core Y^β.

Lemma 6.2 *We have*

$$\bullet_X = \bullet_{X^\beta} : H_1(X) \times H_1(X) \to \mathbb{Z}, \tag{6.7}$$

$$[-,-]_X = [-,-]_{X^\beta} : M \times M \to M, \tag{6.8}$$

$$\nu_X = \nu_{X^\beta} : M \to M \otimes M. \tag{6.9}$$

Proof The gates of X^β are the gates of X and additional gates associated with the components $\{\beta_l\}_l$ of β. Namely, each β_l gives rise to two gates of X^β which are proper segments in Σ' running "parallel" to β_l on different sides of β_l in Σ'. For each l, fix an orientation of β_l and orient the associated gates so that they look in the same direction as β_l. Then every gate orientation ω of X determines a gate orientation ω^β of X^β. Formulas (6.7)–(6.9) are consequences of the following stronger claim: for any ω, we have

$$\bullet_{X,\omega} = \bullet_{X^\beta,\omega^\beta} : H_1(X) \times H_1(X) \to \mathbb{Z}. \tag{6.10}$$

$$[-,-]_{X,\omega} = [-,-]_{X^\beta,\omega^\beta} : M \times M \to M, \tag{6.11}$$

$$\nu_{X,\omega} = \nu_{X^\beta,\omega^\beta} : M \to M \otimes M. \tag{6.12}$$

To prove (6.10), pick an ω-admissible pair a, b of loops in X representing $x, y \in H_1(X)$. Deforming if necessary a, b near β we can assume that a, b are transversal to β, all crossings of a with β lie near the tails of the components, and all crossings of b with β lie near the heads of the components. Then the pair a, b is ω^β-admissible. The integer $x \bullet_{X,\omega} y$ is the algebraic number of intersections of a, b in Σ'. Since all these intersections lie away from β and from the gates of X, they bijectively correspond to the intersections of a, b in the reduced surface core of X^β (and have the same signs). Therefore $x \bullet_{X,\omega} y = x \bullet_{X^\beta,\omega^\beta} y$. The proofs of the equalities (6.11) and (6.12) are similar. □

7 Stars in Quasi-Surfaces

We state and prove analogues of Theorems 1.1–1.4 for quasi-surfaces.

7.1 Stars in X

By a *star* in the quasi-surface X we mean a star s embedded in the surface core Σ of X so that $\partial s = s \cap \partial\Sigma \subset \partial\Sigma \setminus \alpha$. We derive from s a new structure of a quasi-surface in the underlying topological space of X. Denote the number of leaves of s by $|s|$. Pick a closed regular neighborhood V of s in $\Sigma \setminus \alpha \subset X$ and provide V with orientation induced by that of Σ. It is clear that V is a 2-disk whose boundary is formed by $|s|$ disjoint segments in $\partial\Sigma \setminus \alpha$ and $|s|$ disjoint proper segments $\beta_1^s, \dots, \beta_{|s|}^s$ in Σ. Set $Y_\star = \overline{X \setminus V} \subset X$ where the overline stands for the closure in the underlying topological space of X. Taking V as the surface core, $Y_\star$ as the singular core and glueing V to $Y_\star$ along the inclusions $\{\beta_i^s \hookrightarrow Y_\star\}_{i=1}^{|s|}$ we obtain a quasi-surface, $X_\star = X_\star(s)$, with the same underlying topological space as X. This allows us to consider the bilinear maps

$$\bullet_s = \bullet_{X_\star} : H_1(X) \times H_1(X) \to \mathbb{Z}, \quad [-,-]_s = [-,-]_{X_\star} : M \times M \to M$$

and the linear map $\nu_s = \nu_{X_\star} : M \to M \otimes M$.

We next compute the form $\bullet_s$ via intersections of loops with s. As in Sect. 1, the orientation of Σ at the center of s determines a cyclic order in the set $\mathrm{Edg}(s)$ of edges of s. For $e \in \mathrm{Edg}(s)$ we let $e^+ \in \mathrm{Edg}(s)$ be the next edge with respect to this order. We say that a loop in X is *s-generic* if it misses the vertices of s, meets all edges of s transversely, and never traverses a point of s more than once. A family of loops in X is *s-generic* if these loops are s-generic and do not meet at points of s. Given an s-generic loop a in X, we let $a \cap s$ be the set of points of s traversed by a. For an edge $e \in \mathrm{Edg}(s)$, set $a \cap e = (a \cap s) \cap e$. For $p \in a \cap e$, the intersection sign of a and e at p is denoted $\mu_p(a)$ (recall that the edges of s are directed from the center of s to the leaves). The integer $a \cdot e = \sum_{p \in a \cap e} \mu_p(a)$ is the algebraic number of intersections of a and e.

Lemma 7.1 *For any s-generic pair of loops a, b in X representing $x, y \in H_1(X)$,*

$$x \bullet_s y = \sum_{e \in Edg(s)} \big((a \cdot e)(b \cdot e^+) - (b \cdot e)(a \cdot e^+)\big). \tag{7.1}$$

Proof Consider the quasi-surface $X_\star = X_\star(s)$ derived as above from a closed regular neighborhood V of s in $\Sigma \setminus \alpha$. For the reduced surface core of $X_\star$ we take a smaller closed regular neighborhood $V' \subset V$ of s. The gates of $X_\star$ are the segments in $\partial V'$ separating V' from the rest of V. Moving along the circle $\partial V'$ in the direction determined by the orientation of V' induced by that of Σ, we meet all gates in a certain cyclic order. For a gate K of $X_\star$, denote the next gate with respect to this cyclic order by K^+. Note that there is a unique edge $e = e_K$ of s such that K is obtained by pushing the segment $e \cup e^+$ into $\Sigma \setminus s$. Clearly, $e_{K^+} = (e_K)^+$.

To proceed, we select the closed regular neighborhood $V' \subset V$ taking into account the loops a, b. We say that V' is *a-adapted* if the set $a(S^1) \cap V'$ is a disjoint union of proper segments $\{f_p\}_{p \in a \cap s}$ in V' such that $p \in f_p$ for all $p \in a \cap s$. We denote the endpoints of the segment f_p by p', p'' so that the pair (the orientation of f_p from p' to p'', the orientation of the edge of s containing p) determines the given orientation of Σ at p. If V' is a-adapted, then a has no self-intersections in V' and crosses the gates precisely at the points $\{p', p''\}_{p \in a \cap s}$. The crossing signs of a with the gates (see Sect. 3.2) are computed by

$$\varepsilon_{p'}(a) = \mu_p(a) \quad \text{and} \quad \varepsilon_{p''}(a) = -\mu_p(a) \tag{7.2}$$

for all $p \in a \cap s$. We select the neighborhood $V' \subset V$ of s so small (= narrow), that it is a-adapted, b-adapted, and the loops a, b do not meet in V'. Then for any gate K of $X_\star$, we have

$$a \cap K = \{p' \,|\, p \in a \cap e_K\} \sqcup \{p'' \,|\, p \in a \cap (e_K)^+\} \tag{7.3}$$

and, similarly,

$$b \cap K = \{q' \,|\, q \in b \cap e_K\} \sqcup \{q'' \,|\, q \in b \cap (e_K)^+\}. \tag{7.4}$$

We compute $x \bullet_{X_\star} y$ via (3.5). The sum on the right hand-side of (3.5) runs over all triples (a gate K of $X_\star$, a point of $a \cap K$, a point of $b \cap K$). By (7.3) and (7.4), the contribution of such triples with fixed K is the sum of the following four expressions:

$$\sigma_K^1 = \sum_{p \in a \cap e_K, q \in b \cap e_K} \delta(p', q')\, \varepsilon_{p'}(a)\, \varepsilon_{q'}(b),$$

$$\sigma_K^2 = \sum_{p \in a \cap e_K, q \in b \cap (e_K)^+} \delta(p', q'')\, \varepsilon_{p'}(a)\, \varepsilon_{q''}(b),$$

$$\sigma_K^3 = \sum_{p \in a \cap (e_K)^+, q \in b \cap e_K} \delta(p'', q')\, \varepsilon_{p''}(a)\, \varepsilon_{q'}(b),$$

$$\sigma_K^4 = \sum_{p \in a \cap (e_K)^+, q \in b \cap (e_K)^+} \delta(p'', q'')\, \varepsilon_{p''}(a)\, \varepsilon_{q''}(b).$$

To simplify these expressions we use the following notation: for distinct points p, q of an edge of s, set $\mu(p, q) = 1$ if p lies between the center of s and q and set

$\mu(p,q) = -1$ otherwise. For any $e \in \text{Edg}(s)$ and any points $p \in a \cap e, q \in b \cap e$, we have $\delta(p', q') = \mu(p,q)$ and $\delta(p'', q'') = -\mu(p,q)$. Using this and (7.2), we get

$$\sigma_K^1 = \sum_{p \in a \cap e_K, q \in b \cap e_K} \mu(p,q)\, \mu_p(a)\, \mu_q(b)$$

and

$$\sigma_K^4 = - \sum_{p \in a \cap (e_K)^+, q \in b \cap (e_K)^+} \mu(p,q)\, \mu_p(a)\, \mu_q(b).$$

When K runs over all gates, both e_K and $(e_K)^+$ run over all edges of s. Therefore $\sum_K (\sigma_K^1 + \sigma_K^4) = 0$. Observe next that $\delta(p', q'') = -1$ for all $p \in a \cap e_K$ and $q \in b \cap (e_K)^+$. Hence,

$$\sigma_K^2 = \sum_{p \in a \cap e_K, q \in b \cap (e_K)^+} \mu_p(a)\, \mu_q(b) = (a \cdot e_K)(b \cdot (e_K)^+).$$

Similarly, $\delta(p'', q') = 1$ for all $p \in a \cap (e_K)^+, q \in b \cap e_K$ and so

$$\sigma_K^3 = - \sum_{p \in a \cap (e_K)^+, q \in b \cap e_K} \mu_p(a)\, \mu_q(b) = -(a \cdot (e_K)^+)(b \cdot e_K).$$

Therefore

$$x \bullet_s y = \sum_K (\sigma_K^1 + \sigma_K^2 + \sigma_K^3 + \sigma_K^4) = \sum_{e \in \text{Edg}(s)} \big((a \cdot e)(b \cdot e^+) - (b \cdot e)(a \cdot e^+) \big).$$

□

Given two s-generic loops a, b in X and points $p \in a \cap s, q \in b \cap s$, we pick a path c from p to q in s and write $a_p b_q$ for the loop $a_p c b_q c^{-1}$. Clearly, the free homotopy class of this loop in X does not depend on the choice of c.

Lemma 7.2 *For any s-generic loops a, b in X, we have*

$$[\langle a \rangle, \langle b \rangle]_s$$

$$= \sum_{e \in Edg(s)} \Big(\sum_{p \in a \cap e, q \in b \cap e^+} \mu_p(a)\, \mu_q(b)\, \langle a_p b_q \rangle - \sum_{p \in a \cap e^+, q \in b \cap e} \mu_p(a)\, \mu_q(b)\, \langle a_p b_q \rangle \Big).$$

Proof We use the same V' as in the proof of Lemma 7.1 and apply Formula (4.7) to compute $[\langle a\rangle, \langle b\rangle]_s$. The expressions $\sigma_K^1, \ldots, \sigma_K^4$ are replaced with the sums

$$\sum_{p\in a\cap e_K, q\in b\cap e_K} \delta(p', q')\, \varepsilon_{p'}(a)\, \varepsilon_{q'}(b)\langle a_p b_q\rangle,$$

$$\sum_{p\in a\cap e_K, q\in b\cap (e_K)^+} \delta(p', q'')\, \varepsilon_{p'}(a)\, \varepsilon_{q''}(b)\langle a_p b_q\rangle,$$

$$\sum_{p\in a\cap (e_K)^+, q\in b\cap e_K} \delta(p'', q')\, \varepsilon_{p''}(a)\, \varepsilon_{q'}(b)\langle a_p b_q\rangle,$$

$$\sum_{p\in a\cap (e_K)^+, q\in b\cap (e_K)^+} \delta(p'', q'')\, \varepsilon_{p''}(a)\, \varepsilon_{q''}(b)\langle a_p b_q\rangle.$$

The rest of the argument goes along the same lines as the proof of Lemma 7.1. □

Given an s-generic loop a in X and points $p_1, p_2 \in a \cap s$, we write a_{p_1,p_2} for the loop going from p_1 to p_2 along a and then going back to p_1 along a path in s. Clearly, the free homotopy class of this loop does not depend on the choice of the latter path.

Lemma 7.3 *For any s-generic loop a in X, we have*

$$\nu_s(\langle a\rangle)$$
$$= \sum_{e\in Edg(s)} \sum_{p_1\in a\cap e, p_2\in a\cap e^+} \mu_{p_1}(a)\, \mu_{p_2}(a) \big(\langle a_{p_1,p_2}\rangle_0 \otimes \langle a_{p_2,p_1}\rangle_0 - \langle a_{p_2,p_1}\rangle_0 \otimes \langle a_{p_1,p_2}\rangle_0\big).$$

Proof We use the same V' as in the proof of Lemma 7.1 and apply Formula (5.1) to compute $\nu_s(\langle a\rangle)$. The rest of the argument uses the same ideas as the proof of Lemma 7.1. □

7.2 *Star-Fillings of X*

A *star-filling* of X is a finite family F of disjoint stars in X such that each component of the set $\Sigma \setminus \cup_{s\in F} s$ is a disk meeting $\partial\Sigma \setminus \alpha$ at one or two open segments.

Lemma 7.4 *If the surface core Σ of X is compact and each component of Σ has a non-void boundary then X has a star-filling.*

Proof Cutting Σ along a finite set of disjoint proper segments $\{\beta_i\}_i$ we can get a disjoint union of closed disks $\{D_j\}_j$. Pushing if necessary the endpoints of the segments $\{\beta_i\}_i$ along $\partial\Sigma$ we can ensure that $\beta_i \cap \alpha = \emptyset$ for all i. For each j, the circle ∂D_j is formed by several, say, n_j segments in $\partial\Sigma \setminus \alpha$ and the same number of segments which are either components of α or copies of some β_i's. If $n_j \geq 2$, then we pick a star $s_j \subset D_j$ with center in $\mathrm{Int}(D_j)$ and with leaves inside the above-mentioned n_j segments in $\partial\Sigma \setminus \alpha$. Composing the inclusion $s_j \subset D_j$ with the natural embedding $D_j \hookrightarrow \Sigma$, we can view each s_j as a star in Σ. Then the family $\{s_j | n_j \geq 2\}_j$ is a star-filling of X. □

Lemma 7.5 *For any star-filling F of X, we have*

$$\bullet_X = \sum_{s\in F} \bullet_s : H_1(X) \times H_1(X) \to \mathbb{Z}, \tag{7.5}$$

$$[-,-]_X = \sum_{s\in F} [-,-]_s : M \times M \to M, \tag{7.6}$$

$$\nu_X = \sum_{s\in F} \nu_s : M \to M \otimes M. \tag{7.7}$$

Proof We prove the first equality, the other two are proven similarly. Without loss of generality we can assume that all stars in the family F lie in the reduced surface core $\Sigma' \subset \Sigma$. We choose closed regular neighborhoods $\{V^s \supset s\}_{s\in F}$ so that they are pairwise disjoint and lie in Σ'. For each $s \in F$, let $\beta_1^s, \ldots, \beta_{|s|}^s$ be the segments in ∂V^s separating V^s from the rest of X, cf. Sect. 7.1. Then the 1-manifold $\beta = \cup_{s\in F} \cup_{i=1}^{|s|} \beta_i^s$ satisfies the requirements needed to apply the transformation C to X. This transformation produces a quasi-surface X^β having the same underlying topological space as X. By (6.7), $\bullet_X = \bullet_{X^\beta}$. Note that the surface core of X^β is a disjoint union of the surfaces $\{V^s \supset s\}_{s\in F}$ and $\overline{\Sigma \setminus \cup_{s\in F} V^s}$. By (6.1), the form $\bullet_{X^\beta}$ is the sum of the forms $\bullet$ associated with the connected components of the surface core of X^β. Each component $V^s \supset s$ contributes the form $\bullet_s$ to this sum. By the definition of a star-filling, all components of the surface $\overline{\Sigma \setminus \cup_{s\in F} V^s}$ are homeomorphic to the quasi-surfaces described in Sect. 5.4. Therefore the associated pairings $\bullet$ are equal to zero and

$$\bullet_X = \bullet_{X^\beta} = \sum_{s\in F} \bullet_s : H_1(X) \times H_1(X) \to \mathbb{Z}.$$

□

8 Proof of Theorems 1.1–1.4 and the Case of Closed Surfaces

8.1 Proof of Theorems 1.1–1.4

Consider the quasi-surface $X = X(\Gamma)$ with surface core Γ, empty singular core, and $\alpha = \emptyset \subset \partial\Gamma$. (Alternatively, one can use in this proof a quasi-surface with the surface core Γ, a 1-point singular core, and a set $\alpha \subset \partial\Gamma$ consisting of a single segment.) Any star s in Γ is a star in X. Lemma 7.1 implies that the skew-symmetric bilinear form $\bullet_s$ in $H_1(\Gamma)$ satisfies the conditions of Theorem 1.1 and we set $\cdot_s = \bullet_s$. Note that $\bullet_X = 2 \cdot_X$ as is clear from the remarks after the statement of Theorem 3.2. Therefore Theorem 1.2 is a direct consequence of Formula (7.5). Similarly, the first claim of Theorem 1.3 follows from Lemma 7.2, and the second claim of Theorem 1.3 follows from Formula (7.6). The claims of Theorem 1.4 follow, respectively, from Lemma 7.3 and Formula (7.7).

8.2 *The Case of Closed Surfaces*

Each closed oriented surface Φ gives rise to the homological intersection form $\cdot_\Phi : H_1(\Phi) \times H_1(\Phi) \to \mathbb{Z}$, to Goldman's bracket $[-,-]_\Phi : M \times M \to M$, and to the cobracket $\nu_\Phi : M \to M \otimes M$. Here M is the free abelian group whose basis is the set of free homotopy classes of loops in Φ. The definitions of $[-,-]_\Phi$ and ν_Φ repeat word for word the definitions given in the introduction in the case of surfaces with boundary. In this setting there seem to be no analogues of the maps $\cdot_s, [-,-]_s, \nu_s$ derived from stars. We compute the maps $\cdot_\Phi, [-,-]_\Phi, \nu_\Phi$ in terms of the so-called filling graphs which we now define.

By a *bipartite graph* we mean a finite graph G provided with a partition of its set of vertices into two disjoint subsets $B(G)$ and $R(G)$ whose elements are called, respectively, the blue vertices and the red vertices so that every edge of G connects a blue vertex to a red vertex. Note that a bipartite graph cannot have edges connecting a vertex to itself.

A *filling graph* of Φ is a bipartite graph G embedded in Φ so that (1) all components of $\Phi \setminus G$ are open disks and (2) going along the boundary of each of these disks one traverses at most 4 edges and 4 vertices of G. For example, any triangulation τ of Φ determines a filling graph of Φ formed by the vertices of τ (declared to be "blue"), the centers of the 2-simplices of τ (declared to be "red"), and the edges connecting the center of each 2-simplex of τ to the vertices of this 2-simplex. Exchanging the colors of the vertices, one derives from any filling graph of Φ the dual filling graph.

We show now how to calculate the maps $\cdot_\Phi, [-,-]_\Phi, \nu_\Phi$ via a filling graph $G \subset \Phi$. Note that the orientation of Φ at any blue vertex $v \in B(G)$ determines a cyclic order in the set Edg_v of the edges of G adjacent to v. For $e \in \mathrm{Edg}_v$ we let $e^+ \in \mathrm{Edg}_v$ be the next edge with respect to this order. The set $\mathrm{Edg}(G)$ of all edges of G is a disjoint union of the sets $\{\mathrm{Edg}_v \mid v \in B(G)\}$. Therefore the map $e \mapsto e^+$ is a permutation in $\mathrm{Edg}(G)$.

We say that a loop in Φ is *G-generic* if it misses the vertices of G, meets all edges of G transversely, and never traverses a point of G more than once. It is clear that any loop in Φ can be made G-generic by a small deformation. We orient each edge e of G from its blue vertex to its red vertex. For a G-generic loop a in Φ, set $a \cap e = a(S^1) \cap e$. The intersection sign of a and e at any point $p \in a \cap e$ is denoted by $\mu_p(a)$. Set $a \cdot e = \sum_{p \in a \cap e} \mu_p(a)$.

Theorem 8.1 *For any $x, y \in H_1(\Phi)$ and any G-generic loops a, b in Φ representing x, y, we have*

$$2\,x \cdot_\Phi y = \sum_{e \in Edg(G)} \big((a \cdot e)(b \cdot e^+) - (b \cdot e)(a \cdot e^+)\big). \tag{8.1}$$

Proof Since the loops a, b miss the vertices of G, each red vertex $w \in R(G)$ has a closed disk neighborhood D_w in $\Phi \setminus (a(S^1) \cup b(S^1))$. We can assume that the disks $\{D_w\}_w$ are disjoint and $D_w \cap G \subset \Phi$ is a star with center w for all w. Then

$$\Gamma = \Phi \setminus \cup_{w \in R(G)} \mathrm{Int}(D_w)$$

is a compact connected subsurface of Φ which we provide with orientation induced by that of Φ. For every blue vertex $v \in B(G)$, the set

$$s_v = \Gamma \cap \cup_{e \in \mathrm{Edg}_v} e$$

is a star in Γ with center v. It is clear that the stars $\{s_v \mid v \in R(G)\}$ are pairwise disjoint. The assumption that G is a filling graph of Φ implies that this family of stars is a star-filling of Γ. Since $a(S^1) \cup b(S^1) \subset \Gamma$, the loops a, b represent certain homology classes $x', y' \in H_1(\Gamma)$. Theorems 1.1 and 1.2 imply that

$$2\,x' \cdot_\Gamma y' = \sum_{e \in \mathrm{Edg}(G)} \big((a \cdot e)(b \cdot e^+) - (b \cdot e)(a \cdot e^+)\big). \tag{8.2}$$

On the other hand, the inclusion homomorphism $H_1(\Gamma) \to H_1(\Phi)$ carries x', y', respectively, to x, y. The usual definition of the intersection forms $\cdot_\Gamma$ and $\cdot_\Phi$ implies that $x' \cdot_\Gamma y' = x \cdot_\Phi y$. Combining this with (8.2), we obtain (8.1). □

To state similar results for the bracket $[-,-]_\Phi$ in the module M, we need more notation. For a loop a in Φ, we let $\langle a \rangle \in M$ be the free homotopy class of a. For a point $p \in \Phi$ traversed by a only once, we let a_p be the loop in Φ starting at p and going along a until the return to p. We say that a pair of loops a, b in Φ is *G-generic* if a, b are G-generic and do not meet at points of G. Consider a *G-generic* pair of loops a, b and consider edges e, e' of G sharing the same blue vertex v. For any points $p \in a \cap e$, $q \in b \cap e'$, we let $c = c_{p,q}$ be the path going from p to v along e and then going from v to q along e'. We write $a_p b_q$ for the loop $a_p c b_q c^{-1}$.

Theorem 8.2 *For any free homotopy classes x, y of loops in Φ and any G-generic pair of loops a, b in Φ representing x, y, we have*

$$2\,[x, y]_\Phi$$

$$= \sum_{e \in Edg(G)} \Big(\sum_{p \in a \cap e,\, q \in b \cap e^+} \mu_p(a)\, \mu_q(b) \langle a_p b_q \rangle - \sum_{p \in a \cap e^+,\, q \in b \cap e} \mu_p(a)\, \mu_q(b) \langle a_p b_q \rangle \Big).$$

This theorem is deduced from Theorem 1.3 using the construction introduced in the proof of Theorem 8.1.

We finally compute the cobracket $\nu_\Phi : M \to M \otimes M$. For a loop a in Φ, we set $\langle a \rangle_0 = \langle a \rangle \in M$ if a is non-contractible and $\langle a \rangle_0 = 0 \in M$ if a is contractible. Consider a G-generic loop a in Φ and edges e, e' of G sharing the same blue vertex v. For points $p \in a \cap e$, $p' \in a \cap e'$, we write $a_{p,p'}$ for the loop going from p to p' along a, then going to v along e', then going back to p along e.

Theorem 8.3 *For any free homotopy class x of loops in Φ and any G-generic loop a in Φ representing x, we have*

$$2\,\nu_\Phi(x)$$

$$= \sum_{e \in Edg(G)} \sum_{p \in a \cap e, p' \in a \cap e^+} \mu_p(a)\,\mu_{p'}(a)\,\big(\langle a_{p,p'}\rangle_0 \otimes \langle a_{p',p}\rangle_0 - \langle a_{p',p}\rangle_0 \otimes \langle a_{p,p'}\rangle_0\big).$$

This theorem is deduced from Theorem 1.4 using the construction in the proof of Theorem 8.1.

8.3 Example

We check Formula (8.1) in a case where all computations can be done explicitly. Set $I = [0, 1]$. Identifying the opposite sides of the square I^2 via $(t, 0) = (t, 1)$ and $(0, t) = (1, t)$ for all $t \in I$ we obtain a 2-torus T. Let $p : I^2 \to T$ be the projection. The vertices $A_1 = (0, 0)$, $A_2 = (1, 0)$, $A_3 = (1, 1)$, $A_4 = (0, 1)$ of I^2 project to the same point of T denoted A. Let $O = (1/2, 1/2)$ be the center of I^2. Consider the graph $G \subset T$ with two vertices $p(O)$, A and four edges $\{e_i = p(OA_i)\}_{i=1}^4$ where OA_i is the straight segment connecting O and A_i in I^2. We let $p(O)$ be the blue vertex of G and let A be the red vertex of G. This turns G into a filling graph of T. Consider the homology classes $x, y \in H_1(T)$ represented, respectively, by the loops

$$a : S^1 \to T, e^{2\pi i t} \mapsto p((t, 1/3)) \quad \text{and} \quad b : S^1 \to T, e^{2\pi i t} \mapsto p((1/4, t)).$$

These loops meet in the point $p((1/4, 1/3))$. We orient T so that $x \cdot_T y = +1$. Then

$$a \cdot e_1 = a \cdot e_2 = -1, \quad a \cdot e_3 = a \cdot e_4 = 0$$

and

$$b \cdot e_1 = b \cdot e_4 = 1, \quad b \cdot e_2 = b \cdot e_3 = 0.$$

Clearly, $(e_i)^+ = e_{i+1}$ for $i = 1, 2, 3$ and $(e_4)^+ = e_1$. Therefore

$$\sum_{e\in \mathrm{Edg}(G)} \big((a\cdot e)(b\cdot e^+) - (b\cdot e)(a\cdot e^+)\big) = -(b\cdot e_1)(a\cdot e_2) - (b\cdot e_4)(a\cdot e_1) = 2 = 2\,x\cdot_T y$$

which checks Formula (8.1) in this case.

References

[AKKN1] A. Alekseev, N. Kawazumi, Y. Kuno, F. Naef, *The Goldman-Turaev Lie bialgebra in genus zero and the Kashiwara-Vergne problem.* Adv. Math. 326 (2018), 1–53.

[AKKN2] A. Alekseev, N. Kawazumi, Y. Kuno, F. Naef, *The Goldman-Turaev Lie bialgebra and the Kashiwara-Vergne problem in higher genera.* arXiv:1804.09566.

[Go1] W. M. Goldman, *The symplectic nature of fundamental groups of surfaces.* Adv. in Math. 54 (1984), no. 2, 200–225.

[Go2] W. M. Goldman, *Invariant functions on Lie groups and Hamiltonian flows of surface group representations.* Invent. Math. 85 (1986), no. 2, 263–302.

[Ha1] R. Hain, *Hodge Theory of the Turaev Cobracket and the Kashiwara–Vergne Problem.* arXiv:1807.09209.

[Ha2] R. Hain, *Johnson homomorphisms.* arXiv:1909.03914.

[Ka] A. Kabiraj, *Center of the Goldman Lie algebra.* Algebr. Geom. Topol. 16 (2016), no. 5, 2839–2849.

[KK1] N. Kawazumi, Y. Kuno, *The logarithms of Dehn twists.* Quantum Topol. 5 (2014), no. 3, 347–423.

[KK2] N. Kawazumi, Y. Kuno, *Intersection of curves on surfaces and their applications to mapping class groups.* Ann. Inst. Fourier 65 (2015), no. 6, 2711–2762.

[LS] D. Li-Bland, P. Severa, *Moduli spaces for quilted surfaces and Poisson structures.* Doc. Math. 20 (2015), 1071–1135.

[Ma] G. Massuyeau, *Formal descriptions of Turaev's loop operations.* Quantum Topol. 9 (2018), no. 1, 39–117.

[MT] G. Massuyeau, V. Turaev, *Quasi-Poisson structures on representation spaces of surfaces.* Int. Math. Res. Not. (2014), no. 1, 1–64.

[Tu1] V. Turaev, *Skein quantization of Poisson algebras of loops on surfaces.* Ann. Sci. École Norm. Sup. (4) 24 (1991), no. 6, 635–704.

[Tu2] V. Turaev, *Topological constructions of tensor fields on moduli spaces.* arXiv:1901.02634.